CARDIOVASCULAR DISEASES

CARDIOVASCULAR DISEASES

From Molecular Pharmacology to Evidence-Based Therapeutics

Y. ROBERT LI
Professor of Pharmacology
Chair of Department of Pharmacology
Campbell University SOM
Buies Creek, North Carolina, USA

Adjunct Professor of Biomedical Engineering and Sciences
Virginia Tech-Wake Forest University
School of Biomedical Engineering and Sciences
Blacksburg, Virginia, USA

Adjunct Professor of Biomedical Sciences and Pathobiology
Department of Biomedical Sciences and Pathobiology
Virginia-Maryland Regional College of Veterinary Medicine
Virginia Polytechnic Institute and State University
Blacksburg, Virginia, USA

Adjunct Professor of Biology
Department of Biology
University of North Carolina College of Arts and Sciences
Greensboro, North Carolina, USA

Published by John Wiley & Sons, Inc., Hoboken, New Jersey
Published simultaneously in Canada

For general information on our other products and services or for technical support, please contact our Customer Care Department within the United States at (800) 762-2974, outside the United States at (317) 572-3993 or fax (317) 572-4002.

Wiley also publishes its books in a variety of electronic formats. Some content that appears in print may not be available in electronic formats. For more information about Wiley products, visit our web site at www.wiley.com.

Library of Congress Cataloging-in-Publication Data:

Li, Yunbo, author.
Cardiovascular diseases : from molecular pharmacology to evidence-based therapeutics / Y. Robert Li.
   p. ;   cm.
  Includes index.
  ISBN 978-0-470-91537-0 (cloth)
I. Title.
[DNLM: 1. Cardiovascular Diseases–drug therapy.   2. Case Reports.   3. Evidence-Based Medicine–methods.
4. Molecular Medicine–methods.   WG 166]
  RM345
  616.1'061–dc23

                                                       2014050157

Set in 10/12pt Times by SPi Publisher Services, Pondicherry, India

Printed in the United States of America

10 9 8 7 6 5 4 3 2 1

1   2015

CONTENTS

PREFACE

Cardiovascular diseases remain the leading cause of death globally, though the mortality associated with these diseases in developed countries has been significantly reduced over the past decades owing to the availability of effective treatment approaches. Among the available therapeutic strategies, drug therapy continues to be an important modality in the management of various forms of cardiovascular diseases. In this context, cardiovascular pharmacology serves as the foundation for pharmacotherapeutics of cardiovascular diseases and has become an increasingly important subject in cardiovascular medicine.

While there are multiple excellent pharmacology books with chapters being devoted to cardiovascular drugs, a book that systematically integrates essentials, advancements, and clinical correlations for cardiovascular drugs would facilitate the learning of knowledge on using these pharmacological agents to prevent and treat cardiovascular diseases. The aim of this book is to provide a comprehensive coverage of molecular pharmacology of various classes of cardiovascular drugs and evidence-based pharmacotherapeutics in the management of common cardiovascular diseases and conditions, including dyslipidemias, hypertension, ischemic heart disease, heart failure, cardiac arrhythmias, and ischemic stroke. As outlined in the following text, the book contains eight units with a total of 28 chapters.

Unit I (Chapters 1 and 2):	General Introduction
Unit II (Chapters 3–5):	Dyslipidemias
Unit III (Chapters 6–12):	Hypertension and Multitasking Cardiovascular Drugs
Unit IV (Chapters 13–15):	Ischemic Heart Disease: Stable Ischemic Heart Disease
Unit V (Chapters 16–19):	Ischemic Heart Disease: Acute Coronary Syndromes
Unit VI (Chapters 20–22):	Heart Failure
Unit VII (Chapters 23–25):	Cardiac Arrhythmias
Unit VIII (Chapters 26–28):	Ischemic Stroke

To set the stage for subsequently discussing the diverse topics of cardiovascular pharmacology and therapeutics, Unit I provides two introductory chapters. Chapter 1 introduces general aspects of cardiovascular diseases, including definition, classification, and epidemiology, as well as the overall strategies for prevention and control. Chapter 2 briefly surveys the general principles of pharmacology and provides an overview of the key and emerging concepts in cardiovascular pharmacology and therapeutics.

Unit II consists of three chapters (Chapters 3–5) devoted to the discussion of pharmacology and therapeutics of dyslipidemias. Chapter 3 reviews lipoprotein metabolism and lipoprotein disorders to lay a basis for understanding how drugs impact diverse lipoprotein metabolic pathways to treat various dyslipidemias. Chapter 4 examines molecular pharmacology of various classes of drugs for treating dyslipidemias, including statins, bile acid sequestrants, cholesterol absorption inhibitors, fibrates, niacin, as well as newly approved drugs for homozygous familial hypercholesterolemia. The chapter also considers phytosterols, phytostanols, omega-3 fatty acids, and emerging drugs for dyslipidemias. The principles and current evidence-based guidelines for management of dyslipidemias in clinical practice are covered in Chapter 5.

Unit III consisting of seven chapters (Chapters 6–12) discusses pharmacology and therapeutics of hypertension, as well as various classes of cardiovascular drugs. Chapter 6 provides an overview of hypertension, including definition, epidemiology, and pathophysiology and drug targeting. Because the different drug classes used for treating hypertension are also commonly employed in the management of other cardiovascular diseases, molecular pharmacology of these multitasking cardiovascular drug classes is considered in separate chapters (Chapters 7–11). These drug classes include diuretics (Chapter 7), sympatholytics (Chapter 8), inhibitors of the renin–angiotensin–aldosterone system (Chapter 9), calcium channel blockers (Chapter 10), and nitrates and other vasodialtors (Chapter 11). Following discussion of these multitasking drug classes, the principles and current evidence-based guidelines for hypertension management in clinical practice are given in Chapter 12, the last chapter of Unit III.

Unit IV contains three chapters (Chapters 13–15) devoted to discussing pharmacology and therapeutics of stable ischemic heart disease. Chapter 13 gives an overview on ischemic heart disease and discusses pathophysiology of stable ischemic heart disease and mechanistically based drug targeting of stable angina. Chapter 14 reviews anti-anginal drugs that have already been discussed in previous chapters and also considers some newly approved and emerging anti-anginal drugs that are not covered elsewhere in the book. The principles and current evidence-based guidelines for management of stable ischemic heart disease/stable angina in clinical practice are given in Chapter 15.

Unit V consisting of four chapters (Chapters 16–19) considers pharmacology and therapeutics of acute coronary syndromes (ACS), including unstable angina (UA), acute non-ST elevation myocardial infarction (NSTEMI), and acute ST elevation myocardial infarction (STEMI). Chapter 16 provides an overview of definition and epidemiology of ACS and discusses current understanding of ACS pathophysiology and the mechanistically based drug targeting as well as related therapeutic modalities. Chapter 17 examines molecular pharmacology of drugs for treating ACS, including anticoagulants, platelet inhibitors, and thrombolytic agents. This lays a basis for understanding general principles and current evidence-based guidelines for the management of UA/NSTEMI and STEMI in clinical practice in Chapters 18 and 19, respectively.

Unit VI has three chapters (Chapters 20–22) that consider pharmacology and therapeutics of heart failure, a common clinical syndrome representing the end stage of a number of different cardiac diseases. Chapter 20 gives an overview of heart failure, including definition, classification, epidemiology, and pathophysiology and drug targeting. The major drug classes for treating heart failure include diuretics (Chapter 7), β-blockers (Chapter 8), inhibitors of the renin–angiotensin–aldosterone system (Chapter 9), vasodilators (Chapter 11), and positive inotropic agents. Chapter 21 discusses pharmacological basis of using the above drug classes in treating heart failure. Since inotropic drugs have not been covered elsewhere in the book, Chapter 21 focuses on molecular pharmacology of this drug class in heart failure treatment. The chapter also introduces emerging therapeutic

modalities for heart failure. The principles and current evidence-based guidelines for management of heart failure in clinical practice are provided in Chapter 22.

Unit VII consisting of three chapters (Chapters 23–25) discusses pharmacology and therapeutics of cardiac arrhythmias. Chapter 23 provides an overview of cardiac arrhythmias, including classification, epidemiology, and pathophysiology and drug targeting. Chapter 24 discusses molecular pharmacology of classical antiarrhythmic drugs (Vaughan–Williams class I–IV drugs) with a focus on those whose efficacy is supported by recent clinical research. The chapter also considers antiarrhythmic agents that do not fall into the Vaughan–Williams classification, as well as emerging drugs with promising results from randomized clinical trials. Chapter 25 describes the general principles and current evidence-based guidelines for management of arrhythmias in clinical practice with an emphasis on pharmacological therapy. Since arrhythmias are a large group of disorders, the chapter focuses on those that have the most significant clinical and public health impact, including atrial fibrillation and certain forms of ventricular tachyarrhythmias.

Unit VIII, the last unit of the book, contains three chapters (Chapters 26–28) devoted to discussing pharmacology and therapeutics of ischemic stroke. Chapter 26 gives an overview of ischemic stroke, including definition, classification, epidemiology, risk factors, and pathophysiology and drug targeting. The preventive and therapeutic intervention of ischemic stroke involves five areas of management: primary prevention, early treatment of acute ischemic stroke, neuroprotection, secondary prevention, and neurorepair. Although neuroprotection and neurorepair are promising strategies, presently, they are primarily experimental approaches. Drug therapy remains as a major component of the effective intervention of ischemic stroke. Chapter 27 discusses the major classes of drugs that are used in each of the above areas. Since most of the drugs discussed in the chapter have been covered in preceding chapters, Chapter 27 only summarizes current evidence-based consensus statements on their clinical efficacy in preventive and therapeutic intervention of ischemic stroke. The principles and current evidence-based guidelines for management of ischemic stroke in clinical practice are provided in Chapter 28, the last chapter of the book.

It is hoped that this book by integrating the most recent advancements in molecular pharmacology and most current evidence-based, guideline-directed therapeutics of cardiovascular diseases will provide the readers a unique approach to understanding the rapidly evolving field of cardiovascular medicine and therapeutics. Because of the rapidly evolving nature of cardiovascular medicine, and medicine as a whole, the information contained in this book is subject to change based on new scientific knowledge and clinical experience. Although the author of the book has checked with sources believed to be reliable and accurate at the time of publication, information included in the book may not be accurate in every respect due to the possibility of human errors and rapid changes in medical sciences. As such, the author of the book does not warrant that the information contained in the work is in every respect accurate and complete. The author disclaims all responsibility for any errors or omissions or for the results obtained from the use of the information contained in this book. The readers are advised to seek independent verification for any data, advice, or recommendations contained in the work.

This book would not have been possible without the assistance of my son Jason Z. Li who drew all of the chemical structures for the whole book, and my wife Hong Zhu, MD, MPH, who critically reviewed the entire book manuscript. I am grateful to those (over 100 scientists worldwide) who provided me reprints of their publications and/or comments on part of the book manuscript. I am thankful for the time and effort made by Mr. Jonathan Rose, Senior Editor, and other editorial personnel, especially Ms. Shiji Sreejish, at Wiley, which made the work possible and of high quality.

Apex, North Carolina   Y. ROBERT LI, MD, MPH, PhD
September 23, 2014

LIST OF ABBREVIATIONS

ABC	ATP-binding cassette		**ANP**	atrial natriuretic peptide
ABCA1	ATP-binding cassette protein A1		**APC**	atrial premature complex
ABCG2	ATP-binding cassette protein G2		**APD**	action potential duration
ABCG5	ATP-binding cassette protein G5		**aPTT**	activated partial thromboplastin time
ABCG8	ATP-binding cassette protein G8		**AR**	aldosterone receptor
ACAT	acyl-CoA:cholesterol acyltransferase		**ARB**	angiotensin receptor blocker
ACC	American College of Cardiology		**ARVC**	arrhythmogenic right ventricular cardiomyopathy
ACCF	American College of Cardiology Foundation		**ASA**	American Stroke Association
ACE	angiotensin-converting enzyme		**ASCVD**	atherosclerotic cardiovascular disease
ACEI	angiotensin-converting enzyme inhibitor		**ASH**	American Society of Hypertension
Ach	acetylcholine		**AST**	aspartate aminotransferase
ACLS	advanced cardiovascular life support		**AT**	atrial tachycardia
ACS	acute coronary syndromes		**AT$_1$**	angiotensin receptor type 1
ADE	adverse drug event		**AT$_2$**	angiotensin receptor type 2
ADHF	acute decompensated heart failure		**AT$_4$**	angiotensin receptor type 4
ADP	adenosine diphosphate		**ATP**	adenosine triphosphate
ADR	adverse drug reaction		**ATP**	Adult Treatment Panel
AF	atrial fibrillation		**ATP III**	Adult Treatment Panel III
AFL	atrial flutter		**AV**	atrioventricular
AHA	American Heart Association		**AVNRT**	Atrioventricular nodal reciprocating tachycardia
AHF	acute heart failure		**AVRT**	atrioventricular reentrant tachycardia
AHFS	acute heart failure syndromes		**b.i.d.**	twice a day
AHRQ	Agency for Healthcare Research and Quality		**BHS**	British Hypertension Society
ALA	alpha-linolenic acid		**BMPR2**	bone morphogenetic protein receptor type 2
ALDH2	mitochondrial aldehyde dehydrogenase-2		**BP**	blood pressure
ALK1	activin receptor-like kinase type 1		**BRMAC**	Biological Response Modifiers Advisory Committee
ALT	alanine aminotransferase		**CABG**	coronary artery bypass grafting
AMA	American Medical Association		**CAD**	coronary artery disease
AMI	acute myocardial infarction		**cAMP**	3'-5'-cyclic adenosine monophosphate

CCB	calcium channel blocker
CCS	Canadian Cardiovascular Society
cCTA	coronary computed tomography angiogram
CDC	Centers for Disease Control and Prevention
CETP	cholesterol ester transfer protein
cGMP	cyclic guanosine monophosphate
cGMP	cyclic-3′,5′-guanosine monophosphate
CHD	coronary heart disease
CHEP	Canadian Hypertension Education Program
CKD	chronic kidney disease
CM	chylomicron
CMR	chylomicron remnant
CNS	central nervous system
COPD	chronic obstructive pulmonary disease
COR	class of recommendation
COX	cyclooxygenase
CPR	cardiopulmonary resuscitation
CPVT	catecholaminergic polymorphic ventricular tachycardia
CQI	continuous quality improvement
CrCl	creatinine clearance
CRT	cardiac-resynchronization therapy
CTEPH	chronic thromboembolic pulmonary hypertension
CYP	cytochrome P450
DAPT	dual antiplatelet therapy
DASH	Dietary Approaches to Stop Hypertension
DBP	diastolic blood pressure
DC	direct current
DES	drug-eluting stent
DGAT-2	acyl-CoA:diacylglycerol acyltransferase-2
DHA	docosahexaenoic acid
DHHS	Department of Health and Human Services
DTI	direct thrombin inhibitor
DVT	deep vein thrombosis
EAS	European Atherosclerosis Society
ECG	electrocardiogram
ECG	electrocardiography
EF	ejection fraction
eGFR	estimated glomerular filtration rate
EMA	European Medicines Agency
EMS	emergency medical services
EPA	eicosapentaenoic acid
EPAD	established peripheral arterial disease
Epi	epinephrine
ERP	effective refractory period
ESC	European Society of Cardiology
ESCs	embryonic stem cells
ESH	European Society of Hypertension
ESO	European Stroke Organisation
ET-1	endothelin-1
ET-2	endothelin-2
ET-3	endothelin-3
ET$_A$	endothelin receptor type A
ET$_B$	endothelin receptor type B
FAT	focal atrial tachycardia
FDA	Food and Drug Administration
FFA	free fatty acid
FMC	first medical contact
FXR	farnesoid X receptor
GAS	genome-wide association study
GBD	global burden of disease
GDMT	guideline-directed medical therapy
GFR	glomerular filtration rate
GP IIb/IIIa	glycoprotein IIb/IIIa
GTP	guanosine triphosphate
GWTG	Get With the Guidelines
HbA1c	glycosylated hemoglobin
HCM	hypertrophic cardiomyopathy
HDL	high-density lipoprotein
HDL-C	high-density lipoprotein cholesterol
HF	heart failure
HF-PEF	heart failure with preserved ejection fraction
HF-REF	heart failure with reduced ejection fraction
HFSA	Heart Failure Society of America
HIT	heparin-induced thrombocytopenia
HITTS	heparin-induced thrombocytopenia and thrombosis syndrome
HIV	human immunodeficiency virus
HL	hepatic lipase
HMG-CoA	3-hydroxy-3-methylglutaryl-coenzyme A
HoFH	homozygous familial hypercholesterolemia
HRE	hormone response element
HRS	Heart Rhythm Society
ICD	implantable cardioverter defibrillator
ICD	International Classification of Diseases and Related Health Problems
ICD-10	International Statistical Classification of Diseases and Related Health Problems—10th Revision
ICH	intracranial hemorrhage
IDL	intermediate-density lipoprotein
IHD	ischemic heart disease
IND	investigational new drug application

INR	international normalized ratio		**NE**	norepinephrine
IOM	Institute of Medicine		**NHANES**	National Health and Nutrition Examination Survey
iPSCs	induced pluripotent stem cells			
ISA	intrinsic sympathomimetic activity		**NHLBI**	National Heart, Lung, and Blood Institute
ISH	International Society of Hypertension		**NICE**	National Institute for Health and Care Excellence
IST	inappropriate sinus tachycardia		**NICE**	National Institute for Health and Clinical Excellence
Iv	intravenous			
JBS	Joint British Societies		**NIHSS**	National Institutes of Health Stroke Scale
JNC	Joint National Committee on Prevention, Detection, Evaluation, and Treatment of High Blood Pressure		**NO**	nitric oxide
			NPC1L1	Niemann–Pick C1-Like 1
			NPR-A	natriuretic peptide receptor-A
JNC7	the Seventh Report of the Joint National Committee on the Prevention Detection, Evaluation, and Treatment of High Blood Pressure		**Nrf2**	nuclear factor E2-related factor 2
			NSAID	nonsteroidal anti-inflammatory drugs
			NSTEMI	non-ST-elevation myocardial infarction
			NSTEMI	non-ST-segment elevation myocardial infarction
JPC	junctional premature complex		**NYHA**	New York Heart Association
LBBB	left bundle-branch block		**OTC**	ornithine transcarboxylase
LCAT	lecithin–cholesterol acyltransferase		**PAF**	platelet-activating factor
LDL	low-density lipoprotein		**PAH**	pulmonary arterial hypertension
LDL-C	low-density lipoprotein cholesterol		**PAR**	protease-activated receptor
LDLR	low-density lipoprotein receptor		**PCI**	percutaneous coronary intervention
LDLRAP	LDL receptor adaptor protein		**PCSK9**	proprotein convertase subtilisin/kexin type 9
LGL	Lown–Ganong–Levine		**PDE**	phosphodiesterase
LM	lifestyle modifications		**PDE3**	phosphodiesterase 3
LMWH	low-molecular-weight heparin		**PDE5**	phosphodiesterase 5
LOE	level of evidence		**PE**	pulmonary embolism
Lp(a)	lipoprotein(a)		**PGD$_2$**	prostaglandin D$_2$
LPL	lipoprotein lipase		**PGs**	prostaglandins
LQTS	long QT syndrome		**PGx**	pharmacogenetics/pharmacogenomics
LV	left ventricular		**PIAs**	positive inotropic agents
LVD	left ventricular dysfunction		**PLA$_2$**	phospholipase A$_2$
LVEF	left ventricular ejection fraction		**PLTP**	phospholipid transfer protein
LXR	liver X receptor		**PON1**	paraoxonase 1
MAO	monoamine oxidase		**PPAR-α**	peroxisome proliferator-activated receptor-α
MAT	multifocal atrial tachycardia		**PPARγ**	peroxisome proliferator-activated receptor-gamma
MCA	middle cerebral artery			
MCS	mechanical circulatory support		**PSVT**	paroxysmal supraventricular tachycardia
METs	metabolic equivalents		**PTCA**	percutaneous transluminal coronary angioplasty
MI	myocardial infarction		**PVC**	premature ventricular complex
MMP	matrix metalloproteinase		**q.i.d.**	4 times a day
MTP	microsomal triglyceride transfer protein		**QTc**	corrected QT
NAD	nicotinamide adenine dinucleotide		**RAAS**	renin–angiotensin–aldosterone system
NADP	nicotinamide adenine dinucleotide phosphate		**ROS**	reactive oxygen species
NCDs	noncommunicable diseases		**ROS/RNS**	reactive oxygen/nitrogen species
NCEP	National Cholesterol Education program		**ROSC**	return of spontaneous circulation
NCHS	National Center for Health Statistics		**SA**	sinoatrial
NDA	new drug application		**SBP**	systolic blood pressure

SCA	sudden cardiac arrest	**t.i.d.**	3 times a day
SCAD	stable coronary artery disease	**TdP**	torsades de pointes
SCD	sudden cardiac death	**TIA**	transient ischemic attack
SFXa	selective factor Xa	**TLC**	therapeutic lifestyle changes
sGC	soluble guanylate cyclase	**TOS**	The Obesity Society
SIHD	stable ischemic heart disease	**tPA**	tissue plasminogen activator
siRNA	RNA interference	**TR**	thyroid hormone receptor
SMC	smooth muscle cell	**TxA$_2$**	thromboxane A$_2$
SND	SA node dysfunction	**UA**	unstable angina
SNP	single nucleotide polymorphism	**UFH**	unfractionated heparin
SNS	sympathetic nervous system	**USDA**	United States Department of Agriculture
SOE	strength of evidence	**VF**	ventricular fibrillation
sPLA$_2$	secretory PLA$_2$	**VKOR**	vitamin K epoxide reductase
SR-B1	scavenger receptor B1	**VKORC1**	the C1 subunit of VKOR
SREBP	sterol regulatory element-binding protein	**VLDL**	very low-density lipoprotein
SREBP1c	sterol regulatory element-binding protein 1c	**VT**	ventricular tachycardia
		WHO	World Health Organization
		WPW	Wolff–Parkinson–White
STEMI	ST-elevation myocardial infarction	**X-SCID**	X-linked severe combined immunodeficiency disease

UNIT I

GENERAL INTRODUCTION

1

INTRODUCTION TO CARDIOVASCULAR DISEASES

1.1 OVERVIEW

The heart has always held a special fascination for humans: it has been the seat of the soul, the home of emotions, and the pump that when beating symbolizes life and when silent signifies death [1]. Perhaps no other organ in the human body has been so closely scrutinized. Hence, the management of cardiovascular diseases, including the use of medications, has always been a focus of medicine. To set a stage for the subsequent discussion of the diverse topics of cardiovascular pharmacology and therapeutics, this chapter provides a brief introduction to various general aspects of cardiovascular diseases. These include definition, classification, and epidemiology of cardiovascular diseases, as well as the overall strategies for their prevention and control. An introduction to the principles of pharmacology in general and cardiovascular pharmacology in particular is given in Chapter 2.

1.2 DEFINITION OF CARDIOVASCULAR DISEASES

In order to define the term cardiovascular diseases, it is imperative to first provide an overview of the cardiovascular system. Briefly, cardiovascular system refers to an integrated organ system consisting of the heart and blood vessels. Blood flows through a network of blood vessels that extend between the heart and the peripheral tissues. These blood vessels are subdivided into a pulmonary circuit, which carries blood to and from the gas exchange surface of the lungs, and a systemic circuit, which transports blood to and from the rest of the body. Each circuit begins and ends at the heart, and the blood vessels and the heart collectively constitute the cardiovascular system. As noted, blood is a central player in the cardiovascular system, and hence, study of the cardiovascular system inevitably involves examination of the blood, including its components and functionality. It should be noted that cardiovascular system and circulatory system are frequently used interchangeably; however, strictly, the circulatory system is composed of the cardiovascular system, which distributes blood, and the lymphatic system, which distributes lymph.

Cardiovascular diseases refer to a group of diseases involving the heart, blood vessels, or the sequelae of poor blood supply due to a decreased vascular supply and include (i) diseases of the heart, (ii) vascular diseases of the brain (also known as cerebrovascular diseases), and (iii) diseases of other blood vessels (Fig. 1.1). Hence, cardiovascular diseases affect the heart, the brain, and other organs or systems of the human body.

1.3 CLASSIFICATION OF CARDIOVASCULAR DISEASES

1.3.1 Classification Based on Anatomical Location

Cardiovascular diseases are classified in various ways. One scheme is based primarily on the anatomical location of the disease pathogenesis and broadly classifies cardiovascular diseases into two categories: (i) diseases of the heart and (ii) vascular diseases (Fig. 1.2).

Cardiovascular Diseases: From Molecular Pharmacology to Evidence-Based Therapeutics, First Edition. Y. Robert Li.
© 2015 John Wiley & Sons, Inc. Published 2015 by John Wiley & Sons, Inc.

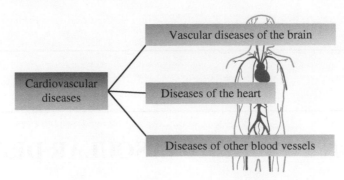

FIGURE 1.1 Definition of cardiovascular diseases. The term cardiovascular diseases refers to a group of diseases, including the diseases of the heart, vascular diseases of the brain, and the diseases of other blood vessels.

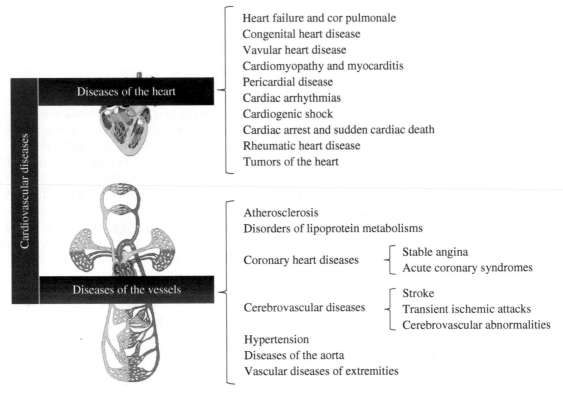

FIGURE 1.2 Classification of cardiovascular diseases. Primarily based on anatomical location, cardiovascular diseases are classified into diseases of the hearts and diseases of the vessels. As illustrated, coronary heart disease and stroke belong to the category of diseases of the vessels.

1.3.2 Classification Based on the Involvement of Atherosclerosis

Another classification scheme emphasizes the primary involvement of atherosclerosis and classifies cardiovascular diseases into (i) cardiovascular diseases due to atherosclerosis (also known as atherosclerotic cardiovascular diseases) and (ii) other cardiovascular diseases (Table 1.1). In this context, atherosclerosis is responsible for ~75% of all deaths due to cardiovascular diseases.

1.3.3 Total Cardiovascular Diseases and ICD-10 Classification

In addition to the aforementioned classification schemes, the American Heart Association (AHA) has recently introduced the concept of total cardiovascular diseases. This category (ICD-10 codes I00–I99, Q20–Q28; see next paragraph for the description of ICD-10) includes rheumatic fever/rheumatic heart disease (I00–I09); hypertensive diseases (I10–I15); ischemic (coronary) heart disease (I20–I25); pulmonary heart

TABLE 1.1   Classification of cardiovascular diseases based on the involvement of atherosclerosis

Classification basis	Disease
Cardiovascular diseases due to atherosclerosis (also known as atherosclerotic cardiovascular diseases)	Coronary heart diseases Cerebrovascular diseases Diseases of the aorta and arteries including hypertension and peripheral vascular diseases
Other cardiovascular diseases	Congenital heart diseases Rheumatic heart diseases Cardiac arrhythmias

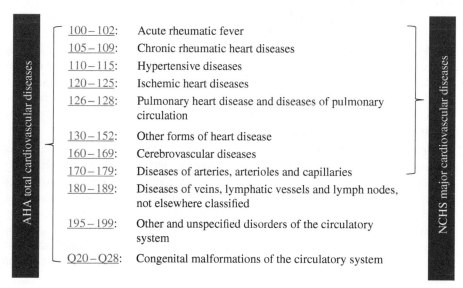

FIGURE 1.3   The ICD-10 disease codes included in the American Heart Association (AHA) total cardiovascular diseases and the US National Center for Health Statistics (NCHS) major cardiovascular diseases. As shown, the AHA total cardiovascular diseases are more comprehensive than the NCHS major cardiovascular diseases.

disease and diseases of pulmonary circulation (I26–I28); other forms of heart disease (I30–I52); cerebrovascular disease (stroke) (I60–I69); atherosclerosis (I70); other diseases of arteries, arterioles, and capillaries (I71–I79); diseases of veins, lymphatics, and lymph nodes not classified elsewhere (I80–I89); and other and unspecified disorders of the circulatory system (I95–I99), as well as congenital cardiovascular defects (Q20–Q28) [2].

ICD denotes International Classification of Diseases. It is the international standard diagnostic classification for all general epidemiological, many health management purposes, and clinical use. The current 10th revision, that is, ICD-10, was endorsed by the Forty-Third World Health Assembly in May 1990 and came into use in the World Health Organization (WHO) member states as from 1994. According to the ICD-10, diseases of the circulatory system are included in I00–I99, whereas the congenital malformations of the circulatory system (Q20–Q28) are included in the disease category of congenital malformations, deformations, and chromosomal abnormalities (Q00–Q99). Hence,

the AHA category of total cardiovascular diseases covers all diseases of the circulatory system, including both cardiovascular and lymphatic systems. On the other hand, the National Center for Health Statistics (NCHS) of the United States employs the term major cardiovascular diseases for reporting mortality data. The NCHS category of major cardiovascular diseases represents ICD codes I00–I78 and hence is less comprehensive than that of the AHA's total cardiovascular diseases (Fig. 1.3).

1.4   PREVALENCE, INCIDENCE, AND TREND OF CARDIOVASCULAR DISEASES

This section provides an overview of some of the major statistical and epidemiological data on cardiovascular diseases in the context of noncommunicable diseases (NCDs) in the globe as well as in selected countries, including the United States and China. The key data are also summarized in tables. Some pertinent terms are provided in Box 1.1.

BOX 1.1   GLOSSARY

- **Disease prevalence:** It is an estimate of how many people have a disease at a given point or period in time. Prevalence is sometimes expressed as a percentage of population.
- **Disease incidence:** An incidence rate refers to the number of new cases of a disease that develop in a population per unit of time. The unit of time for incidence is not necessarily 1 year although we often discuss incidence in terms of 1 year.
- **Mortality:** It refers to the total number of deaths attributable to a given disease in a population during a specific interval of time, usually a year.
- **Death rate or mortality rate:** It refers to the relative frequency with which death occurs within some specified interval of time in a population. Mortality rate is typically expressed as number of deaths per 100,000 individuals per year.
- **The World Health Organization (WHO):** The WHO is the directing and coordinating authority for health within the United Nations system. WHO was established in 1948 with headquarters in Geneva of Switzerland. It is responsible for providing leadership on global health matters, shaping the health research agenda, setting norms and standards, articulating evidence-based policy options, providing technical support to countries, and monitoring and assessing health trends.
- **Epidemiology:** Epidemiology is the study of the distribution and determinants of health-related states or events in specified populations and the application of this study to control of health problems. The objectives of epidemiology include (i) identification of the etiology or cause of a disease and the relevant risk factors; (ii) determination of the extent of disease found in the community; (iii) study of the natural history and prognosis of disease; (iv) evaluation of both existing and newly developed preventive and therapeutic measures and modes of healthcare delivery; and (v) providing the foundation for developing public policy relating to environmental problems, genetic issues, and other considerations regarding disease prevention and health promotion.
- **Global burden of disease:** Global burden of disease analysis provides a comprehensive and comparable assessment of mortality and loss of health due to diseases, injuries, and risk factors for all regions of the world. The overall burden of disease is assessed using the disability-adjusted life year (DALY), a time-based measure that combines years of life lost due to premature mortality and years of life lost due to time lived in states of less than full health. The original Global Burden of Disease Study (GBD 1990 Study) was commissioned by the World Bank in 1991 to provide a comprehensive assessment of the burden of 107 diseases and injuries and 10 selected risk factors for the world and 8 major regions in 1990. The methods of the GBD 1990 Study created a common metric to estimate the health loss associated with morbidity and mortality. It generated widely published findings and comparable information on disease and injury incidence and prevalence for all world regions. It also stimulated numerous national studies of burden of disease. These results have been used by governments and nongovernmental agencies to inform priorities for research, development, policies, and funding. The new Global Burden of Diseases, Injuries, and Risk Factors Study (GBD 2010 Study) commenced in the spring of 2007 and is the first major effort since the original GBD 1990 Study to carry out a complete systematic assessment of global data on all diseases and injuries. The GBD 2010 Study constitutes an unprecedented collaboration of 488 scientists from 303 institutions in 50 countries, focusing on describing the state of health around the world using a uniform method. The GBD 2010 Study results for the world and 21 regions have recently been reported for 291 diseases and injuries, 1160 sequelae of these causes, and 67 risk factors or clusters of risk factors [3]. This project is funded by the Bill and Melinda Gates Foundation.
- **Statistics:** Statistics is the study of the collection, organization, analysis, and interpretation of data.

BOX 1.2   THE WHO DEFINITION OF NONCOMMUNICABLE DISEASES

Noncommunicable diseases are identified by the WHO as "group II diseases," a category that aggregates (based on ICD-10 code; see Section 1.3.3 for ICD-10) the following conditions/causes of death: malignant neoplasms, other neoplasms, diabetes mellitus, endocrine disorders, neuropsychiatric conditions, sense organ diseases, cardiovascular diseases, respiratory diseases (e.g., chronic obstructive pulmonary disease, asthma, others), digestive diseases, genitourinary diseases, skin diseases, musculoskeletal diseases (e.g., rheumatoid arthritis), congenital anomalies (e.g., cleft palate, Down syndrome), and oral conditions (e.g., dental caries). These are distinguished from group I diseases (communicable, maternal, perinatal, and nutritional conditions) and group III diseases (unintentional and intentional injuries).

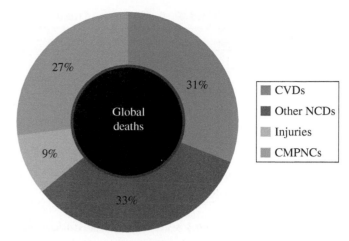

FIGURE 1.4   Global deaths caused by cardiovascular diseases (CVDs). As illustrated, CVDs are responsible for ~30% of all global deaths. NCDs denote noncommunicable diseases; CMPNCs denote communicable, maternal, perinatal, and nutritional conditions.

1.4.1   NCDs and Cardiovascular Diseases: The Global Status

According to the WHO, NCDs, including chiefly cardiovascular diseases (heart disease and stroke), cancer, chronic respiratory diseases, and diabetes, are the leading cause of mortality in the world, responsible for 36 million (or 63%) of the 57 million of the global deaths in 2008. The WHO definition for NCDs is given in Box 1.2. The burden of these diseases is rising disproportionately among lower-income countries and populations. In 2008, nearly 80% of NCD deaths (i.e., 29 million) occurred in low- and middle-income countries with about 29% of deaths occurring before the age of 60 years in these countries, dispelling the myth that such conditions are mainly a problem of affluent societies. Without action, the NCD epidemic is projected to kill 52 million people annually by 2030. A report by the World Economic Forum and the Harvard School of Public Health in September 2011 showed that the estimated cumulative output loss due to NCDs over the next 20 years represents ~4% of annual global gross domestic product (GDP) and will be $47 trillion by 2030. The increasing global crisis in NCDs is a barrier to development goals including poverty reduction, health equity, economic stability, and human security. The above staggering numbers and issues convinced the United Nations (UN) to convene its second-ever high-level general assembly meeting on health in September 2011 in New York, United States. This UN high-level meeting on NCDs along with its political declaration is an unprecedented opportunity to create a sustained global movement against premature death and preventable morbidity and disability from NCDs [4–6].

Among the 36 million NCD deaths in 2008, cardiovascular diseases caused 17.3 million deaths (or 48% of all NCD deaths) followed by cancers (7.6 million or 21% of all

NCD deaths), respiratory diseases (4.2 million or 11.7% of all NCD deaths), and diabetes (1.3 million or 3.6% of all NCD deaths). These four groups of diseases account for around 80% of all NCD deaths. Globally, NCD deaths are projected to increase by 15% between 2010 and 2020.

As shown in Figure 1.4, cardiovascular diseases remain the number one global killer of the human population, accounting for about 30% of all deaths (including communicable, noncommunicable, and other disease deaths) in the world. Notably, based on the WHO 2011 Global Atlas on Cardiovascular Disease Prevention and Control, out of the 17.3 million cardiovascular deaths in 2008, ischemic heart diseases (myocardial infarction) were responsible for 7.3 million deaths, and strokes were responsible for 6.2 million deaths. This figure remained largely unchanged in 2010 based on a report from the Global Burden of Disease 2010 Study [7]. Together, ischemic heart diseases and strokes account for nearly 80% of all cardiovascular deaths in the world and are the top two killers of the human population (Table 1.2), making them globally the two most pressing cardiovascular diseases for prevention and control.

1.4.2   The Status of Cardiovascular Diseases in the United States

1.4.2.1   Statistics   In the United States, currently, more than 82 million adults (more than one in three) have one or more types of cardiovascular diseases. Mortality data show that cardiovascular diseases, as the underlying causes of death, accounted for 31.9% (787,650) of all 2,468,435 deaths in 2010, or approximately one of every 3 deaths in the United States. The 2010 overall death rate from cardiovascular diseases in the United States was 235.5 per 100,000. On the

TABLE 1.2   Top 10 causes of the death in the world in 2011

Rank	Disease	Deaths in millions	% deaths
1	Ischemic heart disease	7.0	11.2
2	Stroke	6.2	10.6
3	Lower respiratory infections	3.2	6.7
4	Chronic obstructive pulmonary disease	3.0	5.8
5	Diarrheal diseases	1.9	4.7
6	HIV/AIDS[a]	1.6	3.0
7	Trachea, bronchus, lung cancers	1.5	2.7
8	Diabetes mellitus	1.4	2.6
9	Road injury	1.3	2.2
10	Prematurity	1.2	1.9

Source: The World Health Organization.
[a] HIV/AIDS denotes human immunodeficiency virus/acquired immunodeficiency syndrome.

TABLE 1.3   Ten leading causes of death in the United States in 2010

Rank	Disease	Deaths[a]	% total deaths
1	Diseases of heart	597,689	24.2
2	Malignant neoplasms	574,743	23.3
3	Chronic lower respiratory diseases	138,080	5.6
4	Cerebrovascular diseases	129,476	5.2
5	Accidents (unintentional injuries)	120,859	4.9
6	Alzheimer's disease	83,494	3.4
7	Diabetes mellitus	69,071	2.8
8	Nephritis, nephrotic syndrome, and nephrosis	50,476	2.0
9	Influenza and pneumonia	50,097	2.0
10	Intentional self-harm (suicide)	38,364	1.6

Source: The United States Centers for Disease Control and Prevention (CDC) National Vital Statistics Report 2013, 61(4).
[a] Total deaths in 2010: 2,468,435.

basis of 2010 mortality rate data, more than 2150 Americans die of cardiovascular diseases each day, an average of one death every 40 s. The total cost of cardiovascular diseases in the United States for 2010 is estimated to be $315.4 billion, accounting for 15% of total health expenditures in 2010, more than any other major diagnostic group [8].

Based on the 2014 update from the AHA [8], the prevalence (incidence) of various types of cardiovascular diseases in adults in the United States is as follows:

- Hypertension: 78,000,000
- Coronary heart disease: 15,400,000
  - Myocardial infarction (also known as heart attack): 7,600,000 (incidence: 720,000)
  - Angina pectoris: 7,800,000 (incidence: 565,000)
- Heart failure: 5,100,000 (incidence: 825,000)
- Stroke: 6,800,000 (incidence: 795,000)
- Congenital cardiovascular defects: 650,000–1,300,000

1.4.2.2 Trend According to the AHA 2014 Update [8], from 2000 to 2010, the overall cardiovascular disease death rates declined 31.0%. However, cardiovascular diseases are

still the leading cause of death in the United States. Declines in stroke death rate (a 35.8% decrease in annual stoke death rate from 2000 to 2010) now rank stroke as the fourth leading cause of death in the nation (as of 2008; Table 1.3). Although the cardiovascular mortality has decreased substantially over the past decades (Fig. 1.5) possibly due to effective prevention and better treatments for heart attacks, congestive heart failure, stroke, and other conditions, the cardiovascular disease prevalence and costs have been growing steadily and are projected to increase substantially in the future. For example, by 2030, 40.5% of the US population is projected to have some form of cardiovascular diseases. Between 2010 and 2030, real total direct medical costs of cardiovascular diseases are projected to triple, from $273 billion to $818 billion. Real indirect costs (due to lost productivity) for all cardiovascular diseases are estimated to increase from $172 billion in 2010 to $276 billion in 2030, an increase of 61% [9].

1.4.3   The Status of Cardiovascular Diseases in China

According to the official data available through the WHO (http://www.who.int), in China, about 230 million people currently have cardiovascular diseases. This translates into

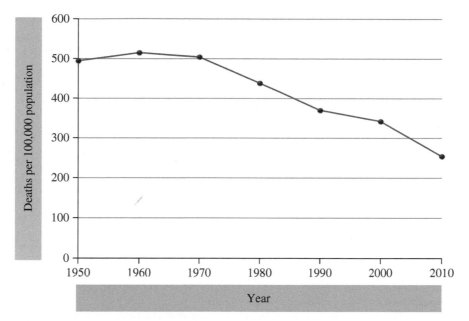

FIGURE 1.5 Cardiovascular disease mortality rates in the United States over the past seven decades. As shown, the past three to four decades have witnessed remarkable decreases in cardiovascular mortality rates. This achievement most likely results from implementation of effective health promotion initiatives and the availability of new effective treatments, including drug therapies.

one in five adults in China having a cardiovascular disease. In 2010, 154.8 per 100,000 deaths per year were estimated to be associated with cardiovascular diseases in urban areas and 163.1 per 100,000 in rural areas. This number accounts for 20.9%/17.9% (urban/rural) of China's total number of deaths per year.

Annual cardiovascular events are predicted to increase by 50% between 2010 and 2030 based on population aging and growth alone in China. Projected trends in blood pressure, total cholesterol, diabetes mellitus (increases), and active smoking (decline) would increase annual cardiovascular disease events by an additional 23%, an increase of ~21.3 million cardiovascular events and 7.7 million cardiovascular deaths.

1.5 RISK FACTORS OF CARDIOVASCULAR DISEASES

1.5.1 Classification of Cardiovascular Disease Risk Factors

It is known that the development of cardiovascular diseases results from the complicated interactions between genes and environmental and dietary factors. The major risk factors for developing cardiovascular diseases are classified by the WHO into (i) behavioral risk factors, (ii) metabolic risk factors, and (iii) other risk factors (Table 1.4). On the other hand, the AHA classifies cardiovascular risk factors into (1) major risk factors and (ii) contributing risk factors. The major risk factors are further divided into modifiable and nonmodifiable major risk factors (Table 1.5).

TABLE 1.4 The WHO classification of cardiovascular disease risk factors

Risk factor category	Risk factor[a]
Behavioral risk factors	Tobacco use
	Physical inactivity
	Unhealthy diet (rich in salt, fat, and calories)
	Harmful use of alcohol
Metabolic risk factors	Hypertension
	Diabetes mellitus
	Dyslipidemia
	Overweight and obesity
	Other metabolic risk factors (e.g., excess homocysteine)
Other risk factors	Advancing age
	Genetic disposition
	Gender
	Psychological factors (e.g., stress, depression, anxiety)
	Poverty and low educational status

Source: The World Health Organization.
[a] The term risk factor is defined as an exposure, behavior, or attribute that, if present and active, clearly increases the probability of a particular disease in a group of people who have the risk factor compared with an otherwise similar group of people who do not.

1.5.2 Major Cardiovascular Disease Risk Factors and Their Impact

As noted earlier, there are many risk factors associated with the development of cardiovascular diseases. The major risk factors, including tobacco use, hypertension, high cholesterol,

TABLE 1.5   The AHA classification of cardiovascular disease risk factors

Risk factor category		Risk factor
Major risk factors (significantly increase the risk of cardiovascular diseases)	Nonmodifiable factors (cannot be changed)	Increasing age Male sex Heredity
	Modifiable factors (can be modified, treated, or controlled by changing your lifestyle or taking medicine)	Tobacco smoke Unhealthy diet[a] High blood cholesterol High blood pressure Physical inactivity Obesity and overweight Diabetes mellitus Chronic kidney disease[b]
Contributing risk factors (other factors are associated with increased risk of cardiovascular disease, but their significance and prevalence haven't yet been precisely determined)		Stress Alcohol

Source: The American Heart Association (http://www.heart.org).

[a] Note that diet and nutrition are classified as contributing factors according to the AHA website listed earlier. This might cause confusion as unhealthy diet represents a major risk factor for cardiovascular diseases. As such, "diet and nutrition" is removed from the contributing risk factors category, and instead, "unhealthy diet" is added to the major risk factors category under "modifiable factors."

[b] Recent evidence suggests that chronic kidney disease is also a major risk factor for cardiovascular diseases. This is a particularly important risk factor considering the high global prevalence (8–16%) of chronic kidney disease [10, 11].

obesity, physical inactivity, and unhealthy diet, have a high prevalence across the world. Of particular significance in developing countries is the fact that while they are grappling with increasing rates of cardiovascular diseases, they still face the scourges of poor nutrition and infectious diseases. Nevertheless, with the exception of sub-Saharan Africa, cardiovascular diseases are also the leading cause of death in the developing world.

You will not necessarily develop cardiovascular diseases if you have a risk factor. But the more risk factors you have, the greater is the likelihood that you will, unless you take actions to modify your risk factors and work to prevent them from compromising your heart health. Table 1.6 summarizes the impact of some of the major risk factors on the development of cardiovascular diseases.

1.6   PREVENTION AND CONTROL OF CARDIOVASCULAR DISEASES

The mortality and morbidity of cardiovascular diseases (with ischemic heart disease and stroke as the major contributors) in the United States have been significantly reduced over the past decades owing to implementation of various health promotion initiatives and the availability of effective surgical procedures and drugs. Regardless of the aforementioned accomplishments, cardiovascular diseases remain a major public health issue in the developed countries including the United States, as well as worldwide. As noted in Section 1.5.2, with the exception of sub-Saharan Africa, cardiovascular diseases are the leading cause of death in the developing world. Globally, cardiovascular diseases (mainly ischemic heart disease and stroke) account for ~30% of all deaths, and

the figure will surely increase in both developing and developed countries as risk factors for the diseases (primarily dyslipidemia, hypertension, obesity, diabetes mellitus, physical inactivity, poor diet, and smoking) continue to increase. In this context, the leading causes of death in the world in 2030 are projected to be ischemic heart disease and stroke.

The globally increasing burden of cardiovascular diseases has prompted various international and national organizations including the WHO, the World Heart Federation, the AHA, as well as many government agencies to take measures to prevent and control these diseases. In this regard, the past several years have witnessed a number of international and national initiatives and activities toward cardiovascular health promotion. These include the UN 2011 High-Level General Assembly Meeting on NCDs, the World Heart Federation Call to Action to Prevent and Control Cardiovascular Diseases, the AHA 2020 Health Impact Goal, and the US Department of Health and Human Services (DHHS) *Million Hearts* initiative. A brief description of these initiatives helps understand the key issues and major measures in the prevention and control of cardiovascular diseases in the world. In essence, the key to prevention and control of the global cardiovascular pandemic is to take measures to control the major modifiable risk factors of cardiovascular diseases. However, enforcement and execution of the effective measures represent a great global challenge.

1.6.1   The UN High-Level Meeting and Tackling Cardiovascular Diseases at the Global Level

The UN high-level meeting (in September 2011) on NCDs and the declaration represents an unprecedented opportunity for those involved in the prevention and treatment of

TABLE 1.6 The impact of some of the major cardiovascular disease risk factors

Risk factor	Impact
Family history	Premature paternal history of a heart attack is associated with a 70% increase in the risk of a heart attack in women and a 100% increase in men [12, 13]. Sibling history of heart disease increases the odds of heart disease by ~50% [14]
Smoking/tobacco use	Cigarette smoking increases cardiovascular disease risk in a "dose"-dependent manner in both men and women. Women smokers have an additional 25% higher risk than men smokers [15]. Nonsmokers who are exposed to secondhand smoke at home or workplace increase their risk of developing cardiovascular diseases by 25–30%. Current smokers have a 2–4 times increased risk of stroke compared with nonsmokers or those who have quit for over 10 years [16]. Although smoking cessation is associated with weight gain, quitting smoking has a net cardiovascular benefit [17]. Hence, every smoker should be encouraged to quit smoking and given support to do so [18]
Physical inactivity	Insufficient physical activity can be defined as <5 times 30 min of moderate activity per week, or <3 times 20 min of vigorous activity per week, or equivalent. People who are insufficiently physically active have a 20–30% increased risk of all-cause mortality compared to those who engage in at least 30 min of moderate-intensity physical activity most days of the week. Physical inactivity is responsible for over 12% of the global burden of myocardial infarction after accounting for other cardiovascular disease risk factors, such as cigarette smoking, diabetes mellitus, hypertension, abdominal obesity, lipid profile, no alcohol intake[a], and psychosocial factors [19]
Unhealthy diet	Dietary habits affect multiple cardiovascular risk factors, including both established risk factors (hypertension, dyslipidemias, glucose levels, and obesity/weight gain) and novel risk factors (e.g., inflammation and endothelial cell function). An unhealthy dietary pattern characterized by higher intake of processed meat, red meat, refined grains, French fries, sweets/desserts, and salt increases cardiovascular mortality by more than 20%. On the other hand, a healthy dietary pattern characterized by higher intake of vegetables, fruits, fish, poultry, and whole grains and lower intake of sodium reduces cardiovascular mortality by >20% [20]
Overweight and obesity	Overweight and obesity increase the risk of developing cardiovascular diseases. Childhood obesity is also a predictor of an increased rate of death, owing primarily to an increased risk of cardiovascular disease. Overweight and obesity are associated with other cardiovascular risk factors, such as hypertension, dyslipidemias, and diabetes mellitus. Interestingly, a recent study reported that those who were overweight or obese as children but who became nonobese as adults had a cardiovascular risk profile that was similar to that of persons who were never obese [21]. This suggests that that childhood obesity does not permanently increase cardiovascular disease risk provided that childhood obesity is successfully treated. Given that atherosclerotic cardiovascular diseases are a major driver of healthcare expenditures in the United States as well as worldwide, the development of more effective strategies for treating and preventing childhood obesity is a cost-effective way of achieving a long-term reduction in global atherosclerotic cardiovascular diseases [22]
Dyslipidemias	Raised blood cholesterol increases the risk of heart disease and stroke. Globally, one third of ischemic heart disease is attributable to high blood cholesterol. For every 30 mg/dl change in low-density lipoprotein cholesterol (LDL-C), the relative risk for coronary artery disease is changed in proportion by about 30% [23]
Hypertension	Nearly 70% of people who have a first heart attack, 77% of those who have a first time stroke, and 74% of those who have congestive heart failure have hypertension [20]
Diabetes mellitus	At least 68% of people >65 years of age with diabetes mellitus die of some form of heart disease; 16% die of stroke. Heart disease death rates among adults with diabetes mellitus are two to four times higher than the rates for adults without diabetes mellitus [20]
Metabolic syndrome[b]	Metabolic syndrome increases the risk of developing cardiovascular diseases by 78–135% [24, 25]
Chronic kidney disease	Cardiovascular mortality is about twice as high in patients with stage 3 chronic kidney disease (estimated glomerular filtration rate 30–59 ml/min per 1.73 m^2) and three times higher at stage 4 (15–29 ml/min per 1.73 m^2) than that in individuals with normal kidney function. The adjusted risk of cardiovascular mortality is more than doubled at the upper end of the microalbuminuria category (30–299 mg/g), compared with the risk in individuals with normal albuminuria [11]

[a] Moderate consumption of alcohol is associated with a decreased risk of developing cardiovascular diseases due, at least partly, to its beneficial effects on high-density lipoprotein cholesterol (HDL-C). Moderation means an average of one to two drinks per day for men and one drink per day for women. A drink (15 ml pure ethanol) is one 12 oz. beer, 4 oz. of wine, 1.5 oz. of 80-proof spirits, or 1 oz. of 100-proof spirits. In contrast, overconsumption of alcohol increases the risk of developing cardiovascular and other diseases.

[b] The term metabolic syndrome (also known as syndrome X, insulin resistance syndrome) refers to a cluster of risk factors for cardiovascular diseases and type 2 diabetes mellitus. Several different definitions for metabolic syndrome are in use; in the United States, the National Cholesterol Education Program (NCEP) Adult Treatment Panel III (ATP III) definition and its two subsequent revisions have been used most commonly. By this definition, metabolic syndrome is diagnosed when three or more of the following five risk factors are present:

1. Fasting plasma glucose ≥100 mg/dl or undergoing drug treatment for elevated glucose
2. HDL-C <40 mg/dl in men or <50 mg/dl in women or undergoing drug treatment for reduced HDL-C
3. Triglycerides ≥150 mg/dl or undergoing drug treatment for elevated triglycerides
4. Waist circumference ≥102 cm in men or >88 cm in women
5. Blood pressure ≥130 mm Hg systolic or ≥85 mm Hg diastolic or undergoing drug treatment for hypertension or antihypertensive drug treatment in a patient with a history of hypertension

TABLE 1.7   The nine challenges and priorities identified by the World Heart Federation

1	Secure an outcomes statement at the United Nations high-level summit on noncommunicable diseases, taking place in September 2011
2	Enhance benefits of smoking cessation and implement affordable smoking cessation programs at the community level
3	Increase access to affordable, quality essential medicines for cardiovascular diseases in low- and middle-income countries
4	Close disparities in cardiovascular disease health
5	Increase the prevalence of workplace-wellness initiatives
6	Integrate cardiovascular disease prevention, detection, and treatment into primary healthcare settings
7	Increase the cardiovascular disease health workforce
8	Strengthen global, regional, and national partnerships
9	Improve data collection and monitoring of care provided to coronary heart disease patients

Source: The World Heart Federation.

cardiovascular diseases and all other concerned parties, including the member nations of the UN and their health ministries, to act and initiate priority programs and interventions that can avert the evolving pandemic of cardiovascular diseases and address the devastating worldwide effects of NCDs [26, 27]. The Lancet NCD Action Group and the NCD Alliance Group have proposed five high-priority interventions that include tobacco control, salt reduction, improved diets and physical activity, reduction in harmful alcohol intake, and access to essential drugs and technologies [28]. It is estimated that the implementation of these interventions (cost/person/year) would be $1.72 in China and $1.52 in India and is generally affordable worldwide. Salt reduction and tobacco control are the two population-directed interventions with the highest health impact. Full implementation of the Framework on Tobacco Control strategies would avert 5.5 million deaths over 10 years in 23 low- and middle-income countries. Reduction of salt intake by only 15% through mass media campaigns and industry reformulation of food products would avert 8.5 million deaths in 23 high-burden countries over 10 years.

1.6.2   The World Heart Federation Call to Action to Prevent and Control Cardiovascular Diseases

The World Heart Federation (http://www.world-heart-federation.org), representing 198 societies of cardiology and heart foundations worldwide, is acting with strong support and involvement from its member societies in developed nations, such as the AHA, the American College of Cardiology, and the European Society of Cardiology, whose expertise and experience with the prevention and treatment of cardiovascular diseases are substantial, to advocate for and assist with the implementation of effective strategies and initiatives that will lessen the global burden of cardiovascular diseases.

In the State of the Heart: Cardiovascular Disease Report (2011), the World Heart Federation and partner organizations call for a sustained worldwide effort to prevent and control cardiovascular diseases and encourage immediate endeavors by international organizations, national governments,

healthcare professionals, and, importantly, the general public. The report identifies nine cardiovascular challenges and priorities for the global community (Table 1.7) to act on to prevent and control the global pandemic of cardiovascular diseases.

1.6.3   The AHA 2010 Health Impact Goal, 2020 Health Impact Goal, and Ideal Cardiovascular Health

1.6.3.1   2010 Impact Goal   The AHA stated mission is "to build healthier lives, free of cardiovascular diseases and stroke." Consistent with that mission, the AHA set a strategic direction in 1998 to provide information and offer solutions for the prevention and treatment of cardiovascular diseases (including stroke) in people of all ages, with special emphasis on those at high risk. The identified goal was to reduce coronary heart disease, stroke, and risk by 25% by 2010, as measured by three key indicators [29, 30] listed below:

- A reduction by 25% in deaths due to coronary heart disease and stroke
- A reduction by 25% in prevalence of smoking, hypercholesterolemia, physical inactivity, and uncontrolled hypertension
- A zero growth rate of obesity and diabetic individuals

Despite the ambitious nature of the 2010 Impact Goal, by 2008, the reduction in the death rate due to coronary artery disease eclipsed 30.7%, and the reduction in the death rate due to stroke reached 29.4% [29]. What is even more provocative, however, is that at least 50% of the reduction in deaths due to coronary artery disease and stroke is attributable to a greater representation of preventive efforts, especially control of blood pressure, treatment of dyslipidemias, and a reduction in smoking. Yet, and ironically, the metric of a 25% risk reduction for smoking and physical inactivity and a zero growth rate for obesity and diabetes were not consistently met and have proven to be more difficult to achieve and will represent major challenges to the even more ambitious 2020 Impact Goal.

*1.6.3.2 2020 Impact Goal and Ideal Cardiovascular
Health* The strategic approaches and progress toward the
2010 Impact Goal pointed to innovations that are required
to define and implement new strategies for cardiovascular
risk prevention, improving cardiovascular health, and preventing disease events and deaths. Accordingly, the AHA
established its 2020 Impact Goal: "By 2020, to improve
the cardiovascular health of all Americans by 20% while
reducing deaths from cardiovascular diseases and stroke
by 20%" [29]. The 2020 Impact Goal for the first time
set an objective improvement in cardiovascular health as a
necessary component of the goal. This was driven by the
need for greater efforts in risk prevention and subsequently
a greater reduction in the burden of cardiovascular diseases
(including stroke).

Several key elements were addressed in the 2020 Impact
Goal, including (i) the definition of cardiovascular health;
(ii) the various attributes to cardiovascular health grouped into
two broad categories, that is, four health behaviors (related
to status of smoking, body mass index, physical activity, and
diet) and three health factors (related to status of total blood
cholesterol, blood pressure, and blood glucose); and (iii) an
algorithm that would not only define health status but also
would promote meaningful changes in cardiovascular health
status for both adults and children.

An aggregation of the above seven health behaviors and
health factors, now referred to as "The Simple 7," was
established and made available to the public on the AHA
website (http://www.heart.org/mylifecheck). "The Simple
7" consists of the following: (i) stop smoking, (ii) lose
weight, (iii) get active, (iv) eat better, (v) control cholesterol,
(vi) manage blood pressure, and (vii) reduce blood sugar.
The definition of ideal cardiovascular health is provided in
Table 1.8. To meet the complete definition of ideal cardiovascular health, an individual would need to meet the ideal
levels of all seven components.

TABLE 1.8 Definition of ideal cardiovascular health: "The Simple 7" [29]

Goal/metric	Ideal cardiovascular health definition
1. Current smoking	
Adults >20 years of age	Never or quit >12 months ago
Children 12–19 years of age	Never tried; never smoked whole cigarette
2. Body mass index	
Adults >20 years of age	<25 kg/m²
Children 12–19 years of age	<85th percentile
3. Physical activity[a]	
Adults >20 years of age	≥150 min/week moderate intensity or ≥75 min/week vigorous intensity or combination
Children 12–19 years of age	≥60 min of moderate- or vigorous-intensity activity every day
4. Healthy diet components[b]	
Adults >20 years of age	4–5 components
Children 12–19 years of age	4–5 components
5. Total cholesterol	
Adults >20 years of age	<200 mg/dl (untreated value)
Children 12–19 years of age	<170 mg/dl (untreated value)
6. Blood pressure	
Adults >20 years of age	<120/<80 mm Hg (untreated values)
Children 12–19 years of age	<90th percentile (untreated value)
7. Fasting plasma glucose	
Adults >20 years of age	<100 mg/dl (untreated value)
Children 12–19 years of age	<100 mg/dl (untreated value)

[a] Intensity of physical activity or exercise intensity can be defined in absolute or relative terms [31]. Absolute intensity reflects the rate of energy expenditure during exercise and is usually expressed in metabolic equivalents (METs), where one MET equals the resting metabolic rate of 3.5 ml O_2/kg body weight/minute. Relative intensity refers to the percent of aerobic power utilized during exercise and is expressed as percent of maximal heart rate or percent of VO_2max. Moderate-intensity activities are those performed at a relative intensity of 40–60% of VO_2max (or absolute intensity of 4–6 METs). Vigorous-intensity activities are those performed at a relative intensity of >60% of VO_2max (or absolute intensity of >6 METs). Moderate activities include such activities as brisk walking at 4 mph (5 METs), stationary cycling at 50 W (3 METs), or bicycling at 8 mph (5 METs). Vigorous activities include jogging 5 mph (8 METs), running 6 mph (10 METs), bicycling 12–13 mph (8 METs), and aerobic stepping 6–8 in. (8.5 METs). Easy or light activities are defined as <3 METs. Examples of light activities are walking 2 mph (2.5 METs), golfing with a cart, or playing piano.
[b] Healthy diet components include the following:
    1. Fruits and vegetables: ≥4.5 cups per day
    2. Fish: ≥two 3.5-oz servings per week (preferably oily fish)
    3. Fiber-rich whole grains (≥1.1 g of fiber per 10 g of carbohydrate): ≥three 1-oz-equivalent servings per day
    4. Sodium: ≤1500 mg per day
    5. Sugar-sweetened beverages: ≤450 kcal (36 oz) per week.

1.6.3.3 Extremely Low Prevalence of Ideal Cardiovascular Health in the Recent "Heart Strategies Concentrating on Risk Evaluation" Study   As stated earlier, the AHA 2020 Impact Goal focuses on promotion of health and control of risk rather than solely on prevention and treatment of specific cardiovascular diseases. As described earlier, this goal includes a new construct of cardiovascular health composed of four health behaviors and three health factors. The prevalence of the new AHA metrics that define ideal cardiovascular health has been addressed in a recent cohort of volunteers participating in a community-based health-screening survey, the Heart Strategies Concentrating on Risk Evaluation (Heart SCORE) [32]. The results of the Heart SCORE study are sobering and the penetration of poor health is alarming. Among the 1933 participants (mean age 59 years; 44% blacks; 66% women), only one person (0.1%) met all seven components of the AHA definition of ideal cardiovascular health. The indices of ideal health behaviors and ideal health factors were only met by 2 and 1.4% of participants, respectively. The large gap between the prevalence of ideal cardiovascular health and the AHA 2020 Impact Goal suggests that the attainment of the stated goals for the next decade may be much more challenging than originally conceived. Targeted efforts will be required at multiple levels (e.g., individual, social, environmental, policies and intervention, and access to quality healthcare) in order to ensure the achievement of this ambitious goal [32, 33].

1.6.4   US DHSS "Million Hearts" Initiative

1.6.4.1 What Is It?   As indicated in Section 1.4.2, at present, more than 14 million Americans are inflicted with a myocardial infarction or stroke with over 1.5 million new cases diagnosed each year. Cardiovascular diseases (notably myocardial infarction and stroke) are the leading cause of death in the United States and the largest cause of lower life expectancy among blacks. Related medical costs and productivity losses approach $450 billion annually, and inflation-adjusted direct medical costs are projected to triple over the next two decades if present trends continue. To reduce this burden, the US DHHS; other federal, state, and local government agencies; and a broad range of private-sector partners including the AHA lunched a "Million Hearts" initiative (http://millionhearts.hhs.gov) on September 13, 2011, to prevent one million heart attacks and strokes over the next 5 years by implementing proven, effective, inexpensive interventions [34]. Building on work already underway thanks to the Affordable Care Act, "Million Hearts" will help improve Americans' health and increase productivity.

1.6.4.2 Two Major Goals   "Million Hearts" is focused on two goals:

- Empowering Americans to make healthy choices such as preventing tobacco use and reducing sodium and trans-fat consumption. This can reduce the number of people who need medical treatment such as blood pressure or cholesterol medications to prevent heart attacks and strokes.
- Improving care for people who do need treatment by encouraging a targeted focus on **a**spirin for people at risk, **b**lood pressure control, **c**holesterol management, and **s**moking cessation ("ABCS")—which address the major risk factors for cardiovascular diseases and can help to prevent heart attacks and strokes.

1.6.4.3 Five Strategies   "Million Hearts" aims to improve heart disease and stroke prevention by:

- Improving access to effective care
- Improving the quality of care
- Focusing more clinical attention on heart attack and stroke prevention
- Increasing public awareness of how to lead a heart-healthy lifestyle
- Increasing the consistent use of high blood pressure and cholesterol medications

1.6.4.4 Six 2017-Specific Goals   By empowering Americans to make healthy choices and improving care, Million Hearts strives to achieve six specific goals by 2017, as listed in Table 1.9.

1.6.4.5 Perspectives   "Million Hearts" makes preventing heart attacks and strokes a top priority for the DHHS, its component agencies, and the broader healthcare system. "Million Hearts" targets improvements in both clinical preventive practice (e.g., reducing uncontrolled blood pressure and cholesterol, increasing aspirin use to prevent and reduce the severity of heart attacks and strokes) and community prevention (e.g., eliminating smoking and exposure to secondhand smoke, decreasing sodium and *trans*-fat intake in the population).

The "Million Hearts" initiative is aligned with the heart disease and stroke targets of the Healthy People 2020 (http://www.healthypeople.gov), which have been set on the basis of achieving a 10–20% improvement in cardiovascular prevention over a 10-year period. By using the diverse platforms of health reform to launch a rigorous effort to achieve successful clinical and community preventive interventions, the campaign is expected to produce a 10% reduction in the rate of acute cardiovascular events each year. There are ~2 million heart attacks and strokes in the United States annually. A 10% reduction would equate to 200,000 prevented cardiovascular events per year. If this rate is achieved and sustained over the 5-year campaign, "Million Hearts" will reach the goal of preventing one million heart attacks and strokes.

"Million Hearts" has the potential to make a significant contribution to the AHA 2020 Impact Goal to prevent 20%

TABLE 1.9 The six 2017-specific goals of the "Million Hearts" initiative

Indicator	Baseline	2017 goal
Aspirin use for people at high risk	47%	65%
Blood pressure control	46%	65%
Effective treatment of high cholesterol (LDL cholesterol)	33%	65%
Smoking prevalence	19%	17%
Sodium intake (average)	3.5 g/day	20% reduction
Artificial trans-fat[a] consumption (average)	1% of calories/day	50% reduction

Source: United States DHHS.

[a]Artificial trans fats (also known as trans-fatty acids or partially hydrogenated oils) are manufactured fats created during an industrial process that adds hydrogen to liquid vegetable oils to stabilize polyunsaturated fatty acids to prevent them from becoming rancid and to keep them solid at room temperature. Natural trans fats are uncommon, and small amounts of trans fats occur naturally in some meat and dairy products, including beef, lamb, and butterfat. The health effects of the naturally occurring trans fats are currently unknown. Unless otherwise specified, the term "trans fats" refers primarily to artificial trans fats. Artificial trans fats impose significant adverse health effects and increase the risk of developing coronary heart disease, stroke, and diabetes mellitus, among others [35, 36]. Many food companies and restaurants have eliminated trans fats over the past decade, in part because of the US Food and Drug Administration (FDA) nutrition label changes enacted in 2006. And some local governments, including New York City, already prohibit the use of trans fats in foods [37]. According to the FDA, these restrictions have helped reduce trans-fat intake among Americans from 4.6 g daily in 2003 to about 1 g a day in 2012. To further reduce the adverse health impact of trans fats, in November 2013, the FDA announced a plan to ban artificial trans fats in foods. According to the FDA estimate, the proposed ban on the use of trans fats could prevent an additional 20,000 heart attacks and 7,000 deaths annually in the United States.

of cardiovascular disease (including stroke) deaths by 2020 by preventing 10% of deaths resulting from myocardial infarction and stroke (which account for one third of all cardiovascular disease deaths) over 5 years. It would be expected that preventing one million heart attacks and strokes would reduce cardiovascular disease deaths even further by also reducing deaths from other cardiovascular disease causes.

To reach the AHA 2020 Impact Goal for cardiovascular health (to improve the cardiovascular health of all Americans by 20% while reducing deaths from cardiovascular diseases including stroke by 20% by 2020), ~4 million cardiovascular events must be prevented in 10 years. The AHA 2020 Impact Goal may be more aggressive than the "Million Hearts" goal, but the data show that with concerted public/private and cross-organizational effort to achieve a range of clinical and community preventive interventions, these goals are achievable [38].

1.7 CARDIOVASCULAR RISK PREDICTION AND EVIDENCE-BASED TREATMENTS

The decrease in the cardiovascular mortality rate in the United States and some developed nations has been hailed as one of the great achievements in public health. The reduction in the cardiovascular mortality rate started before powerful modern medical treatments entered mainstream medical practice, signifying that improvements in risk factors (primarily smoking, total cholesterol, and blood pressure) were key milestones to initiate decline. Nevertheless, analyses suggest that, more recently, both propitious changes in risk factors and the introduction of effective treatments have contributed greatly to reducing cardiovascular mortality rates in the United States and some developed countries, although the balance of these two contributors varies among countries

[39]. It has been suggested that further reductions in cardiovascular mortality can be realized if more aggressive targets for improving the distribution of risk factors in the population can be met and if compliance with evidence-based treatments can be increased [39, 40]. Hence, risk management based on cardiovascular risk prediction and evidence-based treatments, including drug therapies, are indispensible components of the armamentarium for combating cardiovascular diseases and reducing their global burden. This section introduces the cardiovascular risk prediction algorithms and briefly describes the status of evidence-based treatments, especially drug therapies, to serve as a prelude to the introduction to principles of cardiovascular pharmacology in Chapter 2.

1.7.1 Cardiovascular Risk Prediction

Primary prevention (Table 1.10) is paramount for the large number of individuals who are at high risk for developing cardiovascular diseases. Given limited resources, finding low-cost prevention strategies is a top priority in both developed and developing regions. Using prediction algorithms or risk scores to identify those at high risk to target specific behavioral or pharmacological interventions is a well-established primary prevention strategy and has proven to be cost-effective.

Various methods to predict cardiovascular risk use information on multiple risk factors, including age, gender, smoking, hypertension, diabetes mellitus, and blood lipids. Of these, perhaps the best known is the Framingham Risk Score [41], with simple online tools readily available to calculate 10-year risk of coronary artery disease-related adverse events (http://www.mecalc.com/framingham-cardiac-risk-score). The predicted risk has been used by certain clinical guidelines to make clinical decisions about treatments with drugs.

Although the original Framingham Risk Score estimates risk of coronary artery disease, the recently developed new Framingham Risk Score/Profile systems are also used to predict 10-year or 30-year risk of general cardiovascular diseases [42, 43] (Fig. 1.6). The simple online tools for calculating the general cardiovascular risks are available on the website (http://www.framinghamheartstudy.org/risk) of the Framingham Heart Study (Box 1.3). In addition to the Framingham Risk Score systems, several other algorithms have been reported for predicting the total cardiovascular risk in various human populations [44]. These include the Systemic Coronary Risk Estimation (SCORE) [45], the Joint British Societies (JBS3) risk calculator [46], and the most recently proposed Pooled Cohort Equations by the American College of Cardiology and American Heart Association (ACC/AHA) [47, 48].

TABLE 1.10   The three levels of prevention of cardiovascular diseases

Level of prevention	Description
Primary prevention	Primary prevention aims to prevent the disease from occurring. Primary prevention reduces both the incidence and prevalence of a disease. Health promotion targeting on avoiding the risk factors of cardiovascular diseases is an example of primary prevention
Secondary prevention	The goal of secondary prevention is to find and treat disease early. In many cases, the disease can be cured if detected early. For example, patients with early stage of coronary artery disease and hypercholesterolemia are treated with a statin drug to prevent the occurrence of myocardial infarction
Tertiary prevention	Tertiary prevention targets the person who already has symptoms of the disease with the goals to slow down the disease, prevent the disease complications, and improve quality of life. For example, patients following myocardial infarction are treated with inhibitors of the renin–angiotensin–aldosterone system to prevent or retard the development of congestive heart failure

(A)

Age	HDL-C	Total C	SBP not treated	SBP treated	Smoker	Diabetic	Points
	60+		<120				−2
	50−59						−1
30−34	45−49	<160	120−129	<120	No	No	0
	35−44	160−199	130−139				1
35−39	<35	200−239	140−159	120−129			2
		240−279	160+	130−139		Yes	3
		280+		140−159	Yes		4
40−44				160+			5
45−49							6
							7
50−54							8
							9
55−59							10
60−64							11
65−69							12
							13
70−74							14
75+							15

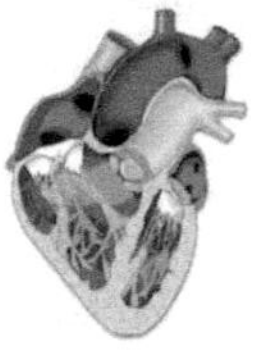

Points	Risk %	CV age	Points	Risk %	CV age
≤−3	<1		8	6.7	48
−2	1.1		9	7.9	51
−1	1.4	<30*	10	9.4	54
0	1.6	30	11	11.2	57
1	1.9	32	12	13.2	60
2	2.3	34	13	15.6	64
3	2.8	36	14	18.4	68
4	3.3	38	15	21.6	72
5	3.9	40	16	25.3	76
6	4.7	42	17	29.4	>80**
7	5.6	45	18+	>30	

FIGURE 1.6   General cardiovascular risk prediction in men (panel A) and women (panel B). This 10-year risk prediction model is based on Ref. [42]. The general cardiovascular diseases include coronary death, myocardial infarction, coronary insufficiency, angina, ischemic stroke, hemorrhagic stroke, transient ischemic attack, peripheral artery disease, and heart failure. As illustrated, six predictors are employed, including (i) age, (ii) diabetes, (iii) smoking, (iv) treated and untreated systolic blood pressure, (v) total cholesterol, and (vi) HDL cholesterol. Body mass index (BMI) can be used to replace lipids in a simpler model. In panel A, *, when the points are <0, the cardiovascular age is <30 years; **, when the points are ≥17, the cardiovascular age is >80 years. In panel B, *, when the points are <1, the cardiovascular age is <30 years.

(B)

Age	HDL-C	Total C	SBP not treated	SBP treated	Smoker	Diabetic	Points
			< 120				−3
	60+						−2
	50−59			< 120			−1
30−34	45−49	<160	120−129		No	No	0
	35−44	160−199	130−139				1
35−39	<35		140−149	120−129			2
		200−239		130−139	Yes		3
40−44		240−279	150−159			Yes	4
45−49		280+	160+	140−149			5
				150−159			6
50−54				160+			7
55−59							8
60−64							9
65−69							10
70−74							11
75+							12

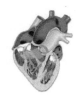

Points	Risk %	CV age		Points	Risk %	CV age
≤−2	<1			10	6.3	59
−1	1.0			11	7.3	64
0	1.2	<30*		12	8.6	68
1	1.5	31		13	10.0	73
2	1.7	34		14	11.7	79
3	2.0	36		15	13.7	>80
4	2.4	39		16	15.9	
5	2.8	42		17	18.5	
6	3.3	45		18	21.5	
7	3.9	48		19	24.8	
8	4.5	51		20	28.5	
9	5.3	55		21+	>30	

FIGURE 1.6 (*Continued*)

BOX 1.3   FRAMINGHAM HEART STUDY

The **Framingham Heart Study** is a long-term, ongoing cardiovascular study on residents of the town of Framingham, Massachusetts. The study under the direction of the National Heart, Lung, and Blood Institute (NHLBI; then known as the National Heart Institute) began in 1948 with 5209 adult subjects from Framingham and is now on its third generation of participants. The origins of the study are closely linked to the cardiovascular health of President Franklin D. Roosevelt and his premature death from hypertensive heart disease and stroke in 1945 at the age of 63 [49]. At the time, little was known about the general causes of heart disease and stroke, but the death rates for cardiovascular diseases had been increasing steadily since the beginning of the century and had become an American epidemic. The objective of the Framingham Heart Study was to identify the common factors or characteristics that contribute to cardiovascular diseases by following its development over a long period of time in a large group of participants who had not yet developed overt symptoms of cardiovascular diseases or suffered a heart attack or stroke. For more than 60 years, the Framingham Heart Study and the residents of Framingham, Massachusetts, have been synonymous with the remarkable advances made in the prevention of heart disease in the United States and throughout the world. More than 60 years of data collected from residents of Framingham have produced over 1000 scientific papers; identified major risk factors associated with heart disease, stroke, and other diseases (and of course the birth of the Framingham Risk Score, as described in Section 1.7.1); paved the way for researchers to undertake singular clinical trials based on Framingham findings; created a revolution in preventive medicine; and forever changed the way the medical community and general public view the genesis of disease. Having spent the past six decades looking at risk factors and lifestyle habits, researchers are now at the forefront of investigating how genes contribute to common metabolic disorders such as obesity, hypertension, diabetes, and even Alzheimer's disease [50, 51].

1.7.2   Evidence-Based Treatments

The entire medical profession strives to deliver care that is safe, timely, evidence based, efficient, equitable, and patient centered. Toward this goal, cardiology probably enjoys the greatest evidence base of any medical specialty [52]. The significant decline in the cardiovascular mortality in the United States and many other developed countries results from both prevention of the risk factor and effective treatment of the diseases. Remarkable progress has recently been made in the evidence-based treatments of cardiovascular diseases with pharmacological management set to assume an increasingly important role. Indeed, pharmacological agents not only play an important part in the treatment of the cardiovascular diseases but also in the management of risk factors of cardiovascular diseases. Pharmacological therapy has become an essential component of the armamentarium of evidence-based medicine for combating cardiovascular diseases. In fact, cardiovascular drugs are among the most widely used prescription drugs in the United States and other developed nations, as well as the developing world. Hence, a thorough understanding on the essentials and advances of cardiovascular pharmacology and therapeutics is of paramount importance for both prevention and treatment of cardiovascular diseases. Chapter 2 introduces the basic principles of pharmacology in general and cardiovascular pharmacology in particular to set a stage for the detailed discussion of the various topics in cardiovascular pharmacology and therapeutics throughout the rest of the book.

1.8   SUMMARY OF CHAPTER KEY POINTS

- The term cardiovascular diseases refers to a group of diseases involving the heart, blood vessels, or the sequelae of poor blood supply due to a decreased vascular supply and include diseases of the heart, vascular diseases of the brain, and diseases of other blood vessels.
- Cardiovascular diseases are classified in various ways, including schemes based on anatomical location and the involvement of atherosclerosis. The World Health Organization ICD-10 category represents the most comprehensive and most widely adopted classification scheme for human diseases, including cardiovascular diseases.
- Cardiovascular diseases are responsible for 48% of all global deaths due to noncommunicable diseases, and as such, they remain as a chief contributor to the global burden of disease.
- Globally, cardiovascular diseases (mainly ischemic heart disease and stroke) account for ~30% of all deaths, and the figure will surely increase in both developing and developed countries as risk factors for the diseases continue to increase.
- The development of cardiovascular diseases results from the complicated interactions between genes and environmental and dietary factors. The major cardio-

vascular disease risk factors include tobacco use, hypertension, high blood cholesterol, obesity, physical inactivity, and unhealthy diet, which have a high prevalence across the world and continue to increase.
- Prevention and control of cardiovascular diseases depend largely on how to effectively identify and manage the risk factors through population-based health promotion programs and initiatives at the community, national, and international levels.
- Risk management based on cardiovascular risk prediction and evidence-based treatments, including drug therapies, are indispensible components of the armamentarium for combating cardiovascular diseases and reducing their global burden.

1.9   SELF-ASSESSMENT QUESTIONS

1.9.1. According to a recent report of the Global Burden of Disease 2010 Study (Lancet 2012; 380:2095–128), there were 52.8 million deaths globally in 2010, of which 34.5 million deaths were caused by noncommunicable diseases, including cardiovascular diseases, cancer, and diabetes, among others. Cardiovascular diseases accounted for which of the following?
   A. ~5% of the total global deaths
   B. ~10% of the total global deaths
   C. ~15% of the total global deaths
   D. ~30% of the total global deaths
   E. ~50% of the total global deaths

1.9.2. According to the World Health Organization 2011 Global Atlas on Cardiovascular Disease Prevention and Control, which of the following is the number one cause of the death in the world?
   A. Congenital heart disease
   B. Heart failure
   C. Hypertension
   D. Ischemic heart disease
   E. Stroke

1.9.3. In a recent community-based Heart Strategies Concentrating on Risk Evaluation (Heart SCORE) study (Circulation 2011; 123:850–7), the AHA construct of cardiovascular health and the AHA ideal health behaviors index and ideal health factors index were evaluated among 1933 participants (mean age 59 years; 44% blacks; 66% women). Out of the 1933 participants in the Heart SCORE study, how many participants met all seven components of the AHA definition of ideal cardiovascular health?
   A. 1
   B. 15
   C. 29
   D. 246
   E. 512

1.9.4. The development of cardiovascular diseases results from the complicated interactions between genes and environmental and dietary factors. There are many risk factors for developing cardiovascular diseases, which are classified into modifiable and nonmodifiable factors. Which of the following is not considered a modifiable risk factor for cardiovascular diseases?

    A. A diet rich in sea salt
    B. Chronic kidney disease
    C. Diabetes mellitus
    D. Moderate consumption of alcohol
    E. Smoking of <20 cigarettes daily

1.9.5. Declines in stroke death rate in the United States now rank stroke as which of the following?

    A. 3rd leading cause of death in the nation
    B. 4th leading cause of death in the nation
    C. 5th leading cause of death in the nation
    D. 6th leading cause of death in the nation
    E. 7th leading cause of death in the nation

REFERENCES

1 Peterson, E.D. and J.M. Gaziano, Cardiology in 2011—amazing opportunities, huge challenges. *JAMA*, 2011. 306(19): p. 2158–9.

2 Roger, V.L., et al., Heart disease and stroke statistics—2012 update: a report from the American Heart Association. *Circulation*, 2012. 125(1): p. e2–e220.

3 Collaboration, G.B.D.C., GBD 2010 country results: a global public good. *Lancet*, 2013. 381(9871): p. 965–70.

4 Rosenbaum, L. and D. Lamas, Facing a "slow-motion disaster"—the UN meeting on noncommunicable diseases. *N Engl J Med*, 2011. 365(25): p. 2345–8.

5 Butler, D., UN targets top killers. *Nature*, 2011. 477(7364): p. 260–1.

6 Beaglehole, R., et al., UN High-Level Meeting on Non-Communicable Diseases: addressing four questions. *Lancet*, 2011. 378(9789): p. 449–55.

7 Lozano, R., et al., Global and regional mortality from 235 causes of death for 20 age groups in 1990 and 2010: a systematic analysis for the Global Burden of Disease Study 2010. *Lancet*, 2012. 380(9859): p. 2095–128.

8 Go, A.S., et al., Heart disease and stroke statistics—2014 update: a report from the American Heart Association. *Circulation*, 2014. 129(3): p. e28–e292.

9 Heidenreich, P.A., et al., Forecasting the future of cardiovascular disease in the United States: a policy statement from the American Heart Association. *Circulation*, 2011. 123(8): p. 933–44.

10 Jha, V., et al., Chronic kidney disease: global dimension and perspectives. *Lancet*, 2013. 382(9888): p. 260–72.

11 Gansevoort, R.T., et al., Chronic kidney disease and cardiovascular risk: epidemiology, mechanisms, and prevention. *Lancet*, 2013. 382(9889): p. 339–52.

12 Lloyd-Jones, D.M., et al., Parental cardiovascular disease as a risk factor for cardiovascular disease in middle-aged adults: a prospective study of parents and offspring. *JAMA*, 2004. 291(18): p. 2204–11.

13 Sesso, H.D., et al., Maternal and paternal history of myocardial infarction and risk of cardiovascular disease in men and women. *Circulation*, 2001. 104(4): p. 393–8.

14 Murabito, J.M., et al., Sibling cardiovascular disease as a risk factor for cardiovascular disease in middle-aged adults. *JAMA*, 2005. 294(24): p. 3117–23.

15 Huxley, R.R. and M. Woodward, Cigarette smoking as a risk factor for coronary heart disease in women compared with men: a systematic review and meta-analysis of prospective cohort studies. *Lancet*, 2011. 378(9799): p. 1297–305.

16 Goldstein, L.B., et al., Guidelines for the primary prevention of stroke: a guideline for healthcare professionals from the American Heart Association/American Stroke Association. *Stroke*, 2011. 42(2): p. 517–84.

17 Clair, C., et al., Association of smoking cessation and weight change with cardiovascular disease among adults with and without diabetes. *JAMA*, 2013. 309(10): p. 1014–21.

18 Fiore, M.C. and T.B. Baker, Should clinicians encourage smoking cessation for every patient who smokes? *JAMA*, 2013. 309(10): p. 1032–3.

19 Yusuf, S., et al., Effect of potentially modifiable risk factors associated with myocardial infarction in 52 countries (the INTERHEART study): case-control study. *Lancet*, 2004. 364(9438): p. 937–52.

20 Go, A.S., et al., Heart disease and stroke statistics—2013 update: a report from the American Heart Association. *Circulation*, 2013. 127(1): p. e6–e245.

21 Juonala, M., et al., Childhood adiposity, adult adiposity, and cardiovascular risk factors. *N Engl J Med*, 2011. 365(20): p. 1876–85.

22 Rocchini, A.P., Childhood obesity and coronary heart disease. *N Engl J Med*, 2011. 365(20): p. 1927–9.

23 Grundy, S.M., et al., Implications of recent clinical trials for the National Cholesterol Education Program Adult Treatment Panel III guidelines. *Circulation*, 2004. 110(2): p. 227–39.

24 Gami, A.S., et al., Metabolic syndrome and risk of incident cardiovascular events and death: a systematic review and meta-analysis of longitudinal studies. *J Am Coll Cardiol*, 2007. 49(4): p. 403–14.

25 Mottillo, S., et al., The metabolic syndrome and cardiovascular risk a systematic review and meta-analysis. *J Am Coll Cardiol*, 2010. 56(14): p. 1113–32.

26 Smith, S.C., Jr., Reducing the global burden of ischemic heart disease and stroke: a challenge for the cardiovascular community and the United Nations. *Circulation*, 2011. 124(3): p. 278–9.

27 Stewart, S., Tackling heart disease at the global level: implications of the United Nations' statement on the prevention and control of noncommunicable disease. *Circ Cardiovasc Qual Outcomes*, 2011. 4(6): p. 667–9.

28 Beaglehole, R., et al., Priority actions for the non-communicable disease crisis. *Lancet*, 2011. 377(9775): p. 1438–47.

29 Lloyd-Jones, D.M., et al., Defining and setting national goals for cardiovascular health promotion and disease reduction: the

American Heart Association's strategic Impact Goal through 2020 and beyond. *Circulation*, 2010. 121(4): p. 586–613.

30 Schwamm, L., et al., Translating evidence into practice: a decade of efforts by the American Heart Association/American Stroke Association to reduce death and disability due to stroke: a presidential advisory from the American Heart Association/ American Stroke Association. *Stroke*, 2010. 41(5): p. 1051–65.

31 Thompson, P.D., et al., Exercise and physical activity in the prevention and treatment of atherosclerotic cardiovascular disease: a statement from the Council on Clinical Cardiology (Subcommittee on Exercise, Rehabilitation, and Prevention) and the Council on Nutrition, Physical Activity, and Metabolism (Subcommittee on Physical Activity). *Circulation*, 2003. 107(24): p. 3109–16.

32 Bambs, C., et al., Low prevalence of "ideal cardiovascular health" in a community-based population: the heart strategies concentrating on risk evaluation (Heart SCORE) study. *Circulation*, 2011. 123(8): p. 850–7.

33 Yancy, C.W., Is ideal cardiovascular health attainable? *Circulation*, 2011. 123(8): p. 835–7.

34 Frieden, T.R. and D.M. Berwick, The "Million Hearts" initiative—preventing heart attacks and strokes. *N Engl J Med*, 2011. 365(13): p. e27.

35 Mozaffarian, D., et al., Trans fatty acids and cardiovascular disease. *N Engl J Med*, 2006. 354(15): p. 1601–13.

36 Micha, R. and D. Mozaffarian, Trans fatty acids: effects on metabolic syndrome, heart disease and diabetes. *Nat Rev Endocrinol*, 2009. 5(6): p. 335–44.

37 Mello, M.M., Nutrition: ejecting trans fat from New York City restaurants. *Nat Rev Endocrinol*, 2012. 8(11): p. 633–5.

38 Tomaselli, G.F., et al., The American Heart Association and the Million Hearts Initiative: a presidential advisory from the American Heart Association. *Circulation*, 2011. 124(16): p. 1795–9.

39 Ford, E.S. and S. Capewell, Proportion of the decline in cardiovascular mortality disease due to prevention versus treatment: public health versus clinical care. *Annu Rev Public Health*, 2011. 32: p. 5–22.

40 Kovacic, J.C. and V. Fuster, From treating complex coronary artery disease to promoting cardiovascular health: therapeutic transitions and challenges, 2010–2020. *Clin Pharmacol Ther*, 2011. 90(4): p. 509–18.

41 Wilson, P.W., et al., Prediction of coronary heart disease using risk factor categories. *Circulation*, 1998. 97(18): p. 1837–47.

42 D'Agostino, R.B., Sr., et al., General cardiovascular risk profile for use in primary care: the Framingham Heart Study. *Circulation*, 2008. 117(6): p. 743–53.

43 Pencina, M.J., et al., Predicting the 30-year risk of cardiovascular disease: the framingham heart study. *Circulation*, 2009. 119(24): p. 3078–84.

44 Cooney, M.T., et al., Total cardiovascular disease risk assessment: a review. *Curr Opin Cardiol*, 2011. 26(5): p. 429–37.

45 Conroy, R.M., et al., Estimation of ten-year risk of fatal cardiovascular disease in Europe: the SCORE project. *Eur Heart J*, 2003. 24(11): p. 987–1003.

46 Board, J.B.S., Joint British Societies' consensus recommendations for the prevention of cardiovascular disease (JBS3). *Heart*, 2014. 100 Suppl 2: p. ii1–ii67.

47 Goff, D.C., Jr., et al., 2013 ACC/AHA guideline on the assessment of cardiovascular risk: a report of the American College of Cardiology/American Heart Association Task Force on Practice Guidelines. *Circulation*, 2014. 129(25): S49–73.

48 Muntner, P., et al., Validation of the atherosclerotic cardiovascular disease Pooled Cohort risk equations. *JAMA*, 2014. 311(14): p. 1406–15.

49 Mahmood, S.S., et al., The Framingham Heart Study and the epidemiology of cardiovascular disease: a historical perspective. *Lancet*, 2014. 383(9921): p. 999–1008.

50 Ehret, G.B., The Contribution of the Framingham Heart Study to Gene Identification for Cardiovascular Risk Factors and Coronary Heart Disease. *Glob Heart*, 2013. 8(1): p. 59–65.

51 Weinstein, G., et al., Risk estimations, risk factors, and genetic variants associated with Alzheimer's disease in selected publications from the Framingham Heart Study. *J Alzheimers Dis*, 2013. 33 Suppl 1: p. S439–45.

52 James, S.K. and J.A. Spertus, Evidence-based treatments for STEMI: are we doing enough? *Lancet*, 2013. 382(9892): p 576–9.

2

INTRODUCTION TO PRINCIPLES OF PHARMACOLOGY

2.1 OVERVIEW

In the foreword to the first edition of *Principles of Pharmacology: The Pathophysiologic Basis of Drug Therapy*, now a widely acclaimed text in its 3rd edition in medical pharmacology, Eugene Braunwald, a world-renowned cardiologist at Harvard Medical School, stated that "Almost every practicing physician prescribes drugs; most write many prescriptions every day. The learning of pharmacology, the science that deals with the action and use of drugs, is among the most important steps in becoming a physician. Rather than reflexly ordering a medication to treat a specific symptom or disease, modern therapeutics requires an understanding of the underlying mechanism of action of a pharmacological agent, how it influences and is influenced by the disease for which it is prescribed, and its capacity for causing both beneficial and harmful clinical effects." Indeed, pharmacology, as a biomedical discipline, focuses on the complex interactions between the drugs and living systems and has contributed substantially to the advances in the management of human diseases in general and cardiovascular diseases in particular. The discovery of new drugs and the rapid development of cardiovascular sciences in the past five decades have helped cardiovascular pharmacology and therapeutics become a major medical subspecialty that plays a central part in cardiovascular medicine. Knowledge in cardiovascular pharmacology and therapeutics is not only essential for the evidence-based treatment of patients with cardiovascular diseases but also important for cardiovascular disease prevention and promotion of public health. In this context, cardiovascular diseases are chief contributors to global burden of disease (Chapter 1). To lay a basis for understanding cardiovascular pharmacology and therapeutics, this chapter examines the general principles of pharmacology and therapeutics, beginning with defining pharmacology and related terms, followed by introducing the pharmacological paradigm as well as drug development and regulation. The chapter ends with a brief survey of the new developments and challenges of cardiovascular pharmacology and therapeutics.

2.2 DEFINITIONS AND HISTORY

2.2.1 What Is Pharmacology?

The term pharmacology is derived from the Greek words pharmakon (meaning a drug or medicine) and logos (meaning the truth about or a rational discussion). In general terms, pharmacology is the science dealing with drug action (including both beneficial and harmful effects) on biological systems. In its entirety, pharmacology embraces knowledge of sources, chemical properties, biological effects, and therapeutic uses of drugs. Pharmacology is a science that is fundamental not only to human clinical medicine but also to pharmacy, nursing, and dentistry, as well as veterinary medicine. Pharmacology taught in medical schools can be defined as a biomedical discipline that deals with the action and use of drugs in the diagnosis, treatment, or prevention of human diseases.

Cardiovascular Diseases: From Molecular Pharmacology to Evidence-Based Therapeutics, First Edition. Y. Robert Li.
© 2015 John Wiley & Sons, Inc. Published 2015 by John Wiley & Sons, Inc.

2.2.2 Definitions of Related Terms

Although medical pharmacology has a focus on mechanisms of drug actions, it also emphasizes the clinical use of the drugs, better known as clinical pharmacology. Hence, medical pharmacology may be considered as a discipline that bridges the basic medical science and clinical medicine. The distinction between medical pharmacology and other related disciplines, such as pharmacotherapeutics, has becoming less obvious. Table 2.1 lists several different terms related to medical pharmacology.

2.2.3 A Brief History of Pharmacology

Historically, the roots of pharmacology go back to the ancient civilizations that used plants and plant extracts both in healing and as poisons. The accumulated total of this empirical knowledge, acquired by mankind through the ages, provided a foundation for the evolution of scientific pharmacology as it exists today. The well-known discovery of the beneficial effects of foxglove extracts for treating heart disease, the use of the bark of the willow and cinchona trees in treating fever, and the effectiveness of extracts of the poppy in the treatment of dysenteries are outstanding examples of such knowledge that have resulted in important advances in pharmacology.

The rise of organic chemistry in the last half of the nineteenth century, together with the development of physiology and, later, biochemistry, allowed empiricism to be discarded in favor of a rational approach, giving birth to modern pharmacology. The first published classic text, "Outline of Pharmacology," written by Oswald Schmiedeberg in 1878, set the momentum for today's pharmacology advancement throughout the world. Table 2.2 summarizes the major historic figures and events in the early development of modern pharmacology. The photos of the major historic figures are shown in Figure 2.1. Table 2.3 lists the Nobel Prize-winning research that has shaped modern pharmacology.

TABLE 2.1   Medical pharmacology and related terms

Term	Definition
Medical pharmacology	Medical pharmacology is the science that deals with the action and use of drugs in the diagnosis, treatment, or prevention of human diseases. Medical pharmacology includes basic pharmacology and clinical pharmacology. Basic pharmacology emphasizes the basic science principles, such as pharmacokinetics and pharmacodynamics, whereas clinical pharmacology is underpinned by the basic science of pharmacology with added focus on the application of pharmacological principles and methods in the clinical management of human diseases
Drug	The term drug is often defined in two ways: (i) the US Federal Food, Drug, and Cosmetic Act (FD&C Act) defines drugs, in part, by their intended use as "articles intended for use in the diagnosis, cure, mitigation, treatment, or prevention of disease" and "articles (other than food) intended to affect the structure or any function of the body of man or other animals" [FD&C Act, sec. 201(g)(1)]; and (ii) in pharmacology, a drug is defined as a natural product, chemical substance, or pharmaceutical preparation intended for administration to a human or animal to diagnose, treat, or prevent a disease
Pharmacotherapeutics	Pharmacotherapeutics is the medical science concerned with the use of drugs in the treatment of diseases, and its essence is clinical use of drugs. Pharmacology provides a rational basis for pharmacotherapeutics by explaining the mechanisms and effects of drugs on the body and the relationship between dose and drug response. Hence, pharmacotherapeutics and pharmacology are closely related and often intertwined. In this context, clinical pharmacology is even more closely intertwined with pharmacotherapeutics. The term pharmacotherapy refers to treatment of disease through the use of drugs
Pharmacy	Pharmacy is the science and profession concerned with the preparation, storage, dispensing, and proper use of drug products
Pharmaceutics	Pharmaceutics is concerned with the formulation and chemical properties of pharmaceutical products, such as tablets, liquid solutions and suspensions, and aerosols. Do not confuse pharmaceutics with pharmacotherapeutics
Pharmacognosy	Pharmacognosy is the study of drugs isolated from natural sources, including plants, microbes, animal tissues, and minerals
Nutraceutical	Nutraceutical is the term used to describe any substance that is considered a food or part of a food, including nutritional supplements that allege to provide health benefits
Functional food	Functional food refers to food that contains physiologically active compounds that provide health benefits beyond their nutrient contributions. The terms nutraceutical and functional food are often used interchangeably
Phytochemical	The term phytochemical refers to any nonnutrient compound in plant-derived foods that possesses biological activity in the body

TABLE 2.2   List of historic figures and events in the early development of modern pharmacology

Figure or event	Major contribution or impact
Rudolf Buchheim (1820–1876) Pioneer of experimental pharmacology	In 1847, as professor of pharmacology at the University of Dorpat in Estonia (then part of Russia), Buchheim established the first laboratory devoted to experimental pharmacology in the basement of his home due to lack of outside funding
Oswald Schmiedeberg (1838–1921) The founder of modern pharmacology	Schmiedeberg, a student of Buchheim, set up an institute of pharmacology in 1872 in Strasbourg, France (Germany at that time), which became a mecca for students who were interested in pharmacological problems. In 1878, he published a classic text, "Outline of Pharmacology." He and his students, including T. Frazer (1841–1921, Scotland), J. Langley (1852–1925, England), and P. Ehrlich (1854–1915, Germany), helped establish the high reputation of pharmacology by introducing fundamental concepts such as structure–activity relationship, drug receptors, and selective toxicity
John J. Abel (1857–1938) The father of American pharmacology	Abel, a student of Schmiedeberg, founded the first pharmacology department in the United States at the University of Michigan in 1891. Abel and 18 of his colleagues founded the American Society for Pharmacology and Experimental Therapeutics in 1908
World War II (1939–1945)	The World War II was the impetus for accelerated research in pharmacology and introduced strong analytical and synthetic chemical approaches

| Rudolph Buchheim
(1820–1876) | Oswald Schmiedeberg
(1838–1921) | Paul Ehrlich
(1854–1915) | John J. Abel
(1857–1938) |

FIGURE 2.1   Photos of major historic figures whose pioneering work has shaped today's pharmacology (From the National Library of Medicine).

TABLE 2.3   List of Nobel Prize-winning research that has shaped modern pharmacology

Nobel laureate	Year of Nobel Prize	Nobel-winning discoveries (and other notable contributions)
Martin Karplus (1930–) Michael Levitt (1947–) Arieh Warshel (1940–)	2013, chemistry	Development of multiscale models for complex chemical systems. Such models of simulating the behavior of molecules at various scales are crucial for modern drug design
Paul Greengard (1925–)	2000, physiology or medicine	Signal transduction in the nervous system
Robert F. Furchgott (1916–2009) Louis J. Ignarro (1941–) Ferid Murad (1938–)	1998, physiology or medicine	Nitric oxide as a signaling molecule in the cardiovascular system
Alfred G. Gilman (1941–) Martin Rodbell (1925–1998)	1994, physiology or medicine	G proteins and the role of these proteins in signal transduction in cells
Edwin G. Krebs (1918–2009)	1992, physiology or medicine	Reversible protein phosphorylation as a biological regulatory mechanism

(continued)

TABLE 2.3   (*Continued*)

Nobel laureate	Year of Nobel Prize	Nobel-winning discoveries (and other notable contributions)
Sir James W. Black (1924–2010) Gertrude B. Elion (1918–1999) George H. Hitchings (1905–1998)	1988, physiology or medicine	Important principles for drug treatment; anticancer agents that block DNA synthesis; Black developed the first beta-blocker, propranolol
Sune K. Bergström (1916–2004) Bengt I. Samuelsson (1934–) John R. Vane (1927–2004)	1982, physiology or medicine	Discoveries concerning prostaglandins and related biologically active substances
Earl W. Sutherland, Jr. (1915–1974)	1971, physiology or medicine	Mechanisms of the action of hormones with regard to cAMP involvement
Julius Axelrod (1912–2004) Sir Bernard Katz (1911–2003) Ulf von Euler (1905–1983)	1970, physiology or medicine	Humoral transmitters in the nerve terminals and the mechanism for their storage, release, and inactivation
Daniel Bovet (1907–1992)	1957, physiology or medicine	Discoveries relating to synthetic compounds (antihistamine agents) that inhibit the action of certain body substances and especially their action on the vascular system and the skeletal muscles
Philip S. Hench (1896–1965) Edward C. Kendall (1886–1972) Tadeus Reichstein (1897–1996)	1950, physiology or medicine	Discoveries relating to the hormones of the adrenal cortex and their structure and biological effects
Linus C. Pauling (1901–1994)	1954, chemistry	Nature of the chemical bond and its application to the elucidation of the structure of complex substances. Pauling also won a Nobel Prize in Peace. He was also known for his studies on high-dose vitamin C for treating cancer
Herbert S. Gasser (1888–1963)	1944, physiology or medicine	The highly differentiated functions of single nerve fibers
Corneille J.F. Heymans (1892–1968)	1938, physiology or medicine	Role played by the sinus and aortic mechanisms in the regulation of respiration
Sir Henry H. Dale (1875–1968) Otto Loewi (1873–1961)	1936, physiology or medicine	Chemical transmission of nerve impulses
Frederick G. Banting (1891–1941)	1923, physiology or medicine	Discovery of insulin
Ernst B. Chain (1906–1979) Sir Alexander Fleming (1881–1955) Sir Howard Florey (1898–1968)	1945, physiology or medicine	Discovery of penicillin and its curative effect in various infectious diseases
Paul Ehrlich (1854–1915)	1908, physiology or medicine	Work on immunity. Ehrlich was also famous for the "magic bullet" concept in treating infectious diseases. He developed the first synthetic antimicrobial agent

2.3 PHARMACOLOGICAL PARADIGM: THE CENTRAL DOGMA IN PHARMACOLOGY

The central dogma is the key to understanding a medical discipline. Figure 2.2 depicts the pharmacological paradigm, the central dogma in pharmacology. This paradigm defines the broad scope of pharmacology, starting from dosing of drugs to pharmacokinetics and pharmacodynamics and finally to clinical responses to drug therapy. As shown, pharmacokinetics precedes pharmacodynamics, which in turn determines the clinical responses. The clinical responses to drug therapy may include the desired responses and the untoward outcomes, which are better known as drug adverse effects or toxicity. This section introduces several key aspects of pharmacology and therapeutics as dictated by the central dogma of pharmacology.

2.3.1 Drug Names, Sources, Preparations, and Administration

2.3.1.1 Drug Names   A particular drug may have up to three different names: chemical name, generic name, and trade name. Trade name is also known as brand name or proprietary name. These different names may be confusing when you first learn pharmacology and pharmacotherapeutics. It is important to know the different names of a drug so that the wrong drug is not prescribed to a patient. Table 2.4 lists the three types of drug names and summarizes their key characteristics.

2.3.1.2 Drug Sources   Drugs can be obtained from natural sources (e.g., plants, microbes, animal tissues, minerals) or synthesized in the laboratories. While in the past,

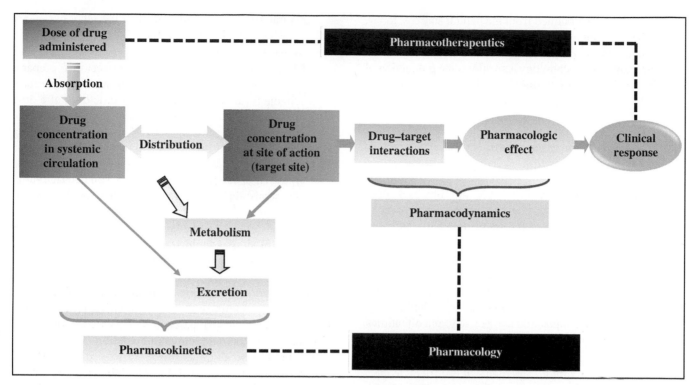

FIGURE 2.2   Pharmacological process. This is considered the central dogma in pharmacology. This paradigm defines the broad scope of pharmacology, starting from dosing of drugs to pharmacokinetics and pharmacodynamics and finally to clinical responses to drug therapy. This paradigm also illustrates the intimate relationship between pharmacology and pharmacotherapeutics.

TABLE 2.4   Drug names

Type of drug name	Characteristics	Example
Chemical name	• It specifies the chemical structure of the drug • It uses standard chemical nomenclature • It is often long and complex	$[R\text{-}(R^*,\,R^*)]$-2-(4-Fluorophenyl)-β, δ-dihydroxy-5-(1-methylethyl)-3-phenyl-4-[(phenylamino) carbonyl]-1H-pyrrole-1-heptanoic acid
Generic name, also known as nonproprietary name	• It is the one recognized internationally • It is the type of drug name most suitable for use by healthcare professionals • It provides some indication of the class to which a particular drug belongs • Using generic drug names is less likely to result in prescribing errors and can give the pharmacist the option of substituting a cheaper generic version	Atorvastatin
Trade name, also known as brand name or proprietary name	• It is the patented exclusive property of the drug manufacturer • It is often designed to be shorter than generic name • It is often not helpful in identifying the pharmacological action or class of drug • Trade names can sometimes be similar yet refer to drugs with entirely different pharmacodynamic actions, increasing the hazard of prescribing error • Many drugs are marketed under two or more brand names, especially after the manufacturer loses patent exclusivity	Lipitor

drugs were usually obtained from the natural sources, today, due to the development of modern synthetic chemistry, drug compounds are mainly synthesized in the laboratories. Recombinant DNA technology also allows the production of protein drugs, such as hormones and antibodies, in large quantity and high purity. Table 2.5 summarizes the different sources of drugs.

2.3.1.3 Drug Preparations
Drug preparations include three types, as outlined below:

- Crude drug preparations: They are obtained from natural sources by drying or pulverizing a plant or animal tissue. Most of the traditional Chinese medicines are crude drug preparations, which contain both active and inactive ingredients. Recently, the green tea extract (Veregen) has been approved by the US Food and Drug Administration (FDA) for treating genital warts. Crude drug preparations, such as herbal remedies, vary in composition and potency, and oftentimes, the exact active component(s) is not known.
- Pure drug compounds: Because crude drug preparations contain multiple ingredients, it is difficult to identify and quantify their pharmacological effects. In this regard, pure drug compounds either isolated from

natural sources or synthesized in the laboratories are obtained to make pharmaceutical preparations suitable for use in humans.

- Pharmaceutical preparations: Pharmaceutical preparations, also known as dosage form, are drug products suitable for administration of a specific dose of a drug to a patient via a particular route of administration. Dosage forms are a mixture of the pure active drug components and nondrug components. Depending on the method of administration, they come in several types. These are liquid dosage form, solid dosage form, and semisolid dosage form, suitable for the various routes of drug administration. Table 2.6 lists the various dosage forms for oral administration, the most common route of drug administration. The dosage forms for other routes of administration are given in Table 2.7.

2.3.1.4 Route of Drug Administration
The major routes of drug administration are classified into five categories: (i) enteral, (ii) parenteral, (iii) transdermal, (iv) inhalational, and (v) topical administration. The definition and a brief description of these various routes of drug administration are given in Table 2.8. Table 2.9 summarizes the different characteristics of some common routes of drug administration.

TABLE 2.5 Drug sources

Drug source	Description and example
Natural sources	• Plants (e.g., digitalis from foxglove) • Microbes (e.g., penicillin from *Penicillium*) • Animal tissues (e.g., insulin from pancreatic tissue) • Minerals (e.g., lithium)
Synthetic drugs	• Small molecules (e.g., aspirin) • Protein therapeutics (e.g., erythropoietin, insulin) • Antisense oligonucleotides and small interference RNAs (siRNAs) (e.g., mipomersen, which is a second-generation antisense oligonucleotide inhibitor of apolipoprotein B-100)

TABLE 2.6 Dosage forms for oral administration (oral medication forms)

Oral dosage form	Description
Tablets	Tablets are dried, and powdered drugs compressed into small shapes that can be swallowed whole
Capsules	Capsules are gelatin containers that hold powder or liquid medicine. Timed-release or sustained-release capsules contain granulates that dissolve at different rates, providing slow and constant release of medications
Suspensions	Suspensions are liquids with solid, insoluble drug particles dispersed throughout. These solid particles tend to settle out in layers, so the medication must be shaken before pouring
Elixirs	Elixirs are liquids made up of drugs dissolved in alcohol and water that may have coloring and flavoring agents added. The alcohol makes the drug more dissolvable than water alone
Emulsions	Emulsions are solutions that have small droplets of water and medication dispersed in oil or oil and medication dispersed in water
Lozenges	Lozenges are medications mixed with hard sugar base to produce small, hard preparation of various sizes and shapes. The drug is released slowly when the lozenge is sucked
Syrups	Syrups are liquids with a high sugar content designed to disguise the bitter taste of a drug. They are usually used for pediatric patients

TABLE 2.7   Dosage forms for nonoral administration

Dosage form	Route of administration	Description
Sterile solutions and suspensions	Parenteral injections	Many drugs are formulated as sterile powders for reconstitution with sterile liquids at the time the drug is to be injected. This is due to the fact that the drug is not stable for long periods of time in solution
Skin patches	Topical	Transdermal skin patches are drug preparations in which the drug is slowly released from the patch for absorption through the skin into systemic circulation
Aerosols	Inhalation	Aerosols are a type of drug preparation administered via inhalation through the nose or mouth
Ointments, creams, lotions	Topical	Ointments and creams are semisolid preparations intended for topical application of a drug to the skin or mucous membranes. Lotion is a low- to medium-viscosity topical preparation intended for application to unbroken skin
Suppositories	Topical (rectal, virginal, urethral)	Suppositories are a type of drug preparation in which the drug is incorporated into a solid base that is inserted into the rectum (rectal suppository), vagina (vaginal suppository), or urethra (urethral suppository), where it dissolves at body temperature. Suppositories are used to deliver both systemically acting and locally acting drugs

TABLE 2.8   Classification of routes of drug administrations

Route of drug administration	Description
Enteral administration	• This term refers to the routes of drug administration that involve absorption of the drug through the gastrointestinal tract • Enteral administration includes sublingual, buccal, oral, and rectal administration, with oral administration being the most common route of drug administration
Parenteral administration	• This term refers to drug administration via a route other than the gastrointestinal tract • The common parenteral administration includes intramuscular, intravenous, and subcutaneous administration • Other less common parenteral routes include intrathecal, epidural, intra-articular, and intradermal administration
Transdermal administration	• This term refers to application of drugs to the skin for absorption into the systemic circulation • Skin patches are commonly used for transdermal administration, and ointments are less commonly used
Inhalational administration	• This term refers to medication being carried through the mouth or nose by inhalation down into the respiratory tract through the use of aerosol nebulizers, metered-dose inhalers, or other apparatus • Inhalational administration can be used to produce either a localized or systemic effect depending on the type of drugs
Topical administration	• This term refers to application of medication to body surfaces such as the skin or mucous membranes such as the vagina, anus, throat, eyes, and ears • It is intended to produce a local effect • This is in contrast to transdermal administration, which is intended to produce a systemic effect

TABLE 2.9   Characteristics of some common routes of drug administration

Route	Advantage	Disadvantage
Oral (po)	• Simple, convenient, and inexpensive • Painless • No risk of infection	• Drugs could be inactivated in the harsh environment of the gut • The drug is subject to first-pass effects • The drug may cause irritation to the gut • It causes slow delivery of the drug to the target site, leading to slow onset of drug action
Intramuscular (im)	• Intermediate onset of action • Suitable for administering oil-based drugs	• Painful • It may cause intramuscular bleeding
Intravenous (iv)	• Rapid onset of action • It allows for rapid titration of dosage	• It may pose more risks for adverse reactions • Skilled personnel are required
Subcutaneous (sc)	• It is suitable for administering suspensions or pellets • It causes slow onset of action	• Only for small volumes of drugs • It cannot be used for drugs that irritate cutaneous tissues

2.3.2 Pharmacokinetics

Following administration, the drug molecules need to reach the target tissue to exert pharmacological effects that lead to clinical responses. The processes of absorption, distribution, metabolism, and excretion collectively constitute the core of pharmacokinetics and determine the concentration of drug delivered to the molecular target in the target tissue. Factors that affect any of the above four processes will inevitably influence drug concentration in the target tissue and hence the pharmacological effects and clinical responses. Pharmacokinetics can be viewed as what the body does to a drug to influence the concentration of the drug at the site of action.

2.3.3 Pharmacodynamics

Once a drug accesses a molecular target of action, it alters the function of that molecular target, with the ultimate result of a drug effect that can be perceived by the patient or healthcare provider. The term pharmacodynamics is used to describe the effects of a drug on the body and the underlying molecular mechanisms. Most drugs interact with specific proteins, such as growth factor receptors or enzymes, to cause pharmacological effects that benefit the patient. These specific proteins or enzymes are termed drug targets or drug receptors.

2.3.4 Drug Toxicity

2.3.4.1 Drug Toxicity and General Mechanistic Aspects
All drugs have the potential to cause unintended consequences, termed toxic effects, side effects, or adverse effects. Such toxic effects are determined by the mechanisms of drug action, the size of the drug dose, the concomitant use of other medications or dietary supplements, as well as the status of the patient's physiology and pathophysiology. In this regard, the genetic variations of the individuals could markedly influence both the desired pharmacological effects and the unwanted toxicity of the medications (Section 2.3.5).

It has been recognized that drug toxicity may result from any of the four general processes:

1. On-target toxic effects, resulting from the drug binding to the intended molecular targets at inappropriate high concentrations or wrong tissues
2. Off-target effects, resulting from the drug binding to the unintended molecular targets
3. Harmful effects mediated by dysregulated immunity
4. Idiosyncratic adverse responses, which are drug reactions that occur rarely and unpredictably among the population and are usually with unclear mechanisms

2.3.4.2 Impact of Drug Toxicity

Definitions of Adverse Drug Reactions, Adverse Drug Events, and Medication Errors In discussion of the impact of drug toxicity, it is necessary to define several commonly used terms, especially adverse drug reactions (ADRs) and adverse drug events (ADEs), as well as medication errors. The World Health Organization (WHO) defines an ADR as any noxious, unintentional, and undesired effect of a drug, which occurs at doses normally used in humans for prophylaxis, diagnosis, or therapy. The WHO definition purposely excludes therapeutic failures, overdoses, drug abuse, noncompliance, and medication errors. In this context, the US FDA defines ADR as any undesirable experience associated with the use of a medical product in a patient. Thus, the FDA definition of ADR is much broader than that of the WHO.

ADE is defined as an injury resulting from the use of a drug. Under this definition, the term ADE includes harm directly caused by the drug (due to adverse drug reactions and overdoses) and harm from the use of the drug (including dose reductions and discontinuations of drug therapy) [1]. ADEs may result from medication errors, but most do not.

Medication errors are mishaps that occur during prescribing, transcribing, dispensing, administering, adhering to, or monitoring a drug. Examples of medication errors include misreading or miswriting a prescription. Medication errors that are stopped before harm can occur are sometimes called "near misses" or "close calls" or, more formally, a potential ADE. Not all medication errors lead to adverse outcomes. Some do not cause harm, while others are caught before harm can occur ("near misses"). Medication errors are more common than ADEs, and about 25% of ADEs are due to medication errors. Medication errors are the most common source of medical errors. Other major sources of medical errors include hospital-acquired infections, falls, handoff errors, diagnostic errors, and surgical errors [2].

Impact of Adverse Drug Reactions Adverse drug reactions are a major health problem to the individuals (including both children and adults) as well as for the society. ADRs are among the leading causes of death in many countries. It is estimated that medical errors, including ADRs, may cost tens of thousands of lives in the United States. The exact number of ADR-caused mortality in the United States remains uncertain. An early meta-analysis in 1998 estimated about 100,000 deaths annually due to ADRs in the United States [3]. However, the reliability of this estimate has been questioned. A 1999 report from the US Institute of Medicine (IOM) estimated 44,000–98,000 deaths due to medical errors, and of this total, an estimated 7000 deaths occurred due to ADRs. Again, this widely quoted statistics has been debated [4–6]. According to a more recent analysis of the ADR death rates reported in the US vital statistics between 1999 and 2006, the annual ADR death rates ranged from 0.08/100,000 to 0.12/100,000 [7]. Regardless of the debate about the estimates, ADRs are increasingly considered as a significant public health problem that is, for a significant part, preventable [8, 9].

Teratogenesis Due to Drug Therapy and the FDA Pregnancy Category Ratings for Drugs   In addition to causing injury to various organs and systems in both children and adults, drugs given to pregnant women may also adversely affect the fetus, causing teratogenesis. Teratogenesis refers to the induction of structural defects in the fetus, and a teratogen is an agent that can induce such defects. Many drugs, including those used for the treatment of cardiovascular diseases, are teratogens or potentially teratogenic. The US FDA has established five categories (A, B, C, D, and X) to indicate the potential of a drug to cause birth defects if used during pregnancy. The categories are determined by the reliability of data from studies in laboratory animals and in human subjects and the risk-to-benefit ratio. Drugs in category A are typically the safest for use in pregnancy, and category X drugs, as the name implies, are contraindicated in pregnancy (Table 2.10).

2.3.5   Pharmacogenetics and Pharmacogenomics

2.3.5.1   *Definitions and Clinical Impact*   As indicated in Section 2.3.4, the pharmacological effects of drugs are influenced by many factors including the patient's genetic makeup. The terms pharmacogenetics and pharmacogenomics are often used interchangeably to describe a field of research focused on how genetic variations affect individual's responses to pharmacological agents. The convergence of recent advances in genomic science and equally striking advances in molecular pharmacology has resulted in the evolution of pharmacogenetics into pharmacogenomics. In this context, pharmacogenomics is generally considered a broader term referring to a large number of genes affecting drug responses, whereas pharmacogenetics refers to a more limited set of genes. However, the difference between the two is largely arbitrary. As stated above, they are oftentimes

used interchangeably. Here, PGx is used to denote pharmacogenetics/pharmacogenomics. Box 2.1 lists several related terms. PGx provides unique methodologies that can lead to DNA-based tests to improve drug selection, identify optimal dosing, maximize drug efficacy, or minimize the risk of toxicity [10]. PGx provides an important path to personalized medicine or patient-centered medicine [11–13].

2.3.5.2   *Mechanisms*   Genetic variations affect the individual's response to drug therapy via four mechanisms. As illustrated in Figure 2.3, genetic changes, primarily single nucleotide polymorphisms, may lead to (i) the alterations of enzymes or proteins involved in drug metabolism and transport, leading to altered pharmacokinetics of drugs; (ii) the changes in drug targets or receptors or downstream signaling cascades, leading to altered pharmacodynamics of drugs; (iii) the alteration of proteins/enzymes involved in both pharmacokinetics and pharmacodynamics of drugs; and (iv) the idiosyncratic reactions, for which there is no apparent alteration in either pharmacokinetics or pharmacodynamics of drugs. Idiosyncratic response to drug therapy is believed to result from the interaction between the drug molecules and a unique aspect of the individual patients. An excellent example of idiosyncratic response is the association between HLA-B*5701 and a serious hypersensitivity reaction to the anti-HIV drug abacavir. The PGx test for screening HLA-B*5701 can be used to prevent the serious hypersensitivity reaction (idiosyncratic response) to the anti-HIV drug abacavir [14].

2.3.5.3   *Impact*   Today, about 10% of labels for the FDA-approved drugs contain PGx information—a substantial increase since the 1990s but hardly the limit of the possibilities for this aspect of personalized medicine. In this context, the FDA has become a proactive and thoughtful advocate of PGx and believes that as a public health agency, it has a responsibility to play a leading role in bringing about the translation of PGx from bench to bedsides [10, 12]. The FDA has provided guidelines on PGx data submissions that

TABLE 2.10   The FDA pregnancy category ratings for drugs

Category	Description
A	Adequate and well-controlled studies have failed to demonstrate a risk to the fetus in the first trimester of pregnancy (and there is no evidence of risk in later trimesters)
B	Animal reproduction studies have failed to demonstrate a risk to the fetus, and there are no adequate and well-controlled studies in pregnant women
C	Animal reproduction studies have shown an adverse effect on the fetus, and there are no adequate and well-controlled studies in humans, but potential benefits may warrant use of the drug in pregnant women despite potential risks
D	There is positive evidence of human fetal risk based on adverse reaction data from investigational or marketing experience or studies in humans, but potential benefits may warrant use of the drug in pregnant women despite potential risks
X	Studies in animals or humans have demonstrated fetal abnormalities, and/or there is positive evidence of human fetal risk based on adverse reaction data from investigational or marketing experience, and the risks involved in use of the drug in pregnant women clearly outweigh potential benefits

---

BOX 2.1   GENETIC VARIATIONS AND TERMS RELATED TO PGx

- **Genetic variations:** The human genome contains 3 billion (3,000,000,000) nucleotides and approximately 25,000 genes, encoding over 100,000 proteins. Any two persons differ at approximately 1 nucleotide in every 1000 or 3 million base pairs per genome. Single nucleotide polymorphisms (SNPs) account for the majority of the genetic variations. The remaining are insertions, deletions, duplications, and reshufflings. The above genetic differences or variations constitute each person's genetic individuality. Some of this genetic individuality affects the way in which each person will respond to drug treatment, which is known as PGx.

- **Polymorphism:** A genetic variation in the DNA sequence with a measurable frequency of detection above 1%. Any polymorphism in the chromosomal DNA (coding and noncoding) can affect mRNA processing, maturation, and translation as a result of changes in conformation. A subset of polymorphisms alters translated proteins. Genetic variants that affect less than 1% of the general population are referred to as mutations.

- **SNP:** A specific location in a DNA sequence at which different people can have a different DNA base. SNPs can occur in coding or noncoding regions of a gene; a coding SNP can be synonymous (the codon encodes the same amino acid), nonsynonymous (an amino acid change), or nonsense (a premature stop results).

- **Allozyme:** Alternate versions of an enzyme determined by genetic variants (alleles) present at a genetic locus.

- **Gene cluster:** Two or more genes in close physical proximity in the genome that encode similar gene products.

- **Genome-wide association study (GAS):** An approach used in genetics research to look for associations between large numbers (typically hundreds of thousands) of specific genetic variations (most commonly SNPs) and particular diseases.

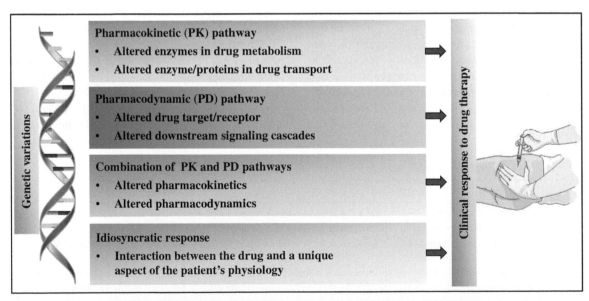

FIGURE 2.3 Mechanisms of pharmacogenetics/genomics (PGx). Genetic variations affect the individual's response to drug therapy via four mechanisms, that is, the alteration of enzymes or proteins involved in drug metabolism and transport, leading to altered pharmacokinetics of drugs; the changes in drug targets or receptors or downstream signaling cascades, leading to altered pharmacodynamics of drugs; the alteration of proteins/enzymes involved in both pharmacokinetics and pharmacodynamics of drugs; and the idiosyncratic reactions, for which there is no apparent alteration in either pharmacokinetics or pharmacodynamics of drugs.

help facilitate the drug development and approval. On the other hand, advances in PGx studies can provide new insights into mechanisms of drug action and as a result can contribute to the development of new therapeutic agents. Section 2.4 provides a brief overview of the principles of drug development and regulation.

2.4 PRINCIPLES OF DRUG DISCOVERY, DEVELOPMENT, AND REGULATION

This section first defines drug discovery and drug development. It then introduces the paradigm of drug creation and survival with an emphasis on the FDA drug approval process.

2.4.1 Definitions

Drug discovery refers to the process that leads to the identification of the potential therapeutic agents for further development into drugs that can be used in humans. In the past, most drugs have been discovered either by identifying the active ingredients from traditional remedies or by serendipitous discoveries. Today, due to our profound understanding of human physiology and disease pathophysiology as well as advances of technologies, useful drugs are rarely discovered as naturally existing things; rather, they are sculpted, designed, and brought into being based on experimentation and optimization of many independent chemical and/or physical properties. As such, the term drug invention has been used to reflect the changes and indicate little serendipity. In this context, computational chemistry has played an important role in modern drug discovery. This is evidenced by the 2013 Nobel Prize in Chemistry for the development and application of methods to simulate the behavior of molecules at various scales, from single molecules to proteins, techniques crucial for drug design.

The term drug development is defined as the process that involves the sequential preclinical and clinical studies of the potential therapeutic agents identified through the drug discovery process. Drug development sometimes is broadly defined to also include the approval of the drugs by regulatory authorities. For clarity, this section considers drug development and drug approval as two separate entities.

2.4.2 The Paradigm of Drug Creation and Survival

Creation and survival of a modern human drug involve four stages, that is, (i) drug discovery, (ii) drug development, (iii) drug approval, and (iv) postapproval regulation of marketed drugs. These four stages collectively form the paradigm of drug creation and survival. As illustrated in

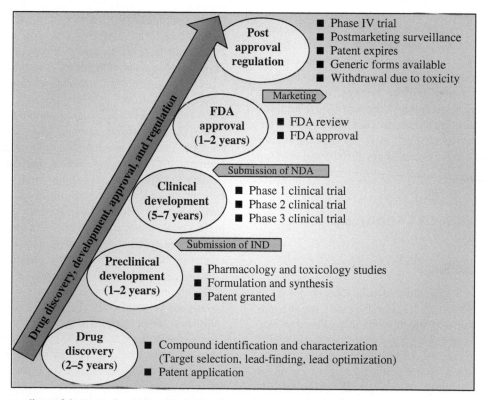

FIGURE 2.4 The paradigm of drug creation and survival. Creation and survival of a modern human drug involves four stages, that is, drug discovery, drug development, drug approval, and postapproval regulation of marketed drugs.

Figure 2.4, the first three stages lead to creation of the drugs for use in humans, and the last stage determines the survival of the approved drugs.

2.4.3   The FDA Drug Review and Approval Process

The FDA is an agency within the US Department of Health and Human Services. The FDA is responsible for protecting the public health by assuring that drugs, vaccines and other biological products, and medical devices intended for human use are safe and effective. Drugs must be approved by the FDA for marketing in the United States. The path a drug travels from a lab to your medicine cabinet is usually long, and every drug takes a unique route. Often, a drug is developed to treat a specific disease. An important use of a drug may also be discovered by accident. Most drugs that undergo preclinical (animal) testing never even make it to human testing and review by the FDA. The drugs that do must undergo the agency's rigorous evaluation process, which scrutinizes everything about the drug—from the design of clinical trials to the severity of side effects and to the conditions under which the drug is manufactured. The major steps involved in the FDA drug review and approval process are summarized in Table 2.11 and also depicted in Figure 2.4.

2.5   PHARMACOLOGY SUBSPECIALTIES

Development of modern pharmacology is closely associated with advances in biomedical sciences and clinical medicine. Indeed, pharmacology is intimately interwoven with the subject matter and experimental techniques of physiology, biochemistry, molecular and cellular biology, microbiology, immunology, genetics, pathology, epidemiology, and the various areas of clinical medicine. As such, modern pharmacology has been divided into many subspecialties or branches. The major subspecialties are briefly summarized in Table 2.12.

2.6   INTRODUCTION TO CARDIOVASCULAR PHARMACOLOGY

2.6.1   Definition and Scope

As defined in Table 2.12, cardiovascular pharmacology is concerned with the effects of drugs on the heart, the vascular system, and those parts of the nervous and endocrine systems that participate in regulating cardiovascular function. Hence, the scope of cardiovascular pharmacology includes studies on (i) drug effects on the heart, (ii) drug effects on the vasculature, and (iii) drug effects on the nervous and endocrine systems that are involved in regulating cardiovascular function. As blood components are intimately involved in vascular disorders, such as coronary artery disease and stroke, many cardiovascular drugs target blood components in treating cardiovascular diseases.

2.6.2   New Developments and Challenges

In the United States, spending for prescription drugs was $234.1 billion in 2008, which was more than double what was spent in 1999. The spending increased to $314.3 billion in 2011 [15]. Notably, cardiovascular drugs are among the most widely used [16]. As new drugs are introduced and new uses for old drugs are found, more patients can have improved health and quality of life with the appropriate use of prescription drugs. Indeed, the past three decades has witnessed dramatic advances in the treatment of cardiovascular diseases due to the development of highly effective and well-tolerated drugs, such as statins, inhibitors of the renin–angiotensin–aldosterone system (RAAS), and beta-blockers. These developments along with the effective nonpharmacological approaches have dramatically improved the survival and life quality of those with major cardiovascular diseases, such as myocardial infarction, heart failure, and stroke. As elderly and chronically ill population increases, the use of cardiovascular drugs is becoming increasingly common. In this context, drug effects are becoming increasingly variable, reflecting the complicated interactions among drugs, underlying cardiovascular disease and disease pathophysiology, and patients' genetic and ethical backgrounds. Hence, increased understanding of the factors influencing drug effects is essential for further improving the drug therapy of cardiovascular diseases. This section provides an overview of recent developments along with a brief discussion of the major challenges facing cardiovascular pharmacology and therapeutics.

2.6.2.1   *Cardiovascular Genomics and Cardiovascular PGx*   Cardiovascular science and medicine have made enormous strides over the past century, beginning with the brilliant elucidation of cardiovascular physiology and leading to molecular and cellular studies, with concomitant epidemiological determinations of risk factors [17]. The cardiovascular field is now primed for genomic medicine to make equal, if not greater, contributions. The field of cardiovascular genomics has two distinct goals: (i) understanding biologic mechanisms and (ii) applying the knowledge to personalized medicine. Knowledge of the molecular pathways can lead to improved therapeutics on a broad base (regardless of the individual genotypes) or at an individual level (targeted specifically to the genotype). During the past several years, the discovery of hundreds of cardiovascular loci is a start. In the years to come, we will require studies of tens of thousands of patients with cardiovascular diseases that combine tests of genome-wide association with sequencing. Discoveries that are provided by genome-wide association studies have strengthened the

TABLE 2.11   Major steps involved in the FDA drug review, approval, and regulatory process

Step	Description	Explanation
1	Preclinical testing	Preclinical testing provides a basis for investigational new drug (IND) application in Step 2
2	IND application	An IND outlines what the sponsor of a new drug proposes for human testing in clinical trials
3	Phase I studies	Phase I clinical trials are done to test a new drug in a small group of people (e.g., 20–80) for the first time to evaluate safety (e.g., determine a safe dosage range and identify side effects)
4	Phase II studies	Phase II clinical trials are done to study the new drug in a larger group of people (several hundred) to determine efficacy and to further evaluate its safety
5	Phase III studies	Phase III clinical trials are done to study the efficacy of the new drug in large groups of human subjects (from several hundred to several thousand) by comparing the new drug to other standard drugs as well as to monitor adverse effects and to collect information that will allow the drug to be used safely
6	The pre-NDA period	It is a period of time just before a new drug application (NDA) is submitted. It is a common time for the FDA and drug sponsors to meet
7	NDA submission	Submission of an NDA is the formal step asking the FDA to consider a drug for marketing approval
8	NDA filing	After an NDA is received, the FDA has 60 days to decide whether to file it so it can be reviewed
9	FDA review of drug safety and efficacy	If the FDA files the NDA, an FDA review team is assigned to evaluate the sponsor's research on the drug's safety and effectiveness
10	FDA review of drug labeling	The FDA reviews information that goes on a drug's professional labeling (information on how to use the drug)
11	FDA inspection of facilities	The FDA inspects the facilities where the drug will be manufactured as part of the approval process
12	FDA decision	• Before August 11, 2008, regarding decision on NDA, the FDA responded to a drug sponsor in one of three types of letters: an "approval" letter, meaning the drug has met agency standards for safety and efficacy and the drug can be marketed for sale in the United States; an "approvable" letter, which generally indicates that the drug can probably be approved at a later date provided that the applicant provides certain additional information or makes specified changes (such as to labeling); or a "not approvable" letter, meaning the application has deficiencies generally requiring the submission of substantial additional data before the application can be approved • Since August 11, 2008, the FDA has used a new way to communicate its decision with drug sponsors. It will either approve the application or issue a complete response letter. A complete response letter is the letter that provides a more consistent and neutral mechanism to convey that the FDA initial review of an application is complete and that the FDA cannot approve the application in its present form. The letter will describe specific deficiencies and, when possible, will outline recommended actions the applicant might take to get the application ready for approval. Hence, a complete response letter is now used to replace the previous "approvable" and "not approvable" letters
13	FDA postmarketing surveillance	Because all possible side effects of a drug cannot be anticipated based on preapproval studies involving only several hundred to several thousand patients, the FDA maintains a system of postmarketing surveillance and risk assessment programs to identify adverse events that did not appear during the drug approval process. The FDA monitors adverse events such as adverse reactions and poisonings. The agency uses this information to update drug labeling and, on rare occasions, to reevaluate the approval or marketing decision. In this context, the term phase IV studies/clinical trials is used to describe studies that are designed to monitor effectiveness of the approved intervention in the general population and to collect information about any adverse effects associated with widespread use

TABLE 2.12 Subspecialties of modern pharmacology

Subspecialty	Description
Neuropharmacology	It refers to the study of drugs that modify the functions of the nervous system, including the brain, the spinal cord, and the nerves that communicate with all parts of the body
Cardiovascular pharmacology	It is concerned with the effects of drugs on the heart, the vascular system, and those parts of the nervous and endocrine systems that participate in regulating cardiovascular function
Respiratory pharmacology	It is concerned with the effects of drugs on the respiratory system, including the airways and lungs
Gastrointestinal pharmacology	It is concerned with the effects of drugs in the gastrointestinal system, including the gastrointestinal tract and liver
Immunopharmacology	It is concerned with the effects of drugs on the immune systems, including both innate and adaptive immune systems
Endocrine pharmacology	It is concerned with the study of actions of drugs that are either hormones or hormone derivatives or drugs that may modify the actions of normally secreted hormones
Chemotherapy	It is the area of pharmacology that deals with drugs used for the treatment of microbial infections and malignancies (cancer)
Molecular pharmacology	It is concerned with the use of the techniques of molecular biology to understand the mechanisms of drug action
Biochemical pharmacology	It is a branch of pharmacology concerned with the biochemical mechanisms responsible for the actions of drugs
Pharmacoepidemiology	It refers to the study of drug effects at the population level. It is concerned with the variability of drug effects between individuals and between populations
Pharmacovigilance	It is the area of pharmacology concerned with the safety of drugs. It involves the characterization, detection, and understanding of adverse effects that arise as a consequence of the short- or long-term use of drugs
Pharmacoeconomics	A branch of health economics aiming to quantify in economic terms the cost and benefit of drug used therapeutically
Clinical pharmacology	It refers to the study of pharmacodynamics and pharmacokinetics in human beings. It is underpinned by the basic science of pharmacology, with added focus on the application of pharmacological principles and methods in the real human world. It has a broad scope, from the discovery of new target molecules to the effects of drug usage in humans
Systems pharmacology	Systems pharmacology is the combination of systems biology and pharmacology and aims to understand both the pharmacological and adverse effects of drugs by considering targets in the context of the biological networks in which they exist. Genome medicine forms the basis on which systems pharmacology can develop. Experimental and computational approaches enable systems pharmacology to obtain holistic, mechanistic information on disease networks and drug responses and to identify new drug targets and specific drug combinations

evidence for PGx interactions in a number of commonly used cardiovascular drugs, including anticoagulants, antiplatelet drugs, statins, beta-blockers, and angiotensin-converting enzyme (ACE) inhibitors [17, 18]. The PGx of cardiovascular drugs will be discussed throughout the various chapters of the book.

2.6.3 Systems Pharmacology in the Management of Cardiovascular Diseases

Systems approaches have long been used in pharmacology to understand drug action at the organ and organismal levels. The application of computational and experimental systems biology approaches to pharmacology allows us to expand the definition of systems pharmacology to include network analyses at multiple scales of biological organization and to explain both therapeutic and adverse effects of drugs [19].

Systems pharmacology analyses rely on experimental "omics" technologies that are capable of measuring changes in large numbers of variables, often at a genome-wide level, to build networks for analyzing drug action. A major use of omics technologies is to relate the genomic status of an individual to the therapeutic efficacy of a drug of interest. In this regard, pharmacogenomics is a component of systems pharmacology. Combining pathway and network analyses, pharmacokinetic and pharmacodynamic models, and a knowledge of polymorphisms in the genome will enable the development of predictive models of therapeutic efficacy and catalyze the practice of personalized precision medicine [20]. Although in its infancy, the field of systems pharmacology has enormous potential to impact both drug development and usage. Systems pharmacology provides holistic, mechanistic information on cardiovascular disease networks and drug responses and helps identify new drug targets and specific drug combinations for the effective management of cardiovascular diseases.

2.6.4 Polypill for the Management of Cardiovascular Diseases

A polypill is a tablet or capsule consisting of a combination of drugs to reduce cardiovascular risk factors simultaneously. This concept was first proposed by cardiologists N.J. Wald and M.R. Law of Queen Mary University of London, and they concluded that a polypill combining low doses of a beta-blocker, diuretic, aspirin, ACE inhibitor, folic acid, and statin would lower cholesterol and blood pressure and decrease the incidence of cardiovascular diseases in at-risk patients by more than 80% [21]. In the past few years, the polypill idea has gained momentum though it still has its critics [22, 23]. Several manufacturers are now making the pill, and a handful of teams are conducting independent clinical trials. Preliminary results suggest that the polypill is safe as well as efficacious in reducing cardiovascular risks [24]. There has been relatively high acceptance of prescribing a polypill for cardiovascular disease prevention [25, 26]. However, evidence from trials on a polypill on cardiovascular outcomes is still lacking. Additional evidence is required to delineate the exact role of such a pill in primary as well as secondary cardiovascular disease prevention [27, 28]. Nevertheless, the polypill should not be considered in isolation, but as an integral part of a comprehensive cardiovascular disease prevention strategy that includes efforts to reduce tobacco use, increase physical activity, and increase consumption of heart-healthy diets.

2.6.5 Protein Therapeutics of Cardiovascular Diseases

Currently, there are over 130 protein- or peptide-based therapeutic agents approved for clinical use. These include drugs that supplement endogenous proteins (e.g., cytokines, growth factors, enzymes, or coagulation factors) and those that block the activity of endogenous proteins (e.g., monoclonal antibodies, soluble receptors, or enzyme inhibitors) [29, 30]. Protein therapeutic agents have been developed for the treatment of diseases in almost every organ system, including the cardiovascular system. A typical example is the antiplatelet drug abciximab, a monoclonal antibody (Chapter 17). Great excitement has now arisen in the field of cardiac regeneration after myocardial infarction (Section 2.6.7). Currently, four classes of potential protein therapeutics have been shown to be effective in improving cardiac function and cardiac regeneration following myocardial infarction in animal models. They are (i) angiogenic growth factors, (ii) proteins that increase recruitment of progenitor cells to damaged myocardium, (iii) proteins that induce mitosis of existing myocytes, and (iv) proteins that increase differentiation and growth of stem cells and

myocytes [31]. As our understanding of cardiac and stem cell biology increases, new proteins that promote cardiac regeneration will be discovered and existing proteins will be more extensively studied for effects on cardiac regeneration. Also, major biotechnological advances, such as PEGylation, have been made that allow for prolonged activity and better pharmacokinetic properties of proteins used as drugs [32, 33]. This will lead to more and better treatment options for patients suffering from myocardial infarction and heart failure, as well as other cardiovascular disorders [31].

2.6.6 Gene Therapy of Cardiovascular Diseases

2.6.6.1 Definition, Approaches, and History The term gene therapy is widely used but poorly defined. Gene therapy can be broadly defined as the introduction, using a vector, of nucleic acids (DNA or RNA) into cells with the intention of altering gene expression to prevent, halt, or reverse a pathophysiological process [34]. The major approaches of gene therapy include (i) replacing a mutated gene that causes a disease with a healthy copy of the gene, (ii) inactivating or knocking down a mutated gene that is functioning improperly, and (iii) introducing a new gene into the body to help fight a disease.

The idea of gene-based therapeutics has been around for some time, but it only received serious attention with the advent of recombinant DNA technology and the ability to transfer and express exogenous genes in mammalian cells. The first federally approved clinical trial of gene therapy was carried out in 1990 in a 4-year-old child with adenosine deaminase deficiency, a rare genetic disease in which children are born with severe immunodeficiency and are prone to repeated serious infections [35]. It was then predicted that gene therapy would become a treatment for serious diseases in just a matter of years. However, during the ensuring two decades, numerous obstacles have tempered the enthusiasm for gene therapy. In particular, the gene therapy suffered a major setback in 1999 with the death of an 18-year-old man during a gene therapy trial for ornithine transcarboxylase (OTC) deficiency. The patient died after an OTC-carrying adenoviral vector triggered a fatal immune response. Another major blow came in January 2003, when the FDA placed a temporary halt on all gene therapy trials using retroviral vectors in blood stem cells. The FDA took this action after it learned that a second child treated in a French gene therapy trial had developed a leukemia-like condition. Both this child and another who had developed a similar condition in August 2002 had been successfully treated by gene therapy for X-linked severe combined immunodeficiency disease (X-SCID), also known as "bubble baby syndrome." The FDA's Biological Response Modifiers Advisory Committee (BRMAC) met at the end of February 2003 to discuss possible measures that could allow a number of retroviral gene therapy trials for treatment of life-threatening diseases to proceed with appropriate safeguards. In April 2003, the FDA eased the ban on gene therapy trials using retroviral vectors in blood stem cells. Over the last several years, some important technical barriers, especially those related to gene delivery, have been overcome to the point where successful examples exist of treating specific diseases, including Leber's congenital amaurosis (an inherited eye disease), hemophilia, β-thalassemia, Wiskott–Aldrich syndrome, chronic lymphoid leukemia, adenosine deaminase deficiency, X-SCID [36, 37], as well as certain forms of cardiovascular disorders (see Section 2.6.6.3 for description of gene therapy of cardiovascular diseases). Although so far there have been over 1700 worldwide clinical trials of gene therapy [38], the US FDA has not yet approved any human gene therapy product for sale.

2.6.6.2 Classification of Gene-Based Therapeutics
Gene-based therapeutics can be broadly classified into two categories: (i) DNA-based gene therapy and (ii) RNA-based gene therapy. While most clinical trials of gene therapy have focused on DNA-based gene therapy, RNA-based therapeutics has recently received increasing attention.

DNA-Based Gene Therapeutics In most DNA-based gene therapy studies, a "normal" gene is inserted into the genome to replace an "abnormal" disease-causing gene. A carrier molecule called a vector must be used to deliver the therapeutic gene to the patient's target cells. Currently, the most common vector is a virus that has been genetically altered to carry normal human DNA. Recently, nonviral vectors, such as plasmid-mediated gene delivery, have also received attention due to their safety profile.

The vector carrying the therapeutic DNA can be injected or given intravenously directly into a specific tissue in the body, where it is taken up by individual cells. Alternately, a sample of the patient's cells can be removed and exposed to the vector in a laboratory setting. The cells containing the vector are then returned to the patient. If the treatment is successful, the new gene delivered by the vector will make a functioning protein.

Researchers must overcome many technical challenges before gene therapy will be a practical approach to treating disease. For example, scientists must find better ways to deliver genes and target them to particular cells. They must also ensure that new genes are precisely controlled by the body.

RNA-Based Gene Therapeutics Since the milestone discoveries of catalytic RNAs in the early 1980s and RNA interference (RNAi) in the late 1990s, the biological understanding of RNA has evolved from simply an intermediate between DNA and protein to a dynamic and versatile molecule that regulates the functions of genes and cells in all living organisms [39]. These and similar breakthroughs have led to the emergence of numerous types of RNA-based

therapeutics that broaden the scope of gene-based therapeutics.

RNA-based therapeutics can be classified by the mechanism of activity and include (i) inhibitors of mRNA translation (antisense), (ii) the agents of RNAi, (iii) catalytically active RNA molecules (ribozymes), and (iv) RNAs that bind proteins and other molecular ligands (aptamers) [40, 41].

Despite a number of hurdles encountered along the way, dozens of RNA-based therapeutics have reached clinical testing for diseases, including inherited genetic disorders, human immunodeficiency virus (HIV) infections, various types of cancer, and certain forms of cardiovascular disorders.

2.6.6.3 Gene-Based Therapeutics in Cardiovascular Diseases
Gene transfer within the cardiovascular system was first demonstrated in 1989 in experimental animals [42]. Despite extensive basic and clinical research, the exact efficacy of gene therapy of cardiovascular diseases remains to be established. Nevertheless, clinical trials over the last several years have shown that gene therapeutics of cardiovascular diseases are relatively safe and may be beneficial for certain forms of cardiovascular disorders, especially heart failure [43]. With a better understanding of the molecular mechanisms of cardiovascular diseases and improved vectors for cardiovascular applications [44], gene therapy can now be considered as a viable adjunctive treatment to mechanical and conventional pharmacological therapies for advanced cardiovascular diseases. Similarly, stem cell therapy has also emerged as a promising approach to combating serious cardiovascular disorders.

2.6.7 Stem Cell Therapy of Cardiovascular Diseases

2.6.7.1 Regenerative Medicine and Stem Cell Therapy
Regenerative medicine may be simply defined as the repair and replacement of damaged tissues and organs. It encompasses a spectrum of technologies and approaches, including cell therapy and tissue engineering (creation of *in vitro* tissues or organs for subsequent transplantation as fully functioning organs or as tissue patches), among others [45]. The paradigm of regenerative medicine includes three Rs: (i) replacement (transplantation of laboratory-grown organs or tissues originated from stem cells), (ii) regeneration (engraftment and differentiation of progenitor cells to restore tissue function through stem cell-based therapy), and (iii) rejuvenation (repair of damaged tissue through activation of endogenous resident stem cells that can simulate biogenesis and replenish functional tissue) [46] (Fig. 2.5). Hence, stem cells are the cornerstone of regenerative medicine [47, 48].

Stem cells can be defined as cells with the ability to divide theoretically for indefinite periods in culture and to give rise to specialized cells. Stem cells have the remarkable potential to develop into many different cell types in the body during early life and growth. In addition, in many tissues, they serve

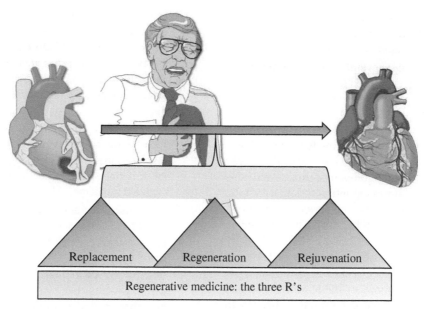

FIGURE 2.5 Regenerative medicine. Regenerative medicine is concerned with the repair and replacement of damaged tissues and organs. The paradigm of regenerative medicine includes three Rs: **r**eplacement (transplantation of laboratory-grown organs or tissues originated from stem cells), **r**egeneration (engraftment and differentiation of progenitor cells to restore tissue function through stem cell-based therapy), and **r**ejuvenation (repair of damaged tissue through activation of endogenous resident stem cells that can simulate biogenesis and replenish functional tissue).

as a sort of internal repair system, dividing essentially without limit to replenish other cells as long as the person or animal is still alive. When a stem cell divides, each new cell has the potential to either remain as a stem cell or become another type of cell with a more specialized function, such as a muscle cell, a red blood cell, or a brain cell.

Stem cells are distinguished from other cell types by two important characteristics. First, they are unspecialized cells capable of renewing themselves through cell division, sometimes after long periods of inactivity. Second, under certain physiological or experimental conditions, they can be induced to become tissue- or organ-specific cells with special functions. In some organs, such as the gut and bone marrow, stem cells regularly divide to repair and replace worn-out or damaged tissues. In other organs, however, such as the pancreas and the heart, stem cells only divide under special conditions.

The most important use of stem cells in medicine is stem cell therapy, which refers to treatment in which stem cells are induced to differentiate into the specific cell type required to repair damaged or destroyed cells or tissues in the human body. The sources of stem cells for stem cell therapy include the following:

1. Embryonic stem cells (ESCs): primitive (undifferentiated) cells derived from a 5-day preimplantation embryo that are capable of dividing without differentiating for a prolonged period in culture and are known to develop into cells and tissues of the three primary germ layers.

2. Adult stem cells: relatively rare undifferentiated cells found in many organs and differentiated tissues with a limited capacity for both self-renewal (in the laboratory) and differentiation. Such cells vary in their differentiation capacity, but it is usually limited to cell types in the organ of origin.

3. Induced pluripotent stem cells: somatic (adult) cells reprogrammed to enter an ESC-like state by being forced to express factors important for maintaining the "stemness" of ESCs.

Section 2.6.7.2 provides a brief overview of the current development in stem cell therapy of cardiovascular diseases, especially ischemic heart disease.

2.6.7.2 Stem Cell Therapy of Cardiovascular Diseases

Preclinical Studies The use of embryonic and adult-derived stem cells for cardiac repair is an active area of research. A number of stem cell types, including ESCs, cardiac stem cells that naturally reside within the heart, myoblasts (muscle stem cells), adult bone marrow-derived cells including mesenchymal cells (bone marrow-derived cells that give rise to tissues such as muscle, bone, tendons, ligaments, and adipose tissue), endothelial progenitor cells (cells that give rise to the endothelium, the interior lining of blood vessels), and umbilical cord blood cells, have been investigated as possible sources for regenerating damaged heart tissue. All have been explored in mouse or rat models, and some have been tested in larger animal models, such as pigs and nonhuman primates [49–52].

Clinical Trials Over the last several years, multiple small-scale clinical trials have been carried out, usually in patients who are undergoing open-heart surgery. Several of these have demonstrated that stem cells that are injected into the circulation or directly into the injured heart tissue appear to improve cardiac function and/or induce the formation of new capillaries. Notably, two recent phase I clinical trials showed that intracoronary infusion of autologous cardiac stem cells was safe and effective in improving left ventricular systolic function and reducing infarct size in patients with heart failure after myocardial infarction [53, 54]. The mechanism for this repair remains controversial, and the stem cells likely regenerate heart tissue through several pathways. The stem cell populations that have been tested in different trials vary widely, as do the conditions of their purification and application. Although much more research is needed to assess the safety and improve the efficacy of this approach, these preliminary clinical experiments show how stem cells may 1 day be used to repair damaged heart tissue, thereby reducing the burden of cardiovascular diseases [55–57].

2.7 SUMMARY OF CHAPTER KEY POINTS

- Medical pharmacology is a biomedical discipline that deals with the action and use of drugs to diagnose, treat, or prevent human diseases. It provides the basis for pharmacotherapeutics. The learning of pharmacology is among the most important steps in becoming a physician.

- In pharmacology, a drug is defined as a natural product, chemical substance, or pharmaceutical preparation intended for administration to a human or animal to diagnose, treat, or prevent a disease.

- Pharmacological paradigm is the central dogma in pharmacology that defines the broad scope of pharmacology, starting from dosing of drugs to pharmacokinetics and pharmacodynamics and finally to clinical responses to drug therapy.

- The major routes of drug administration are classified into five classes: enteral, parenteral, transdermal, inhalational, and topical administration, with oral administration being the most common and convenient route.

- All drugs have the potential to cause unintended consequences, termed toxic effects, side effects, or adverse effects. Drug toxicity is determined by the mechanisms of drug action, the size of the drug dose, the concomitant use of other medications or dietary supplements, as well as the status of the patient's physiology and pathophysiology.

- Many drugs, including those used for the treatment of cardiovascular diseases, are teratogens or potentially teratogenic. The US FDA has established five categories (A, B, C, D, and X) to indicate the potential of a drug to cause birth defects if used during pregnancy. Drugs in category X are contraindicated during pregnancy.

- The terms pharmacogenetics and pharmacogenomics are often used interchangeably to describe a field of research focused on how genetic variations affect individual's responses to drugs. Pharmacogenetics/genomics provides an important path to personalized medicine or patient-centered medicine.

- Drug discovery and drug development are two closely related but distinct terms. Drug discovery refers to the process that leads to the identification of the potential therapeutic agents for further development into drugs that can be used in humans. Drug development is defined as the process that involves the sequential preclinical and clinical studies of the potential therapeutic agents identified through the drug discovery process. Drug development sometimes is broadly defined to also include the approval of the drugs by regulatory authorities.

- The paradigm of creation and survival of a modern human drug involves four stages, that is, drug discovery, drug development, drug approval, and postapproval regulation of marketed drugs.

- The US FDA is responsible for protecting the public health by assuring that drugs, vaccines and other biological products, and medical devices intended for human use are safe and effective. Drugs must be approved by the FDA for marketing in the United States.

- Development of modern pharmacology is closely associated with advances in biomedical sciences and clinical medicine. Modern pharmacology has evolved to include many subspecialties, among which is cardiovascular pharmacology.

- Cardiovascular pharmacology is concerned with the effects of drugs on the heart, the vascular system, and those parts of the nervous and endocrine systems that participate in regulating cardiovascular function. As such, cardiovascular drugs comprise many different classes of pharmacological agents.

- The past three decades has witnessed dramatic advances in the treatment of cardiovascular diseases due to the development of highly effective and well-tolerated drugs, such as statins, inhibitors of the renin–angiotensin–aldosterone system (RAAS), and beta-blockers. Over the past few years, integration of genomics, systems biology, polypill concept, and novel therapeutics (e.g., gene therapy, stem cell therapy) into cardiovascular pharmacology has greatly advanced the field and significantly contributed to the improved management of human cardiovascular diseases.

2.8 SELF-ASSESSMENT QUESTIONS

2.8.1 The term pharmacology is derived from the Greek words pharmakon (meaning a drug or medicine) and logos (meaning the truth about or a rational discussion). Medical pharmacology is defined as which of the following?

A. The medical science concerned with the clinical use of drugs in the treatment of human diseases

B. The science and profession concerned with the preparation, storage, dispensing, and proper use of drug products

C. The science concerned with the formulation and chemical properties of pharmaceutical products, such as tablets, liquid solutions and suspensions, and aerosols

D. The science that deals with the action and use of drugs in the diagnosis, treatment, or prevention of human diseases

E. The study of drugs isolated from natural sources, including plants, microbes, animal tissues, and minerals

2.8.2 The first published classic text, "Outline of Pharmacology," written in 1878, set the momentum for today's pharmacology advancement throughout the world. The author of this classic text is which of the following?

A. Alexander Fleming

B. John J. Abel

C. Oswald Schmiedeberg

D. Paul Ehrlich

E. Rudolf Buchheim

2.8.3 A particular drug may have up to three different names: chemical name, generic name, and trade name. Which of the following is a characteristic of generic name?

A. Country specific and not recognized internationally

B. Often designed to be longer than trade name

C. Often not helpful in identifying the pharmacological action or class of drug

D. The patented exclusive property of the drug manufacturer

E. The type of drug name most suitable for use by healthcare professionals

2.8.4 Azilsartan is an angiotensin receptor blocker, approved in 2011 by the US FDA for hypertension management. This drug is classified into the X category of the FDA pregnancy category ratings for drugs. Which of the following best describes the potential of azilsartan regarding its potential teratogenicity?

A. Adequate and well-controlled studies have failed to demonstrate a risk for this drug to the fetus in the first trimester of pregnancy (and there is no evidence of risk in later trimesters).

B. Animal reproduction studies have failed to demonstrate a risk for this drug to the fetus and there are no adequate and well-controlled studies in pregnant women.

C. Animal reproduction studies have shown an adverse effect of this drug on the fetus and there are no adequate and well-controlled studies in humans, but potential benefits may warrant use of the drug in pregnant women despite potential risks.

D. Studies in animals or humans have demonstrated fetal abnormalities and/or there is positive evidence of human fetal risk based on adverse reaction data from investigational or marketing experience, and the risks involved in use of the drug in pregnant women clearly outweigh potential benefits.

E. There is positive evidence of human fetal risk based on adverse reaction data from investigational or marketing experience or studies in humans, but potential benefits may warrant use of the drug in pregnant women despite potential risks.

2.8.5 A 55-year-old man is being treated with atorvastatin (a generic statin drug) for his hypercholesterolemia. He is concerned with statin-induced rhabdomyolysis (a severe form of statin-induced myopathy). You tell him that the risk of statin-induced myopathy may be elevated in those with polymorphism(s) in a gene that encodes SLCO1B1. This is an example of which of the following?

A. Genome-wide association study

B. Pharmacodynamic consideration

C. Pharmacogenetic consideration

D. Pharmacokinetic consideration

E. Postmarketing surveillance

2.8.6 Professor T.M. and a DO/PhD candidate propose a clinical study involving 22 healthy volunteers to determine if "BrocoProtector," a novel phytochemical recently isolated by Professor T.M. from broccoli, is safe for oral intake at a dose of 0.5 g daily for 7 days. This clinical study falls into which of the following phases?

A. Phase I

B. Phase II

C. Phase III

D. Phase IV

E. Phase V

2.8.7 Several hospitals are participating in a study to test the efficacy of a newly developed drug prior to its release. This drug is designed to lower cholesterol levels. Of the 1000 patients who are involved in this study, half receive the drug and half receive a placebo. Neither the physicians in charge of the study nor the patients were permitted to know what the patients have received. Which of the following steps in the drug development process does this scenario most closely describe?

A. Phase 0 trial
B. Phase I trial
C. Phase II trial
D. Phase III trial
E. Phase IV trial

2.8.8 Multiple hospitals are participating in a well-designed phase III trial of a new analgesic drug for mild pain. Which of the following would probably not be included?

A. A group of 1000–5000 subjects with a clinical condition requiring analgesia
B. A negative control (placebo)
C. A positive control (current standard therapy)
D. Double-blind protocol
E. Prior submission of a new drug application (NDA) to the FDA

2.8.9 A recent clinical study reported in the New England Journal of Medicine (2011; 365:2357–85) demonstrated that infusion of a single dose of a serotype-8-pseudotyped, self-complementary adenovirus-associated virus (AAV) vector expressing a codon-optimized human factor IX (FIX) transgene (scAAV2/8-LP1-hFIXco) in a peripheral vein in six patients with severe hemophilia B (FIX activity, <1% of normal values) improved the bleeding phenotype, with few side effects. Which of the following best describes this study?

A. Gene therapy
B. Polypill therapy
C. Protein therapeutics
D. Regenerative medicine
E. Stem cell therapy

2.8.10 The idea of gene-based therapeutics has been around for some time, but it only received serious attention with the advent of recombinant DNA technology and the ability to transfer and express exogenous genes in mammalian cells. The first US federally approved clinical trial of gene therapy was carried out in 1990 in a 4-year-old child with adenosine deaminase deficiency, a rare genetic disease in which children are born with severe immunodeficiency and are prone to repeated serious infections. Up to date, how many gene therapy products has the US FDA approved for sale?

A. 0
B. 4
C. 13
D. 37
E. 68

REFERENCES

1 Nebeker, J.R., P. Barach, and M.H. Samore, Clarifying adverse drug events: a clinician's guide to terminology, documentation, and reporting. *Ann Intern Med*, 2004. 140(10): p. 795–801.

2 Pham, J.C., et al., Reducing medical errors and adverse events. *Annu Rev Med*, 2012. 63: p. 447–63.

3 Lazarou, J., B.H. Pomeranz, and P.N. Corey, Incidence of adverse drug reactions in hospitalized patients: a meta-analysis of prospective studies. *JAMA*, 1998. 279(15): p. 1200–5.

4 McDonald, C.J., M. Weiner, and S.L. Hui, Deaths due to medical errors are exaggerated in Institute of Medicine report. *JAMA*, 2000. 284(1): p. 93–5.

5 Leape, L.L., Institute of Medicine medical error figures are not exaggerated. *JAMA*, 2000. 284(1): p. 95–7.

6 Hayward, R.A. and T.P. Hofer, Estimating hospital deaths due to medical errors: preventability is in the eye of the reviewer. *JAMA*, 2001. 286(4): p. 415–20.

7 Shepherd, G., et al., Adverse drug reaction deaths reported in United States vital statistics, 1999–2006. *Ann Pharmacother*, 2012. 46(2): p. 169–75.

8 Hakkarainen, K.M., et al., Percentage of patients with preventable adverse drug reactions and preventability of adverse drug reactions—a meta-analysis. *PLoS One*, 2012. 7(3): p. e33236.

9 Smyth, R.M., et al., Adverse drug reactions in children-a systematic review. *PLoS One*, 2012. 7(3): p. e24061.

10 Lesko, L.J. and J. Woodcock, Translation of pharmacogenomics and pharmacogenetics: a regulatory perspective. *Nat Rev Drug Discov*, 2004. 3(9): p. 763–9.

11 Hamburg, M.A. and F.S. Collins, The path to personalized medicine. *N Engl J Med*, 2010. 363(4): p. 301–4.

12 Wang, L., H.L. McLeod, and R.M. Weinshilboum, Genomics and drug response. *N Engl J Med*, 2011. 364(12): p. 1144–53.

13 Johnson, J.A. and L.H. Cavallari, Pharmacogenetics and cardiovascular disease—implications for personalized medicine. *Pharmacol Rev*, 2013. 65(3): p. 987–1009.

14 Mallal, S., et al., HLA-B*5701 screening for hypersensitivity to abacavir. *N Engl J Med*, 2008. 358(6): p. 568–79.

15 Hoffman, J.M., et al., Projecting future drug expenditures—2012. *Am J Health Syst Pharm*, 2012. 69(5): p. 405–21.

16 Roden, D.M., et al., Cardiovascular pharmacogenomics. *Circ Res*, 2011. 109(7): p. 807–20.

17 O'Donnell, C.J. and E.G. Nabel, Genomics of cardiovascular disease. *N Engl J Med*, 2011. 365(22): p. 2098–109.

18 Pereira, N.L. and R.M. Weinshilboum, Cardiovascular pharmacogenomics and individualized drug therapy. *Nat Rev Cardiol*, 2009. 6(10): p. 632–8.

19 Zhao, S. and R. Iyengar, Systems pharmacology: network analysis to identify multiscale mechanisms of drug action. *Annu Rev Pharmacol Toxicol*, 2012. 52: p. 505–21.

20 Iyengar, R., et al., Merging systems biology with pharmacodynamics. *Sci Transl Med*, 2012. 4(126): p. 126ps7.

21 Wald, N.J. and M.R. Law, A strategy to reduce cardiovascular disease by more than 80%. *BMJ*, 2003. 326(7404): p. 1419.

22 Reardon, S., Public health. Experts debate polypill: a single pill for global health. *Science*, 2011. 333(6051): p. 1813.

23 Muntner, P., et al., Projected impact of polypill use among US adults: medication use, cardiovascular risk reduction, and side effects. *Am Heart J*, 2011. 161(4): p. 719–25.

24 Yusuf, S., et al., Effects of a polypill (Polycap) on risk factors in middle-aged individuals without cardiovascular disease (TIPS): a phase II, double-blind, randomised trial. *Lancet*, 2009. 373(9672): p. 1341–51.

25 Viera, A.J., et al., Acceptance of a polypill approach to prevent cardiovascular disease among a sample of U.S. physicians. *Prev Med*, 2011. 52(1): p. 10–5.

26 Viera, A.J., The polypill to prevent cardiovascular disease: physicians' perspectives. *Curr Opin Cardiol*, 2011. 26(5): p. 438–42.

27 Dabhadkar, K.C., et al., Prospects for a cardiovascular disease prevention polypill. *Annu Rev Public Health*, 2011. 32: p. 23–38.

28 Lonn, E., et al., The polypill in the prevention of cardiovascular diseases: key concepts, current status, challenges, and future directions. *Circulation*, 2010. 122(20): p. 2078–88.

29 Krejsa, C., M. Rogge, and W. Sadee, Protein therapeutics: new applications for pharmacogenetics. *Nat Rev Drug Discov*, 2006. 5(6): p. 507–21.

30 Leader, B., Q.J. Baca, and D.E. Golan, Protein therapeutics: a summary and pharmacological classification. *Nat Rev Drug Discov*, 2008. 7(1): p. 21–39.

31 Segers, V.F. and R.T. Lee, Protein therapeutics for cardiac regeneration after myocardial infarction. *J Cardiovasc Transl Res*, 2010. 3(5): p. 469–77.

32 Parrott, M.C. and J.M. DeSimone, Drug delivery: relieving PEGylation. *Nat Chem*, 2012. 4(1): p. 13–4.

33 Carter, P.J., Introduction to current and future protein therapeutics: a protein engineering perspective. *Exp Cell Res*, 2011. 317(9): p. 1261–9.

34 Kay, M.A., State-of-the-art gene-based therapies: the road ahead. *Nat Rev Genet*, 2011. 12(5): p. 316–28.

35 Anderson, W.F., Human gene therapy. *Science*, 1992. 256(5058): p. 808–13.

36 Mullard, A., Gene therapies advance towards finish line. *Nat Rev Drug Discov*, 2011. 10(10): p. 719–20.

37 Nathwani, A.C., et al., Adenovirus-associated virus vector-mediated gene transfer in hemophilia B. *N Engl J Med*, 2011. 365(25): p. 2357–65.

38 Lee, B. and B.L. Davidson, Gene therapy grows into young adulthood: special review issue. *Hum Mol Genet*, 2011. 20(R1): p. R1.

39 Esteller, M., Non-coding RNAs in human disease. *Nat Rev Genet*, 2011. 12(12): p. 861–74.

40 Burnett, J.C. and J.J. Rossi, RNA-based therapeutics: current progress and future prospects. *Chem Biol*, 2012. 19(1): p. 60–71.

41 Kole, R., A.R. Krainer, and S. Altman, RNA therapeutics: beyond RNA interference and antisense oligonucleotides. *Nat Rev Drug Discov*, 2012. 11(2): p. 125–40.

42 Nabel, E.G., et al., Recombinant gene expression in vivo within endothelial cells of the arterial wall. *Science*, 1989. 244(4910): p. 1342–4.

43 Tilemann, L., et al., Gene therapy for heart failure. *Circ Res*, 2012. 110(5): p. 777–93.

44 Katz, M.G., et al., Gene delivery technologies for cardiac applications. *Gene Ther*, 2012. 19(6):659–69.

45 Corona, B.T., et al., Regenerative medicine: basic concepts, current status, and future applications. *J Investig Med*, 2010. 58(7): p. 849–58.

46 Nelson, T.J., A. Behfar, and A. Terzic, Strategies for therapeutic repair: the "R(3)" regenerative medicine paradigm. *Clin Transl Sci*, 2008. 1(2): p. 168–71.

47 Nelson, T.J., et al., Stem cell platforms for regenerative medicine. *Clin Transl Sci*, 2009. 2(3): p. 222–7.

48 Teo, A.K. and L. Vallier, Emerging use of stem cells in regenerative medicine. *Biochem J*, 2010. 428(1): p. 11–23.

49 Nelson, T.J., A. Martinez-Fernandez, and A. Terzic, Induced pluripotent stem cells: developmental biology to regenerative medicine. *Nat Rev Cardiol*, 2010. 7(12): p. 700–10.

50 Laflamme, M.A. and C.E. Murry, Heart regeneration. *Nature*, 2011. 473(7347): p. 326–35.

51 Smart, N., et al., De novo cardiomyocytes from within the activated adult heart after injury. *Nature*, 2011. 474(7353): p. 640–4.

52 Sturzu, A.C. and S.M. Wu, Developmental and regenerative biology of multipotent cardiovascular progenitor cells. *Circ Res*, 2011. 108(3): p. 353–64.

53 Bolli, R., et al., Cardiac stem cells in patients with ischaemic cardiomyopathy (SCIPIO): initial results of a randomised phase 1 trial. *Lancet*, 2011. 378(9806): p. 1847–57.

54 Makkar, R.R., et al., Intracoronary cardiosphere-derived cells for heart regeneration after myocardial infarction (CADUCEUS): a prospective, randomised phase 1 trial. *Lancet*, 2012. 379(9819): p. 895–904.

55 Ptaszek, L.M., et al., Towards regenerative therapy for cardiac disease. *Lancet*, 2012. 379(9819): p. 933–42.

56 Donndorf, P., B.E. Strauer, and G. Steinhoff, Update on cardiac stem cell therapy in heart failure. *Curr Opin Cardiol*, 2012. 27(2): p. 154–60.

57 Mordwinkin, N.M., A.S. Lee, and J.C. Wu, Patient-specific stem cells and cardiovascular drug discovery. *JAMA*, 2013. 310(19): p. 2039–40.

UNIT II

DYSLIPIDEMIAS

3

OVERVIEW OF DYSLIPIDEMIAS AND DRUG THERAPY

3.1 INTRODUCTION

Dyslipidemias, including hyperlipidemia (e.g., hypercholesterolemia) and low levels of high-density lipoprotein (HDL) cholesterol (HDL-C), constitute a chief cause of atherosclerosis and atherosclerosis-associated conditions, such as coronary heart disease, ischemic cerebrovascular disease, and peripheral vascular disease. Recognition of dyslipidemias as a major risk factor of cardiovascular diseases has led to the development of drugs that modify lipid metabolism and correct the dyslipidemias. Currently, there are five groups of drugs for treating dyslipidemias, including (i) statins, (ii) bile acid sequestrants, (iii) cholesterol absorption inhibitors, (iv) fibrates, and (v) niacin. The use of these drugs, especially the statins, has revolutionized the management of dyslipidemias, particularly high levels of low-density lipoprotein (LDL) cholesterol. Indeed, statins have gained a reputation of being a major player in preventive cardiology. Since statins and other lipid-lowering drugs target the pathways of lipoprotein metabolism, this chapter first reviews the major metabolic pathways of lipoproteins to lay a basis for understanding how drugs impact lipoprotein metabolism to treat dyslipidemias. The chapter then introduces the major types of dyslipidemias, focusing on discussing their molecular etiologies. To set a stage for the subsequent discussion of drugs for dyslipidemias in Chapter 4, the chapter ends with a brief survey of the various targeting sites of the major drug classes for treating dyslipidemias.

3.2 LIPOPROTEIN METABOLISM

3.2.1 Definition, Structure, and Classification of Lipoproteins

3.2.1.1 Definition of Lipoproteins
To understand lipoprotein metabolism, we need to first define lipoproteins. As the name indicates, lipoproteins are large macromolecular complexes of lipids and proteins that are essential for transporting hydrophobic lipids (primarily triglycerides and cholesterol) and lipid-soluble vitamins. They are responsible for (i) the absorption of dietary cholesterol, long-chain fatty acids, and fat-soluble vitamins; (ii) the transport of triglycerides, cholesterol, and fat-soluble vitamins from the liver to peripheral tissues; and (iii) the transport of cholesterol from peripheral tissues to the liver, the so-called reverse cholesterol transport, primarily mediated by HDL (see Section 3.2.2.3 for description of reverse cholesterol transport).

3.2.1.2 Structure of Lipoproteins
Structurally, lipoproteins are microscopic spherical particles ranging from 7 to 100 nm in diameter. Each lipoprotein particle contains a core of hydrophobic lipids in the form of triglycerides and cholesteryl esters surrounded by relatively hydrophilic lipids (phospholipids, free unesterified cholesterol) and proteins that interact with the body fluids (Fig. 3.1). The proteins in the lipoproteins are called apolipoproteins. Each lipoprotein particle contains one or more types of apolipoproteins. Apolipoproteins are required for the proper assembly, structure, and functions of lipoproteins.

Cardiovascular Diseases: From Molecular Pharmacology to Evidence-Based Therapeutics, First Edition. Y. Robert Li.
© 2015 John Wiley & Sons, Inc. Published 2015 by John Wiley & Sons, Inc.

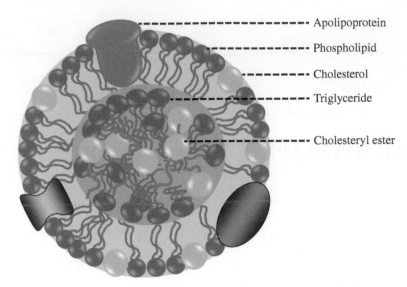

— — — — — — — — — — Apolipoprotein

— — — — — — — — — — Phospholipid

— — — — — — — — — — Cholesterol

— — — — — — — — — — Triglyceride

— — — — — — — — — — Cholesteryl ester

FIGURE 3.1   Lipoprotein structure and composition. See text (Section 3.2.1.2) for description.

TABLE 3.1   Composition and characteristics of human plasma lipoproteins

	CM	CMR	VLDL	IDL	LDL	HDL
Total protein (%)	2	4	10	18	25	40
Total lipid[a] (%)	98	96	90	82	75	60
Density (g/ml)	<0.95	0.95–1.006	<1.006	1.006–1.019	1.019–1.063	1.063–1.210
Diameter (nm)	75–1200	30–80	40–50	25–35	20–25	5–10
Electrophoretic motility[b]	Origin	Slow pre-β	Pre-β	Slow pre-β	β	α
Major apolipoproteins	B48	B48, E	B100	B100, E	B100	A-I, A-II
Other apolipoproteins	A-I, A-IV, A-V, C-I, C-II, C-III, E	A-I, A-IV, C-I, C-II, C-III	A-I, A-II, A-V, C-I, C-II, C-III, E	C-I, C-II, C-III		A-IV, A-V, C-III, E
Other constitutes	Retinyl esters	Retinyl esters	Vitamin E	Vitamin E	Vitamin E	LCAT, CETP, PON1
Origin	Intestine	CM metabolism	Liver	VLDL	IDL	Liver, intestine
Cholesterol in plasma[c]	0.0	0.0	0.1–0.4 mM or 4–15 mg/dl	0.1–0.3 mM or 4–12 mg/dl	1.5–3.5 mM or 58–135 mg/dl	0.9–1.6 mM or 35–62 mg/dl
Triglyceride in fasting plasma[d]	0.0	0.0	0.2–1.2 mM or 18–106 mg/dl	0.1–0.3 mM or 9–27 mg/dl	0.2–0.4 mM or 18–36 mg/dl	0.1–0.2 mM or 9–18 mg/dl

[a] The ratios of triglycerides to cholesterol (both unesterified cholesterol and cholesteryl esters) for CM, VLDL, IDL, LDL, and HDL are 90/10, 70/30, 50/50, 20/80, and 20/80, respectively.

[b] Based on electrophoretic motility, lipoproteins can be classified into alpha-lipoprotein (i.e., HDL), beta-lipoprotein (i.e., LDL), and pre-beta-lipoproteins (i.e., CM, CMR, VLDL, and IDL).

[c] For cholesterol, mM multiplied by 38.67 gives rise to mg/dl.

[d] For triglyceride, mM multiplied by 88.5 gives rise to mg/dl.

CETP, cholesteryl ester transfer protein; CM, chylomicron; CMR, chylomicron remnant; LCAT, lecithin–cholesterol acyltransferase; PON1, paraoxonase 1.

3.2.1.3   Classification of Lipoproteins

The plasma lipoproteins are classified into five major classes based on their relative density, size, and protein content: (i) chylomicrons, (ii) very-low-density lipoproteins (VLDL), (iii) intermediate-density lipoproteins (IDL), (iv) LDL, and (v) HDL (Table 3.1). As a general rule, larger, less dense lipoproteins have a greater percentage composition of lipids. In this regard, chylomicrons are the largest and least dense lipoproteins containing the highest lipid content and the lowest proportion of proteins, whereas HDL particles are the smallest lipoproteins containing the lowest lipid content and the highest percentage of protein composition.

In addition to the above five major classes of lipoproteins, an LDL-like lipoprotein, known as lipoprotein(a) with the abbreviation of Lp(a) has recently received great attention. Like LDL, Lp(a) is composed of a central core

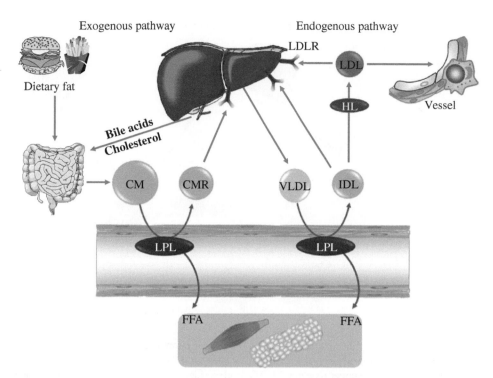

FIGURE 3.2 Endogenous and exogenous pathways of lipoprotein metabolism. See text (Sections 3.2.2.1 and 3.2.2.2) for description. CM, chylomicron; CMR, chylomicron remnant; HL, hepatic lipase; IDL, intermediate-density lipoprotein; LDL, low-density lipoprotein; LDLR, LDL receptor; FFA, free fatty acid; LPL, lipoprotein lipase; VLDL, very-low-density lipoprotein. For color details, please see color plate section.

of cholesteryl esters and triglycerides surrounded by phospholipids and free cholesterol. Lp(a) is produced by the liver and each Lp(a) particle contains a single molecule of ApoB-100 and an additional protein, called apolipoprotein(a) [Apo(a)]. ApoB-100 and Apo(a) are attached via a disulfide bond.

As described in Section 3.3.1, dysregulated expression of Apo(a)-encoding gene may cause elevated plasma levels of Lp(a), also known as hyperlipoproteinemia(a) or Lp(a) hypercholesterolemia. Hyperlipoproteinemia(a) is a risk factor for cardiovascular diseases [1–3]. Elevated Lp(a) increases the risk of cardiovascular diseases via two possible mechanisms: (i) prothrombotic and antifibrinolytic effects as Apo(a) possesses structural homology with plasminogen and plasmin but has no fibrinolytic activity and (ii) accelerated atherogenesis as a result of intimal deposition of the Lp(a) particles and Lp(a)-provoked inflammatory responses.

3.2.2 Metabolic Pathways of Lipoproteins and Drug Therapy

The metabolism of lipoproteins involves three pathways: (i) the exogenous pathway (transport of dietary lipids), (ii) the endogenous pathway (transport of hepatic lipids), and (iii) the reverse cholesterol transport. The schematic illustration of the exogenous and endogenous pathways is shown in Figure 3.2 and that for the reverse cholesterol transport is given in Figure 3.3. As described later in the chapter, the metabolic pathways of lipoproteins serve as the targeting sites of the drugs for treating dyslipidemias.

3.2.2.1 Exogenous Pathway

Solubilization of Dietary Fat The exogenous pathway of lipoprotein metabolism deals with the transport of dietary lipids. Before ingested dietary triglycerides (also known as triacylglycerols) can be absorbed through the intestinal wall, they must be converted from insoluble macroscopic fat particles to finely dispersed microscopic micelles. This solubilization process is carried out by bile salts, which are synthesized from cholesterol in the liver, stored in the gallbladder, and released into the small intestine after ingestion of a fatty meal.

Digestion by Pancreatic Lipase and Formation of Chylomicrons Micelle formation enormously increases the fraction of lipid molecules accessible to the action of water-soluble pancreatic lipases in the intestine. The lipase action converts triglycerides to monoglycerides and diglycerides and glycerol. These products of lipase action diffuse into the enterocytes, where they are converted to triglycerides and packed with ApoB-48 as well as diet-derived cholesterol and cholesteryl esters and phospholipids to form chylomicrons.

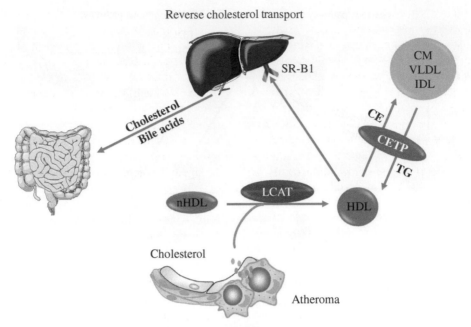

FIGURE 3.3   Reverse cholesterol transport. See text (Section 3.2.2.3) for description. CE, cholesteryl ester; CETP, cholesterol ester transporting protein; CM, chylomicron; HDL, high-density lipoprotein; IDL, intermediate-density lipoprotein; LCAT, lecithin–cholesterol acyltransferase; nHDL, nascent HDL; SR-B1, scavenger receptor B1; TG, triglyceride; VLDL, very-low-density lipoprotein.

Dietary cholesterol is transported into enterocytes by selective cholesterol transporting protein, which is the molecular target of the lipid-lowering drug, ezetimibe (Chapter 4). Inside the enterocytes, cholesterol is esterified to form cholesteryl esters, which, as described earlier, together with the unesterified cholesterol and other constituents, forms chylomicrons (Table 3.1).

Hydrolysis of Chylomicrons by Vascular Lipoprotein Lipase   Nascent chylomicrons are secreted into the intestinal lymph and delivered via the thoracic duct directly to the systemic circulation, where they are extensively processed by peripheral tissues before reaching the liver. In circulation, the chylomicron particles encounter lipoprotein lipase, which is anchored to proteoglycans that decorate the capillary endothelial surfaces of the adipose tissue, heart, and skeletal muscle. The triglycerides of chylomicrons are hydrolyzed by lipoprotein lipase to release free fatty acids. ApoC-II, which is transferred to circulating chylomicrons from HDL, serves as an important cofactor for this reaction. As discussed later, both ApoC-II and lipoprotein lipase are molecular targets of the lipid-lowering drugs, fibrates (Chapter 4).

Metabolism of Free Fatty Acids and Formation of Chylomicron Remnants   The free fatty acids released from the lipoprotein lipase-catalyzed reaction are taken up by adjacent myocytes or adipocytes and either oxidized to generate energy or re-esterified and stored as triglycerides. Some of the released free fatty acids bind albumin and are

transported to other tissues, especially the liver. The chylomicron particles progressively shrink in size as triglycerides are hydrolyzed by the lipoprotein lipase, resulting in the formation of chylomicron remnants.

Chylomicron remnants are rapidly removed by the liver via a process that requires ApoE of the remnants as a ligand for receptors in the plasma membrane of hepatocytes. These receptors are known as ApoE-containing lipoprotein receptors (ApoE-R). It should be noted that the lipoprotein lipase discussed here and the aforementioned lipases in the intestine, as well as the hormone-sensitive lipase (a molecular target of the lipid-lowering drug niacin covered in Chapter 4) in adipose tissue are distinct enzymes.

Enterohepatic Circulation   As noted earlier, bile salts are synthesized from cholesterol in the liver. After being released into the small intestine, the majority of the bile salts is reabsorbed and transported into the liver. This cycle is known as enterohepatic circulation or recycling of bile salts. As a result of this recycling, the de novo synthesis of bile salts is downregulated. Disruption of this recycling by bile acid sequestering drugs results in enhanced de novo synthesis of bile salts, leading to decreased levels of cholesterol in hepatocytes (see Chapter 4).

3.2.2.2   *Endogenous Pathway*

Hepatic Secretion of VLDL   The endogenous pathway of lipid metabolism deals with the transport of hepatic lipids synthesized endogenously. This pathway primarily refers to the hepatic secretion of ApoB-containing VLDL particles and their metabolism. VLDL particles are similar to chylomicrons in protein composition but contain ApoB-100 rather than ApoB-48 and have a higher ratio of cholesterol to triglycerides (Table 3.1). The triglycerides in VLDL are derived mainly from the esterification of free fatty acids in the liver. The assembly of hepatic triglycerides with the other major components of the nascent VLDL particles (ApoB-100, cholesteryl esters, phospholipids, and alpha-tocopherol; alpha-tocopherol is one of the eight forms of vitamin E) requires the action of the enzyme microsomal triglyceride transfer protein.

Conversion of VLDL to IDL After secretion into the circulation, VLDL particles acquire multiple copies of ApoE and the apolipoproteins of the C series via transfer from HDL. As with chylomicrons, the triglycerides of VLDL particles are hydrolyzed by the lipoprotein lipase, especially in the muscle and adipose tissues. After the VLDL remnants dissociate from the lipoprotein lipase, they are referred to as IDL particles, which contain similar amounts of cholesterol and triglycerides (Table 3.1).

Conversion of IDL to LDL About 50% of the IDL particles are removed by the liver via LDL receptor-mediated endocytosis with IDL-associated ApoE as the ligand. The remaining 50% of IDL particles are remodeled by hepatic lipase to form LDL particles. Hepatic lipase is synthesized and secreted by the liver and binds to heparan sulfate proteoglycans on the cell surface of hepatocytes and endothelial cells [4]. During hepatic lipase-catalyzed lipolysis, most of the triglycerides in the IDL particles are hydrolyzed and all apolipoproteins except ApoB-100 are transferred to other lipoproteins. As a result, LDL is a distinct, cholesteryl ester-enriched lipoprotein with ApoB-100 as its only apolipoprotein. The cholesterol in LDL accounts for 65–75% of the total plasma cholesterol in most individuals.

Receptor-Mediated Endocytosis of LDL in the Liver Approximately, 70% of circulating LDL is cleared by LDL receptor-mediated endocytosis in the liver. The LDL receptor is the only receptor capable of clearing significant amounts of LDL from the plasma. The LDL receptor is expressed on the surface of hepatocytes as well as several other types of cells (e.g., macrophages, smooth muscle cells). Due to the lack of ApoE, an important ligand for LDL receptor, LDL particles have relatively weak affinity to LDL receptor, and as a result, the half-life of LDL in the circulation is relatively long (about 2–4 days). This explains why LDL cholesterol accounts for a major portion of the total plasma cholesterol (see section "Conversion of IDL to LDL").

Interaction of ApoB-100 with LDL receptors facilitates receptor-mediated endocytosis of LDL particles and the subsequent vesicle fusion with lysosomes [5]. The LDL receptor is recycled back to the cell surface, while the LDL particles are hydrolyzed to release the unesterified cholesterol. The increase in the levels of the free cholesterol impacts three biochemical pathways: (i) inhibition of 3-hydroxy-3-methyl coenzyme A (HMG-CoA) reductase, the key enzyme involved in de novo synthesis of cholesterol (this enzyme is the major targeting molecule of statin drugs; see Chapter 4); (ii) activation of acetyl coenzyme A cholesterol acyltransferase (ACAT) to increase esterification and storage of cholesterol in the cells; and (iii) downregulation of LDL receptors, reducing further uptake of LDL particles into the cells [5]. Since approximately 70% of circulating LDL is cleared by LDL receptor-mediated endocytosis in the liver, upregulation of hepatocyte LDL receptors is an important mechanism by which many lipid-lowering drugs decrease the levels of plasma LDL cholesterol (Chapter 4).

Receptor-Mediated Uptake of LDL in Nonliver Cells and Atherogenesis As mentioned earlier, LDL receptors are also expressed in several other types of cells, including macrophages, lymphocytes, adrenocortical cells, gonadal cells, and smooth muscle cells. Elevated circulating LDL is a major risk factor for the development of atherosclerosis. Uptake of LDL by macrophages via receptor-mediated mechanisms in vascular tissue is a critical process underlying atherosclerosis. While macrophages can take up native LDL via LDL receptor-mediated endocytosis, the scavenger receptors, which are expressed predominantly in mononuclear phagocytic cells, play a major role in the formation of foam cells (see the following text for description of foam cells). In this regard, native LDL particles that migrate into the subendothelial space of vascular tissue can be oxidatively modified to form oxidized LDL. Oxidized LDL provokes inflammatory responses, leading to recruitment of blood monocytes to the subendothelial space to become macrophages. Expression of scavenger receptors on macrophages greatly facilitates the uptake of oxidized LDL, resulting in the formation of lipid-laden macrophages, known as foam cells.

Foam cells, as a major constituent of atherosclerotic lesion, may exacerbate vascular injury via releasing reactive oxygen species and digestive enzymes, as well as proinflammatory cytokines, thereby forming a vicious cycle. This, along with other molecular pathophysiological events, contributes to the progression of atherosclerosis. On the other hand, mechanisms that facilitate the cholesterol efflux from vascular tissues may counteract the earlier vicious cycle and afford vascular protection. Indeed, as discussed next, HDL-mediated reverse cholesterol transport serves as an important protective mechanism against atherosclerosis.

3.2.2.3 Reverse Cholesterol Transport Another important aspect of lipid metabolism is cholesterol efflux from cells of peripheral tissues, especially blood vessels. This is primarily mediated by HDL particles. Therefore, this section first considers the formation and maturation of HDL particles and then defines the reverse cholesterol transport.

Formation and Maturation of Nascent HDL   Nascent HDL particles are synthesized by the liver and intestine. The earliest events occur when lipid-poor ApoA-I is secreted by the hepatocytes or enterocytes or dissociated from lipoprotein particles in the plasma. These amphipathic ApoA-I molecules acquire phospholipids and unesterified cholesterol from the membranes of hepatocytes and enterocytes via the action of the membrane protein ATP-binding cassette protein A1 (ABCA1). The resulting small, disk-shaped particles, which consist of mainly ApoA-I and phospholipids, are referred to as nascent or pre-β-HDL particles due to their characteristic migration on agarose gels.

As the nascent discoidal HDL particles are relatively inefficient in removing excess cholesterol from cell membranes, they must mature into spherical particles in the plasma. HDL maturation occurs via the action of two circulating proteins: (i) lecithin–cholesterol acyltransferase (LCAT) and (ii) phospholipid transfer protein (PLTP). LCAT converts the cholesterol molecules in the nascent LDL particles to cholesteryl esters. This is accomplished by transesterification of a fatty acid from a phosphatidylcholine molecule on the surface of the nascent HDL to the hydroxyl group of a cholesterol molecule. This reaction also creates a lysophosphatidylcholine molecule, which dissociates from the HDL and binds to plasma albumin. Due to high insolubility, cholesteryl esters migrate into the core of the HDL particles, resulting in the development of a hydrophobic core. This converts the nascent HDL (or pre-β-HDL) to spherical mature HDL particles. Mature HDL is also known as α-HDL due to its characteristic electrophoretic motility (Table 3.1).

The PLTP transfers phospholipids from the surface coat of ApoB-containing remnant particles to the surface coat of HDL. As noted earlier, during lipoprotein lipase-catalyzed hydrolysis of ApoB-containing lipoproteins (e.g., VLDL, chylomicrons), the particles become smaller as triglycerides are removed from the core. This leaves a relative excess of phospholipids on the surface of the ApoB-containing particles. Because phospholipids cannot be spontaneously dissociated from the particles, PLTP removes excess phospholipids and thereby maintains the appropriate surface phospholipid concentration for a shrinking core of the ApoB-containing particles. By transferring phospholipids from the ApoB-containing particles to the surface of HDL, PLTP also replaces the molecules that are consumed by the LCAT-catalyzed reaction. This allows the core of HDL to continue to grow. During the maturation, HDL particles also acquire additional apolipoproteins from other lipoprotein particles (see Table 3.1 for HDL composition).

Definition and Process of Reverse Cholesterol Transport   Reverse cholesterol transport refers to the transport of cholesterol from the plasma membranes of peripheral cells (such as macrophages in vascular tissues) to the liver (Fig. 3.3). This occurs when unesterified cholesterol is transferred from plasma membranes of the peripheral cells to the HDL particles. When mature HDL particles circulate to the liver, they interact with the scavenger receptor class B1 (SR-B1) on the surface of hepatocytes. SR-B1 serves as a principal receptor for HDL. This receptor promotes the selective uptake of lipids. In this process, the cholesterol and cholesteryl esters of HDL particles are taken up into the hepatocytes in the absence of the uptake of the apolipoproteins. The hepatic removal of cholesteryl esters from HDL particles increases the capacity of these particles to take on additional cholesterol molecules from peripheral cells. Here, peripheral cells refer to cells outside the liver, especially cells in vascular tissues.

HDL-C may also be transported into hepatocytes via an indirect pathway. In this process, HDL cholesteryl esters are transferred to ApoB-containing remnant lipoproteins in exchange for triglycerides via the action of the cholesteryl ester transfer protein (CETP). The physiological significance of this pathway remains unclear. On the one hand, transfer of the cholesteryl esters from HDL to ApoB-containing remnant lipoproteins would theoretically increase the capacity of HDL particles to take on additional cholesterol from peripheral cells. On the other hand, this process would also lead to increased cholesterol levels of ApoB-containing lipoproteins, thereby potentially increasing the risk of atherosclerosis. Hence, the transfer of cholesteryl esters from HDL to ApoB-containing remnant lipoproteins might not be physiologically advantageous. Indeed, CETP deficiency in humans is associated with an increase in the cholesteryl ester content of HDL and a reduction in plasma levels of LDL cholesterol. CETP genotypes causing moderate inhibition of CETP activity (and, therefore, modestly higher HDL-C levels) are associated with a reduced coronary risk, which is consistent with the notion that increased HDL-C is cardiovascular protective [6]. Pharmacological agents have also been recently developed to inhibit CETP to augment HDL-C and achieve cardiovascular protection, though the exact clinical efficacy of such a strategy remains to be established [7–9] (see Chapter 4 for more description of CETP inhibitors).

In addition to CETP, microRNAs also control HDL abundance and function by regulating gene networks that control HDL biogenesis and uptake, as well as discrete steps in the reverse cholesterol transport pathway [10]. Notably HDL is found to transport miRNAs selectively in health and disease, offering new possibilities of how this lipoprotein may alter gene expression in distal target cells and tissues and impact the development of atherosclerosis [10, 11]. These novel findings may open new avenues for the development of

HDL-targeted therapeutics for treating dyslipidemias and controlling atherosclerotic cardiovascular diseases.

3.3 DYSLIPIDEMIAS AND GENETIC LIPOPROTEIN DISORDERS

3.3.1 Classification and Molecular Etiologies

Dyslipidemia and lipoprotein disorder (also known as disorder of lipoprotein metabolism) are probably among the most commonly encountered terms in cardiovascular medicine. The term lipoprotein disorder emphasizes the molecular etiology and pathophysiology, whereas the term dyslipidemia focuses on the plasma lipid profile (primarily cholesterol and triglycerides) of the lipoprotein disorder. Dyslipidemias are usually classified into four categories: (i) hypercholesterolemia, (ii) hypertriglyceridemia, (iii) mixed hyperlipidemia, and (iv) hypoalphalipoproteinemia (i.e., low HDL levels). These dyslipidemias may result from genetic lipoprotein disorders and/or secondary causes. The secondary causes may include hormonal and metabolic disorders, renal and liver diseases, as well as unhealthy lifestyle and the use of certain drugs. This section primarily considers genetic lipoprotein disorders. Table 3.2 lists the various types of genetic lipoprotein disorders and their molecular etiologies.

3.3.2 The Four Types of Dyslipidemias and Their Underlying Genetic Lipoprotein Disorders

As noted earlier, dyslipidemias may stem from lipoprotein disorders caused by genetic alterations in the genes encoding the various components of the lipoproteins, the receptors of the lipoproteins, or the enzymes involved in the metabolism of lipoproteins. As described in the following text, a particular type of dyslipidemias may result from more than one lipoprotein disorders. Dyslipidemias resulting from genetic lipoprotein disorders are frequently called primary dyslipidemias. Most cases of primary dyslipidemias do not result from a single gene defect but instead stem from the cumulative burden of multiple genes (polygenic) that predisposes the individuals to milder forms of dyslipidemias, particularly in the presence of unhealthy lifestyle, including excessive dietary intake of saturated fats and lack of physical activity. On the other hand, dyslipidemias may also result from a variety of other diseases and conditions, including obesity, diabetes mellitus, lipodystrophy, thyroid disease, kidney disorders, hepatic disorders, lysosomal storage diseases, Cushing's syndrome, alcohol intake, and use of estrogens and many other medications (e.g., thiazide diuretics, beta-antagonists, anti-HIV protease inhibitors, and antischizophrenic drugs). Dyslipidemias resulting from the above diseases and conditions are known as secondary

dyslipidemias. This section describes the four types of primary dyslipidemias and their underlying genetic lipoprotein disorders, which provide a foundation for understanding drug therapy of dyslipidemias.

3.3.2.1 Hypercholesterolemia Primary hypercholesterolemia is characterized by increased levels of total plasma cholesterol and LDL cholesterol with normal plasma concentrations of triglycerides. The causes of primary hypercholesterolemia include familial hypercholesterolemia, familial defective ApoB-100, autosomal dominant hypercholesterolemia, and, more commonly, polygenic hypercholesterolemia (Table 3.2).

3.3.2.2 Hypertriglyceridemia Primary hypertriglyceridemia is characterized by high fasting plasma levels of triglycerides (200–500 mg/dl or higher) and frequently normal plasma levels of LDL cholesterol. The major causes of primary hypertriglyceridemia include familial hypertriglyceridemia, lipoprotein lipase deficiency, and ApoC-II deficiency (Table 3.2).

3.3.2.3 Mixed Hyperlipidemia Patients with primary mixed hyperlipidemia exhibit complex lipid profiles, which are frequently characterized by increased plasma levels of total cholesterol, LDL cholesterol, and triglycerides. HDL-C is often reduced. Mixed hyperlipidemia mainly results from familial combined hyperlipidemia and dysbetalipoproteinemia (Table 3.2).

3.3.2.4 Hypoalphalipoproteinemia Low plasma of HDL-C is referred to as hypoalphalipoproteinemia because HDL is the "alpha-lipoprotein" (Table 3.1). Decreased HDL-C is an independent risk factor of atherosclerotic cardiovascular diseases [13]. The causes of decreased HDL-C include ApoA-I deficiency, Tangier disease, familial HDL deficiency, and LCAT deficiency (Table 3.2).

3.4 MECHANISTICALLY BASED DRUG THERAPY OF DYSLIPIDEMIAS

The in-depth understanding of lipoprotein metabolism and lipoprotein disorders at the molecular and genetic levels over the past decades has dramatically advanced the development of mechanistically based pharmacological modalities for treating dyslipidemias to reduce the risk of atherosclerotic cardiovascular diseases. These, as detailed in Chapter 4, include (i) HMG-CoA reductase inhibitors (also known as statins) that inhibit the de novo cholesterol synthesis in the liver; (ii) bile acid sequestrants that block the enterohepatic circulation of bile acids, thereby causing increased synthesis of bile acids from cholesterol; (iii) cholesterol absorption inhibitors that block the absorption of

TABLE 3.2   Various types of genetic lipoprotein disorders

	Name	Genetic etiology	Incidence
LDL disorders (increased levels of LDL cholesterol)	Familial hypercholesterolemia	LDL receptor gene mutations, causing decreased expression of LDL receptors, leading to reduced uptake of LDL by hepatocytes	1/500 (heterozygous); 1/1,000,000 (homozygous)
	Familial defective ApoB-100	Mutations of the ApoB-encoding gene, causing disruption of LDL receptor-binding domain of ApoB-100 and the consequent decreased hepatic uptake of LDL	<1/1,000 (in individuals of German descent)
	Autosomal dominant hypercholesterolemia	Gain-of-function mutations in PCSK9-encoding gene, leading to increased LDL receptor degradation. PCSK9 binds to LDL receptors, causing receptor degradation	<1/1,000,000
	Autosomal recessive hypercholesterolemia	Mutations in LDL receptor adaptor protein (LDLRAP)-encoding gene. LDLRAP is a protein involved in LDL receptor-mediated endocytosis	<1/1,000,000
	Sitosterolemia (also known as phytosterolemia)	Mutations in either ABCG5- or ABCG8-encoding gene, leading to decreased pumping out of cholesterol from enterocytes to the gut lumen and from the hepatocytes into the bile. ABCG5 and ABCG8 are two ATP-binding cassette (ABC) proteins that are expressed in enterocytes and hepatocytes. Heterodimerization of the two proteins forms a functional complex that pumps both plant steroids and animal steroids (mainly cholesterol) into the gut lumen and into the bile	<1/1,000,000
	Familial lipoprotein(a) hyperlipoproteinemia	Variation in the Apo(a)-encoding gene	
Triglyceride-rich lipoprotein disorders (increased levels of chylomicrons/VLDL)	Familial hypertriglyceridemia (mainly increased VLDL levels)	Polygenic (the disorder results from the cumulative burden of multiple genes)	1/500
	Familial combined hyperlipidemia (elevated levels of both VLDL triglycerides an LDL cholesterol)	Polygenic (see above for definition)	
	Lipoprotein lipase deficiency (mainly chylomicronemia, also elevated plasma levels of VLDL)	Mutations in lipoprotein lipase-encoding gene, leading to lipoprotein lipase deficiency and thereby decreased catabolism of VLDL and chylomicrons	1/1,000,000
	Familial ApoC-II deficiency (mainly chylomicronemia, also elevated plasma levels of VLDL)	Mutations in ApoC-II-encoding gene, leading to ApoC-II deficiency. ApoC-II is a cofactor of lipoprotein lipase, and deficiency of ApoC-II disables lipoprotein lipase	<1/1,000,000
	ApoA-V deficiency (chylomicronemia)	Mutations in ApoA-V-encoding gene, leading to ApoA-V deficiency. ApoA-V is present in chylomicrons, VLDL, and HDL and is involved in triglyceride metabolism. Deficiency of ApoA-V results in elevated levels of triglyceride-rich lipoproteins [12]	<1/1,000,000

TABLE 3.2 (*Continued*)

	Name	Genetic etiology	Incidence
Remnant lipoprotein disorders (increased levels of chylomicron and VLDL remnants)	Familial hepatic lipase deficiency	Mutations in hepatic lipase-encoding gene, leading to hepatic lipase deficiency. Hepatic lipase is involved in the conversion of IDL to LDL, and deficiency of this enzyme results in increased levels of IDL particles, also called VLDL remnants	<1/1,000,000
	Familial dysbetalipoproteinemia	Genetic variation in ApoE-encoding gene, leading to ApoE dysfunction. ApoE is present on the surface of chylomicron remnants and is essential for the receptor-mediated endocytosis of the particles via hepatic receptors. Several types of mutations of the ApoE can cause a deficiency in the clearance of these remnant particles	1/10,000
HDL disorders (decreased levels of HDL-C)	ApoA-I deficiency	Deletion of ApoA-I-encoding gene, leading to ApoA-I deficiency. ApoA-I is a major apolipoprotein of HDL and required for LCAT activity	
	ABCA1 deficiency, also known as Tangier disease	Mutations in ABCA1-encoding gene, leading to ABCA1 deficiency. ABCA1 is a cellular transporting protein that facilitates efflux of unesterified cholesterol and phospholipids from peripheral cells to ApoA-I. Hence, ABCA1 deficiency affects HDL maturation	
	Familial HDL deficiency	Mutations in ABCA1 or ApoA-I-encoding gene, leading to decreased formation of ABCA1 or ApoA-I	
	LCAT deficiency (there are two forms of LCAT deficiency: complete deficiency, also called classic LCAT deficiency, and partial deficiency, also known as fish-eye disease)	Mutations in LCAT-encoding gene, leading to LCAT deficiency. LCAT is activated by ApoA-I and catalyzes the esterification of cholesterol to form cholesteryl esters. LCAT deficiency results in increased proportion of free cholesterol in circulating lipoproteins and impairs the functional maturation of HDL, leading to the rapid catabolism of circulating ApoA-I	

cholesterol from the intestine; (iv) fibrate drugs that act to enhance the activity of lipoprotein lipase, thereby promoting the catabolism of triglyceride-enriched lipoproteins; (v) niacin that acts to decrease free fatty acids used for triglyceride synthesis and that acts to increase HDL; and (vi) emerging pharmacologic modalities, including agents that inhibit CETP to augment HDL-C [9, 14], as well as drugs that inhibit PCSK9 to increase LDL receptors [15–17]. As our understanding of the molecular etiology and pathophysiology of lipoprotein disorders increases, more effective drugs will continue to emerge.

3.5 SUMMARY OF CHAPTER KEY POINTS

- Dyslipidemias, including hyperlipidemia and low levels of HDL cholesterol, constitute a chief cause of atherosclerosis and atherosclerotic cardiovascular diseases.

- Dyslipidemias are typically classified into hypercholesterolemia, hypertriglyceridemia, mixed hyperlipidemia, and hypoalphalipoproteinemia (i.e., low HDL levels). Dyslipidemias are frequently caused by genetic lipoprotein disorders. They may also result from other diseases and conditions, such as obesity, diabetes, and use of certain drugs.

- Lipoproteins are large macromolecular complexes of lipids and proteins that are essential for transporting hydrophobic lipids (primarily triglycerides and cholesterol) and lipid-soluble vitamins. The plasma lipoproteins are classified into five major classes, that is, chylomicrons, VLDL, IDL, LDL, and HDL.

- The metabolism of lipoproteins involves three pathways: the exogenous pathway (transport of dietary lipids), the endogenous pathway (transport of hepatic lipids), and the reverse cholesterol transport. Dysregulation of any of the above pathways by either genetic disorders or dietary factors (frequently via combination of both genetic and dietary factors) leads to dyslipidemias.

- Advancement in the molecular medicine of dyslipidemias has resulted in the development of different classes of drugs that target various aspects of lipoprotein metabolism to treat dyslipidemias. As our understanding of the molecular mechanisms of lipoprotein disorders increases, novel therapeutic modalities will continue to emerge for more effectively controlling dyslipidemias. The subsequent Chapter 4 considers the molecular pharmacology of both currently available and emerging drugs for treating dyslipidemias.

3.6   SELF-ASSESSMENT QUESTIONS

3.6.1 A 35-year-old male presents to his doctor's office for an annual checkup. Lab tests show a severely elevated level of plasma LDL cholesterol. He is diagnosed with familial hypercholesterolemia. Mutation in which of the following genes is most likely responsible for the elevated LDL cholesterol?
A. ABCG5
B. ApoB-100
C. ApoC-II
D. LDL receptor
E. Scavenger receptor B1

3.6.2 A 16-year-old female presents with tendon xanthomas on the extensor surface of her hands and on her Achilles tendons. Blood test reveals remarkably elevated levels in sitosterol (18.5 mg/dl) and LDL cholesterol (485 mg/dl). Genetic sequencing for ABCG5 reveals compound heterozygous null mutations. Which of the following is the patient most likely having?
A. ApoA-V deficiency
B. Chylomicronemia
C. Hepatic lipase deficiency
D. Phytosterolemia
E. Tangier disease

3.6.3 A 46-year-old male is diagnosed with familial ApoC-II deficiency. Which of the following is most likely experienced by the patient?
A. Severely elevated plasma HDL-cholesterol levels
B. Severely elevated plasma LDL-cholesterol levels
C. Severely elevated plasma Lp(a) levels
D. Severely elevated plasma plant steroid levels
E. Severely elevated plasma triglyceride levels

3.6.4 A 10-year-old boy is diagnosed with a gain-of-function mutation in PCSK9-encoding gene. Which of the following is most likely to be found on a blood lipid test?
A. Elevated Lp(a)
B. Elevated HDL cholesterol
C. Elevated LDL cholesterol
D. Elevated plant steroids
E. Elevated triglycerides

3.6.5 Substantial studies demonstrate that low HDL cholesterol is a significant risk factor for developing atherosclerotic cardiovascular diseases. As such, increasing HDL cholesterol with therapeutic agents has been considered an attractive strategy for controlling atherosclerotic cardiovascular diseases. Inhibition of which of the following proteins would result in elevated HDL-cholesterol levels?
A. ABCG8
B. CETP
C. LCAT
D. LDLRAP
E. PCSK9

REFERENCES

1 Tsimikas, S. and J.L. Hall, Lipoprotein(a) as a potential causal genetic risk factor of cardiovascular disease: a rationale for increased efforts to understand its pathophysiology and develop targeted therapies. *J Am Coll Cardiol*, 2012. 60(8): p. 716–21.

2 O'Donoghue, M.L., et al., Lipoprotein(a) for risk assessment in patients with established coronary artery disease. *J Am Coll Cardiol*, 2014. 63(6): p. 520–7.

3 Thanassoulis, G., et al., Genetic associations with valvular calcification and aortic stenosis. *N Engl J Med*, 2013. 368(6): p. 503–12.

4 Chatterjee, C. and D.L. Sparks, Hepatic lipase, high density lipoproteins, and hypertriglyceridemia. *Am J Pathol*, 2011. 178(4): p. 1429–33.

5 Schekman, R., Discovery of the cellular and molecular basis of cholesterol control. *Proc Natl Acad Sci U S A*, 2013. 110(37): p. 14833–6.

6 Thompson, A., et al., Association of cholesteryl ester transfer protein genotypes with CETP mass and activity, lipid levels, and coronary risk. *JAMA*, 2008. 299(23): p. 2777–88.

7 Schaefer, E.J., Effects of cholesteryl ester transfer protein inhibitors on human lipoprotein metabolism: why have they

failed in lowering coronary heart disease risk? *Curr Opin Lipidol*, 2013. 24(3): p. 259–64.

8   Rader, D.J. and E.M. deGoma, Future of cholesteryl ester transfer protein inhibitors. *Annu Rev Med*, 2014. 65: p. 385–403.

9   van Capelleveen, J.C., et al., Novel therapies focused on the high-density lipoprotein particle. *Circ Res*, 2014. 114(1): p. 193–204.

10 Rayner, K.J. and K.J. Moore, MicroRNA control of high-density lipoprotein metabolism and function. *Circ Res*, 2014. 114(1): p. 183–92.

11 Fernandez-Hernando, C., Antiatherogenic Properties of High-Density Lipoprotein-Enriched MicroRNAs. *Arterioscler Thromb Vasc Biol*, 2014. 34(6):e13–4.

12 Sharma, V., T.M. Forte, and R.O. Ryan, Influence of apolipo-protein A-V on the metabolic fate of triacylglycerol. *Curr Opin Lipidol*, 2013. 24(2): p. 153–9.

13 Feig, J.E., et al., High-density lipoprotein and atherosclerosis regression: evidence from preclinical and clinical studies. *Circ Res*, 2014. 114(1): p. 205–13.

14 Schwartz, G.G., et al., Effects of dalcetrapib in patients with a recent acute coronary syndrome. *N Engl J Med*, 2012. 367(22): p. 2089–99.

15 Stein, E.A., et al., Effect of a monoclonal antibody to PCSK9 on LDL cholesterol. *N Engl J Med*, 2012. 366(12): p. 1108–18.

16 Fitzgerald, K., et al., Effect of an RNA interference drug on the synthesis of proprotein convertase subtilisin/kexin type 9 (PCSK9) and the concentration of serum LDL cholesterol in healthy volunteers: a randomised, single-blind, placebo-controlled, phase 1 trial. *Lancet*, 2014. 383(9911): p. 60–8.

Raal, F.J., et al., Reduction in lipoprotein(a) with PCSK9 monoclonal antibody evolocumab (AMG 145): a pooled analysis of more than 1,300 patients in 4 phase II trials. *J Am Coll Cardiol*, 2014. 63(13): p. 1278–88.

4

DRUGS FOR DYSLIPIDEMIAS

4.1 OVERVIEW

Chapter 3 has discussed lipoprotein metabolism, dyslipidemias, and lipoprotein disorders and outlined the established and emerging molecular targets that drugs act to treat dyslipidemias. This chapter examines the molecular pharmacology of the various classes of drugs for treating dyslipidemias, including statins, bile acid sequestrants, cholesterol absorption inhibitors, fibrates, and niacin, as well as the new drugs for homozygous familial hypercholesterolemia. The chapter also discusses the pharmacological basis of phytosterols, phytostanols, and omega-3 fatty acids in the management of dyslipidemias. The recently emerging drugs for dyslipidemias are considered at the end of the chapter. Following the discussion of the molecular pharmacology of drugs for dyslipidemias in this chapter, mechanistically and evidence-based guidelines for the management of dyslipidemias are considered in Chapter 5.

4.2 STATINS

Listed below are the seven US Food and Drug Administration (FDA)-approved statin drugs for clinical use:

- Atorvastatin (Lipitor)
- Fluvastatin (Lescol)
- Lovastatin (Mevacor)
- Pitavastatin (Livalo)
- Pravastatin (Pravachol)
- Rosuvastatin (Crestor)
- Simvastatin (Zocor)

4.2.1 General Introduction to Drug Class

4.2.1.1 Nomenclature Statins (also called statin drugs) are formally known as 3-hydroxy-3-methylglutaryl-coenzyme A (HMG-CoA) reductase inhibitors. They are called statins because all members of this drug class end with "statin." However, not every drug that ends with "statin" is a statin drug. For example, nystatin is an antifungal drug, not a statin for treating dyslipidemias.

HMG-CoA reductase catalyzes the rate-limiting step of cholesterol biosynthesis (more in Section 4.2.3.1). As such, statin drugs potently inhibit the de novo synthesis of cholesterol. Statins are highly effective and well tolerated and the most commonly prescribed drugs for treating dyslipidemias. Multiple well-controlled clinical trials have demonstrated the efficacy and safety of statins in reducing atherosclerosis-associated fatal and nonfatal coronary heart disease events, strokes, and total mortality.

Currently, there are seven statin drugs approved by the US FDA for clinical use. They are atorvastatin, fluvastatin, lovastatin, pitavastatin, pravastatin, rosuvastatin, and simvastatin. Among them, pitavastatin is the newest member, approved by the FDA in 2009 [1].

4.2.1.2 History of Discovery The first statin compound was initially isolated from a fungal culture and identified as an inhibitor of cholesterol biosynthesis in 1976 by the Japanese biochemist Akira Endo, working at the Sankyo Company in Tokyo [2]. A. Endo had been interested in cholesterol metabolism for some time and proposed in 1971 that the fungi being screened at that time for new antibiotics might also possess an inhibitor of cholesterol synthesis.

Cardiovascular Diseases: From Molecular Pharmacology to Evidence-Based Therapeutics, First Edition. Y. Robert Li.
© 2015 John Wiley & Sons, Inc. Published 2015 by John Wiley & Sons, Inc.

Over a period of multiple years, A. Endo screened more than 6000 fungal cultures until a positive result was observed. The compound isolated from the culture was first named compactin and then renamed mevastatin.

Compactin subsequently proved effective in decreasing cholesterol levels in dogs and monkeys, and the work came to the attention of Michael Brown and Joseph Goldstein at the University of Texas Southwestern Medical School. M. Brown and J. Goldstein began to work with A. Endo and confirmed his results. The positive results from the first limited clinical trials convinced several pharmaceutical companies to join the hunt for statins. A team at Merck led by Alfred Alberts and P. Roy Vagelos began screening fungal cultures and found a positive result after

screening just 18 cultures. The new statin was eventually called lovastatin and approved by the US FDA for use in humans. Several newer generations of statin drugs have been subsequently approved for clinical use. Pitavastatin, pravastatin, and simvastatin are chemically modified derivatives of lovastatin. Atorvastatin, fluvastatin, and rosuvastatin are structurally distinct synthetic compounds (see Section 4.2.2.1).

4.2.2 Chemistry and Pharmacokinetics

4.2.2.1 *Chemistry* Among the seven statins (structures shown in Fig. 4.1), lovastatin and simvastatin are prodrugs that need to undergo metabolism to yield the active forms

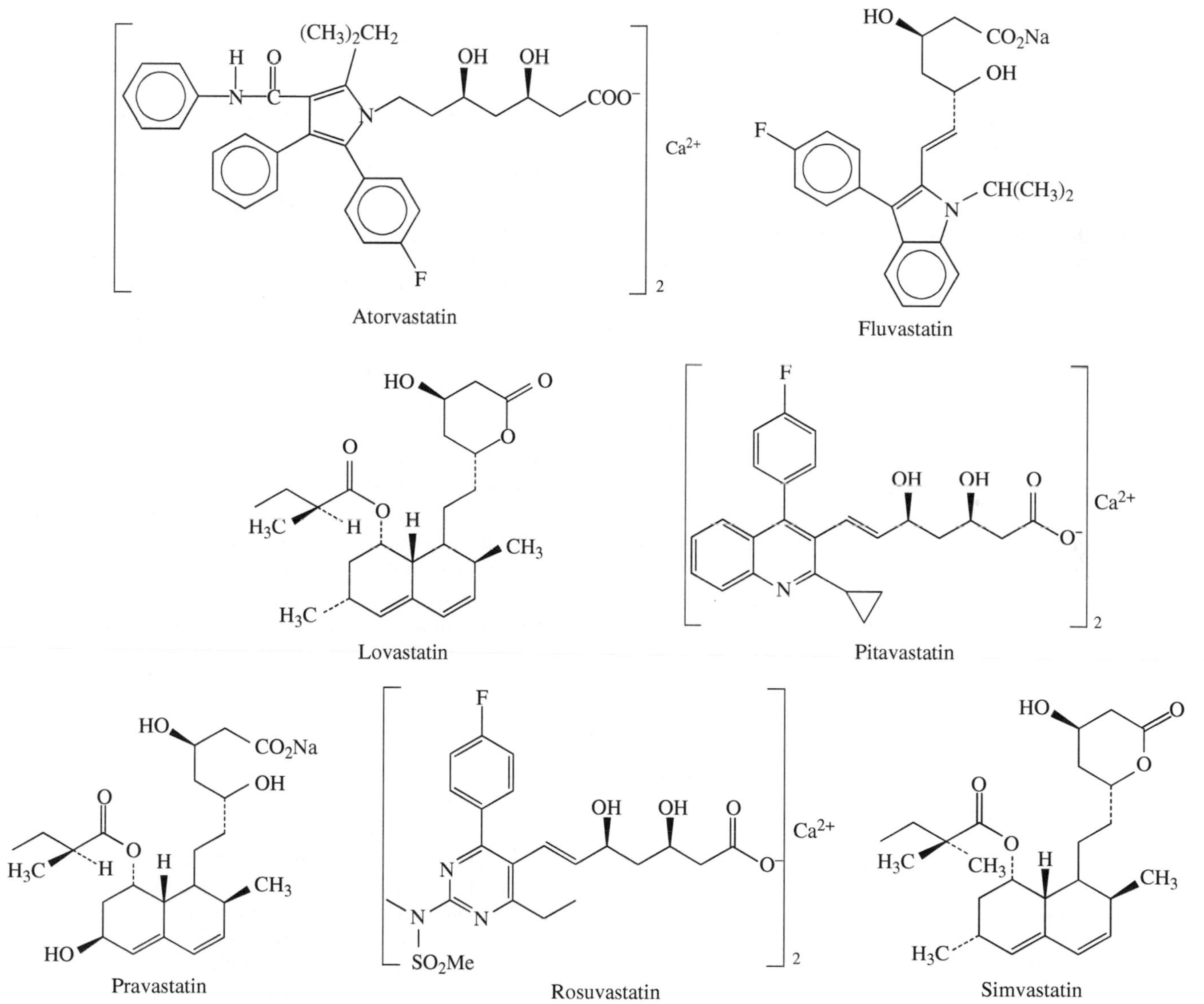

FIGURE 4.1 Chemical structures of the seven US FDA-approved statins. Lovastatin and simvastatin are prodrugs that need to undergo hydrolysis in the gut and liver to form the active drugs that inhibit HMG-CoA reductase. Atorvastatin, pitavastatin, and rosuvastatin exist as calcium salts, whereas fluvastatin and pravastatin are sodium salts.

that inhibit HMG-CoA reductase. These two statins are inactive lactone compounds that are hydrolyzed in the gut and liver to form the active β-hydroxyl derivatives that inhibit the reductase. The other five statins directly inhibit HMG-CoA reductase.

4.2.2.2 Pharmacokinetics

After oral administration, intestinal absorption of statins varies from 40 to 75% with the exception of fluvastatin, which is almost completely absorbed. There is extensive first-pass hepatic uptake of all statin drugs, mediated primarily by the organic anion transport OATP1B1, which is selectively expressed in the liver [3]. All statins undergo extensive metabolism in the liver, and the oral bioavailability varies between 5 and 30% of administered doses. The metabolism of atorvastatin, lovastatin, and simvastatin occurs mainly via the action of cytochrome P450 3A4 (CYP3A4), whereas that of fluvastatin and rosuvastatin is largely mediated by CYP2C9. Pravastatin is metabolized through other pathways, including sulfation.

The principal route of metabolism of the recently approved pitavastatin is glucuronidation via liver UDP-glucuronosyltransferases (UGT1A3 and UGT2B7) with subsequent formation of pitavastatin lactone. There is only minimal metabolism of this new statin by the CYP system. Because of this, pitavastatin is thought to have a low potential for drug interactions.

More than 70% of statin metabolites are excreted into bile and subsequently eliminated in feces. The rest is excreted via the renal route. After an oral dose, plasma concentrations of statin peak in 1–4 h. The plasma half-lives of the parent compounds for atorvastatin, pitavastatin, and rosuvastatin are 14, 12, and 19 h, respectively, and, for the other four statins, range from 1 to 4 h. The longer half-lives of atorvastatin and rosuvastatin may contribute to their greater cholesterol-lowering efficiency. The major pharmacokinetic parameters of the seven statin drugs are listed in Table 4.1.

4.2.3 Molecular Mechanisms and Pharmacological Effects

The mechanisms and pharmacological effects of statins are considered from three general aspects: (i) decrease of cholesterol synthesis by inhibiting HMG-CoA reductase and impact on lipoprotein metabolism, (ii) decreased prenylation of signaling molecules resulting from the initial inhibition of HMG-CoA reductase, and (iii) other novel effects independent of inhibition of HMG-CoA reductase. Among these mechanistic pathways, reduction of cholesterol biosynthesis via inhibiting HMG-CoA reductase has been most extensively studied and firmly established. As such, statin drugs are also known as HMG-CoA reductase inhibitors. The reduction of plasma cholesterol is crucial in reducing cardiovascular events and has led to practice guidelines that emphasize the importance of achieving target goals for LDL levels (see Chapter 5).

4.2.3.1 Inhibition of Cholesterol Biosynthesis and Impact on Lipoprotein Metabolism

As noted earlier, HMG-CoA reductase catalyzes the conversion of HMG-CoA to mevalonic acid and is the rate-limiting enzyme in cholesterol biosynthesis (Fig. 4.2). The active forms of the statin drugs contain a structural moiety that is similar to the structure of HMG-CoA, and as such, they competitively inhibit HMG-CoA reductase, leading to decreased biosynthesis of cholesterol in the liver. Statins primarily impact hepatic cholesterol biosynthesis due to two reasons: (i) following oral administration, statins are extracted by the liver, and as such, the liver receives the most statin molecules; and (ii) the liver is a major organ in de novo cholesterol biosynthesis.

Treatment with statin drugs not only impacts plasma LDL-cholesterol levels but also affects plasma concentrations of triglycerides, as well as HDL cholesterol. The magnitude of LDL cholesterol lowering depends on the efficacy and dosage of the statin that is administered. In general, statin drugs reduce plasma LDL-cholesterol concentrations

TABLE 4.1   Major pharmacokinetic properties of statin drugs

Statin	Oral bioavailability (%)	Elimination half-life (h)	Metabolism and elimination
Atorvastatin	12	14	CYP3A4; eliminated in the bile/feces
Fluvastatin	25	1	CYP2C9; eliminated in the bile/feces
Lovastatin	5	3.5	CYP3A4; eliminated in the bile/feces
Pitavastatin	51	12	Glucuronidation; eliminated in the bile/feces (major) and the urine (minor)
Pravastatin	17	2	Sulfation; eliminated in the bile/feces (major) and the urine (minor)
Rosuvastatin	20	19	CYP2C9; eliminated in the bile/feces
Simvastatin	<5	3	CYP3A4; eliminated in the bile/feces (major) and the urine (minor)

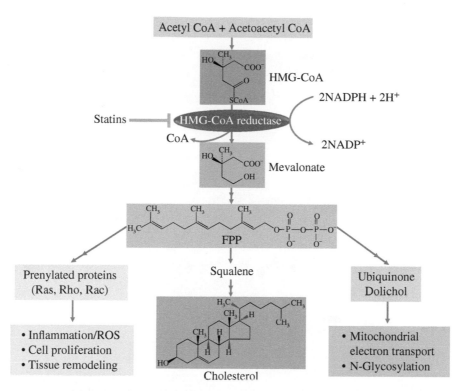

FIGURE 4.2 Molecular mechanism by which statins inhibit cholesterol synthesis. Statins competitively inhibit HMG-CoA reductase, a key enzyme in cholesterol biosynthesis. Inhibition of HMG-CoA reductase by statins also decreases protein prenylation, a process that leads to inflammation, reactive oxygen species (ROS) formation, cell proliferation, and tissue remodeling. These events play an important part in atherosclerosis. Decreased protein prenylation of small G proteins, including Ras, Rho, and Rac, may hence account for the lipid-lowering-independent beneficial effects of statin therapy. On the other hand, statin drugs decrease ubiquinone formation, which might partly be responsible for the development of myopathy associated with statin treatment. FPP denotes farnesyl pyrophosphate. For color details, please see color plate section.

by 25–55%. Statins reduce plasma triglyceride levels by 10–35% depending on the type of statins, the doses, and the degree of hypertriglyceridemia. On average, treatment with stains results in a 5–10% increase in plasma HDL-cholesterol levels. The molecular mechanisms underlying the effects of statins on lipoproteins are described next.

Effects on Plasma LDL-Cholesterol Levels Decreased de novo synthesis of cholesterol in hepatocytes results in reduced amounts of free cholesterol, thereby leading to augmented expression of LDL receptor gene. As illustrated in Figure 4.3, decreased levels of free cholesterol in hepatocytes lead to proteinase activation and cleavage of the sterol regulatory element-binding protein (SREBP), which is a transcription factor that normally resides in the cytoplasm. The cleaved SREBP undergoes nuclear translocation and binds to sterol response element of the LDL receptor gene, leading to upregulation of LDL receptor transcription and the subsequent increased synthesis and surface expression of LDL receptors.

The increased cell surface expression of LDL receptors promotes uptake of LDL particles from plasma into hepatocytes, resulting in reduced LDL-cholesterol

concentrations in plasma. This is a major mechanism by which statins reduce plasma LDL-cholesterol levels. There is evidence showing that the plasma LDL-cholesterol-lowering effects of statins may also occur via several minor pathways, including (i) enhanced removal of LDL precursors, such as VLDL and IDL, from plasma via LDL receptor-mediated endocytosis, and (ii) decreased production of VLDL particles in hepatocytes due to decreased levels of free cholesterol within the hepatocytes. Cholesterol is a major building block for VLDL synthesis in the hepatocytes. Reduced synthesis of hepatic VLDL causes decreased release of VLDL to plasma, which contributes to decreased levels of LDL cholesterol as VLDL is a precursor of LDL.

The effects of statin drugs on plasma LDL-cholesterol levels vary among the various members. Of the seven statins, fluvastatin is the least potent, and pitavastatin is the most potent statin in reducing plasma LDL cholesterol, followed by atorvastatin and rosuvastatin. Indeed, before the approval of pitavastatin, atorvastatin and rosuvastatin were the most potent LDL-cholesterol-lowering drugs.

With regard to LDL cholesterol lowering, the dose–response relationship of statins is nonlinear: the largest

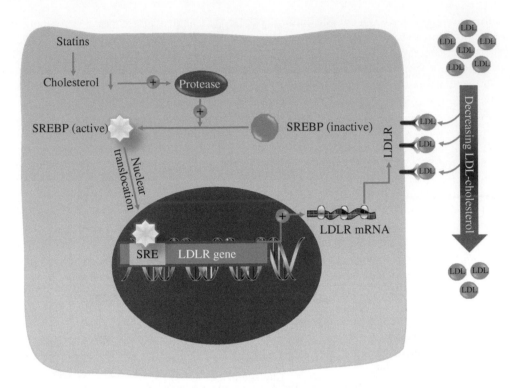

FIGURE 4.3   Molecular mechanism by which statins reduce LDL cholesterol. Statin-induced decreases in cholesterol concentration in hepatocytes result in protease activation. Protease activation causes the activation and nuclear translocation of sterol regulatory element-binding protein (SREBP), which binds to the sterol regulatory element of the LDL receptor (LDLR) gene, leading to increased expression of LDLR. Increased LDLR expression on the surface of hepatocytes promotes LDL uptake from plasma, thereby reducing plasma LDL cholesterol. For color details, please see color plate section.

reduction occurs with the starting dose. Each subsequent doubling of the dose produces, on average, an additional 6% LDL reduction. This is known as the "role of 6's" in statin therapy [4].

Effects on Plasma Triglyceride Levels   Statin treatment also decreases plasma triglycerides in patients with hyper-triglyceridemia. Hypertriglyceridemic patients taking the highest doses of the most potent statins experience a 35–45% reduction in fasting triglyceride levels, and the magnitude of reduction in triglyceride levels is similar to that of LDL-cholesterol reduction. The reducing effects of statins on plasma triglyceride levels result from two mechanisms: (i) the decreased VLDL production by the liver due to decreased levels of free cholesterol within hepatocytes (since VLDL is rich in triglycerides, decreased release of VLDL from the liver to plasma leads to reduced levels of plasma triglycerides) and (ii) increased clearance of remnant lipoproteins by the liver due to upregulation of LDL receptors on the surface of hepatocytes. As remnant lipoproteins carry high amounts of triglycerides, augmented hepatic uptake of these particles results in decreased plasma triglyceride concentration. Here, remnant lipoproteins primarily refer to chylomicron remnants and VLDL remnants. VLDL remnants are also known as IDL particles (see Chapter 3).

Effects on Plasma HDL-Cholesterol Levels   Statin treatment of patients with high LDL-cholesterol and normal HDL-cholesterol levels leads to an average 5–10% increase in HDL cholesterol. In patients with lower HDL cholesterol, the magnitude of HDL-cholesterol induction can be as much as 15–20% for certain potent statin drugs. The detailed mechanisms underlying statin-induced HDL-cholesterol elevation remain poorly understood. Statins increase ApoA-I levels, which may serve as a potential mechanism for the observed elevation of HDL cholesterol following statin therapy.

4.2.3.2   Decreased Prenylation of Signaling Molecules and Impact on Vascular Biology   In addition to their impact on lipoprotein metabolism, statins also exert many other pharmacological effects independent of their lipid-lowering action [5–7]. These include (i) improvement of endothelial dysfunction, (ii) decreased coagulation and thrombogenesis, (iii) decreased inflammation, (iv) increased stability of atherosclerotic plaques, and (v) decreased production of reactive oxygen species (ROS). These diverse actions are also known as the pleiotropic effects of statins.

Although the pleiotropic effects of statins are independent of the changes in the plasma levels of lipoproteins, such effects do stem from the initial inhibition of the HMG-CoA reductase. The synthesis of cholesterol is a complex multistep process. Isoprenoids are important intermediates in the cholesterol biosynthetic pathway that serve as lipid attachments for a wide variety of signaling molecules, such as heterotrimeric G proteins and the small GTP-binding protein Ras and its related Ras-like proteins (e.g., Rho, Rac, Rab, Rap, and Ral) (see Fig. 4.2).

In response to a number of growth factors and cytokines, the aforementioned signaling molecules are activated and thereby exert a diverse array of cellular effects. Although their functions are overlapping to some degree, Ras has been associated with dysregulated cell proliferation, Rac with generation of ROS, and Rho with activation of proinflammatory pathways. All of these contribute to vascular dysfunction.

By inhibiting HMG-CoA reductase, statins decrease the biosynthesis of isoprenoids and protein prenylation. As a consequence, vascular inflammation, dysregulated cell proliferation, and overproduction of ROS are ameliorated. Substantial studies suggest that the pleiotropic effects of statins may significantly contribute to their clinical efficacy in decreasing cardiovascular mobility and mortality. This notion is supported by several lines of evidence, as outlined below [5–8]:

- Results of multiple clinical trials suggest that the higher than expected degree of cardiovascular benefit of statins does not correlate fully with the magnitude of cholesterol lowering.
- Statins reduce the risk of transplant-associated arteriopathy and stroke, disease processes that are not strongly associated with elevated lipid levels.
- Reduction in new coronary events by statin treatment occurs more rapidly than changes in morphology of arterial plaques. Indeed, some statin trials have reported benefits with just a few weeks of statin therapy.
- Lipid lowering by alternative means, such as ileal bypass, requires more time to manifest cardiovascular clinical benefits than statin therapy.
- Inflammation plays a crucial role in cardiovascular disease processes. Statins possess potent anti-inflammatory activities, and treatment with statins decreases inflammatory biomarkers, such as C-reactive protein. Clinical studies demonstrate that the magnitude of benefits associated with a low C-reactive protein level may be as large as that associated with a low LDL-cholesterol level, and the benefits of decreasing C-reactive protein are additive and independent of the lipid lowering.

4.2.3.3 Other Novel Effects Independent of Inhibition of HMG-CoA Reductase
The pharmacological effects of statins described earlier are primarily derived from the potent inhibition of HMG-CoA reductase by these drugs. Evidence also exists, showing that statins may directly interact with various cellular signaling pathways involved in inflamma-

tion and antioxidant defenses [9, 10]. For example, statins can directly interact with lymphocyte function-associated antigen-1 and reduce adhesion of lymphocytes to the endothelium [11]. Statins also cause activation of nuclear factor E2-related factor 2 (Nrf2), a transcription factor playing a central role in regulating the expression of cellular antioxidant and other cytoprotective genes. Statins may directly scavenge free radicals and protect against oxidation of LDL. The contribution of these novel effects to statin-mediated cardiovascular protection, however, remains to be elucidated in clinical studies.

4.2.4   Clinical Uses

The clinical uses of statins are related to the following indications and considerations: (i) treatment of dyslipidemias in adults, (ii) treatment of dyslipidemias in children, (iii) treatment of dyslipidemias in women, and (iv) other potential indications.

4.2.4.1   Treatment of Dyslipidemias in Adults

Hypercholesterolemia and Hypertriglyceridemia Statin drugs are used alone or in combination with other lipid-lowering drugs (such as bile acid sequestrants, ezetimibe, or niacin) for reducing plasma LDL cholesterol in patients with hypercholesterolemia. Hypercholesterolemia is the primary indication of statin treatment, and statins are considered first-line therapy for increased LDL cholesterol. As discussed earlier, statins also increase HDL cholesterol. An early study demonstrated that statin therapy was associated with regression of coronary atherosclerosis when LDL cholesterol was substantially reduced and HDL cholesterol was increased by more than 7.5% [12]. A recent study also showed that maximal doses of rosuvastatin and atorvastatin decreased LDL cholesterol and increased HDL cholesterol and caused significant regression of coronary atherosclerosis [13]. These findings suggest that statin benefits may be derived from both reductions in atherogenic lipoprotein levels and increases in HDL cholesterol. However, it remains to be determined whether the atherosclerotic regression associated with these changes in lipid levels will translate to meaningful reductions in clinical events and improved clinical outcomes.

Statins, especially the most potent ones, such as atorvastatin and rosuvastatin, are also effective in reducing plasma triglycerides. These drugs are useful in treating patients with mixed hyperlipidemia (i.e., having both hypercholesterolemia and hypertriglyceridemia). It should be noted that although statin treatment lowers plasma triglyceride levels, fibrates and niacin (see Sections 4.5 and 4.6) are more effective in treating hypertriglyceridemia.

Combination Therapy with Other Lipid-Lowering Drugs Listed below are the typical combination options:
- A statin plus a bile acid sequestrant
- A statin plus ezetimibe

- A statin plus niacin
- A statin plus a fibrate drug

If a statin alone is insufficient to lower LDL cholesterol to target levels, the statin drug can be effectively combined with other lipid-lowering drugs. The combination of a statin with a bile acid sequestrant (Section 4.3) or the cholesterol absorption inhibitor ezetimibe (Section 4.4) results in additive decreases in LDL-cholesterol levels. A fixed-dose combination of simvastatin and ezetimibe (Vytorin) is available. The US FDA also recently approved a fixed-dose combination of atorvastatin and ezetimibe (Liptruzet).

The combination of a statin with niacin may be most useful in patients with high levels of LDL cholesterol and low levels of HDL cholesterol as niacin is the most effective agent in raising HDL cholesterol (see Section 4.6 for niacin). Fibrates and statins have also been reported to be efficacious in combination; however, such a combination may increase the risk of myopathy, a significant adverse effect associated with statin treatment (see Section 4.2.6 for adverse effects of statin therapy).

4.2.4.2   Treatment of Dyslipidemias in Children   The use of statins in children is restricted to selected patients with heterozygous familial hypercholesterolemia. In this regard, atorvastatin, lovastatin, and simvastatin are indicated for children ≥11 years, and pravastatin for children ≥8 years.

4.2.4.3   Treatment of Dyslipidemias in Women   The safety of statin drugs during pregnancy has not been well established. The US FDA has classified statins as category X drugs, and pregnant women are advised to stop statins during the preconception period and in pregnancy (see Chapter 2 for discussion of the FDA's pregnancy category of drugs). Accordingly, women with hyperlipidemia who are pregnant or likely to become pregnant should not take statins. Nursing mothers also are advised to avoid taking statins.

Due to the widespread use of statins in the management of dyslipidemias and cardiovascular diseases, the safety of these drugs in pregnancy has recently been revisited. A recent review of existing studies concluded that there was no evidence to demonstrate that use of statins in pregnancy increased the risk of fetal abnormalities [14]. However, the existing clinical studies on the safety of statins in pregnancy are limited by study quality, sample size, and confounding factors involved.

4.2.4.4   Other Potential Indications   The pleiotropic effects of statins imply that these drugs may exert benefits in other disease conditions in addition to atherosclerotic cardiovascular diseases. Clinical studies suggest that statins may be useful in improving the outcomes of ventricular arrhythmias, sudden cardiac death, cardiac transplant rejection, chronic obstructive pulmonary disease, sepsis,

and, most recently, periodontal inflammation [15]. However, for these conditions, the level of evidence was inadequate to recommend statin use. The evidence for improving outcomes in atrial fibrillation, mortality in heart failure, contrast-induced nephropathy, cataract, age-related macular degeneration, subarachnoid hemorrhage, osteoporosis, dementia, and cancer incidence was conflicting and inconclusive. The currently ongoing and future large-scale randomized, double-blind, placebo-controlled trials will certainly provide important insights into the potential clinical use of statins in diverse disease conditions other than dyslipidemias and atherosclerotic cardiovascular diseases.

4.2.5   Therapeutic Dosages

Listed below are the dosage forms and strengths of the seven statin drugs:

- Atorvastatin (Lipitor): Oral, 10, 20, 40, and 80 mg tablets
- Fluvastatin (Lescol): Oral, 20 and 40 mg capsules; extended release (Lescol XL), 80 mg capsules
- Lovastatin (Mevacor): Oral, 10, 20, and 40 mg tablets; extended release (Altoprev), 20, 40, and 60 mg tablets
- Pitavastatin (Livalo): Oral, 1, 2, and 4 mg tablets
- Pravastatin (Pravachol): Oral, 10, 20, 40, and 80 mg tablets
- Rosuvastatin (Crestor): Oral, 5, 10, 20, and 40 mg tablets
- Simvastatin (Zocor): Oral, 5, 10, 20, 40, and 80 mg tablets

As statins are primarily used to treat hypercholesterolemia, this section describes the therapeutic dosage of various statins with regard to their efficacy in reducing plasma LDL-cholesterol levels. Table 4.2 provides the dosages of statins required to attain various reductions in plasma LDL cholesterol from baseline levels. As shown in the table, before the recently approved pitavastatin, atorvastatin and rosuvastatin were the two most potent statins for reducing LDL-cholesterol levels.

Each statin drug has a low recommended starting dose that reduces LDL cholesterol by 20–30%. It is recommended that a high dose of statin be used to achieve the patient's target goal for lowering LDL cholesterol. Because hepatic cholesterol synthesis occurs predominantly at night, statins with half-lives <4 h (all but atorvastatin, pitavastatin, and rosuvastatin) should be given in the evening if a single daily dose is used.

The choice of statin drugs should be based on efficacy in reducing LDL cholesterol as well as the cost of the drug. Among the seven approved statins, atorvastatin, lovastatin, pravastatin, and simvastatin have been used safely in clinical trials involving thousands of patients for multiple years. The

TABLE 4.2 The once daily dosages of currently available statins required to attain various reductions in plasma LDL-cholesterol levels

Statin	20–25%	26–30%	31–35%	36–40%	41–50%	51–55%
Atorvastatin			10 mg	20 mg	40 mg	80 mg
Fluvastatin	20 mg	40 mg	80 mg			
Lovastatin	10 mg	20 mg	40 mg	80 mg		
Pitavastatin			1 mg	2 mg	4 mg	
Pravastatin	10 mg	20 mg	40 mg			
Rosuvastatin				5 mg	10 mg	20 mg, 40 mg
Simvastatin		10 mg	20 mg	40 mg	80 mg	

documented safety record should be considered, especially when initiating therapy in younger patients. In this context, although the recently approved pitavastatin is the most potent statin drug, the safety profile of this drug is less well established than that of the old members of the statin class. The therapeutic strategies and regimens of using statins along with other modalities for treating patients with high LDL-cholesterol levels are further discussed in Chapter 5.

4.2.6 Adverse Effects and Drug Interactions

Outlined below are key points regarding the adverse effects of statins and drug interactions, which are discussed individually in this section. In addition, as stated earlier, statins should be avoided during pregnancy or in nursing mothers:

- Statins are generally well-tolerated drugs.
- Less frequently, statins may cause liver toxicity.
- Myopathy is the major adverse effect of statins.
- Possible diabetogenic effects are far outweighed by benefits.
- Drug interactions for statins can be significant.

4.2.6.1 Statins as Generally Well-Tolerated Drugs
Statins appear to be remarkably safe. The potential common mild adverse effects include dyspepsia, headache, fatigue, and muscle and joint pain.

4.2.6.2 Hepatotoxicity Associated with Statins Less frequently, statins may cause liver toxicity as indicated by elevations of serum hepatic aminotransferase activity. Serum aminotransferases including alanine aminotransferase (ALT) and aspartate aminotransferase (AST) should be checked before starting therapy, 1–2 months after therapy, and then every 6–12 months if stable. Substantial (>3 times upper limit of normal) elevations in serum hepatic aminotransferases are relatively rare in patients treated with statins. Mild to moderate (1–3 times normal) elevations in serum hepatic aminotransferases in the absence of symptoms need not mandate discontinuing the medication.

In some patients who may have an underlying liver disorder, serum levels of aminotransferases may exceed three

times normal following statin therapy, and these patients may present with symptoms, such as malaise and anorexia. Statin therapy should be immediately discontinued in such patients as well as in asymptomatic patients whose serum aminotransferase activity is persistently elevated to greater than three times the upper limit of normal. The statin-associated elevations in serum hepatic aminotransferases usually resolve upon discontinuation of the medication. Due to the potential liver toxicity, statins should be used with caution or in reduced dose in patients with concomitant liver disorders.

4.2.6.3 Myopathy Associated with Statins Muscle and joint pain may not be uncommon in patients treated with statins [16]. Minor increases in serum creatine kinase activity (a biomarker of myopathy) are observed in some patients receiving statin therapy, frequently associated with heavy physical activity. Significant myopathy including the most severe form, namely, rhabdomyolysis (see definition in the following text), occurs rarely with statin therapy. Although rarely occurring, myopathy, especially rhabdomyolysis, represents a major adverse effect of statin drugs. Patients with severe myopathy present with marked elevations in serum creatine kinase, frequently accompanied by generalized discomfort or weakness in skeletal muscles. If the medication is not discontinued, myoglobinuria can occur, leading to renal injury, which is a characteristic of rhabdomyolysis. Use of statins must be discontinued if myopathy occurs.

Here, the term rhabdomyolysis refers to the dissociation and damage of skeletal muscle fibers and the release of myoglobin into the circulation, an event that results in myoglobinuria and acute renal failure. The US FDA recorded 42 deaths from statin-induced rhabdomyolysis between 1987 and 2001, corresponding to an approximate rate of one death per million prescriptions (30-day supply). Although the exact mechanisms of statin-induced myopathy including rhabdomyolysis remain to be elucidated, alteration of the plasma membrane composition and electrophysiological properties due primarily to inhibition of cholesterol biosynthesis may play a significant part in the pathophysiology [17]. The risk of statin-associated myopathy is increased by the presence of advanced age, frailty, renal and hepatic dysfunction, as well as coadministration of drugs that interfere with the metabolism of statins (see Section 4.2.6.5 for more discussion of drug interactions). A genome-wide study demonstrated that genetic variations in SLCO1B1, a gene encoding the organic anion-transporting polypeptide OATP1B1, are strongly associated with an increased risk of statin-induced myopathy [18]. OATP1B1 has been shown to regulate the hepatic uptake of statins (see Section 4.2.2.2).

4.2.6.4 Possible Diabetogenic Effects

Early meta-analysis studies suggested that statin therapy, especially at high doses, might be associated with slightly increased risk of new-onset type 2 diabetes mellitus (diabetes for short) [19, 20]. Reexamination of the early studies revealed possible flaws in the study design among others, which may render the conclusion inaccurate and misleading [21]. Furthermore, even if statin therapy is diabetogenic, the small risk of developing diabetes with statin therapy is far outweighed by the potential of statins to reduce cardiovascular events [22].

4.2.6.5 Drug Interactions in Statin Therapy

As statins (except for pitavastatin and pravastatin) are primarily metabolized by hepatic CYP system, especially CYP3A4 (atorvastatin, lovastatin, and simvastatin) and CYP2C9 (fluvastatin and rosuvastatin), drugs or conditions that alter the activity of these CYP enzymes may have significant impact on statin metabolism and thereby the adverse effects of statins [23].

Drug Interactions due to Altered CYP3A4 Activity Many drugs are able to inhibit or compete for CYP3A4 and thereby increase the risk of statin-induced myopathy. These include macrolide antibiotics (e.g., erythromycin), azole antifungal drugs (e.g., itraconazole), anti-HIV agents (e.g., indinavir), and fibric acid derivatives (particularly gemfibrozil; see Section 4.5). As CYP3A4 is primarily responsible for the metabolism of atorvastatin, lovastatin, and simvastatin, these three statins most likely interact with the aforementioned CYP3A4-inhibiting drugs, causing increased risk of myopathy. Because grapefruit juice contains compounds that inhibit CYP3A4, plasma levels of the above three statins may be elevated in patients ingesting large amounts of grapefruit juice.

Conversely, drugs that induce CYP3A4 may increase the hepatic metabolism and thereby decrease the plasma levels of the CYP3A4-dependent statins (atorvastatin, lovastatin, and simvastatin). The CYP3A4-inducing drugs include phenytoin, barbiturates, rifamycin, and thiazolidinediones, among many others.

Drug Interactions due to Altered CYP2C9 Activity Fluvastatin and rosuvastatin are CYP2C9-dependent drugs. Hence, the adverse effects of these two statins may be potentiated by drugs that inhibit CYP2C9, such as ketoconazole (an antifungal agent), metronidazole (an antimicrobial agent), sulfinpyrazone (an antigout drug), amiodarone

(an antiarrhythmic agent), and cimetidine (an antihistamine drug).

Other Drugs Other drugs that may interact with statins and increase statin-induced adverse effects include immunosuppressive agents (e.g., cyclosporine A), anticoagulants (e.g., warfarin), and calcium channel blockers (e.g., verapamil).

4.2.6.6 Contraindications and Pregnancy Category

- Active liver disease, which may include unexplained persistent elevations in hepatic transaminase levels.
- Hypersensitivity to any component of the medication.
- Nursing mothers: women who require statin treatment should not breastfeed their infants.
- Pregnancy category: X.

4.2.7 Summary of Statin Drugs

- Statins are the first-line drugs for treating patients with high LDL cholesterol. They are also useful for reducing plasma triglycerides in patients with mixed dyslipidemia (hypercholesterolemia and hypertriglyceridemia).
- Statins reduce cholesterol biosynthesis by inhibiting HMG-CoA reductase, the key enzyme in cholesterol synthesis. In addition to lipid lowering, statins also exert many other beneficial activities, collectively known as the pleiotropic effects.
- Statins reduce atherosclerosis-associated fatal and nonfatal coronary heart disease events, strokes, and total mortality and have become an important component of preventive cardiology. Statins may also be potentially useful for many noncardiovascular disease conditions.
- Statins are generally well tolerated. Rarely, significant liver toxicity and myopathy may occur that mandate the dosage adjustment or discontinuation of the medication.

4.3 BILE ACID SEQUESTRANTS

Listed below are the three US FDA-approved bile acid sequestrants for treating dyslipidemias:

- Cholestyramine (Questran)
- Colesevelam (Welchol)
- Colestipol (Colestid)

4.3.1 General Introduction to Drug Class

Bile acid sequestrants, also known as bile acid-binding resins, are moderately effective drugs for treating

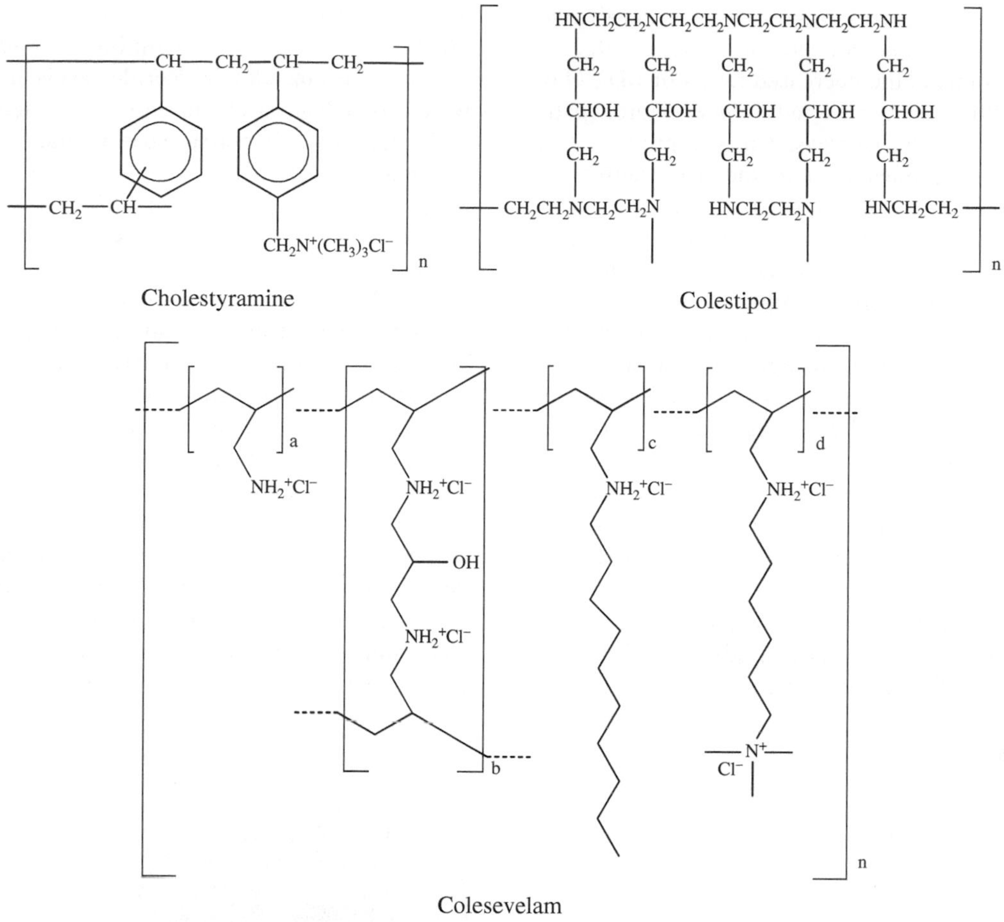

Cholestyramine

Colestipol

Colesevelam

FIGURE 4.4 Chemical structures of the three US FDA-approved bile acid sequestrants for treating dyslipidemias. All three bile acid sequestrants are water-insoluble high-molecular-mass resins in the form of hydrochloride. Colestipol hydrochloride is a copolymer of diethylenetriamine and 1-chloro-2,3-epoxypropane, with approximately 1 out of 5 amine nitrogens protonated (chloride form). Colesevelam hydrochloride is poly(allylamine hydrochloride) cross-linked with epichlorohydrin and alkylated with 1-bromodecane and (6-bromohexyl)-trimethylammonium bromide.

hypercholesterolemia. They cause 10–25% reduction in LDL cholesterol at therapeutic dosages. Currently, three bile acid sequestrants are available: cholestyramine, colesevelam, and colestipol. Because they are not absorbed from the gut, these drugs have an excellent safety record and are especially valuable for patients who cannot tolerate other lipid-lowering drugs.

Of the three bile acid sequestrants, cholestyramine and colestipol are among the oldest drugs for dyslipidemias and probably also the safest lipid-lowering agents. As such, both cholestyramine and colestipol are recommended for treating hypercholesterolemia in children and young patients (11–20 years of age). Colesevelam is a relatively newer bile acid sequestrant approved by the US FDA in 2000. It is also used for treating hypercholesterolemia in children. Recent clinical studies demonstrate an efficacy for colesevelam in treating type 2 diabetes when added to antidiabetic drugs [24].

4.3.2 Chemistry and Pharmacokinetics

The bile acid sequestrants (structures shown in Fig. 4.4) for clinical use are large molecular weight cationic polymer resins that bind noncovalently to negatively charged bile acids in the intestinal lumen. The resin-bound bile acids cannot be reabsorbed in the small intestine and are thus excreted in the feces.

4.3.3 Molecular Mechanisms and Pharmacological Effects

The pharmacological effects of bile acid sequestrants and the underlying mechanisms are generally discussed in the context of modulation of lipoprotein metabolism. In addition, recent work shows that bile acid sequestrants, especially colesevelam exert favorable pharmacological effects in treating diabetes.

4.3.3.1 Effects on Plasma LDL-Cholesterol Levels and Mechanisms

The most notable pharmacological effect of bile acid sequestrants is the decreased levels of LDL cholesterol. Bile acid sequestrants reduce the enterohepatic circulation of bile acids. Decreased reabsorption of bile acids from the small intestine into the liver causes the upregulation of cholesterol 7-α-hydroxylase in hepatocytes, the key enzyme in bile acid biosynthesis (Fig. 4.5). The increased biosynthesis of bile acids reduces the concentrations of free cholesterol in hepatocytes, which then lead to the concurrent upregulation of LDL receptors. The augmented expression of LDL receptors on the surface of hepatocytes causes enhanced uptake of LDL particles from the blood plasma, thereby leading to decreased levels of plasma LDL cholesterol.

As noted earlier, at therapeutic doses, all three bile acid sequestrants have similar efficacy in reducing plasma LDL-cholesterol levels. A 20–25% reduction of plasma LDL cholesterol occurs with maximal doses of the bile acid sequestrants. However, the maximal doses are associated with significant gastrointestinal adverse effects that limit patient's compliance (see Section 4.3.6).

4.3.3.2 Effects on Plasma Triglyceride Levels and Mechanisms

Treatment with bile acid sequestrants may increase the plasma triglyceride levels in some patients, which could lead to adverse clinical consequences in those with significant hypertriglyceridemia. In such patients, bile acid sequestrants may cause further marked increases in plasma triglyceride levels. Hence, use of bile acid sequestrants in those patients should be avoided or with great caution.

How bile acid sequestrants increase plasma triglyceride levels remains elusive, and the most plausible mechanism may be associated with bile acid receptors. Bile acids, including cholic acid and chenodeoxycholic acid, are natural ligands for the farnesoid X receptor (FXR), which is involved in the regulation of lipid and glucose metabolism via controlling the activity of various downstream molecules. Reduction of bile acid flux in the portal vein by bile acid sequestrants decreases FXR activity in the liver, and this decreased FXR activity may induce an increase in the activity of another receptor, known as liver X receptor (LXR). Activation of LXR increases circulating triglyceride levels by promoting the production of a transcription factor, known

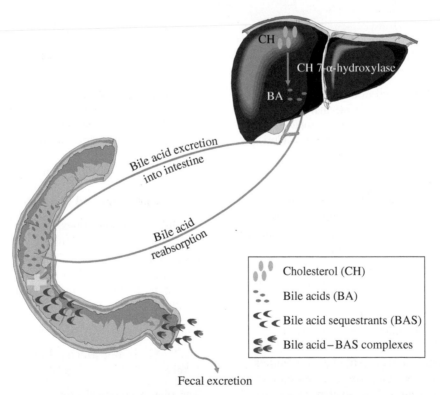

CH

CH 7-α-hydroxylase

BA

Bile acid excretion into intestine

Bile acid reabsorption

Cholesterol (CH)

Bile acids (BA)

Bile acid sequestrants (BAS)

Bile acid–BAS complexes

Fecal excretion

FIGURE 4.5 Molecular mechanism by which bile acid sequestrants reduce cholesterol. A significant portion of bile acids (bile salts) secreted into the intestine is reabsorbed and delivered into the liver via portal vein, hence forming enterohepatic circulation of bile acids (BA). Bile acid sequestrants (BAS) bind BA forming complexes that are eliminated in feces. The decreased return of BA to the liver causes upregulation of cholesterol (CH) 7-α-hydroxylase, the key enzyme in BA synthesis, thereby leading to decreases in CH concentration in hepatocytes. As described in Figure 4.3 legend, decreased cholesterol causes increased expression of LDL receptors on the surface of hepatocytes and the subsequent reduction in plasma LDL cholesterol. For color details, please see color plate section.

as stimulating sterol regulatory element-binding protein 1c (SREBP1c), in the liver. SREBP1c plays an important role in regulating the biosynthesis of triglycerides.

4.3.3.3   Effects on Type 2 Diabetes and Mechanisms

History and Approval   The beneficial effects of bile acid sequestrants in diabetes were first observed in a short-term (6 weeks), double-blind, crossover trial in patients with both type 2 diabetes and dyslipidemia [25]. In the trial, cholestyramine therapy not only lowered LDL-cholesterol levels but also improved glycemic control as evidenced by a 13% decrease in mean plasma glucose values. In addition, a tendency to lower glycosylated hemoglobin (HbA1c) concentration was noted. In a short-term observational study with another bile acid sequestrant, colestimide (this is not a US FDA-approved drug) in Japan, an absolute HbA1c reduction of 0.9% (from 7.7 to 6.8%) and a 0.8 mM reduction (from 8.5 mm to 7.7 mM) in fasting plasma glucose were reported in patients with type 2 diabetes and hypercholesterolemia [26]. These pilot studies suggested a potential value for bile acid sequestrants in glycemic control in type 2 diabetes.

The effects of addition of colesevelam to established antidiabetes monotherapy or combination therapy with metformin, sulfonylureas, or insulin in patients with type 2 diabetes have been recently evaluated in three randomized, double-blind, placebo-controlled clinical trials [27–29]. These trials collectively enrolled a total of 1054 patients with type 2 diabetes with inadequate glycemic control on their current antidiabetes treatment regimen. The addition of colesevelam resulted in a consistent reduction in HbA1c (ranging from 0.50 to 0.54%) across all three trials, regardless of the background antidiabetes medication used, which was accompanied by reduction in fasting glucose. Based on these trials, colesevelam has been approved by the US FDA as an adjunct to antidiabetic therapy for improving glycemic control in adults with type 2 diabetes.

Mechanisms   The exact mechanisms underlying the glycemic effects of bile acid sequestrants remain unclear and studies to elucidate the mechanisms are ongoing. Two potential mechanisms have been proposed, which are summarized below:

- FXR- and LXR-dependent mechanism [30]: As described in Section 4.3.3.2, reduction of bile acid flux in the portal vein by bile acid sequestrants decreases the activity of FXR in the liver, and this decreased FXR activity in turn induces an increase in the activity of LXR. On the one hand, activation of LXR is responsible for the elevated plasma levels of triglycerides. On the other hand, LXR activation in the liver results in improved glucose metabolism by causing inhibition of gluconeogenesis. LXR activation may also improve

glucose metabolism by promoting the expression of glucokinase and glucose transporter 4 in adipose tissue or by promoting insulin secretion by the pancreatic β-cells.
- Incretin hormone-dependent mechanism [31]: Bile acid sequestrants may increase the secretion of incretin hormones, especially glucagon-like peptide-1 and glucose-dependent insulinotropic polypeptide. These hormones play a crucial role in regulating glucose metabolism.

4.3.4   Clinical Uses

Bile acid sequestrants have two major clinical applications: (i) treatment of hypercholesterolemia and (ii) as an adjunctive therapy for type 2 diabetes. In addition, they are also useful in several other conditions.

4.3.4.1   Hypercholesterolemia   As noted earlier, bile acid sequestrants were the drugs of choice for LDL-cholesterol reduction in the prestatin era, reducing LDL cholesterol by up to 20–30%. Due to their well-documented safety profiles, these drugs are still commonly used in the management of patients with hypercholesterolemia, especially for those who cannot tolerate other drugs, such as high doses of statins. In this context, combination of a bile acid sequestrant with a statin drug can have additive effects on lowering LDL-cholesterol levels. This allows the use of relatively low doses of statins to avoid significant adverse effects. Due to long-term safety record, bile acid sequestrants may be particularly useful for treating hypercholesterolemia in children, such as those with heterozygous familial hypercholesterolemia. They can also be used in combination with statins and other lipid-lowering drugs for treating patients with homozygous familial hypercholesterolemia (HoFH).

4.3.4.2   Type 2 Diabetes   Due to the documented benefits of colesevelam in improving glycemic control in multiple clinical trials, this drug is the only bile acid sequestrant approved by the US FDA as an adjunctive in the treatment of hyperglycemia in adults with type 2 diabetes. The dual efficacy in glycemic control and LDL-cholesterol reduction as well as the well-documented safety make colesevelam a drug of interest for individuals with both hyperglycemia and hypercholesterolemia. In this regard, hyperglycemia and hypercholesterolemia frequently coexist in adults with type 2 diabetes.

4.3.4.3   Others Indications   Bile acid sequestrants have been used to relieve diarrhea and pruritus (itching) in patients who have cholestasis and bile salt accumulation. Because they bind to digitalis glycosides, bile acid sequestrants may also be useful in alleviating digitalis toxicity.

4.3.5 Therapeutic Dosages

Listed below are the dosage forms and strengths of the three bile acid sequestrants:

- Cholestyramine (Questran): Oral, 4 g packets granules
- Colestipol (Colestid): Oral, 5 g packets granules; 1 g tablets
- Colesevelam (Welchol): Oral, 625 mg tablets; 1.875 and 3.75 packets granules

Cholestyramine and colestipol are available in powder (or granular) in the individual packets of 4 and 5 g, respectively, for mixing with water or juice before administration. Ideally, the resins should be taken shortly before or with breakfast or supper (the drugs should be present in the small intestine during gallbladder empting to achieve maximal binding). A gradual increase of dosage from 4 g or 5–20 g/day is recommended. Total daily dosages of 30–32 g may be needed to attain maximal LDL-cholesterol reduction; however, such large dosages are frequently associated with significant gut disturbances (bloating and constipation) that reduce patients' compliance. The usual daily dosages in children range from 10 to 20 g. Colestipol is also available in 1 g tablets, with a maximal daily dosage of 16 g.

The newer resin colesevelam hydrochloride is available as tablets of 625 mg or as a powder in packets of 1.875 or 3.75 g. The recommended dose is six tablets once daily or three tablets twice daily. The tablets should be taken with meals and liquid. For the powder formulation, the recommended dose is 3.75 g packet once daily or 1.875 g packet twice daily. The powder is first suspended in 4–8 oz of water and then taken with meals.

4.3.6 Adverse Effects and Drug Interactions

4.3.6.1 Adverse Effects As stated before, bile acid sequestrants are probably the safest drugs for lowering LDL cholesterol. This is largely due to the fact that they are not absorbed systematically and thereby have little potential for systemic adverse effects. Common complaints include gastrointestinal disturbance. In addition, bile acid sequestrants may interfere with the absorption of dietary vitamins. Another adverse effect is the elevation of plasma triglyceride levels (see Section 4.3.3.2).

Gut Disturbance Gut disturbance is manifested as bloating, dyspepsia, and constipation, which can be relieved by increasing dietary fiber intake. Steatorrhea may also occur, particularly in patients with preexisting bowel disease or cholestasis. Many of these adverse effects are due to the increased delivery of fat and bile acids to the large intestine. Bile acid sequestrants should be avoided in patients with diverticulitis and individuals with a history of bowel obstruction.

Interference with Vitamin Absorption Bile acid sequestrants can reduce the absorption of fat-soluble vitamins, such as vitamin K. Bleeding due to vitamin K deficiency has occasionally been reported. As noted in the following, bile acid sequestrants also affect the absorption of many drugs (Section 4.3.6.2).

Elevation of Plasma Triglyceride Levels As bile acid sequestrants increase plasma triglyceride levels, these drugs should be avoided in patients with severe hypertriglyceridemia (>500 mg/dl). They are also contraindicated in patients with a history of hypertriglyceridemia-induced pancreatitis. Use of bile acid sequestrants to lower LDL cholesterol in individuals with less severe hypertriglyceridemia (>250 and <500 mg/dl) should be accompanied by frequent monitoring of fasting plasma triglyceride levels.

4.3.6.2 Drug Interactions In the intestine, cholestyramine and colestipol bind and reduce the absorption of many drugs, including warfarin, digoxin, furosemide, some thiazides, propranolol, thyroxin, and some statins. In contrast, colesevelam does not appear to interfere with the gut absorption of warfarin, digoxin, or lovastatin, but it reduces the gut absorption of verapamil. The effects of bile acid sequestrants on the gut absorption of most other drugs are unknown. Hence, it is wise to administer all drugs either at least 1–2 h before or 3–4 h after the resin to avoid potential drug interactions.

4.3.6.3 Contraindications and Pregnancy Category

- Patients with a history of bowel obstruction.
- Patients with serum triglyceride concentrations >500 mg/dl.
- Patients with a history of hypertriglyceridemia-induced pancreatitis.
- Hypersensitivity to the drug products.
- Pregnancy category: B.

4.3.7 Summary of Bile Acid Sequestrants

- Cholestyramine, colestipol, and colesevelam are three US FDA-approved bile acid sequestrants that inhibit the reabsorption of bile acids from the small intestine.
- Reduced bile acid reabsorption leads to decreased concentrations of free cholesterol in hepatocytes. This is due to increased biosynthesis of bile acids, thereby consuming cholesterol in hepatocytes. As a result, hepatocyte LDL receptors are upregulated, resulting in enhanced clearance of LDL cholesterol from plasma.
- Bile acid sequestrants are probably the safest lipid-lowering drugs. They are useful in individuals who cannot tolerate other lipid-lowering drugs or used in combination with other drugs to achieve the goal of

LDL-cholesterol reduction. Due to an excellent safety record, these drugs are also valuable in treating hypercholesterolemia in children.

- Bile acid sequestrants are also effective in improving glycemic control in type 2 diabetes. Colesevelam is a US FAD-approved drug (as an adjunctive) for the management of adult type 2 diabetes. The exact mechanisms underlying the favorable effects of bile acid sequestrants on glycemic control remain elusive.

4.4 CHOLESTEROL ABSORPTION INHIBITORS

Listed below is the only US FDA-approved cholesterol absorption inhibitor for clinical use:

- Ezetimibe (Zetia)

4.4.1 General Introduction to Drug Class

Cholesterol absorption inhibitors reduce cholesterol (including dietary and biliary cholesterol) absorption by the intestine. Ezetimibe is the first drug approved by the US FDA for lowering total and LDL cholesterol that inhibits cholesterol absorption by enterocytes in the small intestine. In addition to ezetimibe, plant sterols and stanols that are naturally present in nuts, vegetables, and fruits also inhibit intestinal cholesterol absorption. Plant sterols and stanols (see Section 4.8 for more discussion) are structurally similar to cholesterol but are much more hydrophilic. As such, they displace cholesterol from micelles, thereby increasing the excretion of cholesterol from feces. As discussed later, ezetimibe inhibits intestinal absorption of cholesterol via a novel, transporter-dependent mechanism.

Ezetimibe was discovered serendipitously during a search for inhibitors of acyl-coenzyme A:cholesterol acyltransferase at Schering-Plough Research Institute (now merged into Merck). The drug was approved by the US FDA in 2002. At the therapeutic dosage, ezetimibe reduces plasma LDL-cholesterol levels by ~15–20% and is employed primarily as an adjunctive therapy with a statin to attain the goal of LDL-cholesterol reduction.

4.4.2 Chemistry and Pharmacokinetics

Ezetimibe (structure shown in Fig. 4.6) is highly water insoluble. The absolute bioavailability of ezetimibe cannot be determined as the compound is virtually insoluble in aqueous media suitable for injection. Absorption of ezetimibe is not significantly affected by food. After oral administration, ezetimibe is absorbed and undergoes extensive glucuronidation in the small intestine and the liver to form ezetimibe-glucuronide. Ezetimibe-glucuronide is believed to be pharmacologically active. Peak plasma levels of ezetimibe and ezetimibe-glucuronide are reached within 4–12 h and between 1 and 2 h, respectively. Ezetimibe undergoes enterohepatic circulation. Both ezetimibe and ezetimibe-glucuronide are eliminated from plasma with a similar half-life of ~22 h. About 80% of the drug is excreted in feces, and ~10% in urine.

4.4.3 Molecular Mechanisms and Pharmacological Effects

The pharmacological effects of ezetimibe can be summarized into two general aspects, as outlined below:

- Reduction of plasma LDL cholesterol
- Other effects

4.4.3.1 Reduction of Plasma LDL Cholesterol Ezetimibe reduces blood cholesterol by inhibiting the absorption of cholesterol by the small intestine. It inhibits intestinal cholesterol absorption by ~50% without reducing the absorption of triglycerides and fat-soluble vitamins. In order to better understand the mechanisms of action and pharmacological effects of ezetimibe, let's briefly look at the sources of cholesterol in the liver. The cholesterol content of the liver is derived predominantly from three sources, as listed below:

- De novo synthesis of cholesterol in hepatocytes
- Uptake of cholesterol from circulating lipoproteins via receptor-mediated endocytosis
- Cholesterol absorbed by the small intestine (intestinal cholesterol is derived primarily from cholesterol secreted in the bile and from dietary cholesterol)

Ezetimibe has a mechanism of action that differs from those of other classes of cholesterol-reducing drugs. The molecular target of ezetimibe has been shown to be the sterol transporter, namely, Niemann–Pick C1-Like 1 (NPC1L1), located on the brush border cells of the small intestine. This transporter is involved in the intestinal uptake of cholesterol and phytosterols. NPC1L1 is also present in the liver. Inhibition of NPC1L1 by ezetimibe leads to a decrease in the delivery of intestinal cholesterol to the liver (Fig. 4.6). This causes a reduction of hepatic cholesterol stores, which results in upregulation of hepatocyte LDL receptors and the consequent augmented clearance of LDL particles from plasma (also see Fig. 4.3).

4.4.3.2 Other Effects In patients with hyperlipidemia, ezetimibe decreases plasma triglyceride levels by ~8% and elevates HDL cholesterol by ~3%; however, the underlying mechanisms remain unclear. As NPC1L1 also mediates the intestinal uptake of phytosterols, ezetimibe reduces the absorption of these plant sterols and decreases their blood levels.

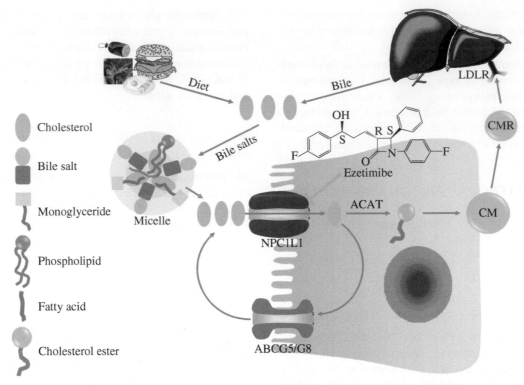

FIGURE 4.6 Chemical structure and molecular mechanism of action of ezetimibe. The molecular target of ezetimibe is the sterol transporter, Niemann–Pick C1-Like 1 (NPC1L1), which is involved in the intestinal uptake of cholesterol and phytosterols (phytosterol transport is not shown here; see Fig. 4.11 for details). Dietary or bile-derived cholesterol in the micelles is transported into the enterocytes via NPC1L1. Cholesterol in enterocytes is converted to cholesterol ester via the action of acyl-CoA:cholesterol acyltransferase (ACAT). Cholesterol ester is then incorporated into chylomicrons and eventually delivered into the liver primarily via LDL receptor-mediated uptake of chylomicron remnants (CMR). Cholesterol in enterocytes can also be transported back into the lumen via ABC transporters G5 and G8. Ezetimibe does not inhibit cholesterol synthesis in the liver or increase bile acid excretion. Instead, ezetimibe localizes at the brush border of the small intestine and inhibits NPC1L1 and the subsequent absorption of cholesterol, leading to a decrease in the delivery of intestinal cholesterol to the liver. This causes a reduction of hepatic cholesterol stores and an increase in clearance of LDL cholesterol from the plasma. This distinct mechanism is complementary to that of statins, and as such, the combination therapy of ezetimibe and a statin results in additive or synergistic effects in lowering plasma LDL cholesterol.

Like statins, ezetimibe has been suggested also to exert pleiotropic effects. When added to statin therapy, ezetimibe improves endothelial dysfunction and lowers blood C-reactive protein levels [32, 33]. Elevation of plasma C-reactive protein is an independent risk factor for cardiovascular diseases [34].

4.4.4 Clinical Uses

The clinical indications of ezetimibe include the following:

- To reduce elevated plasma total and LDL cholesterol in patients with hyperlipidemia
- To reduce elevated plasma levels of sitosterol and campesterol in patients with homozygous sitosterolemia, also known as phytosterolemia

4.4.4.1 Hypercholesterolemia Treatment of hypercholesterolemia may involve ezetimibe monotherapy but more commonly employs combination therapy with a statin or another lipid-lowering agent (e.g., fenofibrate).

Monotherapy versus Combination Therapy Ezetimibe monotherapy reduces plasma LDL cholesterol maximally by ~15–20%. However, inhibition of intestinal cholesterol absorption by ezetimibe results in compensatory augmentation of hepatic cholesterol biosynthesis. Hence, ezetimibe monotherapy has limited efficacy in reducing plasma LDL cholesterol upon long-term treatment. Likewise, inhibition of hepatic cholesterol synthesis by statins also causes increased absorption of cholesterol from the small intestine, which limits the effectiveness of statins in lowering plasma LDL cholesterol. The combination of ezetimibe and a statin has additive or synergistic effects, leading to marked reduction of LDL cholesterol.

A fixed-dose combination formulation (Vytorin) containing 10 mg ezetimibe and various doses of simvastatin (10, 20, 40, and 80 mg) has been approved by the US FDA.

Combination of 10 mg of ezetimibe and 80 mg of simvastatin can reduce LDL cholesterol by 60%, which is greater than what can be achieved with any statin as monotherapy. Although the effects of the combination of ezetimibe and simvastatin on lipid profiles and C-reactive protein have been documented by clinical trials, the impact of this combination therapy on cardiovascular morbidity and mortality remains to be determined. The following section summarizes the completed and ongoing clinical trials in this regard. More recently, another fixed-dose combination formulation (Liptruzet) containing 10 mg ezetimibe and various doses of atorvastatin (10, 20, 40, and 80 mg) was approved by the US FDA in 2013 [35].

Clinical Trials on Combination Therapy of Ezetimibe and a Statin   The combination therapy of ezetimibe with a statin (simvastatin) has been evaluated in several trials, including (i) the Ezetimibe and Simvastatin in Hypercholesterolemia Enhances Atherosclerosis Regression (ENHANCE) trial, (ii) the Simvastatin and Ezetimibe in Aortic Stenosis (SEAS) trial, and (iii) the Arterial Biology for the Investigation of the Treatment Effects of Reducing Cholesterol 6—HDL and LDL Treatment Strategies in Atherosclerosis (ARBITER 6-HALTS) trial.

The ENHANCE trial found that although the addition of ezetimibe (10 mg daily) to high-dose simvastatin (80 mg daily) in patients with heterozygous familial hypercholesterolemia caused an additional 16.5% reduction in LDL cholesterol, it did not significantly affect the mean change in the carotid intima-media thickness (a surrogate for atherosclerosis) compared with simvastatin monotherapy. However, addition of ezetimibe caused a 6.6% and 25.7% greater reduction in plasma triglycerides and C-reactive protein, respectively, than simvastatin alone [33].

The subsequent SEAS study showed that treatment with ezetimibe (10 mg daily) plus simvastatin (40 mg daily) significantly reduced plasma LDL-cholesterol levels in patients with aortic stenosis compared with placebo, but did not reduce the composite outcome of combined aortic valve events and ischemic events [36]. The SEAS trial was not designed to compare the ezetimibe/simvastatin combination therapy with simvastatin monotherapy, but rather to determine if intensive lipid lowering with ezetimibe and simvastatin affected the composite primary outcome of aortic valve events and ischemic events in the patients compared with placebo. Notably, the SEAS study found that cancer occurred more frequently in the ezetimibe/simvastatin group compared with the placebo group (see section "Effect of Ezetimibe on Cancer Rate" for more discussion of ezetimibe and cancer risk).

The ARBITER 6-HALTS trial found that whereas niacin-extended release added to statin therapy caused a smaller LDL-cholesterol reduction compared with ezetimibe added to statin therapy in patients with coronary heart disease or a coronary heart disease equivalent, niacin had

a greater efficacy than ezetimibe when combined with a statin regarding the induction of regression of carotid intima-media thickness [37].

Although the aforementioned trials primarily using carotid intima-media thickness as a surrogate marker of atherosclerosis did not find a beneficial effect of ezetimibe when added to statin therapy, they didn't disprove the effects of ezetimibe on cardiovascular events. In this context, the ongoing large-scale Improved Reduction of Outcomes: Vytorin Efficacy International Trail (IMPROVE-IT) was designed to compare the cardiovascular outcome (death from cardiovascular or cerebrovascular events) in 18,000 patients with high-risk acute coronary syndromes treated with ezetimibe (10 mg/day) plus simvastatin (40 mg/day) versus placebo plus simvastatin (40 mg/day) [38]. The trial is projected to conclude in late 2014. Until then, the exact value of ezetimibe in combination with statin therapy in reducing cardiovascular events remains unknown.

Effect of Ezetimibe on Cancer Rate   As mentioned earlier, the SEAS trial suggested a possible increased rate of cancer in patients treated with ezetimibe plus simvastatin. However, analyses of cancer data from three ezetimibe trials, including the SEAS trial, did not provide credible evidence of any adverse effects of ezetimibe on cancer rate [39]. In contrast, when the cancer mortality data from these three trials were combined, there was an increase in cancer mortality risk in the ezetimibe plus simvastatin groups as compared with controls [40]. The potential cancer hazard of ezetimibe is currently inconclusive and remains to be further determined by follow-up of longer duration.

Current Recommendations on Using Ezetimibe   In view of the currently available data from ezetimibe clinical trials, especially the ENHANCE trial, until the results from IMPROVE-IT are available, it seems prudent to encourage patients whose LDL-cholesterol levels remain elevated despite treatment with an optimal dose of a statin to redouble their efforts at dietary control and regular exercise. Addition of niacin, fibrates, and bile acid sequestrants should be considered when diet, exercise, and a statin have failed to achieve the target level, with ezetimibe reserved for patients who cannot tolerate these agents [41, 42].

4.4.4.2 Sitosterolemia

Introduction to Sitosterolemia   Plant sterols and stanols are also known as phytosterols and phytostanols, respectively. Phytostanols are much less abundant than phytosterols in nature. As such, phytosterols have received more attention. Over 40 phytosterols have been identified; of those, β-sitosterol, stigmasterol, and campesterol account for over 95% of total phytosterol dietary intake [43]. Phytosterols are poorly absorbed in the intestine with an absorption rate of 0.4–3.5%. Phytosterols absorbed into the enterocytes are excreted back to the intestinal lumen by ATP-binding cassette (ABC) transporters, such as ABCG5/ABCG8.

Sitosterolemia (also known as phytosterolemia) is a rare autosomal recessively inherited disease caused by mutations affecting ABCG5 and ABCG8 and characterized by elevated plasma and tissue levels of phytosterols (also see Table 3.2). In contrast to healthy individuals in whom total plasma phytosterols are <1 mg/dl, patients with sitosterolemia have plasma phytosterol concentrations ranging from 12 to 40 mg/dl. Plasma cholesterol levels may be normal or elevated. The disease is characterized clinically by tendon and tuberous xanthomas and by a strong predisposition to premature coronary atherosclerosis.

Treatment with Ezetimibe As ezetimibe inhibits NPC1L1-mediated uptake of phytosterols from the gut lumen into enterocytes, this drug is indicated as adjunctive therapy to diet for the reduction of elevated sitosterol and campesterol levels in patients with homozygous familial sitosterolemia.

4.4.5 Therapeutic Dosages

The dosage form and strength of ezetimibe are listed below:

- Ezetimibe (Zetia): Oral, 10 mg tablets

To reduce plasma LDL-cholesterol levels, ezetimibe is given at one 10 mg tablet once daily, with or without food. The same dosage is also for treating sitosterolemia. As stated earlier, ezetimibe (10 mg daily) is usually used in combination with other lipid-lowering drugs for treating hypercholesterolemia.

Dosing of ezetimibe should occur either ≥2 h before or ≥4 h after administration of a bile acid sequestrant. This is because bile acid sequestrants bind to ezetimibe and reduce its intestinal absorption (see Section 4.3 for bile acid sequestrants).

4.4.6 Adverse Effects and Drug Interactions

4.4.6.1 Adverse Effects Ezetimibe is generally well tolerated. Based on clinical trial database, the commonly reported adverse effects in ezetimibe monotherapy or combination therapy with a statin include upper respiratory infection, sinusitis, diarrhea, arthralgia, and pain in the extremity. It may rarely also cause elevated serum hepatic aminotransferase activity. As with statins, ezetimibe should be avoided in pregnant women or nursing mothers. Ezetimibe may also rarely cause allergic reactions.

4.4.6.2 Drug Interactions Except for the aforementioned bile acid sequestrants that reduce ezetimibe absorption from the intestine, drug interactions with other medications seem to happen rarely. Nevertheless, interaction of ezetimibe

with several drugs, such as cyclosporine and coumarin anticoagulants, may occur. In this regard, caution should be exercised when using ezetimibe and cyclosporine concomitantly due to increased exposure to both ezetimibe and cyclosporine. Cyclosporine concentrations should be monitored in patients receiving ezetimibe and cyclosporine. If ezetimibe is added to warfarin, a coumarin anticoagulant, the international normalized ratio (INR) should be appropriately monitored.

4.4.6.3 Contraindications and Pregnancy Category

- Active liver disease, which may include unexplained persistent elevations in serum hepatic aminotransferase levels.
- Women who are pregnant or may become pregnant.
- Nursing mothers.
- Patients with a known hypersensitivity to the drug.
- Pregnancy category: X.

4.4.7 Summary of Ezetimibe

- Ezetimibe inhibits the absorption of dietary and biliary cholesterol from the intestine, thereby reducing plasma LDL-cholesterol levels. The molecular target of ezetimibe is the cholesterol (and also phytosterol) transporter NPC1L1 located on the brush border enterocytes.
- Ezetimibe is used in monotherapy and more frequently in combination therapy with a statin to attain the target LDL-cholesterol levels in patients with hypercholesterolemia.
- Ezetimibe also inhibits intestinal absorption of phytosterols and, as such, is indicated for treating familial homozygous sitosterolemia.
- The effects of ezetimibe (either monotherapy or in combination with other lipid-lowering agents) on cardiovascular morbidity and mortality remain to be determined.
- Ezetimibe is generally well tolerated. It is contraindicated in individuals with active liver disease, pregnant women, nursing mothers, and patients with a known hypersensitivity to the drug.

4.5 FIBRATES

Listed below are the two US FDA-approved fibrates for treating dyslipidemias:

- Fenofibrate (Tricor, Antara, Lofibra)
- Gemfibrozil (Lopid)

4.5.1   General Introduction to Drug Class

Fibrates have been used as lipid (primarily triglyceride)-lowering drugs for more than 40 years. The first lipid-lowering fibrate, clofibrate, was discovered when its use as a vehicle for androgen administration resulted in the finding that the vehicle alone (used as a control) was as effective in reducing plasma lipids as the combination [44]. In 1967, clofibrate was approved for clinical use in the United States and became the most widely prescribed lipid-lowering drug for effectively reducing plasma triglyceride levels.

In the 1970s, the effects of clofibrate on cardiovascular disease risk were evaluated in clinical studies. In the Coronary Drug Project and later in a World Health Organization (WHO) study in 1978, clofibrate therapy in patients with histories of coronary heart disease produced significant reduction in cardiovascular events. However, in the WHO study, the total mortality was reported to be significantly greater with clofibrate treatment than with placebo. The increased mortality was due to multiple reasons, including cholelithiasis associated with the long-term use of clofibrate.

The negative results of the 1978 WHO study raised widespread concern on the safety of clofibrate as a lipid-lowering drug, which led to virtual abandonment of the drug. A series of analogs have been developed and have essentially replaced clofibrate in most countries. These include gemfibrozil, fenofibrate, bezafibrate, and ciprofibrate. Among them, gemfibrozil and fenofibrate are available in the United States. Hence, this section focuses on these two fibrates.

Fibrates can decrease plasma triglyceride levels by up to 40–50% in patients with hypertriglyceridemia. They also reduce plasma LDL cholesterol by about 10% and raise HDL cholesterol by 10–20%.

4.5.2   Chemistry and Pharmacokinetics

As shown in Figure 4.7, the fibrate drugs are derivatives of a branched-chain carboxylic acid, known as fibric acid. In contrast to other fibrate drugs, gemfibrozil is not halogenated.

Both fenofibrate and gemfibrozil are rapidly and efficiently absorbed from the intestine, with an oral bioavailability for both drugs of >90%. More than 95% of the fibrate drugs bind to albumin in plasma. The half-life for fenofibrate and gemfibrozil is 20 and 1.5 h, respectively. Gemfibrozil also crosses the blood–placenta barrier. Most of the drugs are eliminated from the urine as glucuronide conjugates, with smaller amounts excreted in the feces.

4.5.3   Molecular Mechanisms and Pharmacological Effects

Listed below are the major pharmacological effects of fibrate drugs. Among them, the triglyceride-lowering activity is of most significance for fibrates:

- Reduction of plasma triglyceride levels by 40–50%
- Reduction of plasma LDL-cholesterol levels by ~10%
- Increase of plasma HDL-cholesterol levels by 10–20%

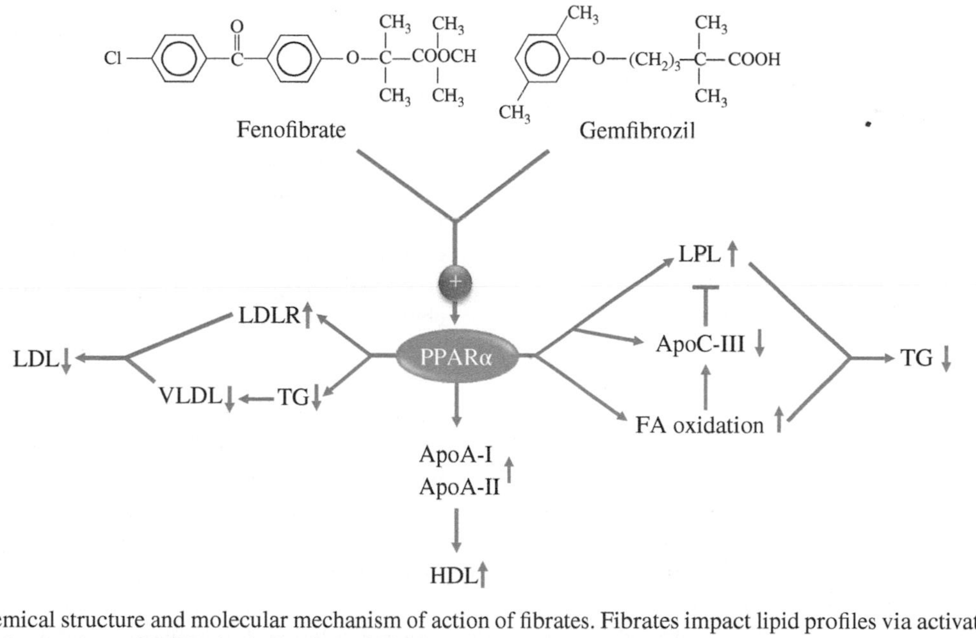

FIGURE 4.7   Chemical structure and molecular mechanism of action of fibrates. Fibrates impact lipid profiles via activation of peroxisome proliferator-activated receptor-α (PPAR-α). Activation of PPAR-α results in upregulation of lipoprotein lipase (LPL); downregulation of ApoC-III, an inhibitor of LPL; and increased oxidation of fatty acids (FA). The first two events augment hydrolysis of triglycerides (TG), whereas the last event leads to the decreased TG synthesis. The LDL-cholesterol-lowering effects result from the upregulation of LDL receptors and decreased formation of VLDL. Upregulation of ApoA-I and ApoA-II, two major components of HDL accounts for the fibrate-induced elevation of HDL cholesterol.

4.5.3.1 Reduction of Plasma Triglyceride Levels

The effects of fibrate drugs on lipoproteins are primarily mediated by peroxisome proliferator-activated receptor-α (PPAR-α). PPAR-α is a transcription factor and its activation by fibrates causes the following three events, which together lead to decreased plasma triglyceride levels (Fig. 4.7):

- Increased expression of lipoprotein lipase
- Decreased expression of apolipoprotein CIII, an inhibitor of lipoprotein lipase
- Increased oxidation of fatty acids due to augmented expression of fatty acid oxidation enzymes

Increased expression of lipoprotein lipase and decreased expression of ApoC-III bring about elevated activity of lipoprotein lipase. This increases the hydrolysis and removal of triglycerides from VLDL and chylomicrons, thereby leading to decreased levels of plasma triglycerides (note: VLDL and chylomicrons are the major contributors to plasma triglycerides).

Increased oxidation of fatty acids reduces the availability of free fatty acids for the biosynthesis of triglycerides. This also contributes to the decreased levels of plasma triglycerides upon treatment with fibrate drugs.

4.5.3.2 Reduction of Plasma LDL-Cholesterol Levels

Fibrate treatment frequently causes an approximately 10% reduction in plasma LDL-cholesterol levels. This effect results from two mechanisms, as outlined below and also depicted in Figure 4.7:

- Activation of PPAR-α also results in increased expression of LDL receptors on liver cells, thereby promoting hepatic uptake of LDL from plasma.
- The increased oxidation of fatty acids, as noted earlier, causes decreased hepatic synthesis of triglycerides. As triglycerides are building blocks for VLDL, the reduced triglyceride synthesis causes decreased formation and release of hepatic VLDL. Since VLDLs are precursors of LDL, decreased plasma levels of VLDL contribute to the LDL-cholesterol-lowering effect of fibrate drugs.

Notably, fibrates also cause a decrease in the proportion of small, dense LDL particles in the LDL density profile. This activity is of potential significance as small, dense LDL particles are more atherogenic. The mechanism underlying this action is currently unclear. In addition to the reduction of small, dense LDL particles, fibrates also improve lipoprotein profiles by causing elevation of HDL cholesterol, as described next.

4.5.3.3 Increase of Plasma HDL-Cholesterol Levels

The exact mechanisms by which fibrates elevate plasma HDL-cholesterol levels remain unclear. It has been suggested that fibrate-mediated activation of PPAR-α results in increased synthesis of ApoA-I and ApoA-II, which are the major components of HDL. This causes increased formation of mature HDL particles and thereby the elevated plasma levels of HDL cholesterol (Fig. 4.7).

4.5.4 Clinical Uses

4.5.4.1 Clinical Indications

- Hypertriglyceridemia
- Hypertriglyceridemia with low HDL-cholesterol levels

Fibrates are indicated for the treatment of hypertriglyceridemia and hypertriglyceridemia with low HDL cholesterol. As fibrates are highly effective in triglyceride lowering and able to decrease plasma triglyceride levels by up to 50%, they are usually the drugs of choice for treating severe hypertriglyceridemia to prevent the occurrence of pancreatitis. In this context, hypertriglyceridemia is one of the most common causes of pancreatitis. On the other hand, in patients with normal or mildly elevated triglyceride levels, use of statin therapy is also associated with a lower risk of pancreatitis [45]. The HDL-cholesterol-raising effects of fibrates also make them useful for treating individuals with both hypertriglyceridemia and low HDL cholesterol.

4.5.4.2 Effects on Cardiovascular Events and Mortality

Fibrates have been used as lipid-lowering drugs for a long time, and the use of these drugs, particularly fenofibrate, in the United States has been increasing over the past decade [46]. The beneficial effects of these drugs in improving lipid profiles have been well documented. However, evidence that fibrates have clinical benefits is mixed. Multiple systemic reviews and meta-analyses of the recent clinical trials demonstrate that fibrates can reduce the risk of major cardiovascular events predominantly by prevention of coronary events in patients with dyslipidemias, especially in those with combined dyslipidemia (high triglycerides and low HDL cholesterol) [47–50]. However, clinical trials overall do not suggest that fibrates substantially affect all-cause mortality.

4.5.4.3 Combination Therapy

Many patients who receive statin therapy for hyperlipidemia (such as patients with diabetes and metabolic syndrome) have residual cardiovascular risk despite intensive LDL-cholesterol reduction. These patients often have other forms of dyslipidemias, including low levels of HDL cholesterol and elevated levels of triglycerides and small, dense LDL particles. For such patients, combination treatment with statins and fibrates is a potentially useful strategy to improve lipid and lipoprotein profiles and reduce cardiovascular risk [51–53]. Combination of fenofibrate with a statin (e.g., simvastatin) has been shown to be well tolerated.

4.5.5 Therapeutic Dosages

Listed below are the dosage forms and strengths of the two US FDA-approved fibrates:

- Fenofibrate (Tricor): Oral, 48 and 145 mg tablets
- Fenofibrate (Fenoglide): Oral, 40 and 120 mg tablets
- Fenofibrate (Lofibra): Oral, 67, 134, and 200 mg capsules
- Fenofibrate (Trilipix): Oral, 45 and 135 mg delayed-release capsules
- Gemfibrozil (Lopid): Oral, 600 mg tablets

Fenofibrate is available in several different formulations. The usual dosage for fenofibrate (as Tricor) is one to three 48 mg tablets daily or a single 145 mg tablet daily. Gemfibrozil is usually taken at a dosage of 600 mg twice daily, preferably 30 min before the morning and evening meals (to maximize its absorption).

4.5.6 Adverse Effects and Drug Interactions

4.5.6.1 Adverse Effects   The adverse effects of fibrates are summarized as follows:

- Gastrointestinal disturbance is the most common adverse effect of fibrates.
- Increases in hepatic transaminases may also occur in ~5% of the patients.
- Rare adverse effects include myopathy, cardiac arrhythmias, hepatotoxicity, and gallstones. Formation of gallstone is probably due to the fibrate-induced increased secretion of cholesterol into the bile.
- Combination of fibrates (especially gemfibrozil) with statins increases the risk of myopathy (see Section 4.5.6.2). Therefore, such combination therapy should be done with caution. In this context, combination of fenofibrate with rosuvastatin or simvastatin appears to have minimal risk of enhanced myopathy.

4.5.6.2 Drug Interactions   Significant drug interactions for fibrates include the following four aspects:

- Gemfibrozil inhibits statin metabolism, increasing their concentrations and hence the risk of myopathy. Fenofibrate has less significant effect on statin metabolism compared with gemfibrozil.
- Fibrate drugs increase the plasma free concentrations of oral coagulants (e.g., warfarin) by displacing them from their binding sites on plasma albumin.
- If given concurrently, bile acid sequestrants reduce the intestinal absorption of fibrates. Hence, fibrates should be taken at least 1 h before or 4–6 h after the administration of a bile acid sequestrant.

- Gemfibrozil markedly increases plasma levels of the antidiabetic drug repaglinide. Coadministration of these two drugs increases the risk of severe hypoglycemia.

4.5.6.3 Contraindications and Pregnancy Category

- Hepatic or renal dysfunction.
- Preexisting gallbladder disease.
- Known hypersensitivity to the drug products.
- Combination therapy of gemfibrozil with repaglinide.
- Pregnancy category: C.

4.5.7 Summary of Fibrates

- Fibrates have been used for treating dyslipidemia for a long time; the beneficial effects on lipid profiles include marked reduction of plasma triglycerides and moderate elevation of HDL cholesterol. They also cause moderate reduction in LDL-cholesterol levels.
- The aforementioned effects in improving lipid profiles occur primarily via activation of PPAR-α and the subsequent increased expression or level of lipoprotein lipase, LDL receptors, and ApoA-I and ApoA-II, among other molecular and biochemical alterations.
- Fibrates are indicated for treating hypertriglyceridemia and hypertriglyceridemia with low HDL-cholesterol levels.
- Treatment with fibrates in patients with dyslipidemias (especially high triglyceridemia and low HDL cholesterol) reduces nonfatal cardiovascular events, but does not substantially affect the mortality.
- Fibrates are generally well tolerated. Rare adverse effects include myopathy, liver injury, and cardiac arrhythmias. Combination of fibrates with statins increases the risk of myopathy, and as such, the combination therapy should be done with caution.
- Fibrates are contraindicated in patients who show hypersensitivity reactions, with liver and renal dysfunction, and with preexisting gallbladder disease. Coadministration of gemfibrozil and the antidiabetic repaglinide is also contraindicated due to the increased risk of severe hypoglycemia.

4.6   NIACIN

Listed below is the only US FDA-approved vitamin drug for treating dyslipidemias:

- Niacin extended release (Niaspan)

4.6.1 General Introduction to Niacin

Niacin, also known as nicotinic acid or vitamin B3, is a water-soluble B-complex vitamin. At physiological concentrations, it is a substrate in the biosynthesis of nicotinamide adenine dinucleotide (NAD) and nicotinamide adenine dinucleotide phosphate (NADP), which are important factors in various metabolic pathways. Niacin functions as a vitamin only after its conversion to NAD or NADP. At pharmacological dosage (which is much larger than that required for its vitamin effects), niacin extended release exerts beneficial effects on lipid profiles.

Niacin was initially discovered in 1873, but its lipid-modulating properties were not discovered until 1955, when R. Altschul and coworkers reported that pharmacological dosage of niacin lowered plasma cholesterol levels in normal as well as hypercholesterolemic subjects [54]. In view of the recognition of plasma cholesterol as an important independent risk factor for coronary artery disease, this discovery by R. Altschul and coworkers formed the basis for the development of lipid-based therapies for coronary artery disease. Multiple subsequent clinical studies established the use of niacin as a broad-spectrum lipid-modifying agent and the first lipid-based intervention to prevent cardiovascular disease and death. The beneficial effects of niacin on lipid profiles include reduction of plasma levels of triglycerides, LDL cholesterol, and Lp(a) and elevation of plasma HDL-cholesterol levels. In addition, niacin reduces the portion of small, dense LDL particles.

4.6.2 Chemistry and Pharmacokinetics

The chemical name of niacin is pyridine-3-carboxylic acid (structure shown in Fig. 4.8). Niacin is readily absorbed from the intestine. Due to extensive and saturable first-pass metabolism, niacin concentrations in the general circulation are dose dependent and highly variable. At smaller pharmacological doses, niacin is extensively metabolized in the liver, and only the major metabolite, nicotinuric acid, is found in the urine. At high pharmacological doses, a greater proportion of the drug is excreted in the urine as unchanged parent compound.

4.6.3 Molecular Mechanisms and Pharmacological Effects

Niacin exerts broad-spectrum beneficial effects on plasma lipid profiles. These effects as outlined below are primarily responsible for its cardiovascular benefits:

- Decrease of plasma triglyceride levels by 30–50%
- Decrease of plasma LDL-cholesterol levels by 15–30%
- Reduction of Lp(a)

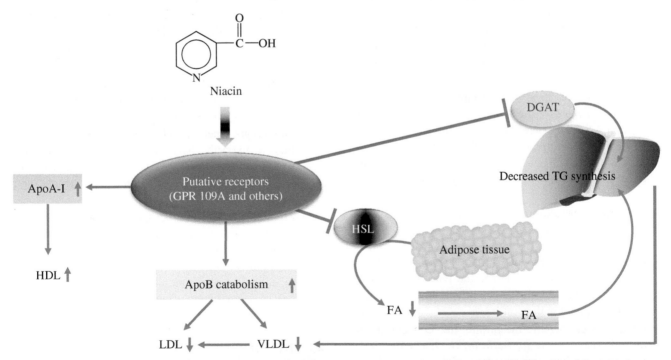

FIGURE 4.8 Chemical structure and molecular mechanism of action of niacin. The impact of niacin on lipid profiles is mediated by putative receptors, primarily the G-protein-coupled receptor (GPR) 109A. Activation of the putative receptors causes inactivation of hormone-sensitive lipase (HSL) and acyl-CoA:diacylglycerol acyltransferase (DGAT), which together lead to the decreased synthesis of triglycerides (TG) in the liver and the subsequent reduction of plasma triglycerides. Increased ApoB catabolism along with the reduced formation accounts for LDL-cholesterol-lowering activity of niacin. Niacin augments HDL cholesterol via attenuating ApoA-I catabolism.

- Reduction in the number of small, dense LDL particles
- Increase of plasma HDL-cholesterol levels by 20–35%

In addition, niacin also exerts pleiotropic effects, including anti-inflammatory, antithrombotic, and antioxidant activities. These pleiotropic actions may also contribute to the favorable cardiovascular outcomes resulting from niacin treatment [55, 56].

Niacin acts by binding to and activating several receptors (Fig. 4.8). The high-affinity G-protein-coupled receptor GPR109A mediates most of niacin's effects on lipid profiles and is also responsible for its adverse effects (e.g., skin flushing). This receptor was discovered independently by three groups in 2003 [57–59]. Niacin also activates the β-chain of ATP synthase in the liver cells. Moreover, niacin is found to directly inhibit enzymes, such as acyl-CoA:diacylglycerol acyltransferase-2 (DGAT-2), an enzyme involved in triglyceride biosynthesis. The roles of the aforementioned pathways in mediating niacin's pharmacological effects are discussed next.

4.6.3.1 Decrease of Plasma Triglyceride Levels
Niacin is as effective as fibrates in lowering plasma triglyceride levels in patients with hypertriglyceridemia. The effects of niacin on triglycerides are thought to be primarily mediated by activation of GPR109A and inhibition of DGAT-2 [55, 60].

Activation of GPR109A   Activation of adipocyte GPR109A by niacin in adipose tissue results in the stimulation of the G_i-adenylyl cyclase pathway, leading to inhibition of cAMP production and decrease of adipocyte hormone-sensitive lipase activity. This causes reduced hydrolysis of peripheral tissue triglycerides to free fatty acids and thereby the decreased flux of free fatty acids to the liver. The reduced flux of free fatty acids decreases the rate of hepatic triglyceride biosynthesis and production of VLDL.

Inhibition of DGAT-2   Niacin inhibits the activity of hepatocyte DGAT-2, an enzyme involved in triglyceride synthesis. This results in decreased triglyceride synthesis and its availability for VLDL assembly. The decreased hepatic triglyceride synthesis and release of VLDL via inhibiting DGAT-2 may play a more significant role than decreased lipolysis of peripheral tissue triglycerides (section "Activation of GPR109A") in niacin's impact on plasma triglyceride levels. In addition, niacin increases ApoB catabolism, which further impairs hepatic VLDL synthesis and secretion.

4.6.3.2 Decrease of Plasma LDL-Cholesterol Levels
The reduced levels of plasma LDL cholesterol following niacin treatment appear to result from two potential pathways: (i) decreased hepatic VLDL assembly and secretion and (ii) increased hepatic catabolism of ApoB. Decreased

VLDL secretion would lead to decreased levels of LDL cholesterol, as LDL is derived from VLDL. Augmented hepatic catabolism of ApoB results in decreased formation of LDL, as ApoB is a major component of LDL. Augmented ApoB catabolism is also a mechanism leading to decreased production of VLDL.

4.6.3.3 Reduction of Lp(a) and Small, Dense LDL Particles
The mechanisms involved in niacin-mediated reduction of Lp(a) and small, dense LDL particles are currently unclear. In addition to niacin, treatment with other lipid-lowering drugs, such as statins also results in decreases in Lp(a) and small, dense LDL particles.

4.6.3.4 Increase of Plasma HDL-Cholesterol Levels
Niacin is the most effective drug for raising plasma HDL-cholesterol levels. The major mechanism by which niacin raises HDL-cholesterol levels is through decreased catabolism of ApoA-I, a major component of HDL particles. Niacin does not increase the de novo synthesis of ApoA-I. In the liver, niacin inhibits the ectopic β-chain of ATP synthase, a protein complex of the mitochondrial inner membrane that, when expressed on the hepatocyte surface, serves as a high-affinity receptor for ApoA-I. Binding of ApoA-I to this receptor triggers the endocytosis of the ApoA-I-containing holo-HDL particles, thereby leading to the augmented catabolism of ApoA-I [61]. Niacin decreases the surface expression of the β-chain of ATP synthase, causing decreased HDL uptake and increased levels of plasma levels of HDL.

By decreasing triglyceride levels in ApoB-containing lipoproteins (VLDL and LDL), niacin also increases HDL cholesterol indirectly. This is because the lower LDL and VLDL levels result in less exchange of cholesterol esters between HDL and LDL/VLDL. In addition, niacin may also increase HDL-cholesterol levels by reducing NAD and NADP. As discussed in Section 4.8, inhibition of cholesterol ester transfer protein (CETP) represents an effective strategy for raising HDL-cholesterol levels although the clinical benefits of pharmacological inhibition of CETP remain to be established.

4.6.4   Clinical Uses

4.6.4.1 Clinical Indications
The clinical indications of niacin include the following:

- Hypertriglyceridemia
- Elevated LDL cholesterol
- Low plasma HDL cholesterol

Niacin is indicated to reduce triglyceride levels in patients with severe hypertriglyceridemia. It is also used (often in combination with a statin or a bile acid sequestrant) to reduce LDL-cholesterol levels in patients with hypercholesterolemia or mixed dyslipidemia. As it is the most effective drug for raising HDL cholesterol, niacin is particularly useful for treating patients with both hypertriglyceridemia and low HDL-cholesterol levels. In fact, over the past decade, niacin use has rapidly increased in the United States and Canada [62].

4.6.4.2 Effects on Cardiovascular Events and Mortality

Early studies suggested that niacin treatment was associated with decreased cardiovascular events and mortality [63]. Small-scale clinical trials also demonstrated that combination therapy of niacin with a statin or a bile acid sequestrant could slow progression or promote regression of atherosclerotic disease in patients with a history of coronary artery disease and hyperlipidemia [64]. However, a recent large-scale cardiovascular outcome trial, namely, the Atherothrombosis Intervention in Metabolic Syndrome with Low HDL/High Triglycerides: Impact on Global Health Outcomes (AIM-HIGH) trial, reported that among patients with atherosclerotic cardiovascular disease and LDL-cholesterol levels of <70 mg/dl (1.81 mmol/l), there was no incremental clinical benefit from the addition of niacin to statin therapy during a 36-month follow-up period, despite significant improvements in HDL-cholesterol and triglyceride levels [65].

More recently, the much larger Heart Protection Study 2: Treatment of HDL to Reduce the Incidence of Vascular Events (HPS2-THRIVE) trial reported that the addition of niacin-extended release and laropiprant (laropiprant, a selective prostaglandin D_2 (PGD_2) receptor 1 antagonist, was used in the trial to prevent niacin-induced flushing) to simvastatin did not reduce cardiovascular events in a high-risk group of patients with existing cardiovascular diseases, despite lowering LDL cholesterol and triglycerides and raising HDL cholesterol. Instead, the risk of myopathy was increased by adding niacin/laropiprant to simvastatin 40 mg daily (with or without ezetimibe), particularly in Chinese patients [66].

The failure of the above two major clinical trials to demonstrate the cardiovascular benefits of niacin has left us uncertain about its utility, particularly in patients with optimal LDL-cholesterol levels on statin therapy. The lack of additional cardiovascular outcome benefits by niacin could be due to the optimal LDL-cholesterol levels achieved by statin therapy as well as the pleiotropic effects of the statin drugs, which collectively make it difficult to further augment the cardiovascular protection by adding niacin to the existing statin therapy. On the contrary, niacin is an excellent adjunctive therapy in familial hypercholesterolemia, where LDL cholesterol remains high despite statin treatment [64, 67].

4.6.4.3 Combination Therapy

Niacin can be used in combination with a statin (e.g., simvastatin, lovastatin) or a bile acid sequestrant. In combination with simvastatin or lovastatin, niacin is used to treat primary hyperlipidemia or mixed dyslipidemia when treatment with niacin, simvastatin, or lovastatin monotherapy is considered inadequate. The combination formulations of niacin with simvastatin and lovastatin are also known by the trade names Simcor and Advicor, respectively.

4.6.5 Therapeutic Dosages

The dosage form and strength of niacin are listed below:

- Niacin (Niaspan): Oral, 500, 750, and 1000 mg tablets

Niacin is available in various formulations, including prescription and dietary supplement forms. Niacin-extended release (Niaspan) is the most widely used formulation, and it is the only niacin formulation approved by the US FDA for treating dyslipidemias and available only by prescription. The immediate- and sustained-release niacin preparations are available without prescription as dietary supplements.

Niacin-extended release (niacin-ER) should be taken at bedtime with a low-fat snack. The dose range is 500–2000 mg once daily. Therapy with niacin-ER must be initiated at 500 mg at bedtime in order to reduce the incidence and severity of adverse effects, which may occur during early therapy. The dosage is increased gradually, and increase should not be more than 500 mg in any 4-week period. The maintenance dose is 1000–2000 mg once daily. Doses >2000 mg daily are not recommended.

For combination therapy with simvastatin, initial dose of simvastatin is 20 mg once a day. The combination therapy with niacin-ER and simvastatin should not exceed doses of 2000 mg and 40 mg daily, respectively. For combination therapy with lovastatin, initial dose of lovastatin is 20 mg once a day, and combination therapy with niacin-ER and lovastatin should not exceed doses of 2000 mg and 40 mg daily, respectively.

4.6.6 Adverse Effects and Drug Interactions

4.6.6.1 Adverse Effects

The adverse effects of niacin can be summarized into two categories: (i) common, less severe adverse effects and (ii) less common but more severe adverse effects.

Common, Less Severe Adverse Effects The most common adverse effects of niacin treatment are skin flushing and associated pruritus and gastric disturbances (e.g., nausea, vomiting, diarrhea), which limit the patient's compliance. Skin flushing results from niacin-induced activation of GPR109A and is mediated by the formation of prostaglandins. Flushing is usually more significant when niacin is started or its dosage increased. Flushing can be minimized by starting the therapy with low doses. Flushing of the skin may be reduced in frequency or severity by pretreatment with aspirin (up to the recommended dose of 325 mg taken 30 min prior to niacin-ER dose). The efficacy of aspirin is due to its inhibition of prostaglandin formation.

In addition to aspirin, flushing can also be managed by a newly developed drug, laropiprant. Laropiprant is a PGD_2 receptor 1 antagonist that selectively blocks the binding of PGD_2 to its receptor, thereby reducing flushing associated with niacin. The combination of niacin-ER and laropiprant has completed successfully phase III development and been recently marketed in many countries, but not yet in the United States. This combination formulation significantly reduces skin flushing caused by niacin [68].

Less Common but More Severe Adverse Effects

- Liver toxicity: Treatment with niacin, especially the immediate-release and sustained-release forms, may cause hepatotoxicity manifested as increased serum levels of hepatic transaminases. The risk for liver injury is reduced with niacin-ER.
- Metabolic disturbances: Niacin may disrupt glucose homeostasis and increase blood glucose levels by causing insulin resistance. Glucose levels should be closely monitored in diabetic and potentially diabetic patients, particularly during the first few months of use or dose adjustment [69]. Although niacin can cause glucose intolerance and potentially aggravate diabetes, clinical trials demonstrate that it can be used in patients whose diabetes is well controlled with little effect on glucose levels. Niacin may also increase blood uric acid levels and occasionally precipitates gout. The antigout drug allopurinol may be given with niacin if needed.
- Myopathy: Myopathy has been reported in patients taking niacin-ER. The risk for myopathy and rhabdomyolysis is increased when lovastatin and simvastatin are coadministered with niacin-ER, especially in elderly patients and patients with diabetes, renal failure, or uncontrolled hypothyroidism.
- Others: The gastric distress caused by niacin may activate peptic ulcer disease. Niacin treatment may also cause reduction in platelet count and increase in prothrombin time, and as such, caution should be taken when niacin is administered concomitantly with anticoagulants.

4.6.6.2 Drug Interactions The major drug interactions for niacin therapy include the following:

- As noted earlier, caution should be used when prescribing niacin with statins as these agents can increase the risk of myopathy and rhabdomyolysis.
- Bile acid sequestrants have a high niacin-binding capacity and should be taken at least 4–6 h before niacin administration.

- Niacin may potentiate the effects of ganglionic blocking agents and vasoactive drugs, resulting in postural hypotension.

4.6.6.3 Contraindications and Pregnancy Category

- Active liver disease, which may include unexplained persistent elevations in hepatic transaminase levels.
- Active peptic ulcer disease.
- Arterial bleeding.
- Known hypersensitivity to the drug product.
- Pregnancy category: C. Niacin, at doses used in humans, has been associated with birth defects in experimental animals. It is not known whether niacin at doses typically used for lipid disorders can cause fetal harm when administered to pregnant women or whether it can affect reproductive capacity. If a woman receiving niacin for primary hypercholesterolemia becomes pregnant, the drug should be discontinued. If a woman being treated with niacin for hypertriglyceridemia conceives, the benefits and risks of continued drug therapy should be assessed on an individual basis.

4.6.7 Summary of Niacin

- Niacin (nicotinic acid, vitamin B3) at pharmacological doses exerts favorable effects on lipid profiles, including reduction of plasma levels of triglycerides, LDL cholesterol, and Lp(a) and elevation of HDL cholesterol. It is the most effective HDL-raising agent currently approved by the US FDA. Niacin also exerts other pleiotropic beneficial effects, including anti-inflammation and antithrombogenesis.
- The beneficial effects of niacin on lipids/lipoproteins are primarily mediated by activation of the G-protein-coupled receptor GPR109A. Niacin also affects other molecular targets to impact lipid profiles as well as to result in pleiotropic effects.
- Niacin is used to treat hypertriglyceridemia and elevated LDL cholesterol. It is a particularly useful agent for treating patients with dyslipidemia and markedly decreased levels of HDL cholesterol when statin or other lipid-lowering monotherapy is inadequate for achieving the target goals of plasma lipoproteins.
- Two recent cardiovascular outcome trials, namely, AIM-HIGH and HPS2-THRIVE, reported that adding niacin failed to further reduce cardiovascular events in high-risk patients with optimal LDL-cholesterol levels on statin therapy despite significant increases in HDL cholesterol and reductions in triglycerides and LDL cholesterol.

- Niacin is generally well tolerated. Common adverse effects include flushing of the skin and pruritus, which can be managed by taking aspirin. Other less common adverse effects may include liver injury, elevation of blood glucose levels, elevation of blood uric acid levels, and activation of peptic ulcer disease.

4.7   NEW DRUGS FOR HoFH

Listed below are two recently US FDA-approved drugs for treating HoFH:

- Lomitapide (Juxtapid).
- Mipomersen (Kynamro).

As described in Chapter 3 (Table 3.2), familial hypercholesterolemia is an inherited lipoprotein disorder caused by defects in the LDL receptor gene that result in very high levels of LDL cholesterol. Patients with HoFH (prevalence: one per million) have extremely high levels of LDL cholesterol; without treatment, cardiovascular events and death can occur in childhood. Standard treatment of patients with HoFH includes dietary fat restriction, high doses of statin therapy combined with ezetimibe, a bile acid sequestrant, and niacin (see Sections 4.2, 4.3, 4.4, and 4.6 for detailed discussion of these drugs).

Standard therapy, however, generally fails to achieve LDL-cholesterol levels anywhere near the goal, and LDL apheresis (a procedure that takes blood outside the body to remove the LDL cholesterol and then returns the blood back to the body) or liver transplantation is usually required to adequately control LDL cholesterol. In fact, LDL apheresis remains the treatment of choice with additional pharmacological therapy. However, the procedure carries high burdens to the patients.

Over the past decade, new pharmacotherapies have been developed to address this patient population (about 300 patients in the United States). The US FDA has recently approved two new drugs (i.e., lomitapide and mipomersen) for reducing LDL cholesterol by inhibiting VLDL production through distinct mechanisms [70] (Fig. 4.9). This section examines the molecular pharmacology of these two new drugs.

4.7.1   Lomitapide

The US FDA approval of lomitapide was based on a single-arm, open-label, phase 3 clinical trial in 29 adult patients with HoFH recruited from 11 centers in four countries (United States, Canada, South Africa, and Italy). Of the 29 enrolled patients, 23 completed both the efficacy phase (26 weeks) and the full study (78 weeks). The median dose of lomitapide was 40 mg a day. LDL cholesterol was reduced by 50% from a baseline of 336 mg/dl to 166 mg/dl after 26 weeks. Concentrations of LDL cholesterol remained

reduced by 44% at week 56 and 38% at week 78. All the reductions were statistically significant [71].

4.7.1.1   Chemistry and Pharmacokinetics   Lomitapide (structure shown in Fig. 4.9) is a synthetic drug. The chemical name of lomitapide mesylate is N-(2,2,2-trifluoroethyl)-9-[4-[4-[[[4′-(trifluoromethyl)[1,1′-biphenyl]-2-yl]carbonyl]amino]-1-piperidinyl]butyl]-9H-fluorene-9-carboxamide, methanesulfonate salt with a molecular mass of 789.8.

Following oral administration, lomitapide is absorbed from the gut with a t_{max} of ~6 h. The drug undergoes extensive metabolism in the liver primarily via CYP3A4, and the oral bioavailability is ~7%. Lomitapide is 99.8% plasma protein bound. The drug is excreted from the kidney (53–60% as metabolites) and feces (33–35%) and has an elimination half-life of 40 h.

4.7.1.2   Molecular Mechanisms and Pharmacological Effects   As depicted in Figure 4.9, lomitapide directly binds and inhibits microsomal triglyceride transfer protein (MTP) (also see section "Hepatic Secretion of VLDL" of Chapter 3), which resides in the lumen of the endoplasmic reticulum, thereby preventing the assembly of ApoB-containing lipoproteins in enterocytes and hepatocytes. This inhibits the synthesis of chylomicrons and VLDL. The inhibition of the synthesis of VLDL leads to reduced levels of plasma LDL cholesterol.

4.7.1.3   Clinical Uses

Clinical Indications   Lomitapide is indicated as an adjunct to a low-fat diet and other lipid-lowering treatments, including LDL apheresis where available, to reduce LDL cholesterol, total cholesterol, ApoB, and non-HDL cholesterol in patients with HoFH.

Limitations of Use   The safety and effectiveness of lomitapide have not been established in patients with hypercholesterolemia who do not have HoFH. The effect of lomitapide on cardiovascular morbidity and mortality has not been determined.

4.7.1.4   Therapeutic Dosages   The dosage forms and strengths of lomitapide are listed below:

- Lomitapide (Juxtapid): Oral, 5, 10, and 20 mg capsules.

Because of the risk of hepatotoxicity, before treatment, plasma levels of ALT, AST, alkaline phosphatase, and total bilirubin should be obtained to establish baseline data. Treatment should be initiated at 5 mg once daily and increased to 10 mg daily after at least 2 weeks and then, at a minimum of 4-week intervals, to 20 mg, to 40 mg, and up

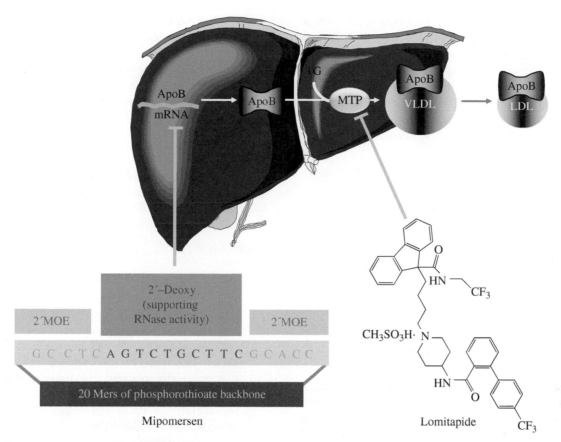

FIGURE 4.9 Chemical structures and molecular mechanism of action of lomitapide and mipomersen. Lomitapide inhibits microsomal triglyceride transfer protein (MTP), an enzyme involved in the assembly of VLDL in the liver. Mipomersen is an antisense that targets ApoB mRNA, thereby reducing ApoB synthesis and the subsequent assembly of VLDL. Decreased formation of VLDL results in reduction of LDL cholesterol. TG denotes triglyceride; 2′MOE denotes 2′-methoxyethyl. For color details, please see color plate section.

to the maximum recommended dose of 60 mg daily. Due to reduced absorption of fat-soluble vitamins/fatty acids, patients should take daily vitamin E, linoleic acid, alpha-linolenic acid (ALA), eicosapentaenoic acid (EPA), and docosahexaenoic acid (DHA) supplements. The drug should be taken without food or at least 2 h after evening meal because taking with food may increase the risk of gastrointestinal adverse effects. Patients with end-stage renal disease on dialysis or with baseline mild hepatic impairment should not take more than 40 mg daily.

4.7.1.5 Adverse Effects and Drug Interactions

Adverse Effects   The adverse effects of lomitapide include the following three aspects:

- Risk of hepatotoxicity (black box warning): Lomitapide may cause elevations in plasma ALT and AST. It may also cause hepatic steatosis, a risk factor for progressive liver disease, including steatohepatitis and cirrhosis.
- Reduced absorption of fat-soluble vitamins and omega-3 fatty acids.

- Gastrointestinal adverse reactions including diarrhea, nausea, vomiting, dyspepsia, and abdominal pain.

Drug Interactions   Outlined below are key aspects of lomitapide drug interactions:

- CYP3A4 inhibitors: Lomitapide is both a substrate and an inhibitor of CYP3A4. CYP3A4 inhibitors increase exposure to lomitapide. Strong and moderate CYP3A4 inhibitors are contraindicated with lomitapide. Patients must avoid grapefruit juice, which contain compounds that inhibit CYP3A4. The daily dose of lomitapide should not exceed 30 mg when used concomitantly with weak CYP3A4 inhibitors, including atorvastatin and oral contraceptives.
- Warfarin: Lomitapide increases plasma concentrations of warfarin. It is necessary to monitor INR regularly, especially with lomitapide dose adjustment.
- Lomitapide increases the plasma levels of simvastatin and lovastatin. The doses of these statins need to be reduced when coadministered with lomitapide due to myopathy risk.

- P-Glycoprotein substrates: Lomitapide is an inhibitor of P-glycoprotein, and dose reduction of drugs that are P-glycoprotein substrates is required because of possible increased absorption of these drugs when coadministered with lomitapide.
- Bile acid sequestrants: Administration of bile acid sequestrants and lomitapide should be separated by at least 4h due to binding of lomitapide by bile acid sequestrants.

Contraindications and Pregnancy Category

- Concomitant use with strong or moderate CYP3A4 inhibitors.
- Moderate or severe hepatic impairment or active liver disease including unexplained persistent abnormal liver function tests.
- Pregnancy category: X.

4.7.2 Mipomersen

The approval of mipomersen by the US FDA was based on a randomized, double-blind, placebo-controlled phase 3 trial in 51 patients aged 12 and older with HoFH reported in Lancet in 2010 [72]. Of the 51 patients, 45 completed the 26-week treatment period (28 mipomersen, 17 placebo). Mean concentrations of LDL cholesterol at baseline were 441 mg/dl (11.4 mmol/l) in the mipomersen group and 402 mg/dl (10.4 mmol/l) in the placebo group. The mean percentage change in LDL-cholesterol concentration was significantly greater with mipomersen (−24.7%) than with placebo (−3.3%) [72]. More recently, multiple clinical trials also demonstrated an efficacy for mipomersen in nonhomozygous familial hypercholesterolemic patients with high LDL cholesterol refractory to conventional lipid-lowering therapy. These include patients with heterozygous familial hypercholesterolemia and severe hypercholesterolemia at high cardiovascular risk [73, 74].

4.7.2.1 Chemistry and Pharmacokinetics
Mipomersen sodium is a synthetic phosphorothioate oligonucleotide sodium salt (20 nucleotides in length). It is a specific antisense inhibitor of ApoB-100 synthesis (Fig. 4.9). Following subcutaneous injection, peak plasma concentrations of mipomersen are typically reached in 3–4 h, and bioavailability ranges from 54 to 78%. Mipomersen is highly bound to human plasma proteins (≥90%). It is not a substrate for CYP metabolism and is metabolized in tissues by endonucleases to form shorter oligonucleotides that are further metabolized by exonucleases. The elimination of mipomersen involves both metabolism in tissues and excretion primarily in the urine. Following subcutaneous administration, elimination half-life for mipomersen is ~1–2 months.

4.7.2.2 Molecular Mechanisms and Pharmacological Effects
Mipomersen is an antisense oligonucleotide targeted to human mRNA for ApoB-100, the principal apolipoprotein of LDL and its metabolic precursor, VLDL. Mipomersen is complementary to the coding region of the mRNA for ApoB-100 and binds by Watson and Crick base pairing. The hybridization of mipomersen to the cognate mRNA results in RNase H-mediated degradation of the cognate mRNA, thus inhibiting translation of the ApoB-100 protein (Fig. 4.9).

4.7.2.3 Clinical Uses

Clinical Indications Mipomersen is indicated as an adjunct to lipid-lowering medications and diet to reduce LDL cholesterol, ApoB, total cholesterol, and non-HDL cholesterol in patients with HoFH.

Limitations of Use

- The safety and effectiveness of mipomersen have not been well established in patients with hypercholesterolemia who do not have HoFH. A planned interim analysis of an ongoing, open-label extension trial in 141 patients with familial hypercholesterolemia receiving a subcutaneous injection of 200 mg mipomersen weekly plus maximally tolerated lipid-lowering therapy for up to 104 weeks suggested that long-term treatment with mipomersen for up to 104 weeks provided sustained reductions in all atherosclerotic lipoproteins measured and a safety profile consistent with prior controlled trials in these high-risk patient populations [75].
- The effect of mipomersen on cardiovascular morbidity and mortality has not been determined.
- The use of mipomersen as an adjunct to LDL apheresis is not recommended.

4.7.2.4 Therapeutic Dosages
The dosage forms and strengths of mipomersen are listed below:

- Single-use vial containing 1 ml of a 200 mg/ml solution
- Single-use prefilled syringe containing 1 ml of a 200 mg/ml solution

The drug is given at 200 mg once weekly as a subcutaneous injection. The injection sites may include the abdomen, thigh region, or outer area of the upper arm. Before treatment, ALT, AST, alkaline phosphatase, and total bilirubin should be measured because the drug may cause hepatotoxicity (see Section 4.7.2.5).

4.7.2.5 Adverse Effects and Drug Interactions

Adverse Effects The most commonly reported adverse reactions (incidence ≥10% and greater than placebo) are injection site reactions, flu-like symptoms, nausea, headache, and elevations in serum transaminases, specifically ALT. The drug may also increase the risk of hepatic steatosis, a risk factor for progressive liver disease, including steatohepatitis and cirrhosis. Risk of hepatotoxicity is a black box warning for mipomersen.

Drug Interactions No significant drug interactions have been reported with mipomersen. This is likely due to its unique pharmacokinetic characteristics, including its inability to affect CYP enzymes as well as the novel mechanism of its metabolism (see Section 4.7.2.1).

Contraindications and Pregnancy Category

- Moderate or severe hepatic impairment, or active liver disease, including unexplained persistent elevations of serum transaminases.
- Known sensitivity to product components.
- Pregnancy category: B.

4.7.3 Summary of New Drugs for HoFH

- HoFH is an inherited lipoprotein disorder caused by defects in the LDL receptor gene that result in very severe hypercholesterolemia.
- The recently approved drugs, lomitapide and mipomersen, act via distinct mechanisms to reduce ApoB-containing lipoprotein levels in patients with HoFH already taking maximum dosages of other lipid-lowering drugs. Lomitapide directly inhibits MTP, thereby preventing the assembly of ApoB-containing lipoproteins in enterocytes and hepatocytes, whereas mipomersen is an antisense oligonucleotide targeting human mRNA for ApoB-100, thereby inhibiting the synthesis of ApoB-100.
- Both lomitapide and mipomersen can cause hepatotoxicity and increase the risk of hepatic steatosis. In addition, both drugs are very expensive with the annual cost ranging from $176,000 to $295,000 [76, 77].

4.8 PHYTOSTEROLS AND PHYTOSTANOLS

4.8.1 Introduction to Phytosterols and Phytostanols

4.8.1.1 Definition and Chemistry Phytosterols, also known as plant sterols, are natural components of cell membranes of plants. The role of phytosterols in plants is similar to that of cholesterol in animals. In fact, the structures of phytosterols and cholesterols are very similar, with a C5 double bond and 3-beta-hydroxyl group, but phytosterols possess structure modifications of the C24 side chains (structures shown in Fig. 4.10). As noted in Section 4.4, most common types of phytosterols in diet are β-sitosterol, campesterol, and stigmasterol. Phytostanols are the saturated form of phytosterols and lack the C5 double bond,

and they are much less abundant than phytosterols. β-Sitostanol and campestanol are the two most common types of dietary stanols.

4.8.1.2 Sources and Dietary Intakes The natural sources of phytosterols/phytostanols include vegetable oils, nuts, seeds, cereals, vegetables, and fruits. The typical Western dietary intake of phytosterols/phytostanols ranges from 200 mg to 400 mg/day, and between the two, phytostanols account for a small portion (about 50 mg/day). Mediterranean diet contains more phytosterols/phytostanols. In contrast to dietary cholesterol, phytosterols are poorly absorbed (0.4–5% of amounts ingested), and the absorption of phytostanols is even less (about 10% of that of phytosterols).

4.8.1.3 History The cholesterol-lowering effects of phytosterols were first demonstrated 60 years ago by D.W. Peterson, who observed that supplementation of soybean sterols in the diet decreased plasma and liver cholesterol levels in chicks [78]. In 1953, O.J. Pollak first reported that administration of crude sitosterol decreased blood cholesterol levels in human subjects [79]. O.J. Pollak also observed that sitosterol was poorly absorbed and its administration inhibited the intestinal absorption of cholesterol [80]. These early observations prompted subsequent clinical studies that collectively demonstrated a therapeutic efficacy for dietary supplementation of phytosterols as well as phytostanols in reducing plasma LDL-cholesterol levels in patients with hypercholesterolemia [81]. The solubility issue associated with phytosterols and phytostanols also stimulated the development of the ester forms with increased solubility and incorporation into fat-based food products, such as margarine (see Section 4.8.2). The cholesterol-lowering effects of the esters were also demonstrated in numerous clinical studies.

4.8.1.4 Esterification and Dietary Supplementation As noted earlier, due to the limited solubility of natural phytosterols and phytostanols, esterification of these compounds is necessary to augment the lipid solubility and facilitate incorporation into fat-based food products and increase the delivery of the compounds to the small intestine. Esterified phytosterols and phytostanols are hydrolyzed in the small intestine to the free forms that inhibit cholesterol absorption. Due to their lipid-lowering effects, phytosterol esters and phytostanol esters added to the foodstuffs are recommended as part of the therapeutic lifestyle change regimen for patients with elevated LDL-cholesterol levels.

4.8.2 Molecular Mechanisms and Pharmacological Effects

Phytosterols and phytostanols reduce the absorption of cholesterol in the small intestine and, as such, result in decreased levels of plasma LDL cholesterol [82]. Two potential

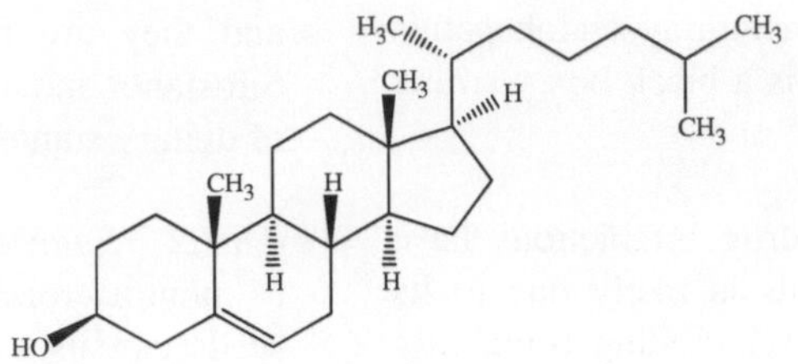

Cholesterol

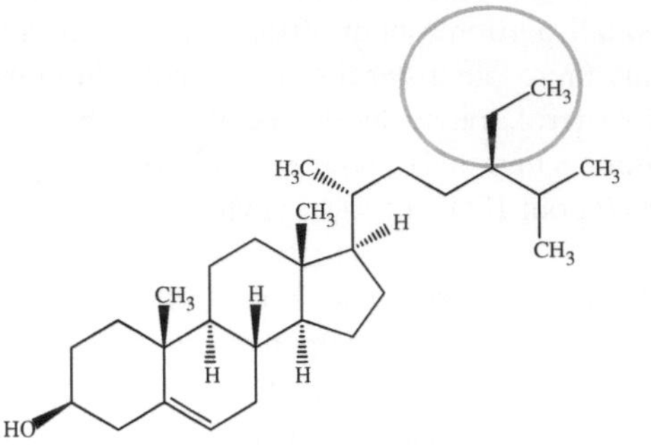

Beta-Sitosterol

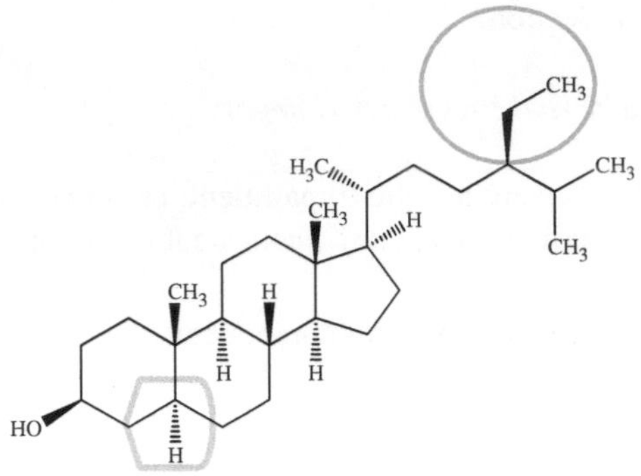

Beta-Sitostanol

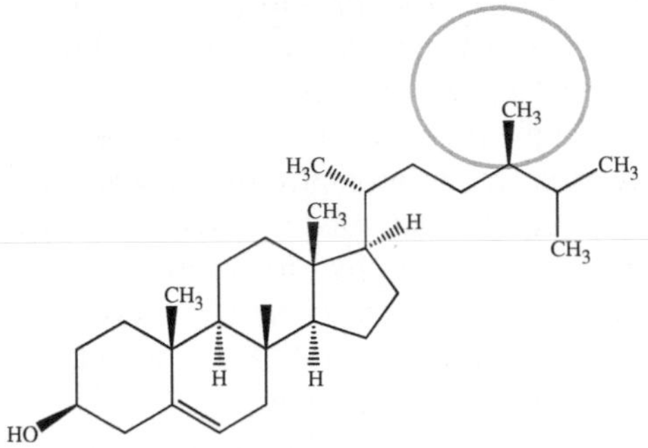

Campesterol

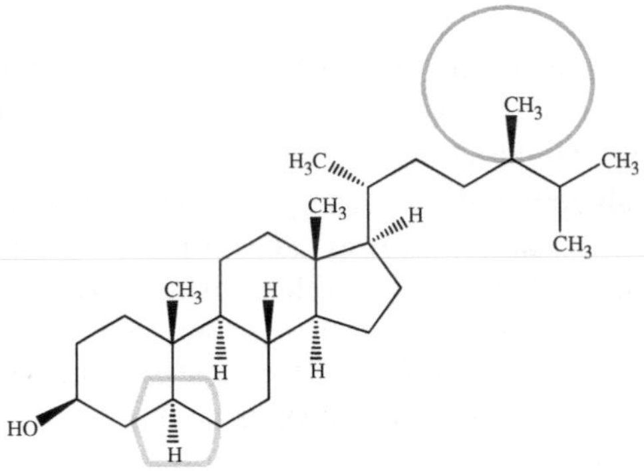

Campestanol

FIGURE 4.10 Chemical structures of common phytosterols and phytostanols. Phytosterols are different from cholesterol only in the side chain (shown in cycle). Phytosterols and phytostanols are different only in the double bond (shown in rectangle).

mechanisms have been proposed to explain the cholesterol-lowering effect of phytosterols/phytostanols (Fig. 4.10):

- Blockage of cholesterol incorporation into micelles: Because phytosterols and phytostanols are more hydrophobic than cholesterol, they have a higher affinity for micelles than has cholesterol. Consequently, they displace cholesterol from micelles. This results in decreased intestinal absorption of cholesterol and hence increased fecal excretion of cholesterol.
- Promotion of cholesterol efflux into the intestinal lumen: Phytosterols/phytostanols may also promote increased efflux of cholesterol molecules that have already entered the enterocytes into the intestinal lumen. This is thought to occur via induction of ABC transport proteins by phytosterols/phytostanols. Induction of these transport proteins augments the efflux of cholesterol molecules.

4.8.3 Clinical Uses

The clinical applications of phytosterols/phytostanols include the following:

- Reduction of plasma LDL-cholesterol levels as part of the therapeutic lifestyle change regimen
- Reduction of plasma LDL cholesterol as an addition to statin therapy

Systemic review and meta-analysis of clinical trials show that supplementation of phytosterols/phytostanols at dosages of 2–3 g/day lowers plasma LDL cholesterol by 5–15% [83, 84]. Clinical trials also demonstrate that addition of phytosterols/phytostanols to statin therapy is associated with a further ~5% reduction in LDL cholesterol per gram of the phytosterols/phytostanols. This additional reduction of LDL cholesterol is equivalent to that caused by doubling of the statin dosage (see section "Effects on Plasma

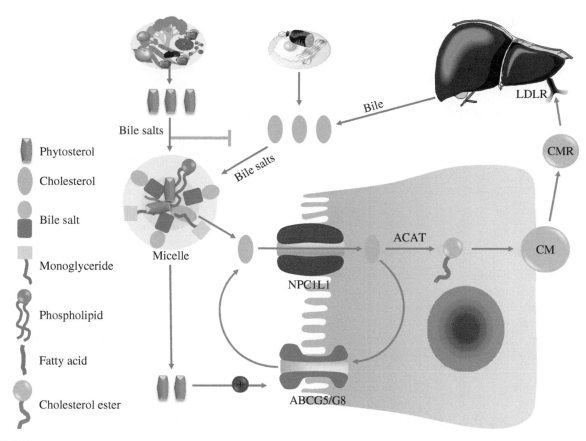

FIGURE 4.11 Molecular mechanism of action of phytosterols and phytostanols. Phytosterols/phytostanols compete with dietary/bile-derived cholesterol for incorporating into micelles, thereby reducing the transport of cholesterol into enterocytes via NPC1L1 (also see Fig. 4.6). Phytosterols/phytostanols cause upregulation of ABCG5/ABCG8, hence promoting efflux of cholesterol from enterocytes to the lumen. See Figure 4.6 legend for more description.

LDL-Cholesterol Levels"). However, it is currently unknown if the LDL-cholesterol-lowering effect persists beyond 2 years as clinical studies on phytosterols/phytostanols have been relatively short term (usually 1–18 months).

4.8.4 Therapeutic Dosages

The Adult Treatment Panel III guidelines recommend a daily intake of 2 g of phytosterol/phytostanol esters as a part of the therapeutic lifestyle change regimen to enhance LDL-cholesterol reduction (see Chapter 5).

Phytosterols are also included in Animi-3® capsules, which also contain folic acid, vitamin B6, vitamin B12, and omega-3 fatty acids. Animi-3® capsules are indicated for improving nutritional status in conditions requiring folic acid, vitamins B6 and B12, and essential fatty acids, as well as for patients whose plasma cholesterol levels are a concern.

4.8.5 Adverse Effects and Drug Interactions

In general, phytosterols/phytostanols are well tolerated and recognized as safe [82]. Commonly reported adverse effects are primarily gastrointestinal in nature (e.g., nausea, dyspepsia, diarrhea, constipation, flatulence, gastroesophageal reflux, appetite changes). These compounds may also result in reduced absorption of carotenoids, especially α-carotene, β-carotene, and lycopene. The long-term safety (>5 years) of phytosterols and phytostanols remains to be established.

4.8.6 Summary of Phytosterols/Phytostanols

- Phytosterols and phytostanols are naturally present in vegetable oils, nuts, seeds, vegetables, and fruits.
- Dietary supplementation of phytosterols/phytostanols leads to inhibition of intestinal absorption of cholesterol and thereby reduced levels of plasma LDL cholesterol. Addition of phytosterols/phytostanols to statin therapy causes additional reduction in plasma LDL-cholesterol levels.
- The Adult Treatment Panel III guidelines recommend a daily intake of 2 g of phytosterol/phytostanol esters as part of the therapeutic lifestyle change regimen to enhance LDL-cholesterol reduction.

- Phytosterols and phytostanols are well tolerated and generally recognized as safe. However, the long-term safety of dietary supplementation of these compounds remains to be established.

4.9   OMEGA-3 FATTY ACIDS

Listed below are the two US FDA-approved drugs of omega-3 fatty acids:

- Omega-3-acid ethyl esters (Lovaza)
- Icosapent ethyl (Vascepa)

4.9.1   Introduction to Omega-3 Fatty Acids

4.9.1.1 Definition   Omega-3 fatty acids, also known as ω-3 fatty acids, constitute a series of essential unsaturated fatty acids that have a final carbon–carbon double bond in the n-3 position (also known as the ω position), that is, the third bond from the methyl end of the fatty acid. As such, omega-3 fatty acids are also referred to as n-3 fatty acids. Nutritionally important omega-3 fatty acids include the plant-derived α-linolenic acid (ALA) and the marine-derived EPA and DHA, all of which are polyunsaturated (structure shown in Fig. 4.12).

4.9.1.2 Preparations   The marine EPA and DHA are present mainly in oily fish and the commercially available supplements, which are available either over the counter (as fish oils) or as concentrated pharmaceutical preparations. Currently, there are two US FDA-approved pharmaceutical preparations of omega-3 fatty acids: Lovaza and Vascepa (Fig. 4.13).

Lovaza (omega-3-acid ethyl esters) is a liquid-filled gel capsule for oral administration. Each 1 g capsule of Lovaza contains at least 900 mg of the ethyl esters of omega-3 fatty acids sourced from fish oils. These are predominantly a combination of ethyl esters of EPA (~465 mg) and DHA (~375 mg). The ethyl esters of EPA and DHA are absorbed into systemic circulation when they are administered orally.

Vascepa is also a liquid-filled gel capsule for oral administration. Each Vascepa capsule contains 1 gram of

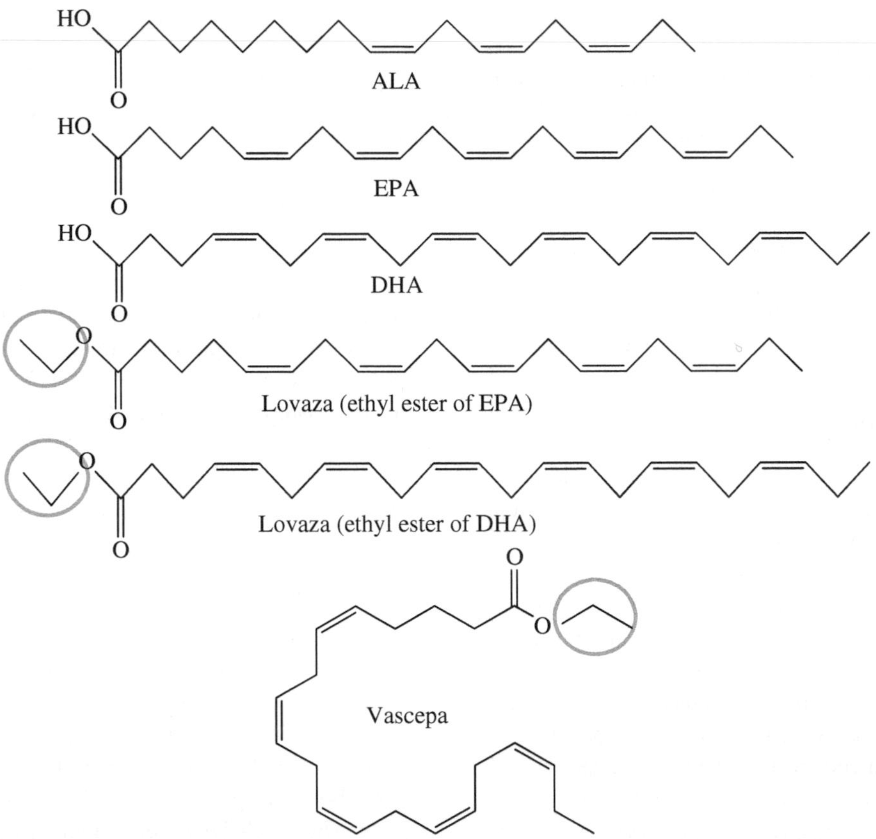

FIGURE 4.12 Chemical structures of omega-3 fatty acids. The three main forms of omega-3 fatty acids are alpha-linolenic acid (ALA), eicosapentaenoic acid (EPA), and docosahexaenoic acid (DHA). Lovaza is a mixture of esters of EPA and DHA, whereas Vascepa is an ester of EPA (ester shown in cycle).

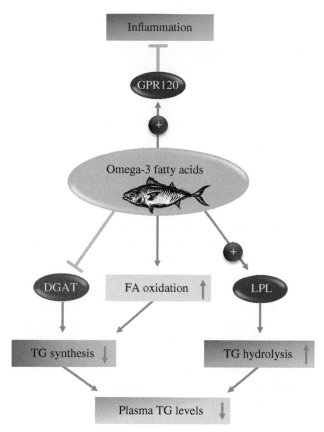

FIGURE 4.13 Molecular mechanism of action of omega-3 fatty acids. The FDA-approved use of omega-3 fatty acids (Lovaza and Vascepa) is to reduce triglycerides (TG) in individuals with severe hypertriglyceridemia. The TG-lowering activity of Lovaza and Vascepa results from inhibition acyl-CoA:diacylglycerol acyltransferase (DGAT) and activation of lipoprotein lipase (LPL), as well as increased oxidation of fatty acids (FA). Omega-3 fatty acids also exert anti-inflammatory effects, likely mediated by activation of the G-protein-coupled receptor 120.

icosapent ethyl. Icosapent ethyl is an ethyl ester of the omega-3 fatty acid EPA.

4.9.1.3 Cardiovascular Effects of Omega-3 Fatty Acids
A number of epidemiological studies and small-scale clinical trials have suggested potential beneficial effects of supplementation of omega-3 fatty acids in various cardiovascular disorders, including coronary artery disease, atrial fibrillation, and heart failure [85, 86]. However, recently, multiple large-scale randomized, double-blind, placebo-controlled trials failed to show an efficacy for omega-3 fatty acids in reducing cardiovascular events and mortality in patients at high risk for cardiovascular events [87–90]. At present, the best characterized and well-documented effect of omega-3 fatty acids is the reduction of plasma triglyceride levels in patients with severe hypertriglyceridemia.

4.9.2 Molecular Mechanisms and Pharmacological Effects

As noted earlier, reduction of triglyceride levels is the best documented effect of omega-3 fatty acids. This section focuses on the discussion of the potential mechanisms underlying omega-3 fatty acid-mediated triglyceride lowering. Accumulating evidence also suggests that omega-3 fatty acids may exert other beneficial effects, such as anti-inflammation.

4.9.2.1 Reduction of Triglyceride Levels
The best documented effect of omega-3 fatty acids is reduction of triglyceride levels in patients with severe hypertriglyceridemia, and this is the only FDA-approved condition for which Lovaza or Vascepa is currently prescribed.

The mechanisms of triglyceride-lowering effects of omega-3 fatty acids are not well understood. Potential pathways include the following:

- Inhibition of acyl-CoA:1,2-diacylglycerol acyltransferase, an enzyme involved in triglyceride biosynthesis.
- Increased mitochondrial and peroxisomal β-oxidation in the liver, leading to increased consumption of fatty acids and thereby reduced triglyceride biosynthesis.
- Increased lipoprotein lipase activity, leading to augmented hydrolysis of triglycerides in VLDLs and chylomicrons.
- Lovaza and Vascepa may also reduce the synthesis of triglycerides in the liver because EPA and DHA are poor substrates for the enzymes responsible for triglyceride synthesis, and EPA and DHA inhibit esterification of other fatty acids.

4.9.2.2 Anti-inflammation and Other Potential Effects
Omega-3 fatty acids have anti-inflammatory and antidiabetic effects in experimental models. It has been shown that the G-protein-coupled receptor GPR120 is a receptor for omega-3 fatty acids, and activation of this receptor by omega-3 fatty acids inhibits multiple inflammation cascades in macrophages and reverses insulin resistance in obese mice [91]. A recent trial has demonstrated that the addition of omega-3-acid ethyl esters to the combination of aspirin and clopidogrel (both aspirin and clopidogrel are inhibitors of platelets) significantly potentiates platelet responses to clopidogrel after percutaneous coronary intervention [92].

4.9.3 Clinical Uses

Both Lovaza and Vascepa are indicated as an adjunct to diet to reduce triglyceride levels in adult patients with severe (≥500 mg/dl) hypertriglyceridemia. Patients should be placed on an appropriate lipid-lowering diet before receiving Lovaza or Vascepa and should continue this diet during treatment with the drug.

4.9.4   Therapeutic Dosages

The dosage forms and strengths of Lovaza and Vascepa are given below:

- Lovaza (omega-3-acid ethyl esters) capsules are supplied as 1 g transparent soft-gelatin capsules filled with light-yellow oil and bearing the designation Lovaza.
- Vascepa (icosapent ethyl) capsules are supplied as 1 g amber-colored soft-gelatin capsules imprinted with Vascepa.

The daily dose of Lovaza is 4 g taken as a single 4 g dose (4 capsules) or as two 2 g doses (2 capsules given twice daily). The daily dose of Vascepa is 4 g/day taken as two capsules twice daily with food.

4.9.5   Adverse Effects and Drug Interactions

4.9.5.1   Adverse Effects   Omega-3 fatty acids are well tolerated and generally regarded as safe. The most common adverse effects of Lovaza are eructation and dyspepsia. Lovaza may increase plasma LDL-cholesterol levels. LDL-cholesterol levels should be monitored periodically during therapy with Lovaza. The most common reported adverse reaction (incidence >2% and greater than placebo) of Vascepa treatment is arthralgia.

4.9.5.2   Drug Interactions   Omega-3 fatty acids may prolong bleeding time due, at least partially, to its potential effects on platelets. Patients taking Lovaza or Vascepa and an anticoagulant or other drug affecting coagulation should be monitored periodically.

4.9.5.3   Contraindications and Pregnancy Category

- Lovaza or Vascepa is contraindicated in patient with known hypersensitivity to the drug or any of its components.
- Pregnancy category: C. It is unknown whether Lovaza or Vascepa can cause fetal harm when administered to pregnant women or can affect reproductive capacity. Lovaza or Vascepa should be used during pregnancy only if the potential benefits to the patients justify the potential risks to the fetus.

4.9.6   Summary of Omega-3 Fatty Acids

- Omega-3 fatty acids mainly refer to the marine EPA and DHA, which are present in oily fish. Lovaza and Vascepa are currently the only two US FDA-approved pharmaceutical formulations of omega-3 fatty acids.
- Recent clinical trials failed to show a beneficial effect of omega-3 fatty acids on cardiovascular events in patients at high cardiovascular risk. Severe hypertriglyceridemia

is the only US FDA-approved indication for Lovaza or Vascepa therapy.
- Lovaza and Vascepa affect multiple potential pathways involved in the synthesis and catabolism of triglycerides to cause reduction of plasma triglyceride levels in patients with severe triglyceridemia.
- Omega-3 fatty acids are well tolerated and generally recognized as safe.

4.10   EMERGING THERAPEUTIC MODALITIES FOR DYSLIPIDEMIAS

The drugs described earlier in this chapter are generally effective for treating various forms of dyslipidemias, including elevated levels of LDL cholesterols and low HDL-cholesterol levels. However, issues continue to exist, including limited maximal effects of the drugs and their intolerable adverse effects, as well as patients who are refractory to the existing drugs. Hence, development of novel therapeutic agents for dyslipidemias is imperative for the effective control of cardiovascular disorders associated with dyslipidemias. This section discusses emerging therapeutic modalities for reducing LDL cholesterol and novel therapies for raising HDL cholesterol.

4.10.1   Emerging Therapeutic Strategies Targeting LDL

There are a number of emerging therapeutic modalities that target LDL cholesterol [93, 94]. They are listed below:

- Antisense oligonucleotides that inhibit the synthesis of ApoB-100, thereby leading to decreased LDL-cholesterol levels
- Squalene synthase inhibitors that inhibit the biosynthesis of cholesterol by blocking the synthesis of squalene
- MTP inhibitors that decrease the levels of LDL cholesterol and triglycerides by decreasing the assembly and secretion of VLDL and chylomicrons
- Thyroid hormone analogs that enhance the expression of hepatic LDL receptor expression, thereby increasing LDL clearance and decreasing LDL-cholesterol levels
- Inhibitors of proprotein convertase subtilisin/kexin type 9 (PCSK9), a protein involved in degradation of LDL receptor

4.10.1.1   Antisense Oligonucleotides Targeting ApoB-100   Among the aforementioned novel therapeutic modalities, antisense oligonucleotides targeting ApoB-100 have received much attention regarding their efficacy in treating refractory hypercholesterolemia. In this context, as discussed earlier, mipomersen, a 20 mer antisense oligonucleotide that specifically binds ApoB-100 mRNA and leads to inhibition of protein translation, reduces the synthesis and secretion of LDL and VLDL that contain ApoB-100. Clinical studies show that mipomersen is effective in lowering LDL-cholesterol levels, causing an ~25% reduction in patients with HoFH [72]. As noted in Section 4.7.2, this antisense agent has recently received the US FDA approval for treating HoFH.

4.10.1.2 Inhibitors of Squalene Synthase

As stated earlier, the most popular drugs for cholesterol reduction are the satins, which are competitive inhibitors of HMG-CoA reductase, the primary rate-limiting enzyme in cholesterol biosynthesis. Although relatively safe and effective, statins can cause liver injury and myopathy owing, at least partly, to their inhibition of the formation of metabolites of the mevalonate pathway, such as coenzyme Q (also known as ubiquinone). Inhibition of cholesterol biosynthesis downstream of these metabolites might theoretically limit statin-associated adverse effects. Squalene synthase catalyzes the conversion of two molecules of farnesyl pyrophosphate to squalene in a two-step reaction that represents the first committed step in the de novo cholesterol biosynthesis (see Fig. 4.2).

Several classes of squalene synthase inhibitors have been developed, and one of the inhibitors, lapaquistat, has undergone extensive clinical trials for safety and efficacy in reducing LDL cholesterol. Systemic analysis of the trials indicates that lapaquistat at 100 mg could significantly reduce LDL cholesterol by ~22% in monotherapy and 18% in combination with a statin. However, the drug caused significant liver injury, which led to the termination of development. The lapaquistat experience illustrates the current challenges in lipid-altering drug development [95].

4.10.1.3 Inhibitors of MTP

MTP is an enzyme expressed in the liver, intestine, and myocardium. It catalyzes the intracellular assembly of chylomicrons in the intestine and VLDLs in the liver by linking triglycerides with ApoB. Functional MTP is absent in subjects with abetalipoproteinemia, a genetic disorder characterized by low levels of circulating cholesterol and triglycerides due to a defect in the assembly and secretion of ApoB-containing lipoproteins [96]. This suggests that inhibition of MTP may be a potential effective strategy for treating mixed dyslipidemia (i.e., elevated levels of LDL cholesterol and triglycerides). To this end, several pharmacological inhibitors of MTP have been investigated in preclinical and clinical studies and are at various stages of development.

As noted in Section 4.7.1, clinical studies show that inhibition of MTP by lomitapide results in the reduction of LDL-cholesterol levels in patients with HoFH, owing to reduced production of ApoB. This led to the recent approval of lomitapide by the US FDA for treating HoFH. However, lomitapide therapy is associated with elevated liver aminotransferase levels and hepatic fat accumulation [97]. To avoid the potential liver toxicity, several enterocyte-specific MTP inhibitors have been developed and tested in preclinical and clinical studies. However, the exact efficacy and safety of these intestine-selective inhibitors remain to be assessed in large-scale clinical trials. Inhibition of MTP by antisense oligonucleotides has also been reported in the literature.

4.10.1.4 Thyroid Hormone Analogs

Thyroid hormones exert their effects via stimulation of thyroid hormone receptors (TR) that have different tissue distribution and metabolic targets. TRβ is predominantly expressed in the liver and mainly responsible for the favorable effects of thyroid hormones on cholesterol and lipoprotein metabolism, whereas TRα is predominantly expressed in the brain, skeletal muscle, and heart and mediates most of the synergism between thyroid hormones and the sympathetic signaling pathway in the heart.

TRβ signaling regulates plasma cholesterol through a number of pathways: (i) activation of TRβ increases LDL receptor expression; (ii) augmentation of 7α-hydroxylase-mediated cholesterol and bile secretion results in a decreased hepatic cholesterol pool (this induces SREBP-2-mediated LDL receptor upregulation (also see Chapter 3)); (iii) TRβ stimulation augments reverse cholesterol transport via upregulating scavenger receptor B1 (SR-B1), a hepatic transporter of cholesteryl esters derived from HDL cholesterol; and (iv) activation of TRβ results in increased synthesis of ApoA-1, a major component of HDL.

Thyroid hormone analogs (also known as thyromimetics) have been developed that selectively activate TRβ in the liver. Such compounds stimulate hepatic LDL receptor expression, increase cholesterol elimination (in the form of bile acids and cholesterol), promote reverse cholesterol transport, and retard atherosclerosis progression in animal models. Phase 1 studies demonstrated that eprotirome, a liver-selective TRβ agonist, exerted favorable lipid-modulating effects while lacking thyroid hormone-related side effects and maintaining normal hypothalamic–pituitary–thyroid feedback. In phase 2 studies, when added to statins, eprotirome reduced LDL and non-HDL cholesterol, ApoB, and triglycerides as well as Lp(a) [98]. Unfortunately, phase 3 studies were terminated due to potential adverse effects on cartilage tissue following long-term treatment with eprotirome. The development of other thyromimetics has also been halted due to similar concerns on long-term adverse effects [99].

4.10.1.5 Inhibitors of PCSK9

PCSK9 is secreted into the plasma primarily by the liver. It binds the LDL receptor at the surface of hepatocytes, thereby preventing its recycling and enhancing its degradation in endosomes/lysosomes, resulting in reduced LDL-cholesterol clearance. As discussed in Chapter 3, rare gain-of-function PCSK9 variants lead to

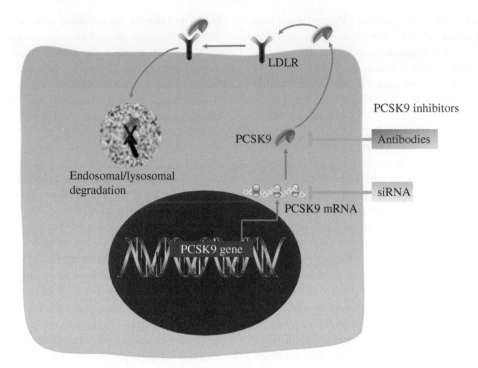

FIGURE 4.14 PCSK9 inhibition as a novel strategy for lowering LDL cholesterol. PCSK9 (proprotein convertase subtilisin/kexin type 9) is primarily released into plasma by the hepatocytes. PCSK9 in the plasma binds to LDL receptor (LDLR), leading to the endosomal/lysosomal degradation of LDL receptors. PCSK inhibitors, including monoclonal antibodies and small RNA interference (siRNA), are currently under clinical studies as a novel strategy for increasing LDL receptor expression on the surface of hepatocytes and reducing plasma LDL cholesterol.

higher levels of LDL cholesterol and increased risk of cardiovascular diseases, and conversely, loss-of-function PCSK9 variants are associated with reductions in both LDL cholesterol and risk of cardiovascular diseases [100]. Because of the critical role of PCSK9 in inducing LDL receptor degradation, several anti-PCSK9 therapeutic strategies have recently been developed and tested in clinical trials, among which monoclonal antibodies are the most advanced PCSK9 inhibitors in development [101–103] (Fig. 4.14).

Two monoclonal antibodies against PCSK9 have been tested in multiple 14-week phase 2 trials, and these PCSK9 inhibitors can decrease LDL cholesterol by 60–70%, even as an add-on therapy to a maximal dose of a statin [102, 103]. Phase 3 studies and larger, event-driven clinical trials are ongoing to determine the long-term safety and efficacy profiles of these anti-PCSK9 antibodies.

In addition to the monoclonal antibody strategy, small molecules and gene-silencing approaches for inhibiting PCSK9 have also recently received attention [103]. In this context, a randomized, single-blind, placebo-controlled phase 1 trial reported that a single intravenous injection of a small RNA interference (siRNA) drug (ALN-PCS) at a maximal dose of 0.4mg/kg in healthy individuals with raised cholesterol was well tolerated and resulted in a 70% reduction in circulating PCSK9 protein and a 40% reduction in LDL cholesterol [104]. The lower efficacy of this siRNA

in reducing plasma LDL cholesterol compared to monoclonal antibodies described earlier might reflect the higher capacity of the latter to block the circulating PCSK9 protein. Nevertheless, this is the first clinical study showing the efficacy of a pharmacological agent able to block intracellular PCSK9 synthesis in reducing LDL cholesterol. Large-scale studies are therefore warranted to further explore the safety and efficacy of this gene-silencing approach.

4.10.2 Novel Therapeutic Strategies Targeting HDL

Treatment of lipid abnormalities has generally focused on LDL-cholesterol reduction. However, it has become increasingly clear that a significant percentage of patients continue to have cardiovascular events despite being on effective LDL-cholesterol-lowering medications, suggesting the influence of other risk factors. In this regard, HDL-cholesterol levels have been shown to be inversely associated with cardiovascular risk in epidemiological studies. Furthermore, this relationship also exists despite reduction of LDL-cholesterol levels to below 70mg/dl [105], suggesting that merely reducing LDL-cholesterol levels with statin therapies does not sufficiently reduce cardiovascular risk.

Elevation of functional HDL cholesterol is cardiovascular protective, and it has been estimated that a 1% elevation of HDL cholesterol may be associated with a 1–3% reduction of cardiovascular risk. The cardiovascular protective action of HDL may be primarily due to its function in reverse cholesterol transport. In addition, HDL also exerts protective effects on oxidative stress, inflammation, thrombogenesis, and endothelial dysfunction [106, 107]. These pleiotropic effects may also contribute to the cardiovascular protective function of HDL.

As discussed earlier in this chapter, current medications that elevate HDL-cholesterol levels include statins, fibrates, and niacin, with niacin being the most effective drug for raising HDL cholesterol. However, augmentation of HDL cholesterol by currently available drugs is limited by their maximal HDL-raising capacity (5–10% for statins, 10–20% for fibrates, and 20–35% for niacin) as well as adverse effects. As such, several novel pharmaceutical agents have been recently developed as potentially effective modalities for raising HDL cholesterol [107–110]. These modalities can be broadly classified into the following three categories:

- CETP inhibitors (Fig. 4.15)
- Administration of HDL-derived proteins and mimetic peptides/lipids
- Inducers of ApoA-I expression

4.10.2.1 CETP Inhibitors

CETP and Its Role in Lipid Metabolism   One of the most promising HDL-cholesterol-raising strategies in development derives from the concept of inhibiting the enzyme CETP. As discussed in Chapter 3, CETP plays a critical role in the transfer of cholesteryl esters from HDL to proatherogenic ApoB-containing lipoproteins, including VLDL and LDL, and of triglycerides from the ApoB-containing lipoproteins to HDL. The transfer of cholesteryl esters from HDL to ApoB-containing lipoproteins leads to reduced HDL-cholesterol levels and ApoA-I content as well as the decreased size of HDL particles. The transfer also results in increased cholesterol content of LDL and VLDL.

CETP deficiency or inhibition results in increased levels of HDL cholesterol and decreased levels of LDL cholesterol, effects that are protective against atherosclerosis. Indeed, substantial experimental and clinical evidence shows an inverse relationship between CETP deficiency or inhibition and increased levels of HDL cholesterol or decreased burden of atherosclerosis. Several CETP inhibitors have been tested for their lipid-modifying effects in humans. These include

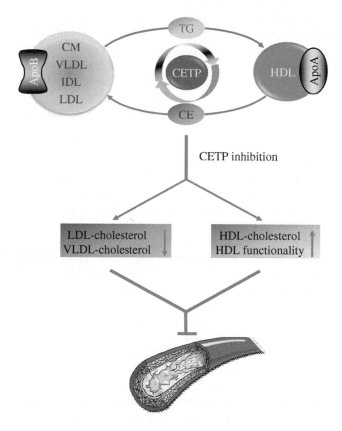

FIGURE 4.15   CETP inhibition as a novel strategy for raising HDL cholesterol. Cholesterol ester transfer protein (CETP) catalyzes the transfer of cholesterol esters (CE) from HDL to ApoB-containing lipoproteins, including LDL and VLDL, and of triglycerides (TG) from the ApoB-containing lipoproteins to HDL. Inhibition of CETP by emerging drugs leads to increased HDL cholesterol and improved functionality of HDL, as well as decreased LDL cholesterol, effects that are protective against atherosclerosis. The cardiovascular benefits of CETP inhibitors are currently under phase 3 clinical trials.

torcetrapib, dalcetrapib, anacetrapib, and evacetrapib. The current status of clinical research on these CETP inhibitors is described next.

Clinical Studies on Torcetrapib Clinical studies demonstrated that in patients with low HDL cholesterol, CETP inhibition with torcetrapib markedly increased HDL-cholesterol levels and also decreased LDL-cholesterol levels both when administered as monotherapy and in combination with a statin [111]. However, clinical trials on torcetrapib evaluating coronary atherosclerosis or carotid atherosclerosis showed that addition of torcetrapib to atorvastatin, as compared with atorvastatin alone, did not result in further reduction of progression of atherosclerosis [112–114]. The combination therapy, however, caused further remarkable elevation of HDL-cholesterol levels and decreased LDL-cholesterol levels as compared with the statin monotherapy.

A large-scale phase III trial (ILLUMINATE) involving 15,076 patients at high cardiovascular risk was conducted to compare the torcetrapib/atorvastatin combination with atorvastatin alone in reducing major cardiovascular events. The combination therapy caused further remarkable elevation and reduction in HDL cholesterol and LDL cholesterol, respectively. However, the trial was stopped due to increased mortality in the combination therapy group [105]. This increased mortality was caused by an off-target effect of torcetrapib, leading to increased aldosterone levels and blood pressure.

Clinical Studies on Dalcetrapib, Anacetrapib, and Evacetrapib Dalcetrapib, anacetrapib, and evacetrapib appear to lack the off-target effect of torcetrapib. Treatment with dalcetrapib, anacetrapib, and evacetrapib, especially the latter two, has robust effects on HDL-cholesterol and LDL-cholesterol levels, as observed with torcetrapib. In contrast to torcetrapib, the above three CETP inhibitors lack the off-target effect on aldosterone levels and blood pressure, and hence, their inhibition of CETP shows promise as a potential strategy to reduce cardiovascular disease risk [115, 116].

Unfortunately, the dal-OUTCOMES trial, the first phase 3 randomized, placebo-controlled trial investigating the effect of dalcetrapib in 15,871 statin-treated patients, was terminated early in May 2012 for futility [110, 117]. Compared with placebo, dalcetrapib treatment resulted in a 30% increase of HDL-cholesterol levels, without a significant effect on LDL-cholesterol levels. There was, however, no significant benefit in overall and cardiovascular disease-related mortality and major cardiovascular events, which formed the basis for the advice of the Data Safety Monitoring Board to terminate the trial [110].

Although the failure of dalcetrapib to reduce cardiovascular events in the dal-OUTCOMES trial has raised serious questions on the validity of CETP inhibition for reducing cardiovascular events in high-risk individuals, the relatively small elevation of HDL cholesterol (30% increase) by dalcetrapib might have rendered the ineffectiveness on cardiovascular outcomes. In this context, anacetrapib and evacetrapib are much more robust in augmenting HDL cholesterol than dalcetrapib. In addition, these two CETP inhibitors, unlike dalcetrapib, also show beneficial effects on other lipid parameters, such as reduction on LDL cholesterol and Lp(a) [115, 118]. More importantly, evidence suggests that these newer CETP inhibitors not only increase HDL-cholesterol levels but also improve functionality of HDL (Fig. 4.15). In this context, dysfunction of HDL is seen in cardiovascular diseases. In view of the aforementioned considerations, anacetrapib and evacetrapib are currently in phase 3 studies to determine their effects on cardiovascular outcomes in patients at high cardiovascular risk, and the trials are expected to be completed in 2015–2017 [119]. Although many questions remain regarding CETP inhibitors, there is optimism that they may prove to be potent HDL-cholesterol-raising medications that will further reduce cardiovascular events in high-risk patients.

4.10.2.2 Administration of HDL-Derived Proteins and Mimetic Peptides/Lipids
ApoA-I Milano is a variant of ApoA-I identified in individuals in rural Italy who exhibit very low levels of HDL cholesterol [120]. However, these patients were shown to have reduced atherosclerosis despite very low levels of HDL cholesterol (10–30 mg/dl), suggesting augmented antiatherosclerotic function of the mutant protein. The recombinant ApoA-I Milano protein differs from the wild-type ApoA-I by a cysteine to arginine substitution at amino acid 173. The antiatherogenic effects of ApoA-I Milano complexed with phospholipids (also known as ETC-216) have been documented in animal models. In a small-scale clinical study, infusion of ETC-216 was shown to produce significant regression of coronary atherosclerosis as measured by intravascular ultrasonography in patients with acute coronary syndromes [121]. ETC-216 is currently under clinical development, and the safety and efficacy of this potential antiatherosclerotic modality remain to be determined by future large-scale clinical trials [122].

Similar to the ApoA-I Milano concept, synthetic ApoA-I mimetic peptides have been developed. Several such mimetics have been tested in animal models and clinical studies for their safety and efficacy in treating atherosclerosis [93, 107, 123].

4.10.2.3 Inducers of ApoA-I Expression
A small synthetic molecule, known as RVX-208, has been shown to increase ApoA-I gene expression, resulting in increased levels of plasma ApoA-I and HDL cholesterol in animal models including monkeys. Clinical studies demonstrated that oral administration of RVX-208 could also cause increased levels of plasma ApoA-I, HDL cholesterol, and larger HDL particles in patients with acute coronary syndromes.

This oral ApoA-I inducer was found to cause increased hepatic transaminases [124]. Clinical studies are currently underway to investigate the efficacy of RVX-208 in reducing coronary atherosclerosis [110].

4.10.3 Summary of Emerging Drugs

- Emerging therapies focus on reducing LDL cholesterol and augmenting HDL cholesterol as well as improving HDL functionality.
- Among the emerging modalities targeting LDL cholesterol, inhibitors of PCSK9 hold great promise due to favorable results from multiple phase 2 trials.
- Phase 3 trials are currently ongoing to determine if augmentation of HDL cholesterol and improvement of HDL functionality by newer CETP inhibitors reduce cardiovascular events and mortality in patients at high cardiovascular risk.
- Other emerging strategies to increase HDL cholesterol include administration of reconstituted HDL and induction of ApoA-I expression by oral inducers.

4.11 SUMMARY OF CHAPTER KEY POINTS

- Commonly used drugs for treating dyslipidemias include statins, bile acid sequestrants, ezetimibe, fibrates, and niacin, among which statins are the most commonly prescribed lipid-lowering agents.
- Statins act via inhibiting HMG-CoA reductase, the key enzyme in cholesterol biosynthesis. Bile acid sequestrants block the enterohepatic circulation of bile acids, thus leading to consumption of cholesterol in hepatocytes. Ezetimibe decreases cholesterol absorption by inhibiting the cholesterol transporting protein, NPC1L1, in enterocytes. Fibrate drugs reduce plasma triglycerides by increasing lipoprotein lipase activity. Fibrates also decrease LDL cholesterol as well as increase HDL cholesterol. Niacin reduces plasma triglycerides and LDL cholesterol and potently augments HDL cholesterol.
- Statins, bile acid sequestrants, and ezetimibe are primarily used for treating hypercholesterolemia, whereas fibrates and niacin are effective agents for treating hypertriglyceridemia.
- Lomitapide and mipomersen are two newly approved drugs for treating severe hypercholesterolemia in patients with HoFH. On the other hand, Lovaza and Vascepa are two recently approved omega-3 fatty acid drugs for treating severe hypertriglyceridemia.
- Emerging drugs for dyslipidemias focus on reducing LDL cholesterol and increasing HDL cholesterol via novel mechanisms. Notably emerging modalities

include inhibitors of PCSK9 and CETP, whose clinical efficacy in reducing cardiovascular events and mortality in patients with high cardiovascular risk is currently being investigated in phase 3 trials.

- In addition to conventional drug therapy, gene therapy has also been developed to treat certain forms of dyslipidemias. In this context, a major milestone in the development of gene therapy for dyslipidemias has been achieved with the approval of alipogene tiparvovec (Glybera) in Europe for the treatment of familial lipoprotein lipase deficiency [125]. Alipogene tiparvovec is a recombinant adeno-associated viral vector of serotype 1 mediating muscle-directed expression of lipoprotein lipase. It is the first gene therapy to receive approval in the Western world.

4.12 SELF-ASSESSMENT QUESTIONS

4.12.1 A 37-year-old male is diagnosed with heterozygous familial hypercholesterolemia. History reveals that a family member has recently died of myocardial infarction at the age of 49 years. The patient is placed on atorvastatin, 20 mg, once daily. To further reduce his LDL-cholesterol levels, the patient is also prescribed a drug that reduces intestinal cholesterol absorption. Which of the following is the molecular target of this added drug?
   A. G-protein-coupled receptor 120
   B. Microsomal triglyceride transfer protein
   C. Niemann–Pick C1-Like 1
   D. Peroxisome proliferator-activated receptor-gamma
   E. Proprotein convertase subtilisin/kexin type 9

4.12.2 A 58-year-old male with hyperlipidemia is treated with a drug that increases the expression of lipoprotein lipase. Which of the following effects would be expected to most likely result from this action?
   A. Decreased plasma fatty acid levels
   B. Decreased plasma HDL-cholesterol levels
   C. Decreased plasma levels of PCSK9
   D. Decreased plasma Lp(a) levels
   E. Decreased plasma triglyceride levels

4.12.3 A 56-year-old female presents to her doctor's office complaining of flushing and pruritus over her upper body after starting a drug for her hypertriglyceridemia. Activation of which of the following is the most likely mechanism of action for this drug to reduce triglycerides?
   A. Cholesterol ester transfer protein
   B. Acyl-CoA:diacylglycerol acyltransferase-2
   C. G-protein-coupled receptor 109A
   D. Lipoprotein lipase
   E. Peroxisome proliferator-activated receptor-alpha

4.12.4   A 50-year-old male presents with elevated plasma levels of total cholesterol (285 mg/dl), LDL cholesterol (147 mg/dl), and triglycerides (164 mg/dl). He has a family history of early cardiovascular disease. Which of the following drugs if prescribed as monotherapy would most likely and most significantly reduce the risk of acute myocardial infarction in this patient?

A. Atorvastatin
B. Colestipol
C. Ezetimibe
D. Gemfibrozil
E. Niacin

4.12.5   The ENHANCE trial involving 720 patients with heterozygous familial hypercholesterolemia showed no significant differences in the primary endpoint [mean change in the intima-media thickness (IMT) measured at three sites in the carotid arteries] between patients treated with ezetimibe/simvastatin (10/80 mg) and patients treated with simvastatin (80 mg) alone over a 2-year period, though ezetimibe/simvastatin group showed a more significant decrease in LDL-C (58% vs. 41%). What is the mechanism of action of ezetimibe in further lowering cholesterol in statin-treated patients?

A. Ezetimibe binds to, and interrupts, bile acid and cholesterol enterohepatic circulation
B. Ezetimibe displaces cholesterol from chylomicrons and facilitates its transport to the liver
C. Ezetimibe displaces dietary cholesterol from micelles and enhances its loss in the feces
D. Ezetimibe inhibits cholesterol incorporation into VLDL in the liver
E. Ezetimibe inhibits cholesterol uptake through a brush border cholesterol transporting protein on enterocytes

4.12.6   In 1976, Dr A. Endo of Japan discovered the first potent and selective inhibitor of HMG-CoA reductase, named ML-236B. Now, HMG-CoA reductase inhibitors approved by the US FDA are the most frequently used and well-studied lipid-lowering drugs. Among the clinically available statins, which of the following is the most potent inhibitor of HMG-CoA reductase?

A. Atorvastatin
B. Fluvastatin
C. Lovastatin
D. Pitavastatin
E. Pravastatin

4.12.7   Statins have become the mainstay of therapy for treating high blood cholesterol to reduce the risk of atherosclerotic cardiovascular diseases. In addition to reducing LDL-cholesterol levels, statins have also been demonstrated to exert other pharmacological effects, which may ameliorate the underlying pathophysiological states associated with the development of atherosclerotic cardiovascular diseases. Which of the following is one of the pleiotropic effects of statin therapy?

A. Decreased expression of ApoB-100
B. Decreased HDL cholesterol
C. Decreased inflammation
D. Decreased size of LDL particles
E. Decreased stability of atherosclerotic plaque

4.12.8   A 12-year-old boy with homozygous familial hypercholesterolemia is treated with a recently US FDA-approved lipid-lowering drug in an effort to reduce the need for LDL apheresis. The patient subsequently develops flu-like symptoms, and a blood test also reveals elevated alanine aminotransferase. Which of the following is the most likely molecular target of this new drug?

A. Apolipoprotein B-100
B. G-protein-coupled receptor 120
C. Microsomal triglyceride transfer protein
D. Niemann–Pick C1-Like 1
E. Proprotein convertase subtilisin/kexin type 9

4.12.9   A recent study reported that treatment with an RNA interference drug targeting specifically on proprotein convertase subtilisin/kexin type 9 (PCSK9) in individuals with elevated LDL cholesterol resulted in a mean 70% reduction in circulating PCSK9 plasma protein and a mean 40% reduction in LDL cholesterol from baseline relative to placebo [104]. Which of the following was most likely responsible for the reduction of LDL cholesterol in the treated group?

A. Increased degradation of ApoB-100 mRNA
B. Increased degradation of Niemann–Pick C1-Like 1
C. Increased deposition of LDL in peripheral tissues
D. Increased expression of lipoprotein lipase
E. Increased LDL receptor density on hepatocytes

4.12.10 Evacetrapib is currently in a phase III trial that was designed to determine its effects on cardiovascular outcomes in patients at high cardiovascular risk. What was the rationale for this phase III study?

A. Evacetrapib activates GPR120 to augment HDL cholesterol
B. Evacetrapib activates PPAR-α to augment HDL cholesterol
C. Evacetrapib induces ApoA-I to augment HDL cholesterol
D. Evacetrapib inhibits CETP to augment HDL cholesterol
E. Evacetrapib inhibits PCSK9 to augment HDL cholesterol

REFERENCES

1 Ahmad, H. and A. Cheng-Lai, Pitavastatin: a new HMG-CoA reductase inhibitor for the treatment of hypercholesterolemia. *Cardiol Rev*, 2010. 18(5): p. 264–7.

2 Steinberg, D., The statins in preventive cardiology. *N Engl J Med*, 2008. 359(14): p. 1426–7.

3 Obaidat, A., M. Roth, and B. Hagenbuch, The expression and function of organic anion transporting polypeptides in normal tissues and in cancer. *Annu Rev Pharmacol Toxicol*, 2012. 52: p. 135–51.

4 Knopp, R.H., Drug treatment of lipid disorders. *N Engl J Med*, 1999. 341(7): p. 498–511.

5 Jain, M.K. and P.M. Ridker, Anti-inflammatory effects of statins: clinical evidence and basic mechanisms. *Nat Rev Drug Discov*, 2005. 4(12): p. 977–87.

6 Antoniades, C., et al., Rapid, direct effects of statin treatment on arterial redox state and nitric oxide bioavailability in human atherosclerosis via tetrahydrobiopterin-mediated endothelial nitric oxide synthase coupling. *Circulation*, 2011. 124(3): p. 335–45.

7 Liao, J.K. and U. Laufs, Pleiotropic effects of statins. *Annu Rev Pharmacol Toxicol*, 2005. 45: p. 89–118.

8 Blaha, M.J. and S.S. Martin, How do statins work?: changing paradigms with implications for statin allocation. *J Am Coll Cardiol*, 2013. 62(25): p. 2392–4.

9 Antoniades, C. and K.M. Channon, Statins: pleiotropic regulators of cardiovascular redox state. *Antioxid Redox Signal*, 2014. 20(8): p. 1195–7.

10 Margaritis, M., K.M. Channon, and C. Antoniades, Statins as regulators of redox state in the vascular endothelium: beyond lipid lowering. *Antioxid Redox Signal*, 2014. 20(8): p. 1198–215.

11 Weitz-Schmidt, G., et al., Statins selectively inhibit leukocyte function antigen-1 by binding to a novel regulatory integrin site. *Nat Med*, 2001. 7(6): p. 687–92.

12 Nicholls, S.J., et al., Statins, high-density lipoprotein cholesterol, and regression of coronary atherosclerosis. *JAMA*, 2007. 297(5): p. 499–508.

13 Nicholls, S.J., et al., Effect of two intensive statin regimens on progression of coronary disease. *N Engl J Med*, 2011. 365(22): p. 2078–87.

14 Morton, S. and S. Thangaratinam, Statins in pregnancy. *Curr Opin Obstet Gynecol*, 2013. 25(6): p. 433–40.

15 Subramanian, S., et al., High-dose atorvastatin reduces periodontal inflammation: a novel pleiotropic effect of statins. *J Am Coll Cardiol*, 2013. 62(25): p. 2382–91.

16 Mansi, I., et al., Statins and musculoskeletal conditions, arthropathies, and injuries. *JAMA Intern Med*, 2013. 173(14): p. 1–10.

17 Reiner, Z., Statins in the primary prevention of cardiovascular disease. *Nat Rev Cardiol*, 2013. 10(8): p. 453–64.

18 Group, S.C., et al., SLCO1B1 variants and statin-induced myopathy—a genomewide study. *N Engl J Med*, 2008. 359(8): p. 789–99.

19 Sattar, N., et al., Statins and risk of incident diabetes: a collaborative meta-analysis of randomised statin trials. *Lancet*, 2010. 375(9716): p. 735–42.

20 Preiss, D., et al., Risk of incident diabetes with intensive-dose compared with moderate-dose statin therapy: a meta-analysis. *JAMA*, 2011. 305(24): p. 2556–64.

21 Mogadam, M., Do statins increase the risk of diabetes or is it guilt by association? *Curr Opin Endocrinol Diabetes Obes*, 2014. 21(2): p. 140–5.

22 Bell, D.S., J.J. Dinicolantonio, and J.H. O'Keefe, Is statin-induced diabetes clinically relevant? A comprehensive review of the literature. *Diabetes Obes Metab*, 2014. 16 (8): p. 689–694.

23 Bellosta, S. and A. Corsini, Statin drug interactions and related adverse reactions. *Expert Opin Drug Saf*, 2012. 11(6): p. 933–46.

24 Ooi, C.P. and S.C. Loke, Colesevelam for type 2 diabetes mellitus. *Cochrane Database Syst Rev*, 2012. 12: p. CD009361.

25 Garg, A. and S.M. Grundy, Cholestyramine therapy for dyslipidemia in non-insulin-dependent diabetes mellitus. A short-term, double-blind, crossover trial. *Ann Intern Med*, 1994. 121(6): p. 416–22.

26 Yamakawa, T., et al., Effect of colestimide therapy for glycemic control in type 2 diabetes mellitus with hypercholesterolemia. *Endocr J*, 2007. 54(1): p. 53–8.

27 Bays, H.E., et al., Colesevelam hydrochloride therapy in patients with type 2 diabetes mellitus treated with metformin: glucose and lipid effects. *Arch Intern Med*, 2008. 168(18): p. 1975–83.

28 Goldberg, R.B., et al., Efficacy and safety of colesevelam in patients with type 2 diabetes mellitus and inadequate glycemic control receiving insulin-based therapy. *Arch Intern Med*, 2008. 168(14): p. 1531–40.

29 Fonseca, V.A., et al., Colesevelam HCl improves glycemic control and reduces LDL cholesterol in patients with inadequately controlled type 2 diabetes on sulfonylurea-based therapy. *Diabetes Care*, 2008. 31(8): p. 1479–84.

30 Takebayashi, K., Y. Aso, and T. Inukai, Role of bile acid sequestrants in the treatment of type 2 diabetes. *World J Diabetes*, 2010. 1(5): p. 146–52.

31 Handelsman, Y., Role of bile acid sequestrants in the treatment of type 2 diabetes. *Diabetes Care*, 2011. 34 Suppl 2: p. S244–50.

32 Yunoki, K., et al., Impact of Hypertriglyceridemia on Endothelial Dysfunction During Statin +/− Ezetimibe Therapy in Patients With Coronary Heart Disease. *Am J Cardiol*, 2011. 108(3): 333–9.

33 Kastelein, J.J., et al., Simvastatin with or without ezetimibe in familial hypercholesterolemia. *N Engl J Med*, 2008. 358(14): p. 1431–43.

34 Emerging Risk Factors Collaboration, et al., C-reactive protein, fibrinogen, and cardiovascular disease prediction. *N Engl J Med*, 2012. 367(14): p. 1310–20.

35 Liptruzet: a combination of ezetimibe and atorvastatin. *Med Lett Drugs Ther*, 2013. 55(1419): p. 49–50.

36 Rossebo, A.B., et al., Intensive lipid lowering with simvastatin and ezetimibe in aortic stenosis. *N Engl J Med*, 2008. 359(13): p. 1343–56.

37 Taylor, A.J., et al., Extended-release niacin or ezetimibe and carotid intima-media thickness. *N Engl J Med*, 2009. 361(22): p. 2113–22.

38 Cannon, C.P., et al., Rationale and design of IMPROVE-IT (IMProved Reduction of Outcomes: Vytorin Efficacy International Trial): comparison of ezetimibe/simvastatin versus simvastatin monotherapy on cardiovascular outcomes in patients with acute coronary syndromes. *Am Heart J*, 2008. 156(5): p. 826–32.

39 Peto, R., et al., Analyses of cancer data from three ezetimibe trials. *N Engl J Med*, 2008. 359(13): p. 1357–66.

40 Drazen, J.M., et al., Ezetimibe and cancer—an uncertain association. *N Engl J Med*, 2008. 359(13): p. 1398–9.

41 Drazen, J.M., et al., Cholesterol lowering and ezetimibe. *N Engl J Med*, 2008. 358(14): p. 1507–8.

42 Brown, B.G. and A.J. Taylor, Does ENHANCE diminish confidence in lowering LDL or in ezetimibe? *N Engl J Med*, 2008. 358(14): p. 1504–7.

43 Calpe-Berdiel, L., J.C. Escola-Gil, and F. Blanco-Vaca, New insights into the molecular actions of plant sterols and stanols in cholesterol metabolism. *Atherosclerosis*, 2009. 203(1): p. 18–31.

44 de Gennes, J.L., et al., Results of atromid or CPIB treatment with and without androsterone in hyperlipemia (110 therapeutic trials in 2 years). *Bull Mem Soc Med Hop Paris*, 1965. 116(8): p. 759–84.

45 Preiss, D., et al., Lipid-modifying therapies and risk of pancreatitis: a meta-analysis. *JAMA*, 2012. 308(8): p. 804–11.

46 Jackevicius, C.A., et al., Use of fibrates in the United States and Canada. *JAMA*, 2011. 305(12): p. 1217–24.

47 Fievet, C. and B. Staels, Combination therapy of statins and fibrates in the management of cardiovascular risk. *Curr Opin Lipidol*, 2009. 20(6): p. 505–11.

48 Abourbih, S., et al., Effect of fibrates on lipid profiles and cardiovascular outcomes: a systematic review. *Am J Med*, 2009. 122(10): p. 962 e1-8.

49 Jun, M., et al., Effects of fibrates on cardiovascular outcomes: a systematic review and meta-analysis. *Lancet*, 2010. 375(9729): p. 1875–84.

50 Lee, M., et al., Efficacy of fibrates for cardiovascular risk reduction in persons with atherogenic dyslipidemia: a meta-analysis. *Atherosclerosis*, 2011. 217(2): p. 492–8.

51 Ginsberg, H.N., et al., Effects of combination lipid therapy in type 2 diabetes mellitus. *N Engl J Med*, 2010. 362(17): p. 1563–74.

52 Tenenbaum, A. and E.Z. Fisman, "If it ain't broke, don't fix it": a commentary on the positive-negative results of the ACCORD Lipid study. *Cardiovasc Diabetol*, 2010. 9: p. 24.

53 Saha, S.A. and R.R. Arora, Hyperlipidaemia and cardiovascular disease: do fibrates have a role? *Curr Opin Lipidol*, 2011. 22(4): p. 270–6.

54 Altschul, R., A. Hoffer, and J.D. Stephen, Influence of nicotinic acid on serum cholesterol in man. *Arch Biochem Biophys*, 1955. 54(2): p. 558–9.

55 Kamanna, V.S. and M.L. Kashyap, Mechanism of action of niacin. *Am J Cardiol*, 2008. 101(8A): p. 20B–6.

56 Florentin, M., et al., Pleiotropic effects of nicotinic acid: beyond high density lipoprotein cholesterol elevation. *Curr Vasc Pharmacol*, 2011. 9(4): p. 385–400.

57 Tunaru, S., et al., PUMA-G and HM74 are receptors for nicotinic acid and mediate its anti-lipolytic effect. *Nat Med*, 2003. 9(3): p. 352–5.

58 Wise, A., et al., Molecular identification of high and low affinity receptors for nicotinic acid. *J Biol Chem*, 2003. 278(11): p. 9869–74.

59 Soga, T., et al., Molecular identification of nicotinic acid receptor. *Biochem Biophys Res Commun*, 2003. 303(1): p. 364–9.

60 Al-Mohaissen, M.A., S.C. Pun, and J.J. Frohlich, Niacin: from mechanisms of action to therapeutic uses. *Mini Rev Med Chem*, 2010. 10(3): p. 204–17.

61 Martinez, L.O., et al., Ectopic beta-chain of ATP synthase is an apolipoprotein A-I receptor in hepatic HDL endocytosis. *Nature*, 2003. 421(6918): p. 75–9.

62 Jackevicius, C.A., et al., Use of niacin in the United States and Canada. *JAMA Intern Med*, 2013. 173(14): p. 1379–81.

63 Giugliano, R.P., Niacin at 56 years of age—time for an early retirement? *N Engl J Med*, 2011. 365(24): p. 2318–20.

64 Michos, E.D., et al., Niacin and statin combination therapy for atherosclerosis regression and prevention of cardiovascular disease events: reconciling the AIM-HIGH (Atherothrombosis Intervention in Metabolic Syndrome With Low HDL/High Triglycerides: Impact on Global Health Outcomes) trial with previous surrogate endpoint trials. *J Am Coll Cardiol*, 2012. 59(23): p. 2058–64.

65 AIM-HIGH Investigators, et al., Niacin in patients with low HDL cholesterol levels receiving intensive statin therapy. *N Engl J Med*, 2011. 365(24): p. 2255–67.

66 HPS2-THRIVE Collaborative Group, HPS2-THRIVE randomized placebo-controlled trial in 25 673 high-risk patients of ER niacin/laropiprant: trial design, pre-specified muscle and liver outcomes, and reasons for stopping study treatment. *Eur Heart J*, 2013. 34(17): p. 1279–91.

67 Ginsberg, H.N. and G. Reyes-Soffer, Niacin: a long history, but a questionable future. *Curr Opin Lipidol*, 2013. 24(6): p. 475–9.

68 Maccubbin, D., et al., Flushing profile of extended-release niacin/laropiprant versus gradually titrated niacin extended-release in patients with dyslipidemia with and without ischemic cardiovascular disease. *Am J Cardiol*, 2009. 104(1): p. 74–81.

69 Goldberg, R.B. and T.A. Jacobson, Effects of niacin on glucose control in patients with dyslipidemia. *Mayo Clin Proc*, 2008. 83(4): p. 470–8.

70 Rader, D.J. and J.J. Kastelein, Lomitapide and mipomersen: two first-in-class drugs for reducing low-density lipoprotein

cholesterol in patients with homozygous familial hypercholesterolemia. *Circulation*, 2014. 129(9): p. 1022–32.

71 Cuchel, M., et al., Efficacy and safety of a microsomal triglyceride transfer protein inhibitor in patients with homozygous familial hypercholesterolaemia: a single-arm, open-label, phase 3 study. *Lancet*, 2013. 381(9860): p. 40–6.

72 Raal, F.J., et al., Mipomersen, an apolipoprotein B synthesis inhibitor, for lowering of LDL cholesterol concentrations in patients with homozygous familial hypercholesterolaemia: a randomised, double-blind, placebo-controlled trial. *Lancet*, 2010. 375(9719): p. 998–1006.

73 Stein, E.A., et al., Apolipoprotein B synthesis inhibition with mipomersen in heterozygous familial hypercholesterolemia: results of a randomized, double-blind, placebo-controlled trial to assess efficacy and safety as add-on therapy in patients with coronary artery disease. *Circulation*, 2012. 126(19): p. 2283–92.

74 Thomas, G.S., et al., Mipomersen, an apolipoprotein B synthesis inhibitor, reduces atherogenic lipoproteins in patients with severe hypercholesterolemia at high cardiovascular risk: a randomized, double-blind, placebo-controlled trial. *J Am Coll Cardiol*, 2013. 62(23): p. 2178–84.

75 Santos, R.D., et al., Long-term efficacy and safety of mipomersen in patients with familial hypercholesterolaemia: 2-year interim results of an open-label extension. Eur Heart J, 2014 (in press).

76 Milani, R.V. and C.J. Lavie, Lipid control in the modern era: an orphan's tale of rags to riches. *J Am Coll Cardiol*, 2013. 62(23): p. 2185–7.

77 Smith, R.J. and W.R. Hiatt, Two new drugs for homozygous familial hypercholesterolemia: managing benefits and risks in a rare disorder. *JAMA Intern Med*, 2013. 173(16): p. 1491–2.

78 Peterson, D.W., Effect of soybean sterols in the diet on plasma and liver cholesterol in chicks. *Proc Soc Exp Biol Med*, 1951. 78(1): p. 143–7.

79 Pollak, O.J., Reduction of blood cholesterol in man. *Circulation*, 1953. 7(5): p. 702–6.

80 Pollak, O.J., Successive prevention of experimental hypercholesteremia and cholesterol atherosclerosis in the rabbit. *Circulation*, 1953. 7(5): p. 696–701.

81 Thompson, G.R. and S.M. Grundy, History and development of plant sterol and stanol esters for cholesterol-lowering purposes. *Am J Cardiol*, 2005. 96(1A): p. 3D–9.

82 Gylling, H., et al., Plant sterols and plant stanols in the management of dyslipidaemia and prevention of cardiovascular disease. *Atherosclerosis*, 2014. 232(2): p. 346–60.

83 Nijjar, P.S., et al., Role of dietary supplements in lowering low-density lipoprotein cholesterol: a review. *J Clin Lipidol*, 2010. 4(4): p. 248–58.

84 Malinowski, J.M. and M.M. Gehret, Phytosterols for dyslipidemia. *Am J Health Syst Pharm*, 2010. 67(14): p. 1165–73.

85 Saravanan, P., et al., Cardiovascular effects of marine omega-3 fatty acids. *Lancet*, 2010. 376(9740): p. 540–50.

86 Watts, G.F. and T.A. Mori, Recent advances in understanding the role and use of marine omega3 polyunsaturated fatty acids in cardiovascular protection. *Curr Opin Lipidol*, 2011. 22(1): p. 70–1.

87 Kromhout, D., et al., n-3 fatty acids and cardiovascular events after myocardial infarction. *N Engl J Med*, 2010. 363(21): p. 2015–26.

88 Investigators, O.T., et al., n-3 fatty acids and cardiovascular outcomes in patients with dysglycemia. *N Engl J Med*, 2012. 367(4): p. 309–18.

89 Risk and Prevention Study Collaborative Group, et al., n-3 fatty acids in patients with multiple cardiovascular risk factors. *N Engl J Med*, 2013. 368(19): p. 1800–8.

90 Rizos, E.C., et al., Association between omega-3 fatty acid supplementation and risk of major cardiovascular disease events: a systematic review and meta-analysis. *JAMA*, 2012. 308(10): p. 1024–33.

91 Oh, D.Y., et al., GPR120 is an omega-3 fatty acid receptor mediating potent anti-inflammatory and insulin-sensitizing effects. *Cell*, 2010. 142(5): p. 687–98.

92 Gajos, G., et al., Effects of polyunsaturated omega-3 fatty acids on responsiveness to dual antiplatelet therapy in patients undergoing percutaneous coronary intervention: the OMEGA-PCI (OMEGA-3 fatty acids after PCI to modify responsiveness to dual antiplatelet therapy) study. *J Am Coll Cardiol*, 2010. 55(16): p. 1671–8.

93 Tavridou, A., G. Ragia, and V.G. Manolopoulos, Emerging targets for the treatment of dyslipidemia. *Curr Med Chem*, 2011. 18(6): p. 909–22.

94 Florentin, M., et al., Emerging options in the treatment of dyslipidemias: a bright future? *Expert Opin Emerg Drugs*, 2011. 16(2): p. 247–70.

95 Stein, E.A., et al., Lapaquistat acetate: development of a squalene synthase inhibitor for the treatment of hypercholesterolemia. *Circulation*, 2011. 123(18): p. 1974–85.

96 Wetterau, J.R., et al., Absence of microsomal triglyceride transfer protein in individuals with abetalipoproteinemia. *Science*, 1992. 258(5084): p. 999–1001.

97 Cuchel, M., et al., Inhibition of microsomal triglyceride transfer protein in familial hypercholesterolemia. *N Engl J Med*, 2007. 356(2): p. 148–56.

98 Angelin, B. and M. Rudling, Lipid lowering with thyroid hormone and thyromimetics. *Curr Opin Lipidol*, 2010. 21(6): p. 499–506.

99 Stoekenbroek, R.M., J.J. Kastelein, and G.K. Hovingh, Recent failures in antiatherosclerotic drug development: examples from the thyroxin receptor agonist, the secretory phospholipase A2 antagonist, and the acyl coenzyme A: cholesterol acyltransferase inhibitor programs. *Curr Opin Lipidol*, 2013. 24(6): p. 459–66.

100 Seidah, N.G., et al., PCSK9: a key modulator of cardiovascular health. *Circ Res*, 2014. 114(6): p. 1022–36.

101 Betteridge, D.J., Cardiovascular endocrinology in 2012: PCSK9-an exciting target for reducing LDL-cholesterol levels. *Nat Rev Endocrinol*, 2013. 9(2): p. 76–8.

102 Farnier, M., PCSK9 inhibitors. *Curr Opin Lipidol*, 2013. 24(3): p. 251–8.

103 Norata, G.D., G. Tibolla, and A.L. Catapano, Targeting PCSK9 for hypercholesterolemia. *Annu Rev Pharmacol Toxicol*, 2014. 54: p. 273–93.

104   Fitzgerald, K., et al., Effect of an RNA interference drug on the synthesis of proprotein convertase subtilisin/kexin type 9 (PCSK9) and the concentration of serum LDL cholesterol in healthy volunteers: a randomised, single-blind, placebo-controlled, phase 1 trial. *Lancet*, 2014. 383(9911): p. 60–8.

105   Barter, P.J., et al., Effects of torcetrapib in patients at high risk for coronary events. *N Engl J Med*, 2007. 357(21): p. 2109–22.

106   Natarajan, P., K.K. Ray, and C.P. Cannon, High-density lipoprotein and coronary heart disease: current and future therapies. *J Am Coll Cardiol*, 2010. 55(13): p. 1283–99.

107   Luscher, T.F., et al., High-density lipoprotein: vascular protective effects, dysfunction, and potential as therapeutic target. *Circ Res*, 2014. 114(1): p. 171–82.

108   Singh, I.M., M.H. Shishehbor, and B.J. Ansell, High-density lipoprotein as a therapeutic target: a systematic review. *JAMA*, 2007. 298(7): p. 786–98.

109   Degoma, E.M. and D.J. Rader, Novel HDL-directed pharmacotherapeutic strategies. *Nat Rev Cardiol*, 2011. 8(5): p. 266–77.

110   van Capelleveen, J.C., et al., Novel therapies focused on the high-density lipoprotein particle. *Circ Res*, 2014. 114(1): p. 193–204.

111   Brousseau, M.E., et al., Effects of an inhibitor of cholesteryl ester transfer protein on HDL cholesterol. *N Engl J Med*, 2004. 350(15): p. 1505–15.

112   Nissen, S.E., et al., Effect of torcetrapib on the progression of coronary atherosclerosis. *N Engl J Med*, 2007. 356(13): p. 1304–16.

113   Kastelein, J.J., et al., Effect of torcetrapib on carotid atherosclerosis in familial hypercholesterolemia. *N Engl J Med*, 2007. 356(16): p. 1620–30.

114   Bots, M.L., et al., Torcetrapib and carotid intima-media thickness in mixed dyslipidaemia (RADIANCE 2 study): a randomised, double-blind trial. *Lancet*, 2007. 370(9582): p. 153–60.

115   Cannon, C.P., et al., Safety of anacetrapib in patients with or at high risk for coronary heart disease. *N Engl J Med*, 2010. 363(25): p. 2406–15.

116   Mitka, M., CETP inhibition shows promise as way to reduce cardiovascular disease risk. *JAMA*, 2011. 305(2): p. 136–7.

117   Schwartz, G.G., et al., Effects of dalcetrapib in patients with a recent acute coronary syndrome. *N Engl J Med*, 2012. 367(22): p. 2089–99.

118   Nicholls, S.J., et al., Effects of the CETP inhibitor evacetrapib administered as monotherapy or in combination with statins on HDL and LDL cholesterol: a randomized controlled trial. *JAMA*, 2011. 306(19): p. 2099–109.

119   Rader, D.J. and E.M. deGoma, Future of cholesteryl ester transfer protein inhibitors. *Annu Rev Med*, 2014. 65: p. 385–403.

120   Sirtori, C.R., et al., Cardiovascular status of carriers of the apolipoprotein A-I(Milano) mutant: the Limone sul Garda study. *Circulation*, 2001. 103(15): p. 1949–54.

121   Nissen, S.E., et al., Effect of recombinant ApoA-I Milano on coronary atherosclerosis in patients with acute coronary syndromes: a randomized controlled trial. *JAMA*, 2003. 290(17): p. 2292–300.

122   Krause, B.R. and A.T. Remaley, Reconstituted HDL for the acute treatment of acute coronary syndrome. *Curr Opin Lipidol*, 2013. 24(6): p. 480–6.

123   Bhatt, K.N., et al., High-density lipoprotein therapy: is there hope? *Curr Treat Options Cardiovasc Med*, 2010. 12(4): p. 315–28.

124   Nicholls, S.J., et al., Efficacy and safety of a novel oral inducer of apolipoprotein a-I synthesis in statin-treated patients with stable coronary artery disease a randomized controlled trial. *J Am Coll Cardiol*, 2011. 57(9): p. 1111–9.

125   Salmon, F., K. Grosios, and H. Petry, Safety profile of recombinant adeno-associated viral vectors: focus on alipogene tiparvovec (Glybera(R)). *Expert Rev Clin Pharmacol*, 2014. 7(1): p. 53–65.

5

MANAGEMENT OF DYSLIPIDEMIAS: PRINCIPLES AND GUIDELINES

5.1 OVERVIEW

As discussed in the preceding two chapters, lipid metabolism can be disturbed in different ways, leading to changes in plasma levels of lipoproteins as well as lipoprotein dysfunction. Such changes by themselves and through interaction with other cardiovascular risk factors may profoundly impact the development of atherosclerosis. In this context, atherosclerosis is responsible for approximately 75% of all cardiovascular deaths. Hence, appropriate management of dyslipidemias is one of the most important steps in controlling the development of atherosclerotic cardiovascular diseases and reducing cardiovascular mortality.

Chapter 4 has discussed the molecular pharmacology of drugs for treating dyslipidemias. This chapter first considers the general principles of the management of dyslipidemias and then examines current evidence-based guidelines on use of pharmacological agents along with other interventional modalities in treating dyslipidemias. The chapter focuses on discussion of the well-known Adult Treatment Panel (ATP) III guideline and the newly released 2013 American College of Cardiology/American Heart Association (ACC/AHA) guideline. This chapter also introduces recent guidelines from the European Society of Cardiology/European Atherosclerosis Society (ESC/EAS), the UK National Institute for Health and Care Excellence (NICE), the Joint British Societies (JBS), and the Canadian Cardiovascular Society (CCS). Due to the controversies facing the 2013 ACC/AHA guideline, the chapter compares and contrasts the above non-ACC/AHA guidelines with the ACC/AHA recommendations to provide a comprehensive and unbiased

discussion of the currently available evidence-based guidelines for the optimal management of dyslipidemias.

5.2 GENERAL PRINCIPLES OF THE MANAGEMENT OF DYSLIPIDEMIAS

Effective management of dyslipidemias requires implementation of a comprehensive approach that addresses all the key aspects of the disorders, ranging from identification of etiologies and molecular pathophysiology, laboratory evaluation of lipid profiles, and assessment of cardiovascular risk to defining treatment targets, implementation of both pharmacological and nonpharmacological treatment approaches, and monitoring treatment effects and ensuring treatment adherence. This section introduces the general principles involved in dyslipidemia management to provide a foundation for understanding the evidence-based guidelines covered in Section 5.3.

5.2.1 Defining Dyslipidemias in the Context of Disease Management

Although dyslipidemias can be manifested as any changes in the plasma lipid profiles, elevation of total cholesterol and LDL cholesterol has received most attention. This is due primarily to two reasons: (i) both total cholesterol and LDL cholesterol can be effectively modified by lifestyle changes and drug therapies, and (ii) reducing total and LDL cholesterol prevents cardiovascular diseases and decreases cardiovascular mortality. Data suggest that for every 30 mg/dl

Cardiovascular Diseases: From Molecular Pharmacology to Evidence-Based Therapeutics, First Edition. Y. Robert Li.
© 2015 John Wiley & Sons, Inc. Published 2015 by John Wiley & Sons, Inc.

change in LDL cholesterol, the relative risk for coronary heart diseases (CHD) is changed in proportion by 30% [1]. Plasma levels of total cholesterol and LDL cholesterol, especially LDL cholesterol, continue to constitute the primary targets of therapy.

In addition to elevated total and LDL cholesterol, several other types of dyslipidemias also predispose the individuals to premature cardiovascular diseases. A particular pattern, termed the atherogenic lipid triad, is more common than others and consists of the coexistence of increased very low-density lipoprotein (VLDL) remnants manifested as mildly elevated triglycerides, increased small dense LDL particles, and reduced HDL-cholesterol levels [2]. However, clinical trial evidence is currently limited on the effectiveness and safety of intervening modalities in this pattern to reduce cardiovascular risk.

5.2.2 Understanding Laboratory Lipid Profiles

Lipid profile, also called lipid panel, lipoprotein panel, or lipoprotein profile, is a term to describe major blood lipid fractions, that is, total cholesterol, LDL cholesterol, HDL cholesterol, and triglycerides. While the levels of plasma total cholesterol, HDL cholesterol, and triglycerides can be easily and reliably measured, direct measurement of LDL-cholesterol levels requires specific techniques. It is for this reason that LDL cholesterol is routinely estimated based on the measurements of total cholesterol, HDL cholesterol, and triglycerides (see the following text for more detail). Lipid profiling of the above four lipid fractions is also known as basic lipid profile. In this context, measurements of other lipid fractions, including apolipoproteins (e.g., ApoB, ApoA-I), Lp(a), and lipoprotein particle size, can also be carried out under certain conditions to provide additional information for cardiovascular risk assessment.

5.2.2.1 Unit Conversion Factors
The units for lipid profiles are in the form of either mg/dl mainly used in the United States or mM (or µM) used in the rest of the world. The conversion factors between these two unit systems are listed below:

1. One mg/dl cholesterol = 38.7 mM cholesterol
2. One mg/dl triglyceride = 88.6 mM triglyceride
3. One mg/dl Lp(a) = 0.0357 µM Lp(a)

5.2.2.2 Total Cholesterol
Total cholesterol is frequently measured in screening programs to estimate total cardiovascular risk by means of the Framingham Risk Score/Profile systems (see Section 1.7). However, measurement of total cholesterol may be misleading in the individual cases. This is particularly true in women who often have high HDL-cholesterol levels and in subjects with diabetes or metabolic syndrome who often have low HDL-cholesterol levels. For an adequate risk analysis, both HDL cholesterol and LDL cholesterol should be measured.

5.2.2.3 LDL Cholesterol
As noted earlier, in most clinical studies, especially those conducted a decade ago, LDL cholesterol is typically calculated using Friedewald's formula listed below when plasma triglyceride level is below 400 mg/dl:

- LDL cholesterol = (total cholesterol) − (HDL cholesterol) − (triglycerides/5) in mg/dl

This formula is based on the fact that total cholesterol is the sum of HDL cholesterol, LDL cholesterol, and VLDL cholesterol. Total cholesterol, HDL cholesterol, and triglycerides are relatively easy and inexpensive to measure, but the same cannot be said of LDL cholesterol and VLDL cholesterol [3]. In 1972, W.T. Friedewald observed that there was a relationship between the level of measured triglycerides and the level of the triglyceride-rich VLDL particles. Thus, VLDL cholesterol can be estimated based on the level of triglycerides. This ratio of triglycerides to VLDL-cholesterol levels on average is 5:1. This observation enabled Friedewald to create the above simple formula for estimating LDL cholesterol [3].

However, Friedewald's formula is not accurate when triglycerides are high. Therefore, it is not used if the triglycerides are above 400 mg/dl, and a more costly direct LDL-cholesterol measurement is necessary. Recently, S.S. Martin and associates developed a novel method to estimate LDL cholesterol. Using a database with approximately 1.3 million lipid profiles, S.S. Martin and associates developed a table that established a more precise but variable triglycerides/VLDL-cholesterol ratio that is dependent on both triglyceride and non-HDL-cholesterol levels. This estimation method provided higher-fidelity estimates than the Friedewald equation or other methods, particularly when classifying LDL-cholesterol levels lower than 70 mg/dl in the presence of high triglyceride levels [4, 5].

If Martin's approach is confirmed, it will create a simple, but important, method to more accurately estimate LDL-cholesterol levels that could be easily implemented at little or no cost at the time that the data are presented to the clinicians. This novel method could reduce the need for more expensive direct LDL-cholesterol measurements, and it also could be particularly useful in parts of the world where direct LDL-cholesterol measurement is not readily available or is too costly to afford [3]. Nevertheless, direct methods for determining LDL cholesterol should be used whenever available and affordable before Martin's method becomes externally validated.

5.2.2.4 Non-HDL Cholesterol Non-HDL cholesterol can be considered as the indicator of total load of atherogenic lipoprotein particles in plasma. Non-HDL cholesterol thus provides a better estimation of cardiovascular risk as compared with LDL cholesterol alone, especially in individuals with hypertriglyceridemia combined with diabetes, metabolic syndrome, or chronic kidney diseases [6, 7]. Non-HDL cholesterol is calculated as the following:

- Non-HDL cholesterol = (total cholesterol) − (HDL cholesterol)

5.2.3 Cardiovascular Risk Assessment

For decades, well-recognized guidelines for the management of cardiovascular diseases, including primary and secondary prevention, have almost exclusively embraced a risk assessment-based paradigm, which links the intensity of preventive therapy with an estimated cardiovascular risk for the individuals. The standard for cardiovascular risk assessment has been the Framingham Risk Score (see Section 1.7). However, weaknesses have emerged over time with the Framingham Risk Score system, including concerns about whether it can be generalized to nonwhite populations and its applicability to women, among others [8]. The new 2013 ACC/AHA Guideline on the Assessment of Cardiovascular Risk makes substantial progress in particular aspects of personalization [9], though it has received criticisms, especially on its potential to overestimate cardiovascular risk [10].

The new 2013 ACC/AHA cardiovascular risk assessment guideline not only provides separate equations for men and women but also offers distinct equations for white and African American individuals, recognizing that the determinants of disease can differ according to ethnicity. The new guideline also broadens the clinical outcome endpoints by inclusion of stroke. As discussed in Section 5.3.3.1, this new cardiovascular risk assessment approach is used in the newly released guideline on the treatment of blood cholesterol to reduce atherosclerotic cardiovascular risk [11].

5.2.4 Treatment Goals

One of the fundamental principles guiding cholesterol-lowering intervention is that the intensity of treatment is directly related to the degree of cardiovascular risk. In nearly all lipid-lowering trials, the LDL-cholesterol levels have been used as a major indicator of response to therapy. LDL-cholesterol levels remain the primary target of therapy in most strategies and guidelines of dyslipidemia management, including the widely adopted US National Cholesterol Education Program (NCEP) Adult Treatment Panel III (ATP III) guideline [12] (see Section 5.3.3.1).

The overall guidelines on cardiovascular disease prevention in clinical practice strongly recommend modulating the intensity of the preventive intervention according to the level of the predicted cardiovascular risk. In this regard, the therapeutic goals for LDL cholesterol should be less demanding when the estimated cardiovascular risk decreases from very high to high or moderate. Setting such treatment goals for LDL cholesterol in dyslipidemia management guidelines is based on the notion that the extent of LDL-cholesterol reduction is proportional to the cardiovascular risk reduction. The most recent Cholesterol Treatment Trialists' Collaboration (CTT) meta-analysis of several trials involving more than 170,000 patients further confirmed the dose-dependent reduction in cardiovascular diseases with LDL cholesterol lowering. Based on this analysis, for every 1.0 mM (∼40 mg/dl) reduction in LDL cholesterol, there is a corresponding 22% reduction in cardiovascular mortality and morbidity [13]. Although most guidelines on dyslipidemia management include treatment goals for LDL cholesterol based on the individual patient's conditions, in the newly released 2013 ACC/AHA guideline, no recommendations are made for or against specific LDL-cholesterol or non-HDL-cholesterol goals for primary or secondary prevention of atherosclerotic cardiovascular diseases (ASCVD) (see Section 5.3.3.1 for more discussion).

5.2.5 General Approaches to Management: Lifestyle Modifications and Drug Therapies

Lifestyle modifications and drug therapies are two essential components of effective management of dyslipidemias. It must be emphasized that lifestyle modification (i.e., adhering to a heart-healthy diet, regular exercise habits, avoidance of tobacco products, and maintenance of a healthy weight) remains the foundation for cardiovascular risk reduction efforts, both prior to and in concert with the use of cholesterol-lowering drug therapies. In this context, lifestyle modifications are considered as background therapy of dyslipidemias. The 2013 AHA/ACC guideline on lifestyle management to reduce cardiovascular risk has recently been released [14], and the reader is suggested to refer to the full document for detailed recommendations.

5.2.6 Management in Specific Clinical Settings

Management of dyslipidemias should not only focus on the general populations but also address specific clinical settings. These include familial dyslipidemias (e.g., familial hypercholesterolemia), dyslipidemias in children, women, the elderly, and different racial and ethical groups. These also include dyslipidemias in patient with comorbid conditions, including diabetes, metabolic syndrome, acute coronary syndromes, heart failure, valvular diseases, peripheral artery disease, stroke, kidney diseases, organ transplantation, autoimmune diseases, and HIV infections [2].

5.2.7 Treatment Monitoring and Adherence

Treatment monitoring and adherence are an integrated part of dyslipidemia management. Lipids and enzymes, such as liver and skeletal muscular enzymes (e.g., alanine aminotransferase and creatine phosphokinase), should be monitored in patients on lipid-lowering drug therapy to assess therapeutic efficacy as well as potential drug adverse effects. Effective strategies should also be considered to improve adherence to lifestyle changes and compliance with drug therapy [2, 12].

5.3 CURRENT EVIDENCE-BASED GUIDELINES ON THE MANAGEMENT OF DYSLIPIDEMIAS

The importance of evidence-based clinical practice guidelines in improving patient's care has been increasingly recognized over the past two decades. This is particularly true for the management of cardiovascular diseases. This section first defines clinical guidelines and then introduces the classification of evidence-based recommendations and strength of evidence. This sets a stage for the discussion of the current guidelines on dyslipidemia management.

5.3.1 Defining Clinical Guidelines in the Context of Dyslipidemia Management

Clinicians and other healthcare providers are oftentimes faced with difficult decisions and considerable uncertainty when treating patients with cardiovascular diseases. They rely on the scientific literature, in addition to their knowledge, skills, experience, and patient preferences, to inform their decisions. According to the US Institute of Medicine (IOM)'s 2011 report on the development of trustworthy clinical practice guidelines, clinical practice guidelines are defined as statements that include recommendations intended to optimize patient care that are informed by a systematic review of evidence and an assessment of the benefits and harms of alternative care options. Rather than dictating a one-size-fits-all approach to patient care, clinical practice guidelines offer an evaluation of the quality of the relevant scientific literature and an assessment of the likely benefits and harms of a particular treatment. This information enables healthcare providers to proceed accordingly, selecting the best care for a unique patient based on his or her preferences. The IOM's definition came about as a result of the US Medicare Improvements for Patients and Providers Act of 2008, in which the US Congress asked the IOM to investigate the best methods for developing clinical practice guidelines [15].

According to the IOM's 2011 report, to be trustworthy, clinical practice guidelines should (i) be based on a systematic review of the existing evidence; (ii) be developed by a knowledgeable, multidisciplinary panel of experts and representatives from key affected groups; (iii) consider important patient subgroups and patient preferences, as appropriate; (iv) be based on an explicit and transparent process that minimizes distortions, biases, and conflicts of interest; (v) provide a clear explanation of the logical relationships between alternative care options and health outcomes and provide ratings of both the quality of evidence and the strength of recommendations; and (vi) be reconsidered and revised as appropriate when important new evidence warrants modifications of recommendations.

In response to the IOM's 2011 report, the US National Heart, Lung, and Blood Institute (NHLBI) Advisory Council recommended that the NHLBI focus specifically on reviewing the highest-quality evidence and partner with other organizations to develop recommendations. Accordingly, in 2013, the NHLBI initiated collaboration with the ACC and AHA to work with other organizations to complete and publish four clinical practice guidelines: (i) 2013 ACC/AHA guideline on the assessment of cardiovascular risk [16], (ii) 2013 AHA/ACC guideline on lifestyle management to reduce cardiovascular risk [14], (iii) 2013 ACC/AHA guideline on the treatment of blood cholesterol to reduce atherosclerotic cardiovascular risk in adults [17] (see Section 5.3.3.1), and (iv) 2013 AHA/ACC/TOS guideline for the management of overweight and obesity in adults [18]. These evidence-based guidelines provide important opportunities for the practice of evidence-based patient care to promote cardiovascular health in the US populations, though the new risk assessment approach and cholesterol treatment guideline have received criticisms. In this context, evidence-based new guidelines on cardiovascular health, including risk assessment and management of dyslipidemias, have also been released from the ESC/EAS, Joint British Societies (JBS), and National Institute for Health and Care Excellence (NICE) to address the European populations and from the CCS to guide patient care in Canada. Each of these guidelines is based on scientific evidence whose strength is classified in different but related manners. The following section introduces the two commonly used classification schemes for evidence-based recommendations and strength of evidence to support the recommendations: one from the ACC/AHA and the other from the NHLBI.

5.3.2 Classification of Evidence-Based Guideline Recommendations and Strength of Evidence

5.3.2.1 *The ACC/AHA Classification Scheme* In the ACC/AHA guidelines, evidence-based clinical practice recommendations are divided into three classes, namely, class I, class II (IIa and IIb), and class III (Table 5.1). Each of the recommendations is also accompanied by one of the three levels of supporting evidence: level A, level B, and level C (Table 5.2).

TABLE 5.1   The ACC/AHA classes of recommendations[a]

Class of recommendation		Definition	Suggested wording to use
Class I		• Benefit >>>risk • Recommendation that the procedure or treatment is useful/effective • Procedure or treatment should be performed or administered	• Should • Is recommended or indicated • Is useful, effective, or beneficial
Class II	IIa	• Benefit ≫ risk • Recommendation in favor of treatment or procedure being useful/effective • Additional studies with focused objectives are needed • It is reasonable to perform procedure or administer treatment	• Is reasonable • Is probably recommended or indicated • Can be useful, effective, or beneficial
	IIb	• Benefit>risk • Usefulness or efficacy is less well established • Additional studies with broad objectives are needed • Additional registry data would be helpful • Procedure/treatment may be considered	• May/might be considered • May/might be reasonable • Usefulness/effectiveness is unknown, unclear, uncertain, or not well established
Class III	No benefit	• Procedure/test is not helpful • Treatment has no proven benefit	• Is not recommended • Is not indicated • Should not be performed/administered • Is not useful/beneficial/effective
	Harm	• Procedure/test results in excess cost without benefit • Procedure/test is harmful • Treatment is harmful to patients	• Potentially harmful • Causes harm associated with excess morbidity/mortality • Should not be performed/administered

[a]Ref. 17.

TABLE 5.2   The ACC/AHA levels of evidence[a]

Level of evidence	Definition
A	• Multiple populations evaluated[b] • Data derived from multiple randomized clinical trials or meta-analysis
B	• Limited populations evaluated[b] • Data derived from a single randomized trial or nonrandomized studies
C	• Very limited populations evaluated[b] • Only consensus opinion of experts, case studies, or standard of care

[a]Ref. 17.
[b]Data available from clinical trials or registries about the usefulness/efficacy in different subpopulations, such as sex, age, history of diabetes, history of prior myocardial infarction, history of heart failure, and prior aspirin use.

5.3.2.2   The NHLBI Grading Format   The NHLBI appoints expert panels to conduct systematic evidence reviews to enable clinical practice guideline development. The IOM 2011 report defines a systematic evidence review as "a scientific investigation that focuses on a specific question and uses explicit, prespecified scientific methods to identify, select, assess, and summarize the findings of similar but separate studies. It may include a quantitative synthesis (meta-analysis), depending on the available data." Systematic evidence reviews of comparative effectiveness research to learn what is known and not known about the potential benefits and harms of alternative drugs, devices, and other healthcare services provide the best evidence to inform clinical decisions. The NHLBI's grading methodologies for clinical practice guideline recommendations are summarized in Tables 5.3 and 5.4.

5.3.3   Current Guidelines on the Management of Dyslipidemias

Table 5.5 summarizes the current guidelines on dyslipidemia management from different nations and organizations. This section focuses on discussing the current guidelines from ACC/AHA and ESC/EAS, as well as NICE, paying particular attention to the strengths and weaknesses of each of the guidelines. Due to the controversies surrounding the newly released 2013 ACC/AHA guideline on cholesterol management (also known informally as ATP IV guideline), the well-known ATP III guideline is discussed here to help the reader understand the major changes in the 2013 ACC/AHA guideline and the possible sources of controversies.

TABLE 5.3   The NHLBI's grading of the strength of recommendation

Grade	Recommendation strength	Definition
A	Strong	There is high certainty based on evidence that net benefit is substantial. Net benefit is defined as benefits minus risks/harms of the service/intervention
B	Moderate	There is moderate certainty based on evidence that net benefit is moderate to substantial or high certainty that net benefit is moderate
C	Weak	There is at least moderate certainty based on evidence that there is a small net benefit
D	Against	There is at least moderate certainty based on evidence that it has no net benefit or that risks/harms outweigh benefits
E	Expert opinion	Net benefit is unclear. Balance of benefits and harms cannot be determined because of no evidence, insufficient evidence, unclear evidence, or conflicting evidence, but the panel thought it was important to provide clinical guidance and make a recommendation. Further research is recommended in this area
N	No recommendation for or against	Net benefit is unclear. Balance of benefits and harms cannot be determined because of no evidence, insufficient evidence, unclear evidence, or conflicting evidence, and the panel thought no recommendation should be made. Further research is recommended in this area

TABLE 5.4   The NHLBI's quality rating of the strength of evidence

Quality rating	Definition	Type of evidence
High	Estimate of effect is unlikely to be impacted and/or changed by further research	• Well-designed, well-executed randomized controlled trials that adequately represent populations to which the results are applied and directly assess effects on health outcomes • Meta-analyses of such studies
Moderate	Estimate of effect may be impacted and/or changed by further research	• Randomized controlled trials with minor limitations affecting confidence in, or applicability of, the results • Well-designed, well-executed nonrandomized controlled studies and well-designed, well-executed observational studies • Meta-analyses of such studies
Low	Estimate of effect is likely to be impacted and/or changed by further research	• Randomized controlled trials with major limitations • Nonrandomized controlled studies and observational studies with major limitations affecting confidence in, or applicability of, the results • Uncontrolled clinical observations without an appropriate comparison group (e.g., case series, case reports) • Physiological studies in humans • Meta-analyses of such studies

5.3.3.1   The ATP III Guideline

Overview   The Third Report of the Expert Panel on Detection, Evaluation, and Treatment of High Blood Cholesterol in Adults (ATP III) constitutes the NCEP's updated clinical guidelines for cholesterol testing and management. The full ATP III document, published in 2002 [12] with a minor update in 2004 [20], is an evidence-based and extensively referenced report for the management of dyslipidemia. This 280-page report with 1121 references cited is probably the most comprehensive guideline, up to date, on dyslipidemia management [12].

The full ATP III report consists of nine major sections: (i) background and introduction, (ii) rationale for intervention, (iii) detection and evaluation, (iv) general approach to treatment goals and thresholds, (v) adopting healthful lifestyle habit to lower LDL cholesterol and reduce CHD risk, (vi) drug therapy, (vii) management of specific dyslipidemias, (viii) special considerations for different population groups, and (ix) adherence [12].

New Features of the ATP III Report   ATP III is constructed on the foundation of ATP I and ATP II, with LDL cholesterol continuing to be identified as the primary target of cholesterol-lowering therapy. Compared with the previous two reports, the new features of ATP III include (i) aggressive treatment of persons who are at relatively high risk for CHD due to multiple risk factors, (ii) use of the lipoprotein profile

TABLE 5.5 Current guidelines on dyslipidemia management from different countries and organizations

Organization	Nation	Year	Title and reference	Note
ACC/AHA	United States	2013	2013 ACC/AHA guideline on the treatment of blood cholesterol to reduce atherosclerotic cardiovascular risk in adults: a report of the American College of Cardiology/American Heart Association Task Force on Practice Guidelines Circulation. 2014, June 24; 129(25 Suppl 2): S1–45. J Am Coll Cardiol. 2014 July 1; 63(25 Pt B): 2889–934.	The blood cholesterol Expert Panel (Expert Panel) was originally convened as the Expert Panel on detection, evaluation, and treatment of high blood cholesterol in adults (Adult Treatment Panel [ATP] IV) appointed by the NHLBI. All 16 members of the NHLBI ATP IV Panel later transitioned to the ACC/AHA guideline Expert Panel, which led to the publication of the long-awaited 2013 ACC/AHA guideline. In this context, the well-known ATP III guideline was published in 2002 [12] and updated in 2004 [1]
ESC/EAS		2011	ESC/EAS Guidelines for the management of dyslipidaemias: the Task Force for the management of dyslipidaemias of the European Society of Cardiology (ESC) and the European Atherosclerosis Society (EAS) Eur Heart J. 2011 July; 32(14):1769–818 Atherosclerosis. 2011 July; 217(1):3–46	Like previous guidelines from ACC/AHA, this ESC/EAS guideline classifies recommendations into three classes, namely, I, II (IIa and IIb), and III, and evidence into three levels, namely A, B, and C
NICE	United Kingdom	2014	Lipid modification: cardiovascular risk assessment and the modification of blood lipids for the primary and secondary prevention of cardiovascular disease. http://www.nice.org.uk/guidance/index.jsp?action=download&o=66547 (accessed March 1, 2014)	The draft guideline is out for consultation with stakeholders. The guideline is expected to be published in July 2014
CCS	Canada	2012	2012 update of the Canadian Cardiovascular Society guidelines for the diagnosis and treatment of dyslipidemia for the prevention of cardiovascular disease in the adult Can J Cardiol. 2013 February; 29(2):151–67	This is an update of the 2009 Canadian Cardiovascular Society (CCS) dyslipidemia guidelines [19]

as the first test for high cholesterol, (iii) a new level at which low HDL becomes a major heart disease risk factor, (iv) a new set of "therapeutic lifestyle changes" (TLC) to improve cholesterol levels, (v) an increased focus on a cluster of heart disease risk factors known as the metabolic syndrome, and (vi) increased attention to the treatment of high triglycerides.

Major Steps of the ATP III Guideline The ATP III guideline includes nine major steps to the management of dyslipidemias, which are briefly described next.

Step 1: Determine lipoprotein levels to obtain complete lipoprotein profile after a 9–12 h fast. The ATP III classification of LDL, total, and HDL cholesterol is given in Table 5.6.

Step 2: Identify the presence of clinical atherosclerotic diseases that confer high risk for CHD events (CHD risk

TABLE 5.6 ATP III classification of plasma LDL, total, and HDL cholesterol

Plasma lipid	Plasma level (mg/dl)	Classification
LDL cholesterol	<100	Optimal
	100–129	Near optimal/above optimal
	130–159	Borderline high
	160–189	High
	≥190	Very high
Total cholesterol	<200	Desirable
	200–239	Borderline high
	≥240	High
HDL cholesterol	<40	Low
	≥60	High

equivalent). CHD risk equivalents include (i) noncoronary forms of atherosclerotic diseases such as peripheral arterial disease, abdominal aortic aneurysm, and carotid artery disease (transient ischemic attacks or stroke of carotid origin or >50% obstruction of a carotid artery), (ii) diabetes, and (iii) ≥2 risk factors with 10-year risk for CHD >20% (see Step 3 for defining risk factors).

Step 3: Determine the presence of major risk factors (other than LDL). The major risk factors (exclusive of LDL cholesterol) that modify LDL-cholesterol goals include (i) cigarette smoking, (ii) hypertension (blood pressure ≥140/90 mm Hg or on antihypertensive medication), (iii) low HDL cholesterol (<40 mg/dl), (iv) family history of premature CHD (CHD in male first-degree relative <55 years of age; CHD in female first-degree relative <65 years of age), and (v) age (men ≥45 years; women ≥55 years). Because HDL is protective against atherosclerosis, HDL cholesterol ≥60 mg/dl counts as a "negative" risk factor, and its presence removes one risk factor from the total count.

Step 4: Assess 10-year (short-term) CHD risk using the Framingham Risk Score system (see Chapter 1) if ≥2 risk factors (other than LDL cholesterol) are present without CHD or CHD risk equivalent. The three levels of 10-year risk are (i) ≥20% (this is considered as CHD equivalent), (ii) 10–20%, and (iii) <10%.

Step 5: Determine the risk category to (i) establish LDL-cholesterol goal of therapy, (ii) determine need for TLC, and determine the LDL-cholesterol level for drug consideration. Listed below are the four risk categories:

- High risk: CHD or CHD risk equivalents (10-year risk >20%)
- Moderate high risk: ≥2 risk factors (10-year risk of 10–20%)
- Moderate risk: ≥2 risk factors (10-year risk <10%)
- Low risk: 0–1 risk factor

The ATP III LDL-cholesterol goals and cutpoints for TLC and drug therapy in different risk categories are given in Table 5.7.

Step 6: Initiate TLC if LDL cholesterol is above goal. The TLC includes three components, as listed below:

1. TLC diet: (i) saturated fat <7% of calories, cholesterol <200 mg/day; (ii) increased viscous (soluble) fiber (10–25 g/day) and plant stanols/sterols (2 g/day) as therapeutic options to enhance LDL cholesterol lowering
2. Weight management
3. Increased physical activity

Step 7: Consider adding drug therapy if LDL cholesterol exceeds recommended levels. This step includes (i) considering drug simultaneously with TLC for CHD and CHD equivalents and (ii) considering adding drug to TLC after 3 months for other risk categories. The drug choices may include statins, bile acid sequestrants, niacin, and fibrates although statins are considered as the most important agents for treating high blood cholesterol.

Step 8: Identify metabolic syndrome and treat, if present, after 3 months of TLC. Metabolic syndrome is defined by the presence of any three of the following five conditions:

1. Central or abdominal obesity (measured by waist circumference): >102 cm for men; >88 cm for women
2. Fasting blood triglycerides: ≥150 mg/dl
3. Blood HDL cholesterol: <40 mg/dl for men; <50 mg/dl for women
4. Blood pressure: ≥130/85 mm Hg
5. Fasting glucose: ≥100 mg/dl

Treatment of metabolic syndrome includes the following two aspects: (i) treating underlying causes (overweight/obesity and physical inactivity) via intensified weight management and increased physical activity and (ii) treating lipid and nonlipid risk factors if they persist despite the above lifestyle therapies, which includes treatment of hypertension, use of aspirin for CHD patients to reduce prothrombotic state, and treatment of elevated triglycerides and/or low HDL cholesterol.

TABLE 5.7   LDL-cholesterol goals and cutpoints for TLC and drug therapy in different risk categories

Risk category	LDL-cholesterol goal	LDL-cholesterol level at which to initiate TLC	LDL-cholesterol level at which to consider drug therapy
High risk	<100 mg/dl (optional goal: <70 mg/dl)	≥100 mg/dl	≥100 mg/dl (<100 mg/dl: consider drug options)
Moderately high risk	<130 mg/dl	≥130 mg/dl	≥130 mg/dl (100–129 mg/dl; consider drug options)
Moderate risk	<130 mg/dl	≥130 mg/dl	≥160 mg/dl
Low risk	<160 mg/dl	≥160 mg/dl	≥190 mg/dl (160–189 mg/dl: LDL-lowering drug optional)

TABLE 5.8   ATP III classification of plasma triglycerides

Triglyceride level (mg/dl)	Classification
<150	Normal
150–199	Borderline high
200–499	High
≥500	Very high

Step 9: Treat elevated triglycerides and low HDL cholesterol. Treatment of elevated triglycerides (see Table 5.8 for ATP III classification of plasma triglycerides) depends on severity of the hypertriglyceridemia. In addition, because patients with elevated triglycerides usually also have increased LDL cholesterol, the primary aim of therapy is to reach the LDL-cholesterol goal unless the triglycerides are severely elevated:

• Treatment of elevated triglycerides (>150 mg/dl): (i) intensify weight management and increase physical activity; (ii) if triglycerides are ≥200 mg/dl after LDL-cholesterol goal is reached, set secondary goal for non-HDL cholesterol at 30 mg/dl higher than LDL-cholesterol goal (i.e., non-HDL-cholesterol goal = LDL-cholesterol goal + 30 mg/dl).

• If triglycerides are between 200 and 499 mg/dl after LDL-cholesterol goal is reached, consider adding drug therapy to reach non-HDL goal via (i) intensifying therapy with LDL-cholesterol-lowering drug or (ii) adding niacin or a fibrate.

• If triglycerides are >500 mg/dl, first lower triglycerides to prevent pancreatitis via (i) very low-fat diet (≤15% of calories from diet), (ii) weight management and physical activity, and (iii) use of a fibrate or niacin. When triglycerides are <500 mg/dl, turn to LDL-lowering therapy.

Similar to the situation described for elevated triglycerides, low HDL cholesterol (<40 mg/dl) is frequently accompanied by high LDL cholesterol and hypertriglyceridemia. Accordingly, the treatment of low HDL cholesterol involves the following aspects: (i) first reach LDL-cholesterol goal and then intensify weight management and increase physical activity (these lifestyle managements can increase HDL cholesterol); (ii) if triglycerides are 200–499 mg/dl, achieve non-HDL goal via intensifying therapy with LDL-cholesterol-lowering drug or adding niacin or a fibrate; and (iii) if triglycerides are <200 mg/dl (isolated low HDL cholesterol) in CHD or CHD equivalent, consider the use of niacin or a fibrate drug to raise HDL cholesterol.

Summary of the ATP III Guideline

• The ATP III report (released in 2001, published in 2002, updated in 2004) has been probably the most extensively referenced evidence-based guideline on the management of dyslipidemias.

• ATP III is constructed on the foundation of ATP I and ATP II, with LDL cholesterol continuing to be identified as the primary target of cholesterol-lowering therapy. Its major new feature is a focus on primary prevention in persons with multiple risk factors and setting specific LDL-cholesterol goals for different risk categories.

• Although the one-decade-old ATP III guideline has been widely accepted and applied with relative consistency, new development in dyslipidemia management over the past decade has necessitated a major revision of the ATP III recommendations to more closely reflect the current evidence. In this regard, as described next, an updated guideline for the management of blood cholesterol (initially known as ATP IV guideline) was recently released by the ACC/AHA.

5.3.3.2   The 2013 ACC/AHA Guideline

Overview   It has been more than a decade since the last full version of NCEP guideline (ATP III) for dyslipidemia management was published. Since then, substantial amounts of new data, especially those from randomized controlled trials, have emerged to justify a new version of the guideline to better target lipid management therapies for more effective control of ASCVD. In this context, in 2008, the NHLBI convened the ATP IV to update the ATP III guideline using a rigorous process to systematically review randomized controlled trials and meta-analyses of such trials that examined cardiovascular outcomes. The panel commissioned independent systematic evidence reviews according to principles recommended by the IOM to answer three questions relevant to clinical care. Two questions focused on the evidence supporting LDL-cholesterol and non-HDL-cholesterol levels as targets of treatment. One question examined the reduction in atherosclerotic cardiovascular events and adverse effects for each cholesterol-lowering drug class.

The panel synthesized the evidence from the aforementioned systemic evidence reviews as well as from Lifestyle Management and Risk Assessment Work Groups reviews [14, 16], and the draft recommendations were transitioned to the ACC/AHA in September 2013. The full guideline was released and in press in November 2013 under the title of "2013 ACC/AHA Guideline on the Treatment of Blood Cholesterol to Reduce Atherosclerotic Cardiovascular Risk in Adults: A Report of the American College of Cardiology/ American Heart Association Task Force on Practice Guidelines." The full guideline is published simultaneously in the *Circulation* [17] and the *Journal of American College of Cardiology* [21], and the synopsis of the full guideline is

published in *Annals of Internal Medicine* [22]. The major recommendations of the 2013 ACC/AHA guideline are summarized next. The reader is advised to refer to the full guideline for a complete list of recommendations and supporting evidence behind each recommendation.

Major Recommendations The full 2013 ACC/AHA guideline includes 10 sections in addition to appendices and references: (i) introduction; (ii) overview of the guideline including lifestyle as the foundation for ASCVD risk reduction efforts and four major statin benefit groups; (iii) critical questions and conclusions; (iv) statin treatment recommendations, including the recommendation on using the new Pooled Cohort Risk Assessment Equations to estimate the 10-year ASCVD risk for the identification of candidates for statin therapy; (v) safety recommendations; (vi) managing statin therapy recommendations; (vii) selected clinical and population subgroups; (viii) limitations; (ix) evidence gaps and future research needs; and (x) conclusion. From the table of contents of the guideline, one can readily notice the emphasis of the new guideline on statin therapy. The major recommendations are summarized in Tables 5.9, 5.10, and 5.11.

Controversies and Potential Impact The release of 2013 ACC/AHA guideline on the treatment of blood cholesterol represents a major event in facilitating evidence-based medicine in the field of cardiovascular diseases. The new guideline is based on rigorous reviews of high-quality evidence according to the IOM's 2011 guideline on developing trustworthy clinical practice guidelines. The new guideline emphasizes prevention of ASCVD, including both CHD and stroke, focuses appropriately on statin therapy rather than alternative unproven therapeutic agents with regard to reducing ASCVD morbidity and mortality, and recognizes that more intensive statin treatment is superior to less intensive treatment for many patient groups. Furthermore, by eliminating emphasis on LDL-cholesterol treatment goals and defining four statin benefit groups, the new guideline greatly simplifies blood cholesterol treatment recommendations as compared with ATP III guideline as well as other guidelines on dyslipidemia management (Table 5.5).

As with all other clinical practice guidelines, the 2013 ACC/AHA cholesterol treatment guideline is not without controversies. Despite the aforementioned notable improvements, the new guideline has received intense criticism immediately since its release in November 2013. Main concerns have focused on methodological issues of the Pooled Cohort Equations for ASCVD risk prediction, omission of LDL-cholesterol treatment goals, and inference that too many people would need statin treatment for primary prevention of ASCVD [10, 23–26]. Continued discussion about the strengths and potential shortfalls of the new guideline along with the continued accumulation of high-quality evidence

will certainly create new momentum for effective implementation and timely updates to maximally benefit the populations throughout the world. In this context, the population and economic impact of the 2013 ACC/AHA guideline as compared with other widely adopted guidelines have been investigated [27–29]. Likewise, a recent study examined the validity of the 2013 ACC/AHA Pooled Cohort Equations for ASCVD risk estimation. It was concluded that in a cohort of US adults ($n = 10,997$) for whom statin initiation is considered based on the ACC/AHA Pooled Cohort risk equations, observed and predicted 5-year ASCVD risks were similar, indicating that these risk equations were well calibrated in the population for which they were designed to be used, and demonstrated moderate to good discrimination [30].

Summary of the 2013 ACC/AHA Guideline

- The 2013 ACC/AHA cholesterol treatment guideline (informally known as the ATP IV guideline) arose from careful consideration of an extensive body of higher-quality evidence derived from randomized controlled trials and systematic reviews and meta-analyses of such trials.

- Rather than defining LDL-cholesterol or non-HDL-cholesterol goals as in the APT III guideline, this new guideline recommends the use of the intensity of statin therapy as the goal of treatment, focusing on those individuals most likely to benefit from evidence-based statin therapy to reduce ASCVD risk.

- The new guideline identifies four groups of individuals as statin benefit groups, for whom an extensive body of randomized controlled trial evidence shows a reduction in ASCVD events with a good margin of safety.

- The new guideline adopts the Pooled Cohort Equations for ASCVD risk prediction and emphasizes lifestyle management as the foundation for ASCVD risk reduction efforts.

- Although the new guideline represents a major step forward in blood cholesterol management, it has received intense criticism, largely for its distinct changes as compared with other widely adopted guidelines, including the ATP III guideline and the 2011 ESC/EAS guideline for the management of dyslipidemias. On the other hand, as discussed next, the 2013 ACC/AHA guideline and the UK 2014 NICE guideline have considerable agreement.

5.3.4 Comparison and Contrast of Current Lipid Guidelines

ASCVD is the most important public health problem of our time in both the United States and the rest of the world, accounting for the greatest expenditure in most healthcare budgets (see Chapter 1). Hence, achieving consistency of

TABLE 5.9 **Major recommendations of the 2013 ACC/AHA guideline on treating blood cholesterol to reduce atherosclerotic cardiovascular disease (ASCVD) risk in adult**[a]

Major recommendation	Description
Encourage adherence to a healthy lifestyle	The guideline emphasizes lifestyle as the foundation for cardiovascular disease risk reduction efforts and endorses the 2013 AHA/ACC Lifestyle Management Guideline [14] to (i) adhere to a diet that is low in saturated fat, trans fat, and sodium; emphasizes vegetables, fruits, whole grains, low-fat dairy products, poultry, fish, legumes, nontropical vegetable oils, and nuts; and limits sweets, sugar-sweetened beverages, and red meats; (ii) engage in regular aerobic physical activity; and (iii) maintain a healthy body weight, avoid smoking, and control hypertension and diabetes when present
Statin therapy is recommended for adults in four groups demonstrated to benefit	Strong randomized controlled trial evidence shows a reduction in atherosclerotic cardiovascular disease (ASCVD) events with a good margin of safety from moderate- or high-intensity statin therapy (see Recommendation 6 for definition). The four statin benefit groups are (i) individuals with clinical ASCVD, (ii) individuals with primary elevations of LDL cholesterol ≥190 mg/dl, (iii) individuals 40–75 years of age with diabetes and LDL cholesterol of 70–189 mg/dl without clinical ASCVD, and (iv) individuals without clinical ASCVD or diabetes who are 40–75 years of age with LDL cholesterol of 70–189 mg/dl and have an estimated 10-year ASCVD risk of ≥7.5%. Individuals in the last group can be identified by using the Pooled Cohort Equations described in the 2013 ACC/AHA Guideline on the Assessment of Cardiovascular Risk [16]
Statins have an acceptable margin of safety when used in properly selected individuals and appropriately monitored	Strong randomized controlled trial evidence supports safety of statins when they are used as directed in conjunction with regular follow-up assessments in properly selected patients. Adjustment of statin intensity is recommended in individuals older than 75 years with a history of statin intolerance or other characteristics or those receiving drug therapy that may increase statin adverse events. Routine monitoring of hepatic aminotransferase level or creatine kinase level is not recommended unless clinically indicated by symptoms suggesting hepatotoxicity or myopathy. Although statin therapy modestly increases the risk for type 2 diabetes, ASCVD risk reduction outweighs the excess risk for diabetes
Engage in a clinician–patient discussion before initiating statin therapy, especially for primary prevention in patients with lower ASCVD risk	Decisions to initiate statin therapy in primary prevention should be based on clinical judgment and preferences of informed patients. In adults without clinical ASCVD or diabetes whose LDL-cholesterol level is <190 mg/dl, calculating the estimated 10-year ASCVD risk should be the start of the clinician–patient discussion and should not automatically lead to statin initiation. As the absolute risk for ASCVD events decreases, so does the net benefit of the intervention. Therefore, discussion of the potential for ASCVD event reduction, adverse effects, drug–drug interactions, and patient preferences is especially important for lower-risk primary prevention. The discussion provides the opportunity to encourage healthy lifestyle habits and control other risk factors
Use the newly developed Pooled Cohort Equations for estimating 10-year ASCVD risk	The Pooled Cohort Equations [16] are currently the best available method for estimating 10-year ASCVD risk to guide statin initiation. The Pooled Cohort Equations were developed using recent data from 5 NHLBI-sponsored, longitudinal, population-based cohorts of African American and white men and women
Initiate the appropriate intensity of statin therapy	The appropriate intensity of statin therapy should be used to reduce ASCVD risk and minimize adverse effects in statin benefit groups. For clarity, the intensity of statin treatment is defined in Table 5.10, and the major statin treatment recommendations are given in Table 5.11
Evidence is inadequate to support treatment to specific LDL-cholesterol or non-HDL-cholesterol goals	Given the absence of randomized controlled trial data on titration of drug therapy to specific goals, no recommendations are made for or against specific LDL-cholesterol or non-HDL-cholesterol goals for primary or secondary prevention of ASCVD. (This is one of the major differences between 2013 ACC/AHA guideline and the ATP III guideline)
Regularly monitor patients for adherence to lifestyle and statin therapy	A fasting lipid panel is needed after initiation of or changes in statin or other drug therapy. Percentage reductions in LDL-cholesterol levels should be used to assess and provide feedback to promote adherence to healthy lifestyle behaviors and statin therapy. Safety measurements should be assessed as clinically indicated. In patients with a less-than-anticipated therapeutic response or intolerance of recommended statin therapy intensity, adherence to healthy lifestyle behaviors and medications should be reemphasized and secondary causes of hyperlipidemia excluded. A nonstatin LDL-cholesterol-lowering drug, preferably one that reduced ASCVD events in randomized controlled trials, can be considered in higher-risk adults, including those with genetic dyslipidemias, such as familial hypercholesterolemia, if the potential for additional ASCVD risk reduction outweighs the potential for adverse effects

ASCVD includes acute coronary syndromes, history of myocardial infarction, stable or unstable angina, coronary or other arterial revascularization, stroke, transient ischemic attack, or peripheral arterial disease presumed to be of atherosclerotic origin.
[a] Refs. 17, 22.

TABLE 5.10   Definition of high-, moderate-, and low-intensity statin therapy

Intensity of statin treatment	Definition	Statin daily dose
High intensity	Daily dose lowers LDL cholesterol on average, by approximately ≥50%	Atorvastatin 40–80 mg Rosuvastatin 20 *(40)* mg
Moderate intensity	Daily dose lowers LDL cholesterol on average, by approximately 30 to <50%	Atorvastatin 10 *(20)* mg Rosuvastatin *(5)* 10 mg Simvastatin 20–40 mg[a] Pravastatin 40 *(80)* mg Lovastatin 40 mg *Fluvastatin XL 80 mg* Fluvastatin 40 mg (twice daily) *Pitavastatin 2–4 mg*
Low intensity	Daily dose lowers LDL cholesterol on average, by <30%	*Simvastatin 10 mg* Pravastatin 10–20 mg Lovastatin 20 mg *Fluvastatin 20–40 mg* *Pitavastatin 1 mg*

Statins and doses that are approved by the US FDA but were not tested in the randomized controlled trials reviewed are listed in italics.

[a] Although simvastatin, 80 mg, was evaluated in randomized controlled trials, initiation of or titration to 80 mg of simvastatin is not recommended by the US FDA due to the increased risk of myopathy, including rhabdomyolysis.

TABLE 5.11   The 2013 ACC/AHA major statin treatment recommendations[a]

Major statin treatment recommendation	COR (LOE)
1. Clinical ASCVD (secondary prevention)	
i. High-intensity statin therapy should be initiated or continued as first-line therapy in women and men ≤75 years of age who have clinical ASCVD, unless contraindicated	I (A)
ii. In individuals with clinical ASCVD in whom high-intensity statin therapy would otherwise be used, when high-intensity statin therapy is contraindicated or when characteristics predisposing to statin-associated adverse effects are present, moderate-intensity statin should be used as the second option if tolerated	I (A)
iii. In individuals with clinical ASCVD >75 years of age, it is reasonable to evaluate the potential for ASCVD risk reduction benefits and for adverse effects and drug–drug interactions and to consider patient preferences when initiating a moderate- or high-intensity statin. It is reasonable to continue statin therapy in those who are tolerating it	IIa (B)
2. Primary prevention in individuals ≥21 years of age with LDL cholesterol ≥190 mg/dl	
i. Individuals with LDL cholesterol ≥190 mg/dl or triglycerides ≥500 mg/dl should be evaluated for secondary causes of hyperlipidemia	I (B)
ii. Adults ≥21 years of age with primary LDL cholesterol ≥190 mg/dl should be treated with statin therapy (10-year ASCVD risk estimation is not required): (i) using high-intensity statin therapy unless contraindicated and (ii), for individuals unable to tolerate high-intensity statin therapy, using the maximum tolerated statin intensity	I (B)
iii. For individuals ≥21 years of age with an untreated primary LDL cholesterol ≥190 mg/dl, it is reasonable to intensify statin therapy to achieve at least a 50% LDL-cholesterol reduction	IIa (B)
iv. For individuals ≥21 years of age with an untreated primary LDL cholesterol ≥190 mg/dl, after the maximum intensity of statin therapy has been achieved, addition of a nonstatin drug may be considered to further lower LDL cholesterol. Evaluate the potential for ASCVD risk reduction benefits, adverse effects, and drug–drug interactions, and consider patient preferences	IIb (C)
3. Primary prevention in individuals with diabetes mellitus and LDL cholesterol 70–189 mg/dl	
i. Moderate-intensity statin therapy should be initiated or continued for adults 40–75 years of age with diabetes mellitus	I (A)
ii. High-intensity statin therapy is reasonable for adults 40–75 years of age with diabetes mellitus with a ≥7.5% estimated 10-year ASCVD risk unless contraindicated	IIa (B)
iii. In adults with diabetes mellitus, who are <40 or >75 years of age, or with LDL cholesterol <70 mg/dl, it is reasonable to evaluate the potential for ASCVD benefits and for adverse effects and for drug–drug interactions and to consider patient preferences when deciding to initiate, continue, or intensify statin therapy	IIa (C)

(Continued)

TABLE 5.11 (*Continued*)

Major statin treatment recommendation	COR (LOE)
4. Primary prevention in individuals without diabetes mellitus and with LDL cholesterol 70–189 mg/dl	
i. The Pooled Cohort Equations should be used to estimate 10-year ASCVD risk for individuals with LDL cholesterol 70–189 mg/dl without clinical ASCVD to guide initiation of statin therapy for the primary prevention of ASCVD	I (B)
ii. (2) Adults 40–75 years of age with LDL cholesterol 70–189 mg/dl without clinical ASCVD or diabetes and an estimated 10-year ASCVD risk ≥7.5% should be treated with moderate- to high-intensity statin therapy	I (A)
iii. It is reasonable to offer treatment with a moderate-intensity statin to adults 40–75 years of age with LDL cholesterol 70–189 mg/dl without clinical ASCVD or diabetes and an estimated 10-year ASCVD risk of 5% to <7.5%	IIa (B)
iv. Before initiating statin therapy for the primary prevention of ASCVD in adults with LDL cholesterol 70–189 mg/dl without clinical ASCVD or diabetes, it is reasonable for clinicians and patients to engage in a discussion that considers the potential for ASCVD risk reduction benefits and for adverse effects and for drug–drug interactions and patient preferences for treatment	IIa (C)
v. In adults with LDL cholesterol < 190 mg/dl who are not otherwise identified in a statin benefit group or for whom after quantitative risk assessment a risk based treatment decision is uncertain, additional factors may be considered to inform treatment decision making. In these individuals, statin therapy for primary prevention may be considered after evaluating the potential for ASCVD risk reduction benefits, adverse effects, drug–drug interactions, and discussion of patient preferences	IIb (C)
5. No recommendations	
i. The guideline makes no recommendations for or against specific LDL-cholesterol or non-HDL-cholesterol targets for the primary or secondary prevention of ASCVD	
ii. The guideline makes no recommendations regarding the initiation or discontinuation of statins in patients with New York Heart Association (NYHA) class II–IV ischemic systolic heart failure or in patients on maintenance hemodialysis	

[a]Refs. 17.

clinical care, incorporating new evidence and their synthesis into practical recommendations for healthcare providers via the coordinated work of various guideline committees throughout the world is crucially important for the effective management of this major public health problem. In this context, any significant changes in a set of clinical practice guidelines would have far-reaching consequences, particularly if they appear to be at variance with the existing widely adopted guidelines. The newly released ACC/AHA guideline on the treatment of blood cholesterol is apparently a good example of such concern. Indeed, as noted earlier, the 2013 ACC/AHA guideline came with intense criticism immediately with its release. This section compares and contrasts the 2013 ACC/AHA guidelines with the widely accepted ATP III guideline and 2011 ESC/EAS guideline, and the newly released draft guideline from NICE to help the reader gain a better understanding of the 2013 ACC/AHA guideline on the treatment of blood cholesterol to reduce ASCVD risk.

5.3.4.1 2013 ACC/AHA Guideline versus ATP III Guideline
A quick comparison of the 2013 ACC/AHA guideline and the prior ATP III report identifies several key features for the new guideline:

- The 2013 ACC/AHA guideline on the treatment of blood cholesterol to reduce ASCVD in adults is not intended to provide for a comprehensive approach to the detection, evaluation, and treatment of lipid disorders as was done in the prior ATP III report.

- The recommendations in the new guideline are exclusively based on rigorous systematic evidence reviews of high-quality randomized controlled trials and meta-analysis of such trials.

- The new guideline focuses on ASCVD risk reduction in four statin benefit groups and focuses efforts to reduce ASCVD events in secondary and primary prevention. The new guideline identifies high-intensity and moderate-intensity statin therapy for use in secondary and primary prevention to maximize the benefit in reducing ASCVD morbidity and mortality.

- The new guideline provides new perspectives on LDL-cholesterol and/or non-HDL-cholesterol treatment goals and states that (i) the Expert Panel was unable to find randomized controlled trial evidence to support continued use of specific LDL-cholesterol and/or non-HDL-cholesterol treatment targets as recommended in the prior ATP III report, (ii) the appropriate intensity of statin therapy should be used to reduce ASCVD risk in those most likely to benefit, and (iii) nonstatin therapies do not provide acceptable ASCVD risk reduction benefits compared to their potential for adverse effects in the routine prevention of ASCVD.

- Instead of using the Framingham Score system used in the prior ATP III report, the new guideline recommends use of the new Pooled Cohort Equations to estimate 10-year ASCVD risk in both white and black men and women. By more accurately identifying higher-risk individuals for statin therapy, the guideline focuses statin therapy on those most likely to benefit.

- The new guideline uses randomized controlled trials to identify important safety considerations in individuals receiving treatment of blood cholesterol to reduce ASCVD risk. Using randomized controlled trials to determine statin adverse effects facilitates understanding of the net benefit from statin therapy. The new guideline provides expert guidance on management of statin-associated adverse effects, including muscle symptoms.

5.3.4.2 The 2013 ACC/AHA Guideline versus the 2011 ESC/EAS Guideline

The 2011 ESC/EAS guideline for the management of dyslipidemias is the most current guideline that provides recommendations to healthcare providers in Europe on effective management of dyslipidemias, including hypercholesterolemia, hypertriglyceridemia, and low HDL cholesterol [2]. Recently, K.K. Ray and coworkers compared the 2013 ACC/AHA guideline with this ESC/EAS guideline and identified several major differences with potential implications, as summarized below:

- The scope of the 2013 ACC/AHA guideline is limited to randomized trials only, which excludes a significant body of data and promotes essentially a statin-centric approach only. On the other hand, the 2011 ESC/EAS guideline considers all available evidence.

- The LDL-cholesterol goals are omitted from the 2013 ACC/AHA guideline in favor of specific statin regimens that produce a 30–50% reduction in LDL cholesterol. The target LDL-cholesterol levels in very high-risk patients with high absolute risk or residual risk factors are also absent in the 2013 ACC/AHA guideline. On the other hand, the 2011 ESC/EAS guideline categorizes people in moderate and very high cardiovascular risk with guidance on specific LDL-cholesterol targets for each level of absolute risk.

- The cardiovascular risk threshold for treatment in primary prevention is reduced from 20 to 7.5% in the 2013 ACC/AHA guideline, which would result in a greater number of patients being prescribed statin therapy. On the other hand, the above 7.5% risk corresponds to 2.5% risk in the Systemic Coronary Risk Estimation (SCORE) (see the following text) system used for risk estimation in the 2011 ESC/EAS guideline.

- The newly published mixed pool risk calculator [16] is used to assess cardiovascular risk in the 2013

ACC/AHA guideline for primary prevention. On the other hand, the SCORE system [31] is used in the 2011 ESC/EAS guideline for cardiovascular risk estimation.

Upon the above comparisons, K.K. Ray and coworkers concluded that the 2011 ESC/EAS guideline on dyslipidemia management appears to be the most appropriate for European countries and that the 2013 ACC/AHA guideline would result in a much larger proportion of patients being treated with statins and especially at higher doses [32]. In this regard, a recent study reported that with application of the 2013 ACC/AHA guideline in a healthy European population-based cohort, nearly all men and the majority of women aged 55 years or older were candidates for drug treatment [28].

5.3.4.3 The 2013 ACC/AHA Guideline versus the 2014 NICE Guideline

The new JBS consensus recommendations for the prevention of cardiovascular disease (JBS3) [33] and its closely linked draft document on cardiovascular risk assessment and modification of blood lipids for the primary and secondary prevention of cardiovascular disease from the NICE were recently released (Table 5.5). The new NICE lipid guideline was published only 4 months after the release of the 2013 ACC/AHA guideline on risk assessment and lipid lowering, which led to comparisons of these two most recent lipid guidelines [34]. It is striking that these two guidelines have considerable agreement with regard to principles of risk calculation, use of statins, and drug treatment versus lifestyle interventions [32]. Of particular note, both guidelines give high priority to use of statins for risk reduction in comparison to other less proven lipid-lowering drugs for cardiovascular risk reduction. This should come as no surprise since they are both based on the same scientific foundation. The fact that totally different committees, in different societies, working separately and independently, came to nearly identical conclusions and recommendations suggests that bias did not greatly play a role in either guideline's conclusions [34].

5.3.4.4 Summary of Comparison and Contrast of Current Lipid Guidelines

- The 2013 ACC/AHA guideline on the treatment of blood cholesterol to reduce ASCVD risk has encountered intense criticism since its release in November 2013. The criticism results primarily from the new recommendations being remarkably different in several key aspects from the existing widely accepted guidelines, such as the prior ATP III report and the 2011 ESC/EAS guideline for the management of dyslipidemias. On the other hand, the newly released 2014

NICE lipid guideline and the 2013 ACC/AHA guideline have much in common.

- It has also been recognized that the 2013 ACC/AHA guideline represents a strong fit with the clinical trial evidence and its implementation would likely maximize the benefit of blood cholesterol treatment in individuals and improve the health of the public. In this context, the goal of therapy is not the achievement of a target level of LDL cholesterol, an approach that previously might have required multiple drugs. The primary aim of drug therapy is the use of a first-line drug (such as a statin) among those likely to benefit, an approach designed to improve the health of the populations [35].

5.4   SUMMARY OF CHAPTER KEY POINTS

- Although dyslipidemias can be manifested as any changes in the plasma lipid profiles, elevation of total cholesterol and LDL cholesterol has received most attention.
- Effective management of dyslipidemias, especially hypercholesterolemia, requires implementation of a comprehensive approach (frequently in the form of clinical practice guidelines) that addresses all the key aspects of the disorders, ranging from risk assessment to selection of drug therapy.
- Clinical practice guidelines are defined as statements that include recommendations intended to optimize patient care that are informed by a systematic review of evidence and an assessment of the benefits and harms of alternative care options.
- Multiple guidelines on blood lipid management have been made available for healthcare providers to consider in treating individuals with dyslipidemias. These include the most widely adopted ATP III guideline (released in 2001, published in 2002, and updated in 2004) on detection, evaluation, and treatment of high blood cholesterol in adults; the 2011 ESC/EAS guideline for the management of dyslipidemias; and the most recently released 2013 ACC/AHA guideline on the treatment of blood cholesterol to reduce atherosclerotic cardiovascular risk in adults.
- The 2013 ACC/AHA guideline (informally known as the ATP IV guideline) arose from careful consideration of an extensive body of higher-quality evidence derived from randomized controlled trials and systematic reviews and meta-analyses of such trials. This new guideline, though facing intense criticism, represents a strong fit with the clinical trial evidence and emphasizes the use of statin therapy in those likely to benefit with regard to reducing cardiovascular events and mortality.

5.5   SELF-ASSESSMENT QUESTIONS

5.5.1. A 55-year-old healthy male presents to the doctor's office for an annual checkup. Lab test reveals that his plasma LDL cholesterol is 95 mg/dl. He has a family history of premature cardiovascular disease. According to the ATP III guideline, which of the following best describes his LDL-cholesterol level?
A. Borderline high
B. High
C. Near optimal
D. Optimal
E. Very high

5.5.2. A 60-year-old diabetic patient presents to the physician's office for a routine checkup. His blood lipid profiles are within normal range, with LDL cholesterol of 89 mg/dl. He denies any history of adverse drug reactions. He is without clinical atherosclerotic cardiovascular disease though one family member died of myocardial infarction at the age of 59 years. According to the 2013 ACC/AHA guideline on treating blood cholesterol to reduce atherosclerotic cardiovascular disease, which of the following should be prescribed to the patient to reduce the risk of cardiovascular events?
A. Atorvastatin 40 mg, once daily
B. Ezetimibe 10 mg, once daily
C. Icosapent ethyl (Vascepa) 1 g, once daily
D. Niacin (Niaspan) 1000 mg, once daily
E. Phytosterol/phytostanol esters 2 g, once daily

5.5.3. A 47-year-old male with clinical atherosclerotic cardiovascular disease presents to the doctor's office asking for advice regarding how to reduce the risk of developing myocardial infarction. He states that his brother recently died of myocardial infarction at the age of 54 years. According to the 2013 ACC/AHA guideline on treating blood cholesterol to reduce atherosclerotic cardiovascular risk in adults in addition to adhering to a healthy lifestyle, which of the following is the most appropriate treatment for the patient?
A. Atorvastatin 10 mg, once daily
B. Fluvastatin 20 mg, once daily
C. Lovastatin 10 mg, once daily
D. Pravastatin 20 mg, once daily
E. Rosuvastatin 20 mg, once daily

5.5.4. The 2013 ACC/AHA guideline (ATP IV) on the treatment of blood cholesterol to reduce atherosclerotic cardiovascular risk in adults has encountered intense criticism since its release in November 2013. The criticism results primarily from the new recommendations being remarkably different in several key aspects from the existing widely accepted ATP III report. Which of the following is a major difference between the ATP IV and ATP III guidelines?

A. Absence of a risk assessment calculator in ATP IV
B. Absence of clinician–patient discussion in ATP IV
C. Absence of LDL-cholesterol treatment goals in ATP IV
D. Absence of lifestyle modifications in ATP IV
E. Absence of primary prevention in ATP IV

5.5.5. A patient with a 5-year history of stable angina is found to have an elevated LDL-cholesterol level of 178 mg/dl. He is prescribed simvastatin (20 mg, once daily) to reduce his LDL cholesterol. According to ATP III guideline, which of the following best describes the LDL-cholesterol goal in this patient?

A. <100 mg/dl
B. <120 mg/dl
C. <130 mg/dl
D. <150 mg/dl
E. <160 mg/dl

REFERENCES

1   Grundy, S.M., et al., Implications of recent clinical trials for the National Cholesterol Education Program Adult Treatment Panel III guidelines. *Circulation*, 2004. 110(2): p. 227–39.

2   European Association for Cardiovascular Prevention & Rehabilitation, et al., ESC/EAS Guidelines for the management of dyslipidaemias: the Task Force for the management of dyslipidaemias of the European Society of Cardiology (ESC) and the European Atherosclerosis Society (EAS). *Eur Heart J*, 2011. 32(14): p. 1769–818.

3   Gaziano, J.M. and T.A. Gaziano, What's new with measuring cholesterol? *JAMA*, 2013. 310(19): p. 2043–4.

4   Martin, S.S., et al., Comparison of a novel method vs the Friedewald equation for estimating low-density lipoprotein cholesterol levels from the standard lipid profile. *JAMA*, 2013. 310(19): p. 2061–8.

5   Martin, S.S., et al., Friedewald-estimated versus directly measured low-density lipoprotein cholesterol and treatment implications. *J Am Coll Cardiol*, 2013. 62(8): p. 732–9.

6   Robinson, J.G., et al., Meta-analysis of the relationship between non-high-density lipoprotein cholesterol reduction and coronary heart disease risk. *J Am Coll Cardiol*, 2009. 53(4): p. 316–22.

7   Boekholdt, S.M., et al., Association of LDL cholesterol, non-HDL cholesterol, and apolipoprotein B levels with risk of cardiovascular events among patients treated with statins: a meta-analysis. *JAMA*, 2012. 307(12): p. 1302–9.

8   Blaha, M.J. and R.S. Blumenthal, Risk factors: new risk-assessment guidelines-more or less personalized? *Nat Rev Cardiol*, 2014. 11(3): p. 136–7.

9   Goff, D.C., Jr., et al., 2013 ACC/AHA guideline on the assessment of cardiovascular risk: a report of the American College of Cardiology/American Heart Association Task Force on Practice Guidelines. *J Am Coll Cardiol*, 2014. 63(25 Pt B): p. 2935–59.

10 Ridker, P.M. and N.R. Cook, Statins: new American guidelines for prevention of cardiovascular disease. *Lancet*, 2013. 382(9907): p. 1762–5.

11 Stone, N.J., et al., 2013 ACC/AHA guideline on the treatment of blood cholesterol to reduce atherosclerotic cardiovascular risk in adults: a report of the American College of Cardiology/American Heart Association Task Force on Practice Guidelines. *J Am Coll Cardiol*, 2014. 63(25 Pt B): p. 2889–934.

12 National Cholesterol Education Program Expert Panel on Detection, E. and A. Treatment of High Blood Cholesterol in, Third Report of the National Cholesterol Education Program (NCEP) Expert Panel on Detection, Evaluation, and Treatment of High Blood Cholesterol in Adults (Adult Treatment Panel III) final report. *Circulation*, 2002. 106(25): p. 3143–421.

13 Cholesterol Treatment Trialists Collaboration, et al., Efficacy and safety of more intensive lowering of LDL cholesterol: a meta-analysis of data from 170,000 participants in 26 randomised trials. *Lancet*, 2010. 376(9753): p. 1670–81.

14 Eckel, R.H., et al., 2013 AHA/ACC guideline on lifestyle management to reduce cardiovascular risk: a report of the American College of Cardiology/American Heart Association Task Force on Practice Guidelines. *J Am Coll Cardiol*, 2014. 63(25 Pt B): p. 2960–84.

15 Mitka, M., Groups aim for trustworthy clinical practice guidelines. *JAMA*, 2014. 311(12): p. 1187–8.

16 Goff, D.C., Jr., et al., 2013 ACC/AHA guideline on the assessment of cardiovascular risk: a report of the American College of Cardiology/American Heart Association Task Force on Practice Guidelines. *Circulation*, 2014. 129(25 Suppl 2): p. 2935–59.

17 Stone, N.J., et al., 2013 ACC/AHA guideline on the treatment of blood cholesterol to reduce atherosclerotic cardiovascular risk in adults: a report of the American College of Cardiology/American Heart Association Task Force on Practice Guidelines. *Circulation*, 2014. 129(25 Suppl 2): p. S1–45.

18 Jensen, M.D., et al., 2013 AHA/ACC/TOS guideline for the management of overweight and obesity in adults: a report of the American College of Cardiology/American Heart Association Task Force on Practice Guidelines and The Obesity Society. *Circulation*, 2014. 129(25 Suppl 2): p. S102–38.

19 Genest, J., et al., 2009 Canadian Cardiovascular Society/Canadian guidelines for the diagnosis and treatment of dyslipidemia and prevention of cardiovascular disease in the adult—2009 recommendations. *Can J Cardiol*, 2009. 25(10): p. 567–79.

20 Brown, D.J., New guidelines for low-density lipoprotein levels from the National Cholesterol Education Program (NCEP): a 2004 update. *Prog Cardiovasc Nurs*, 2004. 19(4): p. 165.

21 Stone, N.J., et al., 2013 ACC/AHA guideline on the treatment of blood cholesterol to reduce atherosclerotic cardiovascular risk in adults: a report of the American College of Cardiology/American Heart Association Task Force on Practice Guidelines. *J Am Coll Cardiol*, 2014. 63(25 pt B): p. 2889–934.

22 Stone, N.J., et al., Treatment of blood cholesterol to reduce atherosclerotic cardiovascular disease risk in adults: synopsis

of the 2013 ACC/AHA cholesterol guideline. *Ann Intern Med*, 2014. 160(5): p. 339–43.

23 Keaney, J.F., Jr., G.D. Curfman, and J.A. Jarcho, A pragmatic view of the new cholesterol treatment guidelines. *N Engl J Med*, 2014. 370(3): p. 275–8.

24 Ioannidis, J.P., et al., Biologic agents in rheumatology: unmet issues after 200 trials and $200 billion sales. *Nat Rev Rheumatol*, 2013. 9(11): p. 665–73.

25 Ginsberg, H.N., The 2013 ACC/AHA guidelines on the treatment of blood cholesterol: questions, questions, questions. *Circ Res*, 2014. 114(5): p. 761–4.

26 Breslow, J.L., Perspective on the 2013 American Heart Association/American College of Cardiology guideline for the use of statins in primary prevention of low-risk individuals. *Circ Res*, 2014. 114(5): p. 758–60.

27 Pencina, M.J., et al., Application of New Cholesterol Guidelines to a Population-Based Sample. *N Engl J Med*, 2014. 370(15): p. 1422–31.

28 Kavousi, M., et al., Comparison of application of the ACC/AHA guidelines, Adult Treatment Panel III Guidelines, and European Society of Cardiology guidelines for cardiovascular disease prevention in a European cohort. *JAMA*, 2014. 311(14): p. 1416–23.

29 Vaucher, J., et al., Population and economic impact of the 2013 ACC/AHA guidelines compared with European guidelines to prevent cardiovascular disease. *Eur Heart J*, 2014. 35(15): p. 958–9.

30 Muntner, P., et al., Validation of the atherosclerotic cardiovascular disease Pooled Cohort risk equations. *JAMA*, 2014. 311(14): p. 1406–15.

31 Conroy, R.M., et al., Estimation of ten-year risk of fatal cardiovascular disease in Europe: the SCORE project. *Eur Heart J*, 2003. 24(11): p. 987–1003.

32 Ray, K.K., et al., The ACC/AHA 2013 guideline on the treatment of blood cholesterol to reduce atherosclerotic cardiovascular disease risk in adults: the good the bad and the uncertain: a comparison with ESC/EAS guidelines for the management of dyslipidaemias 2011. *Eur Heart J*, 2014. 35(15): p. 960–8.

33 Board, J.B.S., Joint British Societies' consensus recommendations for the prevention of cardiovascular disease (JBS3). *Heart*, 2014. 100(Suppl 2): p. ii1–ii67.

34 Greenland, P., British and American prevention guidelines: different committees, same science, considerable agreement. *Heart*, 2014. 100(9): p. 678–9.

35 Psaty, B.M. and N.S. Weiss, 2013 ACC/AHA guideline on the treatment of blood cholesterol: a fresh interpretation of old evidence. *JAMA*, 2014. 311(5): p. 461–2.

UNIT III

HYPERTENSION AND MULTITASKING CARDIOVASCULAR DRUGS

6

OVERVIEW OF HYPERTENSION AND DRUG THERAPY

6.1 INTRODUCTION

Hypertension is the most common condition seen in primary care settings and leads to myocardial infarction, stroke, renal failure, and death if not detected early and treated appropriately. Advances in risk factor identification, molecular pathophysiology, and mechanistically based therapies, especially pharmacological agents, have led to significant improvement of hypertension management. To set a stage for the subsequent discussion of the molecular pharmacology of antihypertensive drugs and evidence-based therapeutics of hypertension in the remaining chapters of Unit III, this chapter provides an overview on several aspects of hypertension, including definition, epidemiology, pathophysiology, and mechanistically based drug therapy. Because the different drug classes used for treating hypertension are also commonly employed in the management of other cardiovascular diseases, these multitasking cardiovascular drugs are considered in separate chapters (Chapters 7–11). Following the discussion of these multitasking drugs, the current guidelines on hypertension management are given in Chapter 12.

6.2 DEFINITIONS, CLASSIFICATIONS, AND EPIDEMIOLOGY OF HYPERTENSION

Hypertension, when not specified, typically refers to increased blood pressure in systemic circulation and hence is also known as systemic arterial hypertension. In addition to systemic circulation, hypertension also occurs in pulmonary and portal circulations. This section defines hypertension and provides an overview of its epidemiology.

6.2.1 Definitions and Classifications

6.2.1.1 The JNC7 Classification of Blood Pressure Hypertension, also known as high blood pressure, is defined as systolic pressure ≥140 mm Hg and/or diastolic pressure ≥90 mm Hg. Based on the Seventh Report of the Joint National Committee on the Prevention, Detection, Evaluation, and Treatment of High Blood Pressure (JNC7), blood pressure is classified into four categories (Table 6.1). According to JNC7, hypertension is further classified into two stages, that is, stage 1 and stage 2.

6.2.1.2 Hypertension Crisis Hypertension crisis is a term that refers to critical clinical conditions of severely elevated blood pressure. Hypertensive crisis can present as hypertensive urgency or as hypertensive emergency.

Hypertensive urgency is a situation where the blood pressure is severely elevated (≥180 mm Hg for systolic pressure or ≥110 mm Hg for diastolic pressure), but there is no associated organ damage. Individuals experiencing hypertensive urgency may or may not experience one or more of these symptoms: severe headache, shortness of breath, nosebleeds, and severe anxiety.

The term hypertensive emergency refers to severely elevated blood pressure levels that are damaging organs, which can be manifested as stroke, myocardial infarction, renal failure, and loss of consciousness. Hypertensive emergencies generally occur at blood pressure levels exceeding 180 mm

Cardiovascular Diseases: From Molecular Pharmacology to Evidence-Based Therapeutics, First Edition. Y. Robert Li.
© 2015 John Wiley & Sons, Inc. Published 2015 by John Wiley & Sons, Inc.

TABLE 6.1 The JNC7 classification of blood pressure

Blood pressure classification	Systolic blood pressure (mm Hg)	Diastolic blood pressure (mm Hg)
Normal	<120	and <80
Prehypertension	120–139	or 80–89
Stage 1 hypertension	140–159	or 90–99
Stage 2 hypertension	≥160	or ≥100

TABLE 6.2 Major mechanisms of secondary hypertension

Cause	Major mechanism
Kidney diseases	Increased renin release due to decreased perfusion
Primary aldosteronism	Water and salt retention due to increased levels of aldosterone
Cushing's syndrome	Water and salt retention due to increased release of corticosteroids
Pheochromocytoma	Release of catecholamines from the tumors of the adrenal chromaffin cells

Hg (systolic) or 120 mm Hg (diastolic) but can also occur at even lower levels in patients whose blood pressure had not been previously high.

6.2.1.3 Primary and Secondary Hypertension

Based on etiology, hypertension is classified into primary and secondary hypertension. In 90–95% of hypertensive patients, a single reversible cause of the elevated blood pressure cannot be identified, and hence, the patients are said to have primary (or essential) hypertension. In the remaining 5–10% of cases, a cause can be identified, and hence, the patients are said to have secondary hypertension. Common causes of secondary hypertension are listed below, and the underlying mechanisms are given in Table 6.2:

- Kidney diseases, especially chronic kidney disease [1]
- Primary aldosteronism [2]
- Cushing's syndrome [3]
- Pheochromocytoma [4]

6.2.1.4 Systemic and Pulmonary Hypertension

The term hypertension is often used synonymously with systemic arterial hypertension. Systemic arterial hypertension is the predominant form of hypertension. It should be kept in mind that hypertension can also occur in pulmonary and portal circulations. Pulmonary hypertension refers to an abnormal elevation in pulmonary artery pressure. It may be the result of left-side heart failure, pulmonary parenchymal or vascular disease, thromboembolism, or a combination of these factors [5]. Pulmonary hypertension is the most common cause of right ventricular enlargement and failure.

6.2.2 Epidemiology and Health Impact of Hypertension

6.2.2.1 Epidemiology

Hypertension currently affects approximately 78 million American adults (one in three adults) and over one billion people worldwide. Many risk factors for development of hypertension have been identified, including age, ethnicity, family history of hypertension and genetic factors, lower education and socioeconomic status, greater weight, lower physical activity, tobacco use, psychosocial stressors, sleep apnea, and dietary factors (including dietary fats, higher sodium intake, lower potassium intake, and excessive alcohol intake) [6]. Notably, the prevalence of hypertension in blacks in the United States is among the highest in the world; 41% of the total black population and 44% of black women have hypertension. Moreover, a recent study reported that blacks, compared with whites, also appear to be more susceptible to stroke, given the same level of elevated blood pressure. Blacks have more hypertension and are less likely to have it controlled, and when it is not controlled, they are at greater risk for incident stroke (relative to whites with the same blood pressure levels) [7].

6.2.2.2 Health Impact

Hypertension remains the most common, readily identifiable, and reversible risk factor for myocardial infarction, stroke, heart failure, and kidney diseases. Hypertension exerts remarkable health impact at both individual and population levels. For example, a 50-year-old man of normal body mass with blood pressure of 146/86 mm Hg has:

- Almost 3 times the risk of dying from a heart attack
- Almost 4 times the risk of dying from a stroke
- About twice the risk of developing heart failure
- About 3 times the risk of developing kidney disease than if he had normal blood pressure (<120/80 mm Hg)

Similarly, a 40-year-old woman of normal body mass with blood pressure of 146/86 mm Hg has:

- More than 3 times the risk of dying from a heart attack
- Almost 4 times the risk of dying from a stroke
- About 3 times the risk of developing heart failure
- About 3 times the risk of developing kidney disease than if she had normal blood pressure (<120/80 mm Hg)

Approximately 69% of people who have a first myocardial infarction, 77% of those who have a first stroke, and 74% of those who have congestive heart failure have blood pressure >140/90 mm Hg. Compared with hypertensive individuals at 50 years of age, people with untreated blood pressure <140/90 mm Hg survive on average 7 years longer without cardiovascular diseases [6].

In addition to the 78 million hypertensive patients in the United States, more than 36% of the American adults have prehypertension. Although prehypertension is not hypertension, it is associated with stroke [8] and estimated to decrease life expectancy by approximately 5 years [9].

6.3 PATHOPHYSIOLOGY OF HYPERTENSION

In order to understand the molecular pathophysiology of hypertension, it is necessary to first examine the physiology of blood pressure regulation.

6.3.1 Physiology of Blood Pressure Regulation

Blood pressure is the product of cardiac output and peripheral resistance (Fig. 6.1). Therefore, factors that affect either the cardiac output or the peripheral resistance will impact blood pressure.

As blood pressure is vital, its regulation is tightly controlled with the coordinated involvement of many organs and pathways. When blood pressure drops, several systems or pathways become activated to cause increased cardiac output and peripheral resistance and the consequent recovery of the blood pressure.

6.3.1.1 *Sympathetic Nervous System*   A decrease in blood pressure activates the baroreceptor reflex, causing increased activities of sympathetic nervous system. The activation of sympathetic nervous system results in recovery of blood pressure via the following three major mechanisms [10]:

1. Increased heart rate and contractility, leading to augmented cardiac output

2. Constriction of blood vessels, leading to increased peripheral resistance
3. Activation of β_1-adrenergic receptors on renal juxtaglomerular cells, causing increased release of renin and the subsequent activation of the renin–angiotensin–aldosterone system (RAAS) (see Section 6.3.1.2)

6.3.1.2 *RAAS*   As noted earlier, activation of the sympathetic nervous system due to blood pressure drop increases the release of renin and the subsequent activation of the RAAS. Hence, activation of sympathetic nervous system is always coupled with the concomitant activation of the RAAS. In addition, increased renin release also occurs as a result of the decreased renal arterial pressure. When the RAAS becomes activated, the following events result, contributing to the recovery of blood pressure:

- Constriction of arterials caused by angiotensin II, leading to increased peripheral resistance
- Increased water and salt retention due to increased levels of aldosterone, leading to increased blood volume and hence the cardiac output

6.3.1.3 *Vasopressin System*   A decrease in arterial pressure causes increased release of vasopressin via baroreflex mechanism. This results in increased water retention and thereby increased blood volume and cardiac output. The increased blood volume and cardiac output lead to elevation of blood pressure.

6.3.1.4 *Fluid Retention by the Kidney*   A decrease in renal arterial pressure also causes less excretion of water and salt by the kidney. This increases blood volume and cardiac output.

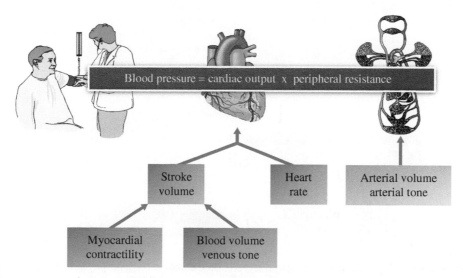

FIGURE 6.1 Physiological determinants of blood pressure. As illustrated, blood pressure is the product of cardiac output and peripheral resistance. Factors that affect either cardiac output or peripheral resistance will inevitably affect blood pressure.

6.3.1.5 Summary of Physiology of Blood Pressure Regulation In summary, a drop of blood pressure activates the aforementioned various systems and pathways, leading to recovery of blood pressure due to increases in cardiac output and peripheral resistance. Hence, the coordinated actions of the various systems and pathways ensure the maintenance of the blood pressure within the physiological ranges. Hypertension develops when these homeostatic processes are disrupted by pathophysiological factors or due to other unknown reasons.

6.3.2 Molecular Pathophysiology of Hypertension Development

6.3.2.1 Overview As noted earlier, 90–95% of hypertension cases are without identifiable causes. Mechanistic studies over the last decades have revealed a critical involvement of both sympathetic nervous system and the RAAS in the genesis and progression of hypertension [10, 11]. Indeed, these two systems serve as targets of many current antihypertensive drugs. In addition, endothelial dysfunction and oxidative stress also serve as important pathophysiological mechanisms of hypertension (Fig. 6.2).

6.3.2.2 Crosstalk Notably, the aforementioned mechanisms are closely related and there is extensive crosstalk between them. For example, as mentioned earlier, activation of sympathetic nervous system causes activation of the RAAS. RAAS activation also increases the activity of the sympathetic nervous system. Activation of the RAAS contributes to endothelial dysfunction and vascular remodeling. Endothelial dysfunction characterized by decreased nitric oxide bioavailability and increased formation of endothelin

plays an important role in the development and progression of hypertension.

6.3.2.3 Oxidative Stress as a Potential Final Common Pathway Activation of sympathetic nervous system and RAAS also results in increased formation of reactive oxygen species (ROS). On the one hand, ROS directly cause damage and dysfunction of the endothelium. On the other hand, ROS, such as superoxide, react with nitric oxide, forming more reactive species, and at the same time decreasing the bioavailability of nitric oxide, leading to endothelial dysfunction. In both experimental models and human subjects, administration of antioxidant compounds has been shown to have antihypertensive effects via decreasing ROS formation or increasing the levels of nitric oxide. Accordingly, antioxidant compounds are emerging therapeutic modalities for hypertension [12].

6.3.2.4 Summary of Molecular Pathophysiology of Hypertension In summary, the major pathophysiological mechanisms of hypertension include the following four pathways and events:

1. Activation of the sympathetic nervous system
2. Activation of the RAAS
3. Endothelial dysfunction
4. Oxidative stress

The aforementioned mechanisms are intertwined with sympathetic and RAAS activation as major pathophysiological components. Indeed, as described next, many current antihypertensive medications mainly act to reduce the activities of the sympathetic nervous system and RAAS.

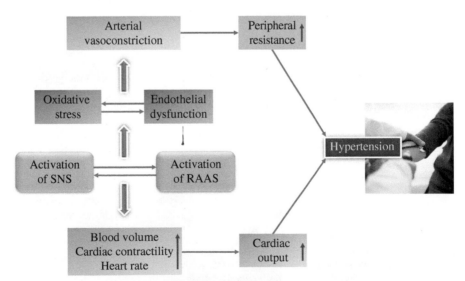

FIGURE 6.2 Pathophysiology of hypertension development. Hypertension develops when the physiological homeostasis of blood pressure is disrupted, leading to uncontrolled activation of sympathetic nervous system and RAAS, endothelial dysfunction, and oxidative stress. RAAS, renin–angiotensin–aldosterone system; SNS, sympathetic nervous system.

6.4 MECHANISTICALLY BASED DRUG THERAPY OF HYPERTENSION: AN OVERVIEW

6.4.1 Historical Overview

The treatment of hypertension has been one of medicine's major successes of the past half century. Since the introduction of the thiazide diuretics in the late 1950s, many classes of antihypertensive drugs have been developed and approved for clinical use (Table 6.3) [13]. Five of these, namely, diuretics, β-blockers, angiotensin-converting enzyme inhibitors, calcium channel blockers, and angiotensin receptor blockers, now represent the primary treatment options for hypertension [14, 15].

6.4.2 Mechanistic Overview

As blood pressure is determined by cardiac output and peripheral resistance, antihypertensive drugs, regardless of their chemical classes, achieve their blood pressure-lowering effects by eventually decreasing either cardiac output or peripheral resistance or both. As listed below and further described in Table 6.4, there are four major categories of antihypertensive agents:

1. Diuretics
2. Sympatholytics
3. RAAS inhibitors
4. Direct vasodilators

Figure 6.3 illustrates the acting sites of these four classes of antihypertensive drugs along the physiological and pathophysiological pathways of blood pressure regulation and hypertension development. The above four categories of antihypertensive agents are also useful in treating various other types of cardiovascular disorders, including ischemic heart disease (see Units IV and V), heart failure (see Unit VI), and cardiac arrhythmias (see Unit VII). Hence, these agents can be considered as multitasking cardiovascular drugs. In addition, many of these drugs are also indicated for disorders other than cardiovascular diseases. The subsequent chapters discuss each of these categories of drugs with a focus on their applications in treating hypertension. Their use in treating other cardiovascular diseases is further covered in other units of the book.

6.5 SUMMARY OF CHAPTER KEY POINTS

- Hypertension, also called high blood pressure, currently affects approximately 78 million American adults (one in three adults) and over one billion people worldwide. It is the most common, readily identifiable, and reversible risk factor for myocardial infarction, stroke, heart failure, and kidney diseases.
- Hypertension is defined by JNC7 as systolic pressure ≥140 mm Hg and/or diastolic pressure ≥90 mm Hg. Based on the extent of blood pressure elevation, hypertension is further divided into stage I and stage II

TABLE 6.3   Historical overview of the development of drug therapy for hypertension[a]

Decade	Drug therapy of hypertension
1940s	• Potassium thiocyanate • Kempner diet (rice and fruits; low in calories, fat, protein, and sodium) • Lumbodorsal sympathectomy
1950s	• *Rauvolfia serpentina* (a plant containing a number of bioactive chemicals, including yohimbine, reserpine, ajmaline, deserpidine, rescinnamine, serpentinine) • Ganglionic blockers • Veratrum alkaloids • Hydralazine • Guanethidine • Thiazide diuretics
1960s	• α_2-Adrenergic receptor agonists • Spironolactone • β-Adrenergic receptor antagonists
1970s	• α_1-Adrenergic receptor antagonists • Angiotensin-converting enzyme inhibitors
1980s	• Calcium channel blockers
1990s	• Angiotensin receptor blockers • Endothelin receptor blockers (for pulmonary hypertension)
2000s	• Renin inhibitors (aliskiren)

[a] Modified from reference [13].

TABLE 6.4 Four major categories of antihypertensive drugs and their mechanisms of action

Drug category	Mechanism of action

Diuretics
  • Thiazides and related diuretics
  • Loop diuretics
  • Potassium-sparing diuretics
Sympatholytics
  • α-Blockers
  • β-Blockers
  • Centrally acting drugs
	• Diuretics decrease plasma volume, thereby leading to the reduced cardiac output
	• Diuretics, especially thiazide diuretics, also reduce peripheral resistance
	• Potassium-sparing diuretics (eplerenone and spironolactone) block aldosterone receptors (also known as mineralocorticoid receptors)
	• Blockage of peripheral α-receptors causes peripheral vasodilation and thereby decreased peripheral resistance
	• Blockage of β_1 receptors decreases heart rate and stroke volume, leading to decreased cardiac output
	• Blockage of β_1 receptors decreases renin release, thereby leading to decreased activation of the RAAS
	• Centrally acting agents decrease the sympathetic outflow from brain to peripheral tissues, leading to decreased peripheral resistance and reduced activation of the RAAS

RAAS inhibitors
  • Angiotensin-converting enzyme inhibitors (ACEIs)
  • Angiotensin receptor blockers (ARBs)
  • Renin inhibitors
  • Aldosterone receptor blockers
Direct vasodilators
  • Calcium channel blockers (CCBs)
  • Other direct vasodilators
	• ACEIs decrease the formation of angiotensin II and thereby reduce the angiotensin II-mediated vasoconstriction and aldosterone production
	• ARBs block the effects of angiotensin II on vasculature by blocking the angiotensin type 1 receptors
	• Renin inhibitors decrease the conversion of angiotensinogen to angiotensin I, thereby leading to decreased activity of the RAAS
	• Calcium channel blockers cause vasodilation, leading to decreased peripheral resistance
	• Some CCBs also suppress cardiac contractility, decreasing cardiac output
	• Other direct vasodilators cause vasodilation, reducing peripheral resistance

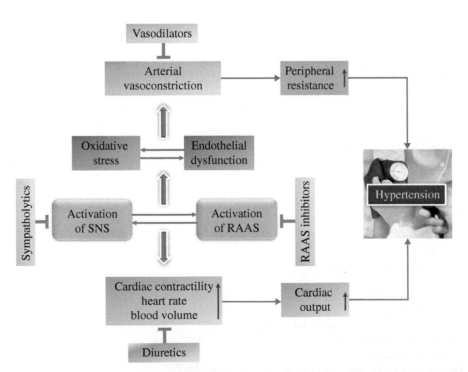

FIGURE 6.3 Mechanistically based drug targeting in hypertension treatment. Depicted are the acting sites of the four categories of antihypertensive drugs along the pathophysiological process of hypertension development. RAAS, renin–angiotensin–aldosterone system; SNS, sympathetic nervous system.

hypertension. The term prehypertension is defined as systolic pressure of 120–139 mm Hg or diastolic pressure of 80–89 mm Hg.

- Based on the part of circulation involved, hypertension is classified as systemic hypertension (also called systemic arterial hypertension) and pulmonary hypertension. Based on etiology, hypertension is classified into essential and secondary hypertension.

- The major pathophysiological mechanisms of hypertension include activation of sympathetic nervous system, activation of the RAAS, endothelial dysfunction, and oxidative stress, with sympathetic and RAAS activation as two key pathophysiological components. Many antihypertensive medications mainly act to reduce the activities of sympathetic nervous system and RAAS.

- Currently, there are four major categories of antihypertensive drugs that act on the various steps of the pathophysiological process of hypertension, which include diuretics, sympatholytics, RAAS inhibitors, and vasodilators. These categories of drugs are also useful for treating other cardiovascular disorders and hence named here multitasking cardiovascular drugs. The molecular pharmacology of these drugs in treating hypertension as well as other cardiovascular diseases is covered in Chapters 7–11.

6.6   SELF-ASSESSMENT QUESTIONS

6.6.1. A 32-year-old female presents to the physician's office complaining of frequent headache and dizziness. Her blood pressure is about 150/90 mm Hg upon repeated measurements. She is otherwise healthy. Which of the following is the best diagnosis for her condition?
   A. Hypertension emergency
   B. Prehypertension
   C. Pulmonary hypertension
   D. State 1 hypertension
   E. State 2 hypertension

6.6.2. Hypertension currently affects about 78 million US adults. Among the US populations, which of the following have the highest prevalence of hypertension?
   A. Asian males
   B. Black females
   C. Black males
   D. Hispanic females
   E. White females

6.6.3. A 48-year-old male is diagnosed with primary aldosteronism. His blood pressure is consistently in the range of 140–150/90–100 mm Hg. He is otherwise healthy. Which of the following is most likely responsible for his high blood pressure?

   A. Increased activation of renal β_1-adrenergic receptors
   B. Increased heart rate
   C. Increased myocardial contraction
   D. Increased renal perfusion
   E. Increased water and salt retention

6.6.4. A 52-year-old male is brought to the emergency department because of severe headache, shortness of breath, and anxiety. Blood pressure is about 190/110 mm Hg upon repeated measurements. He is otherwise normal without evidence of organ damage. Which of the following best describes the patient's condition?
   A. Hypertension emergency
   B. Hypertension urgency
   C. Pulmonary hypertension
   D. Secondary hypertension
   E. State 1 hypertension

6.6.5. A recent meta-analysis of prospective studies (762,393 participants from 19 prospective cohort studies) shows a clear association between blood pressure ≥120/80 mm Hg and stroke, with significantly increased risk even in the 120–129/80–84 mm Hg range. Which of the following best describes the conclusion of this meta-analysis?
   A. Control of prehypertension is unlikely associated with decreased risk of stroke.
   B. Hypertension or high blood pressure increases the risk of stroke.
   C. Increase of blood pressure within the normal range is associated with increased risk of stroke.
   D. Prehypertension is associated with increased risk of stroke.
   E. Stroke increases the risk of prehypertension.

REFERENCES

1 Razzak, M., Hypertension: understanding baroreflex dysfunction in chronic kidney disease. *Nat Rev Nephrol*, 2014. 10(3): p. 124.

2 Tomaschitz, A., et al., Aldosterone and arterial hypertension. *Nat Rev Endocrinol*, 2010. 6(2): p. 83–93.

3 Cicala, M.V. and F. Mantero, Hypertension in Cushing's syndrome: from pathogenesis to treatment. *Neuroendocrinology*, 2010. 92(Suppl 1): p. 44–9.

4 Lenders, J.W., et al., Phaeochromocytoma. *Lancet*, 2005. 366(9486): p. 665–75.

5 Shah, S.J., Pulmonary hypertension. *JAMA*, 2012. 308(13): p. 1366–74.

6 Go, A.S., et al., Heart disease and stroke statistics—2014 update: a report from the American Heart Association. *Circulation*, 2014. 129(3): p. e28–292.

7 Howard, G., et al., Racial differences in the impact of elevated systolic blood pressure on stroke risk. *JAMA Intern Med*, 2013. 173(1): p. 46–51.

8   Huang, Y., et al., Prehypertension and the risk of stroke: a meta-analysis. *Neurology*, 2014. 82(13): p. 1153–61.

9   Schunkert, H., Pharmacotherapy for prehypertension—mission accomplished? *N Engl J Med*, 2006. 354(16): p. 1742–4.

10 Guyenet, P.G., The sympathetic control of blood pressure. *Nat Rev Neurosci*, 2006. 7(5): p. 335–46.

11 Paulis, L. and T. Unger, Novel therapeutic targets for hypertension. *Nat Rev Cardiol*, 2010. 7(8): p. 431–41.

12 Drummond, G.R., et al., Combating oxidative stress in vascular disease: NADPH oxidases as therapeutic targets. *Nat Rev Drug Discov*, 2011. 10(6): p. 453–71.

13 Chobanian, A.V., Shattuck Lecture. The hypertension paradox—more uncontrolled disease despite improved therapy. *N Engl J Med*, 2009. 361(9): p. 878–87.

14 Chobanian, A.V., et al., The Seventh Report of the Joint National Committee on prevention, detection, evaluation, and treatment of high blood pressure: the JNC 7 report. *JAMA*, 2003. 289(19): p. 2560–72.

15 James, P.A., et al., 2014 evidence-based guideline for the management of high blood pressure in adults: report from the panel members appointed to the Eighth Joint National Committee (JNC 8). *JAMA*, 2014. 311(5): p. 507–20.

7

DIURETICS

7.1 OVERVIEW

Diuretic agents, also known as diuretics, refer to drugs that cause increased production of urine (diuresis). Most diuretics act at the various sites of the tubular system of the renal nephrons to inhibit the reabsorption of sodium from the tubular fluid into the circulation, thereby leading to increased excretion of sodium in the urine. The increased excretion of sodium in the urine is referred to as natriuresis, and the diuretics that cause natriuresis are also called natriuretics. Increased sodium excretion is accompanied with increased water excretion and hence the increased production of urine. Diuretics also affect the excretion of other ions, including potassium (K^+), calcium (Ca^{2+}), magnesium (Mg^{2+}), chloride (Cl^-), and bicarbonate (HCO_3^-).

As diuretics decrease plasma volume and volume of tissue fluid by increasing urine production, they are commonly used in the management of edema associated with cardiovascular disorders, renal diseases, hepatic disorders, and endocrine abnormalities [1]. Some diuretics are also used for treating hypertension, glaucoma, and other clinical conditions. This chapter discusses the molecular pharmacology of various classes of diuretics with an emphasis on their mechanistically based application in treating cardiovascular diseases, including hypertension and congestive heart failure. The guidelines on using diuretics to treat hypertension are discussed in Chapter 12. Use of diuretics in the management of congestive heart failure is covered in Unit VI.

7.2 VOLUME REGULATION AND DRUG TARGETING

As mentioned earlier, most diuretics act at various sites of the renal tubular system to inhibit reabsorption of tubular sodium into circulation. Hence, background knowledge on kidney physiology and volume regulation is necessary for understanding the molecular pharmacology of diuretics.

7.2.1 Renal Physiology and Volume Regulation

The kidney is one of the most highly differentiated organs in the body that has evolved to carry out a number of important physiological activities. The primary function of the kidney is regulation of the extracellular fluid (plasma and interstitial fluid) environment in the body. This is accomplished through the formation of urine, which is a modified filtrate of plasma. In the process of urine formation, the kidney regulates important physiological events as listed below:

- The volume of plasma
- The concentration of waste products in the blood
- The concentration of electrolytes (Na^+, K^+, HCO_3^-, and other ions) in the plasma
- The pH of plasma

The basic functional unit of the kidney is the nephron, which consists of the renal corpuscle (glomerulus and

Cardiovascular Diseases: From Molecular Pharmacology to Evidence-Based Therapeutics, First Edition. Y. Robert Li.
© 2015 John Wiley & Sons, Inc. Published 2015 by John Wiley & Sons, Inc.

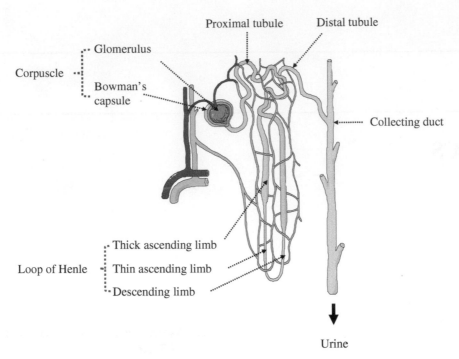

FIGURE 7.1 Schematic illustration of a nephron. Nephron is the basic structural and functional unit of the kidney and consists of the corpuscle and tubular system. The corpuscle further comprises the glomerulus and Bowman's capsule. The glomerulus produces filtrate of plasma that flows through the tubular system. The tubular system includes the proximal tubule, loop of Henle, distal tubule, and collecting duct. This tubular system is responsible for reabsorption of the solutes and water from the filtrate into the circulation as well as for excretion of metabolic waste products and xenobiotics, including drugs. Diuretics act on different segments of the tubular system to affect the reabsorption of solutes and water (see Fig. 7.2).

Bowman's capsule) and the tubular system (Fig. 7.1). The glomerulus produces an ultrafiltrate (also called filtrate) of plasma that flows through the tubular system. The tubular system includes the proximal tubule, loop of Henle, distal tubule, and collecting duct. This tubular system is responsible for reabsorption of the solutes and water from the filtrate into the circulation as well as for excretion of metabolic waste products and xenobiotics, including drugs.

In a single day, the renal glomeruli generate approximately 180 l of filtrate, and 99% of it is reabsorbed, with the rest 1% excreted as the urine. Hence, agents that affect the reabsorption of the filtrate will have significant impact on the production of the urine. For example, if the reabsorption rate were decreased from the initial 99 to 98% by a drug, the production of urine would be doubled. This would lead to a significant reduction in the volume of plasma and interstitial fluid as regulation of extracellular fluid volume by the kidney occurs via the formation of urine.

7.2.2 Drug Class and Drug Targeting

7.2.2.1 Drug Class   There are different ways for classification of diuretics. One commonly used scheme classifies clinical useful diuretics into the following six groups,

with the first three groups being commonly used in cardiovascular diseases:

1. Thiazide and thiazide-type diuretics
2. Loop diuretics
3. Potassium-sparing diuretics
4. Carbonic anhydrase inhibitors
5. Osmotic diuretics
6. Antidiuretic hormone antagonists

7.2.2.2 Drug Targeting   The most important process of renal reabsorption of the filtrate is the reabsorption of Na^+ as well as an accompanying anion, usually Cl^- from the tubular fluid into the circulation. NaCl in the body is the major determinant of the extracellular fluid volume. Hence, most diuretic agents (e.g., thiazide and thiazide-type diuretics, loop diuretics, potassium-sparing diuretics) primarily affect the reabsorption of Na^+. As noted before, diuretic drugs also affect the reabsorption of other ions in the filtrate or the secretion of ions into the tubular fluid and may thus potentially cause electrolyte disturbance. Figure 7.2 illustrates the reabsorption of Na^+ and water as well as other solutes at the various segments of the tubular system. The figure also depicts the acting sites of the various

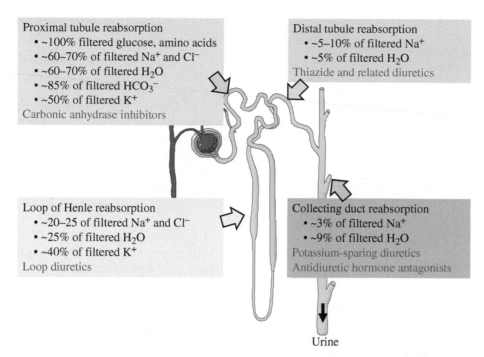

Proximal tubule reabsorption
- ~100% filtered glucose, amino acids
- ~60–70% of filtered Na^+ and Cl^-
- ~60–70% of filtered H_2O
- ~85% of filtered HCO_3^-
- ~50% of filtered K^+

Carbonic anhydrase inhibitors

Distal tubule reabsorption
- ~5–10% of filtered Na^+
- ~5% of filtered H_2O

Thiazide and related diuretics

Loop of Henle reabsorption
- ~20–25 of filtered Na^+ and Cl^-
- ~25% of filtered H_2O
- ~40% of filtered K^+

Loop diuretics

Collecting duct reabsorption
- ~3% of filtered Na^+
- ~9% of filtered H_2O

Potassium-sparing diuretics
Antidiuretic hormone antagonists

Urine

FIGURE 7.2   Schematic illustration of the reabsorption percentage of solutes and water at the various segments of the renal tubular system and the targeting sites of diuretics. Carbonic anhydrase inhibitors act on proximal tubule. As the name indicates, loop diuretics act on the thick ascending limb of the loop of Henle. The distal tubule is the acting site of thiazide and thiazide-type diuretics. Potassium-sparing diuretics and antidiuretic hormone antagonists act on the collecting duct.

types of diuretic agents. The detailed mechanisms of action as well as the clinical use of the various classes of diuretics are described next.

7.3   THIAZIDE AND THIAZIDE-TYPE DIURETICS

Listed below are the US FDA-approved thiazide and thiazide-type diuretics, with chlorthalidone and hydrochlorothiazide being the most commonly used members of the drug class:

- Bendroflumethiazide (Naturetin)
- Chlorothiazide (Diuril)
- Chlorthalidone (Hygroton)
- Cyclothiazide (Anhydron)
- Hydrochlorothiazide (Esidrix, Hydrodiuril)
- Hydroflumethiazide (Diucardin, Saluron)
- Indapamide (Lozol)
- Methyclothiazide (Enduron)
- Metolazone (Diulo, Mykrox, Zaroxolyn)
- Polythiazide (Renese)
- Quinethazone (Hydromox)
- Trichlormethiazide (Metahydrin, Naqua)

7.3.1   General Introduction to Drug Class

Thiazide and thiazide-type diuretics are the most commonly used diuretics and particularly useful in the management of hypertension. These agents are orally efficacious, have a moderate diuretic effect, and are generally well tolerated. Among them, hydrochlorothiazide is the most commonly prescribed. Recently, the use of chlorthalidone, a thiazide-type diuretic, in the treatment of hypertension has increased substantially due to favorable clinical outcome associated with its use [2]. Structures of chlorthalidone and hydrochlorothiazide are given in Figure 7.3.

7.3.2   Chemistry and Pharmacokinetics

Thiazides are sulfonamide compounds that contain a benzothiadiazide (thiazide) moiety. Many of the initially developed thiazide diuretics are actually analogs of 1,2,4-benzothiadiazine-1,2-dioxide. Subsequently, drugs that are pharmacologically similar to thiazide diuretics, but chemically are not thiazides, were developed and called thiazide-type or thiazide-like diuretics. In the literature, the term "thiazide diuretics" is used loosely to refer to both the thiazide and thiazide-type diuretics. But, strictly, this is not accurate as the thiazide-type diuretics are chemically not thiazide compounds. Because of this, in this book, the compound term "thiazide and thiazide-type diuretics" is used.

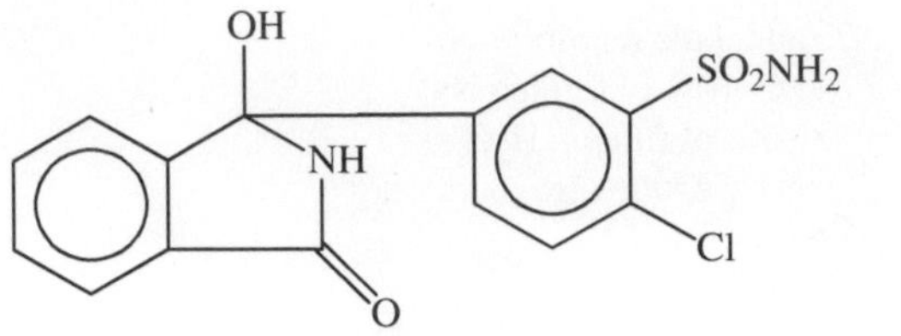

Chlorthalidone

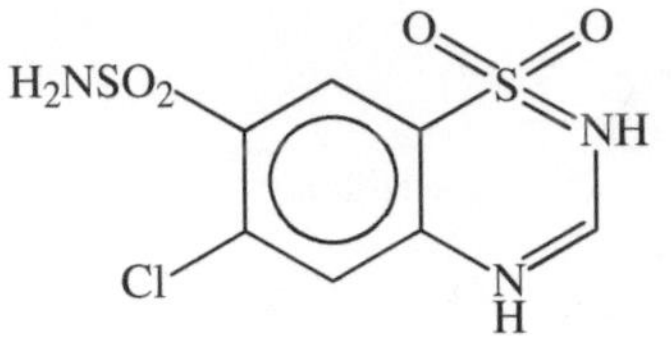

Hydrochlorothiazide

FIGURE 7.3 Structures of hydrochlorothiazide and chlorthalidone. Hydrochlorothiazide is a thiazide compound, whereas chlorthalidone is a thiazide-like drug.

TABLE 7.1 Major pharmacokinetic properties of hydrochlorothiazide and chlorthalidone

Drug	Oral bioavailability	Half-life (h)	Duration of action (h)
Hydrochlorothiazide	~70%	~2.5	~12
Chlorthalidone	~65%	~40–60	~72

Thiazide and thiazide-type diuretics exhibit good oral bioavailability, ranging from 50 to >90%. These drugs are actively secreted into the tubular fluid by the organic acid secretory system in the proximal tubular cells and travel through the tubular fluid to reach their site of action in the distal tubule. As uric acid is also secreted by the same organic acid secretory system, use of the thiazide and thiazide-type diuretics may decrease uric acid secretion by the proximal tubule and thereby cause increased levels of uric acid in plasma. This explains the increased risk of gout associated with the use of these diuretics. Thiazide and thiazide-type diuretics undergo metabolism to varying degrees and a major portion of the drug is excreted intact in urine. Table 7.1 summarizes the key pharmacokinetic properties of the two commonly used thiazide and thiazide-type diuretics, hydrochlorothiazide and chlorthalidone. As shown, chlorthalidone has a much longer half-life and duration of action than hydrochlorothiazide. This may explain the better efficacy of chlorthalidone in treating hypertension (see Chapter 12).

7.3.3 Molecular Mechanisms and Pharmacological Effects

The pharmacological effects of thiazide and thiazide-type diuretics are outlined below:

- Diuresis due to increased excretion of Na^+ and Cl^-
- Altered excretion of other ions
- Reduced peripheral vascular resistance
- Other potential novel effects

7.3.3.1 Diuresis due to Increased Excretion of Na^+ and Cl^- The key mechanism of action of thiazide and thiazide-type diuretics is the inhibition of the Na^+/Cl^- cotransporter at the early portion of the distal tubule (Fig. 7.4). This cotransporter is responsible for the absorption of Na^+ and Cl^- by the distal tubular cells. Inhibition of this cotransporter results in increased excretion of Na^+ and Cl^- in urine.

7.3.3.2 Altered Excretion of Other Ions Thiazide and thiazide-type diuretics also affect the excretion of other ions, including (i) increased excretion of K^+, (ii) decreased excretion of Ca^{2+}, and (iii) increased excretion of Mg^{2+}.

Increased Excretion of K^+ The mechanism of increased K^+ excretion is as follows: inhibition of the Na^+/Cl^- cotransporter at the early portion of the distal tubule leads to the delivery of a greater volume of NaCl-enriched tubular fluid to the late distal tubule and collecting duct, which in turn stimulates the exchange of Na^+ and K^+ at these sites. In the process, a small amount of Na^+ is reabsorbed as K^+ is secreted into the tubular fluid. Because of this increased K^+ excretion in urine, thiazide and thiazide-type diuretics have a kaliuretic effect (increased K^+ excretion in urine) that may result in hypokalemia in some patients (see Section 7.3.6).

Decreased Excretion of Urinary Ca^{2+} Thiazide and thiazide-type diuretics enhance Ca^{2+} reabsorption, thereby decreasing the Ca^{2+} concentration in tubular fluid. Because of this effect, thiazide and thiazide-type diuretics are used in the management of nephrolithiasis (kidney stones) due to idiopathic hypercalciuria (see Section 7.3.4).

The exact mechanisms of the enhanced reabsorption of Ca^{2+} by thiazide and thiazide-type diuretics are unclear. Three potential mechanisms have been suggested [3–6]: (i) in the proximal tubule, thiazide and thiazide-type diuretic-induced volume depletion leads to enhanced passive Ca^{2+} reabsorption; (ii) in the distal tubule, lowering of the intracellular Na^+ by thiazide and thiazide-type drug-induced blockage of Na^+ entry results in the enhanced Na^+/Ca^{2+} exchange in the basolateral membrane; and (iii) thiazide and thiazide-type diuretics induce enhanced expression of apical calcium channels.

Increased Urinary Excretion of Mg^{2+} Thiazide and thiazide-type diuretics increase urinary excretion of Mg^{2+} via a poorly understood mechanism possibly involving the reduced Mg^{2+} channel abundance [5]. This effect may lead to a mild magnesuria, and long-term use of the drugs may cause magnesium deficiency, particularly in the elderly.

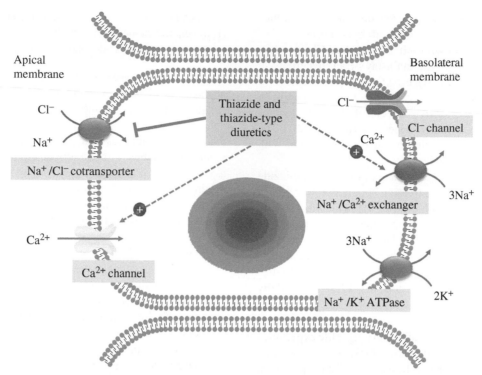

FIGURE 7.4   Molecular mechanisms of action of thiazide and thiazide-type diuretics. Thiazide and thiazide-type diuretics inhibit Na^+/Cl^- cotransporter at the early portion of the distal tubule. This cotransporter is responsible for the absorption of Na^+ and Cl^- by the distal tubular cells. Inhibition of this cotransporter results in increased excretion of Na^+ and Cl^- in urine. Thiazide and thiazide-type diuretics also enhance Ca^{2+} reabsorption via three possible mechanisms: (1) in the proximal tubule, these drugs induce volume depletion resulting in enhanced passive Ca^{2+} reabsorption; (2) in the distal tubule, lowering of the intracellular Na^+ by the drug-induced blockage of Na^+ entry results in the enhanced Na^+/Ca^{2+} exchange in the basolateral membrane; and (3) these drugs induce enhanced expression of apical calcium channels.

7.3.3.3   *Reduced Peripheral Vascular Resistance*   Initially, thiazide and thiazide-type diuretics lower blood pressure by decreasing plasma volume and cardiac output. Cardiac output eventually returns to normal and plasma and extracellular fluid volumes return to slightly less than normal, but peripheral vascular resistance is reduced, resulting in lower blood pressure. The exact mechanisms responsible for the lowered peripheral resistance upon chronic treatment with thiazide and thiazide-type diuretics remain unclear. Two possible pathways have been hypothesized: (i) direct vasodilatory effects and (ii) indirect vasodilatory effects [7]. For the former mechanism, both endothelium and vascular smooth muscle have been suggested to be the source of vasodilatory effects, likely involving calcium and Rho signaling. For the later mechanism, it is thought that blood vessels adapt to the initial thiazide-induced plasma volume loss and decrease in cardiac output by undergoing vasoconstriction. Then, over time, blood vessels dilate to increase cardiac output back toward baseline levels (cardiac output increases as arterials dilate). According to this hypothesis, thiazide and thiazide-type diuretic-induced vasodilation would still originate from sodium-induced fluid loss via inhibition of the Na^+/Cl^- cotransporter [7].

7.3.3.4   *Other Potential Novel Effects*   The thiazide-type diuretics, including chlorthalidone and indapamide, have antiplatelet activity *in vitro*. Recently, chlorthalidone has been shown to decrease platelet aggregation and vascular permeability, and promote angiogenesis [8]. These novel effects may explain the cardiovascular benefits in patients treated with chlorthalidone [2, 9, 10] (see Section 7.3.4.1). Thiazide and thiazide-type diuretics may increase prostaglandin synthesis. As prostaglandins cause vasodilation, the aforementioned effect may be responsible partly for the antihypertensive activity of these diuretic drugs.

7.3.4   Clinical Uses

Outlined below are the major clinical indications of thiazide and thiazide-type diuretics:

- Hypertension
- Edema associated with congestive heart failure and other conditions
- Nephrolithiasis due to idiopathic hypercalciuria
- Nephrogenic diabetes insipidus

7.3.4.1 Hypertension

Thiazide and thiazide-type diuretics are widely used in the treatment of hypertension either alone or in combination with other drugs (see Chapter 12). These drugs have become the mainstay of antihypertensive therapy because they are (i) effective, (ii) convenient to administer, (iii) inexpensive, and (iv) well tolerated. In addition, the antihypertensive effects of the inhibitors of the renin–angiotensin–aldosterone system (RAAS) (see Chapter 9) are enhanced when given in combination with thiazide and thiazide-type diuretics.

It should be noted that the efficacy of the various thiazide and thiazide-type diuretics in treating hypertension varies. In this regard, although hydrochlorothiazide is the most widely used diuretic drug for hypertension, the thiazide-type drug chlorthalidone has been shown to be more effective because of its longer half-life as well as other potential novel effects (see Section 7.3.3.4). Indeed, there is no evidence that hydrochlorothiazide at the usually dose of 12.5–50 mg daily reduces myocardial infarction, stroke, or death. In contrast, treatment of hypertensive patients with chlorthalidone is associated with decreased mortality and longer life expectancy [11]. It has been becoming increasingly recognized that chlorthalidone appears to be the best diuretic for the treatment of hypertension, with regard to both blood pressure-lowering efficacy and, most importantly, prevention of hypertension-related morbidity and mortality.

7.3.4.2 Edema Associated with Congestive Heart Failure and Other Conditions

Due to their ability to reduce extracellular fluid via diuresis, thiazide and thiazide-type diuretics are useful for treating edema associated with mild to moderate congestive heart failure. For severe edema associated with congestive heart failure, loop diuretics are usually preferred due to their high efficacy (see Section 7.4).

Thiazide and thiazide-type diuretics are also useful for the management of edema associated with liver and renal diseases. However, with the possible exception of metolazone and indapamide, most of these drugs are ineffective when the glomerular filtration rate (GFR) is <30–40 ml/min.

7.3.4.3 Nephrolithiasis due to Idiopathic Hypercalciuria

Thiazide and thiazide-type diuretics decrease renal excretion of Ca^{2+} and as such are useful in the management of idiopathic nephrolithiasis, especially for prevention of stone recurrence [12]. However, there is a paucity of clinical data on the dose-dependent effects of these drugs in preventing the recurrence of calcium-containing kidney stones. It has been suggested that the doses for treating nephrolithiasis should be higher than those used for treating hypertension. In this regard, the best available evidence for prevention of kidney stone recurrence suggests the use of chlorthalidone (25–50 mg, daily), indapamide (2.5 mg daily), or hydrochlorothiazide (25 mg twice daily or 50 mg daily) [12].

Due to their inhibition of renal calcium excretion and the consequent positive calcium balance, thiazide and thiazide-type diuretics may be useful for the management of osteoporosis. However, the clinical efficacy in this regard remains to be established via large-scale randomized clinical trials [13].

7.3.4.4 Nephrogenic Diabetes Insipidus

Patients with nephrogenic diabetes insipidus do not respond to desmopressin (a synthetic replacement for vasopressin, the hormone that reduces urine production). However, thiazide and thiazide-type diuretics can paradoxically reduce the urine volume in patients with nephrogenic diabetes insipidus, and as such, are the mainstay for the treatment of patients with this syndrome. As described below, two mechanisms have been proposed to explain the seemly paradoxical effects of these drugs in treating nephrogenic diabetes insipidus [14–16]:

- Decreased GFR hypothesis: This widely accepted hypothesis suggests that the antidiuretic action of thiazide and thiazide-type diuretics in nephrogenic diabetes insipidus is secondary to increased renal sodium excretion. The renal sodium loss causes extracellular volume contraction leading to reduced GFR and increased proximal tubular sodium and water reabsorption. Hence, less water and solutes are delivered to the distal tubule and collecting duct and are lost as urine.

- Upregulation of aquaporin and Na^+ transporters hypothesis: In a lithium-induced nephrogenic diabetes insipidus model, G.H. Kim and associates showed that hydrochlorothiazide could upregulate aquaporin-2 and distal renal Na^+ transporters [15]. These effects might account, at least partly, for the antidiuretic action of this thiazide drug in nephrogenic diabetes insipidus.

7.3.5 Therapeutic Dosages in Cardiovascular Applications

The major use of thiazide and thiazide-type diuretics in cardiovascular medicine is for treating hypertension. These drugs can be used alone but more frequently are used in combination with other antihypertensive drugs for effective control of blood pressure. Fixed-dose combination products of thiazide and thiazide-type diuretics with angiotensin-converting enzyme inhibitors, angiotensin receptor blockers, β-blockers, or the renin inhibitor aliskiren are available. Table 7.2 lists the dosages of commonly used thiazide and thiazide-type diuretics in treating hypertension.

TABLE 7.2 Therapeutic dosages of thiazide and thiazide-type diuretics in treating hypertension

Thiazide and thiazide-type diuretic drug	Oral once daily dose range (mg)
Chlorthalidone	6.25–25
Hydrochlorothiazide	12.5–25
Indapamide	1.25–2.5
Metolazone	2.5–10

7.3.6 Adverse Effects and Drug Interactions

7.3.6.1 Adverse Effects   In general, thiazide and thiazide-type diuretics are well tolerated. However, use of these drugs, especially upon prolonged use, may cause several adverse effects, which are classified into two categories: (i) hypokalemia and other electrolyte disturbance and (ii) metabolic abnormalities.

Hypokalemia and Other Electrolyte Disturbance   As with loop diuretics (see Section 7.4), the most significant adverse effects associated with thiazide and thiazide-type diuretics are abnormalities of fluid and electrolyte balance, including hypokalemia, hyponatremia, hypomagnesemia, and hypercalcemia. The hypokalemic effect, however, makes these drugs useful in combination with K^+-sparing diuretics to avoid disruption of the K^+ balance (see Section 7.5).

Metabolic Abnormalities   Thiazide and thiazide-type diuretics may cause three metabolic abnormalities as described below:

1. Hyperlipidemias: Thiazide and thiazide-type diuretics may cause a small (5–10%) increase in the levels of plasma total cholesterol, LDL cholesterol, and triglycerides. These changes may return toward baseline after prolonged use. There is currently no evidence showing that such lipid effects result in a negation of the beneficial effects of these drugs on blood pressure and cardiovascular events. The mechanisms behind thiazide and thiazide-type diuretic-induced dyslipidemias remain unclear. Two possible mechanisms have been proposed [17, 18]. One is related to the decreased insulin sensitivity due to hypokalemia induced by these drugs. The decreased insulin sensitivity promotes hepatic production of cholesterol. The other mechanism is related to the compensatory activation of the renin–angiotensin–aldosterone axis and the sympathetic nervous system due to the drug-induced loss of blood volume and sodium. The compensatory increase of the overall adrenergic tone may mediate the lipid effects of the diuretics by increasing adipocyte lipolysis, leading to increased levels of free fatty acids and the subsequent elevated synthesis of hepatic very-low-density lipoprotein (VLDL). Increased VLDL levels then result in increases in LDL cholesterol as well as triglycerides in plasma.

2. Hyperglycemia: Use of thiazide and thiazide-type diuretics is associated with an increased risk of diabetes mellitus [19]. This hyperglycemic effect is likely due to hypokalemia, resulting in decreased insulin release from the pancreatic β-cells. In this regard, release of insulin is dependent partially on K^+ levels. Although these diuretic drugs adversely affect glucose homeostasis, clinical findings suggest that thiazide and thiazide-type diuretic (e.g., chlorthalidone)-related incident diabetes has less adverse long-term cardiovascular disease impact than incident diabetes that develops while on other antihypertensive medications (e.g., the calcium channel blocker amlodipine, the angiotensin-converting enzyme inhibitor lisinopril) [20]. Therefore, concerns regarding potential adverse diabetic effects associated with thiazide and thiazide-type diuretic therapy should not discourage its use. In this context, a recent pooled analysis of five statin studies shows that incident diabetes mellitus is more common in persons treated with intensive-dose therapy versus moderate-dose therapy [21]. Nonetheless, the benefits of reduced cholesterol by statin therapy are deemed to outweigh any possible deleterious effects of incident diabetes mellitus on cardiovascular disease outcomes (see Chapter 4). Similarly, thiazide and thiazide-type diuretics, especially chlorthalidone have been shown to be highly effective for preventing cardiovascular outcomes through decades of rigorously controlled clinical trials.

3. Hyperuricemia: Thiazide and thiazide-type diuretics increase plasma uric acid levels by inhibiting the renal tubular secretion of uric acid. Use of these drugs may be associated with an increased risk of gout attacks [22–24], and this association is mediated by increases in plasma uric acid levels [25]. Hence, an important question arises: should thiazide or thiazide-type diuretics be used in hypertensive patients with gout? For patients with gout and untreated hypertension, it would be prudent to avoid thiazide and thiazide-type diuretics. However, frequently, clinicians are faced with patients with gout who are already taking thiazide and thiazide-type diuretics. For patients taking a stable dose of a thiazide diuretic, intermittent use, as compared with consistent daily use, confers an increased risk of gout attacks [26]. Given the cardiovascular risk associated with hypertension and the high prevalence of inadequate blood pressure control, alteration of a regimen that is appropriately controlling a patient's hypertension is not recommended. Since poorly controlled gout is most commonly related to underdosing of urate-lowering therapy and poor adherence, it is recommended to optimize urate-lowering therapy before making changes that might adversely affect blood pressure control. For patients who are refractory to appropriate maximal urate-lowering therapy, switching to an alternate antihypertensive agent (e.g., the uricosuric losartan; see Chapter 9) with close monitoring of blood pressure control would be appropriate. Decision analysis and cost-effectiveness studies are warranted to guide optimal management [23].

7.3.6.2 Drug Interactions Thiazide and thiazide-type diuretics may add to or potentiate the action of other antihypertensive drugs. These drugs may also decrease the renal clearance of lithium (an antipsychotic drug), thereby increasing the risk of lithium toxicity. Because prostaglandins are involved in the antihypertensive activity of thiazide and thiazide-type diuretics (Section 7.3.3.4), nonsteroidal anti-inflammatory drugs may reduce the antihypertensive efficacy of these diuretic drugs in monotherapy [27]. However, the antihypertensive effect of the combination of hydrochlorothiazide and the angiotensin-converting enzyme inhibitor fosinopril was shown not to be affected by nonsteroidal anti-inflammatory drugs [28].

7.3.6.3 Contraindications and Pregnancy Category

- Thiazide and thiazide-type diuretics are contraindicated in patients with anuria.
- Hypersensitivity to these agents or other sulfonamide-derived drugs is also a contraindication.
- Pregnancy category: B.

7.3.7 Summary of Thiazide and Thiazide-Type Diuretics

Thiazide and thiazide-type diuretics inhibit Na^+/Cl^- cotransporter at the early portion of the distal tubule, resulting in increased excretion of Na^+ and Cl^- in urine. These drugs, especially hydrochlorothiazide and chlorthalidone are among the commonly used drugs for treating hypertension. The antihypertensive activity appears to result from the decreased blood volume at the early phase of treatment, and later from vasodilation through unknown mechanisms. Thiazide and thiazide-type diuretics are generally well tolerated though long-term use of these drugs may cause metabolic disturbances.

7.4 LOOP DIURETICS

There are currently four loop diuretics approved by the US FDA for clinical use. Furosemide is the most commonly prescribed loop diuretic drug:

1. Bumetanide (Bumex)
2. Ethacrynic acid (Edecrin)
3. Furosemide (Lasix)
4. Torsemide (Demadex)

7.4.1 General Introduction to Drug Class

Loop diuretics are the most efficacious diuretic agents currently available. They are so called because they act on the thick ascending limb of the loop of Henle (see Section 7.4.3).

It should be noted that loop diuretics such as furosemide may also inhibit the reabsorption of Na^+ and Cl^- in both the proximal and distal tubules. In contrast to thiazide and thiazide-type diuretics, loop diuretics exhibit dose-dependent diuresis throughout their clinical dosage range. As such, these drugs are also known as high-ceiling diuretics.

7.4.2 Chemistry and Pharmacokinetics

Loop diuretics are a chemically diverse group of drugs (Fig. 7.5). Bumetanide and furosemide contain a sulfonamide moiety. Ethacrynic acid is a phenoxyacetic acid derivative and electrophilic. Likely due to its electrophilic activity, ethacrynic acid is a potent activator of Nrf2 [29], a transcription factor involved in regulation of cytoprotective genes. Torsemide is a sulfonylurea.

All four loop diuretics are available in oral and parenteral formulations. They are rapidly absorbed and eliminated by the kidney via proximal tubular secretion (major route) and glomerular filtration (minor route due to the high binding of the drugs to plasma proteins). Drugs secreted by the proximal tubular cells into the lumen travel to the thick ascending limb of the loop of Henle to act on the drug target there (Section 7.4.3). Table 7.3 lists the major pharmacokinetic properties of the loop diuretics.

7.4.3 Molecular Mechanisms and Pharmacological Effects

The pharmacological effects of loop diuretics are outlined below:

- Diuresis due to increased excretion of Na^+ and Cl^-
- Altered excretion of other ions
- Reduced peripheral vascular resistance
- Other potential novel effects

7.4.3.1 Diuresis due to Increased Excretion of Na^+ and Cl^- Loop diuretics act from within the lumen of the thick ascending portion of the loop of Henle, where they inhibit the $Na^+/K^+/2Cl^-$ cotransporter. As illustrated in Figure 7.6, inhibition of $Na^+/K^+/2Cl^-$ cotransporter on the apical membrane of the cells of the loop of Henle results in decreased reabsorption of Na^+ and Cl^- and thereby the increased excretion of these solutes in urine. As depicted in the figure, inhibition of $Na^+/K^+/2Cl^-$ cotransporter also leads to disruption of the positive transepithelial potential, thereby causing increased urinary excretion of Ca^{2+} and Mg^{2+} (see Section 7.4.3.2).

7.4.3.2 Altered Excretion of Other Ions Loop diuretics also affect the excretion of other ions, including (i) increased excretion of K^+ and (ii) increased excretion of both Ca^{2+} and Mg^{2+}.

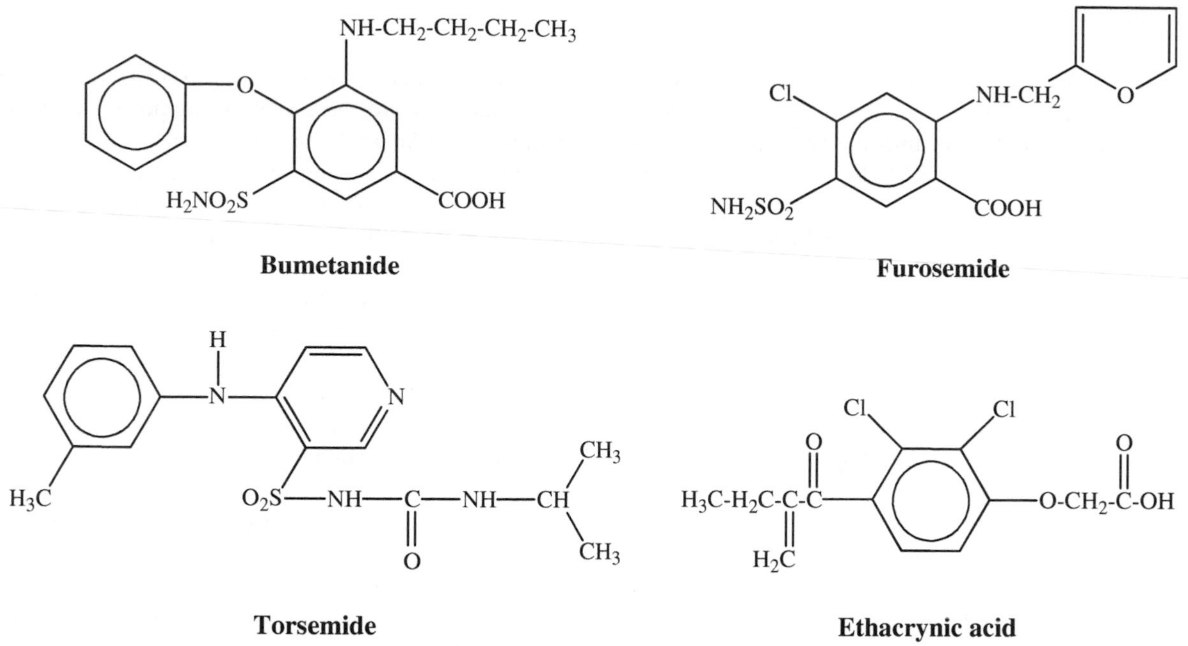

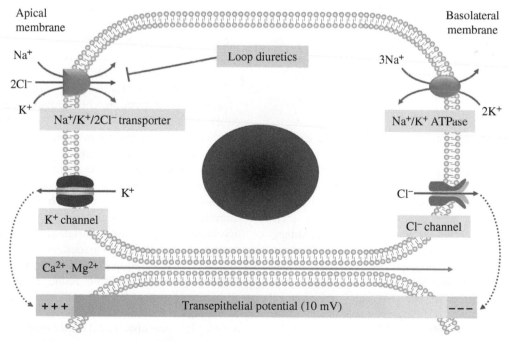

Bumetanide

Furosemide

Torsemide

Ethacrynic acid

FIGURE 7.5 Structures of loop diuretics. Loop diuretics are a chemically diverse group of drugs. Bumetanide and furosemide contain a sulfonamide moiety. Ethacrynic acid is a phenoxyacetic acid derivative and electrophilic. Torsemide is a sulfonylurea.

TABLE 7.3 Major pharmacokinetic properties of loop diuretics

Loop diuretic	Oral bioavailability (%)	Half-life (h)	Duration of action after oral administration (h)
Bumetanide	60	1	4–6
Ethacrynic acid	80	1	6–8
Furosemide	100	1–2	6–8
Torsemide	80	3–4	6–8

FIGURE 7.6 Molecular mechanisms of action of loop diuretics. Loop diuretics inhibit $Na^+/K^+/2Cl^-$ cotransporter on the apical membrane of the cells of the loop of Henle, thereby resulting in decreased reabsorption of Na^+ and Cl^- and the consequent increased excretion of these solutes in urine. As illustrated, the positive transepithelial potential serves as the driving force for the reabsorption of Ca^{2+} and Mg^{2+}. Hence, reduction of the positive transepithelial potential by loop diuretics causes marked increases in Ca^{2+} and Mg^{2+} secretion, predisposing to hypocalcemia and hypomagnesemia.

Increased Excretion of K⁺   The mechanism involved in the loop diuretic-induced increased excretion of K^+ is similar to that discussed for thiazide and thiazide-type diuretics (see section "Increased Excretion of K^+"). In brief, the delivery of Na^+-rich fluid to the collecting duct causes increased secretion of K^+ and H^+, predisposing to hypokalemia and alkalosis. In addition, activation of the renin–angiotensin–aldosterone axis due to fluid reduction caused by the diuretics may also contribute to the increased loss of K^+ in the urine (aldosterone augments Na^+ accumulation and K^+ excretion).

Increased Excretion of Both Ca²⁺ and Mg²⁺   As illustrated in Figure 7.6, the positive transepithelial potential serves as the driving force for the reabsorption of Ca^{2+} and Mg^{2+}. Hence, reduction of the positive transepithelial potential by loop diuretics causes marked increases in Ca^{2+} and Mg^{2+} excretion, predisposing to hypocalcemia and hypomagnesemia.

7.4.3.3   *Reduced Peripheral Vascular Resistance*   Studies in both animal models and human subjects suggest that loop diuretics, including furosemide and torsemide, may cause direct venodilation via a prostaglandin and nitric oxide-dependent mechanism [30, 31]. However, the clinical significance of this venodilatory effect remains unclear.

7.4.3.4   *Other Potential Novel Effects*   Animal studies demonstrate that loop diuretics, including bumetanide and furosemide, may modulate GABAergic signaling through their antagonism of cation–chloride cotransporters in the central nervous system, leading to potential antiepileptic and anxiolytic activities [32, 33]. Bumetanide is also shown to have neuroprotective effects in traumatic brain injury likely via its modulation of GABAergic signaling [34]. The clinical implications of these novel effects warrant further investigation.

7.4.4   Clinical Uses

The major clinical indications for loop diuretics are pulmonary edema and peripheral edema associated with congestive heart failure. They are also widely used for treating other edematous conditions, including liver cirrhosis and nephritic syndrome. Other uses of loop diuretics include hypercalcemia and hyperkalemia due to their enhancement of both Ca^{2+} and K^+ excretion from the kidney (Section 7.4.3.2). Compared with thiazide and thiazide-type diuretics, loop diuretics are less useful in treating hypertension due to two primary reasons: (i) limited efficacy and short half-life and (ii) availability of many other more effective and better-tolerated antihypertensive agents. This section primarily considers the use of loop diuretics in the management of heart failure. Additional discussion of the role of loop diuretics in heart failure management is provided in Unit VI.

7.4.4.1   *Acute Decompensated Heart Failure*   Intravenous loop diuretics are an essential component of therapy for patients with acute decompensated heart failure because of their efficacy in reducing cardiac preload, thereby improving pulmonary congestion. Among the loop diuretics, furosemide has been the most widely used and studied in this clinical setting. It is believed that in addition to sodium and fluid reduction due to diuresis, the direct venodilatory activity may also contribute to the preload reduction (see Unit VI).

Although intravenous diuretics are an essential component of current treatment and are administered to approximately 90% of patients who are hospitalized with heart failure, prospective data to guide the use of these drugs are sparse and current guidelines are based primarily on expert opinions [35] (also see Chapter 22). As a result, clinical practice varies widely with regard to both the mode of administration (i.e., bolus vs. continuous infusion) and the dosage. In this context, a recent prospective, double-blind, randomized controlled trial demonstrated that among patients with acute decompensated heart failure, there were no significant differences in patients' global assessment of symptoms or in the change in renal function when furosemide diuretic therapy was administered by bolus as compared with continuous infusion or at a high dose as compared with a low dose [36]. These findings may change current practice. Since a high-dose regimen may relieve dyspnea more quickly without adverse effects on renal function, such a regimen is preferable to a low-dose regimen. Administration of boluses may be more convenient than continuous infusion and equally effective [37, 38].

7.4.4.2   *Clinically Stable Patients with Congestive Heart Failure*   Diuretic treatment is recommended in all patients with congestive heart failure and clinical evidence of volume overload, including those with preserved left ventricular ejection fraction [39]. Treatment may begin with either a thiazide or loop diuretic. In more severe volume overload or if response to a thiazide is inadequate, treatment with a loop diuretic should be implemented. Excessive diuresis, which may lead to orthostatic changes in blood pressure and worsening of renal function, should be avoided. In this context, use of high doses of oral loop diuretics in clinically stable patients with heart failure is associated with increased mortality. Notably, a recent study suggested that the risk associated with high-dose loop diuretic use is strongly dependent on blood urea nitrogen (BUN) concentrations with reduced survival in patients with an elevated BUN level and improved survival in patients with a normal BUN level. These data suggest a role for neurohormonal activation in loop diuretic-associated mortality as blood BUN is a surrogate marker for activation of the sympathetic nervous system and renin–angiotensin–aldosterone axis [40].

7.4.4.3 Loop Diuretic Resistance

Both acute and long-term use of loop diuretics may lead to drug resistance via multiple mechanisms involving interactions between the pathophysiology of sodium retention in the heart failure and the renal response to the diuretics [41]. Strategies to improve patients' responsiveness to these agents include fluid and salt restriction, switching from oral to intravenous loop diuretics, increasing diuretic dosage, continuous infusion, and combination therapy with thiazide and thiazide-type diuretics [42].

7.4.5 Therapeutic Dosages in Heart Failure Treatment

Table 7.4 lists the dosage for loop diuretics in treating congestive heart failure. Loop diuretics should be initiated in low doses and then carefully titrated upward to relieve signs and symptoms of fluid overload in patients with congestive heart failure. Intravenous loop diuretics are necessary for acute decompensated heart failure to relieve congestion acutely. Once congestion has been relived, treatment with diuretics should be continued to prevent the recurrence of sodium and water retention.

7.4.6 Adverse Effects and Drug Interactions

7.4.6.1 Adverse Effects

Loop diuretics are generally well tolerated, and the adverse effects are mainly due to disruption of fluid and electrolyte balance. These drugs may also cause metabolic abnormalities. Use of loop diuretics, especially ethacrynic acid is associated with ototoxicity.

Abnormalities of Fluid and Electrolyte Balance Loop diuretics cause increased excretion of K^+, Ca^{2+}, and Mg^{2+}, which may lead to hypokalemia, hypocalcemia, and hypomagnesemia, particularly in patients with dietary deficiency of these electrolytes. Overzealous use of loop diuretics can cause severe fluid and sodium depletion and the development of hypotension.

Metabolic Abnormalities Similar to thiazide and thiazide-type diuretics, loop diuretics can also cause hyperuricemia, predisposing to acute gouty attacks. Increased levels of blood glucose, total cholesterol, and LDL cholesterol also occur in patients treated with loop diuretics. The mechanisms underlying these metabolic abnormalities are similar to those described for thiazide and thiazide-type diuretics (see Section 7.3).

Ototoxicity Use of loop diuretics may cause ototoxicity especially in patients with concurrent use of other potential ototoxic drugs, such as aminoglycoside antimicrobials. The ototoxicity may include tinnitus, vertigo, a sense of fullness in the ears, impairment of hearing, and deafness. Ethacrynic acid causes more severe ototoxicity than do other loop diuretics, and as such, it is less frequently used. Ethacrynic acid is used when patients are hypersensitive to other loop diuretics.

The loop diuretics are direct inhibitors of the $Na^+/K^+/2Cl^-$ cotransport system, which also exists in the marginal and dark cells of the stria vascularis, which are responsible for endolymph secretion. The ototoxicity of these agents may be indirect, due to changes in ionic composition and fluid volume within the endolymph [43, 44].

Allergic Reactions and Other Rare Adverse Effects As loop diuretics with the exception of ethacrynic acid are sulfonamides, allergic reactions, such as skin rashes, may occur. Allergic reactions are less common with ethacrynic acid. Other rare adverse effects may include hematotoxicity and bone marrow suppression.

7.4.6.2 Drug Interactions

Similar to what was discussed for thiazide and thiazide-type diuretics, the diuretic effects of loop diuretics can be reduced by concurrent treatment with nonsteroidal anti-inflammatory drugs and probenecid. Loop diuretics may also increase the blood levels of lithium, predisposing to lithium toxicity via a similar mechanism described in Section 7.3.6.2. Notably, the ototoxicity of loop diuretics can be potentiated by other ototoxic drugs, including aminoglycoside antimicrobials, cisplatin, and carboplatin.

7.4.6.3 Contraindications and Pregnancy Category

- Severe sodium and water depletion.
- Anuria.
- Hypersensitivity to the drugs or sulfonamides or sulfonylureas.
- Pregnancy category: B (ethacrynic acid and torsemide) and C (bumetanide and furosemide).

7.4.7 Summary of Loop Diuretics

Loop diuretics are the most efficacious diuretic agents currently available. These drugs inhibit $Na^+/K^+/2Cl^-$ cotransporter on the apical membrane of the cells of the loop of

TABLE 7.4 Dosages of loop diuretics in treating congestive heart failure

Loop diuretic	Initial dose	Maximal dose (mg/day)
Bumetanide	0.5–1 mg, once or twice daily	10
Ethacrynic acid	50–100 mg, once daily	200
Furosemide	20–80 mg, once daily	400
Torsemide	10–20 mg, once daily	200

Henle, resulting in decreased reabsorption of Na$^+$ and Cl$^-$ and thereby the increased excretion of these solutes in urine. Loop diuretics are primarily used to treat edematous conditions, such as pulmonary edema and peripheral edema, associated with congestive heart failure. They are not useful for treating hypertension. Loop diuretics are generally well tolerated and the adverse effects are mainly due to disruption of fluid and electrolyte balance. Use of loop diuretics, especially ethacrynic acid may also cause ototoxicity.

7.5   POTASSIUM-SPARING DIURETICS

Listed below are the US FDA-approved potassium-sparing diuretics. Amiloride and triamterene are sodium channel blockers, whereas eplerenone and spironolactone are aldosterone receptor antagonists:

- Amiloride (Midamor)
- Triamterene (Dyrenium)
- Eplerenone (Inspra)
- Spironolactone (Aldactone)

7.5.1   General Introduction to Drug Class

Potassium-sparing diuretics act on the principal cells of late distal tubule and the collecting duct to decrease Na$^+$ reabsorption and K$^+$ excretion. The ability to decrease K$^+$ excretion differentiates potassium-sparing diuretics from thiazide and loop diuretics, which cause increased K$^+$ excretion (see Sections 7.3 and 7.4). Based on the mechanisms involved, potassium-sparing diuretics are classified into two subgroups: (i) direct Na$^+$ channel blockers (amiloride and triamterene) and (ii) aldosterone receptor antagonists (eplerenone and spironolactone).

7.5.2   Chemistry and Pharmacokinetics

The chemical structures of the four potassium-sparing diuretics are given in Figure 7.7. The major pharmacokinetic properties of these drugs are summarized in Table 7.5.

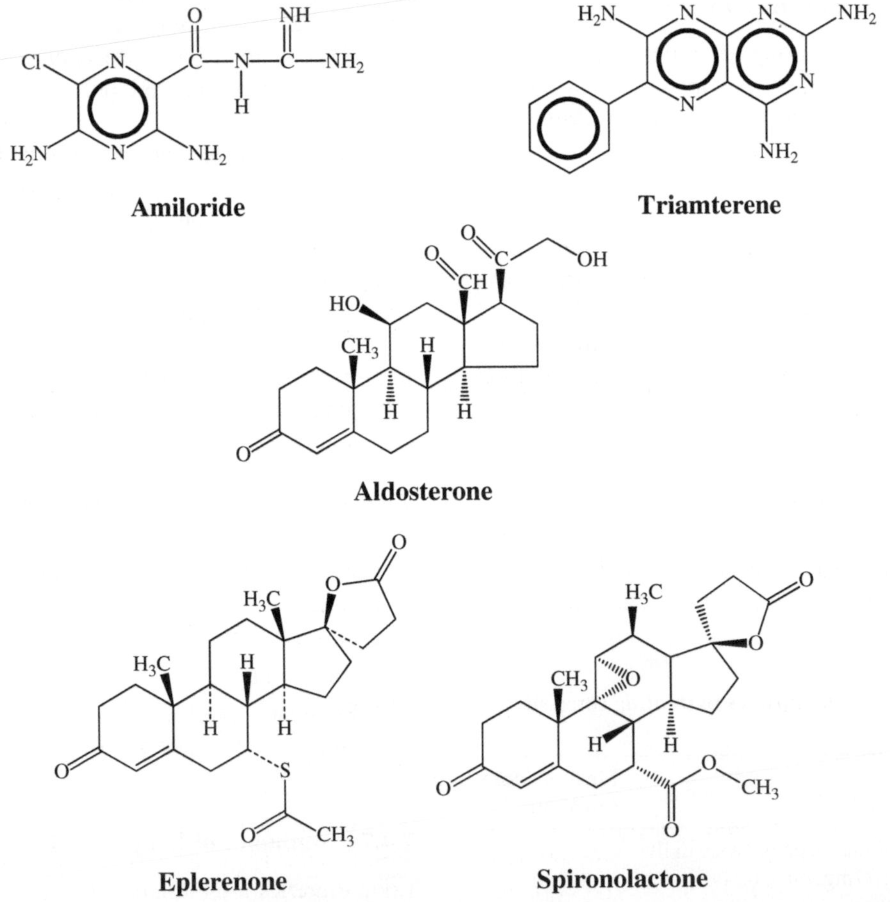

FIGURE 7.7   Structures of potassium-sparing diuretics. The structure of aldosterone is included to show the structural similarity between this hormone and its receptor antagonists eplerenone and spironolactone.

TABLE 7.5   Major pharmacokinetic properties of potassium-sparing diuretics

Potassium-sparing diuretic	Oral bioavailability (%)	Elimination half-life (h)	Duration of action (h)	Route of elimination
Amiloride	20	8	24	Renal excretion of intact drug
Triamterene	50	4	8	Renal excretion of sulfate conjugate of hydroxytriamterene
Eplerenone	70	5	Insufficient data	Urinary and biliary excretion of CYP3A4 metabolites
Spironolactone	65	1.5	60	Urinary and biliary excretion of sulfur-containing metabolites

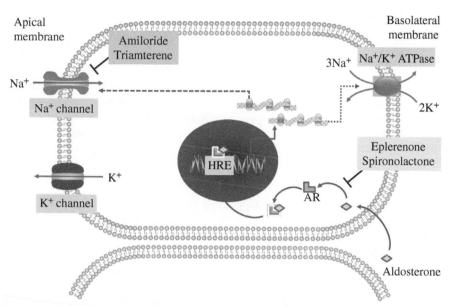

FIGURE 7.8   Molecular mechanisms of action of potassium-sparing diuretics. Amiloride and triamterene block the activity of the Na^+ channels in the apical membranes of the principal cells, whereas eplerenone and spironolactone decrease the biosynthesis of new Na^+ channels in these cells via blocking the aldosterone receptors (AR). As illustrated, aldosterone–AR complex binds to the hormone response element (HRE) of the gene encoding the Na^+ channel protein, thereby leading to enhanced expression of the Na^+ channels. Eplerenone and spironolactone competitively block the binding of aldosterone to AR, thereby inhibiting the transcription of the Na^+ channel-encoding gene. Since Na^+ reabsorption in the principal cells is coupled with K^+ secretion, inhibition of the Na^+ reabsorption by these diuretic drugs causes decreased K^+ excretion. For color details, please see color plate section.

7.5.3   Molecular Mechanisms and Pharmacological Effects

The pharmacological effects of potassium-sparing diuretics include the following three aspects:

1. Increased excretion of Na^+ and decreased excretion of K^+
2. Altered excretion of other ions
3. Inhibition of aldosterone-mediated deleterious effects

7.5.3.1   *Increased Excretion of Na^+ and Cl^- and Decreased Excretion of K^+*   Drugs in this class act via one of two mechanisms: amiloride and triamterene block the activity of the Na^+ channels in the apical membranes of the principal cells, whereas eplerenone and spironolactone decrease the biosynthesis of new Na^+ channels in these cells via blocking

the aldosterone receptors (also known as mineralocorticoid receptors). Since Na^+ reabsorption in the principal cells is coupled with K^+ excretion, inhibition of the Na^+ reabsorption by these diuretic drugs causes decreased K^+ secretion (Fig. 7.8). The detailed mechanism underlying the decreased K^+ excretion is as follows: decreased Na^+ entry through the apical membrane Na^+ channels results in decreased lumen negative transepithelial potential, thereby decreasing the driving force for K^+ secretion as well as the secretion of other cations, including H^+, Ca^{2+}, and Mg^{2+} (see Section 7.5.3.2).

Since the late distal tubule and collecting duct have a limited capacity to reabsorb Na^+ and other solutes, potassium-sparing diuretics only mildly increase Na^+ and Cl^- excretion rate (about 2–3% of filtered load). As such, they are relatively weak diuretics.

TABLE 7.6   Deleterious effects resulting from dysregulated aldosterone

Organ/system	Deleterious effects of dysregulated aldosterone
Heart	• Coronary atherosclerosis • Decreased natriuretic peptide synthesis • Interstitial fibrosis • Myocyte hypertrophy • Oxidative and inflammatory stress • Reduced norepinephrine uptake
Vasculature	• Atherosclerosis • Endothelial cell hypertrophy • Oxidative and inflammatory stress • Platelet aggregation • Reduced nitric oxide bioavailability • Vascular smooth muscle cell hypertrophy • Vasomotor dysfunction
Kidney	• Glomerulosclerosis • Oxidative and inflammatory stress • Podocyte apoptosis and proteinuria • Potassium and magnesium wasting • Sodium and water retention • Tubulointerstitial fibrosis

7.5.3.2 Altered Excretion of Other Ions

As described earlier, decreased lumen negative transepithelial potential by potassium-sparing diuretics also leads to the decreased excretion of H^+, Ca^{2+}, and Mg^{2+}. The decreased H^+ excretion may lead to metabolic acidosis.

7.5.3.3 Inhibition of Aldosterone-Mediated Deleterious Effects

In addition to the weak diuresis, the potassium-sparing diuretics eplerenone and spironolactone exert beneficial activities via their inhibition of the deleterious effects of aldosterone. This mechanism is largely responsible for the clinical efficacy of these two drugs in the management of heart failure as well as resistant hypertension (see Chapters 12, 21, and 22) and renal diseases. In this context, dysregulated aldosterone elicits diverse deleterious effects in the cardiovascular and renal systems (Table 7.6), contributing to the pathophysiology of heart failure, hypertension, and other disorders [45].

7.5.4 Clinical Uses

7.5.4.1 Clinical Uses of Amiloride and Triamterene

Due to their weak natriuresis, amiloride and triamterene are rarely used as monotherapy for treating edematous conditions and hypertension. Instead, they are primarily used as adjunctive treatment with thiazide diuretics or other kaliuretic diuretic agents in congestive heart failure or hypertension to (i) help restore normal plasma potassium levels in patients who develop hypokalemia on the kaliuretic diuretic and (ii) prevent development of hypokalemia in patients who

would be exposed to particular risk if hypokalemia were to develop, for example, patients treated with digitalis or patients with significant cardiac arrhythmias.

Fixed-dose combination with a thiazide diuretic is available for amiloride and triamterene:

- Hydrochlorothiazide/amiloride (Moduretic)
- Hydrochlorothiazide/triamterene (Dyazide)

7.5.4.2 Clinical Uses of Eplerenone and Spironolactone

The major clinical uses of eplerenone and spironolactone include treatment of (i) hypertension and heart failure, and (ii) hyperaldosteronism. Spironolactone is also used for the treatment of other edematous conditions, including liver cirrhosis accompanied by edema and/or ascites, and nephrotic syndrome, as well as for the treatment of patients with hypokalemia when other measures are considered inappropriate or inadequate. It is worth mentioning that spironolactone is a much older drug with more available clinical research data than eplerenone, and as such, it has more FDA-approved indications than eplerenone.

Hypertension and Heart Failure Like amiloride and triamterene, eplerenone and spironolactone are often used in combination with other drugs, such as thiazide or loop diuretics, to treat hypertension and ameliorate edema associated with heart failure. Such a combination use results in increased natriuresis without significant disturbance of K^+ homeostasis. Eplerenone and spironolactone are also useful drugs for treating resistant hypertension because dysregulated aldosterone is involved in the development of resistant hypertension (see Chapter 12).

Clinical trials demonstrate that addition of eplerenone or spironolactone to standard therapy results in attenuation of disease progress and improvement of survival in congestive heart failure patients with either severe, moderate, or mild symptoms [35, 45, 46]. In addition, a recent study suggested that in patients with systolic heart failure and mild symptoms, eplerenone also reduced the incidence of new-onset atrial fibrillation [47]. However, a recent trial reported that in patients with heart failure and a preserved ejection fraction, treatment with spironolactone did not significantly reduce the incidence of the primary composite outcome of death from cardiovascular causes, aborted cardiac arrest, or hospitalization for the management of heart failure [48]. This suggests that the survival benefit of aldosterone antagonists may depend on a reduced ventricular ejection fraction in heart failure patients.

The survival benefits of aldosterone antagonists in patients with heart failure and reduced ventricular ejection fraction are independent of their diuretic activity and most likely stem from the inhibition of the deleterious effects of dysregulated aldosterone in the cardiovascular and renal systems (see Table 7.6) [49, 50].

Hyperaldosteronism Spironolactone is used for (i) establishing the diagnosis of primary hyperaldosteronism by therapeutic trial, (ii) short-term preoperative treatment of patients with primary hyperaldosteronism, (iii) long-term maintenance therapy for patients with discrete aldosterone-producing adrenal adenomas who are judged to have high operative risks or who decline surgery, and (iv) long-term maintenance therapy for patients with bilateral micro- or macronodular adrenal hyperplasia (idiopathic hyperaldosteronism).

7.5.5 Therapeutic Dosages

The therapeutic dosage regimens of potassium-sparing diuretics are given in Table 7.7.

7.5.6 Adverse Effects and Drug Interactions

7.5.6.1 Adverse Effects

Hyperkalemia The most characteristic and also principal adverse effect of potassium-sparing diuretics is hyperkalemia, especially in patients who also ingest potassium supplements or take other drugs that increase plasma K^+ levels (e.g., angiotensin-converting enzyme inhibitors, angiotensin receptor blockers) or in patients with renal disorders that predispose them to hyperkalemia. Hyperkalemia may predispose the individuals to life-threatening conditions, such as cardiac arrhythmias. As such, plasma K^+ levels need to be measured before initiation of the drug therapy and monitored during the therapy.

Endocrine Abnormalities Owing to its ability to modulate other steroid receptors, such as androgen receptors, spironolactone may cause endocrine abnormalities, including gynecomastia, erectile dysfunction, and menstrual irregularities.

TABLE 7.7 Major therapeutic dosage regimens of potassium-sparing diuretics

Potassium-sparing diuretic	Dosage regimen
Amiloride	• 5 mg, once daily, added to other drugs; may be increased to 10 mg/day
Triamterene	• 50–100 mg, twice daily, added to other drugs
Eplerenone	• For heart failure, 25 mg, once daily, titrated to the recommended dose of 50 mg, once daily • For hypertension, 50 mg, once daily; may be increased to 50 mg, twice daily
Spironolactone	• For severe heart failure, 25–50 mg, once daily • For hypertension, 50–100 mg, once daily

Eplerenone is a new aldosterone antagonist that produces less endocrine adverse effects than does spironolactone.

Other Adverse Effects Other adverse effects associated with potassium-sparing diuretics may include gut disturbance, allergic reactions, headache, and weakness.

7.5.6.2 Drug Interactions

Interactions with Drugs That also Increase Plasma K^+ Levels The most significant drug interaction is the concomitant use of other drugs that increase plasma K^+ levels, including angiotensin-converting enzyme inhibitors and angiotensin receptor blockers. The risk of hyperkalemia is also increased with coadministration of nonsteroidal anti-inflammatory drugs. Such combinations increase the risk of life-threatening hyperkalemia. Potassium-sparing diuretics also decrease lithium excretion, thereby predisposing to lithium toxicity.

Interactions with Drugs That Modulate CYP3A4 Activity Because eplerenone is primarily metabolized by CYP3A4 (Table 7.5), strong inhibitors of CYP3A4 (e.g., ketoconazole, itraconazole, nefazodone, troleandomycin, clarithromycin, ritonavir, and nelfinavir) may significantly reduce the metabolism of eplerenone and markedly increase its plasma concentrations. Such drugs should not be taken with eplerenone. Eplerenone dosage needs to be reduced when patients also take the drugs that are moderate inhibitors of CYP3A4 (e.g., verapamil, erythromycin, saquinavir, fluconazole).

7.5.6.3 Contraindications and Pregnancy Category

- Patients with hyperkalemia or impaired renal function (anuria, acute or chronic renal insufficiency).
- For eplerenone, concomitant treatment with strong CYP3A4 inhibitors.
- Pregnancy category: B (eplerenone) and C (amiloride, triamterene, spironolactone).

7.5.6.4 Summary of Potassium-Sparing Diuretics

Potassium-sparing diuretics act on the principal cells of late distal tubule and the collecting duct to decrease Na^+ reabsorption and K^+ excretion. Because the late distal tubule and collecting duct have a limited capacity to reabsorb Na^+ and other solutes, potassium-sparing diuretics only mildly increase Na^+ and Cl^- excretion rate (about 2–3% of filtered load). As such, they are relatively weak diuretics. Potassium-sparing diuretics are mainly used in combination with other drugs for the treatment of hypertension and congestive heart failure. Aldosterone receptor antagonists attenuate disease progression and prolong the survival of patients with congestive heart failure, and such benefits are independent of their diuretic activity and most likely stem from antagonizing aldosterone's deleterious effects on the cardiovascular and renal systems. The most characteristic and also major adverse effect of potassium-sparing diuretics is hyperkalemia, especially in individuals at risk of developing hyperkalemia. Spironolactone may also cause endocrine abnormalities.

7.6 OTHER DIURETICS

As introduced earlier, diuretics represent a large category of drugs that are used in the management of cardiovascular diseases as well as other disorders. With regard to cardiovascular disease management, thiazide and thiazide-type diuretics, loop diuretics, and potassium-sparing diuretics are among the commonly used diuretics, and as such, the molecular pharmacology and cardiovascular indications of these drugs are the focus of this chapter. The remaining three classes of diuretics, namely, carbonic anhydrase inhibitors, osmotic diuretics, and antidiuretic hormone antagonists are used primarily for the management of noncardiovascular conditions, such as mountain sickness, glaucoma, cerebral edema, acute renal failure, and hyponatremia. Table 7.8 summarizes the major mechanisms of action and clinical indications of these drugs.

TABLE 7.8   Other diuretic agents

Diuretic class	Mechanisms of action	Clinical uses	Adverse effects
Carbonic anhydrase inhibitors • Acetazolamide (Diamox) • Dichlorphenamide (Daranide) • Dorzolamide (Trusopt) • Methazolamide (Neptazane)	Inhibition of carbonic anhydrase activity of proximal tubular cells results in decreased Na^+ and HCO_3^- reabsorption The drugs are very weak diuretics because the increased tubular sodium ions are reabsorbed at other parts of the tubular system	Mountain sickness: these drugs cause metabolic acidosis by promoting HCO_3^- excretion, thereby counteracting the alkalosis in mountain sickness Glaucoma: carbonic anhydrase is involved in the formation of aqueous humor Inhibition of this enzyme by the drugs leads to decreased aqueous humor secretion and intraocular pressure	Paresthesias, gut disturbances, and metabolic disturbances
Osmotic diuretics • Mannitol (Osmitrol) • Glycerol • Isosorbide • Urea	These drugs increase osmotic pressure in plasma, which attracts water from interstitial fluid, resulting in increased renal blood flow and washout of medullary tonicity. The drugs also increase the osmotic pressure on the tubular fluid, thereby retarding water reabsorption	Cerebral edema Acute glaucoma Acute renal failure	Excessive plasma volume expansion
Antidiuretic hormone antagonists • Conivaptan (Vaprisol) • Tolvaptan (Samsca)	Conivaptan is a nonselective antagonist of antidiuretic hormone V1A and V2 receptors, whereas tolvaptan is a selective V2 receptor antagonist. V2 receptors are coupled with insertion of aquaporin channels in the apical membrane of the renal collecting duct, leading to reabsorption of water. By activating these receptors, antidiuretic hormone helps maintain plasma osmolality in the normal range. Antagonism of V2 receptors by conivaptan or tolvaptan causes free water excretion or aquaresis, and the drugs are thus called aquaretics	Euvolemic and hypervolemic hyponatremia	Conivaptan: infusion site reactions (including phlebitis), pyrexia, hypokalemia, headache, and orthostatic hypotension Tolvaptan: thirst, dry mouth, asthenia, constipation, pollakiuria or polyuria, and hyperglycemia

7.7   SUMMARY OF CHAPTER KEY POINTS

- Diuretics refer to drugs that cause increased production of urine. These drugs are classified into six classes including thiazide and thiazide-type diuretics, loop diuretics, potassium-sparing diuretics, carbonic anhydrase inhibitors, osmotic diuretics, and antidiuretic hormone antagonists. The first three classes are commonly used in treating cardiovascular diseases, including hypertension and heart failure, and the remaining three classes are primarily for noncardiovascular diseases.

- Thiazide and thiazide-type diuretics inhibit Na^+/Cl^- cotransporter at the early portion of the distal tubule, resulting in increased excretion of Na^+ and Cl^- in urine. These drugs, especially hydrochlorothiazide and chlorthalidone, are among the commonly used drugs for treating hypertension. They are also used to treat mild edema associated with heart failure.

- Loop diuretics, with furosemide being the most commonly prescribed member, are the most efficacious diuretic agents. They inhibit $Na^+/K^+/2Cl^-$ cotransporter on the apical membrane of the cells of the loop of Henle, resulting in decreased reabsorption of Na^+ and Cl^-. Loop diuretics are primarily used to treat edematous conditions, such as pulmonary edema and peripheral edema associated with congestive heart failure. These drugs are not useful for treating hypertension.

- Potassium-sparing diuretics, including direct sodium channel blockers and aldosterone receptor antagonists, act on the principal cells of late distal tubule and the collecting duct to decrease Na^+ reabsorption and K^+ excretion via either directly inhibiting the sodium channels or inhibiting the expression of the channels. As weak diuretics, they are mainly used in combination with potassium-wasting diuretics for the treatment of hypertension and congestive heart failure and for the prevention of potassium disturbance. Aldosterone receptor antagonists improve the survival of heart failure patients via antagonizing the deleterious effects of aldosterone, rather than through diuresis.

- Other diuretics, including carbonic anhydrase inhibitors, osmotic diuretics, and antidiuretic hormone antagonists, are used primarily for treating noncardiovascular conditions, such as mountain sickness, glaucoma, cerebral edema, acute renal failure, and hyponatremia.

7.8   SELF-ASSESSMENT QUESTIONS

7.8.1. A 48-year-old man presents with a blood pressure of 146/95 mmHg. He is treated with a diuretic at a dose of 25 mg once daily. While his blood pressure is under control, he later develops hypokalemia. Which of the following diuretics has been most likely prescribed to the patient?
   - A. Acetazolamide
   - B. Amiloride
   - C. Conivaptan
   - D. Eplerenone
   - E. Hydrochlorothiazide

7.8.2. A 50-year-old male presents with a blood pressure of 145/90 mmHg. History reveals that a family member recently develops a non-ST-elevation acute coronary syndrome at the age of 55 years. A decision is made to put the patient on a diuretic therapy. Which of the following is most appropriate for this patient in order to effectively reduce the risk of cardiovascular events and mortality?
   - A. Amiloride
   - B. Chlorthalidone
   - C. Hydrochlorothiazide
   - D. Methazolamide
   - E. Torsemide

7.8.3. A 65-year-old male is brought to the emergency department due to shortness of breath. He is diagnosed with pulmonary edema due to acute decompensated heart failure. He is started on a diuretic to improve his congestion. While his symptoms improve quickly upon the diuretic therapy, he develops hypokalemia and hypocalcemia. If the diuretic therapy is the cause of the patient's electrolyte disturbance, which of the following is most likely prescribed to the patient?
   - A. Chlorthalidone
   - B. Conivaptan
   - C. Furosemide
   - D. Mannitol
   - E. Triamterene

7.8.4. A 35-year-old female is diagnosed with stage 1 hypertension and hypokalemia due to Liddle syndrome, an extremely rare disorder caused by a genetic defect leading to excessive expression of the apical sodium channels in the principal cells of the renal collecting duct. Which of the following is the most appropriate treatment for the patient's condition?
   - A. Dorzolamide
   - B. Eplerenone
   - C. Ethacrynic acid
   - D. Hydrochlorothiazide
   - E. Triamterene

7.8.5. A 29-year-old male is brought to the emergency department following an apparent traumatic brain injury in an automobile accident. Computerized tomographic image reveals an increased intracranial pressure. A decision is made to start the patient immediately on a diuretic to reduce the elevated intracranial pressure. Which of the following is most appropriate to administer?

A. Conivaptan
B. Furosemide
C. Indapamide
D. Mannitol
E. Methazolamide

7.8.6. A 58-year-old female complains of paresthesia, drowsiness, and occasional nausea associated with one of the drugs she is taking. She is found to have hyperchloremic metabolic acidosis. Which of the following drugs she is most likely taking?

A. Acetazolamide for glaucoma
B. Clopidogrel for post-myocardial infarction
C. Hydrochlorothiazide for hypertension
D. Mannitol for cerebral edema
E. Torsemide for heart failure

7.8.7. A 68-year-old female with a history of heart disease is brought into the emergency room because of severe difficulty in breathing. Examination reveals that she has pulmonary edema. Which of the following drugs should be given to improve her pulmonary congestion?

A. Amiloride
B. Chlorthalidone
C. Furosemide
D. Hydrochlorothiazide
E. Spironolactone

7.8.8. A 60-year-old male has been placed on a diuretic to control his high blood pressure. However, after a few weeks, he develops an acute gouty attack. If the gouty attack is precipitated by the diuretic therapy, which of the following drugs has he most likely been taking?

A. Dorzolamide
B. Eplerenone
C. Hydrochlorothiazide
D. Torsemide
E. Triamterene

7.8.9. A 55-year-old male has a history of frequent episodes of renal colic with calcium-containing renal stones. A careful workup indicates that he has a defect in proximal tubular calcium reabsorption, which results in high concentrations of calcium salts in the tubular fluid. Which of the following is the most useful diuretic agent in treating his condition?

A. Acetazolamide
B. Chlorthalidone
C. Furosemide
D. Spironolactone
E. Triamterene

7.8.10. A 71-year-old female is admitted to the emergency department because of a "fainting spell" at home. She does not appear to have suffered trauma from her fall, but her blood pressure is 120/60 mmHg when she lies down and 80/40 mmHg when she sits up. Neurologic examination and ECG are within normal limits when she is lying down. Medical history reveals that she has recently started taking "water pill" for a heart condition. Which of the following drugs is the most likely cause of her fainting spell?

A. Acetazolamide
B. Chlorthalidone
C. Eplerenone
D. Furosemide
E. Hydrochlorothiazide

REFERENCES

1   Clark, A.L. and J.G. Cleland, Causes and treatment of oedema in patients with heart failure. *Nat Rev Cardiol*, 2013. 10(3): p. 156–70.

2   Roush, G.C., V. Buddharaju, and M.E. Ernst, Is chlorthalidone better than hydrochlorothiazide in reducing cardiovascular events in hypertensives?. *Curr Opin Cardiol*, 2013. 28(4): p. 426–32.

3   Gesek, F.A. and P.A. Friedman, Mechanism of calcium transport stimulated by chlorothiazide in mouse distal convoluted tubule cells. *J Clin Invest*, 1992. 90(2): p. 429–38.

4   Lee, C.T., et al., Effect of thiazide on renal gene expression of apical calcium channels and calbindins. *Am J Physiol Renal Physiol*, 2004. 287(6): p. F1164–70.

5   Nijenhuis, T., et al., Enhanced passive Ca2+ reabsorption and reduced Mg2+ channel abundance explains thiazide-induced hypocalciuria and hypomagnesemia. *J Clin Invest*, 2005. 115(6): p. 1651–8.

6   Reilly, R.F. and C.L. Huang, The mechanism of hypocalciuria with NaCl cotransporter inhibition. *Nat Rev Nephrol*, 2011. 7(11): p. 669–74.

7   Duarte, J.D. and R.M. Cooper-DeHoff, Mechanisms for blood pressure lowering and metabolic effects of thiazide and thiazide-like diuretics. *Expert Rev Cardiovasc Ther*, 2010. 8(6): p. 793–802.

8   Woodman, R., C. Brown, and W. Lockette, Chlorthalidone decreases platelet aggregation and vascular permeability and promotes angiogenesis. *Hypertension*, 2010. 56(3): p. 463–70.

9   Kurtz, T.W., Chlorthalidone: don't call it "thiazide-like" anymore. *Hypertension*, 2010. 56(3): p. 335–7.

10 Kaplan, N.M., Chlorthalidone versus hydrochlorothiazide: a tale of tortoises and a hare. *Hypertension*, 2011. 58(6): p. 994–5.

11 Kostis, J.B., et al., Association between chlorthalidone treatment of systolic hypertension and long-term survival. *JAMA*, 2011. 306(23): p. 2588–93.

12 Reilly, R.F., A.J. Peixoto, and G.V. Desir, The evidence-based use of thiazide diuretics in hypertension and nephrolithiasis. *Clin J Am Soc Nephrol*, 2010. 5(10): p. 1893–903.

13 Walsh, J.S., C. Newman, and R. Eastell, Heart drugs that affect bone. *Trends Endocrinol Metab*, 2012. 23(4): p. 163–8.

14 Magaldi, A.J., New insights into the paradoxical effect of thiazides in diabetes insipidus therapy. *Nephrol Dial Transplant*, 2000. 15(12): p. 1903–5.

15 Kim, G.H., et al., Antidiuretic effect of hydrochlorothiazide in lithium-induced nephrogenic diabetes insipidus is associated with upregulation of aquaporin-2, Na-Cl co-transporter, and epithelial sodium channel. *J Am Soc Nephrol*, 2004. 15(11): p. 2836–43.

16 Loffing, J., Paradoxical antidiuretic effect of thiazides in diabetes insipidus: another piece in the puzzle. *J Am Soc Nephrol*, 2004. 15(11): p. 2948–50.

17 Weir, M.R. and M. Moser, Diuretics and beta-blockers: is there a risk for dyslipidemia? *Am Heart J*, 2000. 139(1 Pt 1): p. 174–83.

18 Brook, R.D., Mechanism of differential effects of antihypertensive agents on serum lipids. *Curr Hypertens Rep*, 2000. 2(4): p. 370–7.

19 Elliott, W.J. and P.M. Meyer, Incident diabetes in clinical trials of antihypertensive drugs: a network meta-analysis. *Lancet*, 2007. 369(9557): p. 201–7.

20 Barzilay, J.I., et al., Long-term effects of incident diabetes mellitus on cardiovascular outcomes in people treated for hypertension: the ALLHAT diabetes extension study. *Circ Cardiovasc Qual Outcomes*, 2012. 5(2): p. 153–162.

21 Preiss, D., et al., Risk of incident diabetes with intensive-dose compared with moderate-dose statin therapy: a meta-analysis. *JAMA*, 2011. 305(24): p. 2556–64.

22 Singh, J.A., S.G. Reddy, and J. Kundukulam, Risk factors for gout and prevention: a systematic review of the literature. *Curr Opin Rheumatol*, 2011. 23(2): p. 192–202.

23 Neogi, T., Clinical practice. Gout. *N Engl J Med*, 2011. 364(5): p. 443–52.

24 Hueskes, B.A., et al., Use of diuretics and the risk of gouty arthritis: a systematic review. *Semin Arthritis Rheum*, 2012. 41(6): p. 879–89.

25 McAdams DeMarco, M.A., et al., Diuretic use, increased serum urate levels, and risk of incident gout in a population-based study of adults with hypertension: the Atherosclerosis Risk in Communities cohort study. *Arthritis Rheum*, 2012. 64(1): p. 121–9.

26 Hunter, D.J., et al., Recent diuretic use and the risk of recurrent gout attacks: the online case-crossover gout study. *J Rheumatol*, 2006. 33(7): p. 1341–5.

27 Gurwitz, J.H., et al., The impact of ibuprofen on the efficacy of antihypertensive treatment with hydrochlorothiazide in elderly persons. *J Gerontol A Biol Sci Med Sci*, 1996. 51(2): p. M74–9.

28 Thakur, V., M.E. Cook, and J.D. Wallin, Antihypertensive effect of the combination of fosinopril and HCTZ is resistant to interference by nonsteroidal antiinflammatory drugs. *Am J Hypertens*, 1999. 12(9 Pt 1): p. 925–8.

29 Wu, R.P., et al., Nrf2 responses and the therapeutic selectivity of electrophilic compounds in chronic lymphocytic leukemia. *Proc Natl Acad Sci U S A*, 2010. 107(16): p. 7479–84.

30 Dormans, T.P., et al., Vascular effects of loop diuretics. *Cardiovasc Res*, 1996. 32(6): p. 988–97.

31 de Berrazueta, J.R., et al., Vasodilatory action of loop diuretics: a plethysmography study of endothelial function in forearm arteries and dorsal hand veins in hypertensive patients and controls. *J Cardiovasc Pharmacol*, 2007. 49(2): p. 90–5.

32 Dzhala, V.I., et al., NKCC1 transporter facilitates seizures in the developing brain. *Nat Med*, 2005. 11(11): p. 1205–13.

33 Krystal, A.D., J. Sutherland, and D.W. Hochman, Loop diuretics have anxiolytic effects in rat models of conditioned anxiety. *PLoS One*, 2012. 7(4): p. e35417.

34 Shulga, A., et al., The loop diuretic bumetanide blocks post-traumatic p75NTR upregulation and rescues injured neurons. *J Neurosci*, 2012. 32(5): p. 1757–70.

35 Jessup, M., et al., 2009 focused update: ACCF/AHA Guidelines for the Diagnosis and Management of Heart Failure in Adults: a report of the American College of Cardiology Foundation/American Heart Association Task Force on Practice Guidelines: developed in collaboration with the International Society for Heart and Lung Transplantation. *Circulation*, 2009. 119(14): p. 1977–2016.

36 Felker, G.M., et al., Diuretic strategies in patients with acute decompensated heart failure. *N Engl J Med*, 2011. 364(9): p. 797–805.

37 Fonarow, G.C., Comparative effectiveness of diuretic regimens. *N Engl J Med*, 2011. 364(9): p. 877–8.

38 Costello-Boerrigter, L.C. and J.C. Burnett, Jr., Heart failure: furosemide—to DOSE or not to DOSE, that is the question. *Nat Rev Cardiol*, 2011. 8(7): p. 365–6.

39 Lindenfeld, J., et al., HFSA 2010 Comprehensive Heart Failure Practice Guideline. *J Card Fail*, 2010. 16(6): p. e1–194.

40 Testani, J.M., et al., Interaction between loop diuretic-associated mortality and blood urea nitrogen concentration in chronic heart failure. *J Am Coll Cardiol*, 2011. 58(4): p. 375–82.

41 Fruscione, F., et al., Regulation of human mesenchymal stem cell functions by an autocrine loop involving NAD+release and P2Y11-mediated signaling. *Stem Cells Dev*, 2011. 20(7): p. 1183–98.

42 Asare, K., Management of loop diuretic resistance in the intensive care unit. *Am J Health Syst Pharm*, 2009. 66(18): p. 1635–40.

43 Delpire, E., et al., Deafness and imbalance associated with inactivation of the secretory Na-K-2Cl co-transporter. *Nat Genet*, 1999. 22(2): p. 192–5.

44 Humes, H.D., Insights into ototoxicity. Analogies to nephrotoxicity. *Ann N Y Acad Sci*, 1999. 884: p. 15–8.

45 Butler, J., et al., Update on aldosterone antagonists use in heart failure with reduced left ventricular ejection fraction heart failure society of America guidelines committee. *J Card Fail*, 2012. 18(4): p. 265–81.

46 Zannad, F., et al., Eplerenone in patients with systolic heart failure and mild symptoms. *N Engl J Med*, 2011. 364(1): p. 11–21.

47 Swedberg, K., et al., Eplerenone and atrial fibrillation in mild systolic heart failure: results from the EMPHASIS-HF

(Eplerenone in Mild Patients Hospitalization And SurvIval Study in Heart Failure) study. *J Am Coll Cardiol*, 2012. 59(18): p. 1598–603.

48 Pitt, B., et al., Spironolactone for heart failure with preserved ejection fraction. *N Engl J Med*, 2014. 370(15): p. 1383–92.

49 Leopold, J.A., Aldosterone, mineralocorticoid receptor activation, and cardiovascular remodeling. *Circulation*, 2011. 124(18): p. e466–8.

50 Rossignol, P., et al., Eplerenone survival benefits in heart failure patients post-myocardial infarction are independent from its diuretic and potassium-sparing effects. Insights from an EPHESUS (Eplerenone Post-Acute Myocardial Infarction Heart Failure Efficacy and Survival Study) substudy. *J Am Coll Cardiol*, 2011. 58(19): p. 1958–66.

8

SYMPATHOLYTICS

8.1 OVERVIEW

The sympathetic nervous system was brought to public awareness in the early decades of the twentieth century by W.B. Cannon, through his research on, and popularization of, the concept of the "fight-or-flight" responses to stress. In the past three decades, the sympathetic nervous system has moved toward the center stage in cardiovascular medicine, with demonstration of the importance of excessive activation of the sympathetic nervous system in various forms of cardiovascular diseases [1]. These include hypertension, coronary heart disease, heart failure, and cardiac arrhythmias.

Drugs that reduce sympathetic stimulation are known as sympatholytic agents or simply sympatholytics. The most important group of sympatholytic drugs consists of the adrenergic receptor antagonists, which further include α-blockers and β-blockers. Other groups of sympatholytics include ganglionic blocking agents and sympathetic neuronal blocking agents. In addition, there are centrally acting sympatholytic agents that are useful for treating particular forms of cardiovascular disorders. This chapter first introduces the sympathetic nervous system and drug targeting and then discusses the molecular pharmacology of α-blockers, β-blockers, and centrally acting sympatholytics with a focus on their mechanisms of action. The evidence-based applications of these drugs in the management of various cardiovascular diseases are covered in other chapters.

8.2 SYMPATHETIC NERVOUS SYSTEM AND DRUG TARGETING

8.2.1 Basic Divisions of the Nervous System

The human nervous system is the most intricate structure known to exist. About one third of the genes encoded in the human genome are expressed in the nervous system. It is responsible for perceiving, processing, and transmitting information throughout the body and generating responses to the information. The nervous system has two anatomical parts: the central nervous system and the peripheral nervous system. The central nervous system consists of the brain and spinal cord and is the integrating and command center of the nervous system. The peripheral nervous system, the part of the nervous system outside the central nervous system, consists mainly of the nerves that extend from and to the brain (i.e., cranial nerves) and spinal cord (i.e., spinal nerves).

The peripheral nervous system includes four subdivisions: (i) somatic sensory, (ii) visceral sensory, (iii) somatic motor, and (iv) visceral motor. The visceral motor system is also known as the autonomic nervous system, which controls the activities of the visceral organs/systems, including the heart and vessels, and the gastrointestinal system. Because we generally have no voluntary control over such activities as the beating of the heart and the movement of food through the digestive tract, the autonomic nervous system is also called the involuntary nervous system.

Cardiovascular Diseases: From Molecular Pharmacology to Evidence-Based Therapeutics, First Edition. Y. Robert Li.
© 2015 John Wiley & Sons, Inc. Published 2015 by John Wiley & Sons, Inc.

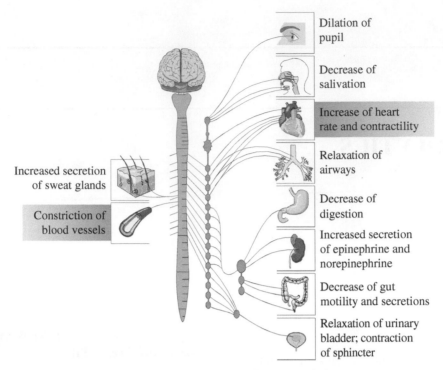

Dilation of pupil

Decrease of salivation

Increase of heart rate and contractility

Relaxation of airways

Decrease of digestion

Increased secretion of epinephrine and norepinephrine

Decrease of gut motility and secretions

Relaxation of urinary bladder; contraction of sphincter

Increased secretion of sweat glands

Constriction of blood vessels

FIGURE 8.1   The sympathetic nervous system. As illustrated, activation of the sympathetic nervous system causes increased heart rate and contractility and constriction of blood vessels.

The autonomic nervous system includes two major divisions: (i) the parasympathetic nervous system and (ii) the sympathetic nervous system. These two systems function in parallel to maintain the physiological homeostasis by regulating bodily functions. Stimulation of the sympathetic nervous system expends energy and leads to the so-called "flight-or-fight" responses, characterized by (i) increased heart rate, blood pressure, and respiration; (ii) increased blood flow to skeletal muscles; and (iii) dilation of the pupil. In contrast, stimulation of the parasympathetic nervous system conserves energy and results in the so-called "rest-and-digest" responses characterized by (i) decreased heart rate, blood pressure, and respiration; (ii) increased secretions and digestion; and (iii) constriction of the pupil (Fig. 8.1).

The enteric nervous system is sometimes called the third division of the autonomic nervous system. It consists of a network of autonomic nerves that are located in the gut wall and regulate gastrointestinal motility and secretion. Although the enteric nervous system is influenced by both parasympathetic and sympathetic nervous systems, it can function independently via a local control mechanism.

8.2.2 Sympathetic Nervous System and Cardiovascular Diseases

As noted earlier, the sympathetic nervous system controls diverse physiological processes, including heart rate, cardiac contractility, and blood pressure, as well as metabolisms of lipids and carbohydrates. These physiological processes are mediated via signaling through adrenergic receptors (Fig. 8.2). Indeed, the sympathetic nervous system via adrenergic signaling has a dominant role in cardiovascular control due to its ability to increase cardiac rate and contractility, cause constriction of arteries and veins, induce release of adrenal catecholamines, and activate the renin–angiotensin–aldosterone system (RAAS). Overactivity of the sympathetic nervous system, which is primarily manifested as augmented adrenergic signaling and abnormal activation of the RAAS, plays a fundamental role in the development of various forms of cardiovascular diseases. This is also evidenced by the demonstrated efficacy of the sympatholytic drugs in treating various types of cardiovascular disorders.

8.2.3 Drug Class and Drug Targeting

Figure 8.3 illustrates the sympathetic flow from the central nervous system to the ganglia and finally to the effector cells/tissues. Sympatholytic drugs target the various steps of the sympathetic flow to reduce the sympathetic activity. Based on the sites and mechanisms involved, sympatholytic drugs are conventionally classified into four groups:

1. Centrally acting sympatholytic drugs: These drugs inhibit central sympathetic discharge via activation of α_2-adrenergic receptors in the brainstem, resulting in decreased sympathetic flow to peripheral tissues and the subsequent reduction in heart rate, cardiac output,

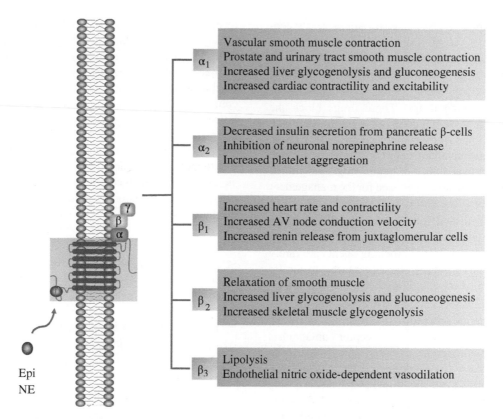

FIGURE 8.2 The physiological processes mediated by adrenergic receptor signaling. As depicted, activation of adrenergic receptors by epinephrine (Epi) and norepinephrine (NE) not only directly impacts the cardiovascular system but also affects other organs/systems and diverse metabolic pathways.

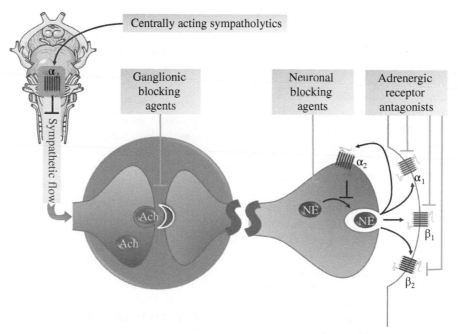

FIGURE 8.3 The sympathetic flow and drug targeting. As illustrated, there are four categories of drugs that inhibit the sympathetic activity: centrally acting sympatholytics inhibit central sympathetic discharge by activating α_2-adrenergic receptors in the brainstem, thereby leading to decreased sympathetic flow to the peripheral tissues and organs; ganglionic blocking agents block the neurotransmission in the ganglia; neuronal blocking agents block the synthesis, storage, or release of norepinephrine from the sympathetic postganglionic nerve terminal; and adrenergic receptor antagonists block α- and β-adrenergic receptors or both located primarily on the membranes of the effector cells. Ach and NE denote acetylcholine and norepinephrine, respectively.

and total peripheral resistance. α_2-Adrenergic receptors are found on both presynaptic neurons and postsynaptic cells. Activation of these receptors on presynaptic neurons results in feedback inhibition of sympathetic transmission.

2. Ganglionic blocking drugs: These drugs block the transmission of nerve impulses between preganglionic and postganglionic neurons in the sympathetic ganglia as well as parasympathetic ganglia. Due to the adverse effects and the availability of better drugs, ganglionic blocking agents are no longer used for the management of cardiovascular diseases.

3. Sympathetic neuronal blocking drugs: These drugs block the synthesis, storage, or release of norepinephrine. Sympathetic neuronal blocking agents are rarely used for treating cardiovascular diseases due to similar reasons described for ganglionic blocking drugs.

4. Adrenergic receptor antagonists: These drugs can block α-adrenergic receptors, β-adrenergic receptors, or both. Hence, adrenergic receptor antagonists include (i) α-adrenergic receptor antagonists, also called α-blockers; (ii) β-adrenergic receptor antagonists, commonly known as β-blockers; and (iii) α- and β-adrenergic receptor antagonists, also called α- and β-blockers. This subclass is commonly included in the β-blocker class.

8.3 α-ADRENERGIC RECEPTOR ANTAGONISTS

α-Adreneryic receptors are G-protein-coupled receptors. There are two types of α-adrenergic reχeptors (α_1 and α_2), and each has three subtypes (α_{1A}, α_{1B}, α_{1D}, α_{2A}, α_{2B}, α_{2C}). There is no α_{1C} receptor. At one time, there was a subtype known as C, βut it was found to be identical to one of the previously discovered subtypes. To avoid confusion, naming was continued with the letter D.

Alpha$_1$ is a Gq-coupled receptor. α_1-Adrenergic receptors are expressed in vascular smooth muscle and genitourinary smooth muscle (prostate, urethra), and activation of the receptors causes smooth muscle contraction. α_1-Adrenergic receptors are also present in the heart and may play an important role in cardiac function. In this context, recent studies demonstrated that cardiac α_1-adrenergic receptors mediate important protective and adaptive functions in the heart, although they are only a minor fraction of total cardiac adrenergic receptors [2]. Cardiac α_1-adrenergic receptors activate pleiotropic downstream signaling to prevent pathological remodeling in heart failure. Mechanisms defined in animal and cell models include activation of adaptive hypertrophy, prevention of cardiac myocyte death, augmentation of contractility, and induction of ischemic preconditioning. Surprisingly, at the molecular level, cardiac α_1-adrenergic

receptors localize to and signal at the nucleus in cardiac myocytes and, unlike most G-protein-coupled receptors, activate "inside-out" signaling to cause cardioprotection. Human clinical studies also showed that α_1-blockade worsened heart failure in hypertension and did not improve outcomes in heart failure, implying a cardioprotective role for human cardiac α_1-adrenergic receptors [2].

In contrast to α_1, α_2 is a Gi-coupled receptor. As noted earlier, α_2-adrenergic receptors work mainly as autoreceptors to mediate feedback inhibition of sympathetic transmission. α_2-Adrenergic receptors are also located on nonneuronal cells, including pancreatic β-cells, and platelets. Activation of these receptors in the β-cells inhibits insulin release. These receptors on platelets mediate platelet aggregation.

α-Blockers can be distinguished on the basis of their selectivity for the types or subtypes of the α-adrenergic receptors and by their reversible or irreversible blockage of these receptors:

1. Nonselective α-blockers: blocking both α_1- and α_2-adrenergic receptors
   - Phenoxybenzamine (Dibenzyline): an irreversible inhibitor
   - Phentolamine (Regitine): a competitive inhibitor
2. Selective α_1-blockers: selectively blocking α_1-adrenergic receptors
   - Alfuzosin (Uroxatral)
   - Doxazosin (Cardura)
   - Prazosin (Minipress)
   - Silodosin (Rapaflo)
   - Tamsulosin (Flomax)
   - Terazosin (Hytrin)

Due to the availability of better drugs, α-blockers are less commonly used in the management of cardiovascular diseases. However, they can be useful for particular cardiovascular conditions, such as the management of pheochromocytoma-associated hypertensive episodes by phenoxybenzamine or phentolamine. Although α_1-selective inhibitors may be used to treat essential hypertension, the chief clinical application of these drugs is to treat benign prostatic hyperplasia via their ability to relax the sphincter of the bladder neck and the smooth muscle of the prostate, thereby improving the lower urinary tract symptoms.

8.4 β-ADRENERGIC RECEPTOR ANTAGONISTS

In contrast to α-blockers, β-blockers are widely used in the management of various forms of cardiovascular diseases, including hypertension, ischemic heart disease, heart failure, and cardiac arrhythmias. In addition, β-blockers are also useful in other clinical conditions, including endocrinological and neurological disorders, glaucoma, and others. As such, this chapter focuses on discussing the pharmacological basis of the use of β-blockers in treating cardiovascular diseases.

8.4.1 General Introduction to Drug Class

β-Adrenergic receptors belong to the G-protein-coupled receptor family and include three types: β_1, β_2, and β_3. All three types of the beta-adrenergic receptors are linked to G_s proteins although β_2 also couples to G_i. Activation of β-adrenergic receptors by catecholamines leads to various physiological responses (Fig. 8.2). Notably, β_3-adrenergic receptors are predominantly expressed in adipose tissue, and stimulation of the receptor signaling results in increased lipolysis. Substantial evidence also demonstrates the presence and physiological function of β_3-adrenergic receptors in the cardiovascular system. In animal models, activation of β_3-adrenergic receptor signaling attenuated pressure overload hypertrophy and heart failure as well as myocardial ischemia–reperfusion injury via a nitric oxide-dependent mechanism [3, 4].

Although β-adrenergic receptors play an important role in diverse physiological processes, dysregulation of these receptor-mediated signaling, frequently manifested as overstimulation, is a central pathophysiological mechanism underlying various cardiovascular disorders. This notion has led to the development of drugs that inhibit these receptors when they become overstimulated. The release of the first clinically useful β-blocker (propranolol) in the early 1960s owing to the efforts of J.W. Black (a Nobel laureate; also see Chapter 1) revolutionized the treatment of angina pectoris. Over the following four decades, numerous members of the β-blockers have been developed and approved for clinical use. These "old dogs" have learned many new tricks over the past decades, from protecting the heart after a heart attack to controlling heart failure. Today, millions of Americans take a β-blocker.

To date, there are 16 US Food and Drug Administration (FDA)-approved β-blockers. Based on their selectivity for the β-adrenergic receptors as well as α-adrenergic receptors and other novel activities, β-blockers are classified into the following three groups (Table 8.1):

- Nonselective β-blockers: These drugs are also known as first-generation β-blockers, and they block both β_1- and β_2-adrenergic receptors.
- Selective β_1-blockers: These drugs are also known as cardioselective β-blockers and second-generation β-blockers, and they selectively block β_1-adrenergic receptors.
- Third-generation β-blockers: These drugs are also known as β-blockers with additional actions.

TABLE 8.1 Classification of β-blockers

Group of β-blockers	Member
Nonselective β-blockers (block both β_1- and β_2-adrenergic receptors)	Carteolol (Ocupress)
	Nadolol (Corgard)
	Penbutolol (Levatol)
	Pindolol (Visken)
	Propranolol (Inderal)
	Sotalol (Betapace, Timolide)
	Timolol (Blocadren)
Selective β_1-blockers (selectively block β_1-adrenergic receptors)	Acebutolol (Sectral)
	Atenolol (Tenormin)
	Betaxolol (Kerlone)
	Bisoprolol (Zebeta)
	Esmolol (Brevibloc)
	Metoprolol (Lopressor, Toprol)
Third-generation β-blockers (block β-adrenergic receptors and possess other novel activities)	Carvedilol (Coreg)
	Labetalol (Normodyne, Trandate)
	Nebivolol (Bystolic)

8.4.1.1 General Pharmacological Properties of Nonselective β-Blockers
Summarized below are the general pharmacological properties of the seven nonselective β-blockers:

- All drugs in this group block both β_1- and β_2-adrenergic receptors.
- Pindolol also has a partial agonist activity (also known as intrinsic sympathomimetic activity, ISA). This means that, apart from blocking β_1- and β_2-adrenergic receptors, it produces some stimulation. Pindolol therefore only slightly influences normal sympathetic drive at rest but effectively reduces the effects of elevated sympathetic activity. In humans, ISA is manifested by a smaller reduction in the resting heart rate (4–8 beats/min) than is seen with drugs lacking ISA. There is also a smaller reduction in resting cardiac output than is seen with drugs lacking ISA. Carteolol, a topical use drug for glaucoma, also exhibits ISA.
- Blockage of cardiac β_1-adrenergic receptors decreases myocardial contractility and heart rate, thereby resulting in decreased cardiac output and blood pressure.
- Blockage of β_1-adrenergic receptors on kidney juxtaglomerular cells decreases renin release, thereby resulting in inhibition of the RAAS.
- Blockage of β_2-adrenergic receptors causes bronchoconstriction. This is why these drugs are contraindicated in patients with asthma or chronic obstructive pulmonary diseases (COPD).
- Blockage of β_2-adrenergic receptors decreases epinephrine-stimulated liver glycogenolysis. This slows the recovery of blood glucose after a hypoglycemic episode caused by insulin injection in patients with diabetes.

Because of this, these drugs should be used with caution in diabetic patients who use insulin.

8.4.1.2 General Pharmacological Properties of Selective β_1-Blockers
The general pharmacological properties of the six selective β_1-blockers are listed below:

- β_1-Adrenergic receptors are primarily located in the heart, and as such, selective β_1-blockers are also called cardioselective β-blockers. However, it should be noted that none of the clinically available β-blockers are absolutely specific for β_1-adrenergic receptors. The selectivity is dose related and it tends to diminish as drug dose increases. In this regard, at low therapeutic doses, selective β_1-blockers do not generally cause bronchoconstriction. However, as doses increase, they may also block β_2-receptors, and as such, they should be used with caution in patients with asthma or COPD.
- As noted earlier, blockage of cardiac β_1-adrenergic receptors decreases myocardial contractility and heart rate, thereby resulting in decreased cardiac output and blood pressure. Blockage of β_1-adrenergic receptors on kidney juxtaglomerular cells decreases renin release, thereby resulting in inhibition of the RAAS.

8.4.1.3 General Pharmacological Properties of Third-Generation β-Blockers
Outlined below are the general pharmacological properties of the 3 third-generation β-blockers:

- Carvedilol and labetalol block β_1- and β_2-adrenergic receptors as well as α_1-receptors, and as such, they are also known as nonselective β- and selective α_1-blockers. Blockage of α_1-adrenergic receptors causes vasodilation, thereby lowering blood pressure. Decreased blood pressure potentially results in reflex tachycardia. Importantly, blockage of cardiac β_1-adrenergic receptors by these drugs suppresses reflex tachycardia caused by α_1 blockage.
- Carvedilol may also have antioxidant activities and increase nitric oxide bioavailability.
- Nebivolol is a β_1-selective blocker, and it has additional actions, such as increasing nitric oxide bioavailability and activating cardiac β_3-adrenergic receptors [5, 6]. As noted earlier, activation of β_3-adrenergic receptors in the cardiovascular system may be a protective mechanism against heart failure and myocardial ischemia–reperfusion injury.

Although blockage of β-adrenergic receptor signaling appears to be largely responsible for the overall pharmacological effects and the clinical efficacy of the β-blockers, other actions, such as increasing nitric oxide bioavailability and antioxidant activities, as noted earlier, may also contribute to the clinical efficacy of some of the β-blockers. Due to the large volume of available information on the various members of the β-blockers, this chapter is intended to focus on discussing those β-blockers that have been documented to be especially useful for treating cardiovascular diseases or exert novel beneficial effects on the cardiovascular system. These β-blockers (in alphabetic order) include atenolol, bisoprolol, carvedilol, labetalol, metoprolol, nebivolol, and propranolol.

8.4.2 Chemistry and Pharmacokinetics

β-Blockers are structurally similar to β-adrenergic receptor agonists. The structures of some β-blockers commonly used in cardiovascular medicine are given in Figure 8.4, and some of their pharmacokinetic properties are provided in Table 8.2.

8.4.3 Molecular Mechanisms and Pharmacological Effects

Blockage of the β-adrenergic receptors is largely responsible for the pharmacological as well as adverse effects of the β-blockers. As stated earlier, some β-blockers also exert beneficial effects independent of the receptor blockage, such as increasing nitric oxide bioavailability and antioxidant activities. As β-adrenergic receptors are widely distributed and the receptor-mediated signaling regulates diverse physiological processes, the pharmacological effects of β-blockers are extended beyond the cardiovascular system. This section examines the major pharmacological effects of β-blockers commonly used in cardiovascular medicine and the underlying molecular mechanisms of action.

8.4.3.1 Pharmacological Effects on Cardiovascular and Renal Systems
The major pharmacological and therapeutic effects of β-blockers are directed at the cardiovascular and renal systems, which occur via three major pathways including effects on myocardial contractility and heart rate, the RAAS, and vascular smooth muscle. Some of the effects are also described in Section 8.4.1.

Effects on Myocardial Contractility and Heart Rate β_1-Adrenergic receptors are predominantly present in the heart and renal juxtaglomerular cells. In the heart, β_1-adrenergic receptors are located in the cardiac conducting system and contracting myocytes. Activation of β_1-adrenergic receptors by norepinephrine released from sympathetic adrenergic nerves, or norepinephrine and epinephrine that circulate in the blood, results in increased heart rate, contractility, conduction velocity, and relaxation rate. They are also known as positive chronotropy, inotropy, dromotropy, and lusitropy, respectively. The increased heart rate results from β_1-adrenergic receptor-mediated increase in the rate of phase 4 depolarization of

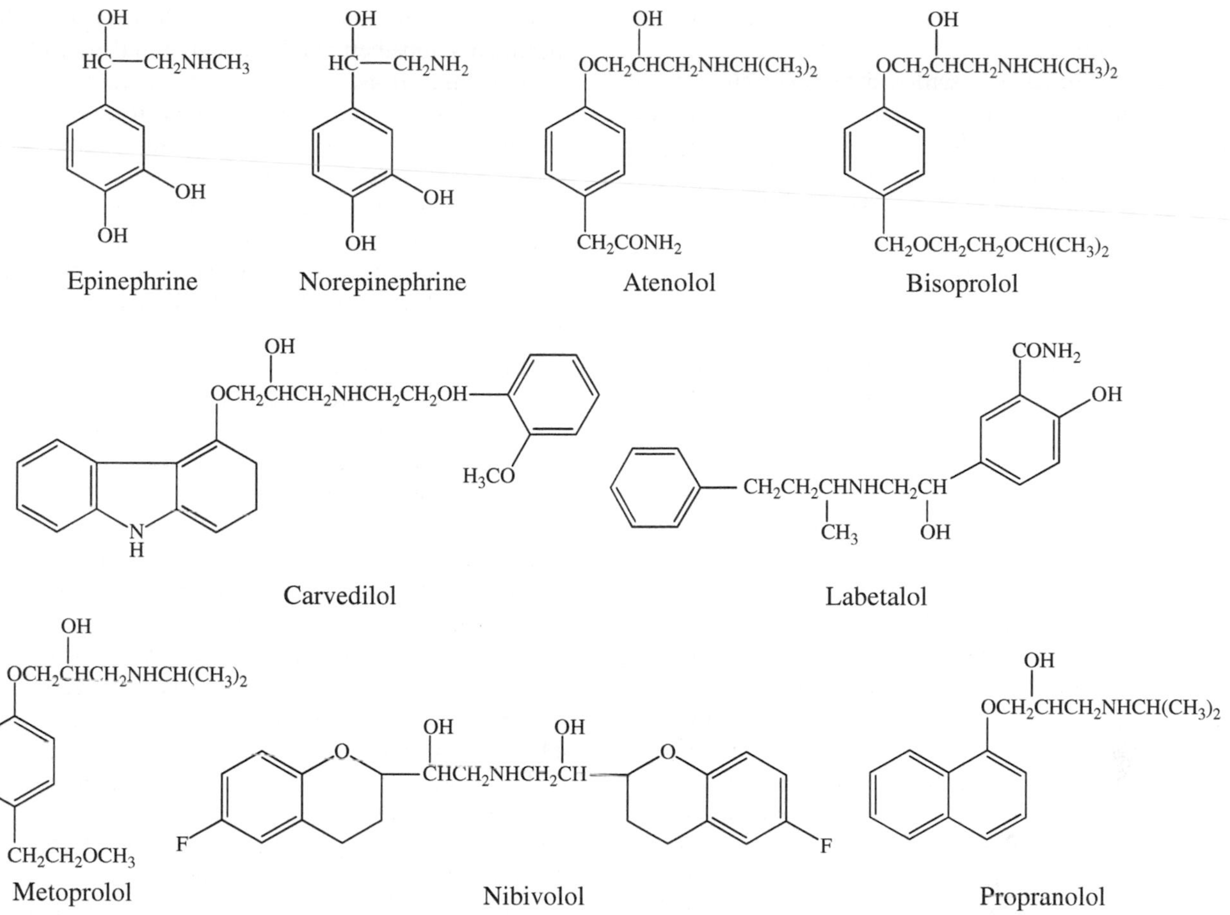

FIGURE 8.4 Structures of β-blockers commonly used to treat cardiovascular diseases. Epinephrine and norepinephrine are included to show the structural similarities between the catecholamines and antagonists of β-adrenergic receptors.

TABLE 8.2 Pharmacokinetic properties of β-blockers commonly used in treating cardiovascular diseases

β-Blocker	Lipid solubility	Oral bioavailability (%)	Elimination half-life (h)	Metabolism and elimination
Atenolol	Low	50	6–7	No metabolism; eliminated in unchanged form in the urine
Bisoprolol	Low	80	9–12	CYP3A4; eliminated in unchanged form (50% of dose) and metabolites in the urine
Carvedilol	Moderate	25–35	7–10	CYP2D6 and CYP2C9 (major); CYP3A4 (minor); eliminated in the bile/feces
Esmolol (iv)	Low		9 min	Esterases; eliminated in the urine
Labetalol	Low	25	6–8	Glucuronidation; eliminated in the urine
Metoprolol	Moderate	50	3–7	CYP2D6; eliminated in the urine
Nebivolol	Low	Unknown	12–19	Glucuronidation; CYP2D6; eliminated in the urine and feces
Propranolol	high	25	3–6	CYP2D6; glucuronidation; eliminated in the urine

sinoatrial node pacemaker cells. The positive inotropic effect is mediated by increased calcium entry into the cells and the subsequent enhanced release of calcium from the sarcoplasmic reticulum in the cardiomyocytes.

Blockage of the β₁-adrenergic receptors of the heart causes decreased heart rate and contractility, thereby reducing cardiac output. This is the basis for using β-blockers to treat hypertension. As described in Chapter 6, blood pressure is the product of cardiac output and peripheral resistance.

Because myocardial contractility determines myocardial oxygen consumption, decreased cardiac contractility caused by β-blockers results in reduction of myocardial oxygen demand. This serves as a major basis for using β-blockers in the management of angina pectoris (see Unit IV).

Effects on RAAS The RAAS plays an important role in regulating blood volume and systemic vascular resistance, which together influence cardiac output and blood pressure. As the name implies, there are three important components in this system: renin, angiotensin, and aldosterone. Renin, which is primarily released by the kidney, stimulates the formation of angiotensin in blood and tissues, which in turn stimulates the release of aldosterone from the adrenal cortex. Hence, factors that affect renin release will influence the activity of the RAAS.

Renin is a proteolytic enzyme that is released into the circulation primarily by the renal juxtaglomerular cells associated with the afferent arteriole entering the renal glomerulus (see Chapter 9). Release of renin is stimulated by three mechanisms, including (i) sympathetic nervous activation (acting via β_1-adrenergic receptors), (ii) renal artery hypotension (caused by systemic hypotension or renal artery stenosis), and (iii) decreased NaCl delivery to the distal tubules of the kidney. Blockage of β_1-adrenergic receptors of kidney juxtaglomerular cells therefore causes decreased renin release and the subsequent attenuation of the renin–angiotensin–aldosterone axis. Reduced RAAS activity by β-blockers contributes to not only decreased fluid retention but also reduction of peripheral resistance. In addition, because of the crosstalk between the sympathetic nervous system and the RAAS (see Chapter 6), inhibition of RAAS also results in attenuation of sympathetic activity.

Effects on Vascular Smooth Muscle Vascular smooth muscle has β_2-adrenergic receptors that have a high binding affinity for circulating epinephrine and a relatively lower affinity for norepinephrine released by sympathetic adrenergic nerves. Vascular smooth muscle β_2-adrenergic receptors, like β_1-adrenergic receptors in the heart, are coupled to a Gs protein, which stimulates the formation of cAMP. Although increased cAMP enhances cardiac myocyte contraction, in vascular smooth muscle, an increase in cAMP leads to smooth muscle relaxation. This is because cAMP inhibits myosin light chain kinase that is responsible for phosphorylating smooth muscle myosin. Hence, activation of β_2-adrenergic receptors in vascular smooth muscle results in less contractile force (i.e., promoting relaxation).

For nonselective β-blockers, blockage of β_2-adrenergic receptors on vascular smooth muscle cells may increase the peripheral resistance. In addition, the initial decrease in blood pressure resulting from the blockage of β_1-adrenergic

receptors by the nonselective β-blockers may lead to increased sympathetic activity and stimulation of vascular smooth muscle α_1-adrenergic receptors. The aforementioned effects, if not counteracted, would result in increased peripheral resistance. However, with long-term use of nonselective β-blockers, the peripheral resistance in hypertensive patients is reduced rather than increased. This net effect is likely due to the combined effects of decreased cardiac output (section "Effects on Myocardial Contractility and Heart Rate") and inhibited RAAS (section "Effects on RAAS") overriding the vasoconstrictive effect of β_2-adrenergic receptor blockage.

For the third-generation β-blockers that are also selective α_1-blockers (i.e., carvedilol and labetalol), the blockage of vascular α_1-adrenergic receptors can lead to a greater fall in peripheral resistance. This may be also true for nebivolol, a third-generation β-blocker with additional vasodilating effect.

In summary, the major pharmacological effects of β-blockers on the cardiovascular system in patients with cardiovascular diseases include (i) decreased blood pressure, (ii) decreased cardiac output and oxygen demand, and (iii) attenuation of the deleterious effects of the dysregulated RAAS (Fig. 8.5). Here, it is imperative to mention that the effects of β-blockers in patients with cardiovascular diseases and in healthy individuals are different. In patients with cardiovascular diseases, such as hypertension and ischemic heart disease, β-blockers have more pronounced inhibitory effects because of the elevated sympathetic activity and high tonic stimulation of the β-adrenergic receptors under the disease conditions. In contrast, in healthy individuals at rest, the tonic stimulation of the β-adrenergic receptors is low, and as such, the inhibitory effects of β-blockers are correspondingly moderate.

8.4.3.2 Effects on Other Organs and Systems

Metabolic and Endocrine Effects β-Adrenergic receptor signaling participates in lipid metabolism and glucose homeostasis. As shown in Figure 8.2, β-adrenergic receptors mediate activation of hormone-sensitive lipase in adipocytes, leading to the increased release of free fatty acids into circulation to serve as an important source of energy for exercising muscle. Nonselective β-blockers may attenuate the sympathetic stimulation of lipolysis, potentially contributing to the unfavorable lipid profiles, including increased levels of plasma triglycerides and LDL cholesterol and decreased levels of HDL cholesterol. In contrast, the third-generation β-blockers, carvedilol and nebivolol, may improve lipid profiles in patients with dyslipidemias. The exact mechanisms underlying the lipid-modifying effects of β-blockers remain to be further elucidated.

Activation of β_2-adrenergic receptor signaling leads to increased muscle and liver glycogenolysis and increased release of glucagon, contributing to glucose recovery from hypoglycemia. Use of nonselective β-blockers may impair the recovery

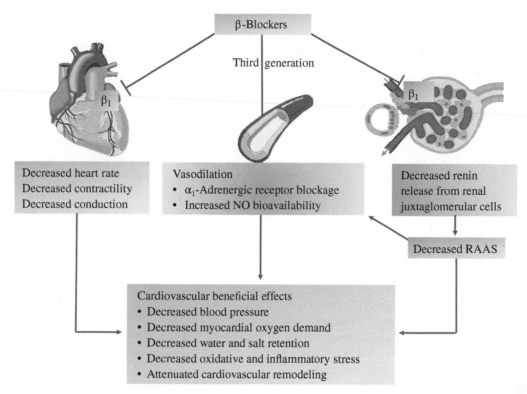

FIGURE 8.5   Pharmacological effects of β-blockers on the cardiovascular system. The beneficial cardiovascular outcomes of β-blockers are primarily attributed to (i) decreased heart rate, contractility, and conduction; (ii) vasodilation; and (iii) inhibition of the renin–angiotensin–aldosterone system (RAAS). Notably, the third-generation β-blockers also possess other novel activities, especially augmentation of nitric oxide (NO) bioavailability. These effects collectively contribute to decreased blood pressure, reduced myocardial oxygen demand, attenuated water and salt retention, inhibition of oxidative and inflammatory stress, and retardation of cardiovascular pathogenic remodeling. Here, the term remodeling refers to alterations in the structure and morphology of the cardiovascular tissues in response to hemodynamic load and/or tissue injury in association with neurohormonal activation, especially activation of the RAAS. Cardiovascular remodeling can be physiological (adaptive) or pathological (maladaptive) in nature. Cardiovascular pathological remodeling is an important mechanism of cardiovascular degeneration under disease conditions, including hypertension, myocardial infarction, and heart failure.

from hypoglycemia in diabetic patients, particularly those who are dependent on insulin therapy. Selective $β_1$-blockers may be less prone to inhibit the recovery from hypoglycemia.

The third-generation β-blockers have been demonstrated to improve glycemic control and insulin resistance in patients with diabetes and metabolic syndrome. The underlying mechanisms are not clear and may be associated with the combination of decreased activation of the RAAS, increased nitric oxide bioavailability, and antioxidant activities.

Effects on the Respiratory Tract   Similar to what was seen with vascular smooth muscle, activation of $β_2$-adrenergic receptors in bronchial smooth muscle causes bronchodilation. Consequently, blockage of the $β_2$-adrenergic receptors on bronchial smooth muscle cells by nonselective β-blockers may lead to bronchoconstriction. As such, nonselective β-blockers should be avoided in patients with asthma or COPD. Although the selective $β_1$-blockers are less prone to cause bronchoconstriction, bronchoconstriction may become significant as doses increase in patients with asthma. In contrast, many patients with COPD may tolerate the use of selective $β_1$-blockers. The use of the selective $β_1$-blockers in COPD patients with concurrent ischemic heart disease, especially acute coronary syndromes, is recommended as the use of the $β_1$-blockers reduces the mortality from the concomitant ischemic heart disease (see Unit V).

Effects on the Eye   β-Adrenergic receptor signaling mediates the production of the aqueous humor. Blockage of β-adrenergic receptors by β-blockers results in the decreased formation of the aqueous humor and the consequent decrease of intraocular pressure in patients with glaucoma. This is the pharmacological basis for using β-blockers to treat glaucoma.

8.4.4   Clinical Uses

Because β-adrenergic receptors are widely distributed and the receptor signaling plays an important role in diverse physiological and pathophysiological processes, β-blockers are useful for treating a wide variety of diseases. This section focuses on discussing the cardiovascular applications of β-blockers. The use of β-blockers in other disease conditions, especially the emerging role of these drugs in cancer treatment, is also introduced.

8.4.4.1 Cardiovascular Diseases   β-Blockers are widely used in the treatment of hypertension, ischemic heart disease, congestive heart failure, and cardiac arrhythmias. They also show beneficial effects in other cardiovascular disorders including hypertrophic obstructive cardiomyopathy and aortic aneurysm. In addition, β-blockers are useful in selected at-risk patients in the prevention of adverse cardiovascular outcomes from noncardiac surgery, which is known as perioperative β-blocker therapy.

Treatment of Hypertension   Over the past four decades, β-blockers have been widely used for the treatment of uncomplicated hypertension as well as for hypertensive patients with concomitant cardiac disorders. They have been recommended as first-line drugs for treating hypertension in both national and international guidelines. However, several recent meta-analyses of the previous studies produced controversies regarding whether β-blockers should be used as first-line drugs in uncomplicated hypertension. The controversies resulted from some meta-analyses suggesting that in patients with uncomplicated hypertension, β-blockers exerted a relatively weak effect in reducing the risk of stroke as compared with other antihypertensive agents [7, 8].

Much of the unfavorable data revealed in the recent meta-analyses were gleaned from studies involving nonvasodilating, traditional β-blockers, primarily the selective β_1-blocker atenolol. Findings with traditional β-blockers may not be extrapolated to other groups of β-blockers, particularly the third-generation β-blockers with vasodilatory activities, including carvedilol and nebivolol [9]. Hence, the use of β-blockers, especially the third-generation β-blockers as first-line drugs for uncomplicated hypertension, might still be an appropriate option though their definitive efficacy in treating uncomplicated hypertension for preventing stroke and cardiovascular events and mortality remains to be determined through vigorous clinical studies. Nevertheless, β-blockers with vasodilating effects, including carvedilol and nebivolol, but not the traditional β-blocker atenolol, should be considered whenever β-blockers are indicated in hypertensive patients. Table 8.3 lists the dosage range of common β-blockers in treating hypertension.

Treatment of Ischemic Heart Disease   β-Blockers are widely used for the management of ischemic heart diseases, including stable angina and unstable angina and non-ST elevation myocardial infarction. The usual dosage range for commonly used β-blockers in treating stable angina is listed in Table 8.4. The dosage information for treating acute coronary syndromes is provided in Unit V.

TABLE 8.3   Usual dosage range of common β-blockers for treating hypertension

β-Blockers	Usual dosage (mg/day) for hypertension
Atenolol	25–100
Bisoprolol	5–20
Carvedilol	10–80
Labetalol	200–800
Metoprolol	25–400
Nebivolol	5–10
Propranolol	40–180

TABLE 8.4   Usual dosage range for common β-blockers in treating stable angina pectoris

β-Blockers	Usual dosage for stable angina
Atenolol	50–200 mg/day
Bisoprolol	10 mg/day
Labetalol	200–600 mg twice daily
Metoprolol succinate	100–400 mg/day
Nebivolol	5–40 mg/day
Propranolol	80–120 mg twice daily

TABLE 8.5   Dosages of β-blockers in treating congestive heart failure

β-Blockers	Initial once daily dose (mg)	Maximal once daily dose (mg)
Bisoprolol	1.25	10
Carvedilol	10	80
Metoprolol	12.5–25	200

Treatment of Heart Failure   β-Blockers have been becoming a major therapy of congestive heart failure with reduced ejection fraction due to their documented ability to prolong survival and reduce mortality. The mechanisms underlying the beneficial effects include (i) inhibition of the sympathetic stimulation of myocardium and cardiac remodeling, (ii) suppression of the RAAS activation by decreasing the release of renin from renal juxtaglomerular cells, and (iii) attenuation of oxidative stress and inflammation in myocardium. Clinical data also suggest potential beneficial effects of β-blockers, especially the third-generation β-blockers on survival and mortality in congestive heart failure patients with concomitant chronic kidney disease or COPD [10–12]. In contrast to the treatment of heart failure with reduced left ventricular ejection fraction, the use of β-blockers in heart failure with preserved left ventricular ejection fraction is controversial [13, 14].

Three β-blockers have been approved by the US FDA for treating congestive heart failure due to the documented efficacy in large clinical trials. They are bisoprolol, carvedilol, and metoprolol succinate. The dosage range of these β-blockers in treating heart failure is listed in Table 8.5.

Treatment of Cardiac Arrhythmias β-Blockers are one of the four classes of antiarrhythmic agents (see Unit VII). The antiarrhythmic effects of β-blockers are related to their ability to inhibit sympathetic stimulation of the cardiac conduction system. Sympathetic stimulation of the cardiac conduction system increases heart rate, conduction velocity, and aberrant pacemaker activity (ectopic beats). These sympathetic influences are mediated primarily through cardiac β_1-adrenergic receptor signaling. Therefore, β-blockers can attenuate these sympathetic effects and thereby reduce sinus rate, decrease conduction velocity, and inhibit aberrant pacemaker activity. β-Blockers also affect nonpacemaker action potentials by increasing action potential duration and the effective refractory period, thereby reducing reentry. Reentry is a key pathophysiological event leading to cardiac arrhythmias (see Chapter 23).

8.4.4.2 Other Diseases and Conditions

Although cardiovascular diseases are the main indications for β-blockers, these drugs also find their use in many other diseases and conditions, including glaucoma, hyperthyroidism, neurological disorders (e.g., migraine headache, tremors, performance anxiety), portal vein high pressure, as well as breast cancer. β-Adrenergic receptor signaling has been demonstrated to promote cell tumorigenesis, and treatment with β-blockers suppresses tumorigenesis and metastasis in laboratory models. Multiple recent population studies also suggest a role for β-blockers in reducing metastases, tumor recurrence, and specific mortality in breast cancer [15–17].

8.4.5 Dosage Forms and Strengths

Listed below are the dosage forms and strengths of some commonly used β-blockers:

- Atenolol (Tenormin): Oral, 25, 50, and 100 mg tablets
- Bisoprolol (Zebeta): Oral, 5 and 10 mg tablets
- Carvedilol (Coreg): Oral, 3.125, 6.25, 12.5, and 25 mg tablets
- Esmolol (Brevibloc): Intravenous, 10 ml (10 mg/ml) vials
- Labetalol (Normodyne, Trandate): Oral, 100, 200, and 300 mg tablets; 5 mg/ml in 20 and 40 ml vials
- Metoprolol (Lopressor, Toprol): Oral, 50 and 100 mg tablets (metoprolol tartrate); 25, 50, 100, and 200 mg extended-release tablets (metoprolol succinate)
- Nebivolol (Bystolic): Oral, 2.5, 5, 10, and 20 mg tablets
- Propranolol (Inderal): Oral, 10, 20, 40, 60, and 80 mg tablets; 60, 80, 120, and 160 mg capsules (extended release); intravenous, 1 mg/ml in 1 ml vials

8.4.6 Adverse Effects and Drug Interactions

8.4.6.1 Adverse Effects

The most common adverse effects of β-blockers are related to the predictable pharmacological consequences of blockage of the β-adrenergic receptors. Significant adverse effects independent of β-adrenergic receptor blockage are rare. Common adverse effects of β-blockers include tiredness, dizziness, shortness of breath, and hypotension. The conventional wisdom that β-blocker therapy is associated with substantial risks of depressive symptoms, fatigue, and sexual dysfunction is not supported by data from clinical trials [18].

Blockage of β_1-adrenergic receptor signaling may aggravate the condition in patients with heart failure. As such, β-blocker therapy should be initiated at lower doses and gradually increased over time with close observation of the patient's conditions. As mentioned earlier, blockage of β_2-adrenergic receptors in bronchial smooth muscle causes bronchoconstriction, and as such, nonselective β-blockers should be generally avoided in patients with asthma or COPD.

8.4.6.2 Drug Interactions

Drug interactions associated with β-blockers may occur both pharmacokinetically and pharmacodynamically. For example, inhibition of CYP enzymes, especially CYP2D6 by cimetidine, results in decreased metabolism of β-blockers that are normally metabolized by the CYP pathway (Table 8.2). Combination of β-blockers with calcium channel blockers may lead to marked bradycardia and syncope. The antihypertensive effects of β-blockers may also be opposed by concomitant use of nonsteroidal anti-inflammatory drugs. The hypoglycemic action of insulin and oral antidiabetic drugs may be enhanced by concomitant treatment with β-blockers, especially the nonselective β-blockers.

8.4.6.3 Contraindications and Pregnancy Category

Asthma or related bronchospastic conditions for nonselective β-blockers

- Severe bradycardia and second- or third-degree atrioventricular (AV) block.
- Patients in cardiogenic shock or decompensated heart failure requiring the use of inotropic therapy.
- Hypersensitivity to the drug.
- Pregnancy category: C (bisoprolol, carvedilol, labetalol, metoprolol, nebivolol, propranolol) and D (atenolol).

8.5 CENTRALLY ACTING SYMPATHOLYTICS

Centrally acting sympatholytic drugs reduce blood pressure mainly by stimulating central α_2-adrenergic receptors in the brainstem, thereby reducing sympathetic

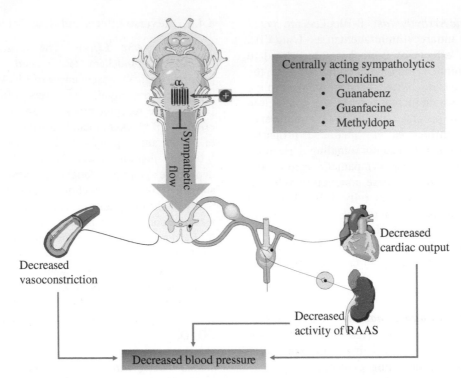

FIGURE 8.6 Molecular mechanisms by which centrally acting sympatholytics reduce blood pressure. Centrally acting sympatholytic agents reduce the sympathetic flow to both heart and blood vessels and cause a reduction in both cardiac output and peripheral resistance, thereby leading to decreased blood pressure. Centrally acting sympatholytic drugs, especially upon prolonged treatment, also attenuate the activation of the renin–angiotensin–aldosterone system (RAAS) by decreasing the release of renin from renal juxtaglomerular cells. This may also contribute to the decreased blood pressure.

nerve activity and neuronal release of norepinephrine. The reduced sympathetic outflow to the heart results in decreased cardiac output by decreasing heart rate and contractility. Reduced sympathetic output to the vasculature decreases sympathetic vascular tone, which causes vasodilation and reduced peripheral resistance. Together, these effects lead to decreased blood pressure (Fig. 8.6). Currently, four centrally acting sympatholytic drugs are approved by the US FDA:

1. Clonidine (Catapres)
2. Guanabenz (Wytensin)
3. Guanfacine (Tenex)
4. Methyldopa

These drugs are infrequently used for hypertension because of adverse effects including drowsiness, fatigue, and dry mouth. Rebound hypertension is also another major concern particularly in patients who are nonadherent to the drug regimen. Centrally acting sympatholytics are primarily used for the treatment of hypertension when other drugs are ineffective or for treatment of hypertensive emergencies (see Chapter 12).

8.6   SUMMARY OF CHAPTER KEY POINTS

- Overactivity of the sympathetic nervous system, which is primarily manifested as augmented adrenergic signaling and abnormal activation of the RAAS, is a fundamental pathophysiological process underlying various forms of cardiovascular diseases. Drugs that reduce sympathetic stimulation are known as sympatholytic agents or, simply, sympatholytics.

- Based on the sites and mechanisms involved, sympatholytic drugs are conventionally classified into four groups, including centrally acting sympatholytic drugs, ganglionic blocking drugs, sympathetic neuronal blocking drugs, and adrenergic receptor antagonists.

- Adrenergic receptor antagonists, including α- and β-blockers, are commonly used drugs, whereas ganglionic blocking drugs and sympathetic neuronal blocking drugs are now rarely used due to significant adverse effects.

- β-Blockers are further classified into nonselective β-blockers, selective β_1-blockers, and third-generation β-blockers. β-Blockers have become a mainstay of therapy for various forms of cardiovascular diseases, including hypertension, ischemic heart disease, heart failure, and arrhythmias. The beneficial effects of β-blockers result from their ability to decrease cardiac output, myocardial oxygen demand, and the activity of the RAAS. The third-generation β-blockers also possess other novel effects, including increased vascular nitric oxide bioavailability, vasodilation, and antioxidant activities.

- Centrally acting sympatholytic drugs are primarily used for the treatment of hypertension when other drugs are ineffective or for treatment of hypertensive emergencies. These drugs activate central α_2-adrenergic receptors in the brainstem. This reduces sympathetic outflow to the heart and vasculature, resulting in decreased cardiac output and reduced peripheral resistance.

8.7   SELF-ASSESSMENT QUESTIONS

8.7.1 A 25-year-old female presents with a blood pressure of 145/95 mmHg. She is prescribed a drug that reduces both cardiac output and peripheral resistance. Which of the following is most likely prescribed?
A. Alfuzosin
B. Carvedilol
C. Phenoxybenzamine
D. Silodosin
E. Tamsulosin

8.7.2 A 35-year-old female presents with episodic severely elevated blood pressure, up to the level of 170/110 mmHg. Urinalysis revolves remarkably elevated levels of epinephrine and norepinephrine metabolites. Computed tomography image shows a "mass" on the right adrenal gland. She is scheduled for a surgery to remove the "mass." However, her blood pressure needs to be controlled pharmacologically before the scheduled surgery. Which of the following drugs would be the most appropriate prescription for her condition?
A. Alfuzosin
B. Carvedilol
C. Chlorthalidone
D. Phenoxybenzamine
E. Propranolol

8.7.3 A 25-year-old type 1 diabetic patient presents with a blood pressure of 145/95 mmHg. He depends on regular insulin injections to control his blood sugar. A decision is made to put him on a beta-blocker to reduce his blood pressure. Which of the following drugs would be most suitable?
A. Atenolol
B. Labetalol
C. Nebivolol
D. Pindolol
E. Propranolol

8.7.4 A 58-year-old male presents to the doctor's office complaining of dizziness and insomnia. History reveals chronic obstructive pulmonary disease (COPD) of 10 years of duration. He smoked cigarettes, 1–2 packs/day, for 20 years until he was diagnosed with COPD a decade ago. His blood pressure is 150/95 mmHg and heart rate 89 beats/min. Lab tests reveal elevated total blood cholesterol and LDL cholesterol. He is prescribed atorvastatin to control his blood cholesterol and a beta-blocker to treat his hypertension. Which of the following drugs is the most appropriate β-blocker to prescribe?
A. Carvedilol
B. Metoprolol
C. Nadolol
D. Pindolol
E. Propranolol

8.7.5 A 55-year-old patient with congestive heart failure and a reduced ejection fraction is put on a beta-blocker therapy to retard pathogenic cardiac remodeling and improve survival. Which of the following drugs is the most appropriate β-blocker to prescribe?
A. Atenolol
B. Carteolol
C. Carvedilol
D. Pindolol
E. Propranolol

REFERENCES

1 Parati, G. and M. Esler, The human sympathetic nervous system: its relevance in hypertension and heart failure. *Eur Heart J*, 2012. 33(9): p. 1058–66.

2 O'Connell, T.D., et al., Cardiac alpha1-adrenergic receptors: novel aspects of expression, signaling mechanisms, physiologic function, and clinical importance. *Pharmacol Rev*, 2014. 66(1): p. 308–33.

3 Niu, X., et al., Cardioprotective effect of beta-3 adrenergic receptor agonism: role of neuronal nitric oxide synthase. *J Am Coll Cardiol*, 2012. 59(22): p. 1979–87.

4 Aragon, J.P., et al., Beta3-adrenoreceptor stimulation ameliorates myocardial ischemia-reperfusion injury via endothelial nitric oxide synthase and neuronal nitric oxide synthase activation. *J Am Coll Cardiol*, 2011. 58(25): p. 2683–91.

5 Dessy, C., et al., Endothelial beta3-adrenoreceptors mediate nitric oxide-dependent vasorelaxation of coronary microvessels in response to the third-generation beta-blocker nebivolol. *Circulation*, 2005. 112(8): p. 1198–205.

6 Rozec, B., et al., Nebivolol, a vasodilating selective beta(1)-blocker, is a beta(3)-adrenoceptor agonist in the nonfailing transplanted human heart. *J Am Coll Cardiol*, 2009. 53(17): p. 1532–8.

7   Lindholm, L.H., B. Carlberg, and O. Samuelsson, Should beta blockers remain first choice in the treatment of primary hypertension? A meta-analysis. *Lancet*, 2005. 366(9496): p. 1545–53.

8   Law, M.R., J.K. Morris, and N.J. Wald, Use of blood pressure lowering drugs in the prevention of cardiovascular disease: meta-analysis of 147 randomised trials in the context of expectations from prospective epidemiological studies. *BMJ*, 2009. 338: p. b1665.

9   Ram, C.V., Beta-blockers in hypertension. *Am J Cardiol*, 2010. 106(12): p. 1819–25.

10 Badve, S.V., et al., Effects of beta-adrenergic antagonists in patients with chronic kidney disease: a systematic review and meta-analysis. *J Am Coll Cardiol*, 2011. 58(11): p. 1152–61.

11 Hawkins, N.M., et al., Heart failure and chronic obstructive pulmonary disease the quandary of beta-blockers and beta-agonists. *J Am Coll Cardiol*, 2011. 57(21): p. 2127–38.

12 Wali, R.K., et al., Efficacy and safety of carvedilol in treatment of heart failure with chronic kidney disease: a meta-analysis of randomized trials. *Circ Heart Fail*, 2011. 4(1): p. 18–26.

13 Waller, D.G. and J.R. Waller, Beta blockers for heart failure with reduced ejection fraction. *BMJ*, 2011. 343: p. d5603.

14 Kubon, C., et al., The role of beta-blockers in the treatment of chronic heart failure. *Trends Pharmacol Sci*, 2011. 32(4): p. 206–12.

15 Powe, D.G. and F. Entschladen, Targeted therapies: using beta-blockers to inhibit breast cancer progression. *Nat Rev Clin Oncol*, 2011. 8(9): p. 511–2.

16 Botteri, E., et al., Therapeutic effect of beta-blockers in triple-negative breast cancer postmenopausal women. *Breast Cancer Res Treat*, 2013. 140(3): p. 567–75.

17 Obeid, E.I. and S.D. Conzen, The role of adrenergic signaling in breast cancer biology. *Cancer Biomark*, 2013. 13(3): p. 161–9.

18 Ko, D.T., et al., Beta-blocker therapy and symptoms of depression, fatigue, and sexual dysfunction. *JAMA*, 2002. 288(3): p. 351–7.

9

INHIBITORS OF THE RENIN–ANGIOTENSIN–ALDOSTERONE SYSTEM

9.1 OVERVIEW

The renin–angiotensin–aldosterone system (RAAS) plays a fundamental role in the control of blood pressure, fluid volume, and sodium balance. However, dysregulation, especially overstimulation of this system, contributes to the pathophysiology of various forms of cardiovascular disorders, including atherosclerosis, hypertension, left ventricular hypertrophy, myocardial infarction, and heart failure. As a result, the dysregulated RAAS represents a logical therapeutic target in the management of cardiovascular diseases. Currently, four classes of drugs are available that target the RAAS axis, including (i) direct renin inhibitors, (ii) angiotensin-converting enzyme (ACE) inhibitors, (iii) angiotensin receptor blockers (ARBs), and (iv) aldosterone receptor antagonists. This chapter examines each of the first 3 classes of the RAAS-inhibiting drugs, focusing on discussing the pharmacological basis of their use in the management of various cardiovascular diseases. The aldosterone receptor antagonists are covered in Chapter 7 under potassium-sparing diuretics.

9.2 THE RAAS AND DRUG TARGETING

9.2.1 History for Discovery of RAAS and Development of the RAAS Inhibitors

Renin, the rate-limiting enzyme of the RAAS, has been discovered for more than a century [1, 2]. In 1898, R. Tigerstedt and P. Bergman first demonstrated that an extract from the renal cortex of rabbits (later named *renin*)

increased blood pressure when it was injected intravenously to recipient rabbits. However, the findings of R. Tigerstedt and P. Bergman could not be reproduced in other studies, and the discovery of renin was once disputed and ignored.

It took another four decades for scientists to realize in the 1940s that renin functioned as an enzyme on a protein substrate to produce a peptide that mediated the vasopressor effect of renin. The protein substrate was later named *angiotensinogen*, and the vasoconstricting peptide known as *angiotensin*. Further work by L.T. Skeggs and coworkers in 1957 demonstrated that angiotensin existed in two distinct forms—angiotensin I and angiotensin II—and angiotensin I was cleaved by ACE to generate the biologically active angiotensin II.

The last molecule of the RAAS axis, aldosterone, was isolated by S.A. Simpson and J.F. Tilt in 1953. The relationship between angiotensin II and aldosterone was hypothesized by F. Gross in 1958 and subsequently confirmed by J.O. Davis in 1959 [1, 2].

As understanding of the pathophysiological role of the RAAS in cardiovascular diseases has unfolded during the past century, so has the interest in developing drugs that could interdict specific components of the RAAS to treat cardiovascular diseases as well as other diseases that involve a mechanism of dysregulated RAAS. The first RAAS-blocking drugs to become commercially available were the aldosterone antagonists in the 1970s, followed by the ACE inhibitors in the 1980s, the angiotensin II receptor blockers in the 1990s, and the direct renin inhibitor aliskiren in 2007 [3].

Cardiovascular Diseases. From Molecular Pharmacology to Evidence Based Therapeutics, First Edition. Y. Robert Li.
© 2015 John Wiley & Sons, Inc. Published 2015 by John Wiley & Sons, Inc.

9.2.2 The RAAS

The RAAS is an ever-evolving endocrine system with considerable checks and balances on the production and catabolism of angiotensin peptides most likely due to the manifold effects of angiotensins [4–6]. To understand the RAAS in drug therapy, a brief review of the components of the RAAS and an introduction to both the classical and contemporary views of the system are imperative.

9.2.2.1 The Classical View of the RAAS

In brief, the classical view of the RAAS focuses on the processing of the single obligate precursor angiotensinogen to angiotensin I by renin, a protease released from renal juxtaglomerular cells (Chapter 8). Human angiotensinogen is a 453-amino-acid-long protein that is produced by different types of cells, including hepatocytes, renal cells, adipocytes, and brain cells, with the liver being the primary source of circulating angiotensinogen. The first 12 amino acids of angiotensinogen are the most important with regard to the formation of angiotensins. Renin cleaves the first 10 amino acids from the N-terminus of the protein to form the 10-amino-acid peptide angiotensin I. Angiotensin I is inactive and is then rapidly hydrolyzed by ACE to form an active peptide of eight amino acids, known as angiotensin II. Interaction of angiotensin II with its receptors, primarily the type 1 receptors (known as AT_1 receptors), leads to the production of aldosterone and diverse physiological and pathophysiological responses. It should be noted that angiotensin II also inhibits renin release, thus providing a negative feedback to the system. Hence, when referring to the RAAS, one should examine not only the process from renin through angiotensin to aldosterone but also its associated negative feedback loop (Fig. 9.1). As discussed later, this feedback loop also affects the pharmacological effects of the RAAS inhibitors.

9.2.2.2 The Contemporary View of the RAAS

In contrast to the aforementioned classical view, the contemporary view of the RAAS is increasingly complex, involving a balance between multiple processing pathways for angiotensin generation and degradation (Fig. 9.2). The contemporary view focuses on a homologue of ACE, known as ACE2, which cleaves angiotensin II to generate angiotensin-(1–7), a peptide fragment that functions through its putative receptor, MAS, to antagonize the adverse cardiovascular actions of angiotensin II. The contemporary view also includes the regulation of renin activity by (pro)renin receptors (receptors that bind to both prorenin and renin) and the concept of intracellular and tissue-based RAAS. In addition, angiotensin II can also be produced by non-ACE pathways.

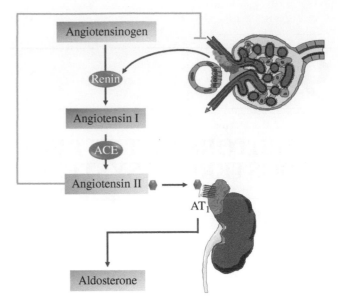

FIGURE 9.1 The classical view of the RAAS. As illustrated, the precursor peptide angiotensinogen is cleaved by renin to form the decapeptide angiotensin I. The dipeptidase angiotensin-converting enzyme (ACE) cleaves angiotensin I to form the octapeptide angiotensin II, the central active component of this system. Angiotensin II activates the AT_1 receptors in the adrenal gland, resulting in the release of aldosterone to the circulation. Of note is the feedback inhibition of renin release by angiotensin II. This feedback inhibition is mediated by activation of AT_1 receptors on the juxtaglomerular cells by angiotensin II. AT_1 denotes angiotensin II receptor type 1.

9.2.3 Drug Class and Drug Targeting

The currently available four classes of RAAS-inhibiting drugs target the classical pathway of the RAAS, that is, (i) inhibition of renin, (ii) inhibition of ACE, (iii) antagonism of AT_1 receptors, and (iv) antagonism of aldosterone receptors (Fig. 9.3). Due to the emerging critical role of the ACE2/angiotensin-(1–7)/MAS receptor axis in protecting against cardiovascular pathophysiology, new drugs may be developed to target this axis to treat cardiovascular diseases [7].

9.3 DIRECT RENIN INHIBITORS

Currently, aliskiren is the only US Food and Drug Administration (FDA)-approved direct renin inhibitor:

- Aliskiren (Tekturna)

9.3.1 General Introduction to Drug Class

The most logical drug target in the RAAS has long been considered to be renin, which is at the top of the enzymatic cascade (Fig. 9.3). However, renin has also proved to be a

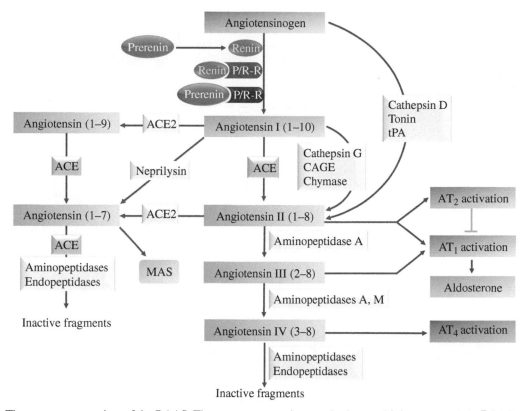

FIGURE 9.2 The contemporary view of the RAAS. The contemporary view emphasizes multiple aspects of the RAAS not recognized in the classical view. For example, the catalytic activity of renin increases when bound to the (pro)renin receptors (P/R-R), and the otherwise inactive prorenin becomes catalytically active when bound to P/R-R. Angiotensin II can be catabolized by angiotensin-converting enzyme 2 (ACE2) to form angiotensin-(1–7), another active peptide of this system, which typically opposes the actions of angiotensin II. Angiotensin-(1–7) can also be derived from angiotensin I with angiotensin-(1–9) as the intermediate. Angiotensin II is cleaved into smaller fragments, such as angiotensin-(2–8) (also called angiotensin III) and angiotensin-(3–8) (also called angiotensin IV) by aminopeptidases. Most effects of angiotensin II are mediated by the angiotensin receptor type 1 (AT$_1$ receptor); however, angiotensin II can also bind to the angiotensin receptor type 2 (AT$_2$ receptor), whose activation generally exhibits opposing effects on the responses caused by the activation of AT$_1$ receptors and may result in cardiovascular protection. Angiotensin-(1–7) acts via its putative MAS receptor to elicit responses that may also counteract those caused by activation of AT$_1$ receptors by angiotensin II. MAS is a proto-oncogene encoding a G-protein-coupled receptor, which has been recently identified as a putative receptor for angiotensin-(1–7). The ACE2/angiotensin-(1–7)/MAS receptor axis may represent new possibilities for developing novel therapeutic strategies for the treatment of cardiovascular diseases. As shown in the scheme, in addition to angiotensin II, angiotensin III also activates AT$_1$ receptors. Angiotensin IV activates AT$_4$ receptors; however, the physiological significance of this receptor activation remains unclear. Likewise, the nomenclature and existence of the AT$_3$ receptor and its relationship to MAS receptor remain controversial. For color details, please see color plate section.

highly challenging enzyme to target. While blockade of the RAAS has become a cornerstone of cardiovascular therapy with ACE inhibition and angiotensin receptor blockade, renin inhibition has lagged. Its late appearance on the therapeutic horizon relates largely to the biochemical features of the enzyme and its early inhibitors. The development of renin inhibitors was hampered by high cost of manufacturing, coupled with low potency and poor bioavailability of early molecules. With the advent of alternative approaches to manufacturing, notably molecular modeling via X-ray crystallography and reconstruction of the active site of renin, potent nonpeptidic oral renin inhibitors are being produced and show promising results in both preclinical and clinical studies [3, 8–10]. Aliskiren is the first in this drug class

approved by the US FDA in 2007 for clinical use in the treatment of hypertension.

9.3.2 Chemistry and Pharmacokinetics

Aliskiren is a nonpeptidic renin inhibitor (structure shown in Fig. 9.4). It is poorly absorbed with a bioavailability of ~2.5%. Absorption of aliskiren is significantly reduced by high-fat meal. About 25% of the absorbed dose appears in the urine as parent drug. Hepatic cytochrome P450 (CYP) 3A4 appears to be the major enzyme responsible for aliskiren metabolism. P-Glycoprotein may mediate the efflux of the drug and be responsible for the decreased absorption as well as drug elimination. The elimination half-life of aliskiren is 20–45 h.

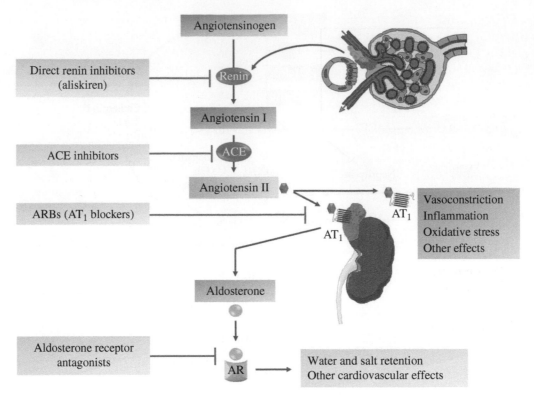

FIGURE 9.3 Targeting sites of the RAAS inhibitors. Shown in the scheme are the four groups of drugs that target the RAAS. AR denotes aldosterone receptor.

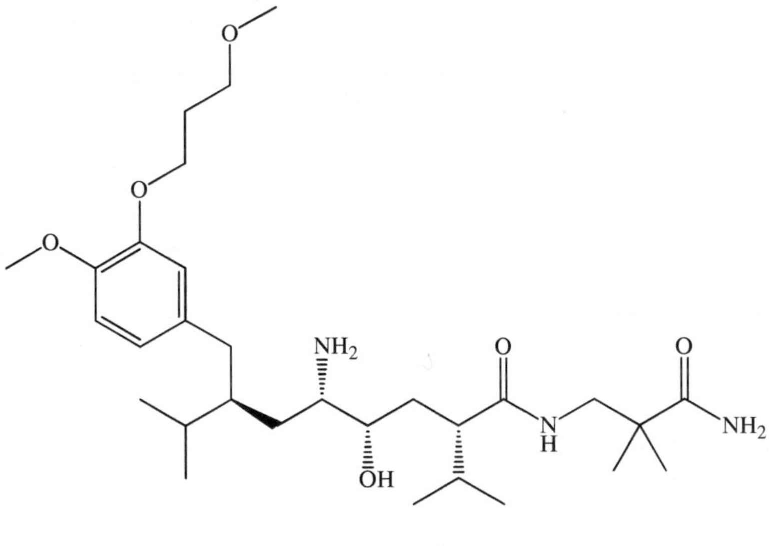

Aliskiren

FIGURE 9.4 Structure of aliskiren. Aliskiren is a nonpeptidic renin inhibitor and is chemically described as (2S,4S,5S,7S)-N-(2-carbamoyl-2-methylpropyl)-5-amino-4-hydroxy-2,7-diisopropyl-8-[4-methoxy-3-(3-methoxypropoxy)phenyl]-octanamide.

9.3.3 Molecular Mechanisms and Pharmacological Effects

Renin inhibitors are unique in their effects on the RAAS. Both ACE inhibition and angiotensin receptor blockade lead to a compensatory rise in plasma renin activity and thus to an increase in the angiotensin peptides, both angiotensin I and angiotensin II in the case of ARBs and angiotensin I with ACE inhibitors. Renin inhibitors, operating at the first and rate-limiting step of the cascade, render the entire pathway quiescent. Because renin is specific for the substrate angiotensinogen, renin inhibitors do not cause stimulation of bradykinin or prostaglandins, as seen with ACE inhibitors (see Section 9.4). In addition, renin inhibitors reduce angiotensin II formed by non-ACE pathways.

Renin inhibitors may also confer benefit by inhibiting activation of the proenzyme prorenin, long thought to be inactive. Recent work shows that prorenin may also contribute to cardiovascular and renal pathophysiology independent of angiotensin production. As such, renin inhibitors may possess an expanded potential for therapy compared with ACE inhibitors or ARBs.

All agents that inhibit the RAAS, including renin inhibitors, suppress the negative feedback loop, leading to a compensatory rise in plasma renin concentration. When this rise occurs during treatment with ACE inhibitors and ARBs, the result is increased levels of plasma renin activity. During treatment with aliskiren, however, the effect of increased renin levels is blocked so that plasma renin activity, angiotensin I, and angiotensin II are all reduced, no matter whether aliskiren is used as monotherapy or in combination with other antihypertensive agents.

9.3.4 Clinical Uses

9.3.4.1 Hypertension Aliskiren is approved by the US FDA for the treatment of hypertension. It may be used alone or in combination with other antihypertensive agents, such as a diuretic, a calcium channel blocker, or an ARB. Combination therapy is additive in lowering blood pressure. Fixed-dose combination formulations for aliskiren with the thiazide diuretic hydrochlorothiazide and the calcium channel blocker amlodipine are available:

- Aliskiren/hydrochlorothiazide (Tekturna)
- Aliskiren/amlodipine (Tekamlo)

The mechanisms underlying the blood pressure-lowering effect of aliskiren are primarily related to the inhibition of renin and the subsequent decreased formation of angiotensin II and aldosterone. Recent studies also suggest that aliskiren may increase the nitric oxide bioavailability and improve endothelial function likely via its inhibition of vascular NADPH oxidase [11, 12]. Inhibition of vascular NADPH oxidase results in decreased formation of superoxide. Because superoxide reacts with nitric oxide to form peroxynitrite, decreased superoxide formation will increase nitric oxide bioavailability.

9.3.4.2 Therapeutic Role in End-Organ Damage The Aliskiren in Left Ventricular Hypertrophy (ALLAY) trial suggested that aliskiren was as effective as the ARB losartan in attenuating left ventricular mass regression, a measure of myocardial end-organ damage, in hypertensive patients with left ventricular hypertrophy [13]. The beneficial effects of aliskiren appeared to be independent of its blood pressure-lowering action. Consistently, the Aliskiren Observation of Heart Failure Treatment (ALOFT) trial reported that addition of aliskiren to an ACE inhibitor (or ARB) and beta-blocker had favorable neurohumoral effects in heart failure and appeared to be well tolerated [14, 15].

Short-term trials suggest that aliskiren may also provide renal protection. In this regard, the Aliskiren in the Evaluation of Proteinuria in Diabetes (AVOID) trial found that aliskiren, when added to recommended treatment with losartan (100 mg daily) and optimal antihypertensive therapy, may have renoprotective effects that are independent of its blood pressure-lowering effect in patients with hypertension, type 2 diabetes, and nephropathy [16, 17]. However, the long-term, randomized, placebo-controlled morbidity/mortality trial Aliskiren Trial in Type 2 Diabetes Using Cardio-Renal Disease Endpoints (ALTITUDE), which included 8600 patients with type 2 diabetes, proteinuria, and a high cardiovascular risk already treated with ACE inhibitors or ARBs, was terminated in December 2011 because of futility and an increased incidence of serious adverse events in the aliskiren 300 mg arm. The serious adverse effects included renal dysfunction, hyperkalemia, hypotension, and possibly an increased risk of nonfatal stroke [10, 18, 19]. In response to these findings, it has been recommended that dual aliskiren and ACE inhibitor/ARB therapy not be used in patients with both hypertension (the current indication for aliskiren) and diabetes or moderate-to-severe renal dysfunction (estimated glomerular filtration rate <60 ml/min/1.73 m^2).

Recently, a randomized controlled trial (ASTRONAUT) involving 1615 heart failure patients reported that among patients hospitalized for heart failure with reduced left ventricular ejection fraction, initiation of aliskiren in addition to standard therapy did not reduce cardiovascular death or heart failure rehospitalization at 6 months or 12 months after discharge [20]. Hence, the results of the ASTRONAUT study do not support the routine administration of aliskiren, in addition to standard therapy, to patients hospitalized for worsening chronic heart failure.

9.3.4.3 Therapeutic Role in Prehypertension and Coronary Artery Disease The Aliskiren Quantitative Atherosclerosis Regression Intravascular Ultrasound Study (AQUARIUS) is a double-blind, randomized, multicenter trial comparing aliskiren with placebo in 613 participants with coronary artery disease, systolic blood pressure between 125 and 139 mmHg (prehypertension range), and two additional cardiovascular risk factors. The results showed that among participants with prehypertension and coronary artery disease, the use of aliskiren compared with placebo did not result in improvement or slowing of progression of coronary atherosclerosis. The finding does not support the use of aliskiren for regression or prevention of progression of coronary atherosclerosis [21]. Therefore, until additional evidence is available, the role of renin inhibition in the aforementioned setting is considered being much limited. Because aliskiren does reduce blood pressure, perhaps this drug could be reserved for use in patients with coronary disease and prehypertension/hypertension who cannot tolerate ACE inhibitors and ARBs [22] (see Sections 9.4 and 9.5 for ACE inhibitors and ARBs, respectively).

9.3.5 Therapeutic Dosages

The dosage form and strength of aliskiren are listed below:

- Aliskiren (Tekturna): Oral, 150 and 300 mg tablets

The usual recommended starting dose of aliskiren is 150 mg once daily. In patients whose blood pressure is not adequately controlled, the daily dose may be increased to 300 mg.

9.3.6 Adverse Effects and Drug Interactions

9.3.6.1 Adverse Effects

Common Adverse Effects Aliskiren is generally well tolerated. The most common adverse effects are mild diarrhea and other gastrointestinal disturbances.

Rare, Serious Adverse Effects Angioedema and renal dysfunction are potential rare adverse events associated with aliskiren therapy. Patients with signs and symptoms of angioedema should stop aliskiren and seek urgent medical assistance. Aliskiren should not be used by patients with a risk of renal impairment.

9.3.6.2 Drug Interactions
The use of nonsteroidal anti-inflammatory drugs may lead to increased risk of renal impairment and loss of antihypertensive effect of aliskiren. Combination treatment with aliskiren and other blockers of the RAAS increases the risk of hyperkalemia [23], which warrants careful monitoring of the plasma potassium levels. As P-glycoprotein and CYP3A4 are involved in the disposition of aliskiren, cotreatment with the p-glycoprotein/CYP3A4 inhibitor, cyclosporine or itraconazole, may lead to increased plasma levels of aliskiren, and as such, concomitant use of these drugs should be avoided.

9.3.6.3 Contraindications and Pregnancy Category

- Aliskiren is contraindicated in patients with diabetes who are receiving ARBs or ACEIs because of the increased risk of renal impairment, hyperkalemia, and hypotension.
- Avoid use of aliskiren with ARBs or ACEI in patients with moderate renal impairment (glomerular filtration rate <60 ml/min/1.73 m^2).

- Pregnancy category: C (first trimester) and D (second and third trimesters). When pregnancy is detected, aliskiren should be discontinued as soon as possible. Drugs that act directly on the RAAS can cause injury and death to the developing fetus.

9.3.7 Summary of Aliskiren

- Aliskiren is the first member of the renin inhibitors approved by the US FDA for clinical use. Renin inhibitors, operating at the first and rate-limiting step of the RAAS cascade, render the entire pathway quiescent.
- Aliskiren is approved by the US FDA for the treatment of hypertension. It may be used alone or in combination with other antihypertensive agents, such as a diuretic, a calcium channel blocker, or an ARB.
- The mechanisms underlying the blood pressure-lowering effect of aliskiren are primarily related to the inhibition of renin and the subsequent decreased formation of angiotensin II and aldosterone. Aliskiren may also reduce blood pressure through other potential mechanisms, such as increased nitric oxide bioavailability.
- Although early studies suggested a beneficial role for aliskiren in heart failure, the results of the recent ASTRONAUT study do not support the routine administration of aliskiren, in addition to standard therapy, to patients hospitalized for worsening chronic heart failure. Similarly, the ALTITUDE trial showed no benefit of dual aliskiren and ACE inhibitor/ARB therapy in patients with both hypertension (the current indication for aliskiren) and diabetes or moderate-to-severe renal dysfunction, and instead, the dual therapy was associated with serious adverse effects. Aliskiren was also found to be ineffective in improvement or slowing of progression of coronary atherosclerosis in patients with prehypertension and coronary artery disease. Aliskiren is well tolerated and the most common adverse effect is diarrhea. Due to its potential fetal toxicity, aliskiren should be discontinued as soon as possible when pregnancy is detected.

9.4 ACE INHIBITORS

Listed below are the 11 US FDA-approved ACE inhibitors:

- Benazepril (Lotensin)
- Captopril (Capoten) (prototype)
- Enalapril (Vasotec)
- Enalaprilat (Vasotec injection)
- Fosinopril (Monopril)
- Lisinopril (Prinivil/Zestril)
- Moexipril (Univasc)
- Perindopril (Aceon)

- Quinapril (Accupril)
- Ramipril (Altace)
- Trandolapril (Mavik)

9.4.1   General Introduction to Drug Class

Drugs that block ACE and the formation of angiotensin II have become mainstays of cardiovascular pharmacotherapeutics. Peptide relatives of modern ACE inhibitors were first identified serendipitously in extracts from *Bothrops* venom. Further pharmacological development culminated in the synthesis in the 1970s of the first oral agent, captopril. Other ACE inhibitors with improved pharmacological properties have since been developed, and most ACE inhibitors, alone and in combination with diuretic or amlodipine, are now available generically. In addition to hypertension, ACE inhibitors are indicated for the treatment of patients at high risk for coronary artery disease, after myocardial infarction, with heart failure or dilated cardiomyopathy, or with chronic kidney disease or diabetes.

9.4.2   Chemistry and Pharmacokinetics

The 11 ACE inhibitors (structures shown in Fig. 9.5) have essentially identical mechanisms of action and pharmacological effects, but they differ in their chemical and pharmacokinetic properties. Captopril is the only ACE inhibitor that contains a sulfhydryl (-SH) group, fosinopril is the only phosphorus-containing ACE inhibitor, and the rest are dicarboxyl-containing compounds. Except for captopril (with a duration of action of 6–12 h), most ACE inhibitors have a duration of action of ~24 h, enabling the once daily dosing regimens. ACE inhibitors are cleared predominantly by the kidneys, and as such, the drug clearance can be significantly reduced in patients with renal impairment. The major pharmacokinetic properties of the 11 ACE inhibitors are given in Table 9.1.

9.4.3   Molecular Mechanisms and Pharmacological Effects

Although evidence suggests multiple biological activities of ACE [24], the primary function of ACE in the RAAS is to catalyze the conversion of angiotensin I to angiotensin II. ACE is also known as kininase II and responsible for breaking down bradykinin, a vasodilating molecule, to inactive metabolites. The mechanisms of action of ACE inhibitors in cardiovascular medicine are illustrated in Figure 9.6. As depicted in the figure, inhibition of ACE causes decreased formation of angiotensin II and reduced catabolism of bradykinin, leading to four pharmacological responses as outlined below:

1. Reduced output of sympathetic nervous system due to decreased levels of angiotensin II
2. Reduced vasoconstriction and vascular remodeling due to decreased levels of angiotensin II

TABLE 9.1   Major pharmacokinetic properties of ACE inhibitors

ACE inhibitor	Oral bioavailability (%)	Elimination half-life (h)	Metabolism and elimination
Benazepril	37	10–11	Hydrolyzed to form active benazeprilat; glucuronidation; mainly excreted in the urine
Captopril	75	<2	40–50% of dose excreted in unchanged form in the urine; the rest is converted to disulfide dimer
Enalapril	60	11	Hydrolyzed to form more potent enalaprilat; no further metabolism; renal excretion of enalapril and enalaprilat
Enalaprilat (iv)		11	Mainly excreted in the urine in unchanged form
Fosinopril	36	12	Hydrolyzed to active fosinoprilat; glucuronidation; excreted in the urine and bile
Lisinopril	25	12	Excreted in the urine in unchanged form; no metabolism
Moexipril	13	2–9	Hydrolyzed to active moexiprilat; no further metabolism; excreted in the urine as moexipril and moexiprilat
Perindopril	75	1	Hydrolyzed to active perindoprilat; extensive metabolism involving hydrolysis, glucuronidation, and cyclization via dehydration
Quinapril[a]	60	2	Hydrolyzed to form active quinaprilat; no further metabolism; excreted in the urine
Ramipril[b]	50–60	9–18	Hydrolyzed to active ramiprilat; glucuronidation; formation of diketopiperazine ester and acid
Trandolapril	10	6	Hydrolyzed to active trandolaprilat; renal and biliary elimination

[a]Quinapril has a prolonged terminal elimination phase with a half-life of 25 h.
[b]Ramipril has a prolonged terminal elimination phase with a half-life of >50 h.

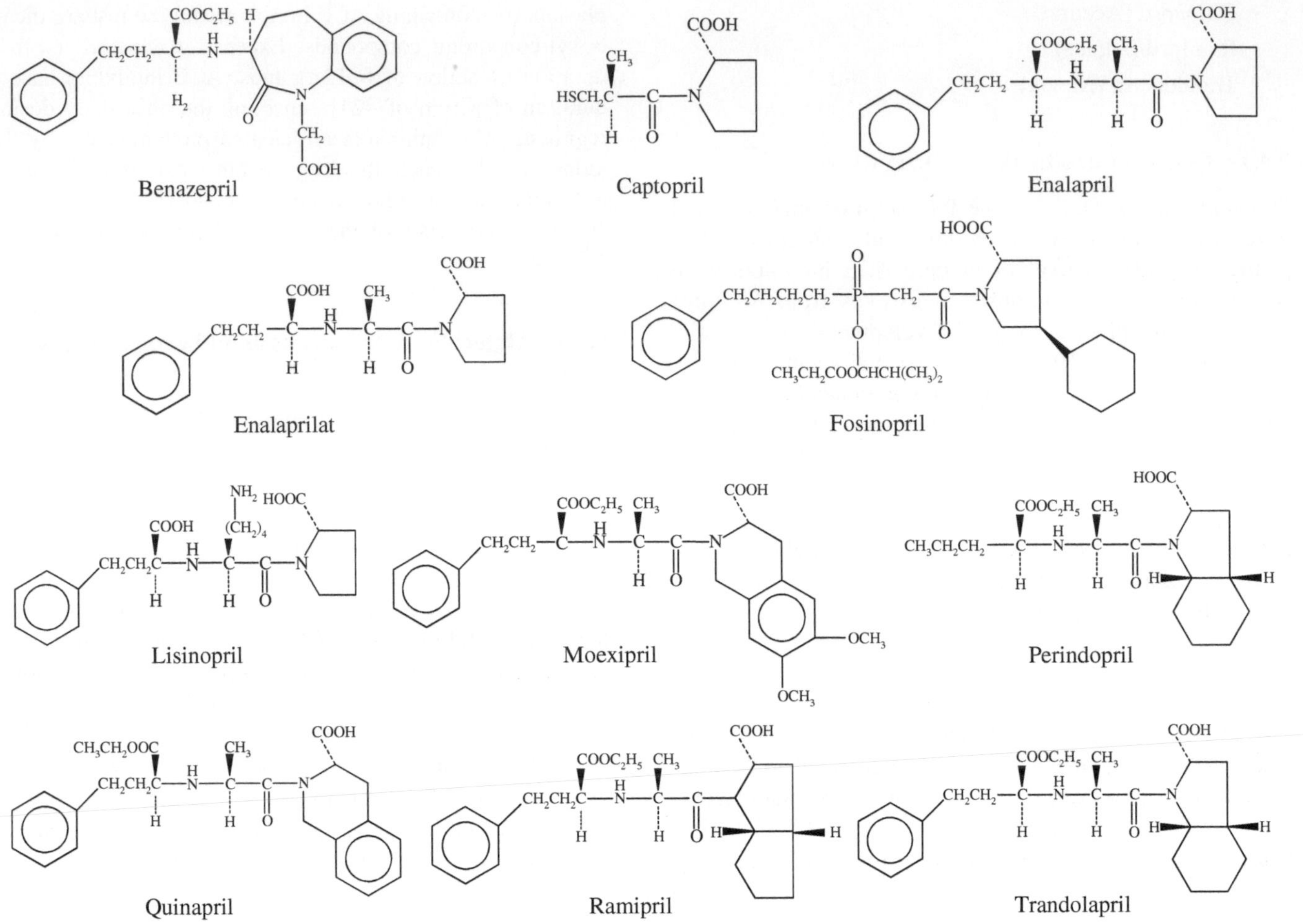

FIGURE 9.5 Structures of angiotensin-converting enzyme (ACE) inhibitors. As shown, captopril is the only ACE inhibitor that contains a sulfhydryl group, fosinopril is the only phosphorus-containing ACE inhibitor, and the rest are dicarboxyl-containing compounds. Most ACE inhibitors (except for captopril, enalaprilat, and lisinopril) are prodrugs that require hydrolysis to form the active forms.

3. Decreased production of aldosterone and the subsequent reduction in sodium and water retention as well as attenuation of aldosterone-mediated cardiovascular remodeling
4. Increased levels of bradykinin due to its decreased catabolism

The above diverse actions serve as the pharmacological basis for the wide applications of ACE inhibitors in treating cardiovascular diseases as well as other disorders involving dysregulated RAAS, such as renal diseases. It should be noted that the pharmacological effects and clinical efficacy of ACE inhibitors are weaker in people of African heritage than in people of other ethical groups. This is likely due to the decreased basal activity of RAAS in people of African heritage.

9.4.4 Clinical Uses

9.4.4.1 Cardiovascular Diseases The cardiovascular applications of ACE inhibitors include treatment of the following diseases and conditions:

- Hypertension
- Heart failure
- Acute coronary syndromes
- High coronary artery disease risk

Hypertension ACE inhibitors are indicated for the treatment of hypertension. It may be used alone as an initial therapy or concomitantly with other classes of antihypertensive drugs (see Chapter 12 for hypertension management). Via the various pathways illustrated in Figure 9.7, ACE inhibitors

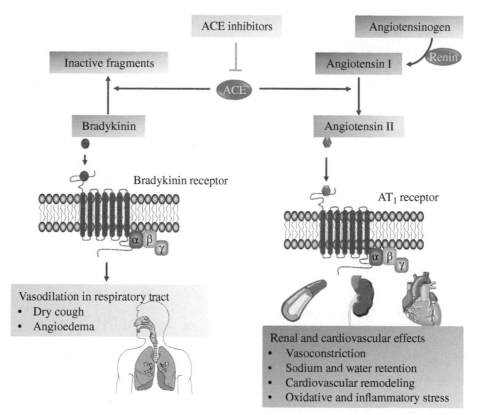

FIGURE 9.6   Molecular mechanisms of action of angiotensin-converting enzyme (ACE) inhibitors in cardiovascular medicine. ACE inhibitors decrease the formation of angiotensin II by inhibiting ACE. As ACE also metabolizes bradykinin, a vasodilating molecule, ACE inhibitors cause bradykinin accumulation. Accumulation of this vasodilating molecule in respiratory tract may cause dry couch and angioedema. As shown, both AT1 receptor and bradykinin receptor are G-protein-coupled receptors.

lower systemic vascular resistance and mean, diastolic, and systolic blood pressures in various hypertensive states except high blood pressure due to primary aldosteronism. Hypertension primarily due to aldosteronism responds well to aldosterone receptor antagonists.

Acute Coronary Syndromes   ACE inhibitors are indicated for the treatment of hemodynamically stable patients within 24 h of acute myocardial infarction, to improve survival (see Unit V for acute coronary syndromes). Patients should receive, as appropriate, the standard recommended treatments such as thrombolytics, aspirin, and beta-blockers. New studies also suggested that a combination of cardiac medications including aspirin, a beta-blocker, a statin, and an ACE inhibitor (or an ARB) at the time of discharge for acute coronary syndromes was strongly associated with decreased long-term mortality in men and women [25].

Heart Failure   When discussing pharmacotherapy of heart failure, it is important to distinguish the heart failure with reduced ejection fraction (also known as systolic heart

failure) from that with preserved ejection fraction (historically known as diastolic heart failure). Each accounts for ~50% of all heart failure cases [26]:

- Heart failure with reduced left ventricular ejection fraction: ACE inhibitors are indicated in the management of systolic heart failure and should be given to all heart failure patients with impaired left ventricular systolic function unless contraindicated. In this clinical setting, use of ACE inhibitors has been extensively documented to prolong survival and reduce mortality. The chief pharmacological basis for this benefit lies in the ability of these drugs to decrease cardiac afterload, improve water and salt retention, and attenuate cardiac remodeling.

- Heart failure with preserved left ventricular ejection fraction: The long-term benefits of ACE inhibitors in heart failure with reduced left ventricular ejection fraction have been well established. However, based on clinical trials on perindopril and enalapril, there is currently no evidence for their benefits in the management of heart failure with preserved left ventricular dysfunction [27, 28]. Similarly, available clinical trials do not suggest an efficacy for ARBs in the management of heart failure with preserved left ventricular ejection fraction [29, 30]. Nevertheless, studies are underway to further determine the clinical effectiveness of ACE inhibitors as well as other inhibitors of the RAAS in the management of heart failure patients with preserved left ventricular ejection fraction [31].

High Coronary Artery Disease Risk   Use of ACE inhibitors in patients at high risk of coronary artery disease has been associated with a decrease in the risk of myocardial infarction, stroke, and death. ACE inhibitors are efficacious in preventing cardiovascular events and death in individuals at high risk of coronary artery disease even when their blood pressure is in the normal range [32]. In general, ACE inhibitors and ARBs are equally effective, and combination of an ACE inhibitor and an ARB doesn't increase benefits, but causes more adverse events [33].

Atrial Fibrillation   The efficacy of ACE inhibitors as well as ARBs in the management of atrial fibrillation remains controversial. While some meta-analyses suggested an efficacy for ACE inhibitors and ARBs for the primary and secondary prevention of atrial fibrillation [34, 35], others suggested an insignificant role for these drugs in the secondary prevention of atrial fibrillation [36].

9.4.4.2   Noncardiovascular Disorders   A meta-analysis of 12 randomized trials suggested that ACE inhibitors may be associated with a reduction in the risk of newly diagnosed type 2 diabetes mellitus [37]. However, none of these trials were designed with reduction in the incidence of type 2 diabetes mellitus as the primary endpoint, and glucose tolerance was not assessed routinely. In this context, the Diabetes Reduction Assessment with Ramipril and Rosiglitazone Medication (DREAM) trial, a randomized trial designed with type 2 diabetes mellitus as a primary outcome, reported that among persons with impaired fasting glucose levels or impaired glucose tolerance, the use of ramipril for 3 years does not significantly reduce the incidence of diabetes or death but does significantly increase regression to normoglycemia [38].

Given the primary findings of the DREAM trial, ramipril cannot be recommended for the prevention of type 2 diabetes mellitus. For patients who take ACE inhibitors for another indication (such as hypertension, congestive heart failure, or high-risk cardiovascular events), improvement in glycemia may turn out to be yet another benefit. For now, ongoing attention to diet and exercise remains the best hope for reducing the rising rate of diabetes [39]. The mechanism by which ACE inhibitors improve glucose metabolism is most likely due to their ability to decrease the formation of angiotensin II as angiotensin II plays an important role in the development of insulin resistance and glucose intolerance.

Despite the controversies on the prevention of diabetes by ACE inhibitors, substantial evidence suggests that these drugs retard the progression of diabetic nephropathy. Two mechanisms underlying ACE inhibitor-mediated renal protection have been proposed:

- Pressure mechanism: Increased glomerular capillary pressure is an important factor leading to glomerular injury. ACE inhibitors reduce glomerular capillary pressure by decreasing the arterial blood pressure and by dilating the renal efferent arterioles.
- Inflammation and remodeling mechanism: Angiotensin II causes inflammation, oxidative stress, mesangial cell proliferation, and renal tissue remodeling, all of which contribute to the development of diabetic nephropathy. ACE inhibitors attenuate the aforementioned deleterious processes by decreasing the formation of angiotensin II.

In addition to the beneficial effects in diabetic nephropathy, ACE inhibitors may also impact the cardiovascular events and mortality in patients with diabetes. In this regard, a recent meta-analysis of 23 randomized trials involving 32,827 diabetic patients comparing ACE inhibitors with placebo or active drugs (amlodipine, atenolol, or a diuretic) reported that ACE inhibitors reduced all-cause mortality, cardiovascular mortality, and major cardiovascular events in patients with diabetes [40]. However, meta-analysis of 13 randomized trials involving 23,867 diabetic patients comparing ARBs with no therapy did not show a beneficial effect on the aforementioned endpoints [40].

9.4.5   Therapeutic Dosages

The dosage forms and strengths of the 11 ACE inhibitors are listed below:

- Benazepril (Lotensin): Oral, 5, 10, 20, and 40 mg tablets
- Captopril (Capoten): Oral, 12.5, 25, 50, and 100 mg tablets
- Enalapril (Vasotec): Oral, 2.5, 5, 10, and 20 mg tablets
- Enalaprilat (Vasotec injection): IV injection, 1.25 mg/ml in vials containing 1 and 2 ml
- Fosinopril (Monopril): Oral, 10, 20, and 40 mg tablets
- Lisinopril (Prinivil/Zestril): Oral, 2.5, 5, 10, 20, 30, and 40 mg tablets
- Moexipril (Univasc): Oral, 7.5 and 15 mg tablets
- Perindopril (Aceon): Oral, 2, 4, and 8 mg tablets
- Quinapril (Accupril): Oral, 5, 10, 20, and 40 mg tablets

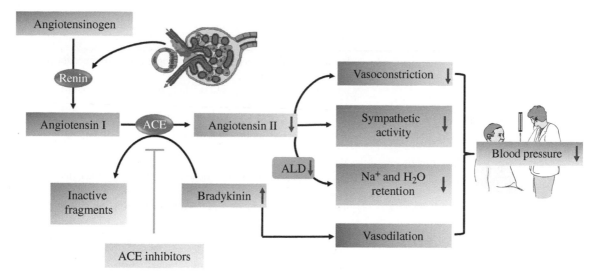

FIGURE 9.7   Molecular mechanisms of action of angiotensin-converting enzyme (ACE) inhibitors in treating hypertension. Angiotensin II via activating AT_1 receptors causes vasoconstriction, increased sympathetic activity, as well as release of aldosterone (ALD) from adrenal gland. Aldosterone in turn causes sodium and water retention. By decreasing angiotensin II formation, ACE inhibitors cause reduction in vasoconstriction, sympathetic activity, and sodium and water retention, thereby leading to decreased blood pressure. In addition, increased bradykinin also causes vasodilation, contributing to blood pressure reduction.

- Ramipril (Altace): Oral, 1.25, 2.5, 5, and 10 mg capsules
- Trandolapril (Mavik): Oral, 1, 2, and 4 mg tablets

As noted earlier, all the 11 ACE inhibitors are approved for treating hypertension. Some are also indicated for other cardiovascular conditions largely based on favorable results from clinical trials. The US FDA-approved indications and the corresponding dosage regimens of the 11 ACE inhibitors are summarized in Table 9.2.

9.4.6   Adverse Effects and Drug Interactions

9.4.6.1   Adverse Effects   ACE inhibitors as a class of drugs are generally well tolerated. Adverse effects are considered to be class specific rather than drug specific, so an intolerable adverse event with one ACE inhibitor generally precludes the use of any other ACE inhibitors. Adverse effects are usually mild and include dry cough, hyperkalemia, decreased blood pressure, headache, weakness, and change of taste. Rare but serious adverse effects include angioedema and impaired renal function:

- Dry cough: Persistent nonproductive cough has been reported with all ACE inhibitors in ~5–20% patients, almost always resolving after discontinuation of therapy. The mechanism may be related to the inhibition of the degradation of endogenous bradykinin and its subsequent accumulation in the lung. Bradykinin induces the formation of prostaglandins and nitric oxide,

which dilate blood vessels and promote cough. ACE inhibitor-induced cough should be considered in the differential diagnosis of cough. If the patient cannot tolerate the dry cough, an ARB should be considered. In this regard, ARBs rarely cause dry cough. This can be explained by the fact that ARBs do not affect the catabolism of bradykinin. The bradykinin accumulation mechanism of ACE inhibitor-induced dry cough is further supported by recent studies showing that genetic polymorphisms in bradykinin B2 receptors are associated with ACE inhibitor-induced dry cough [41, 42].

- Hyperkalemia: Significant hyperkalemia rarely occurs in patients with normal renal function. However, ACE inhibitors may cause hyperkalemia in patients with renal insufficiency or in patients who concomitantly take drugs that cause potassium retention (e.g., potassium-sparing diuretics) or potassium supplements.

- Hypotension: Excessive hypotension is rare in patients with uncomplicated hypertension treated with ACE inhibitors alone. Hypotension may occur after initiation of ACE inhibitor treatment in patients on other antihypertensive drugs (e.g., diuretic agents) or in patients with congestive heart failure.

- Angioedema: Angioedema is a potentially life-threatening but rare adverse event in patients treated with ACE inhibitors. It is manifested by swelling of the face, extremities, eyes, lips, and tongue and difficulty in swallowing or breathing. If angioedema of the face, extremities, lips, tongue, glottis, and/or larynx occurs, treatment with ACE inhibitors should be discontinued

TABLE 9.2 Dosage regimens of ACE inhibitors

ACE inhibitor	Indication	Dosage
Benazepril	Hypertension	Initial dose, 10 mg once daily; dose range, 20–40 mg once daily; maximum daily dose, 80 mg
Captopril	Hypertension	Initial dose, 25 mg b.i.d. or t.i.d.; daily dose, 50 mg b.i.d. or t.i.d.; maximum total daily dose, 450 mg
	Heart failure	Initial dose, 25 mg t.i.d.; daily dose, 50–100 mg t.i.d.
	Myocardial infarction	Initial dose, 6.25 mg single dose, then 12.5 mg t.i.d.; target maintenance dose, 50 mg t.i.d.
	Diabetic nephropathy	25 mg t.i.d.
Enalapril	Hypertension	Initial dose, 5 mg once daily; dose range, 10–40 mg once daily
	Heart failure	Initial dose, 2.5 mg; dose range, 2.5–20 mg b.i.d.
	Asymptomatic left ventricular dysfunction	Initial dose, 2.5 mg b.i.d.; target daily dose, 20 mg in divided doses
Enalaprilat	Hypertension	Given when oral therapy is not practical; IV injection only; 1.25 mg every 6 h
Fosinopril	Hypertension	Initial dose, 10 mg once daily; dose range, 20–40 mg once daily
	Heart failure	Initial dose, 10 mg once daily; dose range, 20–40 mg once daily
Lisinopril	Hypertension	Initial dose, 10 mg once daily; dose range, 20–40 mg once daily
	Heart failure	Initial dose, 5 mg once daily; dose range, 5–20 mg once daily
	Myocardial infarction	First dose 5 mg followed by 5 mg after 24 h, 10 mg after 48 h, and then 10 mg once daily
Moexipril	Hypertension	Initial dose, 7.5 mg once daily; dose range, 7.5–30 mg once daily
Perindopril	Hypertension	Initial dose, 4 mg once daily; dose range, 4–8 mg once daily; maximum daily dose, 16 mg
	Stable CAD	Initial dose, 4 mg once daily for 2 weeks; then maintenance dose, 8 mg once daily
Quinapril	Hypertension	Initial dose, 10–20 mg once daily; dose range, 20–80 mg once daily
	Heart failure	Initial dose, 5 mg b.i.d.; maintenance dose, 20–40 mg/day given in two equally divided doses
Ramipril	Hypertension	Initial dose, 2.5–20 mg once daily; maintenance dose, 2.5–20 mg/day given as single or two divided doses
	Heart failure	Initial dose, 2.5 mg twice daily; maintenance dose, 5 mg twice daily
	Risk reduction	For reduction of risk of myocardial infarction, stroke, or death from cardiovascular causes, initial dose, 2.5 mg once daily for 2–4 weeks; maintenance dose, 10 mg once daily
Trandolapril	Hypertension	Initial dose, 1 mg once daily in nonblacks and 2 mg once daily in blacks; dose range, 2–4 mg once daily

and appropriate therapy instituted immediately. Angioedema occurs in 0.1–0.5% of patients taking these drugs and appears to be more common in patients of African heritage (a meta-analysis showed that African Americans were twice more likely to develop angioedema on ACE inhibitors compared to Caucasians). African Americans have been shown to exhibit an increased sensitivity to bradykinin compared to Caucasian subjects. Thus, the increased risk of angioedema should be taken into account in this population and proactively discussed with the patient. Antihypertensive efficacy of ACE inhibitors is decreased in African American patients compared to other ethnic groups, but the risk of angioedema is increased. This brings up the question as to whether ACE inhibitors should still be considered first-line therapy for hypertension in black patients. When

pondering this issue, we should remember that ACE inhibitor-based treatment was shown to be superior to β-blocker or calcium channel blocker treatment in preventing progression of kidney disease in African American patients [43].

As compared with the incidence for ACE inhibitors, the risk for developing angioedema is considerably lower with ARBs or direct renin inhibitors. A recent meta-analysis of randomized trials reported that incidence of angioedema with ARBs and direct renin inhibitors was less than a half of that with ACE inhibitors and not significantly different from placebo [44].

Although the exact mechanism of angioedema remains unknown, accumulation of bradykinin is believed to be a potential culprit in the development of angioedema. Increased accumulation of bradykinin results in the regional vasodilation and increased vascular permeability that are characteristics of angioedema. In this context, patients with heart failure were found to have significantly higher rates of angioedema with ACE inhibitors and ARBs. This may be related to the observation that heart failure per se increases bradykinin levels.

Data regarding the safety of ARBs in patients with a history of angioedema associated with ACE inhibitors are relatively sparse. In a recent meta-analysis of trials, incidence of cross-reactivity of angioedema in patients who received an ARB after developing ACE inhibitor-associated angioedema was 2.5% with a conservative estimate of <10% [45]. Angioedema related to ARBs is reported to be less severe and occur earlier compared to angioedema that develops during ACE inhibitor therapy. ARBs may be an alternative for patients who develop angioedema while using an ACE inhibitor but should be reserved for patients with high therapeutic need for angiotensin inhibition. Treatment should be started with observation, patients should be educated on the signs of angioedema, and proper emergency management should be emphasized to patients and care providers. There are no data regarding the safety of direct renin inhibitors in patients who have developed ACE inhibitor-associated angioedema. Further studies should be conducted with larger numbers of patients to evaluate the absolute risk of angioedema with ARBs or direct renin inhibitors after ACE inhibitor-associated angioedema.

• Impaired renal function: As a consequence of inhibiting the RAAS, changes in renal function may be anticipated in susceptible individuals on ACE inhibitors. Angiotensin II preferentially constricts the renal efferent arterioles to help maintain adequate glomerular filtration rate when renal perfusion pressure is low. Consequently, decreased formation of angiotensin II by ACE inhibitors may be associated with oliguria and/or progressive azotemia and rarely with acute renal failure and/or death in patients whose renal function may depend on the activity of the RAAS. These may include patients with severe congestive heart failure, bilateral renal arterial stenosis, stenosis of the artery to a single remaining kidney, or volume depletion due to diarrhea or the use of diuretics.

9.4.6.2 Drug Interactions

As stated earlier, ACE inhibitors may pharmacodynamically interact with other antihypertensive agents (e.g., diuretic agents) to cause hypotension and interact with potassium-sparing diuretics to induce hyperkalemia. In some patients with comprised renal function who are being treated with nonsteroidal anti-inflammatory drugs, the coadministration of ACE inhibitors may result in further deterioration of renal function. In addition, the concomitant use of a nonsteroidal anti-inflammatory drug may also reduce the pharmacological effects of ACE inhibitors. This is because prostaglandins partially mediate the vasodilating effects of ACE inhibitors.

ACE inhibitors may increase plasma lithium level, precipitating lithium intoxication. Lithium toxicity is usually reversible upon discontinuation of lithium and the ACE inhibitor. It is recommended that serum lithium levels be monitored frequently if the ACE inhibitor is administered concomitantly with lithium.

9.4.6.3 Contraindications

• Hypersensitive to the drug.
• A history of angioedema related to previous treatment with an ACE inhibitor.
• Patients with hereditary or idiopathic angioedema.
• Pregnancy category: D. When they are used in pregnancy during the second and third trimesters, ACE inhibitors can cause injury and even death to the developing fetus. When pregnancy is detected, ACE inhibitors should be discontinued as soon as possible. Due to the wide use of ACE inhibitors, the potential impact of these drugs on fetus is further discussed next.

As noted earlier, it is believed that ACE inhibitors can cause fetal and neonatal morbidity and death when administered to pregnant women. Consequently, when pregnancy is detected, ACE inhibitors should be discontinued as soon as possible. However, a recent large retrospective cohort study including 465,754 mother–infant pairs demonstrated that maternal use of ACE inhibitors in the first trimester has a risk profile similar to the use of other antihypertensives regarding malformations in live born offspring. The apparent increased risk of malformations associated with use of ACE inhibitors (and other antihypertensives) in the first trimester is likely due to the underlying hypertension rather than the medications [46, 47]. A recent meta-analysis also suggested that first trimester exposure to ACE inhibitors and ARBs was not associated with an elevated risk of major malformations compared with other antihypertensives [48].

Although studies suggest that inadvertent exposure to ACE inhibitors (or ARBs) in the first trimester of pregnancy may not present significant risk for malformations in live births, use of these drugs during the first trimester may be associated with an increased risk of miscarriage [49]. These new findings are important in counseling women who used ACE inhibitors before realizing that they have conceived and also likely to encourage the medical and scientific communities to refocus on the potential fetal risks of uncontrolled hypertension. Nevertheless, given the evidence for teratogenicity beyond the first trimester and the availability of other safer effective antihypertensive agents in pregnancy, it is imperative that women on ACE inhibitors receive prompt attention in the early part of pregnancy so that their antihypertensive medication can be appropriately adjusted.

9.4.7   Summary of ACE Inhibitors

- Currently, there are 11 US FDA-approved ACE inhibitors. Inhibition of ACE causes decreased formation of angiotensin II and reduced catabolism of bradykinin. As such, ACE inhibitors are more frequently associated with dry cough as compared with other RAAS inhibitors.

- ACE inhibitors are widely used in treating hypertension, heart failure, and acute coronary syndromes. ACE inhibitors reduce mortality and prolong survival in the aforementioned settings. Use of ACE inhibitors in patients at high risk of coronary artery disease is associated with decreased risk of myocardial infarction, stroke, and death.

- ACE inhibitors are generally well tolerated. Adverse effects are usually mild and include dry cough and hyperkalemia. Rare but serious adverse effects include angioedema and impaired renal function.

9.5   ARBs

9.5.1   General Introduction to Drug Class

Listed below are the eight US FDA-approved ARBs with azilsartan being the newest member, which was approved in 2011:

1. Azilsartan (Edarbi)
2. Candesartan (Atacand)
3. Eprosartan (Teveten)
4. Irbesartan (Avapro)
5. Losartan (Cozaar) (prototype)
6. Olmesartan (Benicar)
7. Telmisartan (Micardis)
8. Valsartan (Diovan)

ARBs are also known as angiotensin receptor antagonists, AT_1 receptor antagonists, or sartans. These drugs selectively block the activation of angiotensin II type 1 receptors (AT_1 receptors) by angiotensin II. Hence, these drugs may be more precisely called angiotensin II type 1 receptor blockers.

Attempts to develop drugs that block angiotensin II receptors for clinical use dated back to the early 1970s. First receptor antagonists for angiotensin II were angiotensin-like peptides with limitations of short duration, lack of oral bioavailability, and partial agonistic activity. However, the breakthrough came in 1982 with the publication of patents from the Takeda Chemical Industries, Ltd., Japan of a nonpeptide imidazole-5-acetic acid derivative that antagonized angiotensin II vasoconstriction. The original Takeda compounds S-8307 and S-8308 had very weak antihypertensive effects but were selective for the angiotensin II receptors.

Based on the work done at Takeda Chemical Industries, Ltd., scientists at DuPont started synthesizing variants of S-8307 without much success. They then moved to computer modeling of the angiotensin II structure and overlapping drugs. A series of changes were made to increase potency and to create an orally active compound, and in 1989, the hard work of the scientists at DuPont eventually resulted in the synthesis of losartan, or DuP-753, which had improved oral absorption and increased potency to 1000 times that of S-8307 [50]. DuPont and Merck signed an agreement to collaborate on the development and marketing of losartan, starting in January 1990, and the drug was approved for clinical use by the US FDA in 1995. Since then, seven additional AT_1 receptor blockers have been approved. As noted earlier, azilsartan is the newest member of ARBs, which was approved by the US FDA in 2011. Losartan is currently the only ARB available generically. Similar to ACE inhibitors, ARBs are a widely used class of drugs in cardiovascular medicine, with established efficacy in the treatment of hypertension, heart failure, and ischemic heart disease.

9.5.2   Chemistry and Pharmacokinetics

The structures of the eight ARBs are shown in Figure 9.8. All of these drugs are highly protein bound (>85%), but they vary widely in their volume of distribution from 101 for candesartan to 5001 for telmisartan. Most of the drugs are dosed once daily, although based on their short plasma elimination half-lives, losartan and eprosartan may need to be administered twice daily to maintain 24 h of efficacy in some patients. Bioavailability also varies widely from 23% for valsartan to 60–80% for irbesartan. These drugs that are eliminated primarily via the renal route, such as eprosartan and candesartan, will have higher plasma concentrations in patients with impaired renal function, and lower doses may be needed to achieve the same pharmacodynamic effect as in patients with normal or modestly impaired renal function [51]. The major pharmacokinetic properties of the eight ARBs are summarized in Table 9.3.

9.5.3   Molecular Mechanisms and Pharmacological Effects

9.5.3.1   Selective Inhibition of AT_1 Receptors and Angiotensin II-Induced Deleterious Effects   All ARBs bind to the AT_1 receptors with high affinity and are over 10,000 times more selective for the AT_1 receptors than for the AT_2 receptors. As illustrated in Figure 9.9, angiotensin II is formed from angiotensin I in a reaction catalyzed by ACE (also known as kininase II). Angiotensin II is the principal pressor agent of the RAAS, with effects that include (i) vasoconstriction, (ii) stimulation of synthesis and release of aldosterone, (iii) renal reabsorption of sodium, and (iv) cardiac and vascular oxidative stress, inflammation, and remodeling,

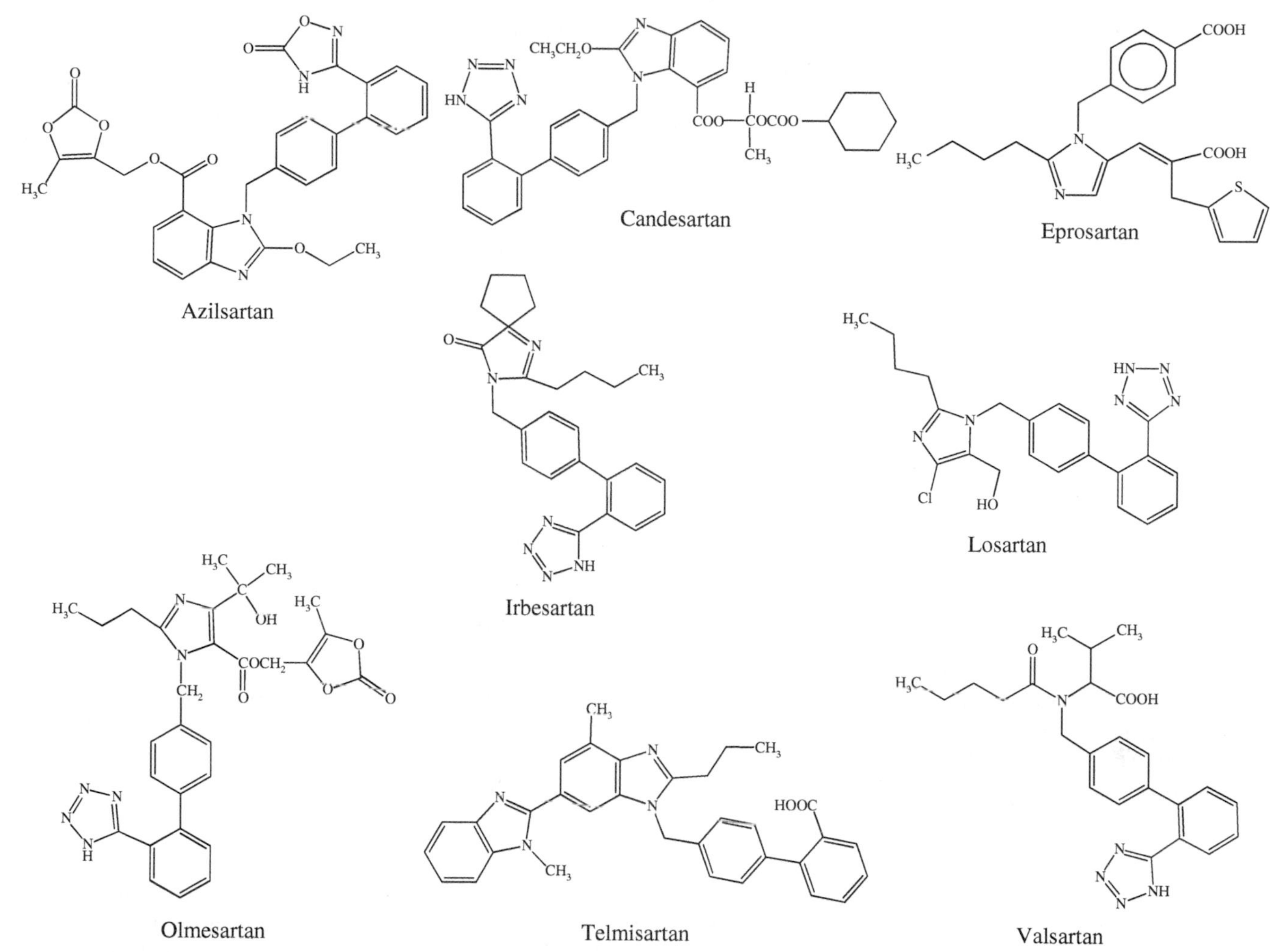

FIGURE 9.8 Structures of angiotensin receptor blockers (ARBs). All ARBs contain a common imidazole-based structural core. The prototype losartan is chemically described as 2-butyl-4-chloro-1-[p-(o-1H-tetrazol-5-ylphenyl)benzyl]imidazole-5-methanol.

TABLE 9.3 Major pharmacokinetic properties of ARBs

ARB	Oral bioavailability (%)	Elimination half-life (h)	Metabolism and elimination
Azilsartan	60	11	CYP2C9; eliminated in the bile/feces and urine
Candesartan	15	9	Eliminated in unchanged form in the urine and bile/feces
Eprosartan	13	20	Eliminated in unchanged form in the urine and bile/feces
Irbesartan	60–80	11–15	CYP2C9; glucuronidation; eliminated in the bile/feces and urine
Losartan	33	2	CYP2C9; CYP3A4; eliminated in the bile/feces and urine
Olmesartan	26	13	Eliminated in unchanged form in the urine and bile/feces
Telmisartan	50	24	Eliminated in unchanged form in the bile/feces
Valsartan	25	6	Eliminated in unchanged form in the bile/feces

among many other potential deleterious effects on the cardiovascular system as well as other organs. ARBs block the aforementioned deleterious effects of angiotensin II by selectively blocking the binding of angiotensin II to the AT$_1$ receptors in cardiovascular and many other tissues. The pharmacological action of ARBs is therefore independent of the pathways for angiotensin II synthesis. In other words,

regardless of the sources of angiotensin II formation, ARBs block the angiotensin II-mediated activation of the AT$_1$ receptors.

9.5.3.2 No Effects on Bradykinin and Reduced Incidence of Adverse Effects
Because ARBs do not inhibit ACE (kininase II), they do not cause accumulation of bradykinin.

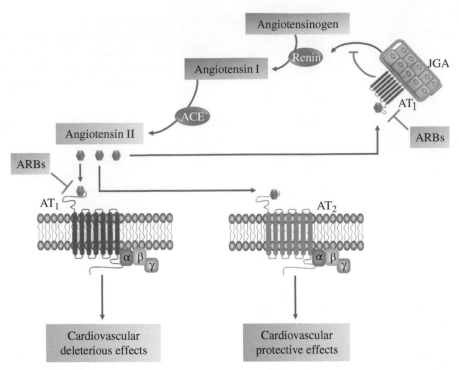

FIGURE 9.9 Molecular mechanisms of action of angiotensin receptor blockers (ARBs). As illustrated, AT_1 receptors mediate the cardiovascular deleterious effects of angiotensin II. Of note is the relatively lower abundance of AT_2 receptors, and as such, activation of AT_1 receptors predominates. ARBs block AT_1 receptors, thereby inhibiting angiotensin II-mediated adverse effects. Since activation of AT_1 receptors in juxtaglomerular apparatus (JGA) inhibits renin release, blocking AT_1 receptors by ARBs in JGA thus increases renin release, which in turn results in increased angiotensin II. This allows the activation of AT_2 receptors, thereby leading to cardiovascular protective effects.

This may explain why use of ARBs is much less frequently associated with dry cough and angioedema as compared with ACE inhibitors (see Section 9.4).

9.5.3.3 Activation of AT_2 Receptors, Potentially Leading to Additional Cardiovascular Protection

Blockade of the angiotensin II receptors by ARBs inhibits the negative regulatory feedback of angiotensin II on renin secretion. However, the resulting increased plasma renin activity and circulating levels of angiotensin II do not overcome the effect of ARBs on blood pressure because the AT_1 receptor-mediated vasoconstriction is blocked by the ARBs. Instead, the increased plasma levels of angiotensin II following AT_1 receptor blockade with ARBs may stimulate the AT_2 receptors, potentially leading to additional cardiovascular protection. In this context, evidence suggests a cardiovascular protective function of AT_2 receptor signaling.

9.5.3.4 Other Novel Effects

Telmisartan, but not the other ARBs, reduces triglyceride levels and improves insulin sensitivity possibly via its activation of peroxisome proliferator-activated receptor-gamma (PPARγ) [52, 53]. In this context, PPARγ-mediated signaling plays an important role in regulating carbohydrate and lipid metabolism. Activation of PPARγ may also be involved in the telmisartan-induced elevation of adiponectin and decrease in C-reactive protein

levels [54, 55]. Both telmisartan and losartan have antiplatelet aggregation activity [51, 56]. However, the exact role of the aforementioned pleiotropic effects in ARB-mediated cardiovascular benefits remains to be established.

9.5.4 Clinical Uses

The clinical uses of ARBs include the treatment of the following diseases:

- Hypertension
- Heart failure
- Stroke
- Diabetic nephropathy

9.5.4.1 Hypertension

All ARBs are approved for the treatment of hypertension, either alone or in combination with other antihypertensive agents. Specific ARBs, as detailed next, are also approved for the management of heart failure, stroke, or diabetic nephropathy based on favorable evidence from clinical trials. It is important to note that because all ARBs share the same mechanisms of action with difference only in pharmacokinetic properties, the benefits of ARBs in cardiovascular diseases as well as other disease conditions are most likely to be class rather than drug specific.

Fixed-dose single-pill combinations of specific ARBs with other antihypertensive drugs have become widely available in recent years. In the United States, 7 of the 8 commercially available ARBs are available as single-pill fixed-dose combinations with the thiazide diuretic hydrochlorothiazide. A combination of the newest ARB, azilsartan, with the thiazide-like diuretic, chlorthalidone, was approved by the FDA in 2011. Fixed-dose combinations of ARBs and the calcium channel blocker amlodipine are also available. Recently, a fixed-dose combination of valsartan with the direct renin inhibitor aliskiren has been approved for the treatment of hypertension not adequately controlled by other medications, patients not likely to respond to a single antihypertensive medication, and patients taking one or the other medications with inadequate blood pressure control. Moreover, two of the ARBs, valsartan and olmesartan, are now available as part of 3-drug combinations that also include amlodipine and hydrochlorothiazide. It is noteworthy that combinations that contain all three drugs are more effective than a combination that includes any two of the three drugs.

The fixed-dose 2-drug combinations of an ARB and a thiazide or thiazide-type diuretic drug are listed below:

- Azilsartan/chlorthalidone
- Candesartan/hydrochlorothiazide
- Eprosartan/hydrochlorothiazide
- Irbesartan/hydrochlorothiazide
- Losartan/hydrochlorothiazide
- Olmesartan/hydrochlorothiazide
- Telmisartan/hydrochlorothiazide
- Valsartan/hydrochlorothiazide

The fixed-dose 2-drug combinations of an ARB and the calcium channel blocker amlodipine are listed below:

- Azilsartan/amlodipine
- Olmesartan/amlodipine
- Valsartan/amlodipine

The fixed-dose 2-drug combination of an ARB and a direct renin inhibitor is listed below:

- Azilsartan/aliskiren

Listed below are fixed-dose 3-drug combinations:

- Olmesartan/amlodipine/hydrochlorothiazide
- Telmisartan/aliskiren/hydrochlorothiazide

9.5.4.2 Heart Failure and Ischemic Heart Diseases

Similar to what found with ACE inhibitors, clinical trials have demonstrated an efficacy for multiple ARBs in the management of heart failure as well as ischemic heart

disease. Valsartan has been approved by the US FDA for treatment of congestive heart failure (NYHA classes II–IV; see Unit VI for classification of heart failure) and for reduction of cardiovascular mortality in clinically stable patients with left ventricular failure or left ventricular dysfunction following myocardial infarction. In addition, the HEAAL trial also firmly established the superiority of full (150 mg daily) over partial (50 mg daily) AT_1 receptor blockade with losartan in decreasing mortality and hospitalizations in heart failure [57].

9.5.4.3 Stroke

Losartan is approved by the US FDA for reducing the risk of stroke in patients with hypertension and left ventricular hypertrophy, but there is evidence that this benefit does not apply to patients of African heritage. A recent meta-analysis of randomized trials among nearly 30,000 participants with a history of symptomatic cerebrovascular disease found that use of ARBs or ACE inhibitors was associated with a 9% relative risk reduction in overall vascular risk and a 7% relative risk reduction in recurrent stroke risk [58]. Although the overall vascular protective effects of ARBs as well as ACE inhibitors among patients with known stroke or transient ischemic attack seem definitive, it is relatively modest when compared with other proven vascular risk-reducing strategies, such as calcium channel blockers [59].

9.5.4.4 Diabetes and Diabetic Nephropathy

Irbesartan and losartan are approved by the US FDA for the treatment of diabetic nephropathy with elevated serum creatinine and proteinuria (>300 mg/day) in patients with type 2 diabetes and hypertension. In this population, both irbesartan and losartan reduce the rate of progression of nephropathy as measured by the occurrence of doubling of serum creatinine or end-stage renal disease (need for dialysis or renal transplantation). A recent trial reported that when added to a lifestyle intervention, valsartan at a daily dose of 160 mg reduced the risk of diabetes mellitus but did not affect the cardiovascular outcomes in patients with glucose intolerance [60]. Recent meta-analyses of randomized controlled trials found that the use of ARBs or ACE inhibitors reduced the incidence of new-onset diabetes mellitus compared with placebo, beta-blockers, and calcium channel blockers and consequently recommended that ARBs or ACE inhibitors should be preferred in patients with clinical conditions that may increase the risk of new-onset diabetes mellitus [61, 62].

9.5.4.5 Combination Therapy of an ARB with an ACE Inhibitor

Heart Failure   The quest for more complete blockade of the RAAS led to the studies combining maximum doses of an ACE inhibitor and an ARB [50]. The Valsartan Heart Failure Trial (Val-HeFT) and CHARM-Added trials represent the only two prospectively designed, adequately sized, and appropriately powered heart failure trials that tested the efficacy of combination of an ARB and an ACE inhibitor [63, 64]. Both showed a statistically significant reduction in mortality–morbidity composite outcomes despite increases in risk of renal dysfunction, hyperkalemia, and hypotension. These results led to strict guidelines for combined use of an ACE inhibitor and an ARB in chronic systolic heart failure from multiple regulatory authorities, including the American College of Cardiology Foundation/American Heart Association (ACCF/AHA) [65]. For instance, the 2009 focused update of ACCF/AHA Guidelines for the Diagnosis and Management of Heart Failure in Adults states that addition of AT_1 receptor blockade to an ACE inhibitor may be considered in patients with reduced left ventricular ejection fraction already treated with conventional therapy [65] (also see Unit VI). The additive efficacy of combined therapy of an ACE inhibitor and an ARB in patients with advanced heart failure suggests that the repetitive cardiac decompensation occurring in patients treated with either an ACE inhibitor or an ARB may be attributable to the incomplete blockade of the RAAS by either an ACE inhibitor or an ARB alone.

Ischemic Heart Disease With regard to reducing mortality and prolonging survival, ARBs are as effective as ACE inhibitors in patients who are at high risk for cardiovascular events after myocardial infarction complicated by heart failure, left ventricular dysfunction, or both. However, the current evidence does not justify the use of an ACE inhibitor/ARB combination in this setting as combination therapy increases the rate of adverse events without improving survival [66]. Similarly, in patients who have vascular disease (coronary artery disease or cerebrovascular disease) or high-risk diabetes mellitus without heart failure, ACE inhibitors and ARBs are equivalent with regard to reducing mortality and morbidity from cardiovascular causes, and the ACE inhibitor/ARB combination therapy is associated with more adverse events without an increase in benefit [67].

Renal Protection The ACE inhibitor/ARB combination therapy may lead to greater reduction in proteinuria in renal disease (mostly associated with diabetes mellitus) than either drug alone [68]. However, meta-analysis of small-scale trials involving diabetic individuals demonstrated that while combination therapy was more effective in reducing proteinuria, there were a decrease in estimated glomerular filtration rate and a trend toward increased creatinine in the combination group [69]. The use of proteinuria as a surrogate maker for kidney outcomes may thus have limitations in assessing the efficacy of combined therapy in diabetic nephropathy. In this context, the Veterans Affairs Nephropathy in Diabetes (VA NEPHRON-D) study, a large-scale randomized trial involving 1448 patients with diabetic nephropathy, was stopped early owing to safety concerns. The trial showed that combination therapy of losartan and lisinopril significantly increased the risk of hyperkalemia and acute kidney injury and failed to provide benefit with respect to mortality or cardiovascular events [70]. Hence, based on the results of the VA NEPHRON-D study, dual RAAS blockade with an ARB and an ACE inhibitor for the treatment of patients with diabetic nephropathy cannot currently be recommended [71].

9.5.5 Therapeutic Dosages

The dosage forms and strengths of the eight ARBs are listed below:

- Azilsartan (Edarbi): Oral, 40 and 80 mg tablets
- Candesartan (Atacand): Oral, 4, 8, 16, and 32 mg tablets
- Eprosartan (Teveten): Oral, 400 and 600 mg tablets
- Irbesartan (Avapro): Oral, 75, 150, and 300 mg tablets
- Losartan (Cozaar): Oral, 25, 50, and 100 mg tablets
- Olmesartan (Benicar): Oral, 5, 20, and 40 mg tablets
- Telmisartan (Micardis): Oral, 20, 40, and 80 mg tablets
- Valsartan (Diovan): Oral, 40, 80, 160, and 320 mg tablets

All ARBs are approved for the treatment of hypertension, either alone or in combination with other antihypertensive agents. Some ARBs are also approved for use in the treatment of heart failure, stroke, or diabetic nephropathy. The FDA-approved indications and the corresponding dosage regimens of the eight ARBs are summarized in Table 9.4.

9.5.6 Adverse Effects and Drug Interactions

Due to the inhibitory effects on the RAAS, ARBs and ACEIs share similar adverse effects, described in Section 9.4.5, except that the incidence of dry cough and angioedema is much lower with ARBs than with ACE inhibitors. The drug interactions, contraindications, and pregnancy categories for ARBs are also similar to those for ACE inhibitors (Section 9.4.5).

9.5.7 Summary of ARBs

- Currently, there are eight US FDA-approved ARBs, with losartan and azilsartan as the prototype and the newest member, respectively.
- ARBs selectively block the activation of angiotensin II type 1 receptors (AT_1 receptors) by angiotensin II and the resulting deleterious effects of angiotensin II.

TABLE 9.4   Dosage regimens of ARBs

ARB	Indication	Dosage
Azilsartan	Hypertension	80 mg once daily
Candesartan	Hypertension	Initial dose, 16 mg once daily; dose range, 8–32 mg once daily
	Heart failure	Initial dose, 4 mg once daily; maintenance dose, 32 mg once daily
Eprosartan	Hypertension	Initial dose, 600 mg once daily; dose range, 400–800 mg once daily or in two divided doses
Irbesartan	Hypertension	Initial dose, 150 mg once daily; dose range, 150–300 mg once daily
	Diabetic nephropathy	Maintenance dose, 300 mg once daily
Losartan	Hypertension	Initial dose, 50 mg once daily; dose range, 25–100 mg once daily or in two divided doses
	Stroke risk reduction	For reducing stroke in patients with hypertension and left ventricular hypertrophy, initial dose, 50 mg once daily; maintenance dose, 100 mg once daily
	Diabetic nephropathy	Initial dose, 50 mg once daily; maintenance dose, 100 mg once daily
Olmesartan	Hypertension	Initial dose, 20 mg once daily; dose range, 20–40 mg once daily
Telmisartan	Hypertension	Initial dose, 40 mg once daily; dose range, 40–80 mg once daily
	Cardiovascular risk reduction	80 mg once daily
Valsartan	Hypertension	Initial dose, 80 or 160 mg once daily; dose range, 80–320 mg once daily
	Heart failure	Initial dose, 40 mg twice daily; maintenance dose, 160 mg twice daily
	Myocardial infarction	To reduce cardiovascular mortality in clinically stable patients with left ventricular dysfunction following myocardial infarction, initial dose, 20 mg twice daily; maintenance dose, 160 mg twice daily

- Because ARBs do not inhibit ACE (kininase II), they do not cause accumulation of bradykinin, and as such, use of these drugs is less frequently associated with dry cough as compared with ACE inhibitors.
- Blockade of AT_1 receptors by ARBs inhibits the negative regulatory feedback of angiotensin II on renin secretion, and the increased plasma levels of angiotensin II may augment the activation of AT_2 receptors, potentially leading to additional cardiovascular protection.
- All ARBs are approved for the treatment of hypertension, either alone or in combination with other antihypertensive agents. Some are also approved for treating heart failure, ischemic heart disease, stroke, and diabetic nephropathy.
- While either ACE inhibitors or ARBs as monotherapy or in combination with other drugs are effective in treating cardiovascular diseases, the combination therapy of an ACE inhibitor and an ARB is generally not recommended due to increased adverse effects.

9.6   COMPARATIVE PHARMACOLOGY OF DIRECT RENIN INHIBITORS, ACE INHIBITORS, AND ARBs

As discussed previously, therapies aimed at modifying the RAAS have been used extensively for the treatment of hypertension, heart failure, myocardial infarction, diabetes, and renal disease. Three classes of drugs that interact with this system are used to inhibit the effects of angiotensin II: (i) ACE inhibitors, (ii) ARBs, and (iii) the direct renin inhibitors. ACE inhibitors block the conversion of angiotensin I into angiotensin II; ARBs selectively inhibit angiotensin II from activating the AT_1 receptors; and direct renin inhibitors block the conversion of angiotensinogen into angiotensin I. This section discusses the comparative pharmacology and therapeutics of these three classes of RAAS inhibitors to guide clinical decision making regarding the use of these drugs in the management of cardiovascular diseases.

9.6.1   Clinical Equivalence

ACE inhibitors and ARBs are frequently considered by clinicians as being equivalent. While both drug classes reduce the downstream effects of angiotensin II, it is not clear that these medications are in fact clinically equivalent. ACE inhibitors, for example, do not entirely block production of angiotensin II because of the presence of unaffected converting enzymes. Also, ACE inhibitors have well-known side effects not shared by ARBs, including cough (estimated incidence 5–20%) and angioedema (estimated incidence 0.1–0.2%, with a lesser reported risk with ARBs). Additional considerations arise with the newer direct renin inhibitors because their side effect profiles and efficacy may differ significantly from ACE inhibitors or ARBs.

9.6.2   Comparative Effectiveness Review

Given the public health importance and widespread use of these agents in treatment of hypertension as well as many other cardiovascular conditions, it is important to understand their comparative effects on clinical outcomes. In this regard, the No. 34 Comparative Effectiveness Review of the Agency for Healthcare Research and Quality (AHRQ) of the US Department of Health and Human Services provides an updated clinical evidence for ACE inhibitors, ARBs, and direct renin inhibitors in the management of hypertension as well as other comorbidities [72].

Consistent with the findings from AHRQ's 2007 report, the 2011 update indicated no overall differences in blood pressure control, mortality rates, and major cardiovascular events in patients treated with ACE inhibitors versus ARBs [72]. With a low strength of evidence, two studies reported a small but significantly greater blood pressure reduction for patients treated with the direct renin inhibitor aliskiren versus the ACE inhibitor ramipril. Studies evaluating aliskiren versus ACE inhibitors and ARBs on mortality and morbidity outcomes were relatively short, and few deaths or cardiovascular events occurred, resulting in insufficient evidence to discern differences. A meta-analysis of 23 randomized controlled trials comparing ACE inhibitors and ARBs found no significant difference in the proportion of patients who achieved successful blood pressure control on a single antihypertensive agent. Compared with ARBs and aliskiren, ACE inhibitors were consistently associated with higher rates of cough. Withdrawals due to adverse events were modestly more frequent for patients receiving ACE inhibitors than those receiving ARBs or aliskiren; this is consistent with the differential rates of cough. There was no evidence of differential effects of ACE inhibitors, ARBs, or direct renin inhibitors on the outcomes of lipids, renal outcomes, carbohydrate metabolism or diabetes, or left ventricular mass; however, there was not a high strength of evidence for any of these outcomes. Regarding the question of whether ACE inhibitors, ARBs, or direct renin inhibitors are associated with better outcomes in specific patient subgroups, the evidence was insufficient to reach firm conclusions. The 2011 AHRQ Review concludes that evidence does not support a meaningful difference between ACE inhibitors and ARBs for any outcomes except medication side effects [72].

9.6.3   Future Research Needs

In a 2012 report from AHRQ [73], seven areas have been identified as the highest priority for future research for the comparative effectiveness of ACEIs, ARBs, or direct renin inhibitors in patients with hypertension. The 2012 AHRQ report also provides recommendations on research design for each of the seven research questions:

1. What is the comparative effectiveness of these medications on cardiovascular and cerebrovascular events measured over several years? Recommended study design: if able to combine with chronic conditions other than hypertension, then a systematic review with broader inclusion criteria could provide additional information not included in the comparative effectiveness review, which was restricted to patients with hypertension. If not, then large long-term clinical trial or observational study would be preferable.

2. What is the impact of comorbidities (such as ischemic heart disease, congestive heart failure, diabetes, peripheral arterial disease, or chronic kidney disease) on ACE inhibitor/ARB/direct renin inhibitor effectiveness or harms in patients with hypertension? Recommended study design: if patient-level data are available from relevant trials, then a patient-level meta-analysis may be the most efficient approach.

3. What is the impact of demographic differences (such as age, race, or sex) on the effectiveness or harms of ACE inhibitor/ARB/direct renin inhibitor in patients with hypertension? Recommended study design: if patient-level data are available from relevant trials, then a patient-level meta-analysis is most appropriate.

4. Do the results differ in practical clinical trials or other external validity-oriented studies that compare these medications in practice settings that better represent real-world practice? Recommended study design: either a large clinical trial with broader inclusion criteria to maximize generalizability or an observational study of patients in typical community practice.

5. What is the impact of ACE inhibitor/ARB/direct renin inhibitor on incidence of new cardiovascular or metabolic diagnoses such as diabetes, atrial fibrillation, or congestive heart failure with or without preserved left ventricular function? Recommended study design: if patients can be combined across clinical conditions (i.e., not exclusively hypertension), then a systematic review of existing studies is most appropriate.

6. What is the impact of ACE inhibitor/ARB/direct renin inhibitor on patient health status including quality of life and functional capacity? Recommended study design: randomized controlled trials with the inclusion of validated quality of life measures as an outcome.

7. Are there important differences in medication adherence and persistence with drug therapy across the different classes of drug? Recommended study design: new observational studies with a focus on the longitudinal measurement of adherence and persistence.

9.7 SUMMARY OF CHAPTER KEY POINTS

- Uncontrolled activation of the RAAS is a chief mechanism of diverse cardiovascular diseases (e.g., hypertension, heart failure, ischemic heart disease) and noncardiovascular diseases, such as diabetic nephropathy. As such, inhibition of the RAAS activity becomes a cornerstone for treating the aforementioned common disorders.

- Currently, four classes of drugs are available that target the RAAS axis, including direct renin inhibitors, ACE inhibitors, ARBs, and aldosterone receptor antagonists. The aldosterone receptor antagonists are covered in Chapter 7 under potassium-sparing diuretics.

- Aliskiren is the first and also the only direct renin inhibitor approved by the US FDA in 2007 for treating hypertension. Aliskiren, operating at the first and rate-limiting step of the RAAS cascade, renders the entire RAAS quiescent. Because renin is specific for the substrate angiotensinogen, aliskiren does not cause stimulation of bradykinin or prostaglandins, as seen with ACE inhibitors. In addition, renin inhibitors reduce the formation of angiotensin II by non-ACE pathways.

- Except for its demonstrated efficacy in treating hypertension, the role of aliskiren in other cardiovascular diseases is not well established. The drug is generally well tolerated, and the most common adverse effects are mild diarrhea and other gut disturbances. Drugs that act directly on the RAAS, including aliskiren, can cause injury and death to the developing fetus and as such should be avoided during pregnancy.

- Drugs that block ACE and the formation of angiotensin II have become mainstays of cardiovascular pharmacotherapeutics. Currently, there are 11 US FDA-approved ACE inhibitors. Inhibition of ACE causes decreased formation of angiotensin II and decreased catabolism of bradykinin. As such, ACE inhibitors are more frequently associated with dry cough as compared with other RAAS inhibitors.

- ACE inhibitors are widely used in treating hypertension, heart failure, and acute coronary syndromes. These drugs reduce mortality and prolong survival in the aforementioned settings. Use of ACE inhibitors in patients at high risk of coronary artery disease has been associated with a decrease in the risk of myocardial infarction, stroke, and death.

- ACE inhibitors as a class of drugs are generally well tolerated. Adverse effects are usually mild and include dry cough and hyperkalemia. Rare but serious adverse effects include angioedema and impaired renal function.

- ARBs selectively block the activation of AT_1 receptors by angiotensin II, allowing the activation of AT_2 receptors, which may lead to additional cardiovascular protection. Because ARBs do not inhibit ACE (kininase II), they do not cause accumulation of bradykinin. As such, use of ARBs is much less frequently associated with dry cough compared with ACE inhibitors.

- All ARBs are approved for treating hypertension, and some are also approved for the treatment of other diseases, including heart failure, stroke, or diabetic nephropathy based on favorable evidence from clinical trials.

- ARBs and ACEIs share similar adverse effects except that the incidence of dry cough and angioedema is much lower with ARBs than with ACE inhibitors.

- Although either ACE inhibitors or ARBs as monotherapy or in combination with other drugs are effective in treating cardiovascular diseases, the combination therapy of an ACE inhibitor and an ARB is generally not recommended due to increased adverse effects.

9.8 SELF-ASSESSMENT QUESTIONS

9.8.1 A 36-year-old female with stage 1 hypertension is planning to become pregnant. In treating her hypertension, which of the following drugs should be avoided?
A. Chlorthalidone
B. Hydralazine
C. Hydrochlorothiazide
D. Irbesartan
E. Methyldopa

9.8.2 A 46-year-old male presents with a blood pressure of 150/95 mmHg. He is prescribed lisinopril. Although his blood pressure is under control, he complains of persistent dry cough. Which of the following drugs acts by a similar but not identical mechanism and is much less likely to cause dry cough in this patient?
A. Captopril
B. Carvedilol
C. Clonidine
D. Hydralazine
E. Losartan

9.8.3 Following treatment with an antihypertensive agent, blood chemistry of a diabetic patient with high blood pressure reveals an increased renin activity and elevated levels of circulating angiotensin II. Which of the following drugs is most likely prescribed for treating his hypertension?
A. Aliskiren
B. Carvedilol
C. Enalaprilat
D. Lisinopril
E. Valsartan

9.8.4 A 38-year-old female presents with a blood pressure of 145/95 mmHg. She denies previous medical problems, but her brother died of myocardial infarction at the age of 51 years and one parent died of diabetic complications. Results of laboratory tests are normal except for a fasting glucose level of 102 mg/dl. Which of the following drugs will be most likely prescribed to control her blood pressure and at the same time also reduce the risk of developing diabetes?

A. Aliskiren
B. Atorvastatin
C. Hydrochlorothiazide
D. Irbesartan
E. Propranolol

9.8.5 A hypertensive patient is being treated with oral lisinopril 20 mg once daily. Which of the following is characteristic of the drug therapy in this patient?

A. Decreased plasma level of angiotensin I
B. Decreased plasma level of angiotensin II
C. Decreased plasma level of angiotensinogen
D. Decreased plasma level of prerenin
E. Decreased plasma level of renin

REFERENCES

1 Ma, T.K., et al., Renin-angiotensin-aldosterone system blockade for cardiovascular diseases: current status. *Br J Pharmacol*, 2010. 160(6): p. 1273–92.

2 Basso, N. and N.A. Terragno, History about the discovery of the renin-angiotensin system. *Hypertension*, 2001. 38(6): p. 1246–9.

3 Jensen, C., P. Herold, and H.R. Brunner, Aliskiren: the first renin inhibitor for clinical treatment. *Nat Rev Drug Discov*, 2008. 7(5): p. 399–410.

4 Schmieder, R.E., et al., Renin-angiotensin system and cardiovascular risk. *Lancet*, 2007. 369(9568): p. 1208–19.

5 Putnam, K., et al., The renin-angiotensin system: a target of and contributor to dyslipidemias, altered glucose homeostasis, and hypertension of the metabolic syndrome. *Am J Physiol Heart Circ Physiol*, 2012. 302(6): p. H1219–30.

6 Nguyen Dinh Cat, A. and R.M. Touyz, A new look at the renin-angiotensin system—focusing on the vascular system. *Peptides*, 2011. 32(10): p. 2141–50.

7 Ferreira, A.J., R.A. Santos, and M.K. Raizada, Angiotensin-(1–7)/angiotensin-converting enzyme 2/mas receptor axis and related mechanisms. *Int J Hypertens*, 2012. 2012: p. 690785.

8 Israili, Z.H., M. Velasco, and V. Bermudez, Direct renin inhibitors as antihypertensive agents. *Am J Ther*, 2010. 17(3): p. 237–54.

9 Fisher, N.D. and E.A. Meagher, Renin inhibitors. *J Clin Hypertens (Greenwich)*, 2011. 13(9): p. 662–6.

10 Azizi, M. and J. Menard, Renin inhibitors and cardiovascular and renal protection: an endless quest? *Cardiovasc Drugs Ther*, 2013. 27(2): p. 145–53.

11 Virdis, A., et al., Effect of aliskiren treatment on endothelium-dependent vasodilation and aortic stiffness in essential hypertensive patients. *Eur Heart J*, 2012. 33(12): p. 1530–8.

12 Schramm, A., et al., Targeting NADPH oxidases in vascular pharmacology. *Vascul Pharmacol*, 2012. 56(5–6): p. 216–31.

13 Solomon, S.D., et al., Effect of the direct Renin inhibitor aliskiren, the Angiotensin receptor blocker losartan, or both on left ventricular mass in patients with hypertension and left ventricular hypertrophy. *Circulation*, 2009. 119(4): p. 530–7.

14 McMurray, J.J., et al., Effects of the oral direct renin inhibitor aliskiren in patients with symptomatic heart failure. *Circ Heart Fail*, 2008. 1(1): p. 17–24.

15 Sidik, N.P., et al., Effect of aliskiren in patients with heart failure according to background dose of ACE inhibitor: a retrospective analysis of the Aliskiren Observation of Heart Failure Treatment (ALOFT) trial. *Cardiovasc Drugs Ther*, 2011. 25(4): p. 315–21.

16 Parving, H.H., et al., Aliskiren combined with losartan in type 2 diabetes and nephropathy. *N Engl J Med*, 2008. 358(23): p. 2433–46.

17 Persson, F., et al., Aliskiren in combination with losartan reduces albuminuria independent of baseline blood pressure in patients with type 2 diabetes and nephropathy. *Clin J Am Soc Nephrol*, 2011. 6(5): p. 1025–31.

18 McMurray, J.J., et al., Aliskiren, ALTITUDE, and the implications for ATMOSPHERE. *Eur J Heart Fail*, 2012. 14(4): p. 341–3.

19 Parving, H.H., et al., Cardiorenal end points in a trial of aliskiren for type 2 diabetes. *N Engl J Med*, 2012. 367(23): p. 2204–13.

20 Gheorghiade, M., et al., Effect of aliskiren on postdischarge mortality and heart failure readmissions among patients hospitalized for heart failure: the ASTRONAUT randomized trial. *JAMA*, 2013. 309(11): p. 1125–35.

21 Nicholls, S.J., et al., Effect of aliskiren on progression of coronary disease in patients with prehypertension: the AQUARIUS randomized clinical trial. *JAMA*, 2013. 310(11): p. 1135–44.

22 Tardif, J.C. and J. Gregoire, Renin-angiotensin system inhibition and secondary cardiovascular prevention. *JAMA*, 2013. 310(11): p. 1130–1.

23 Harel, Z., et al., The effect of combination treatment with aliskiren and blockers of the renin-angiotensin system on hyperkalaemia and acute kidney injury: systematic review and meta-analysis. *BMJ*, 2012. 344: p. e42.

24 Shen, X.Z., et al., Nontraditional roles of angiotensin-converting enzyme. *Hypertension*, 2012. 59(4): p. 763–8.

25 Lahoud, R., et al., Effect of use of combination evidence-based medical therapy after acute coronary syndromes on long-term outcomes. *Am J Cardiol*, 2012. 109(2): p. 159–64.

26 Udelson, J.E., Heart failure with preserved ejection fraction. *Circulation*, 2011. 124(21): p. e540–3.

27 Cleland, J.G., et al., The perindopril in elderly people with chronic heart failure (PEP-CHF) study. *Eur Heart J*, 2006. 27(19): p. 2338–45.

28 Kitzman, D.W., et al., A randomized double-blind trial of enalapril in older patients with heart failure and preserved

ejection fraction: effects on exercise tolerance and arterial distensibility. *Circ Heart Fail*, 2010. 3(4): p. 477–85.

29 Shah, R.V., A.S. Desai, and M.M. Givertz, The effect of renin-angiotensin system inhibitors on mortality and heart failure hospitalization in patients with heart failure and preserved ejection fraction: a systematic review and meta-analysis. *J Card Fail*, 2010. 16(3): p. 260–7.

30 Oghlakian, G.O., I. Sipahi, and J.C. Fang, Treatment of heart failure with preserved ejection fraction: have we been pursuing the wrong paradigm? *Mayo Clin Proc*, 2011. 86(6): p. 531–9.

31 Zhang, Y., et al., Design and rationale of studies of neurohormonal blockade and outcomes in diastolic heart failure using OPTIMIZE-HF registry linked to Medicare data. *Int J Cardiol*, 2013. 166(1): p. 230–5.

32 McAlister, F.A., Angiotensin-converting enzyme inhibitors or angiotensin receptor blockers are beneficial in normotensive atherosclerotic patients: a collaborative meta-analysis of randomized trials. *Eur Heart J*, 2012. 33(4): p. 505–14.

33 Dagenais, G.R., et al., Effects of ramipril and rosiglitazone on cardiovascular and renal outcomes in people with impaired glucose tolerance or impaired fasting glucose: results of the Diabetes REduction Assessment with ramipril and rosiglitazone Medication (DREAM) trial. *Diabetes Care*, 2008. 31(5): p. 1007–14.

34 Bhuriya, R., et al., Prevention of recurrent atrial fibrillation with angiotensin-converting enzyme inhibitors or angiotensin receptor blockers: a systematic review and meta-analysis of randomized trials. *J Cardiovasc Pharmacol Ther*, 2011. 16(2): p. 178–84.

35 Huang, G., et al., Angiotensin-converting enzyme inhibitors and angiotensin receptor blockers decrease the incidence of atrial fibrillation: a meta-analysis. *Eur J Clin Invest*, 2011. 41(7): p. 719–33.

36 Disertori, M., et al., Systematic review and meta-analysis: renin-Angiotensin system inhibitors in the prevention of atrial fibrillation recurrences: an unfulfilled hope. *Cardiovasc Drugs Ther*, 2012. 26(1): p. 47–54.

37 Abuissa, H., et al., Angiotensin-converting enzyme inhibitors or angiotensin receptor blockers for prevention of type 2 diabetes: a meta-analysis of randomized clinical trials. *J Am Coll Cardiol*, 2005. 46(5): p. 821–6.

38 Bosch, J., et al., Effect of ramipril on the incidence of diabetes. *N Engl J Med*, 2006. 355(15): p. 1551–62.

39 Ingelfinger, J.R. and C.G. Solomon, Angiotensin-converting-enzyme inhibitors for impaired glucose tolerance—is there still hope? *N Engl J Med*, 2006. 355(15): p. 1608–10.

40 Cheng, J., et al., Effect of angiotensin-converting enzyme inhibitors and angiotensin II receptor blockers on all-cause mortality, cardiovascular deaths, and cardiovascular events in patients with diabetes mellitus: a meta-analysis. *JAMA Intern Med*, 2014. 174(5): p. 773–85.

41 Mas, S., et al., Pharmacogenetic predictors of angiotensin-converting enzyme inhibitor-induced cough: the role of ACE, ABO, and BDKRB2 genes. *Pharmacogenet Genomics*, 2011. 21(9): p. 531–8.

42 Nishio, K., et al., Angiotensin-converting enzyme and bradykinin gene polymorphisms and cough: a meta-analysis. *World J Cardiol*, 2011. 3(10): p. 329–36.

43 Wright, J.T., Jr., et al., Effect of blood pressure lowering and antihypertensive drug class on progression of hypertensive kidney disease: results from the AASK trial. *JAMA*, 2002. 288(19): p. 2421–31.

44 Makani, H., et al., Meta-analysis of randomized trials of angioedema as an adverse event of renin-angiotensin system inhibitors. *Am J Cardiol*, 2012. 110(3): p. 383–91.

45 Beavers, C.J., S.P. Dunn, and T.E. Macaulay, The role of angiotensin receptor blockers in patients with angiotensin-converting enzyme inhibitor-induced angioedema. *Ann Pharmacother*, 2011. 45(4): p. 520–4.

46 Li, D.K., et al., Maternal exposure to angiotensin converting enzyme inhibitors in the first trimester and risk of malformations in offspring: a retrospective cohort study. *BMJ*, 2011. 343: p. d5931.

47 Koren, G., Hypertension: ACE inhibitor use in pregnancy—setting the record straight. *Nat Rev Cardiol*, 2012. 9(1): p. 7–8.

48 Walfisch, A., et al., Teratogenicity of angiotensin converting enzyme inhibitors or receptor blockers. *J Obstet Gynaecol*, 2011. 31(6): p. 465–72.

49 Moretti, M.E., et al., The fetal safety of angiotensin converting enzyme inhibitors and angiotensin II receptor blockers. *Obstet Gynecol Int*, 2012. 2012: p. 658310.

50 Dell'Italia, L.J., Translational success stories: angiotensin receptor 1 antagonists in heart failure. *Circ Res*, 2011. 109(4): p. 437–52.

51 Taylor, A.A., H. Siragy, and S. Nesbitt, Angiotensin receptor blockers: pharmacology, efficacy, and safety. *J Clin Hypertens (Greenwich)*, 2011. 13(9): p. 677–86.

52 Takagi, H. and T. Umemoto, Telmisartan reduces triglyceride levels over other angiotensin II receptor blockers: a meta-analysis of randomized head-to-head trials. *Int J Cardiol*, 2012. 157(3): p. 403–7.

53 Takagi, H. and T. Umemoto, Telmisartan improves insulin sensitivity: a meta-analysis of randomized head-to-head trials. *Int J Cardiol*, 2012. 156(1): p. 92–6.

54 Takagi, H. and T. Umemoto, Telmisartan increases adiponectin levels: a meta-analysis and meta-regression of randomized head-to-head trials. *Int J Cardiol*, 2012. 155(3): p. 448–51.

55 Takagi, H., et al., Effects of telmisartan on C-reactive protein levels: a meta-analysis of randomized controlled trials. *Int J Cardiol*, 2012. 156(2): p. 238–41.

56 Yamada, K., T. Hirayama, and Y. Hasegawa, Antiplatelet effect of losartan and telmisartan in patients with ischemic stroke. *J Stroke Cerebrovasc Dis*, 2007. 16(5): p. 225–31.

57 Konstam, M.A., et al., Effects of high-dose versus low-dose losartan on clinical outcomes in patients with heart failure (HEAAL study): a randomised, double-blind trial. *Lancet*, 2009. 374(9704): p. 1840–8.

58 Lee, M., et al., Renin-Angiotensin system modulators modestly reduce vascular risk in persons with prior stroke. *Stroke*, 2012. 43(1): p. 113–9.

59 Webb, A.J., et al., Effects of antihypertensive-drug class on interindividual variation in blood pressure and risk of stroke: a systematic review and meta-analysis. *Lancet*, 2010. 375(9718): p. 906–15.

60 McMurray, J.J., et al., Effect of valsartan on the incidence of diabetes and cardiovascular events. *N Engl J Med*, 2010. 362(16): p. 1477–90.

61 Tocci, G., et al., Angiotensin-converting enzyme inhibitors, angiotensin II receptor blockers and diabetes: a meta-analysis of placebo-controlled clinical trials. *Am J Hypertens*, 2011. 24(5): p. 582–90.

62 Geng, D.F., et al., Angiotensin receptor blockers for prevention of new-onset type 2 diabetes: a meta-analysis of 59,862 patients. *Int J Cardiol*, 2012. 155(2): p. 236–42.

63 McMurray, J.J., et al., Effects of candesartan in patients with chronic heart failure and reduced left-ventricular systolic function taking angiotensin-converting-enzyme inhibitors: the CHARM-Added trial. *Lancet*, 2003. 362(9386): p. 767–71.

64 Cohn, J.N. and G. Tognoni, A randomized trial of the angiotensin-receptor blocker valsartan in chronic heart failure. *N Engl J Med*, 2001. 345(23): p. 1667–75.

65 Jessup, M., et al., 2009 focused update: ACCF/AHA guidelines for the diagnosis and management of heart failure in adults: a report of the American College of Cardiology Foundation/American Heart Association Task Force on Practice Guidelines: developed in collaboration with the International Society for Heart and Lung Transplantation. *Circulation*, 2009. 119(14): p. 1977–2016.

66 Pfeffer, M.A., et al., Valsartan, captopril, or both in myocardial infarction complicated by heart failure, left ventricular dysfunction, or both. *N Engl J Med*, 2003. 349(20): p. 1893–906.

67 Yusuf, S., et al., Telmisartan, ramipril, or both in patients at high risk for vascular events. *N Engl J Med*, 2008. 358(15): p. 1547–59.

68 Kunz, R., et al., Meta-analysis: effect of monotherapy and combination therapy with inhibitors of the renin angiotensin system on proteinuria in renal disease. *Ann Intern Med*, 2008. 148(1): p. 30–48.

69 Jennings, D.L., et al., Combination therapy with an ACE inhibitor and an angiotensin receptor blocker for diabetic nephropathy: a meta-analysis. *Diabet Med*, 2007. 24(5): p. 486–93.

70 Fried, L.F., et al., Combined angiotensin inhibition for the treatment of diabetic nephropathy. *N Engl J Med*, 2013. 369(20): p. 1892–903.

71 de Zeeuw, D., The end of dual therapy with renin-angiotensin-aldosterone system blockade? *N Engl J Med*, 2013. 369(20): p. 1960–2.

72 Powers, B., L. Greene, and L.M. Balfe, Updates on the treatment of essential hypertension: a summary of AHRQ's comparative effectiveness review of angiotensin- converting enzyme inhibitors, angiotensin II receptor blockers, and direct renin inhibitors. *J Manag Care Pharm*, 2011. 17(8 Suppl): p. S1–14.

73 Powers, B.J., et al., *Future Research Needs for Angiotensin-Converting Enzyme Inhibitors (ACEIs), Angiotensin II Receptor Antagonists (ARBs), or Direct Renin Inhibitors (DRIs) for Treating Hypertension: Identification of Future Research Needs From Comparative Effectiveness Review No. 34*. 2012: Rockville (MD): Agency for Healthcare Research and Quality (US).

10

CALCIUM CHANNEL BLOCKERS

10.1 OVERVIEW

Calcium channel blockers (CCBs), also known as calcium channel antagonists or calcium antagonists, were introduced into clinical medicine in the 1960s and are now among the most frequently prescribed drugs for the treatment of cardiovascular diseases. Although currently available CCBs for treating cardiovascular diseases are chemically diverse, they share the common property of blocking the transmembrane flow of calcium ions through voltage-gated L-type channels in humans. When inward calcium ion flux is inhibited, vascular smooth muscle cells relax, resulting in vasodilation and a lowering of blood pressure. In cardiac muscle, when calcium ion influx is inhibited, contractility is reduced, and the sinus pacemaker and atrioventricular conduction velocity is slowed. These drugs have proved effective in patients with hypertension and angina pectoris. Some are also effective in treating cardiac arrhythmias and may be beneficial in patients with left ventricular diastolic dysfunction. In addition, CCBs may also be useful in the treatment of Raynaud's phenomenon, migraine, preterm labor, esophageal spasm, and bipolar depressive disorders. In addition to the L-type CCBs, recently, an N-type calcium channel-selective blocker, known as ziconotide, has been approved by the US Food and Drug Administration (FDA) for the management of severe chronic pain refractory to conventional therapy. This chapter discusses the molecular pharmacology of the clinically available L-type CCBs with a focus on the pharmacological basis of their use in the treatment of cardiovascular diseases.

10.2 CALCIUM CHANNELS AND DRUG TARGETING

10.2.1 Calcium Channels

A calcium channel is an ion channel that displays selective permeability to calcium ions. There are a wide variety of calcium channels, which, based on their dependence on voltage or ligand, are classified into voltage-gated calcium channels and ligand-gated calcium channels. Several subtypes of voltage-gated calcium channels have been identified based on the recorded calcium currents, which include L-, N-, P/Q-, R-, and T-type calcium channels [1]. These subtypes differ in their electrochemical and biophysical properties and in their tissue distribution patterns. For instance, the L-type channels are located in tissues including skeletal muscle, cardiac tissue (myocardium and pacemaker cells in sinus and atrioventricular nodes), vascular smooth muscle, endocrine glands, and nervous tissue. L-type channels are the main calcium currents recorded in muscle and endocrine cells, where they initiate contraction and secretion. On the other hand, N-type channels are expressed primarily in nerve terminals and dendrites and are involved in initiating neurotransmitter release and mediating calcium entry into cell bodies and dendrites [1, 2]. Hence, N-type CCBs are used for pain management.

10.2.2 Drug Class and Drug Targeting

As noted earlier, the clinically available CCBs are exclusively L-type channel selective except for a recently approved N-channel-selective blocker, ziconotide [3]. Ziconotide

Cardiovascular Diseases: From Molecular Pharmacology to Evidence-Based Therapeutics, First Edition. Y. Robert Li.
© 2015 John Wiley & Sons, Inc. Published 2015 by John Wiley & Sons, Inc.

(Prialt) is a synthetic equivalent of a naturally occurring conopeptide found in the piscivorous marine snail, *Conus magus*. Ziconotide is a 25-amino-acid polybasic peptide containing three disulfide bridges with a molecular weight of 2639 daltons. Intrathecal infusion of ziconotide is indicated for the management of severe chronic pain in adult patients for whom intrathecal therapy is warranted, and who are intolerant of or refractory to other treatments, such as systemic analgesics, adjunctive therapies, or intrathecal morphine.

Ziconotide binds to N-type calcium channels located on the primary nociceptive (A-δ and C) afferent nerves in the superficial layers (Rexed laminae I and II) of the dorsal horn in the spinal cord. Although the mechanism of action of ziconotide has not been established in humans, results in experimental animals suggest that its binding blocks N-type calcium channels, which leads to a blockade of excitatory neurotransmitter release from the primary afferent nerve terminals and antinociception. Because ziconotide has no approved use in treating cardiovascular diseases, it is not further discussed in this chapter.

10.3 L-TYPE CCBs

Listed below are the 10 L-type CCBs approved by the US FDA. They are classified into dihydropyridine CCBs and nondihydropyridine CCBs:

Dihydropyridine CCBs
- Amlodipine (Amvaz/Norvasc)
- Clevidipine (Cleviprex)
- Felodipine (Plendil)
- Isradipine (DynaCirc)
- Nicardipine (Cardene)
- Nifedipine (Adalat/Procardia)
- Nimodipine (Nimotop)
- Nisoldipine (Sular)

Nondihydropyridine CCBs
- Verapamil (Calan/Isoptin)
- Diltiazem (Cardizem)

10.3.1 General Introduction to Drug Class

As mentioned earlier, based on chemical structures, CCBs are classified into two groups: (1) dihydropyridine CCBs with eight members and (2) nondihydropyridine CCBs consisting of two members. Drugs in these two groups bind to separate sites on the L-type calcium channels. The most important difference between the two groups is that nondihydropyridine CCBs can slow down the heart rate, while dihydropyridine CCBs do not. Because nondihydropyridine CCBs slow down the heart rate, they can be useful for certain types of cardiac arrhythmias.

The dihydropyridine CCBs are potent vasodilators that have little or no negative effect on cardiac contractility or conduction. Nifedipine, the first-generation CCB, is a short-acting drug, and the rest are second-generation longer-acting CCBs. Clevidipine is a recently approved CCB indicated for intravenous use only. Amlodipine is the most commonly prescribed CCB, accounting for approximately 70% of all CCB market, followed by diltiazem (~11%) and verapamil (~8%) [4].

10.3.2 Chemistry and Pharmacokinetics

The chemical structures of the eight dihydropyridine CCBs and the two nondihydropyridine CCBs are shown in Figure 10.1. Verapamil and diltiazem belong to the chemical classes of phenylalkylamines and benzothiazepines, respectively. Although CCBs differ in pharmacokinetic properties, many are metabolized by cytochrome P450 (CYP) 3A4 (Table 10.1). Hence, drugs that affect CYP3A4 activity (e.g., inducers or inhibitors) can cause significant drug–drug interactions with CCBs. Notably, amlodipine has the longest elimination half-life (about 30–50 h) among all CCBs.

10.3.3 Molecular Mechanisms and Pharmacological Effects

10.3.3.1 Effect on Vascular Smooth Muscle, Myocardium, and Cardiac Conduction System Calcium influx into the myocyte initiates a series of events essential for contractility (Fig. 10.2). Calcium entry into the myocyte first triggers intracellular calcium release; the released calcium then binds the regulatory protein troponin, resulting in a calcium–troponin complex that allows actin and myosin to interact and contract. The sequence of events is the same in vascular smooth muscle cells, except that a calcium–calmodulin complex instead of calcium–troponin complex permits the interaction between actin and myosin of smooth muscle.

CCBs work by blocking the initial calcium influx into myocytes and vascular smooth muscle cells, preventing the cascade of events detailed earlier, and thereby causing vasodilation and a negative inotropic effect. In addition, some CCBs also decrease the conduction velocity of both sinus and atrioventricular nodes, causing negative chronotropic effects. In this context, the phase zero of action potential in pacemaker cells is caused by calcium influx through L-type calcium channels in these cells. Inhibition of the calcium ion influx reduces phase zero of the action potential, leading to decreased conduction velocity. It is worth mentioning that conduction velocity is determined by the phase zero of action potential.

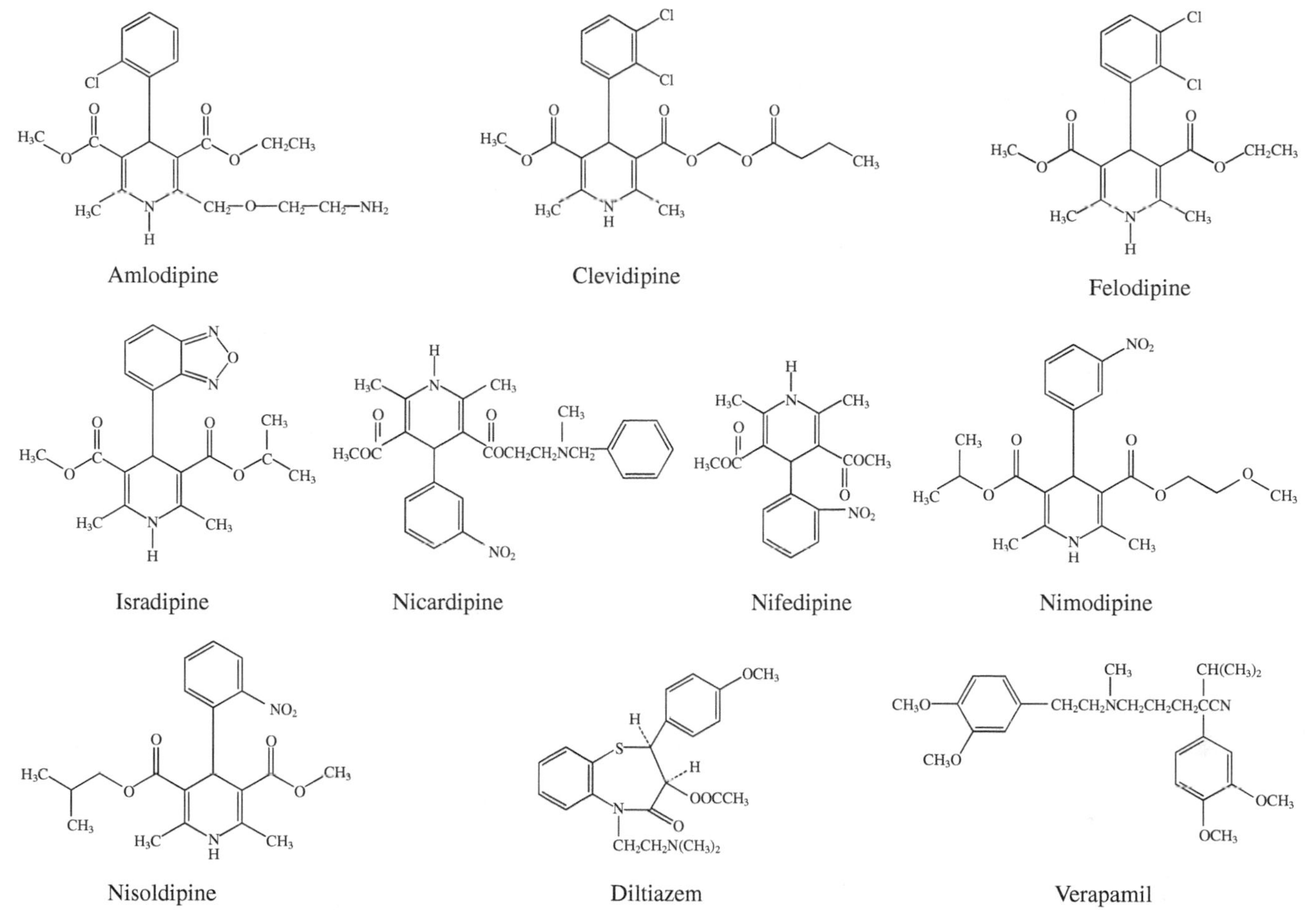

FIGURE 10.1 Structures of calcium channel blockers (CCBs). Among the 10 CCBs, verapamil and diltiazem belong to the chemical classes of phenylalkylamines and benzothiazepines, respectively, and the rest are dihydropyridines.

TABLE 10.1 Major pharmacokinetic properties of CCBs

CCB	Oral bioavailability (%)	Elimination half-life	Metabolism and excretion
Amlodipine	64–90	30–50h	CYP3A4; renal excretion
Clevidipine		Initial phase, 1 min; terminal phase, 15 min	Hydrolysis by esterases; renal and biliary excretion
Felodipine	20	11–16h	CYP3A4; renal (major) and biliary excretion
Isradipine	15–24	Early phase, 1.5–2h; terminal phase, 8h	CYP3A4; renal (main) and biliary excretion
Nicardipine	30	2–4h	CYP3A4; renal and biliary excretion
Nifedipine	50	2h	CYP3A4; renal (main) and biliary excretion
Nimodipine	13	Early phase, 0.5–2h; terminal phase, 8–9h	CYP3A4; renal excretion
Nisoldipine	5	14h	CYP3A4; renal (major) and biliary excretion
Diltiazem	40	3.0–4.5h	CYP3A4; biliary excretion
Verapamil	20–35	4.5–12h	CYP3A4; renal (main) and biliary excretion

Dihydropyridine CCBs have a greater selectivity for vascular smooth muscle than for the myocardium, making them more potent vasodilators than the nondihydropyridine CCBs verapamil and diltiazem. The greater degree of arterial vasodilation caused by dihydropyridine CCBs is accompanied by a sufficient baroreflex-mediated increase in sympathetic tone to overcome the negative inotropic and chronotropic effects. This may explain why the dihydropyridine CCBs have minimal net effect on the heart and are not useful in treating cardiac arrhythmias. On the other hand, the nondihydropyridine CCBs verapamil and diltiazem exert significant negative chronotropic effects (diltiazem is more

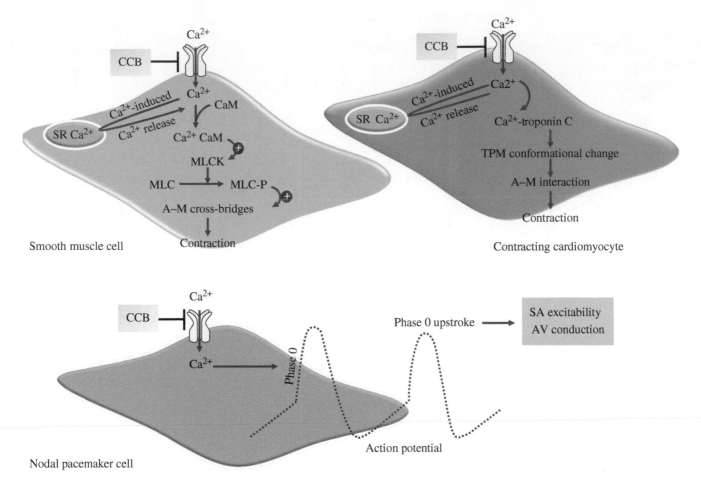

FIGURE 10.2   Molecular mechanisms of action of calcium channel blockers (CCBs). In vascular smooth muscle cells, influx of extracellular calcium ions through L-type voltage-dependent calcium channels (VDCC) causes further release of calcium ions from sarcoplasmic reticulum (SR), a phenomenon known as calcium-induced calcium release. The increased cytosolic calcium ions bind to calmodulin (CaM) to form a calcium–CaM complex. The calcium–CaM complex then activates myosin light chain kinase (MLCK), which in turn causes phosphorylation of myosin light chain (MLC). The phosphorylated myosin light chain (MLC-P) interacts with actin to form actin–myosin (A–M) crossbridges, thereby leading to smooth muscle cell contraction. Hence, blockage of L-type VDCC in vascular smooth muscle cells by CCBs results in smooth muscle relaxation and the subsequent vasodilation. In contracting cardiomyocytes, influx of extracellular calcium ions through L-type VDCC causes further release of calcium ions from the SR. The increased cytosolic calcium ions bind to troponin-C, which in turn causes conformational change of tropomyosin (TPM). This conformational change allows myosin to form an active complex with actin, leading to myocardial contraction. Therefore, blockage of the L-type VDCC in cardiomyocytes by CCBs causes decreased myocardial contractility. In sinoatrial (SA) and atrioventricular (AV) nodal cells, calcium influx through the L-type VDCC causes the upstroke (0 phase) of the action potential in these pacemaker cells. This calcium influx is responsible for the excitability of the SA node and the conduction velocity of the AV node. As such, blockage of the L-type VDCC by CCBs reduces the excitability of SA node and decreases the AV node conduction.

effective than verapamil), making them useful for treating cardiac arrhythmias. Both drugs also cause negative inotropic effects, with verapamil being more effective than diltiazem.

10.3.3.2 Novel Effects Nimodipine and other dihydropyridine CCBs can increase fibrinolytic activity, which may partly contribute to the efficacy in preventing ischemic events in patients with hypertension and subarachnoid hemorrhage [5]. Recently, a large population-based case–control study reported that long-term use of CCBs was associated with a greater than twofold increase in the risk of breast cancer in postmenopausal women [6]. However, the causal relationship and the underlying biological mechanisms remain to be established.

10.3.4   Clinical Uses

10.3.4.1 Cardiovascular Diseases CCBs are among the most commonly used drugs for treating hypertension and angina pectoris. The nondihydropyridine CCBs are also useful for treating certain types of cardiac arrhythmias as well as diastolic heart failure (see Unit VI).

Hypertension With the exception of nimodipine (which was originally developed for treating hypertension but is approved for subarachnoid hemorrhage) and the short-acting nifedipine (i.e., the immediate-release formulation), all CCBs are approved by the US FDA for lowering blood pressure either alone or in combination with other antihypertensive agents. Among the CCBs for treating hypertension, amlodipine is the one most commonly used. For dihydropyridine CCBs, blood pressure-lowering effects result from arterial vasodilation and the consequent reduction in peripheral resistance (blood pressure = cardiac output × peripheral resistance). For nondihydropyridine CCBs, in addition to arterial vasodilation, the negative inotropic effects also contribute to blood pressure lowering.

As noted earlier, CCBs can be used either alone or in combination with other antihypertensive agents in treating hypertension. The fixed-dose combinations of CCBs with other antihypertensive drugs are listed below. In addition, a fixed-dose combination of amlodipine and atorvastatin is also available.

Two drug combinations include the following:

- CCB + diuretic
- CCB + angiotensin-converting enzyme (ACE) inhibitor
- CCB + angiotensin receptor blocker (ARB)
- CCB + direct renin inhibitor (aliskiren)

Today, triple combinations are available as antihypertensive therapy and are frequently used in many countries. Triple combinations at any dose seem to decrease blood pressure more effectively than dual combinations of the same molecules without any remarkable risk elevation for adverse events [7]. Three drug combinations include the following:

- CCB + diuretic + ACE inhibitor
- CCB + diuretic + ARB

Stable Angina CCBs are indicated for the symptomatic treatment of chronic stable angina alone or in combination with other antianginal drugs (see Chapter 14). The antianginal effects of CCBs result from either dilation of the coronary artery and the subsequent increase of oxygen supply or decreased oxygen demand (secondary to a decrease in arterial blood pressure, myocardial contractility, or heart rate), or both. Owing to their coronary dilating effects, CCBs are also useful for treating variant angina (also known as Prinzmetal angina, which is caused by transient localized coronary artery spasm).

All CCBs are effective in treating chronic stable angina. CCBs approved by the US FDA for treating chronic stable angina include amlodipine, nifedipine, and diltiazem. Amlodipine and nifedipine are also approved by the FDA for treating variant angina. In patients with recently documented coronary artery disease by angiography and without heart failure, amlodipine is also indicated to reduce the risk of hospitalization due to angina and to reduce the risk of a coronary revascularization procedure.

Acute Coronary Syndromes Available evidence suggests that CCBs have a limited role in the management of patients with acute coronary syndromes (ACS), which include unstable angina, non-ST-segment elevation myocardial infarction (STEMI), and STEMI (see Unit V). No CCB has been shown to reduce mortality, and in certain patients with ACS, the short-acting CCBs, such as nifedipine, may even be harmful (see the following text).

Because coronary vasospasm occurs in some patients with unstable angina, CCBs may be useful in alleviating the symptoms. However, evidence for clinical outcomes of using CCBs in this setting is lacking. Regarding the use of CCBs in acute myocardial infarction (MI), the excess early mortality with short-acting nifedipine in some patients with acute MI has led to the recommendation that this drug should be avoided. The increase in risk is thought to be mediated at least in part by repeated episodes of hypotension and reflex sympathetic activation, changes that are not generally seen with long-acting CCBs.

Use of diltiazem and verapamil in acute MI patients with heart failure, left ventricular systolic dysfunction, or significant atrioventricular block has been associated with worse outcomes. It has been suggested that verapamil or diltiazem may be used, in addition to other beneficial interventions, in patients with MI who have ongoing or recurring ischemia and who do not have left ventricular dysfunction, heart failure, or atrioventricular block when β-blockers are absolutely contraindicated (e.g., active bronchospasm or allergy). Verapamil or diltiazem may also be used in patients with STEMI for the short-term control of rapid ventricular response in atrial fibrillation or atrial flutter, in the absence of left ventricular dysfunction or heart failure, when β-blockers are absolutely contraindicated or ineffective.

With regard to the impact of CCBs on clinical outcomes in patients with coronary artery disease, a meta-analysis of randomized trials involving 47,694 patients with coronary artery disease demonstrated that long-acting CCBs (either dihydropyridines or nondihydropyridines) were not associated with any excess cardiovascular events and were associated with a reduction in the risk of stroke, angina pectoris, and heart failure compared with the comparison group. The authors suggested that long-acting CCBs should therefore be considered in the armamentarium of treatment for patients with coronary artery disease [8]. However, given the lack of data in each trial, this meta-analysis did not adjust analyses for degree of blood pressure control, dose of medications used, add-on therapy used, and no compliance with assigned therapy. In addition, the authors were unable to adjust the analysis for other comorbidities. The definitions of coronary artery disease and the diagnostic criteria used in the studies were also heterogeneous.

Heart Failure   Available evidence points to the lack of benefit of CCBs in patients with systolic heart failure. These agents do not improve exercise tolerance, quality of life, or survival. The negative inotropic effects of CCBs, especially verapamil, may worsen the symptoms in patients with severe left ventricular dysfunction. Based upon these observations, the use of CCBs in patients with systolic heart failure was **not recommended** in the current guidelines (see Chapter 22). Nevertheless, amlodipine and felodipine appear to be safe and well tolerated in patients with systolic heart failure and can be used for the treatment of hypertension or angina in such patients. Although CCBs can be used in the aforementioned setting, ACE inhibitors and β-blockers are preferred due to their ability to reduce mortality and improve survival of the patients.

CCBs may be useful in treating diastolic heart failure. Evidence from small clinical trials suggested that verapamil, compared to placebo, significantly reduced the signs and symptoms of diastolic heart failure and increased left ventricular diastolic filling rate and treadmill exercise time. In patients with hypertrophic cardiomyopathy, verapamil also improved left ventricular diastolic function and may prolong long-term survival.

Cardiac Arrhythmias   Diltiazem and verapamil are useful antiarrhythmic agents in the management of certain arrhythmias, especially supraventricular tachyarrhythmias (see Unit VII). These drugs preferentially affect slow-response myocardial tissues (i.e., sinoatrial and atrioventricular nodes, which depend on calcium currents to generate slowly propagating action potentials), in contrast to fast-response myocardial tissues (i.e., the atria, specialized intranodal conducting system, the ventricles, and accessory pathways), which rely on sodium channels. As discussed in Unit VII, in addition to adenosine, diltiazem and verapamil are also treatments of choice for the termination of supraventricular tachycardia. Diltiazem and verapamil can be used both acutely (via the intravenous route) and chronically (via the oral route) to slow the ventricular response in atrial fibrillation and atrial flutter.

10.3.4.2   *Noncardiovascular Diseases*   In addition to the cardiovascular indications, CCBs may be beneficial in other disease conditions though the evidence supporting their efficacy is relatively weak. These include symptom relief of Raynaud's disease due to their peripheral vasodilating effect, prevention of migraine headache due to improvement of vasoconstriction, prevention of preterm labor likely due to relaxation of the myometrium, and treatment of Peyronie's disease possibly by altering penile fibroblasts at several

levels, including cell proliferation, extracellular matrix protein synthesis and secretion, as well as collagen degradation.

10.3.5   Therapeutic Dosages

Listed below are the dosage forms and strengths of the 10 US FDA-approved CCBs:

1. Amlodipine (Amvaz/Norvasc): Oral, 2.5, 5, and 10 mg tablets
2. Clevidipine (Cleviprex): Intravenous, 50 and 100 ml vials containing 0.5 mg/ml
3. Felodipine (Plendil): Oral, 2.5, 5, and 10 mg extended-release tablets
4. Isradipine (DynaCirc): Oral, 2.5 and 5 mg capsules
5. Nicardipine (Cardene): Oral, 20 and 30 mg capsules; intravenous, 10 ml ampules containing 2.5 mg/ml
6. Nifedipine (Adalat/Procardia): Oral, 10 mg capsules; oral, 30, 60, and 90 mg extended-release tablets
7. Nimodipine (Nimotop): Oral, 30 mg capsules; oral, solution 60 mg/20 ml
8. Nisoldipine (Sular): Oral, 8.5, 17, 25.5, and 34 mg extended-release tablets
9. Diltiazem (Cardizem): Oral, 30, 60, 90, and 120 mg tablets; 120, 180, 240, 300, 360, and 420 mg extended-release tablets; 120, 180, 240, 300, and 360 mg capsules; 120, 180, 240, 300, and 360 mg extended-release capsules; intravenous, 0.5% (5 mg/ml) in 5, 10, and 25 ml vials
10. Verapamil (Calan/Isoptin): Oral, 80 and 120 mg tablets; 120, 180, and 240 mg extended-release tablets; 100, 200, and 300 mg extended-release capsules; 240 and 360 mg delayed-release capsules; intravenous, 2.5 mg/ml in single use containers containing 2 and 4 ml

The dosage regimens of CCBs in treating hypertension and other cardiovascular diseases are given in Table 10.2.

10.3.6   Adverse Effects and Drug Interactions

10.3.6.1   *Adverse Effects*   CCBs are generally well tolerated. The adverse effects seen with CCBs vary with the drug that is used. The potent vasodilators can, in 10–20% of patients, lead to one or more of the following: headache, dizziness or lightheadedness, flushing, and peripheral edema. The major adverse effect with verapamil is constipation, occurring in over 25% patients. This is due to its potent relaxation of intestinal smooth muscle, thereby inhibiting peristalsis. Use of short-acting nifedipine may increase the risk for acute MI in hypertensive patients and increase the mortality in patients with acute MI.

TABLE 10.2 Dosage regimens of CCBs in treating cardiovascular diseases

CCB	Indication	Dosage regimen
Amlodipine	Hypertension	Initial dose, 5 mg once daily; maximum dose, 10 mg once daily
	Coronary artery disease (CAD)	For chronic stable angina, coronary vasospasm angina, or angiographically documented CAD in patients without heart failure or without an ejection fraction of <40%, dose range: 5–10 mg once daily
Clevidipine	Hypertension	Given intravenously (iv) when oral therapy is not possible, feasible, or not desirable; initial iv infusion,1–2 mg/h; maintenance dose, 4–6 mg/h; maximum dose, 16 mg/h
Felodipine (extended release)	Hypertension	Initial dose, 5 mg once daily; dose range, 2.5–10 mg once daily
Isradipine	Hypertension	Initial dose, 2.5 mg, b.i.d.; maximum daily dose, 20 mg/day
Nicardipine (oral)	Hypertension	Initial dose, 20 mg t.i.d.; dose range, 20–40 mg t.i.d.
	CAD	For stable angina, initial dose, 20 mg t.i.d.; dose range, 20–40 mg t.i.d.
Nicardipine (iv)	Hypertension	For short-term treatment of hypertension when oral therapy is not feasible or not desirable; initial dose, 5 mg/h; maximum dose, 15 mg/h
Nifedipine (extended release)	Hypertension	Initial dose, 30 or 60 mg once daily; maximum dose, 120 mg once daily
	CAD	For chronic stable angina or vasospasm angina, initial dose, 30 or 60 mg once daily; maximum dose, 120 mg once daily
Nifedipine	CAD	For chronic stable angina or vasospasm angina, initial dose, 10 mg t.i.d.; dose range, 10–20 mg t.i.d.
Nimodipine	Stroke	Nimodipine is indicated for the improvement of neurological outcome by reducing the incidence and severity of ischemic deficits in patients with subarachnoid hemorrhage from ruptured intracranial berry aneurysms regardless of their postictus neurological condition (i.e., Hunt and Hess grades I–V). Dose regimen, 60 mg every 4 h for 21 consecutive days
Nisoldipine (extended release)	Hypertension	Initial dose, 17 mg once daily; dose range, 17–34 mg once daily
Diltiazem (extended release)	Hypertension	Initial dose, 180–240 mg once daily; dose range, 240–360 mg once daily
	CAD	For chronic stable angina or vasospasm angina, initial dose, 120 or 180 mg once daily; maximum dose, 480 mg once daily
Diltiazem (iv)	Arrhythmias	It is used for temporary control of rapid ventricular rate in atrial fibrillation or atrial flutter associated with an accessory bypass tract such as in Wolff–Parkinson–White (WPW) syndrome or short PR syndrome [it should not be used in patients with atrial fibrillation or atrial flutter associated with an accessory bypass tract such as in Wolff–Parkinson–White (WPW) syndrome or short PR syndrome]. It is also indicated for rapid conversion of paroxysmal supraventricular tachycardias (PSVT) to sinus rhythm. This includes AV nodal reentrant tachycardias and reciprocating tachycardias associated with an extranodal accessory pathway such as the WPW syndrome or short PR syndrome. The initial dose should be 0.25 mg/kg body weight as a bolus administered over 2 min. If response is inadequate, a second dose (0.35 mg/kg body weight) may be administered after 15 min. Subsequent iv bolus doses should be individualized for each patient
		For continued reduction of the heart rate (up to 24 h) in patients with atrial fibrillation or atrial flutter, an iv infusion may be initiated at 10 mg/h and titrated to 15 mg/h for up to 24 h
Verapamil	Hypertension	Initial dose, 80 mg t.i.d.; maximum dose, 120 mg t.i.d.
	CAD	For chronic stable angina, vasospasm angina, or unstable angina, 80–120 mg t.i.d.
	Arrhythmias	It is used in association with digitalis for the control of ventricular rate at rest and during stress in patients with chronic atrial flutter and/or atrial fibrillation. It is also used for prophylaxis of repetitive PSVT. The dosage in digitalized patients with chronic atrial fibrillation ranges from 240 to 320 mg/day in divided (t.i.d. or q.i.d.) doses. The dosage for prophylaxis of PSVT (nondigitalized patients) ranges from 240 to 480 mg/day in divided (t.i.d. or q.i.d.) doses
Verapamil (extended-release capsules)	Hypertension	Initial dose, 200 mg once daily; dose range, 200–400 mg once daily
Verapamil (iv)	Arrhythmias	iv verapamil is used for rapid conversion to sinus rhythm of PSVT, including those associated with accessory bypass tracts. It is also used for temporary control of rapid ventricular rate in atrial flutter or atrial fibrillation except when the atrial flutter and/or atrial fibrillation is associated with accessory bypass tracts. Initial dose, 5–10 mg (0.075–0.15 mg/kg body weight) given as an iv bolus over at least 2 min; repeat dose, 10 mg (0.15 mg/kg body weight) 30 min after the first dose if the initial response is not adequate

10.3.6.2 Drug Interactions CCBs may exhibit many significant drug interactions. Verapamil and diltiazem increase digoxin levels. Verapamil, diltiazem, and nicardipine increase plasma levels of and decrease the dosing requirement for cyclosporine. Verapamil and diltiazem are metabolized by CYP3A4; therefore, inducers (e.g., rifampin) and inhibitors (e.g., erythromycin, cimetidine) are likely to result in decreased and increased plasma levels of these two CCBs, respectively. Because of their shared negative effects on heart rate and myocardial contractility, β-blockers and verapamil are not used simultaneously.

10.3.6.3 Contraindications and Pregnancy Category

- All CCBs are contraindicated in patients who are allergic to any component of a given preparation.
- Verapamil and diltiazem are contraindicated in patients with hypotension, sick sinus syndrome (unless a permanent pacemaker is in place), or second- or third-degree atrioventricular block. In addition, verapamil is contraindicated in patients with severe left ventricular dysfunction, whereas diltiazem is contraindicated in patients with acute MI and pulmonary congestion on X-ray. Verapamil is also contraindicated in patients with atrial flutter or atrial fibrillation and an accessory bypass tract (e.g., Wolff–Parkinson–White or Lown–Ganong–Levine syndromes).
- Pregnancy category: C.

10.4 SUMMARY OF CHAPTER KEY POINTS

- Clinically used CCBs, when not specified, include both L-type and N-type CCBs. L-type CCBs, which are covered in this chapter, are among the most frequently prescribed drugs for treating cardiovascular diseases, including hypertension and coronary artery disease.
- Currently, there are 10 US FDA-approved L-type CCBs. They are further classified, based on chemical structures, into dihydropyridines (with eight CCBs belonging to this class) and nondihydropyridines, which consist of two members (i.e., diltiazem and verapamil).
- Calcium ion influx through the L-type voltage-dependent channels in vascular smooth muscle cells and contracting cardiomyocytes is a critical initial event that leads to contraction of these muscular cells, resulting in vasoconstriction and increased myocardial contraction, respectively. Blockage of the calcium channels by CCBs thus causes vasodilation and decreased myocardial contractility. Calcium influx in nodal pacemaker cells causes the zero-phase upstroke of the action potential, which determines the excitability of the SA node and the conduction velocity of the AV

node. Hence, blockage of the calcium channels in these nodal tissues suppresses SA nodal excitability and AV nodal conduction.
- Dihydropyridine CCBs have a greater selectivity for vascular smooth muscle than for the myocardium, making them more potent vasodilators than the nondihydropyridine CCBs verapamil and diltiazem. The greater degree of arterial vasodilation caused by dihydropyridine CCBs is accompanied by a sufficient baroreflex-mediated increase in sympathetic tone to overcome the negative inotropic and chronotropic effects. This may account for the minimal net effects of dihydropyridines on the heart. In contrast, diltiazem and verapamil can significantly decrease myocardial contractility and cardiac conduction.
- With the exception of nimodipine and the short-acting nifedipine (i.e., the immediate-release formulation), all CCBs are approved by the US FDA for treating hypertension either alone or in combination with other antihypertensive agents. Some CCBs are also approved for treating coronary artery disease (chronic stable angina and vasospasm angina). Diltiazem and verapamil are used for treating certain forms of cardiac arrhythmias as well.
- CCBs are generally well tolerated. The major adverse effects typically result from vasodilation and may include headache, dizziness, and peripheral edema. Many CCBs are metabolized by CYP enzymes, especially CYP3A4, and as such may cause significant drug interactions.

10.5 SELF-ASSESSMENT QUESTIONS

10.5.1. A 60-year-old male with obstructive pulmonary disease requires therapy to prevent vasospasm angina attacks. Which of the following drugs would be most appropriate for this patient?
A. Aliskiren
B. Amlodipine
C. Losartan
D. Nimodipine
E. Propranolol

10.5.2. A 55-year-old male presents to the physician's office, complaining of constipation since taking a "new" cardiovascular pill. History reveals that the patient is recently diagnosed with hypertension and stable angina. He has been on a cardiovascular drug for 3 weeks, which he was told can control both his hypertension and angina. Which of the following is most likely the cardiovascular drug?
A. Amlodipine
B. Diltiazem
C. Isradipine
D. Nimodipine
E. Verapamil

10.5.3. A 60-year-old female is diagnosed with subarachnoid hemorrhage from ruptured intracranial berry aneurysms. Which of the following may be prescribed to improve the neurological outcome by reducing the incidence and severity of ischemic deficits in the patient?
   A. Amlodipine
   B. Clevidipine
   C. Nicardipine
   D. Nimodipine
   E. Nisoldipine

10.5.4. A 25-year-old hospitalized patient develops paroxysmal supraventricular tachycardia (PSVT). A calcium channel blocker is given at 0.25 mg/kg body weight as a bolus administered over 2 min. Immediately after the injection, the patient's PSVT is converted to sinus rhythm. Which of the following is most likely prescribed?
   A. Amlodipine
   B. Clevidipine
   C. Diltiazem
   D. Nifedipine
   E. Nisoldipine

10.5.5. In a recent issue of JAMA Internal Medicine, a report by Li and associates from a large population-based case–control study suggested that long-term use of a class of cardiovascular drugs may be associated with a greater than twofold increase in the risk of breast cancer in postmenopausal women. Which of the following classes of cardiovascular drugs upon long-term use may be most likely associated with an increased risk of breast cancer?
   A. Angiotensin receptor blockers
   B. Angiotensin-converting enzyme inhibitors
   C. Beta-blockers
   D. Calcium channel blockers
   E. Thiazide diuretics

REFERENCES

1 Catterall, W.A., et al., International union of pharmacology. XL. Compendium of voltage-gated ion channels: calcium channels. *Pharmacol Rev*, 2003. 55(4): p. 579–81.

2 Catterall, W.A., Voltage-gated calcium channels. *Cold Spring Harb Perspect Biol*, 2011. 3(8): p. a003947.

3 Schmidtko, A., et al., Ziconotide for treatment of severe chronic pain. *Lancet*, 2010. 375(9725): p. 1569–77.

4 Elliott, W.J. and C.V. Ram, Calcium channel blockers. *J Clin Hypertens (Greenwich)*, 2011. 13(9): p. 687–9.

5 Vergouwen, M.D., et al., Dihydropyridine calcium antagonists increase fibrinolytic activity: a systematic review. *J Cereb Blood Flow Metab*, 2007. 27(7): p. 1293–308.

6 Li, C.I., et al., Use of antihypertensive medications and breast cancer risk among women aged 55 to 74 years. *JAMA Intern Med*, 2013. 173(17): p. 1629–37.

7 Kizilirmak, P., et al., The efficacy and safety of triple vs dual combination of angiotensin II receptor blocker and calcium channel blocker and diuretic: a systematic review and meta-analysis. *J Clin Hypertens (Greenwich)*, 2013. 15(3): p. 193–200.

8 Bangalore, S., S. Parkar, and F.H. Messerli, Long-acting calcium antagonists in patients with coronary artery disease: a meta-analysis. *Am J Med*, 2009. 122(4): p. 356–65.

11

NITRATES AND OTHER VASODILATORS

11.1 OVERVIEW

As described in Chapter 10, calcium channel blockers (CCBs) cause vascular smooth muscle relaxation, leading to decreased vascular tone and thereby vasodilation. The dynamics of vascular tone, determined by the degree of vascular smooth muscle contraction, is an essential physiological process controlling organ and tissue perfusion. Various pathways including both systemic (neurohormonal) factors and local mechanisms are involved in the physiological regulation of vascular tone. Dysregulated vascular tone (usually manifested as increased vascular constriction) under pathophysiological conditions contributes to the development of cardiovascular disease conditions, including hypertension, ischemic heart disease, and heart failure. Hence, drugs that cause vasodilation (also known as vasodilators), including CCBs covered in the preceding chapter, are commonly used in the management of the aforementioned cardiovascular conditions.

Vasodilators can be broadly defined as any agents capable of causing vasodilation either through systemic or local mechanisms or both. Vasodilators used in treating cardiovascular diseases are typically classified into the following nine categories:

1. Sympatholytics, including α-blockers, β-blockers, and centrally acting drugs (see Chapter 8)
2. Inhibitors of the renin–angiotensin–aldosterone system (RAAS), including direct renin inhibitors, angiotensin-converting enzyme (ACE) inhibitors, and angiotensin receptor blockers (see Chapter 9)
3. CCBs (see Chapter 10)
4. Organic nitrates and sodium nitroprusside (nitric oxide-releasing vasodilators)
5. Endothelin (ET) receptor antagonists
6. Phosphodiesterase 5 inhibitors
7. Soluble guanylate cyclase (sGC) stimulators
8. K^+_{ATP} channel openers
9. Others (hydralazine, fenoldopam)

Chapters 8–10 have discussed the sympatholytics, inhibitors of RAAS, and CCBs, respectively. As you recall, though acting via different mechanisms, these classes of drugs are all able to cause vasodilation in addition to their other pharmacological effects. This chapter examines the molecular pharmacology of the remaining six classes of vasodilators, focusing on discussing the mechanistic basis of their use in the management of hypertension and other cardiovascular diseases.

11.2 DRUG CLASS AND DRUG TARGETING

To understand how nitrates and other vasodilators act to treat cardiovascular diseases, it is imperative to first review the molecular regulation of vascular tone.

11.2.1 Molecular Regulation of Vascular Tone

Vascular tone refers to the degree of constriction experienced by a blood vessel relative to its maximally dilated state. All arterial and venous vessels under basal conditions

Cardiovascular Diseases: From Molecular Pharmacology to Evidence-Based Therapeutics, First Edition. Y. Robert Li.
© 2015 John Wiley & Sons, Inc. Published 2015 by John Wiley & Sons, Inc.

exhibit some degree of smooth muscle contraction that determines the diameter and hence the tone of the vessels. Vascular tone is dictated by many different vasoconstricting and vasodilating factors acting on the blood vessels. These factors are divided into two categories: (1) systemic factors that originate from outside of the organ or tissue, in which the blood vessel is located, and (2) localized factors that originate from the vessel itself or the surrounding tissue. The primary function of the systemic factors is to regulate arterial blood pressure by altering systemic vascular resistance, whereas local factors are important for local blood flow regulation within an organ or tissue. Vascular tone at any given time is determined by the balance of vasoconstricting and vasodilating mechanisms. Disruption of this balance may lead to sustained vasoconstriction, contributing to hypertension, ischemic heart disease, and heart failure.

11.2.1.1 Systemic Factors in Regulating Vascular Tone
The major systemic factors that regulate vascular tone include the following:

- Sympathetic flow: Activation of α_1-adrenergic receptors on vascular smooth muscle cells leads to vasoconstriction, whereas activation of β_2-adrenergic receptors on these cells causes vasodilation. Epinephrine and norepinephrine preferably cause α_1-adrenergic receptor activation in vascular smooth muscle. Hence, sympathetic activation causes vasoconstriction.

- Circulating angiotensin II: Angiotensin II causes vasoconstriction. Activation of AT_1 receptors on smooth muscle cells by angiotensin II induces contraction via intracellular calcium ions and JAK2-dependent pathway [1, 2].

- Atrial natriuretic peptide: Atrial natriuretic peptide (ANP) is a hormone that is released from myocardial cells in the atria, and in some cases, the ventricles in response to volume expansion and possibly increased wall stress. ANP can directly cause vasodilation and it also increases water and sodium excretion. ANP activates its receptor, namely, natriuretic peptide receptor-A (NPR-A). NPR-A is a guanylyl cyclase, which catalyzes the synthesis of cyclic guanosine monophosphate (cGMP) from guanosine triphosphate (GTP). In vascular smooth muscle cells, the intracellular effects of NPR-A activation are mediated through protein kinase G (also known as cGMP-dependent kinase), which phosphorylates downstream targets, such as phospholamban. Phospholamban is an important regulator of intracellular Ca^{2+} concentrations by mediating Ca^{2+} sequestration in the endoplasmic or sarcoplasmic reticulum and downregulation of the L-type Ca^{2+} channels located in the cell membrane [3].

11.2.1.2 Localized Mechanisms in Regulating Vascular Tone
Multiple localized mechanisms are involved in the regulation of vascular tone. These include (i) myogenic mechanism, (ii) formation of endothelial factors, and (iii) presence of local vasoactive substances.

Myogenic Mechanism Myogenic mechanisms are intrinsic to the smooth muscle of blood vessels, particularly in small arteries and arterioles. When the pressure within a vessel is suddenly increased, the vessel responds by constricting. This occurs because mechanical stretching of the smooth muscle cells causes membrane depolarization and the consequent activation of voltage-dependent calcium channels. Influx of extracellular calcium ions provokes smooth muscle cell contraction and hence vasoconstriction. On the other hand, diminishing pressure within the vessel causes smooth muscle relaxation and vasodilation. Notably, evidence suggests that the AT_1 receptor may act as a mechanosensor in smooth muscle cells and is responsible, at least partly, for myogenic vasoconstriction [4].

Endothelial Factors Endothelium plays a critical role in regulating vascular tone. In this regard, nitric oxide and ET are among the most important endothelium-derived factors that impact smooth muscle contractility. As such, nitric oxide and ET serve as important targets for some vasodilating drugs, namely, nitrates and ET receptor antagonists, respectively. Nitric oxide causes vasodilation through activating guanylyl cyclase (the detailed signaling mechanism is discussed in the following section on nitrate drugs), whereas ET induces vasoconstriction via activation of ET receptors on smooth muscle cells. Activation of the ET receptors on smooth muscle cells causes release of calcium ions from intracellular stores, leading to smooth muscle cell contraction and vasoconstriction (also see ET Receptor Antagonists section for more detail).

Local Vasoactive Substances Local vasoactive substances refer to molecules generated locally in vasculature or adjacent tissues that cause vasodilation or constriction. These include prostaglandins (PGs), leukotrienes, thromboxanes, bradykinin, and histamine.

Smooth muscle cells produce both PGE_2 and $PGF_{2\alpha}$, with the former acting as a vasodilator and the latter as a vasoconstrictor. On the other hand, PGI_2 is the major PG produced by endothelial cells. PGI_2 acts as a potent vasodilator and inhibitor of platelet adhesion and, as such, plays an important role in protecting against thrombogenesis. In contrast, thromboxane A_2 produced by platelets causes vasoconstriction in addition to platelet aggregation and activation. Leukotrienes are produced by leukocytes during inflammation, and like thromboxane A_2, they cause vasoconstriction. Histamine released during inflammation causes vasodilation, increased vascular permeability, and tissue edema. Similar to histamine, bradykinin is also a vasodilator (also see Chapter 9).

Bradykinin induces the formation of nitric oxide and PGI_2 from vascular endothelium, which is responsible for its vasodilating activity.

In addition to the aforementioned local vasoactive substances, changes in the levels of tissue metabolites and ions (e.g., adenosine, lactic acid, carbon dioxide, hydrogen ions, and potassium ions), especially during hypoxia, also play a role in regulating local vascular tone. For example, increased formation of adenosine during hypoxia may be an important mechanism for regulating coronary blood flow as adenosine is a potent vasodilator [5, 6].

11.2.2   Drug Class

Although numerous mechanisms as discussed earlier are involved in the regulation of vascular tone, calcium ions and nitric oxide appear to be the two key players. They may be viewed as the final common pathways leading to vasoconstriction and vasodilation, respectively. Indeed, the clinically available vasodilators affect these two factors either directly or indirectly to cause vasodilation. The detailed molecular pharmacology of these drug classes is discussed next.

11.3   ORGANIC NITRATES AND SODIUM NITROPRUSSIDE (NITRIC OXIDE-RELEASING VASODILATORS)

Organic nitrates and sodium nitroprusside are nitric oxide-releasing drugs. Listed below are the US Food and Drug Administration (FDA)-approved organic nitrates (first three drugs) and sodium nitroprusside. It should be noted that sodium nitroprusside is not an organic nitrate compound:

- Isosorbide dinitrate (Isordil)
- Isosorbide mononitrate (Ismo)
- Nitroglycerin
- Sodium nitroprusside (Nitropress)

11.3.1   General Introduction to Drug Class

Organic nitrates are the second oldest cardiovascular drug class after digitalis glycosides (see Chapter 21 for digitalis) and have been employed continuously for one and a half centuries in cardiovascular therapy. Despite the advent of modern cardiovascular therapeutic agents (which are the focus of this book), nitrates remain one of the major cardiovascular drug classes in the management of ischemic heart disease and congestive heart failure [7].

As listed earlier, organic nitrate drugs include nitroglycerin, isosorbide dinitrate, and isosorbide mononitrate. Both organic nitrates and sodium nitroprusside cause vasodilation via releasing nitric oxide, and as such, they are collectively called nitrovasodilators or nitric oxide-releasing vasodilators. While organic nitrates are widely used drugs for treating ischemic heart disease as well as heart failure, sodium nitroprusside is mainly indicated for the immediate reduction of blood pressure of patients in hypertensive crises. It is also useful for the treatment of acute congestive heart failure.

11.3.2   Chemistry and Pharmacokinetics

The chemical structures of organic nitrates and sodium nitroprusside are given in Figure 11.1. The major pharmacokinetic properties of these drugs are summarized in Table 11.1. Notably, infused sodium nitroprusside is rapidly

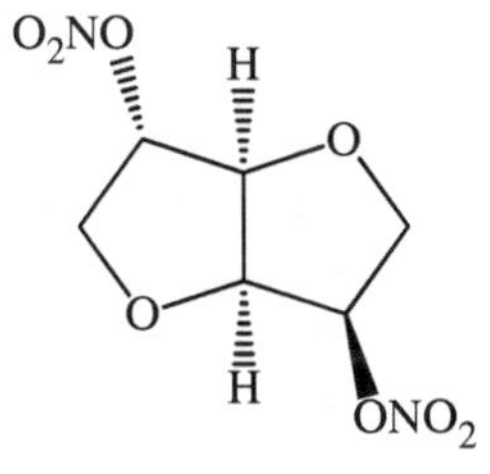

Isosorbide dinitrate

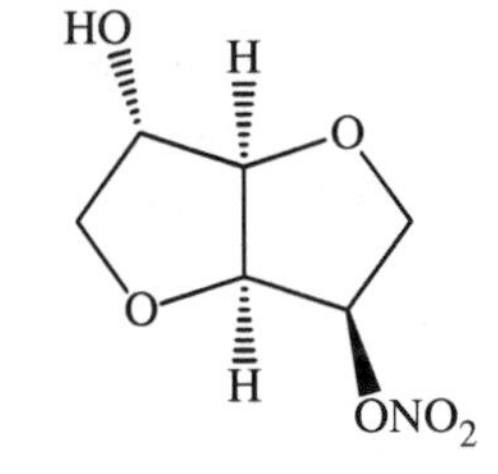

Isosorbide mononitrate

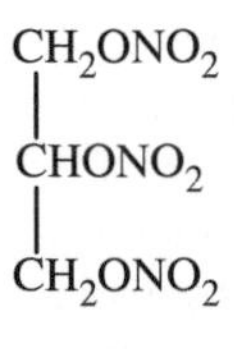

Nitroglycerin

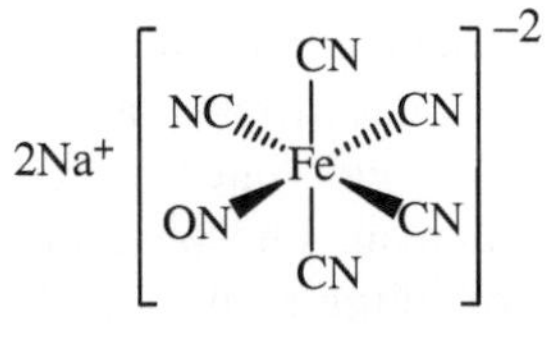

Sodium nitroprusside

FIGURE 11.1   Structure of nitrates and sodium nitroprusside. As shown, sodium nitroprusside is not a nitrate compound. Both nitrates and sodium nitroprusside give rise to nitric oxide (NO), and these drugs are hence called NO-releasing vasodilators.

TABLE 11.1   Major pharmacokinetic properties of organic nitrates and sodium nitroprusside

Drug	Oral bioavailability (%)	Elimination half-life	Metabolism and excretion
Isosorbide dinitrate	25	1 h; 5 h for the active metabolite, 5-mononitrate	Hepatic denitration followed by glucuronidation; the metabolite 5-mononitrate also has biological activities with an elimination half-life of 5 h; renal excretion
Isosorbide mononitrate	~100	5 h	No significant first-pass effects; hepatic denitration and renal excretion
Nitroglycerin	~40	2–3 min	Hepatic denitration to dinitrates and mononitrates; renal excretion
Sodium nitroprusside		2 min	Reaction with hemoglobin to form cyanmethemoglobin and cyanide ions; renal excretion

distributed to a volume that is approximately coexisting with the extracellular space. The drug is cleared from this volume by readily reacting with hemoglobin, and as such, the circulatory half-life of sodium nitroprusside is about 2 min. The products of the nitroprusside/hemoglobin reaction are cyanmethemoglobin and cyanide ions. Some cyanide is eliminated from the body as expired hydrogen cyanide, but most is enzymatically converted to thiocyanate by thiosulfate–cyanide sulfur transferase (also known as rhodanese), a mitochondrial enzyme. The enzyme is normally present in great excess, so the reaction is rate limited by the availability of sulfur donors, especially thiosulfate, cystine, and cysteine. Thiocyanate is also a normal physiological constituent of serum, with normal levels typically in the range of 50–250 μM (3–15 mg/l). Clearance of thiocyanate is done primarily by renal excretion, with an elimination half-life of about 3 days. In renal failure, the half-life can be doubled or tripled, leading to cyanide poisoning.

11.3.3   Molecular Mechanisms and Pharmacological Effects

11.3.3.1   Molecular Mechanisms and Pharmacological Effects of Organic Nitrates

Molecular Mechanisms of Nitric Oxide Formation and Smooth Muscle Relaxation   Nitrates dilate veins, arteries, and coronary arteries by relaxing vascular smooth muscle. They produce these effects by entering vascular smooth muscle cells where they are metabolized to 1,2-glyceryl dinitrate and nitrite, via mitochondrial aldehyde dehydrogenase-2 (ALDH2 or mtALDH), and then nitric oxide and S-nitrosothiols. Sulfhydryl groups on ALDH2 are required for activity, which can explain the known sulfhydryl requirement for vascular smooth muscle relaxation by nitrates. In vascular smooth muscle cells, nitric oxide activates guanylate cyclase, which increases cGMP leading to dephosphorylation of myosin light chain and thereby smooth muscle relaxation. Nitric oxide may also activate calcium-dependent potassium channels, leading to membrane hyperpolarization and thereby smooth muscle cell relaxation (Fig. 11.2).

Nitrates cause vasodilating effects on both peripheral veins and arteries but with more prominent effects on the veins. Nitrates primarily reduce cardiac oxygen demand by decreasing preload (left ventricular end-diastolic pressure). They may modestly reduce afterload, dilate coronary arteries, and improve collateral flow to ischemic regions. For use in rectal fissures, intra-anal administration of nitrates results in decreased sphincter tone and intra-anal pressure.

Pharmacological Effects   The pharmacological actions of nitrates include effects on (i) systemic hemodynamics, (ii) coronary hemodynamics, and (iii) platelets and thrombogenesis. This section examines the aforementioned effects and also considers the tolerance issue of nitrate drug therapy:

- *Effects on systemic hemodynamics*: The principal pharmacological action of nitroglycerin is relaxation of vascular smooth muscle and consequent dilation of peripheral arteries and veins, especially the latter. Dilation of the veins promotes peripheral pooling of blood and decreases venous return to the heart, thereby reducing left ventricular end-diastolic pressure and pulmonary capillary wedge pressure (preload). Arteriolar relaxation reduces systemic vascular resistance, systolic arterial pressure, and mean arterial pressure (afterload). As noted in the following text, nitrates also dilate coronary arteries. However, the relative importance of preload reduction, afterload reduction, and coronary dilation with regard to nitrate therapy of ischemic heart disease and heart failure remains to be further defined.

- *Effects on coronary hemodynamics*: A nitrate-induced increase in coronary blood flow has been proposed as a potential mechanism for relieving ischemia. Animal and human studies have shown that nitrates dilate both normal and abnormal coronary arteries; this response is preserved in saphenous vein grafts. The clinical importance of this effect is uncertain because the coronary

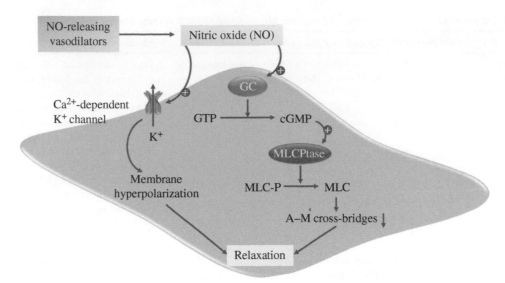

FIGURE 11.2 Molecular mechanisms of action of nitric oxide (NO)-releasing vasodilators. As illustrated, NO released from the NO-releasing drugs activates guanylate cyclase (GC), which in turn catalyzes the formation of cGMP from GTP. cGMP then activates myosin light chain phosphatase (MLCPtase), which causes dephosphorylation of the phosphorylated myosin light chain (MLC-P). The dephosphorylation of MLC-P diminishes the formation of actin–myosin (A–M) crossbridges, thereby leading to smooth muscle cell relaxation. In addition, NO may directly activate calcium ion-dependent K^+ channels, leading to efflux of intracellular K^+ and membrane hyperpolarization, which may also contribute to smooth muscle cell relaxation and vasodilation.

arterioles in patients with a flow-limiting coronary stenosis are already dilated to maintain resting blood flow, making further coronary dilation during ischemia difficult. There are, however, settings in which a direct effect on coronary hemodynamics may be beneficial. Nitrates can reduce or reverse coronary vasospasm. Thus, patients with primarily vasospastic angina or a large vasoconstrictor component to their angina can benefit from the direct coronary action of nitrate therapy. Nitrates also indirectly improve subendocardial blood flow as the reduction in left ventricular end-diastolic pressure induced by systemic venous dilation decreases the resistance to coronary blood flow from the epicardium to the endocardium. In addition, nitrates may lower the resistance to collateral vessel blood flow.

- *Effects on platelets and thrombogenesis*: Nitrates have significant antiplatelet and antithrombotic properties; however, the clinical importance of these potentially beneficial effects is unclear [8]. Stimulation of platelet guanylate cyclase by nitrates prevents fibrinogen binding to platelet IIb/IIIa receptors, which is essential for platelet aggregation (see Chapter 17). Transdermal nitroglycerin has been shown to inhibit platelet aggregation and thrombus formation in patients with angina pectoris. The antiplatelet activity of nitrates is independent of hemodynamic tolerance. In addition, intravenous administration of nitroglycerin causes further inhibition of platelets in patients with stable coronary artery disease on dual antiplatelet therapy following percutaneous coronary intervention [9].

- *Nitrate tolerance*: Tolerance has been a major problem with the use of nitrates as chronic antianginal therapy. At least three, not mutually exclusive, mechanisms have been proposed to explain the development of nitrate tolerance [10–12]. These include (i) impaired nitroglycerin bioconversion to 1,2-glyceryl dinitrate with decreased formation of nitric oxide, (ii) reduced bioactivity of nitric oxide, and (iii) activation of the RAAS and sympathetic nervous system in response to nitrate-induced vasodilation.

11.3.3.2 Molecular Mechanisms and Pharmacological Effects of Sodium Nitroprusside
In contrast to organic nitrates that need biotransformation to release nitric oxide, sodium nitroprusside spontaneously releases nitric oxide at physiological pH. The principal pharmacological action of sodium nitroprusside is relaxation of vascular smooth muscle and consequent dilation of peripheral arteries and veins. Other smooth muscles (e.g., uterus, duodenum) are not affected. Sodium nitroprusside is more active on veins than on arteries, but this selectivity is much less marked than that of nitrates. Dilation of the veins promotes peripheral pooling of blood and decreases venous return to the heart, thereby reducing left ventricular end-diastolic pressure and pulmonary capillary wedge pressure (preload). Arteriolar relaxation reduces systemic vascular resistance, systolic arterial pressure, and mean arterial pressure (afterload). Dilation of the coronary arteries also occurs. The hypotensive effect of sodium nitroprusside is seen within a minute or two after the start of an adequate infusion, and it dissipates almost as rapidly after an infusion is discontinued.

11.3.4 Clinical Uses

11.3.4.1 *Clinical Uses of Nitrates*

Nitrates are approved by the US FDA for the management of chronic stable angina. Nitroglycerin is indicated for either the acute relief of an attack or prophylaxis of angina pectoris due to coronary artery disease. Isosorbide dinitrate and isosorbide mononitrate are indicated only for the prevention of angina pectoris due to coronary artery disease. This is because the onset of action of isosorbide dinitrate or isosorbide mononitrate is not sufficiently rapid for them to be useful in aborting an acute anginal episode. This is also true for the extended-release formulations of nitroglycerin, which are only indicated for prophylaxis of angina pectoris. In addition to the approved use in angina pectoris, nitrates may also be used in the management of acute coronary syndromes (see Unit V) and heart failure (see Unit VI).

Stable Angina   T.L. Brunton, a British physician, first used the organic nitrite amyl nitrite in the treatment of angina pectoris in 1867. As a medical student, T.L. Brunton had become aware of prior clinical findings of B.W. Richardson that inhaled amyl nitrite rapidly increased the action of the heart, and also the unpublished observations of A. Gamgee demonstrating that amyl nitrite greatly lessened "arterial tension" in both animals and humans. During the same period in which T.L. Brunton used amyl nitrite, another British physician, W. Murrell, began using the organic nitrate, nitroglycerin, for treating angina pectoris and reported the antianginal efficacy in 1879. With nitroglycerin therapy, patients would obtain relief from angina with some patients also reporting that their angina could be aborted by taking the drug at the onset of symptoms. W. Murrell also worked with F. Barnes and compared the effects of amyl nitrite and nitroglycerin, where it was observed that the actions of these drugs differed in the time of onset and duration. It was concluded in the early 1880s that nitroglycerin would be more clinically useful than amyl nitrite [13]. Use of nitroglycerin, the world's first synthesized drug, in treating angina pectoris is still happening today and in fact remains the first-line drug therapy for chronic stable angina.

While they act as venodilators, coronary vasodilators, and modest arteriolar dilators, the primary anti-ischemic effect of nitrates seems to be the decreased myocardial oxygen demand by producing systemic vasodilation more than coronary vasodilation. This systemic vasodilation reduces left ventricular systolic wall stress. In patients with exertional stable angina, nitrates improve exercise tolerance, time to onset of angina, and ST-segment depression during exercise testing. In combination with β-blockers or CCBs, nitrates produce greater antianginal and anti-ischemic effects.

Acute Coronary Syndromes   Sublingual, intravenous, and oral nitrate preparations may also be used in the management of acute coronary syndromes, despite the absence of a mortality benefit. They can be of value in reducing or potentially eliminating pain (either initial or recurrent) due to myocardial ischemia, improving symptoms of pulmonary congestion, lowering blood pressure in hypertensive patients, and aiding in the diagnosis and management of the rare patients who present with variant angina (also known as coronary vasospasm angina or Prinzmetal angina).

Heart Failure   Nitrates have been used in the management of both acute and chronic heart failure. Although frequently used, the clinical efficacy of nitrates in treating acute heart failure remains unclear due to a lack of high-quality studies [14]. On the other hand, combination therapy of isosorbide dinitrate and hydralazine provides symptomatic and mortality benefit in patients of African heritage with chronic congestive heart failure due to systolic dysfunction [15, 16] (see Unit VI). Furthermore, hydralazine in the combination therapy reduces nitrate tolerance likely due to the ability of hydralazine to decrease superoxide formation. Despite its proven benefits, combination hydralazine and nitrate therapy is not commonly used in heart failure likely due to efficacy limited to a specific ethical group (people of African heritage) and adverse effects associated with hydralazine. Since ACE inhibitors also provide afterload reduction along with nitrate tolerance benefits seen with hydralazine, it is suggested that demonstrating benefit with nitrates and ACE inhibitors without concurrent hydralazine use in patients with congestive heart failure may represent a major opportunity to improve nitrate therapy in this cardiac disorder [17].

11.3.4.2 *Clinical Uses of Sodium Nitroprusside*

The US FDA-approved uses of sodium nitroprusside include blood pressure control in hypertensive crises as well as in surgery and treatment of acute congestive heart failure.

Blood Pressure Control   Sodium nitroprusside is indicated for the immediate reduction of blood pressure of adult and pediatric patients in hypertensive crises. Concomitant longer-acting antihypertensive medication should be administered so that the duration of treatment with sodium nitroprusside can be minimized. Sodium nitroprusside is also indicated for producing controlled hypotension in order to reduce bleeding during surgery.

Acute Congestive Heart Failure   Sodium nitroprusside is indicated for the treatment of acute congestive heart failure. It improves patients' symptoms and hemodynamics in this setting though its use does not improve patients' survival [18]. Sodium nitroprusside should not be used for the treatment of acute congestive heart failure associated with reduced peripheral vascular resistance such as high-output heart failure that may be seen in endotoxic sepsis.

11.3.5 Therapeutic Dosages

Listed below are the dosage forms and strengths of the US FDA-approved organic nitrates and sodium nitroprusside. The dosage regimens of these drugs in treating cardiovascular diseases are given in Table 11.2:

- Isosorbide dinitrate (Isordil): Oral, 5 and 40 mg tablets; 40 mg extended-release tablets and capsules
- Isosorbide mononitrate (Ismo): Oral, 10 mg tablets; 30 and 60 mg extended-release tablets
- Nitroglycerin: Oral, 0.3, 0.4, and 0.6 mg tablets; 2.5, 6.5, and 9 mg extended-release capsules; 2.5 and 6.5 mg sustained-release capsules; transdermal, 7 cm^2 (37.3 mg, 0.2 mg/h), 14 cm^2 (74.6 mg, 0.4 mg/h), and 21 cm^2 (111.9 mg, 0.6 mg/h) patches; injection, 50 mg/10 ml single-dose vials (5 mg/ml)

- Sodium nitroprusside (Nitropress): Injection, 50 mg/2 ml single-dose vials (25 mg/ml)

11.3.6 Adverse Effects and Drug Interactions

11.3.6.1 Adverse Effects and Drug Interactions of Nitrates

Adverse Effects Adverse reactions to organic nitrates are generally dose related, and most, if not all, of these reactions are the result of their action as vasodilators. Headache, which may be severe, is the most commonly reported side effect. Headache may be recurrent with each daily dose, especially at higher doses. Transient episodes of lightheadedness, occasionally related to blood pressure changes, may also occur. Hypotension occurs infrequently, but in some patients, it may be severe enough to warrant discontinuation of therapy. Reflex tachycardia occurs, which may be alleviated by simultaneous use of β-blockers if indicated. Syncope,

TABLE 11.2 Dosage regimens of organic nitrates and sodium nitroprusside

Drug	Indication	Dose regimen
Isosorbide dinitrate (tablets)	Angina pectoris (prevention only)	For prevention of angina pectoris due to coronary artery disease (CAD), initial dose, 5–20 mg, 2–3 times/day; maintenance dose, 10–40 mg, 2–3 times/day; daily dose-free interval of at least 14 h (8:00 A.M., 2:00 P.M., 6:00 P.M.) to minimize tolerance
Isosorbide dinitrate (extended-release tablets or capsules)	Angina pectoris (prevention only)	For prevention of angina pectoris due to CAD, total daily dose, 40–160 mg; dose-free interval >18 h
Isosorbide mononitrate (tablets)	Angina pectoris (prevention only)	For prevention of angina pectoris due to CAD, 20 mg twice daily with doses 7 h apart
Isosorbide mononitrate (extended-release tablets)	Angina pectoris (prevention only)	For prevention of angina pectoris due to CAD, initial dose, 30 or 60 mg once daily; after 7 days, dose may be increased to 120 mg once daily
Nitroglycerin (tablets)	Angina pectoris (prevention and treatment)	For prevention of angina pectoris due to CAD, one tablet should be dissolved under the tongue or in the buccal pouch 5–10 min prior to engaging activities that might precipitate an acute anginal attack; for treatment, one tablet should be dissolved under the tongue or in the buccal pouch at the first sign of an acute angina attack
Nitroglycerin (extended-release capsules)	Angina pectoris (prevention only)	For prevention of angina pectoris due to CAD, initial dose, 2.5–6.5 mg, 3–4 times/day with subsequent upward dose adjustment guided by symptoms and adverse effects
Nitroglycerin (transdermal patches)	Angina pectoris (prevention only)	For prevention of angina pectoris due to CAD, initial dose, 0.2–0.4 mg/h; usual dose range, 0.4–0.8 mg/h with daily patch-off period of 10–12 h
Nitroglycerin (injection)	Perioperative hypertension; congestive heart failure; angina pectoris refractory to sublingual nitroglycerin and β-blockers; induction of intraoperative hypertension	Initial intravenous infusion rate, 5 μg (mcg)/min; initial titration should be in 5 μg (mcg)/min increments with increases every 3–5 min until some response is noted
Sodium nitroprusside (injection)	Hypertension crisis; producing controlled hypotension to reduce bleeding during surgery; acute congestive heart failure	Initial dose, 0.3 μg (mcg)/kg/min with upward titration every few minutes until the desired effect is obtained; maximum infusion rate, 10 μg (mcg)/kg/min

crescendo angina, and rebound hypertension have been reported but are uncommon. Extremely rarely, ordinary doses of organic nitrates may cause methemoglobinemia.

Drug Interactions The vasodilating effects of nitroglycerin may be additive with those of other vasodilators. Alcohol, in particular, has been found to exhibit this type of additive effects. Marked symptomatic orthostatic hypotension has been reported when CCBs and organic nitrates were used in combination. Dose adjustments of either class of agents may be necessary. Nitrates and PDE5 inhibitors synergize to cause severe hypotension, and the concomitant use should thus be avoided.

Contraindications and Pregnancy Category

- Allergic reactions to organic nitrates are extremely rare, but they do occur. Nitrates are contraindicated in patients who are allergic to them.
- Sublingual nitroglycerin therapy is contraindicated in patients with early myocardial infarction, severe anemia, and increased intracranial pressure.
- Nitrates and PDE5 inhibitors (e.g., sildenafil for erectile dysfunction) synergize to cause severe hypotension, and as such, combination should be avoided.
- Pregnancy category: C.

11.3.6.2 *Adverse Effects and Drug Interactions of Sodium Nitroprusside*

Adverse Effects The major adverse effects of sodium nitroprusside administration are excessive hypotension and excessive accumulation of cyanide. Other less common effects may include methemoglobinemia, tachycardia, flushing, and increased intracranial pressure. Small transient excesses in the infusion rate of sodium nitroprusside can result in dramatic decreases in blood pressure. In patients not properly monitored, the excessive blood pressure reduction can lead to irreversible ischemic injuries or death. Infused sodium nitroprusside reacts with hemoglobin to produce cyanmethemoglobin with the release of cyanide ion, whose accumulation may lead to cyanide toxicity and death. The usual dose rate is 0.5–10 µg (mcg)/kg body weight/min, but infusion at the maximum dose rate should never last >10 min.

Drug Interactions The blood pressure-lowering effect of sodium nitroprusside is augmented by drugs that decrease blood pressure, including inhaled anesthetics.

Contraindications and Pregnancy Category

- Sodium nitroprusside should not be used in the treatment of compensatory hypertension, where the primary hemodynamic lesion is aortic coarctation or arteriovenous shunting.

- Sodium nitroprusside should not be used to produce hypotension during surgery in patients with known inadequate cerebral circulation or in moribund patients (ASA class 5E) coming to emergency surgery.
- Patients with congenital (Leber's) optic atrophy or with tobacco amblyopia have unusually high cyanide/thiocyanate ratios. These rare conditions are probably associated with defective or absent rhodanese, and sodium nitroprusside should be avoided in these patients.
- Sodium nitroprusside should not be used for the treatment of acute congestive heart failure associated with reduced peripheral vascular resistance such as high-output heart failure that may be seen in endotoxic sepsis.
- Pregnancy category: C.

11.4 ET RECEPTOR ANTAGONISTS

Listed below are the three US FDA-approved ET receptor antagonists, with bosentan and macitentan as nonselective ET receptor type A (ET_A) and type B (ET_B) antagonists and ambrisentan as a selective ET_A antagonist. Macitentan is the newest member of the drug class, which was approved by the US FDA in 2013:

1. Ambrisentan (Letairis): ET_A-selective antagonist
2. Bosentan (Tracleer): ET_A/ET_B antagonist
3. Macitentan (Opsumit): ET_A/ET_B antagonist

11.4.1 General Introduction to ET and Drug Class

As discussed earlier, ET is an important factor involved in the regulation of vascular tone as well as other vascular processes. Three ET, ET-1, ET-2, and ET-3, are expressed in humans. They are all 21-amino-acid-long peptides. ET-1 is the main ET secreted by endothelial cells. It is also the ET that has been shown to play an important role in cardiovascular physiology and pathophysiology. ET-2 is produced mainly by the kidney and the intestine, whereas ET-3 is primarily localized in the brain, the intestine, and the kidney tubule cells.

ET-1 acts in an autocrine/paracrine manner to activate G-protein-coupled receptors ET type A (ET_A) and type B (ET_B) to produce its physiological effects on vessels. Vascular smooth muscle cells express both ET_A and ET_B receptors, whereas endothelial cells express only ET_B receptors. Binding of ET to ET_A and ET_B receptors on smooth muscle causes vasoconstriction, as well as inflammatory responses in the vascular wall. The vasoconstriction results from inositol triphosphate-induced release of calcium ions from the sarcoplasmic reticulum. Both vasoconstriction and vascular inflammation are involved in the detrimental actions of ET-1 on blood vessels in hypertension and other vascular disorders [19, 20]. Evidence suggests that binding of ET to ET_B receptors on endothelial cells may lead to vasodilation via a nitric oxide-dependent mechanism. ET_B receptors are also involved in the clearance of circulating ET-1 that has a half-life of approximately 1 min.

Ambrisentan has a high selectivity for the ET_A receptor compared to the ET_B receptor, whereas bosentan and macitentan block both receptors. The clinical significance of high selectivity for the ET_A receptor is not known. All three drugs are approved by the US FDA for the treatment of pulmonary hypertension.

11.4.2 Chemistry and Pharmacokinetics

The structures of ambrisentan, bosentan, and macitentan are shown in Figure 11.3. Ambrisentan is a propionic acid derivative, whereas bosentan belongs to a class of highly substituted pyrimidine derivatives. The newly approved macitentan is a bosentan derivative. Table 11.3 summarizes the major pharmacokinetic properties of the three drugs.

11.4.3 Molecular Mechanisms and Pharmacological Effects

As noted earlier, ET-1 is a potent autocrine and paracrine peptide. Two receptor subtypes, ET_A and ET_B, mediate its effects in the vascular smooth muscle and endothelium. The primary consequences of ET_A and ET_B activation in smooth muscle cells are vasoconstriction, cell proliferation, and vascular inflammation, while the predominant actions of ET_B activation in endothelial cells are vasodilation, antiproliferation, and ET-1 clearance. In patients with pulmonary hypertension, plasma ET-1 levels are elevated as much as 10-fold and correlate with increased mean right atrial pressure and disease severity. Notably, both protein and mRNA expression are increased as much as ninefold in the lung tissue of patients with pulmonary hypertension, primarily in the endothelium of pulmonary arteries. These findings suggest that ET-1 plays a critical role in the pathogenesis and progression of pulmonary hypertension, and indeed, ET receptor antagonists via blocking ET-1-mediated receptor activation are effective therapy for pulmonary hypertension.

Ambrisentan is a high-affinity ET_A receptor antagonist with a high selectivity for the ET_A versus ET_B receptor (>4000-fold). Bosentan is a specific and competitive antagonist at ET_A and ET_B receptors, with a slightly higher affinity for ET_A than for ET_B receptors. Macitentan displays high

affinity and sustained occupancy of the ET_A and ET_B receptors in human pulmonary arterial smooth muscle cells. Although ET_B activation in endothelial cells appears to be vascular protective, the clinical impact of selective ET_A inhibition or dual receptor (ET_A/ET_B) blockage is currently unknown. The three ET receptor antagonists are similar with regard to reducing pulmonary artery pressure, improving exercise ability, and slowing clinical worsening (Fig. 11.4).

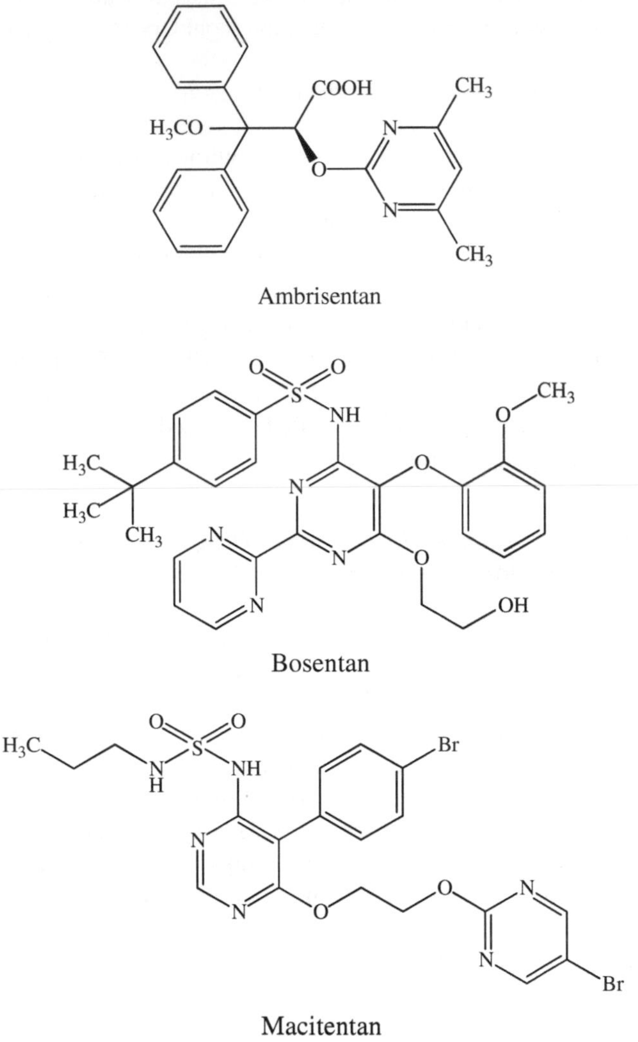

FIGURE 11.3 Structures of endothelin receptor antagonists. The recently approved macitentan is a derivative of bosentan.

TABLE 11.3 Major pharmacokinetic properties of endothelin receptor antagonists

Drug	Oral bioavailability (%)	Elimination half-life (h)	Metabolism and excretion
Ambrisentan	Unknown	9	Hepatic CYP3A4 and CYP2C19; glucuronidation; biliary excretion
Bosentan	50	5	Hepatic CYP2C9 and CYP3A4; also inducer of CYP2C9 and CYP3A4; biliary excretion
Macitentan	Unknown	16	Hepatic CYP3A4; CYP2C19 (minor); urinary and biliary excretion

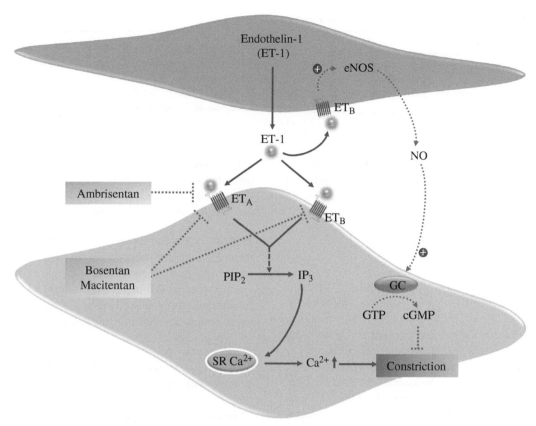

FIGURE 11.4 Molecular mechanisms of action of endothelin receptor antagonists. Endothelin (ET-1) is produced by endothelial cells, and the released ET-1 activates endothelin receptors type A and type B (ET_A and ET_B) on smooth muscle cells, leading to the subsequent formation of inositol 1,4,5-triphosphate (IP_3) through phospholipase C-catalyzed hydrolysis of phosphatidylinositol 4,5-bisphosphate (PIP_2). IP_3 then causes calcium ion release from sarcoplasmic reticulum (SR), thereby causing smooth muscle cell constriction. Abnormal activation of ET_A and ET_B on smooth muscle cells may also cause cell proliferation and vascular inflammation, which together with vasoconstriction contribute to the pathogenesis of hypertension (notably pulmonary arterial hypertension) and other cardiovascular disorders. On the other hand, activation of ET_B receptors by ET-1 on endothelial cells results in increased production of nitric oxide (NO), and the released NO activates guanylate cyclase (GC), leading to increased formation of cGMP, and the subsequent smooth muscle relaxation (or counteraction of smooth muscle cell constriction). However, the significance of this vasodilating effect of activation of ET_B receptors on endothelial cells in cardiovascular physiology and pathophysiology remains to be elucidated.

11.4.4 Clinical Uses

All three ET receptor antagonists are approved for treating pulmonary arterial hypertension (WHO group 1) to improve exercise ability and delay clinical worsening (also see Chapter 12). The newly approved macitentan also reduces mortality due to pulmonary arterial hypertension [21].

11.4.5 Therapeutic Dosages

The dosage forms and strengths of the three ET receptor antagonists are listed below, and the dosage regimens of these drugs in treating pulmonary hypertension are summarized in Table 11.4:

1. Ambrisentan (Letairis): Oral, 5 and 10 mg tablets
2. Bosentan (Tracleer): Oral, 62.5 and 125 mg tablets
3. Macitentan (Opsumit): Oral, 10 mg tablets

11.4.6 Adverse Effects and Drug Interactions

11.4.6.1 Adverse Effects The most common adverse effects of ET receptor antagonists are upper respiratory infections, peripheral edema, and anemia. Use of bosentan and macitentan is also associated with liver toxicity. Peripheral edema is the most common adverse effect that requires attention. Mild cases can be managed with diuretics, but more severe cases warrant discontinuation of the medication. ET receptor antagonists are also potent teratogens, requiring meticulous contraception if used by women who have childbearing potential, and are contraindicated during pregnancy.

11.4.6.2 Drug Interactions

Ambrisentan Drug interactions for ambrisentan at therapeutic doses are less dramatic than bosentan and macitentan. This is because ambrisentan does not

TABLE 11.4   Dosage regimens of endothelin receptor antagonists

Drug	Indication	Dosage regimen
Ambrisentan	Pulmonary arterial hypertension	Initial dose, 5 mg once daily; the dose may be increased to 10 mg once daily if 5 mg once daily is tolerated
Bosentan	Pulmonary arterial hypertension	Initial dose, 62.5 mg twice daily for 4 weeks; the dose may then be increased to maintenance dose of 125 mg twice daily
Macitentan	Pulmonary arterial hypertension	10 mg once daily

significantly affect drug metabolism enzymes at therapeutic doses. The only clinically significant drug interaction is that cotreatment with ambrisentan and cyclosporine results in an approximately twofold increase in ambrisentan exposure, and as such, the dose of ambrisentan should be reduced when coadministered with cyclosporine.

Bosentan   Bosentan may cause significant drug interactions due to its ability to alter CYP enzymes. Bosentan is metabolized by CYP2C9 and CYP3A4. Inhibition of these enzymes may increase the plasma concentrations of bosentan. Concomitant administration of both a CYP2C9 inhibitor (such as fluconazole or amiodarone) and a strong CYP3A4 inhibitor (e.g., ketoconazole, itraconazole) or a moderate CYP3A4 inhibitor (e.g., amprenavir, erythromycin, fluconazole, diltiazem) with bosentan will likely lead to large increases in plasma concentrations of bosentan. Coadministration of such combinations of a CYP2C9 inhibitor plus a strong or moderate CYP3A4 inhibitor with bosentan is not recommended. Bosentan is also an inducer of CYP3A4 and CYP2C9. Consequently, plasma concentrations of drugs metabolized by these two isozymes will be decreased when bosentan is coadministered. In this regard, combination use of bosentan decreases the bioavailability of simvastatin and other CYP3A4-metabolized statins, resulting in decreased efficacy in treating dyslipidemia and atherosclerotic cardiovascular diseases (see Chapter 4 for statin drugs).

Notably, initial concomitant administration of cyclosporine A may increase the plasma concentrations of bosentan by about 30-fold. The mechanism of this interaction is most likely inhibition of transport protein-mediated uptake of bosentan into hepatocytes by cyclosporine A. Steady-state bosentan plasma concentrations are three- to fourfold higher than in the absence of cyclosporine A. As such, concomitant administration of bosentan and cyclosporine A is contraindicated. Coadministration of bosentan also decreases the plasma concentrations of cyclosporine A (a CYP3A4 substrate) by approximately 50%.

Macitentan   Macitentan is primarily metabolized by CYP3A4, and as such, drugs that affect CYP3A4 may significantly alter metabolism of macitentan. Strong inducers of CYP3A4, such as rifampin, significantly reduce macitentan exposure. Hence, concomitant use of macitentan with strong CYP3A4 inducers should be avoided. On the other hand, concomitant use of strong CYP3A4 inhibitors like ketoconazole and many anti-HIV drugs may markedly increase the plasma concentrations of macitentan.

11.4.6.3   Contraindications and Pregnancy Category   All three ET receptor antagonists are contraindicated in pregnancy (pregnancy category: X) because of their potential to cause fetal harm. While pregnancy is the only contraindication for macitentan, as described in the following, ambrisentan and bosentan also have other contraindications.

Ambrisentan   Ambrisentan is contraindicated in patients with idiopathic pulmonary fibrosis (IPF), including IPF patients with pulmonary hypertension (WHO group 3). A randomized controlled study in patients with IPF, with or without pulmonary hypertension (WHO group 3), comparing ambrisentan ($n=329$) to placebo ($n=163$) was terminated after 34 weeks for lack of efficacy and a greater risk of disease progression or death on ambrisentan [22].

Bosentan

- Use with cyclosporine A: As noted earlier, coadministration of cyclosporine A and bosentan results in markedly increased plasma concentrations of bosentan. As such, concomitant use of bosentan and cyclosporine A is contraindicated.
- Use with glyburide: An increased risk of liver enzyme elevations has been observed in patients receiving glyburide concomitantly with bosentan. Therefore, coadministration of glyburide and bosentan is contraindicated.
- Hypersensitivity: Bosentan is contraindicated in patients who are hypersensitive to bosentan or any component of the product. Observed hypersensitivity reactions include rash and angioedema.

11.5   PHOSPHODIESTERASE 5 INHIBITORS

Listed below are the four US FDA-approved phosphodiesterase 5 inhibitors, with avanafil being the newest member, which was approved in 2012. All four drugs are indicated for treating erectile dysfunction, and sildenafil and tadalafil are also approved for the treatment of pulmonary arterial hypertension. Hence, this section focuses on sildenafil and tadalafil:

- Avanafil (Stendra): For treating erectile dysfunction
- Sildenafil (Viagra): For treating erectile dysfunction
- Sildenafil (Revatio): For treating pulmonary arterial hypertension
- Tadalafil (Cialis): For treating erectile dysfunction
- Tadalafil (Adcirca): For treating pulmonary arterial hypertension
- Vardenafil (Levitra): For treating erectile dysfunction

11.5.1 General Introduction to Drug Class

Cyclic nucleotide phosphodiesterases (PDEs) catalyze the hydrolysis of cyclic nucleotides adenosine 3′5′-cyclic monophosphate (cAMP) and guanosine 3′5′-cyclic monophosphate (cGMP) to their inactive 5′ monophosphates, thereby regulating the intracellular concentrations of these cyclic nucleotides, their signaling pathways and, consequently, myriad biological responses in health and disease [23]. Mammalian PDEs are composed of 21 genes and are classified into 11 families based on sequence homology, enzymatic properties, and sensitivity to inhibitors. These 11 families are denoted as PDE1 to PDE11. Each family has various members.

At the molecular level, PDE families contain many splice variants that mostly are unique in tissue-expression patterns, gene regulation, enzymatic regulation by phosphorylation and regulatory proteins, subcellular localization, and interaction with association proteins. Each unique variant is closely related to the regulation of a specific cellular signaling. Thus, multiple PDEs function as a particular modulator of each cardiovascular function and regulate physiological homeostasis [23, 24]. Inhibitors of PDEs have been developed over the past decades to treat cardiovascular diseases. Among them are drugs that inhibit the type-5 phosphodiesterase (PDE5). These drugs are thus known as PDE5 inhibitors.

The PDE5 is found primarily in the penis and lung tissues although they are also expressed in other tissues/organs, including the visceral smooth muscle, skeletal muscle, platelets, kidney, cerebellum, and pancreas. PDE5 inhibitors cause vasodilation in the penis and lung by blocking the breakdown of cGMP, which results in prolongation of the action of vasodilating mediators including nitric oxide. For these reasons, the two major actions of the PDE5 inhibitors are to prolong penile erection and decrease pulmonary vascular pressure. They have relatively less pronounced effects on the systemic vasculature.

As noted earlier, all the four PDE5 inhibitors are currently approved for the treatment of erectile dysfunction. Sildenafil and tadalafil, under the trade name of Revatio and Adcirca, respectively, are also approved for treating pulmonary arterial hypertension.

11.5.2 Chemistry and Pharmacokinetics

The chemical structures of PDE5 inhibitors are provided in Figure 11.5. Table 11.5 summarizes the major pharmacokinetic properties of PDE5 inhibitors.

11.5.3 Mechanisms and Pharmacological Effects

As aforementioned, PDE5 inhibitors are primarily used for treating erectile dysfunction. Some are also approved for treating pulmonary arterial hypertension. The physiologic mechanism of erection of the penis involves release of nitric oxide in the corpus cavernosum during sexual stimulation. Nitric oxide then activates the enzyme guanylate cyclase, which results in increased levels of cGMP, producing smooth muscle relaxation in the corpus cavernosum and allowing inflow of blood. PDE5 inhibitors have no direct relaxant effect on isolated human corpus cavernosum. When sexual stimulation causes local release of nitric oxide, inhibition of PDE5 by PDE5 inhibitors causes increased levels of cGMP in the corpus cavernosum, resulting in smooth muscle relaxation and inflow of blood to the corpus cavernosum. Hence, PDE5 inhibitors at recommended doses have no significant effect in the absence of sexual stimulation.

Pulmonary arterial hypertension is associated with impaired release of nitric oxide by the vascular endothelium and consequent reduction of cGMP concentrations in the pulmonary vascular smooth muscle. PDE5 is the predominant phosphodiesterase in the pulmonary vasculature. Inhibition of PDE5 by sildenafil or tadalafil increases the concentrations of cGMP, resulting in relaxation of pulmonary vascular smooth muscle cells and vasodilation of the pulmonary vascular bed and, to a lesser degree, vasodilation in the systemic circulation.

The 1,000–10,000-fold selectivity of PDE5 inhibitors for PDE5 versus PDE3 is important because PDE3 is involved in control of cardiac contractility. Sildenafil is only about 10-fold as potent for PDE5 compared to PDE6, an enzyme found in the retina and involved in the phototransduction pathway of the retina. This relatively lower selectivity is thought to be the basis for abnormalities related to color vision observed with higher doses or plasma levels of sildenafil. In contrast, tadalafil is 700-fold more potent for PDE5 than for PDE6, which may explain the decreased risk of color vision abnormalities as compared with sildenafil.

In addition to pulmonary vascular smooth muscle and the corpus cavernosum, PDE5 is also found in other tissues including vascular and visceral smooth muscle and in platelets. The inhibition of PDE5 in these tissues by PDE5 inhibitors may be the basis for the enhanced platelet antiaggregatory activity of nitric oxide observed *in vitro* and the mild peripheral arterial–venous dilation *in vivo*.

11.5.4 Clinical Uses

PDE5 inhibitors are approved by the US FDA for the treatment of two clinical conditions: erectile dysfunction (all four PDE5

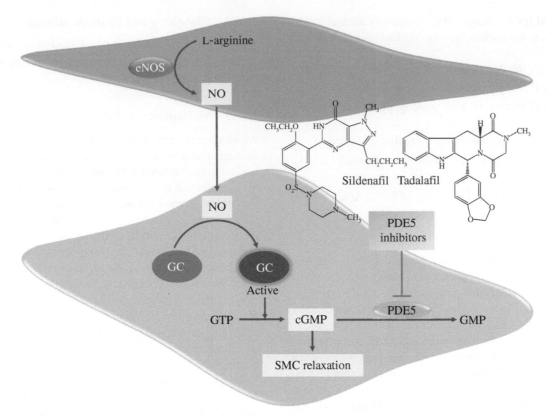

FIGURE 11.5   Molecular mechanisms of action of PDE5 inhibitors. As depicted, endothelial cells produce nitric oxide (NO) via the action of endothelia nitric oxide synthase (eNOS). The NO released then activates guanylate cyclase (GC) in smooth muscle cells, which in turn catalyzes the formation of cGMP from GTP. cGMP is metabolized to GMP by PDE5. Inhibition of PDE5 by sildenafil or tadalafil causes increased accumulation of cGMP, thereby resulting in smooth muscle cell (SMC) relaxation and vasodilation.

TABLE 11.5   Major pharmacokinetic properties of PDE5 inhibitors

Drug	Oral bioavailability (%)	Elimination half-life (h)	Metabolism and excretion
Sildenafil	41	4	Hepatic CYP3A (major) and CYP2C9 (minor); biliary (major) and urinary (minor) excretion
Tadalafil	Unknown	15	Hepatic CYP3A; biliary (major) and urinary (minor) excretion

TABLE 11.6   Dose regimens of PDE5 inhibitors in treating pulmonary arterial hypertension

Drug	Indication	Dosage regimen
Sildenafil (tablet and oral suspension)	Pulmonary arterial hypertension	5 or 20 mg 3 times/day
Sildenafil (injection)	Pulmonary arterial hypertension	2.5 or 10 mg administered as an intravenous bolus injection twice a day
Tadalafil (tablet)	Pulmonary arterial hypertension	40 mg once daily

inhibitors) to improve sexual performance and pulmonary arterial hypertension (WHO group I) (sildenafil and tadalafil) to improve exercise ability and delay clinical worsening.

11.5.5   Therapeutic Dosages

The dosage forms and strengths of PDE5 inhibitors for treating pulmonary arterial hypertension are listed below, and the typical dose regimens are given in Table 11.6:

- Sildenafil (Revatio): Oral, 20 mg tablets; 10 mg/ml oral suspension (when reconstituted); injection, 10 mg/12.5 ml in single-use vials
- Tadalafil (Adcirca): Oral, 20 mg tablets

11.5.6   Adverse Effects and Drug Interactions

11.5.6.1   Adverse Effects   The most common adverse effects of sildenafil include epistaxis, headache, dyspepsia, flushing, insomnia, erythema, dyspnea, and rhinitis. Headache is the most common adverse effect of tadalafil therapy. Other less common adverse effects for both sildenafil and tadalafil include hypotension, visual disturbance (especially for sildenafil), hearing impairment, and priapism.

11.5.6.2 Drug Interactions
Coadministration of PDE5 inhibitors with nitrates causes hypotension and such a combination should be avoided. Since CYP3A4 is involved in metabolism of sildenafil and tadalafil, concomitant use of these PDE5 inhibitors with ritonavir and other potent CYP3A4 inhibitors is not recommended. Since PDE5 inhibitors have mild systemic blood pressure-lowering effects, coadministration of other blood pressure-lowering drugs, including α-blockers (e.g., doxazosin) and CCBs (e.g., amlodipine), may cause significant decreases in blood pressure due to additive blood pressure-lowering effects.

11.5.6.3 Contraindications and Pregnancy Category

- Coadministration of nitrate: Administration of sildenafil or tadalafil with nitrates (either regularly or intermittently) or other nitric oxide donors is contraindicated. The PDE5 inhibitors potentiate the hypotensive effect of nitrates. This potentiation is thought to result from the combined effects of nitrates and PDE5 inhibitors on the nitric oxide/cGMP pathway. A suitable time interval following dosing of PDE5 inhibitors for the safe administration of nitrates or other nitric oxide donors has not been determined.
- Hypersensitivity reactions: Tadalafil is contraindicated in patients with a known serious hypersensitivity to the drug. Hypersensitivity reactions have been reported, including Stevens–Johnson syndrome and exfoliative dermatitis.
- Pregnancy category: B.

11.6 sGC STIMULATORS

Listed below is the first and only soluble guanylate cyclase stimulator approved by the US FDA for clinical use:

- Riociguat (Adempas)

11.6.1 General Introduction to Drug Class

sGC is activated by nitric oxide and catalyzes the formation of cGMP from GTT. It plays an important role in a wide variety of physiological and pathophysiological processes, particularly in the cardiovascular system. As such, sGC is a potential therapeutic target in cardiovascular disorders such as myocardial infarction, stroke, and pulmonary hypertension. Multiple drug compounds have been developed to stimulate sGC to treat pulmonary hypertension. Among them, riociguat has recently been approved by the US FDA and is the first in its drug class for treating pulmonary hypertension.

11.6.2 Chemistry and Pharmacokinetics

Riociguat is methyl 4,6-diamino-2-[1-(2-fluorobenzyl)-1*H*-pyrazolo [3,4-*b*]pyridin-3-yl]-5-pyrimidinyl(methyl)carbamate and its structure is shown in Figure 11.6. Riociguat is readily absorbed through the gut and its oral bioavailability is about 94%. It is also a substrate of P-glycoprotein and ATP-binding cassette protein G2 (ABCG2). Riociguat is metabolized by CYP1A1, CYP3A, CYP2C8, and CYP2J2 and eliminated via urinary and biliary excretion. Its elimination half-life is 7–12 h.

11.6.3 Molecular Mechanisms and Pharmacological Effects

When nitric oxide binds to sGC, the enzyme is activated and catalyzes synthesis of the signaling molecule cGMP. Intracellular cGMP plays an important role in regulating processes that influence vascular tone, proliferation, fibrosis, and inflammation. Pulmonary hypertension is associated with endothelial dysfunction, impaired synthesis of nitric oxide, and insufficient stimulation of the nitric oxide–sGC–cGMP pathway. Riociguat stimulates the nitric oxide–sGC–cGMP pathway and leads to increased generation of intracellular cGMP with subsequent smooth muscle relaxation and vasodilation. Riociguat has a dual mode of action (Fig. 11.6). It sensitizes sGC to endogenous nitric oxide by stabilizing the nitric oxide–sGC binding. Riociguat also directly stimulates sGC via a different binding site, independently of nitric oxide. Riociguat treatment results in decreases in systemic vascular resistance, systolic blood pressure, pulmonary vascular resistance, and pulmonary arterial pressure.

11.6.4 Clinical Uses

Riociguat is approved for treating two clinical conditions—chronic thromboembolic pulmonary hypertension (CTEPH) and pulmonary arterial hypertension:

1. CTEPH: Riociguat is indicated for the treatment of adults with persistent/recurrent CTEPH (WHO group 4) after surgical treatment, or inoperable CTEPH, to improve exercise capacity and WHO functional class.
2. Pulmonary arterial hypertension: Riociguat is indicated for the treatment of adults with pulmonary arterial hypertension (WHO group 1), to improve exercise capacity and WHO functional class and to delay clinical worsening.

11.6.5 Therapeutic Dosages

The dosage forms and strengths of riociguat are listed below:

- Riociguat (Adempas): Oral, 0.5, 1, 1.5, 2, and 2.5 mg tablets

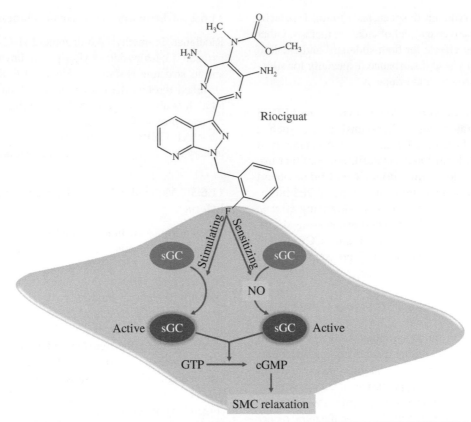

FIGURE 11.6   Molecular mechanisms of action of riociguat. Riociguat directly stimulates soluble guanylate cyclase (sGC), augmenting its activity. The drug also sensitizes sGC to nitric oxide (NO)-mediated activation. Together, the dual actions of riociguat on sGC result in increased formation of cGMP, thereby causing smooth muscle cell (SMC) relaxation and vasodilation.

The recommended starting dosage is 1 mg taken three times a day. For patients who may not tolerate the hypotensive effect of riociguat, a starting dose of 0.5 mg taken three times a day may be considered. If systolic blood pressure remains >95 mmHg and the patient has no signs or symptoms of hypotension, the dose can be uptitrated by 0.5 mg taken three times a day. Dose increases should be no sooner than 2 weeks apart. The dose can be increased to the highest tolerated dosage, up to a maximum of 2.5 mg taken three times a day. If, at any time, the patient has symptoms of hypotension, decrease the dosage by 0.5 mg taken three times a day.

11.6.6   Adverse Effects and Drug Interactions

11.6.6.1   Adverse Effects   Adverse effects of riociguat may include headache, dyspepsia/gastritis, dizziness, nausea, diarrhea, hypotension, vomiting, anemia, gastroesophageal reflux, and constipation. Riociguat may also cause bleeding, pulmonary edema in patients with pulmonary veno-occlusive disease, and fetal harm.

11.6.6.2   Drug Interactions   Due to the involvement of multiple pathways in the deposition of riociguat, drug interactions can be significant. For example, cigarette

smoking (inducer of CYP1A1) augments metabolism of riociguat, and the plasma concentrations of the drug in smokers are reduced by 50–60% compared to nonsmokers. Concomitant use of riociguat with strong CYP inhibitors and P-glycoprotein/ABCG2 inhibitors such as azole antimycotics (e.g., ketoconazole, itraconazole) or HIV protease inhibitors (such as ritonavir) increases riociguat exposure and may result in hypotension. Strong inducers of CYP3A (e.g., rifampin, phenytoin, carbamazepine, phenobarbital, or St. John's Wort) may significantly reduce riociguat exposure. Finally, antacids such as aluminum hydroxide and magnesium hydroxide decrease riociguat absorption and should not be taken within 1 h of taking riociguat.

11.6.6.3   Contraindications and Pregnancy Category

- Nitrates and nitric oxide donors: Coadministration of riociguat with nitrates or nitric oxide donors in any form is contraindicated because of additive blood pressure-lowering effects.
- Phosphodiesterase inhibiting agents: Concomitant administration of riociguat with specific PDE5 inhibitors (such as sildenafil, tadalafil, or vardenafil) or nonspecific PDE inhibitors (such as dipyridamole or

theophylline) is contraindicated due to additive blood pressure-lowering effects.

- Pregnancy: Riociguat may cause fetal harm when administered to a pregnant woman. It is thus contraindicated in pregnancy (pregnancy category: X).

11.7   K+$_{ATP}$ CHANNEL OPENERS

While several K+$_{ATP}$ channel openers are available for research use or clinical use in other nations, minoxidil listed below is currently the only US FDA-approved K+$_{ATP}$ channel opener for treating hypertension:

- Minoxidil (Loniten)

11.7.1   General Introduction to Drug Class

K+$_{ATP}$ channel openers are drugs that activate ATP-sensitive potassium ion channels in vascular smooth muscle. Opening of these channels hyperpolarizes the smooth muscle, which closes voltage-gated calcium channels and decreases intracellular calcium concentrations (see Chapter 10 for

discussion of voltage-dependent calcium channels). With less calcium available to combine with calmodulin, there is less activation of myosin light chain kinase and phosphorylation of myosin light chains (see Fig. 10.2 of Chapter 10). This results in smooth muscle cell relaxation and vasodilation. Because small arteries and arterioles normally have a high degree of smooth muscle tone, K+$_{ATP}$ channel openers are particular effective in dilating these resistance vessels, decreasing systemic vascular resistance, and lowering arterial pressure.

K+$_{ATP}$ channel openers include cromakalim, minoxidil, nicorandil, and pinacidil. As noted earlier, minoxidil is the only K+$_{ATP}$ channel opener approved by the US FDA for treating hypertension. It should be noted that the recently US FDA-approved ezogabine is an activator of the voltage-gated K(V)7 potassium channels primarily expressed in neurons and is indicated as an adjunctive treatment of partial-onset seizures in adult patients [25]. Hence, when not specified, potassium channel openers may include K+$_{ATP}$ channel openers (e.g., minoxidil) and voltage-gated potassium channel openers (e.g., ezogabine). Since K+$_{ATP}$ channel openers are used in cardiovascular medicine, this section focuses on discussing the molecular pharmacology and clinical use of minoxidil.

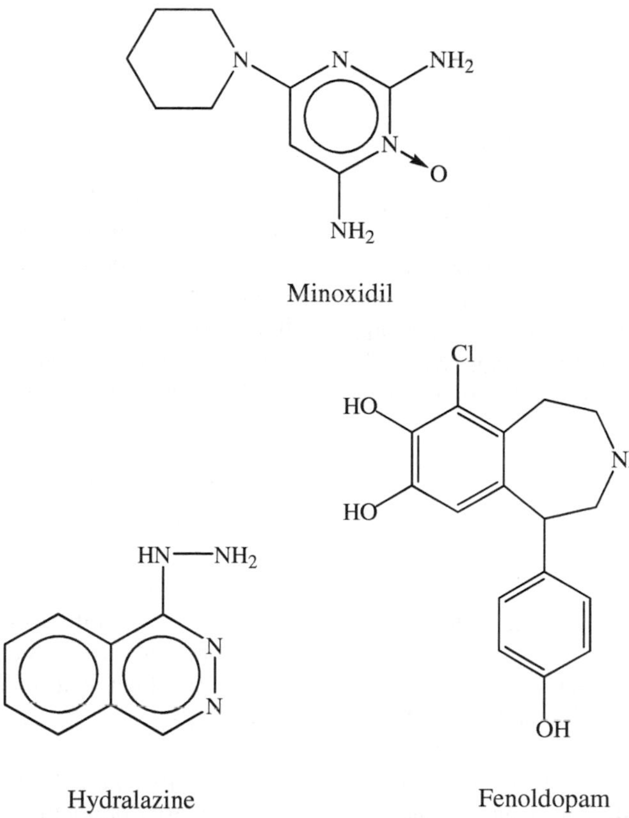

FIGURE 11.7   Structures of minoxidil, hydralazine, and fenoldopam. The chemical names for minoxidil, hydralazine, and fenoldopam are 2,4-pyrimidinediamine 6-(1-piperidinyl)-3-oxide, 1-hydrazinophthalazine, and 6-chloro-2,3,4,5-tetrahydro-1-(4-hydroxyphenyl)-[1H]-3-benzazepine-7,8-diol, respectively.

11.7.2   Chemistry and Pharmacokinetics

The chemical name for minoxidil is 2,4-pyrimidinediamine 6-(1-piperidinyl)-3-oxide (structure shown in Fig. 11.7). Minoxidil is readily absorbed from the gut. Approximately 90% of the administered drug is metabolized, predominantly by conjugation with glucuronic acid at the N-oxide position in the pyrimidine ring, and excreted principally in the urine. The elimination half-life is approximately 4 h.

11.7.3   Molecular Mechanisms and Pharmacological Effects

Minoxidil directly opens K^+_{ATP} channels on vascular smooth muscle cells, thereby causing vasodilation and reducing elevated systolic and diastolic blood pressure by decreasing peripheral vascular resistance. Because it causes peripheral vasodilation, minoxidil elicits a number of predictable reactions. Reduction of peripheral arteriolar resistance and the associated fall in blood pressure trigger sympathetic, vagal inhibitory, and renal homeostatic mechanisms, including an increase in renin secretion, that lead to increased cardiac rate and output and salt and water retention. These adverse effects can usually be minimized by concomitant administration of a diuretic and a β-blocker or other sympathetic nervous system suppressant.

11.7.4   Clinical Uses

Because of the potential for serious adverse effects, minoxidil is indicated only in the treatment of hypertension that is symptomatic or associated with target organ damage and is not manageable with maximum therapeutic doses of a diuretic plus two other antihypertensive drugs. At the present time, use in milder degrees of hypertension is not recommended because the benefit–risk relationship in such patients has not been defined. Because of its ability to cause hypertrichosis (excessive hair growth), minoxidil is also used to treat baldness.

11.7.5   Therapeutic Dosages

The dosage forms and strengths of minoxidil are listed below:

- Minoxidil (Loniten): Oral, 2.5 and 10 mg tablets

For treating hypertension, the recommended initial dosage of minoxidil is 5 mg given as a single daily dose. Daily dosage can be increased to 10 and 20 and then to 40 mg in single or divided doses if required for optimum blood pressure control. The effective dosage range is usually 10–40 mg/day. The maximum recommended dosage is 100 mg/day.

11.7.6   Adverse Effects and Drug Interactions

11.7.6.1   Adverse   Effects   Minoxidil   may   cause significant adverse effects, including fluid and salt retention, cardiovascular abnormalities (e.g., pericarditis, pericardial effusion, and tamponade), and hypertrichosis.

11.7.6.2   Drug   Interactions   Although minoxidil itself does not cause orthostatic hypotension, its administration to patients already receiving guanethidine can result in profound orthostatic effects. If at all possible, guanethidine should be discontinued well before minoxidil is begun.

11.7.6.3   Contraindications and Pregnancy Category

- Minoxidil is contraindicated in pheochromocytoma, because it may stimulate secretion of catecholamines from the tumor through its antihypertensive action.
- Minoxidil is contraindicated in those patients with a history of hypersensitivity to any of the components of the preparation. Reactions may include rashes, and rarely bullous eruptions, and Stevens–Johnson syndrome.
- Pregnancy category: C.

11.8   OTHER VASODILATORS

- Hydralazine (Apresoline)
- Fenoldopam (Corlopam)

In addition to the aforementioned different classes of vasodilators, several other drugs also possess vasodilating activity and are useful for treating hypertension and other disorders in certain patients. These drugs, as listed earlier, include hydralazine and fenoldopam (structure shown in Fig. 11.7). This section briefly introduces their pharmacological basis of use in cardiovascular medicine.

11.8.1   Hydralazine

Hydralazine is an old drug used for treating hypertension. It is not commonly used today due to its adverse effects, limited efficacy, and in particular the availability of other more effective antihypertensive agents. It is mainly used (i) in combination with nitrates for treating heart failure, and (ii) in the treatment of severe hypertension, including preeclampsia.

11.8.1.1   Pharmacokinetics   Hydralazine   is   rapidly absorbed after oral administration, and peak plasma levels are reached at 1–2 h. Hydralazine is subject to polymorphic acetylation; slow acetylators generally have higher plasma levels of hydralazine and require lower doses to maintain control of blood pressure. Hydralazine undergoes extensive hepatic metabolism; it is excreted mainly in the form of metabolites in the urine. The elimination half-life is 3–7 h.

11.8.1.2 Molecular Mechanisms and Pharmacological Effects

Although the precise mechanism of action of hydralazine is not fully understood, the major effects are on the cardiovascular system. Hydralazine apparently lowers blood pressure by exerting a peripheral vasodilating effect through a direct relaxation of vascular smooth muscle. Proposed mechanisms include potassium channel activation and membrane hyperpolarization of smooth muscle cells and inhibition of inositol 1,4,5-triphosphate-induced calcium release from sarcoplasmic reticulum. Hydralazine, by altering cellular calcium metabolism, interferes with the calcium movements within the vascular smooth muscle that are responsible for initiating or maintaining the contractile state.

The peripheral vasodilating effects of hydralazine result in (i) decreased arterial blood pressure (diastolic more than systolic); (ii) decreased peripheral vascular resistance; and (iii) an increased heart rate, stroke volume, and cardiac output. The preferential dilation of arterioles, as compared to veins, minimizes postural hypotension and promotes the increase in cardiac output.

11.8.1.3 Clinical Uses

Hydralazine is approved for treating hypertension, acute severe hypertension, and preeclampsia/eclampsia. It is combined with a nitrate for treating congestive heart failure in patients of African heritage. In this ethical group, hydralazine/nitrate combination reduces mortality of congestive heart failure.

11.8.1.4 Therapeutic Dosages

The dosage forms and strengths of hydralazine are listed below, and the dose regimens are given in Table 11.7:

- Hydralazine: Oral, 10, 25, 50, and 100 mg tablets; injection, 20 mg/ml, 1 ml fill, in single-use vials

11.8.1.5 Adverse Effects and Drug Interactions

Adverse Effects Adverse effects are mainly related to vasodilation. Common adverse effects include headache, anorexia, nausea, vomiting, diarrhea, palpitations, tachycardia, and angina pectoris. Others may include lupus and mild disturbances of the gut and central nervous system.

Drug Interactions Monoamine oxidase inhibitors should be used with caution in patients receiving hydralazine. When other potent parenteral antihypertensive drugs, such as diazoxide, are used in combination with hydralazine, patients should be continuously observed for several hours for any

excessive fall in blood pressure. Profound hypotensive episodes may occur when diazoxide injection and hydralazine are used concomitantly.

Contraindications and Pregnancy Category

- Hypersensitivity to hydralazine.
- Coronary artery disease: Myocardial stimulation produced by hydralazine can cause anginal attacks and ECG changes of myocardial ischemia. The drug has been implicated in the production of myocardial infarction.
- Mitral valvular rheumatic heart disease: The "hyperdynamic" circulation (e.g., tachycardia) caused by hydralazine may accentuate specific cardiovascular inadequacies. In this regard, hydralazine may increase pulmonary artery pressure in patients with mitral valvular disease.
- Pregnancy category: C.

11.8.2 Fenoldopam

Fenoldopam is a rapid-acting vasodilator and given via intravenous injection. Fenoldopam has vasodilating effects in coronary, renal, mesenteric, and peripheral arteries and is indicated for the management of severe hypertension.

11.8.2.1 Pharmacokinetics

Following intravenous administration, the elimination half-life is about 5 min. Elimination is largely by conjugation, without participation of CYP enzymes. The principal routes of conjugation are methylation, glucuronidation, and sulfation.

TABLE 11.7 Dose regimens of hydralazine

Indication	Dose regimen
Hypertension	Oral administration: initial dose, 10 mg 4 times/day; usual dose range, 25–100 mg/day in two divided doses
Severe acute hypertension	Intramuscular or intravenous injection: initial dose, 10–20 mg/dose every 4–6 h as needed; the dose may be increased to 40 mg/dose; change to oral therapy as soon as possible
Preeclampsia/ eclampsia	Intramuscular or intravenous injection: 5 mg/dose and then 5–10 mg every 20–30 min as needed
Congestive heart failure	Oral administration: initial dose, 10–25 mg 3–4 times/day; dosage must be adjusted based on individual response; target dose, 225–300 mg/day in divided doses; use in combination with isosorbide dinitrate

11.8.2.2 Molecular Mechanisms and Pharmacological Effects

Fenoldopam is an agonist for D1-like dopamine receptors and binds with moderate affinity to α_2-adrenergic receptors. It is a selective postsynaptic dopamine agonist (activating D1-like dopamine receptors), which exerts hypotensive effects by decreasing peripheral vasculature resistance with increased renal blood flow, diuresis, and natriuresis. It is six times as potent as dopamine in producing renal vasodilation.

11.8.2.3 Clinical Uses

Fenoldopam is indicated for the in-hospital, short-term (up to 48 h) management of severe hypertension when rapid, but quickly reversible, emergent reduction of blood pressure is clinically indicated, including malignant hypertension with deteriorating end-organ function. Transition to oral therapy with another agent can begin at any time after blood pressure is stable during fenoldopam infusion.

11.8.2.4 Therapeutic Dosages

The dosage forms and strengths of fenoldopam are listed below:

- Fenoldopam (Corlopam): Intravenous injection, 10 mg/ml, 1 ml fill, single-use ampules; 10 mg/ml, 2 ml fill, single-use ampules

Fenoldopam should be administered by continuous intravenous infusion. A bolus dose should not be used. The initial infusion rate may be between 0.03 and 0.1 µg (mcg)/kg/min to minimize reflex tachycardia. The initial dose should be titrated, no more frequently than every 15 min to achieve the desired therapeutic effect. The recommended increments for titration are 0.05–0.1 µg (mcg)/kg/min.

11.8.2.5 Adverse Effects and Drug Interactions

The most common events associated with fenoldopam use are headache, cutaneous dilation (flushing), nausea, and hypotension, each reported in >5% of patients. Fenoldopam does not cause significant drug interactions. It belongs to the pregnancy category B.

11.9   SUMMARY OF CHAPTER KEY POINTS

- Abnormal vasoconstriction is an important pathophysiological mechanism underlying hypertension, ischemic heart disease, and heart failure. Vasodilators are drugs that decrease vascular smooth muscle contractility and thereby cause vasodilation, and these drugs are commonly used in the management of the aforementioned cardiovascular disorders.

- Vasodilators used in cardiovascular medicine include nine groups of drugs: (1) sympatholytics, (2) inhibitors of the RAAS, (3) CCBs, (4) nitric oxide-releasing vasodilators (nitrates and sodium nitroprusside), (5) endothelin receptor antagonists, (6) PDE5 inhibitors, (7) soluble guanylate cyclase stimulators, (8) K^+_{ATP} channel openers, and (9) other vasodilators.

- Although numerous mechanisms are involved in regulating vascular tone, calcium ions and nitric oxide appear to the two key players. They may be viewed as the final common pathways leading to vasoconstriction and vasodilation, respectively. The clinically available vasodilators affect these two factors either directly or indirectly to cause vasodilation.

- Nitrates, including isosorbide dinitrate, isosorbide mononitrate, and nitroglycerin, are primarily used for the management of stable angina due to CAD, whereas sodium nitroprusside is used to treat hypertensive crises. Nitrates and sodium nitroprusside are also indicated in the management of heart failure.

- Endothelin receptor antagonists, including ambrisentan, bosentan, and macitentan, cause pulmonary arterial smooth muscle relaxation and decrease pulmonary arterial pressure. They are approved for treating pulmonary arterial hypertension (WHO group 1) to improve exercise ability and delay clinical worsening.

- The PDE5 inhibitors, sildenafil and tadalafil, are not only used for treating erectile dysfunction, but also approved for the treatment of pulmonary arterial hypertension. These drugs inhibit PDE5, thereby increasing cGMP levels and causing smooth muscle cell relaxation and vasodilation. Likewise, the recently approved riociguat is a stimulator of soluble guanylate cyclase, which also leads to increased cGMP levels and the subsequent vasodilation. This drug is approved for treating chronic thromboembolic pulmonary hypertension and pulmonary arterial hypertension.

- Minoxidil is a K^+_{ATP} channel opener that causes vascular smooth muscle membrane hyperpolarization and thereby vasodilation. The drug is used to treat hypertension refractory to common antihypertensives.

- Other vasodilators, including hydralazine and fenoldopam, are primarily used to treat severe hypertensive conditions.

11.10   SELF-ASSESSMENT QUESTIONS

11.10.1 A 48-year-old male presents with angina pectoris. He uses one tablet (0.4 mg) of nitroglycerin sublingually when he experiences chest pain. Which of the following drugs may interact to cause serious hypotension if taken together?
   A. Diltiazem
   B. Lisinopril

C. Losartan

D. Metoprolol

E. Tadalafil

11.10.2 A 35-year-old female is diagnosed with essential hypertension. She has started on an antihypertensive medication, and after a few weeks, she notices rash over her face, joint aches, and fever. A serum antinuclear antibody test is positive. Which of the following is the most likely drug that the patient takes for her hypertension?

A. Bosentan

B. Hydralazine

C. Irbesartan

D. Propranolol

E. Sildenafil

11.10.3 A 65-year-old male presents with chronic thromboembolic pulmonary hypertension. Following surgical treatment, he is prescribed a drug that sensitizes soluble guanylate cyclase to nitric oxide-mediated activation. Which of the following is most likely prescribed?

A. Ambrisentan

B. Isosorbide dinitrate

C. Macitentan

D. Minoxidil

E. Riociguat

11.10.4 A 45-year-old female is diagnosed with chronic stable angina. She is prescribed a vasodilator that increases cGMP levels in vascular smooth muscle cells for the prophylaxis of acute anginal attack upon exertion. Which of the following is most likely prescribed?

A. Amlodipine

B. Isosorbide dinitrate

C. Macitentan

D. Riociguat

E. Sildenafil

11.10.5 A 45-year-old female with severe hypertension refractory to diuretic and beta-blocker therapy, is started on a drug that acts via a mechanism different from that of many other antihypertensives. After a period of therapy, the patient has noticed excessive hair growth. Which of the following drugs is most likely used for treating her hypertension?

A. Captopril

B. Fenoldopam

C. Hydralazine

D. Minoxidil

E. Sodium nitroprusside

REFERENCES

1 Guilluy, C., et al., The Rho exchange factor Arhgef1 mediates the effects of angiotensin II on vascular tone and blood pressure. *Nat Med*, 2010. 16(2): p. 183–90.

2 Bernstein, K.E. and S. Fuchs, Angiotensin II and JAK2 put on the pressure. *Nat Med*, 2010. 16(2): p. 165–6.

3 Zois, N.E., et al., Natriuretic peptides in cardiometabolic regulation and disease. *Nat Rev Cardiol*, 2014.11: p. 403–12

4 Mederos y Schnitzler, M., U. Storch, and T. Gudermann, AT1 receptors as mechanosensors. *Curr Opin Pharmacol*, 2011. 11(2): p. 112–6.

5 Lopez-Palop, R., et al., Comparison of effectiveness of high-dose intracoronary adenosine versus intravenous administration on the assessment of fractional flow reserve in patients with coronary heart disease. *Am J Cardiol*, 2013. 111(9): p. 1277–83.

6 Layland, J., et al., Adenosine: physiology, pharmacology, and clinical applications. *JACC Cardiovasc Interv*, 2014.7: p. 581–91.

7 Iachini Bellisarii, F., et al., Nitrates and other nitric oxide donors in cardiology: current positioning and perspectives. *Cardiovasc Drugs Ther*, 2012. 26(1): p. 55–69.

8 Zhou, R.H. and W.H. Frishman, The antiplatelet effects of nitrates: is it of clinical significance in patients with cardiovascular disease? *Cardiol Rev*, 2010. 18(4): p. 198–203.

9 Davlouros, P., et al., Platelet inhibition by IV glyceryl trinitrate in patients with stable coronary artery disease on dual antiplatelet therapy subjected to PCI. *Int J Cardiol*, 2013. 168(3): p. 3069–70.

10 Munzel, T., A. Daiber, and T. Gori, Nitrate therapy: new aspects concerning molecular action and tolerance. *Circulation*, 2011. 123(19): p. 2132–44.

11 Munzel, T., A. Daiber, and T. Gori, More answers to the still unresolved question of nitrate tolerance. *Eur Heart J*, 2013. 34(34): p. 2666–73.

12 Munzel, T. and T. Gori, Nitrate therapy and nitrate tolerance in patients with coronary artery disease. *Curr Opin Pharmacol*, 2013. 13(2): p. 251–9.

13 Nossaman, V.E., B.D. Nossaman, and P.J. Kadowitz, Nitrates and nitrites in the treatment of ischemic cardiac disease. *Cardiol Rev*, 2010. 18(4): p. 190–7.

14 Wakai, A., et al., Nitrates for acute heart failure syndromes. *Cochrane Database Syst Rev*, 2013. 8: p. CD005151.

15 Cole, R.T., et al., Hydralazine and isosorbide dinitrate in heart failure: historical perspective, mechanisms, and future directions. *Circulation*, 2011. 123(21): p. 2414–22.

16 Ferdinand, K.C., et al., Use of isosorbide dinitrate and hydralazine in African-Americans with heart failure 9 years after the African-American heart failure trial. *Am J Cardiol*, 2014. 114: p. 151–9.

17 Gupta, D., et al., Nitrate therapy for heart failure: benefits and strategies to overcome tolerance. *JACC Heart Fail*, 2013. 1(3): p. 183–91.

18 Carlson, M.D. and P.M. Eckman, Review of vasodilators in acute decompensated heart failure: the old and the new. *J Card Fail*, 2013. 19(7): p. 478–93.

19 Jandeleit-Dahm, K.A. and A.M. Watson, The endothelin system and endothelin receptor antagonists. *Curr Opin Nephrol Hypertens*, 2012. 21(1): p. 66–71.

20 Rautureau, Y. and E.L. Schiffrin, Endothelin in hypertension: an update. *Curr Opin Nephrol Hypertens*, 2012. 21(2): p. 128–36.

21 Pulido, T., et al., Macitentan and morbidity and mortality in pulmonary arterial hypertension. *N Engl J Med*, 2013. 369(9): p. 809–18.

22 Raghu, G., et al., Treatment of idiopathic pulmonary fibrosis with ambrisentan: a parallel, randomized trial. *Ann Intern Med*, 2013. 158(9): p. 641–9.

23 Maurice, D.H., et al., Advances in targeting cyclic nucleotide phosphodiesterases. *Nat Rev Drug Discov*, 2014. 13(4): p. 290–314.

24 Omori, K. and J. Kotera, Overview of PDEs and their regulation. *Circ Res*, 2007. 100(3): p. 309–27.

25 Stafstrom, C.E., S. Grippon, and P. Kirkpatrick, Ezogabine (retigabine). *Nat Rev Drug Discov*, 2011. 10(10): p. 729–30.

12

MANAGEMENT OF HYPERTENSION: PRINCIPLES AND GUIDELINES

12.1 OVERVIEW

The treatment of hypertension has been one of medicine's major successes of the past half century. The preceding chapters have discussed the molecular pharmacology of various classes of drugs that are effective for treating hypertension as well as other forms of cardiovascular diseases. This lays a basis for understanding the principles of the management of hypertension. This chapter emphasizes systemic arterial hypertension. The chapter begins with an introduction to the current guidelines for the management of hypertension. Then, it focuses on discussing the JNC7 guideline, which is widely considered as a standard of management of hypertension, as well as the newly released JNC8 guideline, which has received criticism. The management of hypertension is discussed in two broad aspects: (i) the principles of management applied to the general patient population and (ii) the principles of management related to specific patient groups or conditions. In addition, the management of pulmonary arterial hypertension, a relative rare form of hypertension, is also covered.

12.2 INTRODUCTION TO CURRENT GUIDELINES FOR THE MANAGEMENT OF SYSTEMIC HYPERTENSION

Management of hypertension involves two components: (i) lifestyle modifications and (ii) drug therapy. Hypertension management continues to evolve as newer antihypertensive

medications, clinical trial data, and evidence-based guidelines become available. Indeed, guidelines from professional organizations have played an important role in the effective management of diseases, including hypertension. In this regard, several current guidelines on hypertension management are available from various organizations, which are listed below and briefly described in this section:

- Joint National Committee on Prevention, Detection, Evaluation, and Treatment of High Blood Pressure (JNC) guidelines
- American Heart Association (AHA) guidelines
- Canadian Hypertension Education Program (CHEP) guidelines
- British Hypertension Society (BHS) guidelines
- British National Institute for Health and Care Excellence (NICE) guidelines
- European Society of Hypertension (ESH)/European Society of Cardiology (ESC) guidelines

12.2.1 The JNC Guidelines

The major guidelines on hypertension management in the United States are issued by the JNC, a group coordinated by the National Heart, Lung, and Blood Institute (NHLBI). The first report, known as JNC1, was published in 1976. The seventh report, known as JNC7, is considered the most comprehensive guideline on hypertension management, which was published in 2003. The long-awaited JNC8 guideline was released in early 2014 [1]. Listed below is the history of the various versions of the JNC guidelines:

- JNC8: published in 2014
- JNC7: published in 2003
- JNC6: published in 1997
- JNC5: published in 1992
- JNC4: published in 1988
- JNC3: published in 1984
- JNC2: published in 1980
- JNC1: published in 1976

12.2.2   The Guidelines from the AHA and Collaborating Organizations

Several recent documents have been developed by the AHA alone or jointly with the American College of Cardiology Foundation (ACCF) or the American Medical Association (AMA). These documents provide recommendations or consensus statements on hypertension management under various conditions. These documents include the following:

- "Dietary Approaches to Prevent and Treat Hypertension," published in 2006 [2]
- "Treatment of Hypertension in the Prevention and Management of Ischemic Heart Disease," published in 2007 (also known as the AHA 2007 guideline) [3]
- "ACCF/AHA/AMA-PCPI 2011 Performance Measures for Adults with Coronary Artery Disease and Hypertension" [4]
- "ACCF/AHA 2011 Expert Consensus Document on Hypertension in the Elderly" [5]
- "AHA/ACC/CDC Science Advisory: An Effective Approach to High Blood Pressure Control," published in 2014 [6]

While the above US-based documents are largely in line with the aforementioned JNC7 guideline published in 2003, they also provide new recommendations based on new evidence from clinical trials on hypertension management.

12.2.3   The American Society of Hypertension/ International Society of Hypertension 2014 Guideline

High blood pressure (BP) affects approximately one in three adults in the Americas, Europe, some Asian countries, and Australia and one billion people worldwide. Because of this epidemic, the American Society of Hypertension (ASH) and the International Society of Hypertension (ISH) have recently jointly created the first-of-their-kind guideline for

the diagnosis and treatment of hypertension, entitled "Clinical Practice Guidelines for the Management of Hypertension in the Community" [7]. This is the first guideline to be usable for medical practitioners in any socioeconomic environment around the globe, from those countries with state-of-the-art equipment to those that lack basic resources. Notably, the guideline is designed with guidance that is easy to implement for doctors and healthcare professionals in even the most impoverished areas. The guideline is divided into 15 sections, as listed below:

1. Introduction
2. Epidemiology
3. Special Issues with Black Patients (African Ancestry)
4. How Is Hypertension Defined?
5. How Is Hypertension Classified?
6. Causes of Hypertension
7. Making the Diagnosis of Hypertension
8. Evaluating the Patient
9. Physical Examination
10. Tests
11. Goals of Treating Hypertension
12. Nonpharmacological Treatment of Hypertension
13. Drug Treatment for Hypertension
14. Brief Comments on Drug Classes
15. Treatment-Resistant Hypertension

12.2.4   The CHEP Guidelines

In an effort to improve the treatment and control of hypertension in Canada, the CHEP was formed in 1999. For the last 15 years, the CHEP annually updates evidence-based hypertension management recommendations, and the 2014 CHEP recommendations have been published [8].

12.2.5   The BHS Guidelines

The BHS has published guidelines for the management of hypertension in 1989, 1993, 1999, and 2004. The BHS 2004 guidelines, also known as BHS IV [9], represent the most recent recommendations for the management of hypertension in the United Kingdom.

12.2.6   The British NICE Guidelines

The British NICE released its current guideline on clinical management of primary hypertension in adults in 2011 (NICE clinical guideline 127). This guideline was developed in collaboration with the BHS and represents the most authoritative source of guide for hypertension management in the United Kingdom [10].

12.2.7 The ESH/ESC Guidelines

First produced in 2003, the original version of the joint ESH/ESC guidelines for the management of arterial hypertension became one of the most highly cited medical papers in the world. The 2013 guideline, which replaces the 2007 edition, gives state-of-the-science recommendations that show how the hypertension landscape has changed and indicates what needs to be done to reduce mortality and morbidity from high BP and associated conditions [11]. A major development of the 2013 guideline is the decision to recommend a single systolic BP target of 140 mmHg for almost all patients. This contrasts with the 2007 version that recommended a 140/90 mmHg target for moderate- to low-risk patients and 130/80 mmHg target for high-risk patients. This new recommendation is based on the lack of enough evidence to justify the above two targets of BP management. This change is also reflected in the JNC8 guideline [1].

12.2.8 Summary

Although the guidelines described earlier have been developed by different organizations and in different countries, the major components of management are similar, which include lifestyle modifications and drug therapy. The rest of this chapter describes these two components of hypertension management with a focus on the US-based guidelines, including the JNC7 report, the AHA 2007 guideline, the AHA/ACC/CDC 2014 Science Advisory, as well as the newly released controversial JNC8 report. Relevant data from the most recent clinical trials on hypertension management and recommendations from non-US-based organizations are also discussed with regard to the consensus and key differences. It should be stressed that effective management of hypertension not only depends on a thorough understanding of both the current guidelines and the results from the most recent clinical trials, but also the responsible physician's judgment.

12.3 KEY RECOMMENDATIONS OF MAJOR GUIDELINES FOR THE MANAGEMENT OF SYSTEMIC HYPERTENSION

Before discussing the lifestyle modifications and drug therapy of hypertension management recommended by JNC7, AHA 2007, and JNC8 guidelines, it is necessary to introduce the key recommendations and algorithms of these three major guidelines on hypertension management in the United States. This sets a stage for the subsequent discussion of the principles of the management of hypertension.

12.3.1 Key Recommendations of the JNC7 Guideline

The key recommendations from the JNC7 report are the following [12]:

- In persons older than 50 years, systolic BP of more than 140 mmHg is a much more important cardiovascular disease risk factor than diastolic blood pressure (DBP).
- The risk of cardiovascular diseases, beginning at 115/75 mmHg, doubles with each increment of 20/10 mmHg; individuals who are normotensive at 55 years of age have a 90% lifetime risk for developing hypertension.
- Individuals with a systolic BP of 120–139 mmHg or a DBP of 80–89 mmHg should be considered as prehypertensive and require health-promoting lifestyle modifications to prevent cardiovascular diseases.
- Thiazide-type diuretics should be used in drug treatment for most patients with uncomplicated hypertension, either alone or combined with drugs from other classes. Certain high-risk conditions are compelling indications for the initial use of other antihypertensive drug classes (angiotensin-converting enzyme inhibitors (ACEIs), angiotensin receptor blockers (ARBs), β-blockers, calcium channel blockers (CCBs)).
- Most patients with hypertension will require two or more antihypertensive medications to achieve goal BP (<140/90 mmHg or <130/80 mmHg for patients with diabetes or chronic kidney disease (CKD)).
- If BP is more than 20/10 mmHg above goal BP, consideration should be given to initiating therapy with two agents, one of which usually should be a thiazide or thiazide-type diuretic.
- The most effective therapy prescribed by the most careful clinician will control hypertension only if patients are motivated. Motivation improves when patients have positive experiences with and trust in the clinician. Empathy builds trust and is a potent motivator. Finally, in presenting these guidelines, the committee recognizes that the responsible physician's judgment remains paramount.

12.3.2 Key Recommendations of the AHA 2007 Guideline

Guidelines from the AHA on hypertension management include the aforementioned "Dietary Approaches to Prevent and Treat Hypertension," published in 2006, and the AHA 2007 guideline "Treatment of Hypertension in the Prevention and Management of Ischemic Heart Disease." Since the lifestyle modifications are addressed separately in the "Dietary Approaches to Prevent and Treat Hypertension," the AHA

2007 guideline mainly discusses drug therapy. The main recommendations of the AHA 2007 guideline [3, 13] are outlined below and also summarized in Table 12.1:

- For most adults with hypertension, the BP goal is <140/90 mmHg but should be <130/80 mmHg in patients with diabetes mellitus, CKD, known coronary artery disease (CAD), CAD equivalents (carotid artery disease, abdominal aortic aneurysm, and peripheral vascular disease), or 10-year Framingham risk score of

≥10%. For those with left ventricular dysfunction, the recommended BP target is <120/80 mmHg.

- For primary CAD prevention, any effective antihypertensive drug or combination is indicated, but preference is given to ACEIs, ARBs, CCBs, and thiazide diuretics.
- For the management of hypertension in patients with established CAD (stable or unstable angina, non-ST-segment elevation myocardial infarction (NSTEMI), ST-segment elevation myocardial infarction (STEMI)),

TABLE 12.1   Summary of the main recommendations of the AHA 2007 guideline[a]

Area of concern	BP target (mmHg)	LM	Specific drug indications	Comments
General CAD prevention	<140/90	Yes	Any effective antihypertensive drug or combination; evidence supports ACEI[c] (or ARB), CCB, or thiazide diuretic as first-line therapy	If SBP ≥160 mmHg or DBP ≥100 mmHg, then start with 2 drugs
High CAD risk[b]	<130/80	Yes	ACEI or ARB or CCB or thiazide diuretic or combination	IF SBP ≥160 mmHg or DBP ≥100 mmHg, then start with 2 drugs
Stable angina	<130/80	Yes	β-blocker and ACEI or ARB	If β-blocker contraindicated or if side effects occur, can substitute diltiazem or verapamil (but not if bradycardia or LVD is present) Can add dihydropyridine CCB (not diltiazem or verapamil) to β-blocker A thiazide diuretic can be added for BP control
UA/NSTEMI	<130/80	Yes	β-Blocker (if patient is hemodynamically stable) and ACEI or ARB (if anterior MI is present, hypertension persists, LVD or heart failure is present, or the patient has diabetes mellitus)	If β-blocker contraindicated or if side effects occur, can substitute diltiazem or verapamil (but not if bradycardia or LVD is present) Can add dihydropyridine CCB (not diltiazem or verapamil) to β-blocker A thiazide diuretic can be added for BP control
STEMI	<130/80	Yes	β-Blocker (if patient is hemodynamically stable) and ACEI or ARB (if anterior MI is present, hypertension persists, LVD or heart failure is present, or the patient has diabetes mellitus)	If β-blocker contraindicated or if side effects occur, can substitute diltiazem or verapamil (but not if bradycardia or LVD is present) Can add dihydropyridine CCB (not diltiazem or verapamil) to β-blocker A thiazide diuretic can be added for BP control
LVD	<120/80	Yes	ACEI or ARB and β-blocker and aldosterone antagonist (if severe heart failure is present) and thiazide or loop diuretic and hydralazine/isosorbide dinitrate (patients of African heritage)	Contraindicated: verapamil, diltiazem, clonidine, moxonidine, β-blocker

[a] Modified from Ref. [3].
[b] High CAD risk includes diabetes mellitus, CKD, known CAD or CAD equivalent (carotid artery disease, peripheral arterial disease, abdominal aortic aneurysm), or 10-year Framingham risk score ≥10%.
[c] ACEI, angiotensin-converting enzyme inhibitor; ARB, angiotensin receptor blocker; BP, blood pressure; CAD, coronary artery disease; CCB, calcium channel blocker; DBP, diastolic blood pressure; LM, lifestyle modifications; LVD, left ventricular dysfunction; MI, myocardial infarction; NSTEMI, non-ST-segment elevation myocardial infarction; SBP, systolic blood pressure; STEMI, ST-segment elevation myocardial infarction; UA, unstable angina.

β-blockers and ACEIs (or ARBs) are the basis of treatment. If further BP lowering is needed, a thiazide diuretic and/or a dihydropyridine CCB (not verapamil or diltiazem) can be added. If a β-blocker is contraindicated or not tolerated, diltiazem or verapamil can be substituted.

- If there is left ventricular dysfunction, recommended therapy consists of an ACEI or ARB, a β-blocker, and either a thiazide or loop diuretic. In patients with more severe heart failure, an aldosterone antagonist and hydralazine/isosorbide dinitrate (in patients of African heritage) should be considered.

12.3.3   Key Points of the 2014 AHA/ACC/CDC Science Advisory

This advisory is intended to provide updated high-quality evidence-based data to complement and support current clinical guidelines on hypertension management, providing clinicians and health systems tools to improve the treatment and control of hypertension. The key points of the advisory include the following:

- The BP goal for an individual is set by utilizing a combination of factors including scientific evidence, clinical judgment, and patient tolerance.
- For most people, the BP goal is <140/90 mmHg; however, lower targets may be appropriate for some populations such as people of African heritage, the elderly, or patients with left ventricular hypertrophy, systolic or diastolic left ventricular dysfunction, diabetes mellitus, or CKD.
- Lifestyle modifications should be initiated in all patients with hypertension, and they should be assessed for target organ damage and existing cardiovascular diseases.
- Self-monitoring is encouraged for most patients throughout their care, and requesting and reviewing readings from home and community settings can help the provider assist the patient in achieving and maintaining good control.
- For patients with hypertension on combination with certain clinical conditions, specific medications should be considered first-line treatments (Table 12.2).

12.3.4   Key Recommendations of the JNC8 Report

The recently released JNC8 report focuses on three main questions: (i) when to begin treatment, (ii) how low to aim for BP control, and (iii) which antihypertensive medications to use. It does not cover many topics that were included in the JNC7 report. Indeed, the JNC7 report is considered the

TABLE 12.2   Suggested medications for the treatment of hypertension in the presence of certain medical conditions [6]

Condition	First-line drugs
Coronary artery disease/ postmyocardial infarction	β-Blocker, ACEI
Systolic heart failure	ACEI or ARB, β-blocker, aldosterone antagonist, thiazide
Diastolic heart failure	ACEI or ARB, β-blocker, thiazide
Diabetes mellitus	ACEI or ARB, thiazide, β-blocker, CCB
Kidney disease	ACEI or ARB
Stroke or transient ischemic attack	Thiazide, ACEI

most comprehensive guideline on hypertension management. Compared with the JNC7 report, the JNC8 guideline is more evidence based and focused and outlines a management strategy that is simpler and, in some instances, less aggressive [1, 14]. The key points of the JNC8 guideline are outlined below, and its nine recommendations are given in Table 12.3:

- In patients aged 60 or older, the JNC8 guideline recommends starting antihypertensive treatment if the BP is ≥150/90 mmHg or higher, with a goal of <150/90 mmHg.
- For everyone else, including people with diabetes or CKD, the threshold is 140/90 mm Hg, and the goal is <140/90 mmHg.
- The recommended classes of drugs for initial therapy in nonblack patients without CKD are thiazide-type diuretics, CCBs, ACEIs, and ARBs, although the last two classes (i.e., ACEIs and ARBs) should not be used in combination. The JNC8 report does not recommend β-blockers for the initial treatment because of increased risk of stroke.
- For black patients, the initial classes of drugs are diuretics and CCBs; patients with CKD should receive an ACEI or ARB.

As noted earlier, the JNC8 guideline is much more focused than the JNC7 report and does not include recommendations on lifestyle modifications in the guideline. Regardless of this omission, the JNC8 report recognizes that the potential benefits of a healthy diet, weight control, and regular exercise cannot be overemphasized, and it supports the 2013 AHA/ACC guideline on lifestyle management to reduce cardiovascular risk [15]. Section 12.4 describes the major recommendations of the JNC7 and the 2013 AHA/ACC on lifestyle modifications with regard to hypertension management.

TABLE 12.3   The nine recommendations of the JNC8 report [1]

	Recommendation	Strength of recommendation[a]
1	In the general population aged ≥60 years, initiate pharmacological treatment to lower blood pressure (BP) at systolic blood pressure (SBP) of ≥150 mmHg or diastolic blood pressure (DBP) of ≥90 mmHg and treat to a goal SBP of <150 mmHg and goal DBP of <90 mmHg	Grade A (strong)
	In the general population aged ≥60 years, if pharmacological treatment for high BP results in lower achieved SBP (e.g., <140 mmHg) and treatment is well tolerated and without adverse effects on health or quality of life, treatment does not need to be adjusted	Grade E (expert opinion)
2	In the general population aged <60 years, initiate pharmacological treatment to lower BP at DBP of ≥90 mmHg and treat to a goal DBP of <90 mmHg	Grade A (strong) for ages 30–59 years); grade E (expert opinion) for ages 18–29
3	In the general population aged <60 years, initiate pharmacological treatment to lower BP at SBP of ≥140 mmHg and treat to a goal SBP of <140 mmHg	Grade E (expert opinion)
4	In the population aged ≥18 years with chronic kidney disease (CKD), initiate pharmacological treatment to lower BP at SBP of ≥140 mmHg or DBP of ≥90 mmHg and treat to goal SBP of <140 mmHg and goal DBP of <90 mmHg	Grade E (expert opinion)
5	In the population aged ≥18 years with diabetes, initiate pharmacological treatment to lower BP at SBP of ≥140 mmHg or DBP of ≥90 mmHg and treat to goal SBP of <140 mmHg and goal DBP of <90 mmHg	Grade E (expert opinion)
6	In the general nonblack population, including those with diabetes, initial antihypertensive treatment should include a thiazide-type diuretic, CCB, ACEI, or ARB	Grade B (moderate)
7	In the general black population, including those with diabetes, initial antihypertensive treatment should include a thiazide-type diuretic or CCB	Grade B (moderate) for general black population; grade C (weak) for black patients with diabetes mellitus
8	In the population aged ≥18 years with CKD and hypertension, initial (or add-on) antihypertensive treatment should include an ACEI or ARB to improve kidney outcomes. This applies to all CKD patients with hypertension regardless of race or diabetes status	Grade B (moderate)
9	The main objective of hypertension treatment is to attain and maintain goal BP. If goal BP is not reached within a month of treatment, increase the dose of the initial drug or add a second drug from one of the classes in recommendation 6 (thiazide-type diuretic, CCB, ACEI, or ARB). The clinician should continue to assess BP and adjust the treatment regimen until goal BP is reached. If goal BP cannot be reached with 2 drugs, add and titrate a third drug from the list provided. Do not use an ACEI and an ARB together in the same patient. If goal BP cannot be reached using the drugs in recommendation 6 because of a contraindication or the need to use more than 3 drugs to reach goal BP, antihypertensive drugs from other classes can be used. Referral to a hypertension specialist may be indicated for patients in whom goal BP cannot be attained using the above strategy or for the management of complicated patients for whom additional clinical consultation is needed	Grade E (expert opinion)

ACEI, angiotensin-converting enzyme inhibitor; ARB, angiotensin receptor blocker; BP, blood pressure; CCB, calcium channel blocker; DBP, diastolic blood pressure; SBP, systolic blood pressure.

[a] Refer to Chapter 5 for description of the grading system of the strength of recommendation.

12.4 LIFESTYLE MODIFICATIONS FOR THE MANAGEMENT OF SYSTEMIC HYPERTENSION

12.4.1 General Considerations

Adoption of healthy lifestyles by all persons is critical for the prevention of high BP and is an indispensable part of the management for those with hypertension. Lifestyle modifications are one of the major components of the various guidelines for the management of hypertension. Lifestyle modifications are also recommended for all individuals with prehypertension. The overall benefits of lifestyle modifications include the following:

- Reduction of BP
- Prevention and delay of the development of hypertension
- Enhancement of the efficacy of the antihypertensive medications
- Reduction of cardiovascular risk

12.4.2 Recommendations

The lifestyle modifications recommended in the JNC7 guideline [16] are summarized in Table 12.4. The dietary approaches to preventing and treating hypertension recommended by the AHA in 2006 are outlined in Table 12.5 [2]. The DASH eating plan is provided in Table 12.6. The newly released 2013 AHA/ACC guideline on lifestyle management to reduce cardiovascular risk (including hypertension) is given in Table 12.7.

Since DASH diet is an important component of lifestyle modifications, here, it is necessary to introduce briefly the DASH eating plan. Scientists supported by the NHLBI conducted two key studies in the late 1990s and early 2000s [18, 19]. Their findings showed that BP was reduced with an eating plan that is low in saturated fat, cholesterol, and total fat and that emphasizes fruits, vegetables, and fat-free or low-fat milk and milk products. This eating plan, known as the DASH eating plan, also includes whole grain products, fish, poultry, and nuts. It is reduced in lean red meat, sweets, added sugars, and sugar-containing beverages compared to

TABLE 12.4 The JNC7 lifestyle modifications to prevent and manage hypertension [16][a]

Modification	Recommendation	Approximate SBP reduction (range)[b]
Weight reduction	Maintain normal body weight (body mass index 18.5–24.9 kg/m²)	5–20 mmHg/10 kg
Adopt DASH eating plan	Consume a diet rich in fruits, vegetables, and low-fat dairy products with a reduced content of saturated and total fat	8–14 mmHg
Dietary sodium reduction	Reduce dietary sodium intake to no more than 100 mmol/day (2.4 g sodium or 6 g sodium chloride)	2–8 mmHg
Physical activity	Engage in regular aerobic physical activity, such as brisk walking (at least 30 min/day, most days of the week)	4–9 mmHg
Moderation of alcohol consumption	Limit consumption to no more than 2 drinks[c] (e.g., 24 oz beer, 10 oz wine, or 3 oz 80-proof whiskey)/day in most men and to no more than 1 drink/day in women and lighter-weight persons	2–4 mmHg

[a] For overall cardiovascular risk reduction, stop smoking.
[b] The effects of implementing these modifications are dose and time dependent and could be greater for some individuals.
[c] One drink is equivalent to 12 oz of beer, 5 oz of wine, or 1.5 oz of 80-proof liquor, each representing on an average 14 g (or 18 ml) of pure ethanol.
DASH, Dietary Approaches to Stop Hypertension; SBP, systolic blood pressure.

TABLE 12.5 The AHA-recommended diet-related lifestyle modifications that effectively lower blood pressure [2]

Lifestyle modification	Recommendation
Weight loss	For overweight or obese persons, lose weight, ideally attaining a BMI ≤25 kg/m²; for nonoverweight persons, maintain desirable BMI ≤25 kg/m²
Reduced salt intake	Lower salt (sodium chloride) intake as much as possible, ideally to ~65 mmol/day sodium (corresponding to 1.5 g/day of sodium or 3.8 g/day sodium chloride)
DASH-type dietary patterns	Consume a diet rich in fruits and vegetables (8–10 servings/day), rich in low-fat dairy products (2–3 servings/day), and reduced in saturated fat and cholesterol
Increased potassium intake	Increase potassium intake to 120 mmol/day (4.7 g/day), which is also the level provided in DASH-type diets
Moderation of alcohol intake	For those who drink alcohol, consume ≤2 alcoholic drinks/day (men) and ≤1 alcoholic drink/day (women)

BMI, body mass index; DASH, Dietary Approaches to Stop Hypertension.

TABLE 12.6 The DASH eating plan developed by the NHLBI[a]

Food group	Daily servings	Serving sizes	Examples and notes	Significance of each food group to DASH eating pattern
Grains[b]	6–8	1 slice bread 1 oz dry cereal[c] 1/2 cup cooked rice, pasta, or cereal	Whole wheat bread and rolls, whole wheat pasta, English muffin, pita bread, bagel, cereals, grits, oatmeal, brown rice, unsalted pretzels and popcorn	Major sources of energy and fiber
Vegetables	4–5	1 cup raw leafy vegetable 1/2 cup cut-up raw or cooked vegetable 1/2 cup vegetable juice	Broccoli, carrots, collards, green beans, green peas, kale, lima beans, potatoes, spinach, squash, sweet potatoes, tomatoes	Rich sources of potassium, magnesium, and fiber
Fruits	4–5	1 medium fruit 1/4 cup dried fruit 1/2 cup fresh, frozen, or canned fruit 1/2 cup fruit juice	Apples, apricots, bananas, dates, grapes, oranges, grapefruit, grapefruit juice, mangoes, melons, peaches, pineapples, raisins, strawberries, tangerines	Important sources of potassium, magnesium, and fiber
Fat-free or low-fat milk and milk products	2–3	1 cup milk or yogurt 11/2 oz cheese	Fat-free (skim) or low-fat (1%) milk or buttermilk; fat-free, low-fat, or reduced-fat cheese; fat-free or low-fat regular or frozen yogurt	Major sources of calcium and protein
Lean meats, poultry, and fish	6 or less	1 oz cooked meats, poultry, or fish 1 egg[d]	Select only lean meat; trim away visible fats; broil, roast, or poach; remove skin from poultry	Rich sources of protein and magnesium
Nuts, seeds, and legumes	4–5 per week	1/3 cup or 11/2 oz nuts 2 Tbsp peanut butter 2 Tbsp or 1/2 oz seeds 1/2 cup cooked legumes (dry beans and peas)	Almonds, hazelnuts, mixed nuts, peanuts, walnuts, sunflower seeds, peanut butter, kidney beans, lentils, split peas	Rich sources of energy, magnesium, protein, and fiber
Fats and oils[e]	2–3	1 tsp soft margarine 1 tsp vegetable oil 1 Tbsp mayonnaise 2 Tbsp salad dressing	Soft margarine, vegetable oil (such as canola, corn, olive, or safflower), low-fat mayonnaise, light salad dressing	The DASH study had 27% of calories as fat, including fat in or added to foods
Sweets and added sugars	5 or less per week	1 Tbsp sugar 1 Tbsp jelly or jam 1/2 cup sorbet, gelatin 1 cup lemonade	Fruit-flavored gelatin, fruit punch, hard candy, jelly, maple syrup, sorbet and ices, sugar	Sweets should be low in fat

[a]The DASH eating plan shown in the table is based on 2000 cal a day. The number of daily servings in a food group may vary from those listed depending on the individual's caloric needs.

[b]Whole grains are recommended for most grain servings as a good source of fiber and nutrients.

[c]Serving sizes vary between 1/2 cup and 11/4 cups, depending on cereal type. Check the product's Nutrition Facts label.

[d]Since eggs are high in cholesterol, limit egg yolk intake to no more than 4 per week; two egg whites have the same protein content as 1 oz of meat.

[e]Fat content changes serving amount for fats and oils. For example, 1 Tbsp of regular salad dressing equals one serving, 1 Tbsp of a low-fat dressing equals one half serving, and 1 Tbsp of a fat-free dressing equals zero servings.

oz, ounce; tsp, teaspoon; Tbsp., tablespoon.

TABLE 12.7   The 2013 AHA/ACC recommendations on lifestyle management to reduce blood pressure [17]

Component	Recommendation	NHLBI grade	AHA/ACC COR, LOE

Diet	1. Consume a dietary pattern that emphasizes intake of vegetables, fruits, and whole grains; includes low-fat dairy products, poultry, fish, legumes, nontropical vegetable oils, and nuts; and limit intake of sweets, sugar-sweetened beverages, and red meats:
	   i. Adapt this dietary pattern to appropriate calorie requirements, personal and cultural food preferences, and nutrition therapy for other medical conditions (including diabetes mellitus)
	   ii. Achieve this pattern by following plans such as the DASH dietary pattern, the USDA Food Patterns[a], or the AHA Diet
		A (strong)	I, A

	2. Lower sodium intake		
	3. Detailed approach to sodium control includes that (1) the individual should consume no more than 2400 mg of sodium/day, (2) further reduction of sodium intake to 1500 mg/day is desirable since it is associated with even greater reduction in blood pressure, and (3) the individual should reduce intake by at least 1000 mg/day since that will lower blood pressure, even if the desired daily sodium is not yet achieved	B (moderate)	IIa, B
	4. Combine the DASH dietary pattern with lower sodium intake	A (strong)	I, A
Physical activity	In general, advise adults to engage in aerobic physical activity to lower blood pressure: 3–4 sessions a week, lasting on average 40 min/session and involving moderate to vigorous intensity physical activity	B (moderate)	IIa, A

[a]The USDA Food Patterns were developed to help individuals carry out dietary guideline recommendations. They identify daily amounts of foods, in nutrient-dense forms, to eat from five major food groups and their subgroups. The patterns also include an allowance for oils and limits on the maximum number of calories that should be consumed from solid fats and added sugars (http://www.cnpp.usda.gov/USDAfoodpatterns.htm; accessed on June 14, 2014).
COR and LOE denote class of recommendation and level of evidence, respectively (see Chapter 5 for description).

the typical American diet. It is rich in potassium, magnesium, and calcium, as well as protein and fiber. See Table 12.6 for the DASH eating plan developed by the NHLBI.

12.4.3   Mechanisms

The role of the lifestyle modifications discussed earlier in reduction of BP and in preventing or delaying the development of essential systemic hypertension has been extensively documented in epidemiological studies and clinical trials. The three cornerstones of lifestyle modifications for the treatment of hypertension, namely, weight loss/reduced body fat, a healthful dietary pattern (DASH-type diet), and reduced sodium intake, influence the pathophysiology of hypertension at many of its points of control, as listed below [20]:

- Weight loss and reduced body fat result in decreased sympathetic nervous system activity and angiotensin II levels.
- Weight loss, low sodium intake, and healthy diet reduce stiffness of larger conduit arteries.
- Weight loss, low sodium intake, and healthy diet improve function of resistance vessels and decrease peripheral resistance.
- Healthy diet improves renal sodium excretion.

12.5   DRUG THERAPY OF SYSTEMIC HYPERTENSION

12.5.1   General Considerations of Drug Therapy

As described previously, the commonly used antihypertensive medications include four classes of drugs, listed below:

1. Diuretics
2. Sympatholytics, especially β-blockers
3. Inhibitors of the renin–angiotensin–aldosterone system (RAAS)
4. Direct vasodilators, especially CCBs

This section discusses the general principles involved in the use of each of the above drug classes for the treatment of systemic hypertension.

12.5.1.1   Diuretics for Treating Hypertension

Mechanistic Basis   Diuretics that can be used to treat hypertension include thiazides and related compounds (collectively, these are conventionally called thiazide diuretics), loop diuretics, and potassium-sparing diuretics. Thiazides and related compounds are the most commonly used diuretics for long-term therapy of hypertension.

Diuretics lower BP by increasing urinary sodium excretion and reducing blood volume as well as other potential mechanisms (e.g., decrease of peripheral arterial resistance). The initial treatment with diuretics results in decreased plasma volume and reduced cardiac output, but these changes return toward normal within 6–8 weeks of treatment. At this point and beyond, the reduction of BP is related to a decline in peripheral arterial resistance.

Clinical Evidence and Recommendations Thiazide diuretics became available in the late 1950s and were the first effective oral antihypertensive agents with an acceptable adverse effect profile. More than a half century later, thiazide diuretics remain important medications for the treatment of hypertension. Extensive clinical studies demonstrate that these drugs reduce BP as monotherapy, enhance the efficacy of other antihypertensive medications, and reduce the hypertension-related morbidity and mortality (see Chapter 7).

The JNC7 guideline recommends that a thiazide diuretic be used for the initial therapy for most patients with stage I hypertension and without compelling indications. According to the JNC7 guideline, an ACEI, an ARB, a β-blocker, a CCB, or combination may also be considered for the initial treatment of stage I hypertension without compelling indications. For most patients with stage II hypertension, a thiazide diuretic is combined with another antihypertensive agent (an ACEI, an ARB, a β-blocker, or a CCB) to achieve more effective control of high BP.

Due to evidence supporting the efficacy of other drug classes in treating hypertension, the AHA 2007 guideline recommends the use of an ACEI, an ARB, a CCB, a thiazide diuretic, or a two-drug combination as first-line therapy for hypertensive patients without compelling indications. It should be noted that for treating hypertensive patients without compelling indications, the use of a β-blocker is recommended by the JNC7 guideline, but not by the AHA 2007 guideline or JNC8 report. This is because new clinical evidence showed limited efficacy of β-blockers in treating hypertensive patients without compelling indications (see Section 12.5.1.2 for more discussion).

New Developments Thiazide diuretics are most commonly prescribed for treating hypertension both as monotherapy and more frequently in combination with other antihypertensive agents. Most patients with mild-to-moderate hypertension respond to the lower doses of the various thiazide diuretics. In this context, hydrochlorothiazide in doses of 12.5–25 mg has been the overwhelming choice. It is also the thiazide diuretic combined with various ACEIs, ARBs, and β-blockers, as well as the renin inhibitor aliskiren, with the exception of three other drugs—atenolol, clonidine, and captopril—that are combined with chlorthalidone. However, hydrochlorothiazide in these doses has not been shown to reduce morbidity and mortality. In contrast, chlorthalidone in doses of 12.5–25 mg has shown benefits in multiple clinical trials.

Due to the favorable clinical trial data, chlorthalidone is now being recommended as an appropriate thiazide-type diuretic for treating hypertension [21]. Pharmacologically, chlorthalidone causes more potent and prolonged BP-lowering effects than hydrochlorothiazide. A recent retrospective cohort analysis concludes that chlorthalidone decreases BP and reduces cardiovascular events more than does hydrochlorothiazide, suggesting that chlorthalidone may be the preferred thiazide-type diuretic for hypertension in patients at high risk of cardiovascular events [22].

Combination of a thiazide diuretic and an ACEI, an ARB, a β-blocker, or a CCB is recommended by the JNC7 guideline should more aggressive treatment of hypertension be desired. However, recently, the ACCOMPLISH trial reported that benazepril (an ACEI)/amlodipine (a CCB) combination was superior to the benazepril/hydrochlorothiazide combination in reducing cardiovascular events in patients with hypertension who were at high risk for such events [23]. It is suggested that due to its superior effects, chlorthalidone should be utilized preferably in combination with other antihypertensive agents for treating hypertension, especially in patients at high risk of cardiovascular events [24].

12.5.1.2 Sympatholytics for Treating Hypertension

Mechanistic Basis Sympatholytics lower BP by reducing peripheral resistance, decreasing cardiac contractility and rate (hence decreased cardiac output), and increasing venous pooling in capacitance vessels (hence decreased cardiac preload, leading to reduced cardiac output) (see Chapter 8). In addition, blockage of β_1-adrenergic receptors decreases the release of renin and the subsequent activity of the RAAS (see Section 12.5.1.3 for more discussion). Among the various types of sympatholytic agents, β-blockers have been most widely used to treat hypertension.

Clinical Evidence and Recommendations β-Blockers are specifically recommended for treating hypertension in patients with concomitant coronary heart disease (particularly after a myocardial infarction), congestive heart failure, or tachyarrhythmias (more discussion in Section 12.4.2). Indeed, β-blocker administration remains a standard of care in patients with angina pectoris, those who have had a myocardial infarction, and those who have left ventricular dysfunction with or without heart failure symptoms, unless contraindicated.

However, in hypertensive patients who do not have symptomatic coronary heart disease, have not had a myocardial infarction, or do not have heart failure, the evidence for β-blocker cardioprotection is weak, especially in the elderly. Systemic analysis of clinical trials suggests that in patients with uncomplicated hypertension, compared with other antihypertensive drugs, such as RAAS inhibitors, CCBs, and thiazide diuretics, β-blockers show evidence of worse outcomes, particularly with regard to stroke [25]. Available evidence does not support the use of β-blockers, especially the traditional β-blockers, such as atenolol, as first-line drugs in the treatment of uncomplicated hypertension. Whether newer β-blockers, such as carvedilol and nebivolol, which show vasodilatory properties and a more favorable hemodynamic and metabolic profile, will be more efficacious in reducing morbidity and mortality in patients with uncomplicated hypertension remains to be determined [26].

In view of the aforementioned, the AHA 2007 guideline states that for uncomplicated hypertension, any effective antihypertensive drug or combination is indicated, but preference is given to ACEIs, ARBs, CCBs, and thiazide-type diuretics [3, 13]. Notably, β-blockers are not among the preferred drugs. Similarly, for the same reason, in the recently released JNC8 guideline, β-blockers are not recommended for the initial treatment of hypertension [1]. In contrast, the 2013 ESH/ESC guideline continues to recommend β-blockers as initial therapy for uncomplicated hypertension [11]. This recommendation is based on a meta-analysis showing that β-blocker-initiated therapy is equally as effective as the other major classes of antihypertensive agents in preventing coronary outcomes in uncomplicated hypertensive patients and highly effective in preventing cardiovascular events in patients with a recent myocardial infarction and those with heart failure [27].

12.5.1.3 The RAAS Inhibitors for Treating Hypertension

Mechanistic Basis The activity of the RAAS may be inhibited in five ways, which can be applied clinically for treating hypertension (see Chapter 9):

1. Inhibition of renin release by β-blockers
2. Inhibition of renin activity by aliskiren
3. Inhibition of ACE activity by ACEIs
4. Blockage of angiotensin receptors by ARBs
5. Blockage of aldosterone receptors by the aldosterone receptor blockers (also known as potassium-sparing diuretics) spironolactone and eplerenone

ACEIs, ARBs, aliskiren, and aldosterone receptor blockers may be collectively called inhibitors of the RAAS. These agents lower BP via various mechanisms including blocking angiotensin II-mediated peripheral vasoconstriction as well as aldosterone-mediated sodium retention. Hence, treatment with RAAS inhibitors decreases both peripheral arterial resistance and plasma volume, leading to decreased BP. Among the four classes of RAAS inhibitors, ACEIs and ARBs have been among the most widely used antihypertensive medications. There is also growing enthusiasm over aliskiren, the only antihypertensive agent introduced in more than a decade. However, the effectiveness of this drug as compared with other antihypertensive agents in reducing morbidity and mortality in hypertensive patients needs to be further determined in clinical trials.

Clinical Evidence and Recommendations The efficacy of ACEIs and ARBs in treating hypertension as either monotherapy or combination with other antihypertensive agents has been extensively documented in clinical trials. In addition to their well-established efficacy in lowering BP, these agents provide benefits in three large groups of patients—those with coronary heart disease, heart failure, or diabetic nephropathy (see Section 12.5.2 for more discussion). They are the drugs of choice for those patients. ACEIs are usually prescribed first and replaced with ARBs if patients cannot tolerate the ACEIs.

Direct renin inhibition by aliskiren reduces all major components of the RAAS, making it an attractive option to combine with other antihypertensive agents for the management of hypertension and its comorbidities. Clinical trials have shown that combination of aliskiren with drugs representing each of the major antihypertensive medications (thiazide-type diuretics, β-blockers, ACEIs, ARBs, and CCBs) reduces BP and improves biomarkers for cardiovascular outcomes. Results of several ongoing clinical trials should provide additional insights into the potential efficacy of therapeutic combinations that include aliskiren to improve cardiovascular morbidity and mortality in patients with hypertension and related comorbidities [28].

The antihypertensive action of the aldosterone receptor blockers (also known as aldosterone antagonists) has long been recognized. As add-on to conventional anti-heart failure therapy, the aldosterone receptor blockers spironolactone and eplerenone lower BP and have a secondary protective effect in patients with severe heart failure, leading to improved survival. As such, aldosterone receptor blockers are indicated in patients with severe heart failure. Recent studies also demonstrate that add-on aldosterone receptor blocker treatment is effective for treating uncontrolled hypertension [29, 30] (also see Section 12.5.2.9). However, there has been no major study of cardiovascular outcomes in individuals treated with aldosterone receptor blockers for hypertension without left ventricular dysfunction.

12.5.1.4 Direct Vasodilators for Treating Hypertension

Mechanistic Basis As described earlier, the antihypertensive agents, ACEIs and ARBs, cause vasodilation indirectly via blocking vasoconstricting angiotensin II. Direct vasodilators reduce BP by directly relaxing vascular smooth muscle, thus dilating resistance vessels. This results in decreased peripheral arterial resistance. The direct vasodilators, to varying degrees, also increase venous pooling in capacitance vessels, causing decreased cardiac preload. The direct vasodilators include CCBs, nitrates and nitroprusside, hydralazine, minoxidil, and fenoldopam, among others (see Chapter 11). Of the direct vasodilators, CCBs have been most commonly used for treating hypertension. In fact, CCBs are among the most popular classes of agents used in the treatment of hypertension [31].

Clinical Evidence and Recommendations CCBs are effective in hypertensive patients of all ages and races. Compared with other classes of antihypertensive agents, CCBs exhibit a similar capacity of reducing the risk of coronary heart disease and a better ability to protect against stroke, but a less effective protection against heart failure [32]. CCBs are particularly effective in the prevention of stroke in older hypertensive patients. They also appear to lower BP more effectively in people of African heritage than in people of other ethical backgrounds. In contrast, people of African heritage show less response to the RAAS inhibitors than to other classes of antihypertensive agents. This is likely due to the fact that people of African heritage tend to have lower activity of RAAS.

12.5.2   Drug Therapy for Special Situations

12.5.2.1   Hypertensive Patients with Diabetes The existence of hypertension in diabetes is particularly pernicious because of the strong linkage of the two conditions with all cardiovascular diseases (including stroke) and progression of kidney diseases. Hence, it is especially important to control BP in diabetic people. Both the JNC7 and the AHA 2007 guidelines, as well as the American Diabetes Association, recommend that BP in diabetics be controlled to levels of 130/80 mmHg or lower. Rigorous control of BP is paramount for reducing the progression of diabetic nephropathy to end-stage renal disease. However, such a BP target has not been supported by high-quality clinical evidence, and as such, the JNC8 guideline does not recommend that 130/80 mmHg or lower be the target BP level for diabetic patients and instead recommends that the goal be <140/90 mmHg as in the general population [1].

The general principles of pharmacological management of hypertension in diabetics are summarized below:

- Pharmacological therapy for patients with diabetes and hypertension should be with a regimen that includes either an ACEI or an ARB. If one class is not tolerated, the other should be substituted. If needed to achieve BP targets, a thiazide diuretic should be added to those with an estimated glomerular filtration rate (GFR) ≥30 ml/min/1.73 m^2 and a loop diuretic for those with an estimated GFR <30 ml/min/1.73 m^2.
- Multidrug therapy (two or more agents at maximal doses) is generally required to achieve BP targets.
- If ACEIs, ARBs, or diuretics are used, kidney function and serum potassium levels should be closely monitored.
- There is currently no high-quality evidence to support a goal BP of ≤130/80 mm Hg in patients with diabetes.

12.5.2.2   Hypertensive Patients with Ischemic Heart Disease

Patients with Chronic Stable Angina Management of hypertension in patients with chronic coronary heart disease and chronic stable angina is directed toward three general goals:

- Prevention of death, myocardial infarction, and stroke
- Reduction in the frequency and duration of myocardial ischemia
- Amelioration of symptoms

Both the JNC7 and the AHA 2007 guidelines recommend that a reasonable BP target for hypertensive patients with demonstrated coronary heart disease or with its risk equivalents (carotid artery disease, peripheral arterial disease, abdominal aortic aneurysm, diabetes mellitus, or chronic renal disease) be <130/80 mmHg. If ventricular dysfunction is present, consideration should be given to lowering the BP even further, to <120/80 mmHg. In addition to lifestyle modifications, pharmacological management of hypertension in patients with coronary heart disease and chronic stable angina is inevitably required. According to the AHA 2007 guideline, the general principles of pharmacological management of hypertension in patients with coronary heart disease and chronic stable angina are as follows [3]:

- Patients with hypertension and chronic stable angina should be treated with a regimen that includes a β-blocker in patients with a history of prior myocardial infarction, an ACEI or ARB if there is diabetes mellitus and/or left ventricular systolic dysfunction, and a thiazide-type diuretic. The combination of a β-blocker, ACEI (or ARB), and a thiazide-type diuretic should also be considered in the absence of a prior myocardial infarction, diabetes mellitus, or left ventricular systolic dysfunction.
- If β-blockers are contraindicated or produce intolerable side effects, a nondihydropyridine CCB (such as diltiazem or verapamil) can be substituted, but not if there is left ventricular dysfunction.
- If either the angina or the hypertension remains uncontrolled, a long-acting dihydropyridine CCB can be added to the basic regimen of β-blocker, ACEI, and thiazide-type diuretic. The combination of a β-blocker and either of the nondihydropyridine CCBs (diltiazem or verapamil) should be used with caution in patients with symptomatic coronary heart disease and hypertension because of the increased risk of significant bradyarrhythmias and heart failure.

Patients with Acute Coronary Syndromes Unstable angina, NSTEMI, and STEMI are collectively known as acute coronary syndromes (see Unit V). Hypertension is highly prevalent in patients with acute coronary syndromes, many of whom are elderly. The majority of patients respond to standard methods of hypertension control.

The cornerstone of the management of acute coronary syndromes is the modification of the balance between myocardial oxygen supply and demand, in addition to the initiation of anticoagulant and platelet inhibitor therapy. Although an elevated BP increases cardiac afterload and thereby the myocardial oxygen demand, rapid and excessive lowering of BP may also result in impairment of coronary blood flow and oxygen supply.

The JNC7 and AHA 2007 guidelines recommend the target BP in patients with acute coronary syndromes be <130/80 mmHg. However, in patients with an elevated DBP and acute coronary syndrome, the BP should be lowered slowly, and caution is advised in inducing falls of DBP below 60 mmHg. In older hypertensive individuals with wide pulse pressures, lowering systolic BP may cause very low DBP values (<60 mmHg). This should alert the clinician to assess carefully any untoward signs or symptoms, especially those due to worsening myocardial ischemia.

The AHA 2007 guideline recommends the following for the pharmacological management of hypertension in patients with acute coronary syndromes:

- In unstable angina, NSTEMI, or STEMI, the initial therapy of hypertension should include short-acting β_1-selective β-blockers without intrinsic sympathomimetic activity, usually intravenously, in addition to nitrates for symptom control. Oral β-blockers can be substituted at a later stage of the hospital stay. Alternatively, oral β-blockers may be started promptly without prior use of intravenous β-blockers. If the patient is hemodynamically unstable, the initiation of β-blocker therapy should be delayed until stabilization of heart failure or shock has been achieved. Diuretics can be added for BP control and for the management of heart failure.

- For patients with unstable angina or NSTEMI, if there is a contraindication to the use of a β-blocker or if the patient develops intolerable side effects of a β-blocker, then a nondihydropyridine CCB, such as verapamil or diltiazem, may be substituted, but not if there is left ventricular dysfunction. If the angina or the hypertension is not controlled with a β-blocker alone, then a longer-acting dihydropyridine CCB may be added. A thiazide diuretic can also be added for BP control.

- For patients with STEMI, CCBs do not reduce mortality rates in the setting of acute STEMI and can increase mortality if there is depressed left ventricular function and/or pulmonary edema. Long-acting dihydropyridine CCBs can be used when β-blockers are contraindicated or inadequate to control angina or as adjunct therapy for BP control.

- If the patient is hemodynamically stable, an ACEI or ARB should be added if the patient has an anterior myocardial infarction, if hypertension persists, if the patient has evidence of left ventricular dysfunction or heart failure, or if the patient has diabetes mellitus.

- Aldosterone antagonists may be useful in the management of STEMI with left ventricular dysfunction and heart failure and may have an additive BP-lowering effect.

12.5.2.3 Hypertension in Patients with Heart Failure

Most patients with heart failure have systemic arterial hypertension. Hypertension is not only a concomitant disorder but also a significant contributor to the pathogenesis of both systolic and diastolic heart failure. Hence, control of high BP is an important component of management of patients with heart failure.

The target BP in patients with heart failure is <130/80 mmHg, but consideration should be given to lowering the BP even further, to <120/80 mmHg. In patients with an elevated DBP who have coronary heart disease and heart failure with evidence of myocardial ischemia, the BP should be lowered slowly, and caution is advised in inducing falls of DBP below 60 mmHg if the patient has diabetes mellitus or is over the age of 60 years. In older hypertensive individuals with wide pulse pressures, lowering systolic BP may cause very low DBP values (<60 mmHg). This should alert the clinician to assess carefully any untoward signs or symptoms, especially those due to myocardial ischemia and worsening heart failure.

The principles of the pharmacological management of hypertension in patients with heart failure are the following:

- In general, drugs that have been shown to improve outcomes for patients with heart failure also lower BP. Patients should be treated with diuretics, ACEIs (or ARBs), β-blockers, and aldosterone receptor antagonists.

- Thiazide diuretics should be used for BP control and to reverse volume overload and associated symptoms. In severe heart failure or in patients with severe renal impairment, loop diuretics should be used for volume control, but these are less effective than thiazide diuretics in lowering BP. Diuretics should be used together with an ACEI or ARB and a β-blocker. However, there is no evidence that diuretics retard disease progression of heart failure.

- Both ACEIs and the ARBs (candesartan or valsartan) are beneficial in heart failure, resulting in decreased morbidity and mortality and improvement of heart failure symptoms. Either class of agents is effective in lowering BP.

- Among the β-blockers, carvedilol, metoprolol succinate, and bisoprolol have been shown to improve outcomes in heart failure and are effective in lowering BP.
- The aldosterone receptor antagonists spironolactone and eplerenone have been shown to be beneficial in heart failure and should be included in the regimen if there is severe heart failure. Spironolactone or eplerenone may be used together with a thiazide diuretic, particularly in patients with refractory hypertension.
- Consider the addition of hydralazine/isosorbide dinitrate to the regimen of diuretic, ACEI or ARB, and β-blocker in patients of African heritage.

12.5.2.4 Hypertension in Patients with CKD

Hypertension and CKD frequently coexist. High BP is a major risk factor of progression of CKD. The National Kidney Foundation Kidney Disease Outcomes Quality Initiative has issued a clinical practice guideline on hypertension and antihypertensive agents in CKD [33]. The recommended BP goal is <130/80 mmHg. This BP goal is also recommended by the US-based Kidney Disease: Improving Global Outcomes (KDIGO) 2012 guideline for CKD patients with proteinuria [34]. The key recommendations of the pharmacological management of the guideline are outlined below and also summarized in Table 12.8.

- All patients with CKD should be considered for pharmacologic therapy to slow progression of kidney disease.
- Patients with diabetic or nondiabetic kidney disease with spot urine total protein-to-creatinine ratio ≥200 mg/g should be treated with a moderate to high dose of an ACEI or an ARB. These agents should also be used even the patients are not hypertensive. Both ACEIs and ARBs protect against progression of CKD.
- For most CKD patients with hypertension, two or more antihypertensive agents will be necessary to achieve the BP goal.
- Selection of other agents to reduce cardiovascular disease risk should follow the recommendations of the JNC7 report.

- Most CKD patients should receive a diuretic; a thiazide-type diuretic is preferable in patients with GFR ≥30 ml/min/1.73 m² and a loop diuretic in patients with GFR <30 ml/min/1.73 m².
- Other antihypertensive agents such as β-blockers or CCBs should be added as necessary to achieve target BP.

12.5.2.5 Hypertension in Elderly

General Considerations Hypertension is common in the elderly (≥65 years of age), and those individuals are more likely to have organ damage or clinical cardiovascular diseases. Approximately 80% of all adults older than 80 years of age in the United States have hypertension. In Framingham (see Chapter 1), 90% of all 65-year-old men and women with normal BP later developed hypertension. The pattern of BP elevation in the US population also changes with age. Before reaching 50 years of age, most people with hypertension have elevated diastolic pressure. After the age of 50 years, as systolic pressure continues to rise and diastolic pressure tends to fall, isolated systolic hypertension predominates [35].

The goal BP in the elderly is debated, but based on current outcome data, a goal BP of 150/80–90 mmHg is reasonable in at least the very elderly [36]. According to the ACCF/AHA 2011 Expert Consensus Document on Hypertension in the Elderly, the target systolic BP in patients aged 55–79 is ≤140 mmHg. But for those aged ≥80, 140–145 mmHg (systolic), if tolerated, can be acceptable [5]. The recently published JNC8 guideline recommends a BP goal of <150/90 mmHg for hypertensive persons aged ≥60 years [1]. Hence, the BP goal for the elderly recommended in the JNC8 report is less aggressive than that recommended in the ACCF/AHA 2011 Expert Consensus Document on Hypertension in the Elderly. As noted earlier, the JNC8 guideline recommends one BP goal of <140/90 mmHg for persons aged <60 years regardless of the moribund diseases, including diabetes and CKD. This recommendation is based on the absence of high-quality randomized controlled trials comparing the outcomes of the current treatment goal of <140/90 mmHg with others (e.g., <130/80 mmHg for persons with diabetes as recommended by the JNC7 and the AHA 2007 guidelines described earlier).

TABLE 12.8   The key recommendations on the management of hypertension in patients with different forms of CKD [33]

Type of CKD	Blood pressure target (mm Hg)	Preferred agents for CDK with (or without) hypertension	Other agents to reduce CVD risk and reach blood pressure target
Diabetic kidney disease	<130/80	ACEI or ARB	Diuretic preferred, then β-blocker or CCB
Nondiabetic kidney disease with spot urine total protein-to-creatinine ratio ≥200 mg/g	<130/80	ACEI or ARB	Diuretic preferred, then β-blocker or CCB
Nondiabetic kidney disease with spot urine total protein-to-creatinine ratio <200 mg/g	<130/80	None preferred	Diuretic preferred, then ACEI, ARB, β-blocker, or CCB
Kidney disease in the transplant recipient	<130/80	None preferred	CCB, diuretic, β-blocker, ACEI, or ARB

Management Use of specific drug classes in older people is largely similar to that recommended in the general algorithm and for individual compelling indications [5, 37]. Combination therapy with two or more drugs in general is needed to achieve optimal BP control. Careful lowering of BP with a diuretic plus an ACEI to reach the goal of 150 mmHg (systolic) can reduce cardiovascular events in patients 80 years of age or older [38]. Overall, CCBs reduce BP and risk of stroke more effectively than ACEIs [32]. Particular care is needed for treating hypertension in the elderly with drugs due to alterations in drug distribution and disposal and changes in homeostatic cardiovascular control. For instance, due to decreased effectiveness of baroreceptor reflex in the elderly, postural hypotension should be carefully sought. All medications should be given in slowly increasing doses to prevent excessive BP lowering.

12.5.2.6 *Hypertension in Children and Adolescents*

Definition and General Considerations According to the JNC7 guideline and the Fourth Report on Diagnosis, Evaluation, and Treatment of High Blood Pressure in Children and Adolescents, hypertension in children and adolescents is defined as elevated BP that persists on repeated measurement at the 95th percentile or greater for age, height, and gender (Table 12.9). For example, for a 12-year-old boy with a 50th percentile for height, if the BP is at or above 123/80 mmHg, he will be diagnosed with hypertension. Prehypertension in children is defined as average systolic BP or DBP levels that are greater than or equal to the 90th percentile but less than the 95th percentile. As with adults, adolescents with BP levels greater than or equal to 120/80 mmHg should be considered prehypertensive.

Hypertension in the young is increasingly being recognized as an emerging critical healthcare problem, not only because of its increasing prevalence in recent years (partly due to the epidemic of childhood obesity) but also because of its significant impact on the health and well-being of children and adolescents and tracking into adult life. It is widely accepted that pediatric hypertension carries an increased risk for future cardiovascular morbidity and mortality [39, 40].

Clinician should be alert to the possibility of identifiable causes of hypertension in younger children. Secondary forms of hypertension are more common in children and in individuals with severe hypertension (>20 mmHg above the 95th percentile). Chronic hypertension is becoming increasingly common in adolescents and is generally associated with obesity, sedentary lifestyle, and a positive family of hypertension and other cardiovascular disorders [41].

Management The JNC7 guideline recommends the following for the management of hypertension in children and adolescents [16]:

- Lifestyle interventions should be recommended for all children with hypertension, with pharmacologic therapy instituted for higher levels of BP or if insufficient response to lifestyle modifications occurs.
- Teenage children with BP below but near the 95th percentile should adopt healthy lifestyles similar to adults with prehypertension.
- Although the recommendations for choice of drugs are generally similar in children and adults, dosages of antihypertensive medication for children should be smaller and adjusted very carefully.
- ACEIs and ARBs should not be used if the patient is pregnant. These agents should be used with extreme caution in sexually active teenage girls and only when careful counseling and effective pregnancy precautions are established.

The major recommendations on the Fourth Report on Diagnosis, Evaluation, and Treatment of High Blood Pressure in Children and Adolescents also include both therapeutic lifestyle changes and pharmacological therapy, and the key messages are summarized below [42].

Therapeutic Lifestyle Changes Weight reduction is the primary therapy for obesity-related hypertension. Prevention of excess or abnormal weight gain will limit future increases in BP.

- Regular physical activity and restriction of sedentary activity will improve efforts at weight management and may prevent an excess increase in BP over time.

TABLE 12.9 **The 95th percentile of BP by selected ages, by the 50th and 75th height percentiles, and by the gender in children and adolescents**

Age	Girls' SBP/DBP		Boys' SBP/DBP	
	50th percentile for height	75th percentile for height	50th percentile for height	75th percentile for height
1	104/58	105/59	102/57	104/58
6	111/73	112/73	114/74	115/75
12	123/80	124/81	123/81	125/82
17	129/84	130/85	136/87	138/88

DBP, diastolic blood pressure; SBP, systolic blood pressure.

- Dietary modification should be strongly encouraged in children and adolescents who have BP levels in the prehypertensive range as well as in those with hypertension.
- Family-based intervention improves success.

Pharmacological Therapy

- Indications for antihypertensive drug therapy in children include secondary hypertension and insufficient response to lifestyle modifications.
- Recent clinical trials have expanded the number of drugs that have pediatric dosing information. Dosing recommendations for many of the newer drugs are provided.
- Pharmacologic therapy, when indicated, should be initiated with a single drug. Acceptable drug classes for use in children include ACEIs, ARBs, β-blockers, CCBs, and diuretics.
- The goal for antihypertensive treatment in children should be reduction of BP to <95th percentile, unless concurrent conditions are present. In that case, BP should be lowered to <90th percentile.
- Severe, symptomatic hypertension should be treated with intravenous antihypertensive drugs.

12.5.2.7 *Hypertension in Pregnancy*

Classification   Hypertensive disorders in pregnancy are a major cause of maternal, fetal, and neonatal morbidity and mortality. Hypertension in pregnancy is classified into five categories (Table 12.10) [16], and it is critical to differentiate preeclampsia, from preexisting hypertension.

Management   Among the various types of hypertension in pregnancy, chronic hypertension and preeclampsia are of the most significant concern, and as such, these two are particularly discussed with regard to their management. This section primarily describes the JNC7 recommendations of pharmacological management of chronic hypertension and preeclampsia in pregnancy [16].

CHRONIC HYPERTENSION   The JNC7-recommended management of chronic hypertension in pregnancy is summarized below:

- Women with stage I hypertension are at low risk for cardiovascular complications during pregnancy and are candidates for lifestyle modification therapy only as there is no evidence that pharmacological treatment improves neonatal outcomes.

TABLE 12.10   Classification of hypertension in pregnancy

Category	Description
Chronic hypertension	• BP ≥140 mmHg systolic or 90 mmHg diastolic prior to pregnancy or before 20 weeks' gestation • Persists >12 weeks postpartum
Preeclampsia	• BP ≥140 mmHg systolic or 90 mmHg diastolic with proteinuria (>300 mg/24 h) after 20 weeks' gestation • Can progress to eclampsia (seizures) • More common in nulliparous women, multiple gestation, women with hypertension for ≥4 years, family history of preeclampsia, hypertension in previous pregnancy, renal disease

Chronic hypertension with superimposed preeclampsia	• New-onset proteinuria after 20 weeks of gestation in a woman with hypertension
	• In a woman with hypertension and proteinuria prior to 20 weeks' gestation:
	    Sudden two- to threefold increase in proteinuria
	    Sudden increase in BP
	    Thrombocytopenia
	    Elevated AST or ALT

Gestational hypertension	• Hypertension without proteinuria occurring after 20 weeks' gestation • Temporary diagnosis • May represent preproteinuric phase of preeclampsia or recurrence of chronic hypertension abated in midpregnancy • May evolve to preeclampsia • If severe, may result in higher rates of premature delivery and growth retardation than mild preeclampsia
Transient hypertension	• Retrospective diagnosis • BP normal by 12 weeks postpartum • May recur in subsequent pregnancies • Predictive of future primary hypertension

ALT, alanine aminotransferase; AST, aspartate aminotransferase; BP, blood pressure.

- The primary goal of treating chronic hypertension in pregnancy is to reduce maternal risk, but the choice of antihypertensive agent(s) is largely driven by the safety of the fetus. In this context, the inhibitors of the RAAS, including aliskiren, ACEIs, and ARBs, are contraindicated.

- Methyldopa is preferred by many as first-line therapy, based on reports of stable uteroplacental blood flow and fetal hemodynamics and the absence of long-term (7.5-year follow-up) adverse effects on development of children exposed to methyldopa *in utero*. Other treatment options are summarized in Table 12.11.

PREECLAMPSIA The key principles of pharmacological management of preeclampsia include the following [16]:

- Antihypertensive therapy should be prescribed only for maternal safety; it does not improve perinatal outcomes and may adversely affect uteroplacental blood flow.

- Selection of antihypertensive agents and route of administration depends on anticipated timing of delivery.

TABLE 12.11 Pharmacological agents for the treatment of chronic hypertension in pregnancy

Drug	Comments
Methyldopa	Preferred on the basis of long-term follow-up studies supporting safety
β-Blockers	Reports of intrauterine growth retardation (atenolol) Generally safe
Labetalol	Increasingly preferred to methyldopa because of reduced side effects
Clonidine	Limited data
Calcium channel blockers	Limited data No increase in major teratogenicity with exposure
Diuretics	Not first-line agents Probably safe

- If delivery is likely more than 48 h off, oral methyldopa is preferred because of its safety record. Oral labetalol is an alternative, and other β-blockers and calcium antagonists are also acceptable on the basis of limited data.

- If delivery is imminent, parenteral agents are practical and effective, with hydralazine as a preferred agent (Table 12.12).

- Antihypertensives are administered before induction of labor for persistent DBPs of 105–110 mmHg or higher, aiming for levels of 95–105 mmHg.

12.5.2.8 Hypertension in Patients of African Heritage

General Considerations Hypertension in African Americans is a major clinical and public health problem because of the high prevalence and premature onset of elevated BP as well as the high burden of comorbid factors that lead to pharmacological treatment resistance (obesity, diabetes mellitus, depressed GFR, and albuminuria). BP control rates are lower in African Americans, especially men, than in other major race/ethnicity–sex groups [43].

Management Since the first International Society on Hypertension in Blacks consensus statement on the "Management of High Blood Pressure in African American" in 2003 [44], data from additional clinical trials have become available. Recently, "Management of High Blood Pressure in Blacks: An Update of the International Society on Hypertension in Blacks Consensus Statement" has been published [43]. The key messages of this update are summarized below:

- In this update, blacks with hypertension are divided into two risk strata: (i) primary prevention, including elevated BP without target organ damage, preclinical cardiovascular disease, or overt cardiovascular disease, and (ii) secondary prevention, including elevated BP with target organ damage, preclinical cardiovascular disease, and/or a history of cardiovascular disease.

TABLE 12.12 Drug treatment of acute severe hypertension in preeclampsia

Drug	Treatment regimen
Hydralazine	• 5 mg iv bolus and then 10 mg every 20–30 min to a maximum of 25 mg, repeat in several hours as necessary
Labetalol (second line)	• 20 mg iv bolus, then 40 mg 10 min later, and 80 mg every 10 min for two additional doses to a maximum of 220 mg
Nifedipine (controversial)	• 10 mg oral, repeat every 20 min to a maximum of 30 mg (controversial) • Caution when using with magnesium sulfate; can see precipitous BP drop • Short-acting nifedipine is not approved by the US FDA for managing hypertension
Sodium nitroprusside (is rarely used; is used when others fail)	• 0.25 μg/kg/min to a maximum of 5 μg/kg/min • Fetal cyanide poisoning may occur if used for more than 4 h

BP, blood pressure; iv, intravenous.

- The recommended target BP for primary prevention is <135/85 mmHg, and for secondary prevention, BP consistently <130/80 mmHg is recommended.
- If BP is ≤10 mmHg above target levels, monotherapy with a diuretic or CCB is preferred.
- When BP is >15/10 mmHg above target, a two-drug therapy is recommended, with either a CCB plus a renin–angiotensin–aldosterone system (RAAS) blocker or, alternatively, in edematous and/or volume-overload states, with a thiazide diuretic plus an RAAS blocker.
- Effective management of BP may require multidrug therapeutic combinations of four drugs.
- Comprehensive lifestyle modifications should be initiated in blacks when BP is ≥115/75 mmHg.
- The updated International Society on Hypertension in Blacks consensus statement on hypertension management in blacks lowers the minimum target BP level for the lowest-risk blacks, emphasizes effective multidrug regimens, and deemphasizes monotherapy.

12.5.2.9 Resistant Hypertension

General Considerations   Resistant hypertension is defined as failure to achieve BP goal (<140/90 mmHg for the overall population and <130/80 mmHg for those with diabetes mellitus or CKD) when a patient adheres to maximum tolerated doses of three antihypertensive drugs including a diuretic [45, 46]. Resistant hypertension is a common clinical problem faced by both primary care clinicians and specialists [47]. While the exact prevalence of resistant hypertension is unknown, clinical trials suggest that it is not rare, involving perhaps 20–30% of study participants. As older age and obesity are two of the strongest risk factors for uncontrolled hypertension, the incidence of resistant hypertension will likely increase as the population becomes more elderly and heavier. The causes of resistant hypertension are multifactorial, with the major factors listed below [47–49]:

- Drug-induced (e.g., nonsteroidal anti-inflammatory drugs, cocaine and amphetamines, oral contraceptive hormones, cyclosporine and tacrolimus, and erythropoietin)
- Excessive alcohol intake
- Volume overload (from excessive sodium intake, volume retention due to kidney disease, or inadequate diuretic therapy)
- Obesity, diabetes mellitus, and older age
- Secondary hypertension (caused by renal diseases, pheochromocytoma, or primary aldosteronism)

The prognosis of resistant hypertension is unknown, but cardiovascular risk is undoubtedly increased as patients often have a history of long-standing, severe hypertension complicated by multiple other cardiovascular risk factors such as obesity, sleep apnea, diabetes, and CKD. The degree to which cardiovascular risk is reduced with treatment of resistant hypertension is unknown. The benefits of successful treatment, however, are likely substantial as suggested by hypertension outcome studies [50].

Management   Recently, the AHA has published a scientific statement, "Resistant Hypertension: Diagnosis, Evaluation, and Treatment" [50], that provides recommendations for the management of resistant hypertension. The key messages from this statement are summarized below:

- Successful treatment of resistant hypertension requires (i) identification and reversal of lifestyle factors contributing to treatment resistance, (ii) diagnosis and appropriate treatment of secondary causes of hypertension, and (iii) use of effective multidrug regimens.
- The pharmacological treatment of resistant hypertension remains largely empiric due to the lack of systematic assessments of 3 or 4 drug combinations.
- The pharmacological treatment strategies include:
  - Maximize diuretic therapy, including possible addition of a mineralocorticoid receptor antagonist.
  - Combine agents with different mechanisms of action.
  - Use loop diuretics in patients with CKD and/or patients receiving potent vasodilators (e.g., minoxidil).

12.5.2.10 Hypertensive Emergencies and Urgencies

Definitions   Hypertensive emergencies and urgencies are collectively known as hypertensive crises. These two terms are defined as follows:

- Hypertensive emergencies are characterized by severe elevations in BP (≥180/120 mmHg) complicated by evidence of impending or progressive target organ dysfunction. They require immediate BP reduction (not necessarily to normal) to prevent or limit target organ damage. Examples include hypertensive encephalopathy, intracerebral hemorrhage, acute myocardial infarction, acute left ventricular failure with pulmonary edema, unstable angina pectoris, dissecting aortic aneurysm, or eclampsia.
- Hypertensive urgencies are characterized by severe elevations in BP without progressive target organ dysfunction. Examples include upper levels of stage II hypertension associated with severe headache, shortness of breath, epistaxis, or severe anxiety. The majority of these patients present as noncompliant or inadequately treated hypertensives, often with little or no evidence of target organ damage.

TABLE 12.13   Parenteral drugs for the treatment of hypertensive emergencies

Drug	Dose	Onset of action	Duration of action	Special indications
Vasodilators				
Nitroprusside	0.25–10 µg/kg/min as iv infusion	Immediate	1–2 min	Most hypertensive emergencies; caution with high intracranial pressure or azotemia
Nicardipine	5–15 mg/h as iv infusion	5–10 min	15–30 min; may exceed 4 h	Most hypertensive emergencies except acute heart failure; caution with coronary ischemia
Clevidipine	1–2 mg iv, rapidly increasing dose to 16 mg maximum	2–4 min	5–15 min	Most hypertensive emergencies
Fenoldopam	0.1–0.3 µg/kg/min as iv infusion	<5 min	30 min	Most hypertensive emergencies; caution with glaucoma
Nitroglycerin	5–100 µg/min as iv infusion	2–5 min	5–10 min	Not preferred, but may be useful for coronary ischemia
Enalaprilat	1.25–5 mg every 6 h iv	15–30 min	6–12 h	Acute left ventricular failure; avoid in acute myocardial infarction
Hydralazine	10–20 mg iv 10–40 mg im	10–20 min iv 20–30 min im	1–4 h iv 4–6 h im	Eclampsia
Adrenergic inhibitors				
Labetalol	20–80 mg iv bolus every 10 min; 0.5–2.0 mg/min iv infusion	5–10 min	3–6 h	Most hypertensive emergencies except acute heart failure
Esmolol	250–500 µg/kg/min iv bolus, then 50–100 µg/kg/min by infusion; may repeat bolus after 5 min or increase infusion to 300 µg/min	1–2 min	10–30 min	Aortic dissection, perioperative
Phentolamine	5–15 mg iv bolus	1–2 min	10–30 min	Catecholamine excess

iv, intravenous; im, intramuscular.

Management of Hypertensive Emergencies The current guidelines, including the JNC7 report, recommend the following for the management of hypertensive emergencies:

- Patients with a hypertensive emergency should be admitted to an intensive care unit for continuous monitoring of BP and parenteral administration of an appropriate agent. Table 12.13 lists the parenteral drugs for the treatment of hypertensive emergencies.

- The potency and rapidity of action of sodium nitroprusside have made it the drug of choice for life-threatening hypertension emergencies. Other parenteral drugs that are also widely used for treating hypertensive emergencies include the β-blocker labetalol and the rapidly acting CCBs nicardipine and clevidipine. With any of these drugs, intravenous furosemide (a loop diuretic) is often needed to lower BP further and prevent retention of sodium and water.

- The initial goal of therapy in hypertensive emergencies is to reduce mean arterial BP by no more than 25% (within minutes to 1 h) and then, if stable, to 160/100–110 mmHg within the next 2–6 h.

- Excessive falls in pressure that may precipitate renal, cerebral, or coronary ischemia should be avoided. For this reason, short-acting nifedipine is no longer considered acceptable in the initial treatment of hypertensive emergencies or urgencies.

- If this level of blood pressure is well tolerated and the patient is clinically stable, further gradual reductions toward a normal BP can be implemented in the next 24–48 h.

Management of Hypertensive Urgencies Some patients with hypertensive urgencies may benefit from treatment with an oral, short-acting agent such as captopril, labetalol, or clonidine followed by several hours of observation. However, there is no evidence to suggest that failure to aggressively lower BP in the emergency room is associated with any increased short-term risk to the patient who presents with severe hypertension. Such a patient may also benefit from adjustment in their antihypertensive therapy, particularly the use of combination drugs, or reinstitution of medications if noncompliance is a problem. Most importantly, patients should not leave the emergency room without a confirmed follow-up visit within one to a few days.

12.6 DRUG THERAPY OF PREHYPERTENSION

12.6.1 General Considerations

As noted in Chapter 6, prehypertension is defined as a systolic BP of 120–139 mmHg or a DBP of 80–89 mmHg and affects nearly 30% of the adult population. The term prehypertension was actually coined in 1939 in the context of early studies that linked high BP recorded during physical examination for life insurance purposes to subsequent morbidity and mortality. These studies demonstrated that individuals with BP >120/80 mmHg but <140/90 mmHg (the accepted value for the lower limit of the hypertensive range) had an increased risk of hypertension, cardiovascular disease, and early death from cardiovascular causes [51].

The prehypertension classification of BP was later used by the JNC7 report to define a group of individuals at increased risk of cardiovascular events because of elevated BP. Indeed, increased cardiovascular disease risk begins at systolic BP levels as low as 115 mmHg. For each 20 mmHg increase in systolic BP or 10 mmHg increase in DBP above 115/75 mmHg, there is a twofold increase in mortality associated with coronary heart disease and stroke.

The reasons for creating a classification of prehypertension include the following [51]:

- To increase the awareness of lifetime risk of hypertension as prehypertension is considered a precursor of stage I hypertension
- To increase the awareness of increased risk of cardiovascular complications as prehypertension carries an elevated risk for cardiovascular disease incidence and mortality
- To identify individuals in whom early intervention by lifestyle modifications could lower BP, decrease the rate of progression to hypertension with age, and prevent hypertension entirely
- To enable insurance coverage for the treatment of prehypertension

12.6.2 Management

The management of prehypertension primarily involves lifestyle modifications. Recently, pharmacological treatment has also been suggested for the management of prehypertension.

12.6.2.1 Lifestyle Modifications   Lifestyle modifications are recommended for all individuals with prehypertension as they effectively reduce the BP, slow or prevent the development of hypertension, and decrease cardiovascular disease risk. Lifestyle modifications are currently the first choice for the treatment of individuals with prehypertension who do not have comorbid conditions (e.g., diabetes mellitus,

CKD, coronary heart disease) that would otherwise mandate BP reduction by use of pharmacological agents.

12.6.2.2 Drug Therapy   Treatment of prehypertensive individuals with antihypertensive agents in addition to lifestyle modifications has been explored in clinical trials. The Trial of Preventing Hypertension (TROPHY) reported in 2006 that over a period of 4 years, stage I hypertension developed in nearly twothirds of patients with untreated prehypertension (the placebo group). Treatment of prehypertension with candesartan appeared to be well tolerated and reduced the risk of incident hypertension during the study period. The TROPHY trial hence suggested that treatment of prehypertension appears to be feasible [52]. Similar to the TROPHY study, the Prevention of Hypertension with the Angiotensin-Converting Enzyme Inhibitor (ACEI) Ramipril in Patients with High-Normal Blood Pressure (PHARAO) trial showed that treatment of patients with prehypertension with the ACEI was well tolerated and significantly reduced the risk of progression to manifest hypertension [53].

More recently, a meta-analysis of 25 trials involving a total of 64,162 participants demonstrated that among patients with clinical history of cardiovascular disease but without hypertension, antihypertensive treatment was associated with decreased risk of stroke, congestive heart failure, composite cardiovascular disease events, and all-cause mortality compared with those not receiving antihypertensive therapy [54]. The antihypertensive agent-treated participants received either an ACEI, an ARB, a β-blocker, a CCB, or an ACEI combined with a diuretic.

The clinical importance of the aforementioned meta-analysis is clear: pharmacological intervention in patients with cardiovascular disease and BP levels less than 140/90 mmHg is associated with a decreased cardiovascular morbidity and mortality. However, this meta-analysis does not determine whether lowering BP levels is the reason for improved clinical outcomes. The antihypertensive agents may improve clinical outcomes through multiple other mechanisms, for example, hemodynamic effects unrelated to BP, neurohormonal effects, and tissue-level effects [55]. Obviously, additional clinical studies are needed to determine the clinical outcomes of antihypertensive treatment in prehypertensive individuals without cardiovascular disease and delineate the contribution of the lowering of BP to the clinical outcomes. Further studies are also needed to determine the safety and cost-effectiveness of pharmacological intervention and to determine whether particular antihypertensive drug classes are more effective than others in the prehypertensive subjects.

12.6.2.3 Controversies on Drug Therapy of Prehypertension   Although available clinical data suggest a benefit of drug therapy in prehypertensive patients, the exact value of pharmacological management of prehypertension is still currently debated. Arguments against

the use of antihypertensive drugs for prehypertension include the following:

- Lack of evidence that antihypertensive agents reduce target organ damage and cardiovascular morbidity and mortality in prehypertensive individuals who do not have comorbid conditions
- Lack of evidence that antihypertensive agents are safe when administered over many decades, as would be required to treat young individuals with prehypertension
- Lack of evidence that antihypertensive agents are cost-effective for the treatment of prehypertensive patients with no comorbid conditions
- Existence of questions that concern appropriate choices and doses of the antihypertensive agents and duration of treatment

On the other hand, arguments in favor of the use of antihypertensive drugs for prehypertension are that:

- The antihypertensive drugs are more convenient and more likely to be adhered to than complex lifestyle-modifying regimens.
- The antihypertensive drugs are already accepted for use in certain high-risk individuals with prehypertension (i.e., in those with BP ≥130/80 mmHg and with diabetes mellitus, CKD, or CAD).

In the absence of further information about these issues and in light of the fact that lifestyle approaches favorably influence global cardiovascular risk as well as BP, lifestyle modification is currently the mainstay of therapy for individuals with prehypertension who do not have comorbid conditions that would mandate BP reduction by antihypertensive agents [51].

12.7 DRUG THERAPY OF PULMONARY HYPERTENSION

12.7.1 Classifications and General Considerations

Pulmonary hypertension may be defined as an abnormal increase in BP in the pulmonary artery, pulmonary vein, or pulmonary capillaries, together known as the lung vasculature, leading to shortness of breath, dizziness, fainting, and other symptoms, all of which are exacerbated by exertion. Pulmonary hypertension can be a severe disease with a markedly decreased exercise tolerance and heart failure. Pulmonary hypertension is a complex, multidisciplinary disorder and classified both clinically and functionally. Table 12.14 shows the updated clinical classification of pulmonary hypertension [56]. The World Health Organization (WHO) has developed a functional classification

scheme to help determine how limited a patient with pulmonary hypertension is in his/her ability to do the activities of daily living. In general, patients with more severe pulmonary hypertension tend to have a higher functional class (Table 12.15).

12.7.2 Management of Pulmonary Arterial Hypertension

Among the various forms of pulmonary hypertension, pulmonary arterial hypertension (PAH) has received the most extensive attention regarding pharmacological treatment. Hence, this section focuses on discussing the pharmacological agents for treating PAH following a brief introduction to the disease pathophysiology.

12.7.2.1 Pathophysiology PAH is defined as a sustained elevation of pulmonary arterial pressure to more than 25 mmHg at rest or to more than 30 mmHg with exercise, with a mean pulmonary capillary wedge pressure and left ventricular end-diastolic pressure of less than 15 mmHg [57]. Our understanding of the pathophysiology of PAH has undergone a paradigm shift in the past decade. Once a condition thought to be dominated by increased vasoconstrictor tone and thrombosis, PAH is now seen as a vasculopathy in which structural changes/vascular remodeling driven by excessive vascular cell growth, inflammation, as well as oxidative stress with recruitment and infiltration of circulating cells play a major role.

Perturbations of a number of molecular mechanisms have been described, including pathways involving growth factors, cytokines, metabolic signaling, elastases, and proteases that may underlie the pathogenesis of the disease. Identifying their contribution to the pathophysiology of PAH could offer new drug targets [58, 59]. Further elucidation of the molecular pathophysiology of PAH would increase our ability to develop more effective therapeutic modalities for combating this disease.

12.7.2.2 Principles of Pharmacological Management Several classes of drugs have been used to treat PAH with the emphasis of ameliorating pulmonary vasoconstriction and thereby the improvement of patients' systems. Although the mainstay of therapy has focused on the use of vasodilators, the management also involves other drugs, such as diuretics, anticoagulants, and digoxin. The pharmacological treatments for PAH are grouped into nonvasodilating and vasodilating therapies (Table 12.16).

Guidelines and treatment algorithms on the management of PAH have become available and provide evidence-based treatment algorithms for PAH [60–63]. The ACCF/AHA 2009 Expert Consensus Document on Pulmonary Hypertension provides detailed recommendations on the

TABLE 12.14   The updated clinical classification of pulmonary hypertension

1. Pulmonary arterial hypertension (PAH)

1.1	Idiopathic PAH
1.2	Heritable
1.2.1	BMPR2
1.2.2	ALK1, endoglin (with or without hereditary hemorrhagic telangiectasia)
1.2.3	Unknown
1.3	Drug and toxin induced
1.4	PAH associated with:
1.4.1	Connective tissue diseases
1.4.2	HIV infection
1.4.3	Portal hypertension
1.4.4	Congenital heart diseases
1.4.5	Schistosomiasis
1.4.6	Chronic hemolytic anemia
1.5	Persistent pulmonary hypertension of the newborn

1′. Pulmonary veno-occlusive disease (PVOD) and/or pulmonary capillary hemangiomatosis (PCH)

2. Pulmonary hypertension owing to left heart disease

2.1	Systolic dysfunction
2.2	Diastolic dysfunction
2.3	Valvular disease

3. Pulmonary hypertension owing to lung diseases and/or hypoxia

3.1	Chronic obstructive pulmonary disease
3.2	Interstitial lung disease
3.3	Other pulmonary diseases with mixed restrictive and obstructive pattern
3.4	Sleep-disordered breathing
3.5	Alveolar hypoventilation disorders
3.6	Chronic exposure to high altitude
3.7	Developmental abnormalities

4. Chronic thromboembolic pulmonary hypertension (CTEPH)

5. Pulmonary hypertension with unclear multifactorial mechanisms

5.1	Hematologic disorders: myeloproliferative disorders, splenectomy
5.2	Systemic disorders: sarcoidosis, pulmonary Langerhans cell histiocytosis, lymphangioleiomyomatosis, neurofibromatosis, vasculitis
5.3	Metabolic disorders: glycogen storage disease, Gaucher disease, thyroid disorders
5.4	Others: tumoral obstruction, fibrosing mediastinitis, chronic renal failure on dialysis

ALK1, activin receptor-like kinase type 1; BMPR2, bone morphogenetic protein receptor type 2; HIV, human immunodeficiency virus.

TABLE 12.15   WHO functional classification of pulmonary hypertension

Class I	Patients with pulmonary hypertension but without resulting limitation of physical activity. Ordinary physical activity does not cause undue dyspnea or fatigue, chest pain, or near syncope
Class II	Patients with pulmonary hypertension resulting in slight limitation of physical activity. They are comfortable at rest. Ordinary physical activity causes undue dyspnea or fatigue, chest pain, or near syncope
Class III	Patients with pulmonary hypertension resulting in marked limitation of physical activity. They are comfortable at rest. Less than ordinary activity causes undue dyspnea or fatigue, chest pain, or near syncope
Class IV	Patients with pulmonary hypertension with inability to carry out any physical activity without symptoms. These patients manifest signs of right heart failure. Dyspnea and/or fatigue may even be present at rest. Discomfort is increased by any physical activity

TABLE 12.16   Pharmacological treatments of pulmonary arterial hypertension (PAH)

Pharmacological treatment	Pharmacological effects and other comments
Nonvasodilating therapies	
Anticoagulants	• Suppression of thrombin, a contributor to disease progression • Improvement of survival
Diuretics	• Relief of symptoms due to PAH-associated right ventricular failure and systemic venous congestion
Oxygen	• Improvement of hypoxemia
Digoxin	• Improvement of PAH-associated right ventricular dysfunction by increasing myocardial contractility
Vasodilating drugs	
Calcium channel blockers (CCBs)	• Reduction of pulmonary artery pressure due to vasodilating activity • The most commonly used CCBs include long-acting nifedipine, diltiazem, and amlodipine • Due to its potential negative inotropic effects, verapamil should be avoided
Prostacyclins (epoprostenol, iloprost, treprostinil)	• Prostacyclin synthesis is reduced in PAH patients, resulting in inadequate production of prostacyclin I_2, a vasodilator with antiproliferative effects • Administration of prostanoids has been a mainstay of PAH therapy for more than a decade • There are currently three commercially available prostanoids: epoprostenol, treprostinil, and iloprost
PDE5 inhibitors (sildenafil, tadalafil)	• Inhibition of PDE5 produces pulmonary vasodilation by promoting an enhanced and sustained level of cGMP, an identical effect to that of inhaled nitric oxide • Both sildenafil and tadalafil have been approved for treating PAH in addition to their use in treating erectile dysfunction
Endothelin (ET) receptor blockers (bosentan, ambrisentan, macitentan)	• ET-1 is a vasoconstrictor and a smooth muscle mitogen that may contribute to the development of PAH • Blockage of ET receptors improves the symptoms of PAH patients, and three blockers have been approved for treating PAH • Ambrisentan is a selective blocker of ET_A, and bosentan and macitentan are blockers of ET_A and ET_B receptors
Soluble guanylate cyclase (sGC) stimulator (riociguat)	• Riociguat is an sGC stimulator. It directly stimulates sGC and also sensitizes nitric oxide-mediated activation of sGC. • Riociguat is the newest vasodilating drug for treating PAH (approved by the US FDA in 2013)

management of PAH as well as other types of hypertension. The major principles of the pharmacological management of PAH include the following:

- Oral anticoagulation is proposed for most patients; diuretic treatment and supplemental oxygen are indicated in cases of fluid retention and hypoxemia, respectively.
- High doses of CCBs are indicated only in the minority of patients who respond to acute vasoreactivity testing.
- Nonresponders to acute vasoreactivity testing or responders who remain in WHO functional class III should be considered as candidates for treatment with either an oral phosphodiesterase-5 inhibitor or an oral endothelin receptor antagonist.
- Continuous intravenous administration of epoprostenol remains the treatment of choice in WHO functional class IV patients.

- Combination therapy is recommended for patients treated with PAH monotherapy who remain in WHO functional class III.
- Atrial septostomy and lung transplantation are indicated for refractory patients or where medical treatment is unavailable.

12.8   SUMMARY OF CHAPTER KEY POINTS

- The treatment of hypertension, especially the systemic primary hypertension, has been one of medicine's major successes of the past half century. Although various guidelines have been developed by different organizations and in different parts of the globe, the major components of hypertension management are essentially the same, which include lifestyle modifications and drug therapy.

TABLE 12.17 Comparison of the JNC7, AHA 2007, and some most recent guidelines on hypertension management

Guideline	BP target in general population	BP targets in patients with comorbid conditions (diabetes, CKD, CAD, HF)	β-Blockers as first-line therapy for uncomplicated hypertension
JNC7 (2003)	<140/90 mmHg	<130/80 mmHg	Yes
AHA 2007	<140/90 mmHg	<130/80 mmHg	No
AHA/ACC/CDC 2014	<140/90 mmHg	<140/90 mmHg and lower targets may be considered	No
JNC8 (2014)	<140/90 mmHg for aged <60 years; <150/90 mmHg for aged ≥60 years	<140/90 mmHg	No
NICE 2011	<140/90 mmHg for aged <80 years; <150/90 mmHg for aged ≥80 years	Not addressed	No
ESH/ESC 2013	<140/90 mmHg; 140–150/<90 mmHg in the elderly	<140/85 mmHg	Yes
ASH/ISH 2014	<140/90 mmHg; <150/90 for aged ≥80 years; consider <130/80 mmHg for young adults if tolerated	<140/90 mmHg	No
CHEP 2014	<140/90 mmHg; SBP <150 mmHg for aged ≥ 80 years	<130/80 mmHg for diabetes; <140/90 mmHg for others	Yes

- Among the various guidelines, the US-based JNC7 report published 10 years ago still represents the most comprehensive guideline on hypertension management. Recently, multiple guidelines have been developed to incorporate new evidence, especially high-quality data from randomized controlled trials to emphasize the importance of evidence-based recommendations. Among these are the US-based AHA 2007 guideline, AHA/ACC/CDC 2014 Science Advisory, and JNC8 report. Guidelines developed outside the United States include the 2011 NICE and 2013 ESH/ESC guidelines, as well as the joint guideline of ASH/ISH published in 2014.

- Although the aforementioned recent guidelines on hypertensive management were developed by different organizations in different nations, they share many similarities regarding key recommendations on drug therapy and lifestyle modifications. Not surprisingly, some also differ with regard to blood pressure targets in the general population and in patients with comorbid disorders as well as selection of certain antihypertensive agents (Table 12.17) [64, 65].

- Regardless of their differences, the key consensus message from the various guidelines remains the same, that is, guidelines are not a substitute for clinical judgment, and decisions about patient care must carefully consider and incorporate the clinical characteristics and circumstances of each individual patient. Specifically, the selection of appropriate blood pressure targets and medications during hypertension therapy for patients at any age should be influenced by the clinical responses of the patients to their treatment and the healthcare

provider's judgment. On the other hand, advances in antihypertensive therapies will continue to provide healthcare providers with new armamentaria to combat hypertension, including PAH [66–68].

12.9 SELF-ASSESSMENT QUESTIONS

12.9.1. A 36-year-old pregnant woman develops pre-eclampsia. Which of the following drugs is most appropriate for treating this patient's hypertension associated with preeclampsia?
  A. Aliskiren
  B. Hydralazine
  C. Lisinopril
  D. Losartan
  E. Macitentan

12.9.2. A 45-year-old male of African heritage with stage I hypertension also develops asthma. Which of the following antihypertensive agents would be most appropriate for treating this patient's hypertension?
  A. Amlodipine
  B. Captopril
  C. Furosemide
  D. Propranolol
  E. Spironolactone

12.9.3. A 30-year-old African American is diagnosed with essential hypertension, and his blood pressure is 150/95 mmHg. He is otherwise healthy. Which of the following drugs is most appropriate for initial therapy?
  A. Aliskiren
  B. Chlorthalidone

C. Lisinopril
D. Losartan
E. Metoprolol

12.9.4. A 56-year-old female presents to the physician's office complaining of dizziness, headache, and tinnitus. Physical exam reviews a blood pressure of 155/100 mmHg upon repeated measurements. Lab test shows elevated fasting blood glucose (8.9 mM). History reveals that her two brothers also have both hypertension and diabetes. According to the JNC8 report, which of the following is the blood pressure target for this patient?
A. <120/80 mmHg
B. <130/80 mmHg
C. <140/85 mmHg
D. <140/90 mmHg
E. <150/90 mmHg

12.9.5. A 48-year-old female with chronic kidney disease is presented to the physician's office for an annual checkup. Her blood pressure is 150/95 mmHg upon repeated measurements. Which of the following is most suitable for treating her high blood pressure?
A. Aliskiren
B. Amlodipine
C. Atenolol
D. Eplerenone
E. Lisinopril

REFERENCES

1 James, P.A., et al., 2014 evidence-based guideline for the management of high blood pressure in adults: report from the panel members appointed to the Eighth Joint National Committee (JNC 8). *JAMA*, 2014. 311(5): p. 507–20.

2 Appel, L.J., et al., Dietary approaches to prevent and treat hypertension: a scientific statement from the American Heart Association. *Hypertension*, 2006. 47(2): p. 296–308.

3 Rosendorff, C., et al., Treatment of hypertension in the prevention and management of ischemic heart disease: a scientific statement from the American Heart Association Council for High Blood Pressure Research and the Councils on Clinical Cardiology and Epidemiology and Prevention. *Circulation*, 2007. 115(21): p. 2761–88.

4 Drozda, J., Jr., et al., ACCF/AHA/AMA-PCPI 2011 performance measures for adults with coronary artery disease and hypertension a report of the American College of Cardiology Foundation/American Heart Association Task Force on Performance Measures and the American Medical Association-Physician Consortium for Performance Improvement. *J Am Coll Cardiol*, 2011. 58(3): p. 316–36.

5 Aronow, W.S., et al., ACCF/AHA 2011 expert consensus document on hypertension in the elderly: a report of the American College of Cardiology Foundation Task Force on Clinical Expert Consensus Documents. *Circulation*, 2011. 123(21): p. 2434–506.

6 Go, A.S., et al., An effective approach to high blood pressure control: a science advisory from the American Heart Association, the American College of Cardiology, and the Centers for Disease Control and Prevention. *J Am Coll Cardiol*, 2014. 63(12): p. 1230–8.

7 Weber, M.A., et al., Clinical practice guidelines for the management of hypertension in the community a statement by the American Society of Hypertension and the International Society of Hypertension. *J Hypertens*, 2014. 32(1): p. 3–15.

8 Dasgupta, K., et al., The 2014 Canadian Hypertension Education Program recommendations for blood pressure measurement, diagnosis, assessment of risk, prevention, and treatment of hypertension. *Can J Cardiol*, 2014. 30(5): p. 485–501.

9 Williams, B., et al., Guidelines for management of hypertension: report of the fourth working party of the British Hypertension Society, 2004-BHS IV. *J Hum Hypertens*, 2004. 18(3): p. 139–85.

10 Krause, T., et al., Management of hypertension: summary of NICE guidance. *BMJ*, 2011. 343: p. d4891.

11 Mancia, G., et al., 2013 ESH/ESC guidelines for the management of arterial hypertension: the Task Force for the management of arterial hypertension of the European Society of Hypertension (ESH) and of the European Society of Cardiology (ESC). *J Hypertens*, 2013. 31(7): p. 1281–357.

12 Chobanian, A.V., et al., The seventh report of the Joint National Committee on Prevention, Detection, Evaluation, and Treatment of High Blood Pressure: the JNC 7 report. *JAMA*, 2003. 289(19): p. 2560–72.

13 Rosendorff, C., Hypertension and coronary artery disease: a summary of the American Heart Association scientific statement. *J Clin Hypertens (Greenwich)*, 2007. 9(10): p. 790–5.

14 Thomas, G., et al., New hypertension guidelines: one size fits most? *Cleve Clin J Med*, 2014. 81(3): p. 178–88.

15 Eckel, R.H., et al., 2013 AHA/ACC guideline on lifestyle management to reduce cardiovascular risk: a report of the American College of Cardiology/American Heart Association Task Force on Practice Guidelines. *J Am Coll Cardiol*, 2014. 63(25 Pt B): p. 2960–84.

16 Chobanian, A.V., et al., Seventh report of the Joint National Committee on Prevention, Detection, Evaluation, and Treatment of High Blood Pressure. *Hypertension*, 2003. 42(6): p. 1206–52.

17 Eckel, R.H., et al., 2013 AHA/ACC guideline on lifestyle management to reduce cardiovascular risk: a report of the American College of Cardiology/American Heart Association Task Force on Practice Guidelines. *Circulation*, 2014. 129(25 Suppl 2): p. S76–S99.

18 Appel, L.J., et al., A clinical trial of the effects of dietary patterns on blood pressure. DASH Collaborative Research Group. *N Engl J Med*, 1997. 336(16): p. 1117–24.

19 Sacks, F.M., et al., Effects on blood pressure of reduced dietary sodium and the Dietary Approaches to Stop Hypertension (DASH) diet. DASH-Sodium Collaborative Research Group. *N Engl J Med*, 2001. 344(1): p. 3–10.

20 Sacks, F.M. and H. Campos, Dietary therapy in hypertension. *N Engl J Med*, 2010. 362(22): p. 2102–12.

21 Ernst, M.E. and B.C. Lund, Renewed interest in chlorthalidone: evidence from the Veterans Health Administration. *J Clin Hypertens (Greenwich)*, 2010. 12(12): p. 927–34.

22 Dorsch, M.P., et al., Chlorthalidone reduces cardiovascular events compared with hydrochlorothiazide: a retrospective cohort analysis. *Hypertension*, 2011. 57(4): p. 689–94.

23 Jamerson, K., et al., Benazepril plus amlodipine or hydrochlorothiazide for hypertension in high-risk patients. *N Engl J Med*, 2008. 359(23): p. 2417–28.

24 Ernst, M.E., B.L. Carter, and J.N. Basile, All thiazide-like diuretics are not chlorthalidone: putting the ACCOMPLISH study into perspective. *J Clin Hypertens (Greenwich)*, 2009. 11(1): p. 5–10.

25 De Caterina, A.R. and A.M. Leone, Why beta-blockers should not be used as first choice in uncomplicated hypertension. *Am J Cardiol*, 2010. 105(10): p. 1433–8.

26 Ram, C.V., Beta-blockers in hypertension. *Am J Cardiol*, 2010. 106(12): p. 1819–25.

27 Law, M.R., J.K. Morris, and N.J. Wald, Use of blood pressure lowering drugs in the prevention of cardiovascular disease: meta-analysis of 147 randomised trials in the context of expectations from prospective epidemiological studies. *BMJ*, 2009. 338: p. b1665.

28 Taylor, A.A. and J.L. Pool, Clinical Role of Direct Renin Inhibition in Hypertension. *Am J Ther*, 2012. 19(3): p. 204–10.

29 Sharabi, Y., et al., Efficacy of add-on aldosterone receptor blocker in uncontrolled hypertension. *Am J Hypertens*, 2006. 19(7): p. 750–5.

30 Jansen, P.M., et al., Long-term use of aldosterone-receptor antagonists in uncontrolled hypertension: a retrospective analysis. *Int J Hypertens*, 2011. 2011: p. 368140.

31 Smith, R.E. and M. Ashiya, Antihypertensive therapies. *Nat Rev Drug Discov*, 2007. 6(8): p. 597–8.

32 Costanzo, P., et al., Calcium channel blockers and cardiovascular outcomes: a meta-analysis of 175,634 patients. *J Hypertens*, 2009. 27(6): p. 1136–51.

33 Kidney Disease Outcomes Quality Initiative (K/DOQI). K/DOQI clinical practice guidelines on hypertension and antihypertensive agents in chronic kidney disease. *Am J Kidney Dis*, 2004. 43(5 Suppl 1): p. S1–290.

34 Taler, S.J., et al., KDOQI US commentary on the 2012 KDIGO clinical practice guideline for management of blood pressure in CKD. *Am J Kidney Dis*, 2013. 62(2): p. 201–13.

35 Chobanian, A.V., Clinical practice. Isolated systolic hypertension in the elderly. *N Engl J Med*, 2007. 357(8): p. 789–96.

36 Cohen, D.L. and R.R. Townsend, Update on pathophysiology and treatment of hypertension in the elderly. *Curr Hypertens Rep*, 2011. 13(5): p. 330–7.

37 Arif, S.A., et al., Treatment of systolic heart failure in the elderly: an evidence-based review. *Ann Pharmacother*, 2010. 44(10): p. 1604–14.

38 Beckett, N.S., et al., Treatment of hypertension in patients 80 years of age or older. *N Engl J Med*, 2008. 358(18): p. 1887–98.

39 Villar, V.A., T. Liu, and P.A. Jose, Recent trends in pediatric hypertension research. *J Med Liban*, 2010. 58(3): p. 179–84.

40 Franks, P.W., et al., Childhood obesity, other cardiovascular risk factors, and premature death. *N Engl J Med*, 2010. 362(6): p. 485–93.

41 Ingelfinger, J.R., Clinical practice. The child or adolescent with elevated blood pressure. *N Engl J Med*, 2014. 370(24): p. 2316–25.

42 Falkner, B. and S.R. Daniels, Summary of the fourth report on the diagnosis, evaluation, and treatment of high blood pressure in children and adolescents. *Hypertension*, 2004. 44(4): p. 387–8.

43 Flack, J.M., S.A. Nasser, and P.D. Levy, Therapy of hypertension in African Americans. *Am J Cardiovasc Drugs*, 2011. 11(2): p. 83–92.

44 Douglas, J.G., et al., Management of high blood pressure in African Americans: consensus statement of the Hypertension in African Americans Working Group of the International Society on Hypertension in Blacks. *Arch Intern Med*, 2003. 163(5): p. 525–41.

45 Moser, M. and J.F. Setaro, Clinical practice. Resistant or difficult-to-control hypertension. *N Engl J Med*, 2006. 355(4): p. 385–92.

46 Sarafidis, P.A. and G.L. Bakris, Resistant hypertension: an overview of evaluation and treatment. *J Am Coll Cardiol*, 2008. 52(22): p. 1749–57.

47 Vongpatanasin, W., Resistant hypertension: a review of diagnosis and management. *JAMA*, 2014. 311(21): p. 2216–24.

48 Calhoun, D.A., Hyperaldosteronism as a common cause of resistant hypertension. *Annu Rev Med*, 2013. 64: p. 233–47.

49 Solini, A. and L.M. Ruilope, How can resistant hypertension be identified and prevented? *Nat Rev Cardiol*, 2013. 10(5): p. 293–6.

50 Calhoun, D.A., et al., Resistant hypertension: diagnosis, evaluation, and treatment: a scientific statement from the American Heart Association Professional Education Committee of the Council for High Blood Pressure Research. *Circulation*, 2008. 117(25): p. e510–26.

51 Pimenta, E. and S. Oparil, Prehypertension: epidemiology, consequences and treatment. *Nat Rev Nephrol*, 2010. 6(1): p. 21–30.

52 Julius, S., et al., Feasibility of treating prehypertension with an angiotensin-receptor blocker. *N Engl J Med*, 2006. 354(16): p. 1685–97.

53 Luders, S., et al., The PHARAO study: prevention of hypertension with the angiotensin-converting enzyme inhibitor ramipril in patients with high-normal blood pressure: a prospective, randomized, controlled prevention trial of the German Hypertension League. *J Hypertens*, 2008. 26(7): p. 1487–96.

54 Thompson, A.M., et al., Antihypertensive treatment and secondary prevention of cardiovascular disease events among persons without hypertension: a meta-analysis. *JAMA*, 2011. 305(9): p. 913–22.

55 Ventura, H.O. and C.J. Lavie, Antihypertensive therapy for prehypertension: relationship with cardiovascular outcomes. *JAMA*, 2011. 305(9): p. 940–1.

56 Simonneau, G., et al., Updated clinical classification of pulmonary hypertension. *J Am Coll Cardiol*, 2009. 54(1 Suppl): p. S43–54.

57 Farber, H.W. and J. Loscalzo, Pulmonary arterial hypertension. *N Engl J Med*, 2004. 351(16): p. 1655–65.

58 Morrell, N.W., et al., Cellular and molecular basis of pulmonary arterial hypertension. *J Am Coll Cardiol*, 2009. 54(1 Suppl): p. S20–31.

59 Schermuly, R.T., et al., Mechanisms of disease: pulmonary arterial hypertension. *Nat Rev Cardiol*, 2011. 8(8): p. 443–55.

60 Barst, R.J., et al., Updated evidence-based treatment algorithm in pulmonary arterial hypertension. *J Am Coll Cardiol*, 2009. 54(1 Suppl): p. S78–84.

61 McLaughlin, V.V., et al., ACCF/AHA 2009 expert consensus document on pulmonary hypertension: a report of the American College of Cardiology Foundation Task Force on Expert Consensus Documents and the American Heart Association: developed in collaboration with the American College of Chest Physicians, American Thoracic Society, Inc., and the Pulmonary Hypertension Association. *Circulation*, 2009. 119(16): p. 2250–94.

62 Galie, N., et al., Updated treatment algorithm of pulmonary arterial hypertension. *J Am Coll Cardiol*, 2013. 62(25 Suppl): p. D60–72.

63 McLaughlin, V.V., et al., Treatment goals of pulmonary hypertension. *J Am Coll Cardiol*, 2013. 62(25 Suppl): p. D73–81.

64 Krakoff, L.R., Recent guidelines for the management of hypertension: what is missing? *Blood Press Monit*, 2014. 19(4): p. 189–91.

65 Weber, M.A., Recently published hypertension guidelines of the JNC 8 panelists, the American Society of Hypertension/International Society of Hypertension and other major organizations: introduction to a focus issue of the Journal of Clinical Hypertension. *J Clin Hypertens (Greenwich)*, 2014. 16(4): p. 241–5.

66 Paulis, L., U.M. Steckelings, and T. Unger, Key advances in antihypertensive treatment. *Nat Rev Cardiol*, 2012. 9(5): p. 276–85.

67 Burnier, M., Y. Vuignier, and G. Wuerzner, State-of-the-art treatment of hypertension: established and new drugs. *Eur Heart J*, 2014. 35(9): p. 557–62.

68 Gomberg-Maitland, M., et al., New trial designs and potential therapies for pulmonary artery hypertension. *J Am Coll Cardiol*, 2013. 62(25 Suppl): p. D82–91.

UNIT IV

ISCHEMIC HEART DISEASE: STABLE ISCHEMIC HEART DISEASE

13

OVERVIEW OF ISCHEMIC HEART DISEASE, STABLE ANGINA, AND DRUG THERAPY

13.1 INTRODUCTION

Ischemic heart disease (IHD) is the single most common cause of death in developed nations as well as in many developing countries [1, 2]. IHD is an umbrella term that encompasses a spectrum of cardiac disorders caused by myocardial ischemia. The notable examples of IHD include stable IHD (SIHD) (with stable angina as its prototypical manifestation) and acute coronary syndromes (ACS), among many others. This chapter provides an overview of IHD and discusses the pathophysiology of SIHD and the mechanistically based drug targeting of stable angina. Chapter 14 reviews antianginal drugs that have already been discussed in previous chapters and also considers some newly approved antianginal drugs that are not covered in previous chapters. The principles and guidelines regarding the management of SIHD/stable angina in clinical practice are given in Chapter 15.

13.2 CLASSIFICATION, EPIDEMIOLOGY, AND PATHOPHYSIOLOGY

13.2.1 Classification

13.2.1.1 Definition of IHD and the International Statistical Classification of Diseases and Related Health Problems–10th Revision Classification As noted earlier, the term IHD refers to a spectrum of diseases of the heart caused by decreased oxygen supply to the myocardium. The International Statistical Classification of Diseases and Related Health Problems–10th Revision (ICD-10) classifies

IHD into the following six categories, and each category consists of multiple disease entities (also in Table 13.1):

1. Angina pectoris
2. Acute myocardial infarction
3. Certain current complications following acute myocardial infarction
4. Subsequent myocardial infarction
5. Other acute IHD
6. Chronic IHD including coronary artery disease (CAD), among others

13.2.1.2 Conventional Classification of IHD ICD-10 classification of IHD is comprehensive and authoritative; however, it is complicated and often times causes confusion. Hence, a simplified classification scheme is frequently used to divide IHD into two general categories (Fig. 13.1): (1) SIHD with stable angina as the prototypical manifestation and (2) ACS that include unstable angina, non-ST-elevation myocardial infarction, and ST-elevation myocardial infarction (see Unit V). SIHD is also frequently known as stable coronary artery disease (SCAD). Regardless of the nomenclature, stable angina is the chief manifestation of SIHD or SCAD. Indeed, the main symptomatic clinical presentations of SIHD include (i) classical chronic stable angina caused by epicardial stenosis; (ii) angina caused by microvascular dysfunction (also known as microvascular angina), (iii) angina caused by vasospasm (vasospastic angina), and (iv) symptomatic ischemic cardiomyopathy [3].

Cardiovascular Diseases: From Molecular Pharmacology to Evidence-Based Therapeutics, First Edition. Y. Robert Li.
© 2015 John Wiley & Sons, Inc. Published 2015 by John Wiley & Sons, Inc.

TABLE 13.1   ICD-10 classification of ischemic heart diseases (I20–I25)[a]

ICD-10 code	ICD-10 subcode and disease description
I20: Angina pectoris	I20.0. Unstable angina Angina: • Crescendo • De novo effort • Worsening effort Intermediate coronary syndrome Preinfarction syndrome I20.1. Angina pectoris with documented spasm Angina: • Angiospastic • Prinzmetal • Spasm induced • Variant I20.8. Other forms of angina pectoris Angina of effort Stenocardia I20.9. Angina pectoris, unspecified Angina: • NOS • Cardiac Anginal syndrome Ischemic chest pain
I21: Acute myocardial infarction	I21.0. Acute transmural myocardial infarction of anterior wall Transmural infarction (acute) (of): • Anterior (wall) NOS • Anteroapical • Anterolateral • Anteroseptal I21.1. Acute transmural myocardial infarction of inferior wall Transmural infarction (acute) (of): • Diaphragmatic wall • Inferior (wall) NOS • Inferolateral • Inferoposterior I21.2. Acute transmural myocardial infarction of other sites Transmural infarction (acute) (of): • Apical–lateral • Basal–lateral • High lateral • Lateral (wall) NOS • Posterior (true) • Posterobasal • Posterolateral • Posteroseptal • Septal NOS I21.3. Acute transmural myocardial infarction of unspecified site Transmural myocardial infarction NOS I21.4. Acute subendocardial myocardial infarction Nontransmural myocardial infarction NOS I21.9. Acute myocardial infarction, unspecified Myocardial infarction (acute) NOS
I22: Subsequent myocardial infarction	I22.0. Subsequent myocardial infarction of anterior wall Subsequent infarction (acute) (of): • Anterior (wall) NOS • Anteroapical • Anterolateral • Anteroseptal

TABLE 13.1   *(Continued)*

ICD-10 code	ICD-10 subcode and disease description
	I22.1. Subsequent myocardial infarction of inferior wall Subsequent infarction (acute) (of): • Diaphragmatic wall • Inferior (wall) NOS • Inferolateral • Inferoposterior I22.8. Subsequent myocardial infarction of other sites Subsequent myocardial infarction (acute) (of): • Apical–lateral • Basal–lateral • High lateral • Lateral (wall) NOS • Posterior (true) • Posterobasal • Posterolateral • Posteroseptal • Septal NOS I22.9. Subsequent myocardial infarction of unspecified site
I23: Certain current complications following acute myocardial infarction	I23.0. Hemopericardium as current complication following acute myocardial infarction
Excl.: the listed conditions, when: • Concurrent with acute myocardial infarction (I21-I22) • Not specified as current complications following acute myocardial infarction (I31.-, I51.-)[b, c]	I23.1. Atrial septal defect as current complication following acute myocardial infarction I23.2. Ventricular septal defect as current complication following acute myocardial infarction I23.3. Rupture of cardiac wall without hemopericardium as current complication following acute myocardial infarction Excl.: with hemopericardium (I23.0) I23.4. Rupture of chordae tendineae as current complication following acute myocardial infarction I23.5. Rupture of papillary muscle as current complication following acute myocardial infarction I23.6. Thrombosis of the atrium, auricular appendage, and ventricle as current complications following acute myocardial infarction I23.8. Other current complications following acute myocardial infarction
I24: Other acute ischemic heart diseases Excl.: angina pectoris (I20.-), transient myocardial ischemia of newborn (P29.4)[d]	I24.0. Coronary thrombosis not resulting in myocardial infarction Coronary (artery)(vein): • Embolism • Occlusion • Thromboembolism Excl.: specified as chronic or with a stated duration of >4 weeks (>28 days) from onset (I25.8) I24.1. Dressler syndrome Postmyocardial infarction syndrome I24.8. Other forms of acute ischemic heart disease Coronary: • Failure • Insufficiency I24.9. Acute ischemic heart disease, unspecified Excl.: ischemic heart disease (chronic) NOS (I25.9)
I25: Chronic ischemic heart disease Excl.: cardiovascular disease NOS (I51.6)[e]	I25.0. Atherosclerotic cardiovascular disease, so described I25.1. Atherosclerotic heart disease Coronary (artery): • Atheroma • Atherosclerosis • Disease • Sclerosis

(Continued)

TABLE 13.1   *(Continued)*

ICD-10 code	ICD-10 subcode and disease description
	I25.2. Old myocardial infarction
	Healed myocardial infarction
	Past myocardial infarction diagnosed by ECG or other special investigation, but currently presenting no symptoms
	I25.3. Aneurysm of the heart
	Aneurysm:
	• Mural
	• Ventricular
	I25.4. Coronary artery aneurysm
	Coronary arteriovenous fistula, acquired
	Excl.: congenital coronary (artery) aneurysm (Q24.5)f
	I25.5. Ischemic cardiomyopathy
	I25.6. Silent myocardial ischemia
	I25.8. Other forms of chronic ischemic heart disease
	Any condition in I21-I22 and I24.- specified as chronic or with a stated duration of >4 weeks (>28 days) from onset
	I25.9. Chronic ischemic heart disease, unspecified
	Ischemic heart disease (chronic) NOS

Excl., excluding; NOS, not otherwise specified.

[a]Adapted from http://apps.who.int/classifications/icd10/browse/2010/en.

[b]I31.-: Other diseases of pericardium under Other forms of heart disease (I30–I52) of Chapter IX: Diseases of the circulatory system (I00–I99).

[c]I51.-: Complications and ill-defined descriptions of heart disease under Other forms of heart disease (I30-I52) of Chapter IX: Diseases of the circulatory system (I00–I99).

[d]P29.4: Transient myocardial ischemia of newborn under Respiratory and cardiovascular disorders specific to the perinatal period (P20–P29) of Chapter XVI: Certain conditions originating in the perinatal period (P00–P96).

[e]I51.6: Cardiovascular disease, unspecified (cardiovascular accident NOS) under Other forms of heart disease (I30–I52) of Chapter IX: Diseases of the circulatory system (I00–I99).

[f]Q24.5: Malformation of coronary vessels under Congenital malformations of the circulatory system (Q20–Q28) of Chapter XVII: Congenital malformations, deformations and chromosomal abnormalities (Q00–Q99).

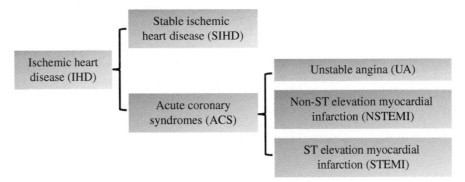

FIGURE 13.1   Conventional classification of ischemic heart disease (IHD). As illustrated, IHD is typically classified into stable IHD (with stable angina as the prototypical clinical manifestation) and acute coronary syndromes (ACS). ACS include unstable angina, non-ST-elevation myocardial infarction, and ST-elevation myocardial infarction.

13.2.1.3   *Definition of CAD and Coronary Heart Disease*

CAD and coronary heart disease (CHD) are two most commonly encountered terms in cardiovascular medicine and frequently used synonymously by healthcare professionals. However, strictly speaking, there are differences between these two terms. CAD is generally used to refer to the pathological process affecting the coronary arteries (usually atherosclerosis). On the other hand, CHD is actually a result of CAD. With CAD, plaque first grows in the coronary arteries until the blood flow to the cardiac muscle is limited. This is also called myocardial ischemia. It may be chronic, caused by narrowing of the coronary artery and limitation of the blood supply to part of the muscle. Or it can be acute, resulting from a sudden plaque rupture. Hence, CHD includes the diagnoses of angina pectoris, myocardial infarction, silent myocardial ischemia, and CHD mortality that result from CAD.

13.2.2 Epidemiology

CHD is a major cause of death and disability in developed countries as well as many developing countries, such as China. Although CHD mortality rates have declined over the past four decades in the United States (and elsewhere), currently, CHD remains responsible for about one sixth of all deaths in the country. The 2014 Heart Disease and Stroke Statistics update of the American Heart Association reported that 15.4 million (or 6.4%) people (age ≥20 years) in the United States have CHD, including 7.6 million with myocardial infarction and 7.8 million with angina pectoris [4]. The reported prevalence increases with age for both women and men. For individuals aged 40 years in the United States, the lifetime risk of developing CHD is 49% in men and 32% in women. Lifetime risk for CHD varies drastically as a function of risk factor profiles. With an optimal risk factor profile, lifetime risk for CHD is 3.6% for men and <1% for women; with ≥2 major risk factors, it is 37.5% for men and 18.3% for women [5]. The key statistical data of CHD in the United States are given in Table 13.2.

Population-based epidemiological data, such as that from the Framingham Heart Study (see Section 1.7.1), provide the best assessment of the risk factors that contribute to the development of CHD and to the way it evolves, progresses, and terminates because these data are less encumbered by the unavoidable selection bias of clinical trials data. In addition, epidemiological data provide critical information regarding targets for the primary and secondary prevention of CHD [7] (see Chapter 1).

13.2.3 Pathophysiology

Atherosclerosis is the fundamental pathophysiological basis of IHD. It is the process that results in the buildup of plaque in the coronary arteries that may subsequently lead to stable angina and ACS. Stable angina is caused by narrowing of the coronary artery and limitation of the blood supply to part of myocardium as a result of gradual buildup of plaque that is stable. On the other hand, the sudden rupture of a plaque and the subsequent thrombosis are responsible for ACS.

TABLE 13.2 Key statistical data of CHD in the United States[a]

CHD	Statistical data
Prevalence	On the basis of data from "NHANES 2007–2010," an estimated 15.4 million US adults (age of ≥20 years) have CHD
	The overall CHD prevalence is 6.4% (7.9% for men and 5.1% for women), and the overall prevalence for myocardial infarction is 2.9% (4.2% for men and 1.7% for women)
	Projections show that by 2030, prevalence of CHD will increase approximately 18% from 2013 estimates
Incidence	The estimated annual incidence of myocardial infarction is 515,000 new attacks and 205,000 recurrent attacks
	Approximately every 44 s, an American will have a myocardial infarction
	Average age at first myocardial infarction is 64.9 years for men and 72.3 years for women
	While some studies suggested an overall decline in the incidence of myocardial infarction over the past decades, others showed no significant changes. Analysis of over 40 years of physician-validated acute myocardial infarction data in the NHLBI's Framingham Heart Study found that acute myocardial infarction rates diagnosed by electrocardiographic criteria declined by approximately 50%, with a concomitant twofold increase in rates of acute myocardial infarction diagnosed by blood markers. These findings may explain the apparently steady national acute myocardial infarction rates in the face of improvements in primary prevention [6]
Mortality	CHD was an underlying cause of death in approximately one of every six deaths in the United States in 2010. In 2010, CHD mortality was 379,559, and among this, myocardial infarction mortality was 122,071
	In 2010, 73% of CHD deaths occurred out of the hospital. According to NCHS mortality data, 278,000 CHD deaths occur out of the hospital or in hospital emergency departments annually
	About 50% of men and >60% of women who die suddenly of CHD have no previous symptoms of this disease
	People who have had a myocardial infarction have a sudden death rate 4–6 times that of the general population
	Within 5 years after a first myocardial infarction, at ≥45 years of age, 36% of men and 47% of women will die
	CHD death rates have fallen from 1968 to the present. From 2000 to 2010, the annual death rate attributable to CHD declined 39.2%, and the actual number of deaths declined 26.3%
	It was estimated that approximately 47% of the decrease in CHD deaths was attributable to treatments (including secondary preventive therapies after myocardial infarction or revascularization, initial treatments for acute myocardial infarction or unstable angina, treatments for heart failure, revascularization for chronic angina, and other therapies such as antihypertensive and lipid-lowering primary prevention therapies) and approximately 44% was attributable to changes in risk factors (lower total cholesterol, lower blood pressure, lower smoking prevalence, and decreased physical inactivity)

NCHS, National Center for Health Statistics; NHANES, National Health and Nutrition Examination Survey; NHLBI, National Heart, Lung, and Blood Institute.
[a]Adapted from Ref. [4].

The term atherosclerosis comes from the Greek words *athero* (meaning gruel or paste) and *sclerosis* (hardness). It refers to the process of fatty substances, cholesterol, cellular waste products, calcium, and fibrin (a clotting material in the blood) building up in the inner lining of an artery. The resulting buildup is called plaque, which, as noted earlier, is responsible for IHD. Plaque may partially or totally block the blood flow through a coronary artery. As listed below, two things can happen where plaque occurs:

1. There may be bleeding (hemorrhage) into the plaque.
2. A blood clot (thrombus) may form on the plaque's surface.

Atherosclerosis is a complex process that begins in childhood. Exactly how atherosclerosis begins or what causes it remains partially understood. It is generally believed that atherosclerosis starts when the endothelium of the artery becomes damaged. Listed below are four possible causes of damage to the arterial wall:

1. Elevated levels of cholesterol and triglycerides in the blood
2. High blood pressure
3. Cigarette smoking
4. Inflammatory and oxidative stress

Cigarette smoking greatly aggravates and speeds up the growth of atherosclerosis in the coronary arteries, the aorta, and the arteries of the extremities. Because of the damage, over time, cholesterol, triglycerides, platelets, cellular debris, and calcium are deposited in the artery wall. These substances may stimulate the cells of the artery wall to produce other molecules, including growth factors and proinflammatory cytokines. This results in more cells accumulating in the innermost layer of the artery wall where the atherosclerotic lesions form. These cells accumulate, and many of them divide. At the same time, fat builds up within and around these cells. They also form connective tissue. The innermost layer of the artery becomes markedly thickened by these accumulating cells and surrounding material. If the wall is thickened sufficiently, the diameter of the artery will be reduced and less blood will flow, thus decreasing the oxygen supply, which may result in ischemia. Often, a blood clot forms and blocks the artery, stopping the flow of blood. If the oxygen supply to the cardiac muscle is reduced, a myocardial infarction can occur. On the other hand, if the oxygen supply to the brain is cut off, an ischemic stroke can occur (see Unit VIII). Additionally, if the oxygen supply to the extremities stops, it may cause gangrene.

13.3   STABLE ANGINA AND DRUG TARGETING

13.3.1   Definition and Classification

Angina, formally known as angina pectoris, is a term used to describe chest pain. Along with chest pain, individuals may also feel a sensation of pressure or tightness in the chest. The term angina is derived from a Latin word, meaning "to choke." The angina pain may be felt in the jaw, arm, neck, back, or shoulder as well. Angina, caused by a reduced amount of oxygen flowing to the heart, is a symptom of IHD and is not a medical condition itself. Angina signifies that the affected individual is at a greater risk of suffering from a heart attack or cardiac arrest.

In addition to the ICD-10 classification described in Table 13.1, there are various other ways of classification. One conventional scheme classifies angina into three categories as listed below (also see Table 13.3):

1. Stable angina
2. Unstable angina
3. Variant angina

Based on the characteristics of chest pain (Table 13.4), angina is also classified into typical and atypical angina.

TABLE 13.3   Conventional classification of angina into stable, unstable, and variant angina

Classification	Characteristics
Stable angina	Stable angina is chest pain that typically occurs when an individual suffering from CAD increases the oxygen demand on the heart. With CAD, the blood vessels that supply blood and oxygen to the heart are weakened or blocked. The relative lack of oxygen delivered to the heart causes the angina. Stable angina is predictable because it is generally provoked by exertion or emotional stress. Stable angina is relieved by rest or nitroglycerin (see Chapter 11)
Unstable angina	Unstable angina occurs when the chest pain begins to last longer than 15 min, comes on without warning, and does not respond well to rest and medication. Unstable angina is associated with an increased risk of an impending heart attack. The pain may change in severity once unstable angina occurs
Variant angina	Variant angina, also referred to as Prinzmetal angina or vasospasm angina, is a rare form of angina. An individual experiences variant angina after a spasm in one of the coronary arteries. If a coronary artery suddenly spasms, the vessel narrows and decreases the blood supply to the heart. Risk factors such as smoking, high blood pressure, high cholesterol, and cold temperatures increase the chance of developing coronary artery spasms. This type of angina pain occurs during rest, usually between the hours of midnight to 8 A.M. and lasts for 5–30 min. Variant angina is relieved by nitroglycerin

TABLE 13.4   Traditional clinical classification of chest pain into typical and atypical angina [3, 8]

Typical angina (definite)	Meets all three of the following characteristics:
	1. Substernal chest discomfort of characteristic quality and duration 2. Provoked by exertion or emotional stress 3. Relieved by rest and/or nitrates within minutes
Atypical angina (probable)	Meets 2 of the three typical anginal characteristics
Nonanginal chest pain	Meets 1 or none of the three typical anginal characteristics

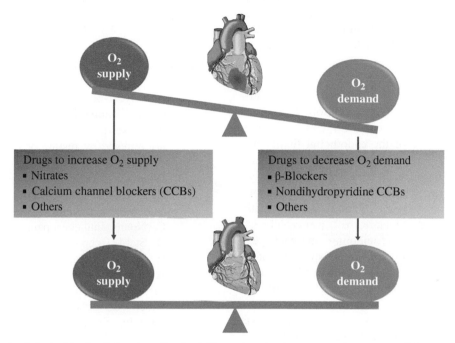

FIGURE 13.2   Pathophysiological basis of drug targeting in stable angina. Angina occurs when myocardial oxygen demand is not met by oxygen supply. Drugs that increase myocardial oxygen supply and/or decrease myocardial oxygen demand are used to treat stable angina.

Typical angina and atypical angina are also known as definite and probable angina, respectively.

13.3.2   Pathophysiology and Drug Targeting

Section 13.2.3 describes the overall pathophysiology of IHD, emphasizing the fundamental causal role of atherosclerosis in disease development. This section discusses the pathophysiology of stable angina, which serves as a basis for understanding antianginal drug targeting.

13.3.2.1   Pathophysiology   The pathophysiology of stable angina can be understood from two different aspects: (1) myocardial oxygen imbalance and (2) histological characteristics of atheroma.

Myocardial Oxygen Imbalance   Under physiological conditions, myocardial oxygen demand and supply are balanced. Angina is caused by disturbance of this balance, characterized by myocardial oxygen demand exceeding oxygen supply or oxygen supply failing to meet the oxygen demand (Fig. 13.2). Hence, an understanding of the pathophysiology of angina first requires a brief review of the determinants of myocardial oxygen demand and supply:

- *Myocardial oxygen demand*: There are four major factors that determine myocardial work and therefore myocardial oxygen demand: (1) heart rate; (2) systolic arterial blood pressure, which determines the afterload of the heart; (3) myocardial wall tension or stress, which is the product of ventricular end-diastolic volume (preload) and myocardial muscle mass; and (4) myocardial contractility. Clinical conditions associated with an increase in myocardial oxygen demand must affect one or more of these parameters. Examples include increased sympathetic activity, as with physical exertion or mental stress, environmental stress (such as cold weather and pollution), tachycardia of any etiology, high blood pressure, and left ventricular hypertrophy.

- *Myocardial oxygen supply*: The major determinants of oxygen supply are the oxygen-carrying capacity of the blood, which is affected by a variety of factors, including (i) oxygen tension and the hemoglobin concentration; (ii) the degree of oxygen unloading from hemoglobin to the tissues; and (iii) the coronary artery blood flow, which is in turn determined by coronary artery diameter and tone, collateral blood flow, coronary perfusion pressure, blood flow within the endocardium (determined by the left ventricular end-diastolic pressure), and heart rate (coronary artery flow primarily occurs during diastole). Although any clinical setting that reduces myocardial oxygen supply can cause ischemia and angina, the predominant cause is the epicardial coronary atherosclerosis that limits the coronary blood flow.

Histological Characteristics of Atheroma In patients with stable angina, the epicardial atherosclerotic lesions, as compared with those of patients with ACS, less commonly show an erosion or rupture of the endothelial lining; the lesions are typically fibrotic, poorly cellular, and with small necrotic cores, thick fibrous caps, and little or no overlying thrombus. In contrast, culprit lesions of ACS patients typically show the rupture or tear of a thin fibrous cap, with exposure toward the lumen of large, soft, prothrombotic, and necrotic core material (containing macrophages, cholesterol clefts, debris, inflammatory cell infiltrates, neovascularization, and/or intraplaque hemorrhage) that can trigger occlusive or subocclusive thrombosis [3, 9].

13.3.2.2 Drug Targeting Various classes of drugs have been used to treat stable angina based on the current understanding of its pathophysiology (Fig. 13.2). Chapter 14 provides a detailed discussion of the major classes of antianginal drugs with an emphasis on the pharmacological basis of their use to treat stable angina. These drugs are also used in the management of unstable angina as well as other cardiovascular disorders. Briefly, drugs used to treat stable angina include those that (i) directly decrease oxygen demand (β-blockers, calcium channel blockers), (ii) directly increase oxygen supply (organic nitrates, calcium channel blockers), and (iii) act via novel mechanisms (e.g., ranolazine).

13.4 SUMMARY OF CHAPTER KEY POINTS

- IHD is the single most important cause of death worldwide. It is an umbrella term that refers to a spectrum of cardiac disorders caused by myocardial ischemia, with notable examples including SIHD/stable angina and ACS. Atherosclerosis is the fundamental pathophysiological basis of IHD.

- Angina occurs when myocardial oxygen demand exceeds oxygen supply (i.e., myocardial ischemia).

Stable angina is the prototypical manifestation of SIHD (also known as SCAD), predominantly caused by the narrowing of coronary arteries due to atherosclerosis.

- Pharmacological therapy for stable angina involves the use of drugs that directly decrease myocardial oxygen demand, directly increase oxygen supply, or act via novel mechanisms. These various classes of drugs in treating stable angina are considered in Chapter 14.

13.5 SELF-ASSESSMENT QUESTIONS

13.5.1 A 56-year-old male presents to the physician's office, complaining of chest tightness and pain upon exertion or emotional stress. The symptoms are relieved by resting or taking sublingual nitroglycerin. The patient most likely has which of the following disorders?
 A. Acute coronary syndrome
 B. Cardiomyopathy
 C. Nonanginal chest pain
 D. Stable ischemic heart disease
 E. Unstable angina

13.5.2 A 65-year-old female presents to the physician's office, complaining of chest pain upon exertion, which can be completely relived by sublingual nitroglycerin. Which of the following is most likely the pathophysiological mechanism underlying the patient chest pain?
 A. Coronary atherosclerosis
 B. Diabetes mellitus
 C. Emphysema
 D. High blood HDL cholesterol
 E. Hypertension

13.5.3 Stable angina occurs as a result of disturbance of the balance between myocardial oxygen supply and oxygen demand. Which of the following conditions would most likely aggravate a person's angina?
 A. Decreased arterial blood pressure
 B. Decreased cardiac preload
 C. Decreased heart rate
 D. Decreased hemoglobin concentration
 E. Decreased myocardial contractility

13.5.4 A 65-year-old man is diagnosed with stable coronary artery disease of 2 years of duration. Which of the following would be the least likely presentation of the patient's condition?
 A. Chronic stable angina
 B. Hypotension
 C. Microvascular angina
 D. Substernal chest discomfort

13.5.5   A 55-year-old male is diagnosed with stable ischemic heart disease. Which of the following is most likely the prototypical manifestation of the patient's condition?
A. Acute coronary syndrome
B. Apnea
C. Hypotension
D. Stable angina
E. Unstable angina

REFERENCES

1 Lozano, R., et al., Global and regional mortality from 235 causes of death for 20 age groups in 1990 and 2010: a systematic analysis for the Global Burden of Disease Study 2010. *Lancet*, 2012. 380(9859): p. 2095–128.

2 Moran, A.E., et al., The global burden of ischemic heart disease in 1990 and 2010: the global burden of disease 2010 study. *Circulation*, 2014. 129(14): p. 1493–501.

3 Montalescot, G, et al., 2013 ESC guidelines on the management of stable coronary artery disease: the Task Force on the management of stable coronary artery disease of the European Society of Cardiology. *Eur Heart J*, 2013. 34(38): p. 2949–3003.

4 Go, A.S., et al., Heart disease and stroke statistics—2014 update: a report from the American Heart Association. *Circulation*, 2014. 129(3): p. e28–292.

5 Berry, J.D., et al., Lifetime risks of cardiovascular disease. *N Engl J Med*, 2012. 366(4): p. 321–9.

6 Parikh, N.I., et al., Long-term trends in myocardial infarction incidence and case fatality in the National Heart, Lung, and Blood Institute's Framingham Heart study. *Circulation*, 2009. 119(9): p. 1203–10.

7 Wong, N.D., Epidemiological studies of CHD and the evolution of preventive cardiology. *Nat Rev Cardiol*, 2014. 11(5): p. 276–89.

8 Fihn, S.D., et al., 2012 ACCF/AHA/ACP/AATS/PCNA/SCAI/STS guideline for the diagnosis and management of patients with stable ischemic heart disease: a report of the American College of Cardiology Foundation/American Heart Association task force on practice guidelines, and the American College of Physicians, American Association for Thoracic Surgery, Preventive Cardiovascular Nurses Association, Society for Cardiovascular Angiography and Interventions, and Society of Thoracic Surgeons. *Circulation*, 2012. 126(25): p. e354–471.

9 Crea, F. and F. Andreotti, The unstable plaque: a broken balance. *Eur Heart J*, 2009. 30(15): p. 1821–3.

14

DRUGS FOR STABLE ANGINA

14.1 OVERVIEW

As outlined in Chapter 13, there are four major classes of antianginal drugs. They are (1) nitrates, (2) β-blockers, (3) calcium channel blockers (CCBs), and (4) novel antianginal agents. The basic pharmacology of the first three classes of antianginal drugs has been covered in previous chapters. This chapter first briefly reviews the use of the nitrates, β-blockers, and CCBs in the management of stable angina and then focuses on discussing novel antianginal drugs. The principles and guidelines regarding the management of stable angina are given in Chapter 15.

14.2 β-BLOCKERS FOR TREATING STABLE ANGINA

β-Adrenergic receptor activation is associated with increases in heart rate, accelerated atrioventricular nodal conduction, and increased contractility, which collectively contribute to increased myocardial oxygen demand (see Chapter 8). β-Blockers are recommended as the initial agents to relieve symptoms in most patients with stable angina. These drugs reduce myocardial oxygen consumption by decreasing heart rate and myocardial contractility with attenuation of cardiovascular remodeling by reducing left ventricular wall tension upon long-term use. The reduction in myocardial oxygen demand is directly proportional to the decreased level of adrenergic tonic stimulation. Furthermore, the reduction in heart rate also shifts the cardiac cycle, permitting more diastolic time and greater coronary perfusion, thereby improving myocardial oxygen supply. These effects collectively contribute to a reduction in angina onset with improvement in the ischemic threshold during exercise and in symptoms [1].

With regard to angina control, β-blockers and CCBs are similar (see next section on CCBs). β-Blockers can be combined with dihydropyridine CCBs to provide more effective control of angina [2]. Combination therapy of β-blockers with cardiac suppressive CCBs (i.e., diltiazem and verapamil) should be avoided because of the risk of bradycardia or atrioventricular block [3].

In addition to the improvement of angina symptoms, long-term β-blocker treatment also improves survival in patients with left ventricular dysfunction or a history of myocardial infarction. When prescribed in combination with agents that block the renin–angiotensin–aldosterone system (see Chapter 9), β-blockers are the preferred agents for the treatment of angina in patients with left ventricular dysfunction after myocardial infarction and in patients with heart failure, on the basis of documented improvements in survival and ventricular performance [1].

In summary, there is substantial evidence for prognostic benefits from the use of β-blockers in postmyocardial infarction patients or in those with heart failure. Extrapolation from these data suggests that β-blockers may be the first-line antianginal therapy in stable anginal patients without contraindications. In this context, the most widely used β-blockers are those with predominant β_1-blockade, such as atenolol, bisoprolol, metoprolol, or nebivolol (in alphabetical order). Carvedilol, a third-generation β-blocker with α_1-blocking activity, is also frequently used.

Cardiovascular Diseases: From Molecular Pharmacology to Evidence-Based Therapeutics, First Edition. Y. Robert Li.
© 2015 John Wiley & Sons, Inc. Published 2015 by John Wiley & Sons, Inc.

14.3 CCBs FOR TREATING STABLE ANGINA

If adverse effects or contraindications limit the use of β-blockers, CCBs are recommended for relief of anginal symptoms. CCBs act chiefly by vasodilation and reduction of the peripheral vascular resistance; some also directly decrease cardiac conduction and contractility (see Chapter 10). CCBs are a heterogeneous group of drugs that can chemically be classified into the dihydropyridines (e.g., amlodipine and nifedipine) and nondihydropyridines. The latter group further includes the phenylalkylamines (e.g., verapamil) and the benzothiazepines (e.g., diltiazem).

All CCBs improve myocardial oxygen supply by decreasing coronary vascular resistance and augmenting epicardial conduit vessel and systemic arterial blood flow. Myocardial oxygen demand is decreased by a reduction in systemic vascular resistance and arterial pressure. Verapamil, and, to a lesser extent, diltiazem also depress cardiac contractility and cardiac pacemaker rate and slow conduction. These depressant effects can cause sinus bradycardia or worsen preexisting conduction defects, leading to heart block.

All CCBs reduce anginal episodes, increase exercise duration, and reduce use of sublingual nitroglycerin in patients with effort-induced angina. Because CCBs also reduce the frequency of Prinzmetal variant angina, they are the drugs of choice, along with nitrates, used alone or in combination, for this specific type of angina.

Because all CCBs seem to be equally efficacious in treating angina, the choice of a particular drug should be based on potential drug interactions and adverse events. For instance, the dihydropyridine class is preferred over the nondihydropyridine drugs in patients with cardiac conduction defects, such as sick sinus syndrome, sinus bradycardia, or significant atrioventricular conduction disturbances (see Unit VII).

14.4 ORGANIC NITRATE FOR TREATING STABLE ANGINA

Nitrates are effective in the treatment of all forms of angina. They relax vascular smooth muscle in the systemic arteries and veins (predominant effect at lower doses) as well as coronary arteries in patients with stable angina (see Chapter 11). Dilation of arterials and decrease of systemic peripheral resistance result in decreased cardiac afterload. On the other hand, dilation of veins causes decreased preload. Decrease of preload leads to reduction in myocardial wall tension and myocardial oxygen demand. Nitrates also contribute to coronary blood flow redistribution by augmenting collateral flow and lowering ventricular diastolic pressure, from areas of normal perfusion to ischemic zones. These effects collectively contribute to the improvement of the disturbance of myocardial oxygen demand and supply in ischemia.

Sublingual nitroglycerin is the standard initial therapy for exertional angina. It is used for both treatment and prevention of stable angina. Long-acting nitrates (e.g., isosorbide dinitrate, isosorbide mononitrate) are not continuously effective if regularly taken over a prolonged period without a nitrate-free or nitrate-low interval of about 8–10 h. This is known as nitrate tolerance. Worsening of endothelial dysfunction is a potential complication of long-acting nitrates, and hence, the common practice of the routine use of long-acting nitrates as first-line therapy for patients with exertional angina needs reevaluation [3, 4]. Nevertheless, in patients with stable exertional angina, long-acting nitrates improve exercise tolerance, increase the time to onset of angina, and decrease ST-segment depression during the treadmill exercise test. In combination with β-blockers or CCBs, nitrates produce greater antianginal and anti-ischemic effects in patients with stable angina. Moreover, combination with β-blockers also reduces the increased heart rate and myocardial contractile state due to reflex sympathetic stimulation that occurs as a result of nitrate-induced vasodilation and hypotension.

14.5 NEW ANTIANGINAL DRUGS: RANOLAZINE

14.5.1 Introduction

Ranolazine (Ranexa) was approved by the US Food and Drug Administration (FDA) for the treatment of chronic angina in 2006. It is the first drug in a new class for treating this condition approved in the United States in about three decades.

14.5.2 Chemistry and Pharmacokinetics

Ranolazine (structure shown in Fig. 14.1) is a racemic mixture, chemically described as 1-piperazineacetamide, N-(2,6-dimethylphenyl)-4-[2-hydroxy-3-(2-methoxyphenoxy) propyl]-,(±)-. The pharmacokinetics of the (+) R- and (−) S-enantiomers of ranolazine are similar in healthy volunteers. Following oral administration, ranolazine is extensively metabolized in the gut and the liver primarily catalyzed by cytochrome P450 (CYP)3A4 and, to a lesser extent, by CYP2D6. The elimination half-life of ranolazine is 7 h, and approximately 75% of the dose is excreted in the urine and 25% in the feces.

14.5.3 Molecular Mechanisms and Pharmacological Effects

Ranolazine at usual therapeutic doses exerts antianginal effects. At high concentrations, the drug may also have an antiarrhythmic activity.

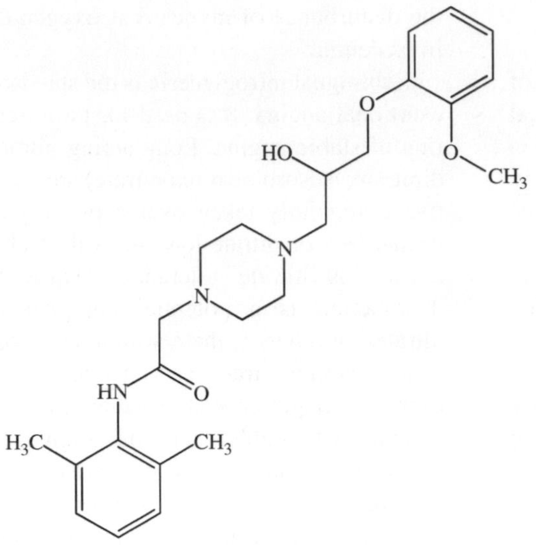

Ranolazine

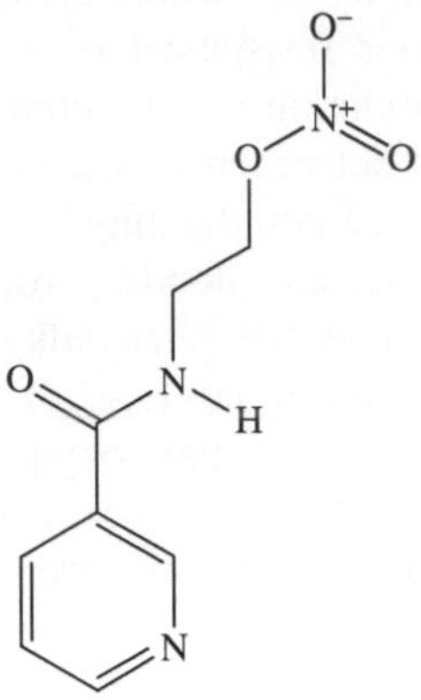

Nicorandil

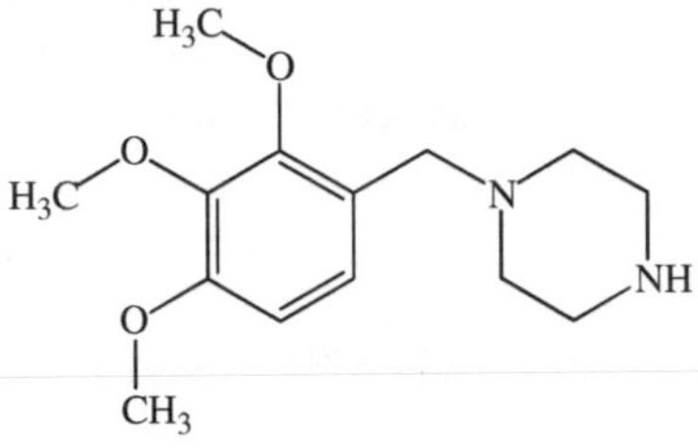

Trimetazidine

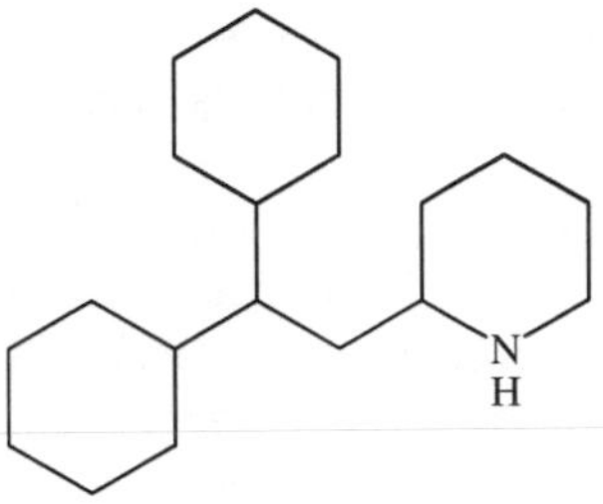

Perhexiline

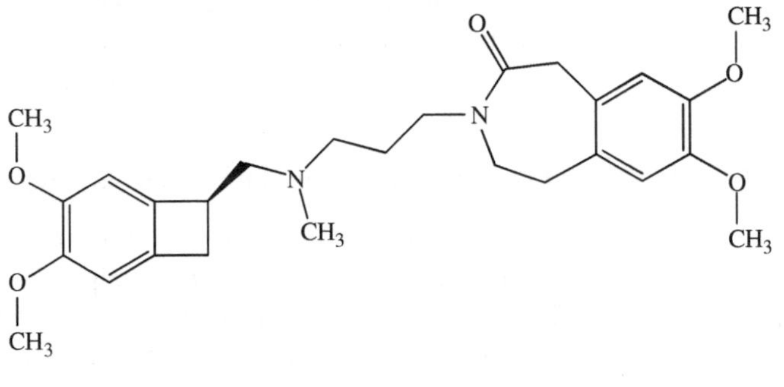

Ivabradine

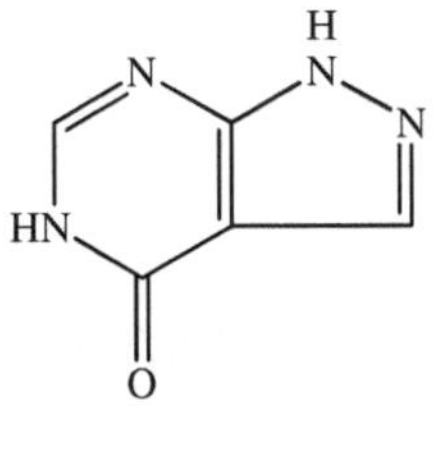

Allopurinol

FIGURE 14.1 Structures of selected antianginal drugs. Among the five drugs, only ranolazine is a US FDA-approved drug for treating angina.

14.5.3.1 Antianginal Effects and Mechanisms The mechanisms underlying the antianginal and anti-ischemic effects of ranolazine are not clear. Although initially thought to act by partial inhibition of fatty acid oxidation (see Section 14.6.1), it was later recognized that ranolazine had that effect only at plasma levels not achieved with the usual dosage.

Recently, investigations of the electrophysiological effects of ranolazine have found that it blocks the late cardiac sodium current (I_{NaL}). Accumulation of intracellular sodium induced by ischemia results in calcium overload in myocardial cells, leading to mechanical dysfunction. It has been suggested that by blocking I_{NaL}, ranolazine might prevent this sodium-induced calcium overload and the subsequent increase in diastolic tension and thereby attenuate ischemia [5] (Fig. 14.2).

Since the I_{NaL} channel frequently fails to inactivate in a number of important myocardial disease states, such as ischemia

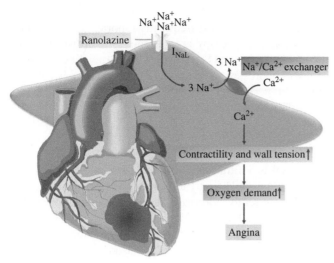

FIGURE 14.2 Molecular mechanism of action of ranolazine in treating angina. Ranolazine inhibits the late cardiac sodium current (I_{NaL}), leading to reduced accumulation of intracellular Ca^{2+} during diastole and thereby decreased wall tension.

and hypertrophy, excess entry of sodium ions leads to activation of the sodium/calcium exchanger, thereby raising intracellular calcium concentrations. Given the normal rapid inactivation of the late inward sodium channels in normal myocytes, ranolazine does not exert a significant effect on the normal myocardium at usual dosages. This potentially increases its therapeutic window.

14.5.3.2 Antiarrhythmic Activity and Mechanisms As noted earlier, ranolazine is believed to alleviate angina through inhibition of late I_{Na}, which is amplified during myocardial ischemia and heart failure. This agent is relatively specific for late I_{Na} and, at the therapeutic doses used to treat angina, has minimal effect on peak I_{Na}. At high concentrations, ranolazine blocks the rapidly activating delayed-rectifier potassium channel current (I_{Kr}), an effect that results in a dose-related prolongation of the QT_c interval (the QT interval of the electrocardiogram corrected for heart rate) by approximately 6 ms at doses of 1000 mg twice daily. The net effect of ranolazine at therapeutic doses on the various cardiac ion currents is a reduction in the frequency of cardiac arrhythmias [6].

14.5.4 Clinical Uses

14.5.4.1 Stable Angina Ranolazine is approved by the US FDA only for the treatment of chronic angina. It may be used with β-blockers, nitrates, CCBs, antiplatelet therapy (see Chapter 17), lipid-lowering therapy (see Chapter 4), angiotensin-converting enzyme inhibitors, and angiotensin receptor blockers (see Chapter 9).

14.5.4.2 Other Potential Applications In the 6560 patients of the Metabolic Efficiency with Ranolazine for Less

Ischemia in Non-ST-Elevation Acute Coronary Syndromes: Thrombolysis in Myocardial Infarction 36 (MERLIN-TIMI 36) trial presenting with recent non-ST-elevation ACS (NSTE-ACS), ranolazine therapy showed no overall benefit [7]. In patients with prior chronic angina enrolled in the MERLIN trial, ranolazine reduced recurrent ischemia [8, 9]. In those studied after the coronary event, ranolazine reduced the incidence of newly increased HbA1c (glycated hemoglobin) by 32%. In the recent Type 2 Diabetes Evaluation of Ranolazine in Subjects with Chronic Stable Angina (TERISA) study, ranolazine reduced episodes of stable angina in 949 diabetic patients already receiving one or two antianginal drugs and led to less use of sublingual nitroglycerin, and the benefits appeared more prominent in patients with higher rather than lower HbA1c levels [10]. These results suggest that ranolazine can be added to other well-established antianginal drugs, in particular in patients with higher HbA1c levels, who may also more often rely on medical management [10].

14.5.5 Therapeutic Dosages

The dosage form and strengths of ranolazine are listed below:

- Ranolazine (Ranexa): Oral, 500 and 1000 mg extended-release tablets

For treating stable angina, the initial dosage of ranolazine is 500 mg twice daily. The dosage may be increased to 1000 mg twice daily, as needed, based on clinical symptoms. The maximum recommended daily dose of ranolazine is 1000 mg twice daily.

14.5.6 Adverse Effects and Drug Interactions

14.5.6.1 Adverse Effects The most common adverse reactions of ranolazine treatment are dizziness, headache, constipation, and nausea. Although ranolazine blocks I_{Kr} and prolongs the QTc interval in a dose-related manner, clinical studies in an ACS population did not show an increased risk of proarrhythmia or sudden death associated with ranolazine treatment [7].

14.5.6.2 Drug Interactions Ranolazine is primarily metabolized by CYP3A4 and is also a substrate for P-glycoprotein. Drugs that affect CYP3A4 or are substrates of P-glycoprotein may interact with ranolazine to cause adverse effects. In this regard, ranolazine should not be used with strong CYP3A inhibitors, including ketoconazole, itraconazole, clarithromycin, nefazodone, nelfinavir, ritonavir, indinavir, and saquinavir. Likewise, ranolazine should not be combined with CYP3A4 inducers, such as rifampin, rifabutin, rifapentine, phenobarbital, phenytoin, carbamazepine, and St. John's wort.

The dose of ranolazine should be limited to 500 mg twice daily in patients on moderate CYP3A inhibitors, including diltiazem, verapamil, erythromycin, fluconazole, and grapefruit juice or grapefruit-containing products. Concomitant use of ranolazine and P-glycoprotein inhibitors, such as cyclosporine, may result in increases in ranolazine concentrations. As such, the dose of ranolazine should be titrated based on clinical response in patients concomitantly treated with P-glycoprotein inhibitors. Ranolazine also inhibits pathways involved in the metabolism of digoxin and simvastatin, and dose reduction of the drugs may be required.

14.5.6.3 *Contraindications and Pregnancy Category*

- Patients taking strong inhibitors of CYP3A4.
- Patients taking inducers of CYP3A4.
- Patients with liver cirrhosis.
- Pregnancy category: C.

14.6 OTHER NEW AND EMERGING DRUGS

14.6.1 Inhibitors of Fatty Acid Oxidation

Modulation of cardiac metabolism via partially inhibiting the oxidation of free fatty acids in the myocardium has been investigated as a novel strategy to alleviate stable angina. Although the heart uses both glucose and fatty acids to provide energy, during periods of stress, the heart uses more fatty acids, which is less oxygen efficient (i.e., as compared with glucose, utilization of fatty acids to produce the same amount of adenosine triphosphate [ATP] would require more oxygen). Inhibition of fatty acid oxidation shifts the equilibrium toward increased use of glucose, improving the efficient use of oxygen and thereby restoring the balance between myocardial oxygen demand and supply.

Trimetazidine and perhexiline (structures shown in Fig. 14.1) were developed as partial inhibitors of fatty acid oxidation in the myocardium, and trimetazidine has become extensively used for treating stable angina throughout Europe and Asia and in >80 countries. However, this agent is not a US FDA-approved drug for use in the United States. As noted earlier, trimetazidine improves cellular tolerance to ischemia by inhibiting fatty acid metabolism and secondarily by stimulating glucose metabolism [11]. In patients with chronic stable angina, this drug increases coronary flow reserve, delaying the onset of ischemia associated with exercise and reducing the number of weekly angina episodes and weekly nitroglycerin consumption. The antiischemic effects are not associated with changes in heart rate or systolic blood pressure. In diabetic persons, trimetazidine also improved HbA1c and glycemia while increasing forearm glucose uptake [3]. Few data exist on the effects of trimetazidine on cardiovascular endpoints, mortality, or quality of life. The most frequently reported adverse events of trimetazidine therapy are gastrointestinal disorders, but the incidence is low [12]. In contrast, perhexiline is associated with more severe adverse effects, including hepatotoxicity and peripheral neuropathy, and as such, it is not commonly used.

14.6.2 K$^+_{ATP}$ Channel Activators

Nicorandil is a nitrate derivative of nicotinamide (structure shown in Fig. 14.1) for treating stable angina in many countries, but currently not available in the United States. The drug has a dual mechanism of action. It activates K$^+_{ATP}$ channels, causing membrane hyperpolarization of vascular smooth muscle cells and thereby vasodilation. Nicorandil also promotes systemic venous and coronary vasodilation through a nitrate moiety (i.e., release of nitric oxide). This dual action increases coronary blood flow, with reductions in afterload, preload, and oxidative injury. Nicorandil does not exhibit effects on cardiac contractility or conduction. The antianginal efficacy and safety of nicorandil are similar to those of oral nitrates, β-blockers, and CCBs [12]. It can be used for the prevention and long-term treatment of stable angina. In the Impact of Nicorandil in Angina (IONA) study of 5126 patients with chronic stable angina, cardiovascular events were reduced by 14% in nicorandil-treated group as compared with placebo group [13]. A recent study also reported that long-term use of oral nicorandil may stabilize coronary plaque in patients with stable angina [14]. Nicorandil is generally well tolerated. Occasional adverse effects include oral, intestinal, and perianal ulceration.

14.6.3 Inhibitors of Sinus Node Pacemaker Current

Ivabradine (structure shown in Fig. 14.1) is a specific inhibitor of the I$_f$ current of pacemaker cells in the sinoatrial node. Treatment with ivabradine results in heart rate reduction, prolonging diastole and thereby improving myocardial oxygen balance. It has no effect on blood pressure, myocardial contractility, or intracardiac conduction. Ivabradine was shown to be as effective as atenolol or amlodipine in patients with stable angina, and adding ivabradine to atenolol therapy gave better control of heart rate and anginal symptoms. Ivabradine was approved by the European Medicines Agency (EMA) for therapy of chronic stable angina in patients intolerant to or inadequately controlled by β-blockers and whose heart rate exceeds 60 beats/min (in sinus rhythm). It is, however, currently not available in the United States.

A randomized controlled trial (BEAUTIFUL) in 10,917 patients with stable coronary artery disease (SCAD) and left ventricular dysfunction reported that ivabradine added to standard treatment had no effect, when compared with placebo, on the composite endpoint of cardiovascular death, admission to the hospital for acute myocardial infarction, and admission to the hospital for new-onset or worsening heart failure [15]. A subsequent subgroup analysis in 1507 patients with prior angina enrolled in the BEAUTIFUL trial showed that ivabradine reduced the composite primary endpoint of cardiovascular death, hospitalization with myocardial infarction and heart failure, and decreased hospitalization for myocardial infarction. The effect was predominant in patients with a heart rate of ≥70 beats/min [16]. Control of heart rate by ivabradine was also shown to improve clinical outcomes in patients with chronic heart failure [17–19].

The most common adverse effect of ivabradine, reported in approximately 15% of patients, is phosphenes, described as a transient enhanced brightness in a limited area of the visual field that typically occurs within the first 2 months of treatment. Most of these luminous visual-field disturbances (77%) resolve without discontinuing treatment.

14.6.4 Emerging Antianginal Drugs and Stem Cell Therapy

14.6.4.1 Xanthine Oxidase Inhibitors Allopurinol (structure shown in Fig. 14.1) inhibits xanthine dehydrogenase/oxidase, an enzyme required in the oxidation of hypoxanthine and xanthine, which produces uric acid as well as reactive oxygen species. A recent small, placebo-controlled, randomized trial of 65 patients showed that high-dose allopurinol (300 mg twice daily) significantly increased total exercise duration, time to onset of angina, and ischemic ST-segment depression [20]. The exact mechanism of action is unknown, but inhibition of xanthine dehydrogenase/oxidase by allopurinol might reduce oxidative stress in ischemic myocardium and improve endothelium-dependent vasodilation [21]. Allopurinol is inexpensive, well tolerated, and safe during long-term use. Hence, the precise place of this drug in the management of stable angina needs to be further investigated in large-scale clinical trials. In addition to stable angina, recent studies also suggested that high-dose allopurinol could regress left ventricular hypertrophy, reduce left ventricular end-systolic volume, and improve endothelial function in patients with ischemic heart disease (IHD) and left ventricular hypertrophy [22]. Similar benefit was also reported in patients with diabetes and left ventricular hypertrophy [23]. These findings raise the possibility that allopurinol might reduce future cardiovascular events and mortality in these patients.

14.6.4.2 Stem Cell Therapy Stem cell therapy has been shown to be a promising option for patients after myocardial infarction [24]. The effectiveness of stem cell therapy in patients with stable angina has been investigated in pilot trials. For instance, a recent study on autotransplantation of mesenchymal stromal cells from the bone marrow to the heart in 31 patients with severe SCAD and refractory angina demonstrated sustained clinical effects, reduced hospital admissions for cardiovascular disease, and excellent long-term safety [25]. Cost-effective analysis also showed promise for stem cell therapy in refractory angina [26]. However, the exact value of stem cell therapy in treating IHD, including stable angina, remains to be established via large-scale randomized trials.

14.7 SUMMARY OF CHAPTER KEY POINTS

- Angina occurs when myocardial oxygen demand exceeds oxygen supply, resulting in myocardial ischemia. Stable angina is the prototypical manifestation of stable IHD (SIHD), also known as SCAD.

- Drugs that improve myocardial oxygen balance and thereby relieve anginal symptoms include β-blockers, CCBs, organic nitrates, and ranolazine, which are approved for use in the United States. Antianginal agents used in other countries, but not in the United States, include the fatty acid oxidation inhibitor trimetazidine, the K^+_{ATP} channel activator nicorandil, and the sinus node pacemaker current inhibitor ivabradine.

- While all of the above drugs improve the symptoms and exercise tolerance in patients with stable angina, they vary with regard to efficacy in reducing cardiovascular outcomes in such patients.

- Notably, long-term β-blocker treatment improves survival in patients with left ventricular dysfunction or a history of myocardial infarction. As such, β-blockers are considered first-line therapy for stable angina.

- Other drugs are used when β-blockers are contraindicated or ineffective, and selection of the different drugs should be based on evidence-based guidelines, individual patient's condition and response to drug therapy, as well as the healthcare provider's judgment. Chapter 15 discusses the principles and guidelines regarding stable angina/SIHD management.

14.8 SELF-ASSESSMENT QUESTIONS

14.8.1 A 55-year-old male presents to the physician's office complaining of constipation since taking a "new" cardiovascular pill. History reveals that the patient is recently diagnosed with hypertension and stable angina. He has been on the "new" cardiovascular drug for 3 weeks, which he was told can control both his hypertension and angina. Which of the following is most likely the "new" cardiovascular drug?
A. Amlodipine
B. Diltiazem

C. Isradipine

D. Nimodipine

E. Verapamil

14.8.2 A 60-year-old male with chronic obstructive pulmonary disease requires therapy to prevent vasospasm angina attacks. Which of the following drugs would be most appropriate for this patient?

A. Aliskiren

B. Amlodipine

C. Losartan

D. Milrinone

E. Propranolol

14.8.3 A 50-year-old male is diagnosed with chronic stable angina. He uses nitroglycerin sublingually when he experiences chest pain. Coadministration of which of the following drugs may cause serious hypotension in this patient?

A. Captopril

B. Diltiazem

C. Losartan

D. Prasugrel

E. Tadalafil

14.8.4 A 60-year-old male has severe chest pain when he walks uphill in cold weather to his store. The pain disappears when he rests. A decision is made to treat him with nitroglycerin. He develops tachycardia following the nitrate therapy. Which of the following can be given to improve his tachycardia and, at the same time, augment the antianginal efficacy?

A. Amlodipine

B. Hydralazine

C. Ibutilide

D. Metoprolol

E. Phentolamine

14.8.5 A 65-year-old male with a history of hypertension, diabetes, and hypercholesterolemia begins to develop episodes of chest pain on exertion. One week after his first episode, a bout of chest pain occurs while he is working in the yard. Five minutes after the onset of this pain, he takes two sublingual nitroglycerin tablets. Within minutes, he feels much better. Which of the following best explains the drug's effect?

A. Increase of cardiac oxygen demand

B. Increase of cardiac wall tension

C. Increase of myocardial contractility

D. Increase of smooth muscle relaxation

E. Increase of venous return

14.8.6 A 45-year-old male with stable ischemic heart disease is prescribed isosorbide mononitrate for prophylaxis of exertional angina. Which of the following adverse effects is the patient likely to experience?

A. Bradycardia

B. Erectile dysfunction

C. Hemolytic anemia

D. Hypertension

E. Throbbing headache

14.8.7 A 65-year-old female is diagnosed with stable coronary artery disease. She is placed on nebivolol and diltiazem for the prevention of angina attack upon exertion. Which of the following effects is caused by both of these drugs?

A. Decreased cGMP levels in cardiomyocytes

B. Decreased heart rate

C. Decreased myocardial oxygen supply

D. Decreased nitric oxide bioavailability

E. Decreased sympathetic activity

14.8.8 A 58-year-old male with stable angina refractory to β-blocker and calcium channel blocker therapy is put on a drug that acts most likely via blocking the late cardiac sodium current. Which of the following is most likely prescribed?

A. Isosorbide dinitrate

B. Ivabradine

C. Ranolazine

D. Sildenafil

E. Trimetazidine

14.8.9 A 60-year-old female presents to the physician's office complaining of substernal chest discomfort and pain upon exertion or emotional stress. She is diagnosed with stable angina/stable coronary artery disease and put on a drug that activates K^+_{ATP} channels of vascular smooth muscle cells. Which of the following is most likely prescribed?

A. Isosorbide dinitrate

B. Ivabradine

C. Nicorandil

D. Ranolazine

E. Trimetazidine

14.8.10 A 65-year-old diabetic patient with stable angina is treated with an anti-ischemic agent that partially suppresses fatty acid oxidation in the myocardium and also likely improves glycemia. Which of the following is most likely prescribed?

A. Ivabradine

B. Nitroglycerin

C. Perhexiline

D. Ranolazine

E. Trimetazidine

REFERENCES

1 Fihn, S.D., et al., 2012 ACCF/AHA/ACP/AATS/PCNA/SCAI/ STS guideline for the diagnosis and management of patients with stable ischemic heart disease: a report of the American College of Cardiology Foundation/American Heart Association task force on practice guidelines, and the American College of Physicians, American Association for Thoracic Surgery,

Preventive Cardiovascular Nurses Association, Society for Cardiovascular Angiography and Interventions, and Society of Thoracic Surgeons. *Circulation*, 2012. 126(25): p. e354–471.

2 Belsey, J., et al., Relative efficacy of antianginal drugs used as add-on therapy in patients with stable angina: a systematic review and meta-analysis. Eur J Prev Cardiol, 2014 (in press).

3 Montalescot, G, et al., 2013 ESC guidelines on the management of stable coronary artery disease: the Task Force on the management of stable coronary artery disease of the European Society of Cardiology. *Eur Heart J*, 2013. 34(38): p. 2949–3003.

4 Henderson, R.A., N. O'Flynn, and Guideline Development Group, Management of stable angina: summary of NICE guidance. *Heart*, 2012. 98(6): p. 500–7.

5 Abrams, J., C.A. Jones, and P. Kirkpatrick, Ranolazine. *Nat Rev Drug Discov*, 2006. 5(6): p. 453–4.

6 Chaitman, B.R. and A.A. Laddu, Stable angina pectoris: anti-anginal therapies and future directions. *Nat Rev Cardiol*, 2012. 9(1): p. 40–52.

7 Morrow, D.A., et al., Effects of ranolazine on recurrent cardio-vascular events in patients with non-ST-elevation acute coronary syndromes: the MERLIN-TIMI 36 randomized trial. *JAMA*, 2007. 297(16): p. 1775–83.

8 Wilson, S.R., et al., Efficacy of ranolazine in patients with chronic angina observations from the randomized, double-blind, placebo-controlled MERLIN-TIMI (metabolic efficiency with ranolazine for less ischemia in non-ST-segment elevation acute coronary syndromes) 36 trial. *J Am Coll Cardiol*, 2009. 53(17): p. 1510–6.

9 Morrow, D.A., et al., Evaluation of the glycometabolic effects of ranolazine in patients with and without diabetes mellitus in the MERLIN-TIMI 36 randomized controlled trial. *Circulation*, 2009. 119(15): p. 2032–9.

10 Kosiborod, M., et al., Evaluation of ranolazine in patients with type 2 diabetes mellitus and chronic stable angina: results from the TERISA randomized clinical trial (type 2 diabetes evaluation of ranolazine in subjects with chronic stable angina). *J Am Coll Cardiol*, 2013. 61(20): p. 2038–45.

11 Kantor, P.F., et al., The antianginal drug trimetazidine shifts cardiac energy metabolism from fatty acid oxidation to glucose oxidation by inhibiting mitochondrial long-chain 3-ketoacyl coenzyme A thiolase. *Circ Res*, 2000. 86(5): p. 580–8.

12 Fihn, S.D., et al., 2012 ACCF/AHA/ACP/AATS/PCNA/SCAI/STS Guideline for the diagnosis and management of patients with stable ischemic heart disease: a report of the American College of Cardiology Foundation/American Heart Association Task Force on Practice Guidelines, and the American College of Physicians, American Association for Thoracic Surgery, Preventive Cardiovascular Nurses Association, Society for Cardiovascular Angiography and

Interventions, and Society of Thoracic Surgeons. *J Am Coll Cardiol*, 2012. 60(24): p. e44–164.

13 The IONA Study Group, Effect of nicorandil on coronary events in patients with stable angina: the Impact Of Nicorandil in Angina (IONA) randomised trial. *Lancet*, 2002. 359(9314): p. 1269–75.

14 Izumiya, Y., et al., Long-term use of oral nicorandil stabilizes coronary plaque in patients with stable angina pectoris. *Atherosclerosis*, 2011. 214(2): p. 415–21.

15 Fox, K., et al., Ivabradine for patients with stable coronary artery disease and left-ventricular systolic dysfunction (BEAUTIFUL): a randomised, double-blind, placebo-controlled trial. *Lancet*, 2008. 372(9641): p. 807–16.

16 Fox, K., et al., Heart rate as a prognostic risk factor in patients with coronary artery disease and left-ventricular systolic dysfunction (BEAUTIFUL): a subgroup analysis of a ran-domised controlled trial. *Lancet*, 2008. 372(9641): p. 817–21.

17 Swedberg, K., et al., Ivabradine and outcomes in chronic heart failure (SHIFT): a randomised placebo-controlled study. *Lancet*, 2010. 376(9744): p. 875–85.

18 Griffiths, A., et al., The cost effectiveness of ivabradine in the treatment of chronic heart failure from the UK National Health Service perspective. *Heart*, 2014. 100(13): p. 1031–6.

19 Borer, J.S., et al., Efficacy and safety of ivabradine in patients with severe chronic systolic heart failure (from the SHIFT study). *Am J Cardiol*, 2014. 113(3): p. 497–503.

20 Noman, A., et al., Effect of high-dose allopurinol on exercise in patients with chronic stable angina: a randomised, placebo con-trolled crossover trial. *Lancet*, 2010. 375(9732): p. 2161–7.

21 Rajendra, N.S., et al., Mechanistic insights into the therapeutic use of high-dose allopurinol in angina pectoris. *J Am Coll Cardiol*, 2011. 58(8): p. 820–8.

22 Rekhraj, S., et al., High-dose allopurinol reduces left ventric-ular mass in patients with ischemic heart disease. *J Am Coll Cardiol*, 2013. 61(9): p. 926–32.

23 Szwejkowski, B.R., et al., Allopurinol reduces left ventricular mass in patients with type 2 diabetes and left ventricular hypertrophy. *J Am Coll Cardiol*, 2013. 62(24): p. 2284–93.

24 Strauer, B.E. and G. Steinhoff, 10 years of intracoronary and intramyocardial bone marrow stem cell therapy of the heart: from the methodological origin to clinical practice. *J Am Coll Cardiol*, 2011. 58(11): p. 1095–104.

25 Mathiasen, A.B., et al., Autotransplantation of mesenchymal stromal cells from bone-marrow to heart in patients with severe stable coronary artery disease and refractory angina--final 3-year follow-up. *Int J Cardiol*, 2013. 170(2): p. 246–51.

26 Hossne, N.A., Jr., et al., Long-term and sustained therapeutic results of a specific promonocyte cell formulation in refractory angina: ReACT (Refractory Angina Cell Therapy) clinical update and cost effective analysis. Cell Transplant, 2014. doi: 10.3727/096368914X681595.

15

MANAGEMENT OF STABLE ANGINA/STABLE ISCHEMIC HEART DISEASE: PRINCIPLES AND GUIDELINES

15.1 OVERVIEW

Management of stable angina involves the use of various classes of medications as discussed in Chapter 14 as well as nonpharmacological procedures, such as revascularization with percutaneous coronary intervention (PCI). This chapter discusses the principles and current guidelines regarding stable angina management. The chapter first introduces current guidelines on the management of stable angina from various organizations, including the American Heart Association (AHA) and its collaborative organizations, the European Society of Cardiology (ESC) and the British National Institute for Health and Clinical Excellence (NICE). The chapter then focuses on discussing the general principles of management and the major recommendations from the US-based guidelines with reference to the ESC and NICE guidelines. Since the terms stable coronary artery disease (SCAD) and stable ischemic heart disease (SIHD) have been evolved to replace stable angina, the compound term "stable angina/SIHD" is used in the chapter to discuss the principles and guidelines regarding disease management.

15.2 INTRODUCTION TO CURRENT GUIDELINES ON MANAGEMENT OF STABLE ANGINA/SIHD

Multiple professional organizations have developed guidelines for the management of stable angina/SIHD. The most notable ones are those from the AHA and its collaborative organizations (US-based guidelines), the ESC and the NICE.

The titles and years of publication of some recent guidelines from these organizations are summarized in Table 15.1.

15.2.1 Guidelines from the AHA and Its Collaborative Organizations

It is important that the medical profession plays a significant role in critically evaluating the use of diagnostic procedures and therapies in the management or prevention of human diseases. Rigorous and expert analysis of the available data documenting relative benefits and risks of those procedures and therapies can produce helpful guidelines that improve the effectiveness of care, optimize patient outcomes, and have a favorable impact on the overall cost of care by focusing resources on the most effective strategies. The American College of Cardiology (ACC) and the AHA have jointly engaged in the production of such guidelines in the area of cardiovascular disease since 1980. This effort is directed by the ACC/AHA Task Force on Practice Guidelines, whose charge is to develop and revise practice guidelines for important cardiovascular diseases and procedures. This section briefly surveys the guideline development for stable angina/SIHD by the AHA and its collaborative organizations over the past decade.

15.2.1.1 The 1999 ACC/AHA/ACP–ASIM Guideline
Recognizing the importance of the management of stable angina, the most common manifestation of IHD, and the absence of national clinical practice guidelines in this area, the ACC/AHA Task Force formed the Committee on

Cardiovascular Diseases: From Molecular Pharmacology to Evidence-Based Therapeutics, First Edition. Y. Robert Li.
© 2015 John Wiley & Sons, Inc. Published 2015 by John Wiley & Sons, Inc.

TABLE 15.1   Some recent guidelines from the AHA and its collaborative organizations, the ESC and the NICE (in chronological order)

Organization	Title of guideline	Journal and date of publication
AHA and its collaborative organizations	ACC/AHA/ACP–ASIM Guidelines for the Management of Patients with Chronic Stable Angina: A Report of the American College of Cardiology/American Heart Association Task Force on Practice Guidelines (Committee on Management of Patients with Chronic Stable Angina)	*J Am Coll Cardiol.* **1999** Jun; 33(7):2092–197
	2002 ACC/AHA Guideline Update for the Management of Patients with Chronic Stable Angina—Summary Article: A Report of the American College of Cardiology/American Heart Association Task Force on Practice Guidelines (Committee on the Management of Patients with Chronic Stable Angina)	*Circulation.* **2003** Jan 7; 107(1):149–58 *J Am Coll Cardiol.* **2003** Jan 1; 41(1):159–68 Full guideline on website (www.americanheart.org)
	2007 Chronic Angina Focused Update of the 2002 ACC/AHA Guidelines for the Management of Patients with Chronic Stable Angina: A Report of the American College of Cardiology/American Heart Association Task Force on Practice Guidelines Writing Group to Develop the Focused Update of the 2002 Guidelines for the Management of Patients with Chronic Stable Angina	*Circulation.* **2007** Dec 4; 116(23):2762–72 *J Am Coll Cardiol.* **2007** Dec 4; 50(23):2264–74
	2012 ACCF/AHA/ACP/AATS/PCNA/SCAI/STS Guideline for the Diagnosis and Management of Patients with Stable Ischemic Heart Disease: A Report of the American College of Cardiology Foundation/American Heart Association Task Force on Practice Guidelines, and the American College of Physicians, American Association for Thoracic Surgery, Preventive Cardiovascular Nurses Association, Society for Cardiovascular Angiography and Interventions, and Society of Thoracic Surgeons	*Circulation.* **2012** Dec 18; 126(25):e354–471 *J Am Coll Cardiol.* **2012** Dec 18; 60(24):e44–164
ESC	Guidelines on the Management of Stable Angina Pectoris: Executive Summary: The Task Force on the Management of Stable Angina Pectoris of the European Society of Cardiology	*Eur Heart J.* **2006** Jun; 27(11):1341–81
	2013 ESC Guidelines on the Management of Stable Coronary Artery Disease: The Task Force on the Management of Stable Coronary Artery Disease of the European Society of Cardiology	*Eur Heart J.* **2013** Oct; 34(38):2949–3003
NICE	Stable Angina (CG126, July 2011) Management of Stable Angina: Summary of NICE Guidance	Summary published in *BMJ.* **2011** Aug 5; 343:d4147. Full guideline available at: http://guidance.nice.org.uk/CG126 (accessed on June 23, 2014)

Management of Patients with Chronic Stable Angina to develop guidelines for the management of stable angina. Because this problem is frequently encountered in the practice of internal medicine, the task force invited the American College of Physicians–American Society of Internal Medicine (ACP–ASIM) to serve as a partner in this effort by identifying three general internists to serve on the committee. The document was released in 1999 and is known as the "1999 ACC/AHA/ACP–ASIM Guidelines for the Management of Patients with Chronic Stable Angina." The guideline document consists of four sections: (i) diagnosis, (ii) risk stratification, (iii) treatment, and (iv) patient follow-up. The full text of the guideline is published in the June 1999 issue of the *Journal of the American College of Cardiology* [1]; the executive summary is published in the June 1, 1999, issue of *Circulation* [2].

15.2.1.2 The 2002 ACC/AHA Guideline Update The ACC/AHA Task Force on Practice Guidelines regularly reviews existing guidelines to determine when an update or a full revision is needed [3, 4]. This process gives priority to areas in which major changes in text, and particularly recommendations, are merited on the basis of new understanding or evidence. In this regard, the committee updated the "1999 ACC/AHA/ACP–ASIM Guidelines for the Management of Patients with Chronic Stable Angina" and released the updated guideline in 2002. The text of the full updated guideline, entitled "ACC/AHA 2002 Guideline Update for the Management of Patients with Chronic Stable Angina: A Report of the American College of Cardiology/ American Heart Association Task Force on Practice Guidelines (Committee to Update the 1999 Guidelines for the Management of Patients with Chronic Stable Angina)," is posted on the websites of the AHA and the ACC, and the summary article is published in the January 1, 2003, issue of the *Journal of the American College of Cardiology* [3] and the January 7/14, 2003, issue of *Circulation* [4]. The 2002 updated guidelines include four important areas of changes, as listed below:

1. Angiotensin-converting enzyme inhibitors (ACEIs)
2. Treatment of risk factors
3. Alternative therapies for chronic stable angina in patients refractory to medical therapy who are not candidates for percutaneous intervention or revascularization
4. Asymptomatic patients with known or suspected CAD

15.2.1.3 The 2007 ACC/AHA Focused Update [5, 6]

A primary challenge in the development of clinical practice guidelines is keeping pace with the stream of new data and evidence upon which recommendations are based. In an effort to respond more quickly to new evidence, the ACC/AHA Task Force on Practice Guidelines has created a new "focused update" process to revise the existing guideline recommendations that are affected by the evolving data or opinions. Prior to the initiation of this focused approach, periodic updates and revisions of existing guidelines required up to 3 years to complete. Now, however, new evidence will be reviewed in an ongoing fashion to more efficiently respond to important science and treatment trends that could have a major impact on patient outcomes and quality of care. In this context, the focused update of the 2002 ACC/AHA guideline for the management of patients with chronic stable angina was released in 2007, which is known as the "2007 Chronic Angina Focused Update of the 2002 ACC/AHA Guidelines for the Management of Patients with Chronic Stable Angina." The text of this focused update is published in the December 4, 2007, issue of the *Journal of the American College of Cardiology* [5] and the December 4, 2007, issue of *Circulation* [6].

15.2.1.4 The 2012 ACCF/AHA/ACP/AATS/PCNA/SCAI/ STS Guideline

The most current US-based guideline on stable angina/SIHD management is the "2012 ACCF/AHA/ACP/AATS/PCNA/SCAI/STS Guideline for the Diagnosis and Management of Patients with Stable Ischemic Heart Disease: A Report of the American College of Cardiology Foundation/American Heart Association Task Force on Practice Guidelines, and the American College of Physicians, American Association for Thoracic Surgery, Preventive Cardiovascular Nurses Association, Society for Cardiovascular Angiography and Interventions, and Society of Thoracic Surgeons." This guideline is published simultaneously in the December 18, 2012, issues of *Circulation* [7] and *Journal of American College of Cardiology* [8]. As

indicated by the title, this most current guideline uses the term "stable ischemic heart disease" instead of stable angina.

This is the first major revision in a decade of the US-based guidelines for patients with SIHD. A major change in focus is a strong emphasis put on the involvement of the patient in his or her own disease management, from making lifestyle changes, such as smoking cessation, weight control, exercise, and stress reduction, to compliance with prescribed medications, such as statins, antiplatelet drugs, and hypertension medications, and to making choices about course of therapy when that decision is viable (e.g., continuing with medical therapy or opting for a revascularization procedure, such as angioplasty, stent placement, or even bypass graft surgery). Another major change is the emphasis on guideline-directed medical therapy (GDMT) that benefits most patients. GDMT was designated by the task force to represent optimal medical therapy as defined by the ACCF/AHA guideline (primarily class I)-recommended evidence-based therapies [8].

15.2.2 The ESC Guidelines

In order to improve clinical practice in Europe, the Committee for Practice Guidelines of the ESC charges groups of European experts with the task of creating recommendations and guidelines for clinical practice. These recommendations and guidelines clarify areas of consensus and disagreement, allowing distribution of the best possible guidance to practicing physicians. Guidelines aim to present all the relevant evidence on a particular clinical issue in order to help physicians weigh the benefits and risks of a particular diagnostic or therapeutic procedure. With regard to management of stable angina, the first major guideline from ESC was published 1997 [10]. Nearly 10 years later, in 2006, ESC released a revised version of the guideline [9]. The most current ESC guideline was released in 2013 [10]. This section briefly introduces the aforementioned two recent ESC guidelines on stable angina/SIHD management so as to provide the reader a source of reference regarding management of stable angina/SIHD in Europe.

15.2.2.1 The 2006 ESC Guideline

The full text of the 2006 ESC guideline on the management of stable angina is posted on the website of the ESC (www.esc.org), and the executive summary is published in the June 2006 issue of the *European Heart Journal* [9]. The 2006 ESC guideline provided a new landmark in the mission to reduce the burden of cardiovascular disease in Europe. Notably, the 2006 ESC guideline is largely consistent with the 2002/2007 ACC/AHA updated guideline with regard to major evidence-based recommendations. Listed below are the areas covered in the 2006 ESC guideline:

1. Definition and pathophysiology
2. Epidemiology

3. Natural history and prognosis
4. Diagnosis and assessment
5. Risk stratification
6. Treatment

15.2.2.2 The 2013 ESC Guideline One major change of the 2013 ESC guideline as compared with the preceding 2006 guideline is the adoption of the term "SCAD" to replace the term stable angina used in previous versions of the guidelines. This change is also in line with that in the 2012 ACCF/AHA/ACP/AATS/PCNA/SCAI/STS guideline described in Section 15.2.1.4, where the term SIHD is used to replace stable angina.

The concept of SCAD or SIHD has evolved over the past decade, and the term is now considered to refer to all different evolutionary phases of stable coronary heart disease with stable angina as a major manifestation. The 2013 ESC guideline defines SCAD as "coronary heart disease generally characterized by episodes of reversible myocardial demand/supply mismatch, related to ischemia or hypoxia, which are usually inducible by exercise, emotion or other stress and reproducible, but, which may also be occurring spontaneously. Such episodes of ischemia/hypoxia are commonly associated with transient chest discomfort (angina pectoris). SCAD also includes the stabilized, often asymptomatic, phases that follow an ACS" [10].

The aforementioned conceptual change will draw more attention from both healthcare providers and patients to the entire phases of the disease instead of only focusing on angina symptoms and will lead to identification of more patients who may benefit from evidence-based intervention. The major components of the 2013 ESC guideline include diagnosis and assessment, lifestyle and pharmacological management, revascularization, and special groups or consideration.

15.2.3 The 2011 NICE Guideline

The NICE is a special health authority of the English National Health Service (NHS), serving both the English NHS and the Welsh NHS. It was set up as the National Institute for Clinical Excellence in 1999 and, on April 1, 2005, joined with the Health Development Agency to become the new NICE.

Clinical guidelines are recommendations by the NICE on the appropriate treatment and care of people with specific diseases and conditions within the NHS. They are based on systematic reviews of the best available evidence and explicit consideration of cost-effectiveness. When minimal evidence is available, recommendations are based on the Guideline Development Group's experience and opinion of what constitutes good practice. Evidence levels for the recommendations are given in italic in square brackets. Notably, the evidence levels used in the NICE guidelines are different from those used in the US-based guidelines and the ESC guidelines.

The recently published NICE clinical guideline (2011; CG126) on the management of stable angina offers advice on treatment of episodes of angina, antianginal drug treatment, secondary prevention, the role of risk scores and noninvasive functional investigation, myocardial revascularization, lifestyle adjustments, and the management of refractory angina. Detailed review of the evidence for the guideline can be found in the full version (http://guidance.nice.org.uk/CG126), and the most important recommendations are summarized in an article published in the August 5, 2011 issue of the *British Medical Journal* [11].

15.3 GENERAL PRINCIPLES OF MANAGEMENT OF STABLE ANGINA/SIHD

The management of stable angina/SIHD focuses on two key aims: (i) relieving symptoms and improving health and function and (ii) improving the prognosis and minimizing the likelihood of death. Therapy directed toward preventing death has the highest priority. When two different therapeutic strategies are equally effective in alleviating symptoms of angina, the therapy with a definite or very likely advantage in preventing death should be recommended. The general principles of stable angina/SIHD management include (i) defining the treatment objectives and (ii) identifying strategies to attain the objectives.

15.3.1 Defining Treatment Objectives

The 2012 ACCF/AHA/ACP/AATS/PCNA/SCAI/STS guideline defines five specific objectives to meet the paramount goals of minimizing the likelihood of death while maximizing health and function [8]. These five objectives are outlined below:

1. Reducing premature cardiovascular death
2. Preventing complications of SIHD that directly or indirectly impair patients' functional well-being, including nonfatal acute myocardial infarction and heart failure
3. Maintaining or restoring a level of activity, functional capacity, and quality of life that is satisfactory to the patient
4. Completely, or nearly completely, eliminating ischemic symptoms
5. Minimizing costs of healthcare, in particular by eliminating avoidable adverse effects of tests and treatments, by preventing hospital admissions, and by eliminating unnecessary tests and treatments

15.3.2 Identifying Strategies to Attain the Treatment Objectives

To pursue the objectives outlined in Section 15.3.1, the 2012 ACCF/AHA/ACP/AATS/PCNA/SCAI/STS guideline provides five fundamental, complementary, and overlapping strategies [8]. They are outlined below:

1. Educating patients about the etiology, clinical manifestations, treatment options, and prognosis of IHD to support active participation of patients in their treatment decisions

2. Identifying and treating conditions that contribute to, worsen, or complicate IHD

3. Effectively modifying risk factors for IHD by both pharmacological and nonpharmacological methods

4. Using evidence-based pharmacological treatments to improve patients' health status and survival, with attention to avoiding drug interactions and adverse effects

5. Using revascularization by percutaneous catheter-based techniques or coronary artery bypass grafting (CABG) when there is clear evidence of the potential to improve patients' health status and survival

Hence, effective management of stable angina/SIHD requires the coordinated efforts into (i) patient education, (ii) GDMT to control risk factors, (iii) GDMT to prevent myocardial infarction and death, (iv) medical therapy to relieve symptoms, and (v) decision for revascularization. Current guidelines from the US- and non-US-based organizations provide recommendations on each of the above five areas.

15.4 CURRENT GUIDELINE RECOMMENDATIONS ON STABLE ANGINA/SIHD MANAGEMENT

This section discusses current guideline recommendations on the management of stable angina/SIHD. It primarily considers the recommendations from the 2012 ACCF/AHA/ACP/AATS/PCNA/SCAI/STS guideline [8]. Pertinent recommendations from the 2013 ESC and NICE 2011 guidelines are also included when necessary. The section focuses on recommendations regarding evidence-based medical therapy to prevent myocardial infarction and death and to relieve symptoms. The reader is advised to refer to the full guidelines [8, 10] for recommendations on patient education, lifestyle modifications, and medical therapy toward risk factor modifications. In addition, the reader may refer to two recently released guidelines on lifestyle management [12] and management of overweight and obesity [13] for updated recommendations.

15.4.1 Drug Therapy to Relieve Symptoms

As discussed in Chapter 14, several classes of drugs can be used to relieve anginal symptoms and improve exercise tolerance in patients with SIHD. These include β-blockers, calcium channel blockers (CCBs), organic nitrates, and ranolazine, as well as several others, which are currently not available in the United States. Table 15.2 summarizes the 2012 ACCF/AHA/ACP/AATS/PCNA/SCAI/STS guideline recommendations on use of anti-ischemic drugs for symptom control in patient with stable angina/SIHD. The corresponding 2013 ESC guideline recommendations are listed in Table 15.3.

TABLE 15.2   The 2012 ACCF/AHA/ACP/AATS/PCNA/SCAI/STS guideline recommendations on use of anti-ischemic drugs for symptom control in patient with stable angina/SIHD

COR	Recommendation	LOE
I	β-Blockers should be prescribed as initial therapy for relief of symptoms in patients with SIHD	B
	CCBs or long-acting nitrates should be prescribed for relief of symptoms when β-blockers are contraindicated or cause unacceptable adverse effects in patients with SIHD	B
	CCBs or long-acting nitrates, in combination with β-blockers, should be prescribed for relief of symptoms when initial treatment with β-blockers is unsuccessful in patients with SIHD	B
	Sublingual nitroglycerin or nitroglycerin spray is recommended for immediate relief of angina in patients with SIHD	B
IIa	Treatment with a long-acting nondihydropyridine CCB (verapamil or diltiazem) instead of a β-blocker as initial therapy for relief of symptoms is reasonable in patients with SIHD	B
	Ranolazine can be useful when prescribed as a substitute for β-blockers for relief of symptoms in patients with SIHD if initial treatment with β-blockers leads to unacceptable adverse effects or is ineffective or if initial treatment with β-blockers is contraindicated	B
	Ranolazine in combination with β-blockers can be useful when prescribed for relief of symptoms when initial treatment with β-blockers is not successful in patients with SIHD	A

COR and LOE denote class of recommendation and level of evidence, respectively (see Chapter 5 for description).

TABLE 15.3   The 2013 ESC guideline recommendations on angina/ischemia relief in patients with stable angina/SCAD

COR	Recommendation	LOE
I	First-line treatment is indicated with β-blockers and/or CCBs to control heart rate and symptoms	A
	Short-acting nitrates are recommended	B
	According to comorbidities/tolerance, it is indicated to use second-line therapies as first-line treatment in selected patients	C
IIa	For second-line treatment, it is recommended to add long-acting nitrates or ivabradine or nicorandil or ranolazine, according to heart rate, blood pressure, and tolerance	B
	In asymptomatic patients with large areas of ischemia (>10%), β-blockers should be considered	C
	In patients with vasospastic angina, CCBs and nitrates should be considered and β-blockers avoided	B
IIb	For second-line treatment, trimetazidine may be considered	B

COR and LOE denote class of recommendation and level of evidence, respectively (see Chapter 5 for description).

15.4.2   Drug Therapy to Prevent Myocardial Infarction and Mortality

Efforts to prevent myocardial infarction and death in IHD focus primarily on reducing the incidence of acute thrombotic events and the development of ventricular dysfunction. These aims are achieved by lifestyle or pharmacological interventions, which (i) reduce plaque progression; (ii) stabilize plaque, by reducing inflammation and preserving endothelial function; and (iii) prevent thrombosis if endothelial dysfunction or plaque rupture occurs. In certain circumstances, such as in patients with severe lesions in coronary arteries supplying a large area of jeopardized myocardium, revascularization offers additional opportunities to improve prognosis by improving existing perfusion or providing alternative route of perfusion.

Pharmacological therapy is not only essential for relieving anginal symptoms but also a central component of medical therapy to prevent myocardial infarction and death in patients with stable angina/SIHD. Drugs that improve the prognosis of stable angina/SIHD include statins, antiplatelet agents, β-blockers, and inhibitors of the renin–angiotensin–aldosterone system (RAAS). This section first reviews the basic pharmacology of these drug classes with regard to their role in preventing cardiovascular events and death in patients with stable angina/SIHD. It then describes the current recommendations on using these agents to improve prognosis. The detailed discussion of the drug classes is provided in other chapters of the book.

15.4.2.1   Basic Pharmacology

Antiplatelet Agents   Platelet activation and aggregation are key events of the thrombotic response to plaque disruption (see Chapter 17). Antiplatelet therapy to prevent coronary thrombosis is indicated due to a favorable ratio between benefit and risk in patients with stable angina/SIHD. Low-dose aspirin is the drug of choice in most cases, and clopidogrel may be considered for some patients. The use of either drug in patients with stable angina/SIHD results in significant reduction (over 30%) of the risk of adverse cardiovascular events.

Aspirin exerts an antithrombotic effect by inhibiting cyclooxygenase and synthesis of platelet thromboxane A_2.

The optimal antithrombotic dosage of aspirin appears to be 75–150 mg/day, as the relative risk reduction afforded by aspirin may decrease both below and above this dose range. Clopidogrel prevents adenosine diphosphate (ADP)-mediated activation of platelets by selectively and irreversibly inhibiting the binding of ADP to its platelet receptors and thereby blocking ADP-dependent activation of the glycoprotein IIb/IIIa complex. Clopidogrel is much more expensive than aspirin but may be considered for aspirin-intolerant patients with significant risks of arterial thrombosis. After coronary stenting or an ACS, clopidogrel may be combined with aspirin during a finite period of time, but such a combination therapy is currently not warranted in chronic stable angina.

Statins   Multiple clinical trials have demonstrated that treatment with statins in patients with documented IHD, including stable angina, results in significant reduction (20–35%) of risk of both mortality rate and major coronary events (see Chapter 4). These clinical trials indicate that in patients with established IHD, including chronic stable angina, statin therapy should be recommended even in the absence of elevation of LDL cholesterol. Statins lower cholesterol effectively, but mechanisms other than cholesterol synthesis inhibition, such as anti-inflammatory and antithrombotic effects, may also contribute to the cardiovascular risk reduction. Notably, in the recently released guideline from the ACC/AHA on the treatment of blood cholesterol to reduce atherosclerotic cardiovascular risk in adults [14], it is recommended that unless contraindicated, high-intensity statin therapy should be initiated or continued as first-line therapy in adults 40–75 years of age who have clinical atherosclerotic cardiovascular disease regardless of LDL-cholesterol levels (see Chapter 5).

β-Blockers   β-Blockers significantly reduce deaths and recurrent myocardial infarctions in patients who have suffered a myocardial infarction and are especially effective when an ST-elevation myocardial infarction (STEMI) is complicated by persistent or recurrent ischemia or tachyarrhythmias early after the onset of infarction (see Chapter 8). However, no large trials have assessed the effects of β-blockers on survival or coronary event rates in patients with stable angina/SIHD [8]. Moreover, recent studies have even questioned the efficacy of β-blocker therapy in patients with STEMI [15, 16].

Inhibitors of the RAAS Inhibitors of the RAAS include ACEIs and angiotensin receptor blockers (ARBs), among others (see Chapter 9). Substantial evidence suggests that ACEIs and ARBs have cardiovascular protective effects, reducing the risks of future ischemic events. These drugs, by inhibiting the deleterious effects of angiotensin II, may contribute to (i) the reductions in left ventricular and vascular hypertrophy, atherosclerosis progression, plaque rupture, and thrombosis; (ii) the favorable changes in cardiac hemodynamics; and (iii) the improved myocardial oxygen supply/demand. Clinical studies have demonstrated significant reduc-

tions in the incidence of acute myocardial infarction and unstable angina and the need for coronary revascularization in patients: (i) after myocardial infarction with left ventricular dysfunction, (ii) with atherosclerotic vascular diseases, or (iii) with diabetes. In addition, as discussed in Chapter 9, ACEIs and ARBs are also renal protective and, as such, indicated in patients with chronic kidney disease (CKD). Hence, it is appropriate to consider ACEIs or ARBs for the treatment of patients with stable angina/SIHD, especially in those with coexisting hypertension, left ventricular dysfunction, diabetes, or CKD, unless contraindicated.

15.4.2.2 Recommendations
Table 15.4 summarizes the 2012 ACCF/AHA/ACP/AATS/PCNA/SCAI/STS guideline recommendations on use of various drug classes for preventing myocardial infarction and death in patients with stable

TABLE 15.4   The 2012 ACCF/AHA/ACP/AATS/PCNA/SCAI/STS guideline recommendations on using drugs to prevent myocardial infarction and death in patients with stable angina/SIHD [8]

Drug class	COR	Recommendation	LOE
Antiplatelet agents	I	Treatment with aspirin (75–162 mg daily) should be continued indefinitely in the absence of contraindications in patients with SIHD	A
		Treatment with clopidogrel is reasonable when aspirin is contraindicated in patients with SIHD	B
	IIb	Treatment with aspirin (75–162 mg daily) and clopidogrel (75 mg daily) might be reasonable in certain high-risk patients with SIHD	B
	III (no benefit)	Dipyridamole is not recommended as antiplatelet therapy for patients with SIHD	B
Statins[a]	I	High-intensity statin therapy should be initiated or continued as first-line therapy in women and men ≤75 years of age who have clinical atherosclerotic cardiovascular disease (ASCVD), unless contraindicated	A
		In individuals with clinical ASCVD in whom high-intensity statin therapy would otherwise be used, when high-intensity statin therapy is contraindicated or when characteristics predisposing to statin-associated adverse effects are present, moderate-intensity statin should be used as the second option if tolerated	A
	IIa	In individuals with clinical ASCVD >75 years of age, it is reasonable to evaluate the potential for ASCVD risk reduction benefits and for adverse effects and drug interactions and to consider patient preferences when initiating a moderate- or high-intensity statin. It is reasonable to continue statin therapy in those who are tolerating it	B
β-Blockers	I	β-Blocker therapy should be started and continued for 3 years in all patients with normal left ventricular function after myocardial infarction or ACS	B
		β-Blocker therapy should be used in all patients with left ventricular systolic dysfunction (ejection fraction ≤40%) with heart failure or prior myocardial infarction, unless contraindicated. Use should be limited to carvedilol, metoprolol succinate, or bisoprolol, which have been shown to reduce risk of death	A
	IIb	β-Blocker may be considered as chronic therapy for all other patients with coronary or other vascular disease	C
RAAS inhibitors	I	ACEIs should be prescribed for all patients with SIHD who also have hypertension, diabetes, left ventricular ejection fraction ≤40%, or CKD, unless contraindicated	A
		ARBs are recommended for patients with SIHD who have hypertension, diabetes, left ventricular systolic dysfunction, or CKD and have indications for, but are intolerant of, ACEIs	A
	IIa	Treatment with an ACEI is reasonable in patients with both SIHD and other vascular diseases	B
		It is reasonable to use ARBs in other patients who are ACEI intolerant	C

[a]In the ACCF/AHA/ACP/AATS/PCNA/SCAI/STS guideline, statins are not discussed in the drug therapy for preventing myocardial infarction and death. In the recently released guideline from the ACC/AHA on the treatment of blood cholesterol to reduce atherosclerotic cardiovascular risk in adults [17], the use of statins is recommended in patients with clinical arteriosclerotic cardiovascular disease (including stable angina/SIHD) regardless of blood cholesterol levels. Hence, statins are included in the table to reflect the most current development on statin therapy.

COR and LOE denote class of recommendation and level of evidence, respectively (see Chapter 5 for description).

angina/SIHD. The corresponding 2013 ESC guideline recommendations are outlined in Table 15.5.

15.4.3 Revascularization

The two well-established approaches to revascularization for treatment of chronic stable angina caused by coronary atherosclerosis are CABG and PCI. Currently, both methods are facing rapid development with the introduction of minimally invasive and off-pump surgery and drug-eluting stents (DES). As in the case of pharmacological therapy, the potential objectives of revascularization are twofold: (i) to improve survival or survival free of infarction and (ii) to diminish or eradicate symptoms. The decision to revascularize a patient and use PCI or CABG should be based on the presence of significant obstructive coronary artery stenosis, the amount of related ischemia, and the expected benefit to prognosis and/or symptoms, as well as technical and environmental factors. Refer to the full guidelines for indications and decisions to perform revascularization in patients with stable angina/SIHD [8, 10]. Tables 15.6 an 15.7 summarize the current ACCF/AHA/ACP/AATS/PCNA/SCAI/STS guideline recommendations on revascularization to improve survival and to alleviate symptoms, respectively.

TABLE 15.5 The 2013 ESC guideline recommendations on using drugs to prevent myocardial infarction and death in patients with stable angina/SIHD [10]

COR	Recommendation	LOE
I	Daily low-dose aspirin is recommended in all stable coronary artery disease (SCAD[a]) patients	A
I	Clopidogrel is indicated as an alternative in case of aspirin intolerance	B
I	Statins are recommended in all SCAD patients	A
I	It is recommended to use ACEIs (or ARBs) if other conditions (e.g., heart failure, hypertension, or diabetes) are present	A

[a] In the ESC guideline, SCAD is used instead of SIHD. SCAD and SICH can be considered interchangeable.
COR and LOE denote class of recommendation and level of evidence, respectively (see Chapter 5 for description).

TABLE 15.6 The 2012 ACCF/AHA/ACP/AATS/PCNA/SCAI/STS guideline recommendations on revascularization to improve survival in patients with stable angina/SIHD

COR	Recommendation	LOE
Left main CAD revascularization		
I	CABG to improve survival is recommended for patients with significant (≥50% diameter stenosis) left main coronary artery stenosis	B
IIa	PCI to improve survival is reasonable as an alternative to CABG in selected stable patients with significant (≥50% diameter stenosis) unprotected left main CAD with (i) anatomic conditions associated with a low risk of PCI procedural complications and a high likelihood of good long-term outcome (e.g., a low SYNTAX score[a] [≤22], ostial or trunk left main CAD) and (ii) clinical characteristics that predict a significantly increased risk of adverse surgical outcomes (e.g., STS-predicted risk of operative mortality ≥5%)[b]	B
	PCI to improve survival is reasonable in patients with unstable angina/non-ST-elevation myocardial infarction (UA/NSTEMI) when an unprotected left main coronary artery is the culprit lesion and the patient is not a candidate for CABG	B
	PCI to improve survival is reasonable in patients with acute ST-elevation myocardial infarction (STEMI) when an unprotected left main coronary artery is the culprit lesion, distal coronary flow is less than thrombolysis in myocardial infarction (TIMI[c]) grade 3, and PCI can be performed more rapidly and safely than CABG	C
IIb	PCI to improve survival may be reasonable as an alternative to CABG in selected stable patients with significant (≥50% diameter stenosis) unprotected left main CAD with (i) anatomic conditions associated with a low to intermediate risk of PCI procedural complications and an intermediate to high likelihood of good long-term outcome (e.g., low-intermediate SYNTAX score of <33, bifurcation left main CAD) and (ii) clinical characteristics that predict an increased risk of adverse surgical outcomes (e.g., moderate–severe chronic obstructive pulmonary disease, disability from previous stroke, or previous cardiac surgery; STS-predicted risk of operative mortality >2%)	B
III (harm)	PCI to improve survival should not be performed in stable patients with significant (≥50% diameter stenosis) unprotected left main CAD who have unfavorable anatomy for PCI and who are good candidates for CABG	B
Non-left main CAD revascularization		
I	CABG to improve survival is beneficial in patients with significant (≥70% diameter) stenoses in three major coronary arteries (with or without involvement of the proximal LAD artery) or in the proximal LAD artery plus 1 other major coronary artery	B
	CABG or PCI to improve survival is beneficial in survivors of sudden cardiac death with presumed ischemia-mediated ventricular tachycardia caused by significant (≥70% diameter) stenosis in a major coronary artery. CABG, LOE: B; PCI, LOE: C	B/C

(Continued)

TABLE 15.6   (*Continued*)

COR	Recommendation	LOE
IIa	CABG to improve survival is reasonable in patients with significant (≥70% diameter) stenoses in two major coronary arteries with severe or extensive myocardial ischemia (e.g., high-risk criteria on stress testing, abnormal intracoronary hemodynamic evaluation, or >20% perfusion defect by myocardial perfusion stress imaging) or target vessels supplying a large area of viable myocardium	B
	CABG to improve survival is reasonable in patients with mild–moderate left ventricular systolic dysfunction (ejection fraction 35–50%) and significant (≥70% diameter stenosis) multivessel CAD or proximal LAD coronary artery stenosis, when viable myocardium is present in the region of intended revascularization	B
	CABG with a left internal mammary artery (LIMA) graft to improve survival is reasonable in patients with significant (≥70% diameter) stenosis in the proximal LAD artery and evidence of extensive ischemia	B
	It is reasonable to choose CABG over PCI to improve survival in patients with complex 3-vessel CAD (e.g., SYNTAX score >22), with or without involvement of the proximal LAD artery who are good candidates for CABG	B
	CABG is probably recommended in preference to PCI to improve survival in patients with multivessel CAD and diabetes mellitus, particularly if a LIMA graft can be anastomosed to the LAD artery	B
IIb	The usefulness of CABG to improve survival is uncertain in patients with significant (70% diameter) stenoses in two major coronary arteries not involving the proximal LAD artery and without extensive ischemia	C
	The usefulness of PCI to improve survival is uncertain in patients with 2- or 3-vessel CAD (with or without involvement of the proximal LAD artery) or 1-vessel proximal LAD disease	B
	CABG might be considered with the primary or sole intent of improving survival in patients with SIHD with severe left ventricular systolic dysfunction (ejection fraction <35%) whether or not viable myocardium is present	B
	The usefulness of CABG or PCI to improve survival is uncertain in patients with previous CABG and extensive anterior wall ischemia on noninvasive testing	B
III (harm)	CABG or PCI should not be performed with the primary or sole intent to improve survival in patients with SIHD with 1 or more coronary stenoses that are not anatomically or functionally significant (e.g., <70% diameter non-left main coronary artery stenosis, fractional flow reserve[d] >0.80, no or only mild ischemia on noninvasive testing), involve only the left circumflex or right coronary artery, or subtend only a small area of viable myocardium	B

[a]SYNTAX score is a measure of the anatomical severity of coronary artery disease and has been arbitrarily classified as low (SYNTAX score 0–22), intermediate (SYNTAX score 23–32), and high severity (SYNTAX score >32), to produce three approximately similar-sized groups. SYNTAX score is an important factor in formulating revascularization recommendations.
[b]The Society of Thoracic Surgeons (STS) risk model predicts the risk of operative mortality and morbidity of adult cardiac surgery on the basis of patient demographic and clinical variables [18].
[c]TIMI grade flow is a grading system for coronary flow developed by the TIMI Study Group (founded by Eugene Braunwald in 1984) (http://www.timi.org). Grade 0, no perfusion; grade 1, penetration without perfusion; grade 2, partial perfusion; grade 3, complete perfusion.
[d]Fractional flow reserve (FFR) refers to the measurement that involves determining the ratio between the maximum achievable blood flow in a diseased coronary artery and the theoretical maximum flow in a normal coronary artery. An FFR of 1.0 is widely accepted as normal. An FFR of <0.75–0.80 is generally considered to be associated with myocardial ischemia.
COR and LOE denote class of recommendation and level of evidence, respectively (see Chapter 5 for description).
LAD denotes left anterior descending.

TABLE 15.7   The 2012 ACCF/AHA/ACP/AATS/PCNA/SCAI/STS guideline recommendations on revascularization to improve symptoms in patients with stable angina/SIHD

COR	Recommendation	LOE
I	CABG or PCI to improve symptoms is beneficial in patients with 1 or more significant (≥70% diameter) coronary artery stenoses amenable to revascularization and unacceptable angina despite GDMT	A
IIa	CABG or PCI to improve symptoms is reasonable in patients with 1 or more significant (≥70% diameter) coronary artery stenoses and unacceptable angina for whom GDMT cannot be implemented because of medication contraindications, adverse effects, or patient preferences	C
	PCI to improve symptoms is reasonable in patients with previous CABG, 1 or more significant (≥70% diameter) coronary artery stenoses associated with ischemia, and unacceptable angina despite GDMT	C
	It is reasonable to choose CABG over PCI to improve symptoms in patients with complex 3-vessel CAD (e.g., SYNTAX score >22), with or without involvement of the proximal LAD artery, who are good candidates for CABG	B
IIb	CABG to improve symptoms might be reasonable for patients with previous CABG, 1 or more significant (≥70% diameter) coronary artery stenoses not amenable to PCI, and unacceptable angina despite GDMT	C
	TMR performed as an adjunct to CABG to improve symptoms may be reasonable in patients with viable ischemic myocardium that is perfused by arteries that are not amenable to grafting	B
III (harm)	CABG or PCI to improve symptoms should not be performed in patients who do not meet the anatomical (≥50% diameter left main or ≥70% non-left main stenosis diameter) or physiological (e.g., abnormal fractional flow reserve) criteria for revascularization	C

COR and LOE denote class of recommendation and level of evidence, respectively (see Chapter 5 for description).

15.5 MANAGEMENT OF SPECIAL TYPES OF STABLE ANGINA

The main symptomatic clinical presentations of SIHD include (i) classical chronic stable angina caused by epicardial stenosis, (ii) angina caused by microvascular dysfunction (microvascular angina), (iii) angina caused by vasospasm (vasospastic angina), and (iv) symptomatic ischemic cardiomyopathy. The preceding sections have discussed the management of classical chronic stable angina. This section considers the management of microvascular angina and vasospastic angina. It also discusses the management of the classical chronic stable angina that is refractory to conventional therapies, which is known as refractory angina.

15.5.1 Microvascular Angina

Microvascular angina may be difficult to distinguish from classical stable angina because both are mainly exercise related. Microvascular angina results from dysfunction of the small coronary arteries. In microvascular angina, coronary flow reserve (CFR) is impaired in the absence of epicardial artery obstruction because of nonhomogeneous metabolic vasodilation that may favor the "steal" phenomenon, by inappropriate prearteriolar/arteriolar vasoconstriction, or by other causes for altered cross-sectional luminal area [19, 20]. Conditions such as ventricular hypertrophy, myocardial ischemia, arterial hypertension, and diabetes can also affect the microcirculation and blunt CFR in the absence of epicardial vessel narrowing, causing microvascular angina.

All patients with microvascular angina should achieve optimal coronary risk factor control as for patients with the classical stable angina [10]. Traditional anti-ischemic drugs are the first step in medical treatment of microvascular angina. Short-acting nitrates can be used to treat anginal attacks, but often, they are only partially effective. β-Blocker therapy is a rational approach because the dominant symptom is effort-related angina. Indeed, β-blockers were found to improve symptoms in several studies and should constitute the first choice of therapy, particularly in patients with evidence of increased adrenergic activity (e.g., high heart rate at rest or during low-workload exercise). CCBs can also be first-line therapy in patients with a significant variable threshold of effort angina. In patients with persisting symptoms despite optimal anti-ischemic drug therapy, other treatments, such as ACEIs and xanthine derivatives (aminophylline, bamiphylline), as well as the nonpharmacological approaches used in treating refractory angina (Section 15.5.3), may also be considered. Table 15.8 summarizes the 2013 ESC guideline recommendations on management of patients with microvascular angina [10].

15.5.2 Vasospastic Angina

Vasospastic angina, in contrast to classical and microvascular angina, is characterized by angina at rest with preserved effort tolerance [10]. Severe focal constriction (spasm) of a normal or atherosclerotic epicardial artery determines vasospastic angina. Spasm can also be multifocal or diffuse and, in the latter case, is most pronounced in the distal coronary arteries. It is predominantly caused by vasoconstricting stimuli acting on hyperreactive vascular smooth muscle cells, although endothelial dysfunction may also be involved. It is currently unclear whether the more common form of diffuse distal vasospasm has the same or different mechanisms. The causes of smooth muscle cell hyperreactivity are unknown, but several possible contributing factors have been suggested, including increased cellular rho-kinase activity and abnormalities in K^+_{ATP} channels and/or membrane Na^+–H^+ countertransport [21, 22]. Other contributing factors may include imbalances in the autonomic nervous system; enhanced intracoronary concentrations of vasoconstricting substances, such as endothelin; and hormonal changes, such as postoophorectomy. Coronary vasospasm, especially the focal occlusive variant, has been found on occasion to cause myocardial infarction.

All patients with vasospastic angina should achieve optimal coronary risk factor control, in particular through smoking cessation and aspirin. A drug-related cause (e.g., cocaine or amphetamines) should be systemically researched and managed if detected. Chronic preventive treatment of vasospastic angina is mainly based on the use of CCBs. Average doses of these drugs (240–360 mg/day of verapamil or diltiazem, 40–60 mg/day of nifedipine) usually prevent spasm in about 90% of patients. Long-acting nitrates can be added in some patients to improve the efficacy of treatment

TABLE 15.8 The 2013 ESC guideline recommendations on management of patients with microvascular angina

COR	Recommendation	LOE
I	It is recommended that all patients receive secondary prevention medications, including aspirin and statins	B
	β-Blockers are recommended as a first-line treatment	B
	CCBs are recommended if β-blockers do not achieve sufficient symptomatic benefit or are not tolerated	B
IIb	ACEIs or nicorandil may be considered in patients with refractory symptoms	B
	Xanthine derivatives or nonpharmacological treatments, such as neurostimulatory techniques, may be considered in patients with symptoms refractory to the aforementioned listed drugs	B

COR and LOE denote class of recommendation and level of evidence, respectively (see Chapter 5 for description).

and should be scheduled to cover the period of the day in which ischemic episodes most frequently occur, in order to prevent nitrate tolerance. β-Blockers should be avoided, as they might favor spasm by leaving α-adrenergic receptor-mediated vasoconstriction unopposed by β-adrenergic receptor-mediated vasodilation [10].

15.5.3 Refractory Stable Angina

Drugs and revascularization procedures (i.e., CABG and PCI) can adequately manage the majority of patients suffering from IHD. However, there are patients who remain severely disabled by angina pectoris in spite of different forms of conventional treatment.

The term "refractory angina" is defined as a chronic condition caused by clinically established SIHD reversible myocardial ischemia in the presence of coronary artery disease, which cannot be adequately controlled by a combination of drug therapy, PCI, or GABC. For this patient group, a number of treatment options have emerged, including some new drugs (see Chapter 14) and nonpharmacological approaches.

Among the nonpharmacological treatments of refractory stable angina are (i) enhanced external counterpulsation (EECP); (ii) neurostimulatory techniques (transcutaneous electrical nerve stimulation [TENS], spinal cord stimulation [SCS]); (iii) angiogenesis through noninvasive techniques (extracorporeal cardiac shock wave therapy) or invasive techniques, such as transmyocardial laser revascularization (TMR) or percutaneous myocardial laser revascularization (PMR); and (iv) emerging investigational stem cell/gene therapy.

The efficacy of these nonconventional approaches in treating refractory angina remains to be established, and some are controversial. In this context, the recent NICE evaluation of TMR and PMR concluded that current evidence on both TMR and PMR for refractory angina shows no efficacy and may pose unacceptable procedure-related risks. Therefore, these procedures should not be used [23].

The stem cell therapy of refractory angina has recently received more attention. A double-blind, randomized, phase II study of 167 patients with refractory angina reported that patients who received intramyocardial

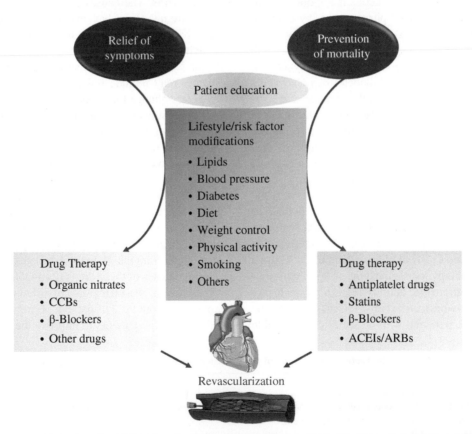

FIGURE 15.1 Schematic illustration of guideline-based management of stable angina/stable ischemic heart disease (SIHD). As illustrated, effective management of stable angina/SIHD requires multiple intertwined efforts, including drug therapy, nonpharmacological approach, lifestyle/risk factor modifications, and patient education. These efforts are aimed to relieve the patients' symptoms and prevent cardiovascular events and mortality, with the latter being the highest priority. ACEIs/ARBs, angiotensin-converting enzyme inhibitors/angiotensin receptor blockers; CCBs, calcium channel blockers. For color details, please see color plate section.

injections of autologous CD34+ cells (10^5 cells/kg body weight) experienced significant improvements in angina frequency and exercise tolerance [17]. Multiple other studies also suggested an efficacy for stem cell therapy in refractory angina [24–26].

15.6 SUMMARY OF CHAPTER KEY POINTS

- Management of stable angina/SIHD has evolved remarkably over the past decade with regard to both drug therapy and nonpharmacological treatments.
- Multiple professional organizations have developed guidelines for the management of stable angina/SIHD. The most notable ones are the 2012 ACCF/AHA/ACP/AATS/PCNA/SCAI/STS guideline and the 2013 ESC guideline.
- Two fundamental goals of management of stable angina/SIHD as described in the various guidelines are improving survival and alleviating symptoms. To attain the above dual goals, the 2012 ACCF/AHA/ACP/AATS/PCNA/SCAI/STS guideline defines five specific objectives and provides five fundamental, complementary, and overlapping strategies.
- The effective management of stable angina/SIHD requires the coordinated efforts in five areas, including (i) patient education, (ii) guideline-directed medical therapy to control risk factors, (iii) guideline-directed medical therapy to prevent myocardial infarction and death, (iv) medical therapy to relieve symptoms, and (v) decision for revascularization.
- Both the 2012 ACCF/AHA/ACP/AATS/PCNA/SCAI/STS guideline and the 2013 ESC guideline provide detailed recommendations on each of the above five areas. Figure 15.1 illustrates the general scheme of current guideline-based management of stable angina/SIHD.

15.7 SELF-ASSESSMENT QUESTIONS

15.7.1 A 54-year-old male is diagnosed with stable angina/stable ischemic heart disease. Which of the following should be prescribed as initial therapy for relief of his symptoms?
   A. A β-blocker
   B. A nitrate
   C. A calcium channel blocker
   D. A platelet inhibitor
   E. An anticoagulant

15.7.2 A 46-year-old male presents to the physician's office complaining of substernal chest tightness and pain upon exertion. History reveals that he has hypertension of 5 years of duration, which has been poorly managed with a thiazide diuretic. He also has asthma and cannot tolerate β-blocker therapy (including β_1-selective drugs). Which of the following should be prescribed for relief of his symptoms?
   A. Amlodipine
   B. Aspirin
   C. Clopidogrel
   D. Ranolazine
   E. Sublingual nitroglycerin

15.7.3 A 45-year-old patient is experiencing angina due to emotional stress. Which of the following is recommended for immediate relief of his angina?
   A. Amlodipine
   B. Isosorbide mononitrate
   C. Metoprolol succinate
   D. Nitroglycerin spray
   E. Ranolazine

15.7.4 A 48-year-old male presents to the emergency room due to severe chest pain while running uphill in his neighborhood. He is diagnosed with typical stable angina. Which of the following drugs should the patient take indefinitely in the absence of contraindication even if he is asymptomatic?
   A. Aspirin
   B. Clopidogrel
   C. Ranolazine
   D. Verapamil
   E. Warfarin

15.7.5 A 56-year-old female is brought to the emergency department because of chest pain. She is diagnosed with stable angina and left ventricular dysfunction (LVEF 39%). History reveals that she had an acute myocardial infarction 14 months ago. A decision is made to put her on β-blocker therapy to prevent recurrent myocardial infarction and death. Which of the following β-blockers should be prescribed?
   A. Atenolol
   B. Carvedilol
   C. Labetalol
   D. Pindolol
   E. Propranolol

REFERENCES

1   Gibbons, R.J., et al., ACC/AHA/ACP-ASIM guidelines for the management of patients with chronic stable angina: a report of the American College of Cardiology/American Heart Association Task Force on Practice Guidelines (Committee on Management of Patients With Chronic Stable Angina). *J Am Coll Cardiol*, 1999. 33(7): p. 2092–197.

2   Gibbons, R.J., et al., ACC/AHA/ACP-ASIM guidelines for the management of patients with chronic stable angina: executive summary and recommendations. A Report of the American

College of Cardiology/American Heart Association Task Force on Practice Guidelines (Committee on Management of Patients with Chronic Stable Angina). *Circulation*, 1999. 99(21): p. 2829–48.

3   Gibbons, R.J., et al., ACC/AHA 2002 guideline update for the management of patients with chronic stable angina—summary article: a report of the American College of Cardiology/American Heart Association Task Force on practice guidelines (Committee on the Management of Patients With Chronic Stable Angina). *J Am Coll Cardiol*, 2003. 41(1): p. 159–68.

4   Gibbons, R.J., et al., ACC/AHA 2002 guideline update for the management of patients with chronic stable angina—summary article: a report of the American College of Cardiology/American Heart Association Task Force on Practice Guidelines (Committee on the Management of Patients With Chronic Stable Angina). *Circulation*, 2003. 107(1): p. 149–58.

5   Fraker, T.D., Jr., et al., 2007 chronic angina focused update of the ACC/AHA 2002 guidelines for the management of patients with chronic stable angina: a report of the American College of Cardiology/American Heart Association Task Force on Practice Guidelines Writing Group to develop the focused update of the 2002 guidelines for the management of patients with chronic stable angina. *J Am Coll Cardiol*, 2007. 50(23): p. 2264–74.

6   Fraker, T.D., Jr., et al., 2007 chronic angina focused update of the ACC/AHA 2002 guidelines for the management of patients with chronic stable angina: a report of the American College of Cardiology/American Heart Association Task Force on Practice Guidelines Writing Group to develop the focused update of the 2002 guidelines for the management of patients with chronic stable angina. *Circulation*, 2007. 116(23): p. 2762–72.

7   Fihn, S.D., et al., 2012 ACCF/AHA/ACP/AATS/PCNA/SCAI/STS guideline for the diagnosis and management of patients with stable ischemic heart disease: a report of the American College of Cardiology Foundation/American Heart Association task force on practice guidelines, and the American College of Physicians, American Association for Thoracic Surgery, Preventive Cardiovascular Nurses Association, Society for Cardiovascular Angiography and Interventions, and Society of Thoracic Surgeons. *Circulation*, 2012. 126(25): p. e354–471.

8   Fihn, S.D., et al., 2012 ACCF/AHA/ACP/AATS/PCNA/SCAI/STS guideline for the diagnosis and management of patients with stable ischemic heart disease: a report of the American College of Cardiology Foundation/American Heart Association Task Force on Practice Guidelines, and the American College of Physicians, American Association for Thoracic Surgery, Preventive Cardiovascular Nurses Association, Society for Cardiovascular Angiography and Interventions, and Society of Thoracic Surgeons. *J Am Coll Cardiol*, 2012. 60(24): p. e44–164.

9   Fox, K., et al., Guidelines on the management of stable angina pectoris: executive summary: The Task Force on the Management of Stable Angina Pectoris of the European Society of Cardiology. *Eur Heart J*, 2006. 27(11): p. 1341–81.

10 Task Force Members, et al., 2013 ESC guidelines on the management of stable coronary artery disease: the Task Force on the management of stable coronary artery disease of the European Society of Cardiology. *Eur Heart J*, 2013. 34(38): p. 2949–3003.

11 O'Flynn, N., et al., Management of stable angina: summary of NICE guidance. *BMJ*, 2011. 343: p. d4147.

12 Eckel, R.H., et al., 2013 AHA/ACC guideline on lifestyle management to reduce cardiovascular risk: a report of the American College of Cardiology/American Heart Association Task Force on Practice Guidelines. *Circulation*, 2014. 129(25 Suppl 2): p. S76–99.

13 Jensen, M.D., et al., 2013 AHA/ACC/TOS guideline for the management of overweight and obesity in adults: a report of the American College of Cardiology/American Heart Association Task Force on Practice Guidelines and The Obesity Society. *Circulation*, 2014. 129(25 Suppl 2): p. S102–38.

14 Stone, N.J., et al., 2013 ACC/AHA guideline on the treatment of blood cholesterol to reduce atherosclerotic cardiovascular risk in adults: a report of the American College of Cardiology/American Heart Association Task Force on Practice Guidelines. *J Am Coll Cardiol*, 2014. 63(25Ptb): p. 3024–5.

15 Park, K.L., et al., Beta-blocker use in ST-segment elevation myocardial infarction in the reperfusion era (GRACE). *Am J Med*, 2014. 127(6): p. 503–11.

16 Bangalore, S., et al., Clinical outcomes with beta-blockers for myocardial infarction a meta-analysis of randomized trials. *Am J Med*, 2014. 127(10): p. 939–53.

17 Stone, N.J., et al., 2013 ACC/AHA guideline on the treatment of blood cholesterol to reduce atherosclerotic cardiovascular risk in adults: a report of the American College of Cardiology/American Heart Association Task Force on Practice Guidelines. *J Am Coll Cardiol*, 2014. 63(25 Pt B): p. 2889–934.

18 Shahian, D.M. and F.H. Edwards, The Society of Thoracic Surgeons 2008 cardiac surgery risk models: introduction. *Ann Thorac Surg*, 2009. 88(1 Suppl): p. S1.

19 Lanza, G.A. and F. Crea, Primary coronary microvascular dysfunction: clinical presentation, pathophysiology, and management. *Circulation*, 2010. 121(21): p. 2317–25.

20 Crea, F., P.G. Camici, and C.N. Bairey Merz, Coronary microvascular dysfunction: an update. *Eur Heart J*, 2014. 35(17): p. 1101–11.

21 Lanza, G.A., G. Careri, and F. Crea, Mechanisms of coronary artery spasm. *Circulation*, 2011. 124(16): p. 1774–82.

22 Kinlay, S., Coronary artery spasm as a cause of angina. *Circulation*, 2014. 129(17): p. 1717–9.

23 Schofield, P.M., et al., NICE evaluation of transmyocardial laser revascularisation and percutaneous laser revascularisation for refractory angina. *Heart*, 2010. 96(4): p. 312–3.

24 Hossne, N.A., Jr, et al., Long-term and sustained therapeutic results of a specific promonocyte cell formulation in refractory angina: react (refractory angina cell therapy) clinical update and cost effective analysis. Cell Transplant, 2014. doi: 10.3727/ 096368914X681595.

25 Haack-Sorensen, M., et al., Direct intramyocardial mesenchymal stromal cell injections in patients with severe refractory angina: one-year follow-up. *Cell Transplant*, 2013. 22(3): p. 521–8.

26 Mathiasen, A.B., et al., Autotransplantation of mesenchymal stromal cells from bone-marrow to heart in patients with severe stable coronary artery disease and refractory angina—final 3-year follow-up. *Int J Cardiol*, 2013. 170(2): p. 246–51.

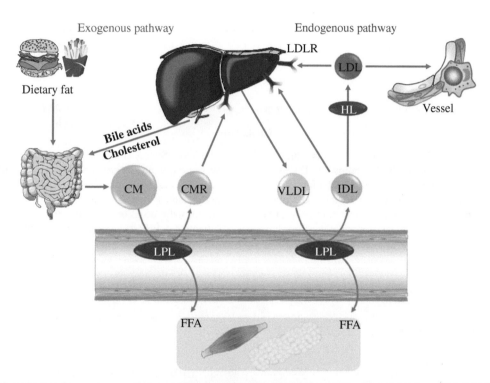

FIGURE 3.2 Endogenous and exogenous pathways of lipoprotein metabolism. See text (Sections 3.2.2.1 and 3.2.2.2) for description. CM, chylomicron; CMR, chylomicron remnant; HL, hepatic lipase; IDL, intermediate-density lipoprotein; LDL, low-density lipoprotein; LDLR, LDL receptor; FFA, free fatty acid; LPL, lipoprotein lipase; VLDL, very-low-density lipoprotein.

Cardiovascular Diseases: From Molecular Pharmacology to Evidence-Based Therapeutics, First Edition. Y. Robert Li.
© 2015 John Wiley & Sons, Inc. Published 2015 by John Wiley & Sons, Inc.

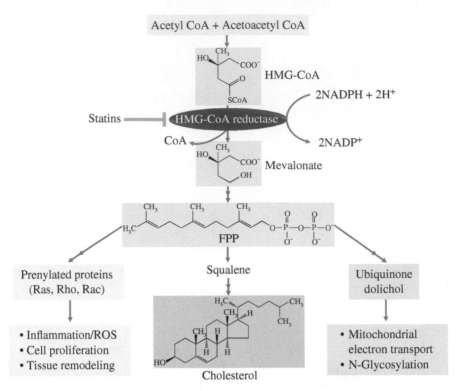

FIGURE 4.2 Molecular mechanism by which statins inhibit cholesterol synthesis. Statins competitively inhibit HMG-CoA reductase, a key enzyme in cholesterol biosynthesis. Inhibition of HMG-CoA reductase by statins also decreases protein prenylation, a process that leads to inflammation, reactive oxygen species (ROS) formation, cell proliferation, and tissue remodeling. These events play an important part in atherosclerosis. Decreased protein prenylation of small G proteins, including Ras, Rho, and Rac, may hence account for the lipid-lowering-independent beneficial effects of statin therapy. On the other hand, statin drugs decrease ubiquinone formation, which might partly be responsible for the development of myopathy associated with statin treatment. FPP denotes farnesyl pyrophosphate.

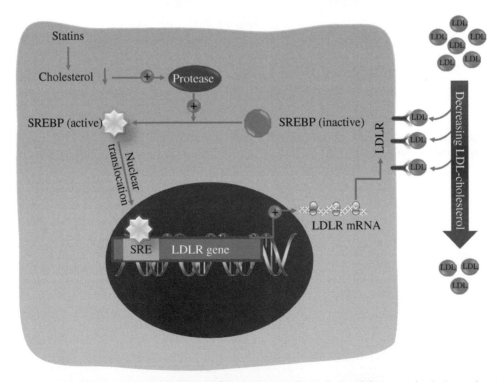

FIGURE 4.3 Molecular mechanism by which statins reduce LDL cholesterol. Statin-induced decreases in cholesterol concentration in hepatocytes result in protease activation. Protease activation causes the activation and nuclear translocation of sterol regulatory element-binding protein (SREBP), which binds to the sterol regulatory element of the LDL receptor (LDLR) gene, leading to increased expression of LDLR. Increased LDLR expression on the surface of hepatocytes promotes LDL uptake from plasma, thereby reducing plasma LDL cholesterol.

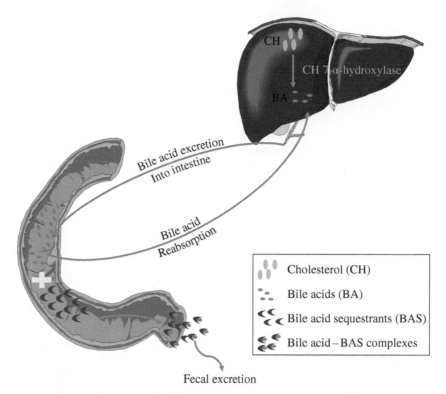

FIGURE 4.5 Molecular mechanism by which bile acid sequestrants reduce cholesterol. A significant portion of bile acids (bile salts) secreted into the intestine is reabsorbed and delivered into the liver via portal vein, hence forming enterohepatic circulation of bile acids (BA). Bile acid sequestrants (BAS) bind BA forming complexes that are eliminated in feces. The decreased return of BA to the liver causes upregulation of cholesterol (CH) 7-α-hydroxylase, the key enzyme in BA synthesis, thereby leading to decreases in CH concentration in hepatocytes. As described in Figure 4.3 legend, decreased cholesterol causes increased expression of LDL receptors on the surface of hepatocytes and the subsequent reduction in plasma LDL cholesterol.

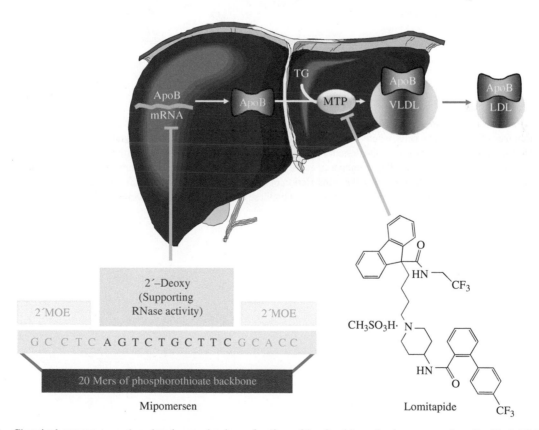

FIGURE 4.9 Chemical structures and molecular mechanism of action of lomitapide and mipomersen. Lomitapide inhibits microsomal triglyceride transfer protein (MTP), an enzyme involved in the assembly of VLDL in the liver. Mipomersen is an antisense that targets ApoB mRNA, thereby reducing ApoB synthesis and the subsequent assembly of VLDL. Decreased formation of VLDL results in reduction of LDL cholesterol. TG denotes triglyceride; 2′MOE denotes 2′-methoxyethyl.

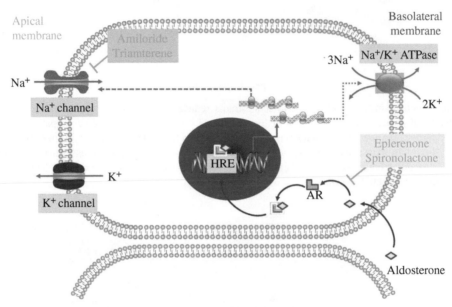

FIGURE 7.8 Molecular mechanisms of action of potassium-sparing diuretics. Amiloride and triamterene block the activity of the Na^+ channels in the apical membranes of the principal cells, whereas eplerenone and spironolactone decrease the biosynthesis of new Na^+ channels in these cells via blocking the aldosterone receptors (AR). As illustrated, aldosterone–AR complex binds to the hormone response element (HRE) of the gene encoding the Na^+ channel protein, thereby leading to enhanced expression of the Na^+ channels. Eplerenone and spironolactone competitively block the binding of aldosterone to AR, thereby inhibiting the transcription of the Na^+ channel-encoding gene. Since Na^+ reabsorption in the principal cells is coupled with K^+ secretion, inhibition of the Na^+ reabsorption by these diuretic drugs causes decreased K^+ excretion.

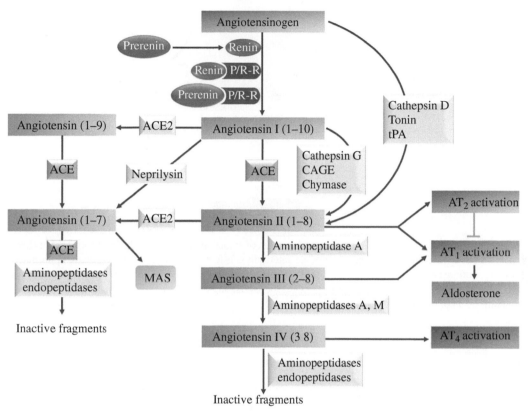

FIGURE 9.2 The contemporary view of the RAAS. The contemporary view emphasizes multiple aspects of the RAAS not recognized in the classical view. For example, the catalytic activity of renin increases when bound to the (pro)renin receptors (P/R-R), and the otherwise inactive prorenin becomes catalytically active when bound to P/R-R. Angiotensin II can be catabolized by angiotensin-converting enzyme 2 (ACE2) to form angiotensin-(1–7), another active peptide of this system, which typically opposes the actions of angiotensin II. Angiotensin-(1–7) can also be derived from angiotensin I with angiotensin-(1–9) as the intermediate. Angiotensin II is cleaved into smaller fragments, such as angiotensin-(2–8) (also called angiotensin III) and angiotensin-(3–8) (also called angiotensin IV) by aminopeptidases. Most effects of angiotensin II are mediated by the angiotensin receptor type 1 (AT$_1$ receptor); however, angiotensin II can also bind to the angiotensin receptor type 2 (AT$_2$ receptor), whose activation generally exhibits opposing effects on the responses caused by the activation of AT$_1$ receptors and may result in cardiovascular protection. Angiotensin-(1–7) acts via its putative MAS receptor to elicit responses that may also counteract those caused by activation of AT$_1$ receptors by angiotensin II. MAS is a proto-oncogene encoding a G-protein-coupled receptor, which has been recently identified as a putative receptor for angiotensin-(1–7). The ACE2/angiotensin-(1–7)/MAS receptor axis may represent new possibilities for developing novel therapeutic strategies for the treatment of cardiovascular diseases. As shown in the scheme, in addition to angiotensin II, angiotensin III also activates AT$_1$ receptors. Angiotensin IV activates AT$_4$ receptors; however, the physiological significance of this receptor activation remains unclear. Likewise, the nomenclature and existence of the AT$_3$ receptor and its relationship to MAS receptor remain controversial.

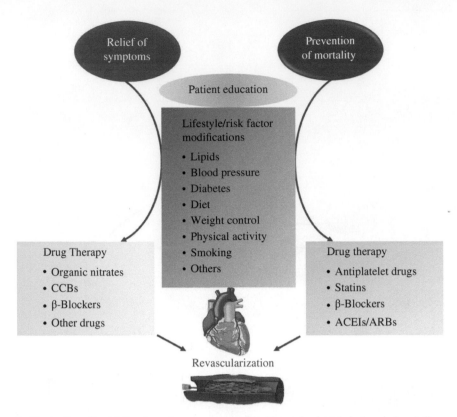

FIGURE 15.1 Schematic illustration of guideline-based management of stable angina/stable ischemic heart disease (SIHD). As illustrated, effective management of stable angina/SIHD requires multiple intertwined efforts, including drug therapy, nonpharmacological approach, lifestyle/risk factor modifications, and patient education. These efforts are aimed to relieve the patients' symptoms and prevent cardiovascular events and mortality, with the latter being the highest priority. ACEIs/ARBs, angiotensin-converting enzyme inhibitors/angiotensin receptor blockers; CCBs, calcium channel blockers.

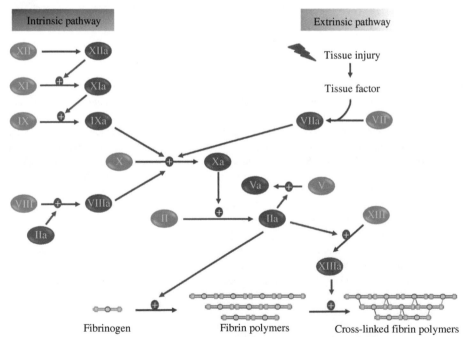

FIGURE 17.2 Blood coagulation cascades and anticoagulants. As illustrated, the intrinsic and extrinsic pathways of coagulation converge with the activation of factor X and the subsequent formation of thrombin (factor IIa), which in turn catalyzes the formation of fibrin, leading to the fibrin clot formation. Vitamin K antagonists inhibit the functional maturation of factors II, VII, IX, and X. On the other hand, heparins, selective factor Xa, and direct thrombin inhibitors either indirectly or directly inhibit factor Xa and/or thrombin.

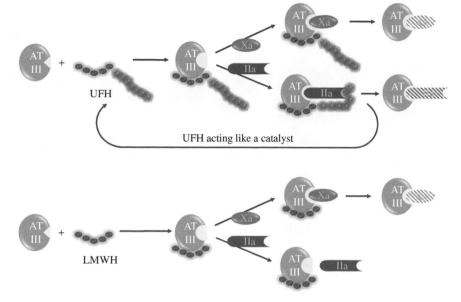

FIGURE 17.4 Molecular mechanisms of action of unfractionated heparin (UFH) and low-molecular-weight heparins (LMWHs). As illustrated, UFH binds to antithrombin III (ATIII) and facilitates the inactivation of both factors Xa and IIa by ATIII. Due to the short chain, LMWHs only primarily facilitate ATIII-mediated inactivation of factor Xa.

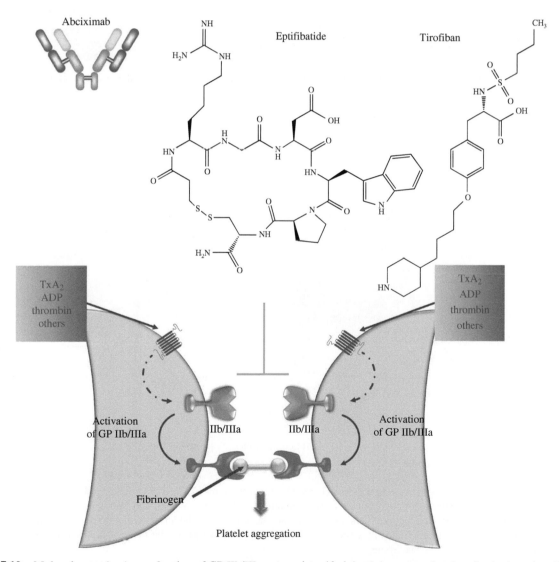

FIGURE 17.10 Molecular mechanisms of action of GP IIb/IIIa antagonists. Abciximab is a monoclonal antibody drug that directly binds (essentially irreversible) to GP IIb/IIIa, preventing fibrinogen-mediated platelet aggregation. Eptifibatide and tirofiban are small molecule drugs that reversibly antagonize fibrinogen binding to the GP IIb/IIIa receptors. Also shown in the scheme is the activation of GP IIb/IIIa receptors by TxA₂, ADP, thrombin, and other platelet activators.

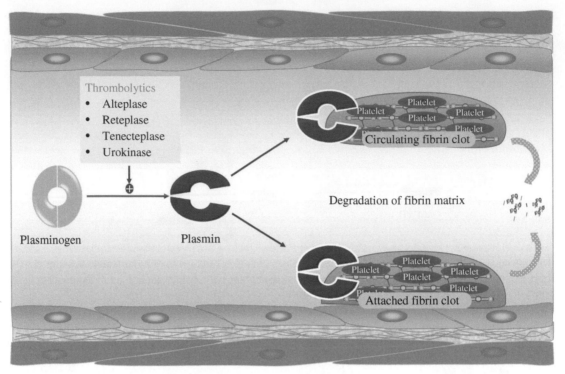

FIGURE 17.11 Molecular mechanism of action of thrombolytic drugs. Thrombolytic drugs stimulate the conversion of plasminogen to plasmin, which then degrades fibrin, leading to thrombus lysis.

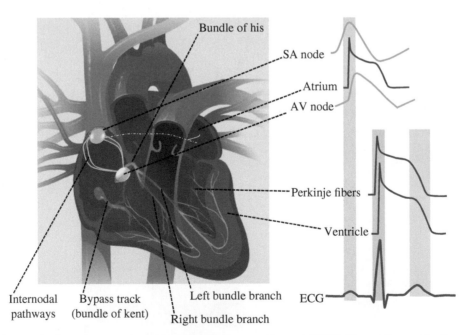

FIGURE 23.1 Cardiac conduction system, action potential, and electrocardiogram (ECG). The cardiac conduction system consists of the sinoatrial (SA) node, atrial ventricular (AV) node, bundle of His, left and right Bundle branches, and Purkinje fibers. In addition, internodal pathways connect SA and AV nodes. Conduction fibers also travel from the SA node to the left ventricle. In some individual, a bypass track connects the right atrium to the right ventricle. This accessory pathway also known as bundle of Kent predisposes the individual to the development of Wolff–Parkinson–White syndrome.

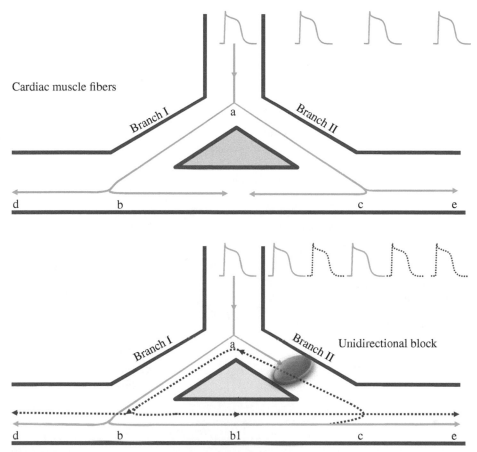

FIGURE 23.5 Normal and reentrant cardiac electrical pathways. Panel A shows a normal impulse (action potential) conduction through a bifurcating pathway to depolarize the different areas of myocardium. Panel B shows formation of a reentrant circuit (dashed lines) due to unidirectional block of one of the branch of the bifurcation pathway leading to tachyarrhythmias. As illustrated in panel A, a normal impulse arrives at point a, where it travels through the bifurcating pathway (branches I and II). When the impulses travel down branches I and II to arrive at points b and c, respectively, the impulses are able to once again travel in two opposite directions. The impulses between b and c canceled each other, whereas as the rest travel to points d and e respectively, they cause excitation of the myocardium in different areas. As depicted in panel B, a unidirectional block occurs in branch II due to pathologic factors, such as ischemia. Hence, in branch II, the impulse from point a cannot travel to point c, whereas impulses are able to travel from point c to a. Because of this unidirectional block in branch II, when the impulse from branch I arrives at point b, it will travel in two directions: one is toward point d as in normal conduction (panel A), and the other toward point b1 because of no impulse from phase II to cancel it. When the impulse from point b1 arrives at point c, it will travel in two directions: one through the retrograde conduction along branch II toward point a (reentry), and the other continues toward point e. Once the reentered impulse arrives at point a, it will continue to travel along branch I to point b, forming a circuit. As illustrated, reentrant conduction results in extra beats, thereby tachyarrhythmias.

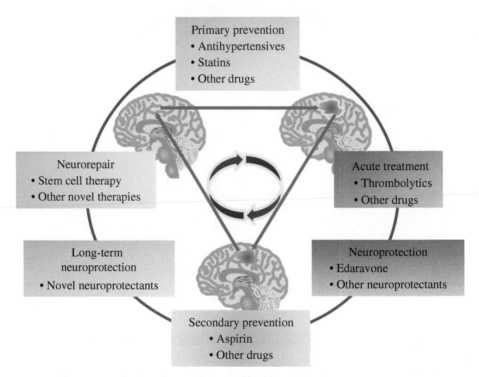

FIGURE 26.3 Management options for the preventive and therapeutic intervention of ischemic stroke. As depicted, pharmacological therapies play an important part in the overall management of ischemic stroke. Of note is the fact that at present neuroprotection and neurorepair are largely experimental approaches whose clinical efficacy remains to be established. Edaravone (structure shown in Fig. 27.1) is a free radical scavenging agent that is approved for treating ischemic stroke in Japan.

UNIT V

ISCHEMIC HEART DISEASE: ACUTE CORONARY SYNDROMES

16

OVERVIEW OF ACUTE CORONARY SYNDROMES AND DRUG THERAPY

16.1 INTRODUCTION

Unstable angina (UA), acute non-ST-elevation myocardial infarction (NSTEMI), and acute ST-elevation myocardial infarction (STEMI) are the three presentations of acute coronary syndromes (ACS). This chapter provides an overview on the definitions and epidemiology of ACS and discusses the current understanding of the pathophysiology of ACS and the mechanistically based drug targeting and related therapeutic modalities. Chapter 17 considers the molecular pharmacology of drugs for treating ACS, including anticoagulants, platelet inhibitors, and thrombolytic agents. This lays a basis for the coverage of the general principles and current guidelines regarding the management of UA/NSTEMI and STEMI in Chapters 18 and 19, respectively.

16.2 DEFINITIONS AND GENERAL CONSIDERATIONS

16.2.1 Definitions

As mentioned earlier, the term ACS refers to a spectrum of clinical presentations ranging from those for STEMI to presentations found in NSTEMI or in UA. In terms of pathogenesis, ACS is almost always associated with rupture (or erosion) of an atherosclerotic plaque and partial or complete thrombosis of the inflicted coronary artery. While UA and NSTEMI are caused by incomplete coronary blockage, STEMI typically results from complete coronary occlusion. UA and NSTEMI are also known as non-ST-elevation ACS (NSTE-ACS) and STEMI as ST-elevation ACS.

UA and NSTEMI differ primarily in whether the ischemia is severe enough to cause sufficient myocardial damage to release detectable quantities of a marker of myocardial injury (e.g., cardiac troponins) (Fig. 16.1). The key diagnostic criteria for UA and NSTEMI are outlined below:

- UA is considered to be present in patients with ischemic symptoms suggestive of an ACS and no elevation in cardiac troponins, with or without electrocardiogram (ECG) changes, indicative of ischemia (e.g., ST-segment depression or transient elevation or new T-wave inversion).
- NSTEMI is considered to be present in patients having the same manifestations as those in UA but in whom an elevation in cardiac troponins is present.

Since an elevation in cardiac troponins may not be detectable for hours after presentation, UA and NSTEMI are frequently indistinguishable at initial evaluation. As a consequence, initial management is the same for these two syndromes. For this reason, and because the pathophysiological mechanisms of the two conditions are similar, UA and NSTEMI are often considered together.

16.2.2 Historical Overview

The condition commonly referred to today as ACS has a long history in the annals of medicine [1]. The dreaded symptoms were eloquently described by the esteemed English physician William Heberden (1710–1801) in 1772 as "a most disagreeable sensation in the breast, which seems

Cardiovascular Diseases: From Molecular Pharmacology to Evidence-Based Therapeutics, First Edition. Y. Robert Li.
© 2015 John Wiley & Sons, Inc. Published 2015 by John Wiley & Sons, Inc.

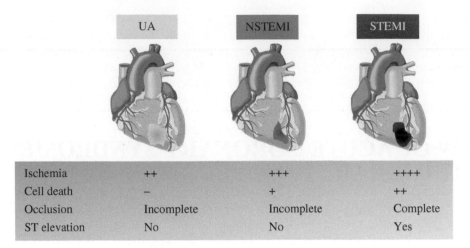

	UA	NSTEMI	STEMI
Ischemia	++	+++	++++
Cell death	–	+	++
Occlusion	Incomplete	Incomplete	Complete
ST elevation	No	No	Yes

FIGURE 16.1   Characteristics of acute coronary syndromes. As illustrated, in unstable angina (UA), ischemia is not severe enough to cause myocardial injury (cell death), whereas in non-ST-elevation myocardial infarction (NSTEMI) and ST-elevation myocardial infarction (STEMI), ischemia causes significant myocardial damage, leading to increased plasma levels of cardiac troponins.

as if it would take their life away, if it were to increase or continue." More than a century later, Sir William Osler (1849–1919) formalized the definition of ACS and highlighted its profound implications for the patient in his famous Lumleian Lecture on "Angina Pectoris" delivered before the Royal College of Physicians in London in 1910: "There are two primary features of the disease, pain and sudden death—pain, paroxysmal, intense, peculiar, usually pectoral, and with well-known lines of radiation—death in a higher percentage than any known disorder, and usually sudden." With the advent of diagnostic tools, such as the ECG and sensitive and specific laboratory tests for myocardial damage, the term ACS has evolved as an umbrella diagnosis to capture the full spectrum of disease severity and protean clinical manifestations of critical coronary atherosclerosis from UA to NSTEMI and STEMI [1].

16.2.3   Global Burden

As discussed in Chapters 1 and 13, the epidemic of ischemic heart disease (IHD) is truly global, with more than 80% of the burden of this disease carried by the developing nations. IHD, and its manifestation as ACS, carries enormous personal, societal, and economic burdens and is a major determinant of morbidity and mortality among all races, ethnic groups, and cultures [1–3]. While prevalence of ACS has reached a pandemic level as a consequence of modernization of the developing world, the demographics of ACS have also evolved, with a precipitous decline in the incidence of STEMI and a progressive rise in the incidence of NSTEMI [1].

16.3   PATHOPHYSIOLOGY AND DRUG TARGETING

16.3.1   Historical Overview

The causative link between arterial thrombosis and ACS has a controversial history [4, 5]. The initial link between coronary thrombosis and acute myocardial infarction was recognized in the late 1800s and gained widespread acceptance within the medical community throughout the early 1900s. But after a series of pathological studies that cast doubt on this association, alternative mechanisms were sought. It was not until the use of coronary angiography in people with an acute myocardial infarction in the late 1970s, in combination with the success of thrombolytic therapies in the 1980s, that the causative role for thrombosis in precipitating UA and acute myocardial infarction was unequivocally established. Since then, there has been dramatic improvement in the understanding and treatment options for ACS, with antithrombotic therapy taking a center stage in the disease management.

16.3.2   Molecular Pathophysiology

ACS represent life-threatening manifestations of coronary atherosclerosis. Atherosclerosis alone may obstruct coronary blood flow and cause stable angina as described in Chapter 13, but this is rarely fatal in the absence of scarring of the myocardium, which can elicit an arrhythmia presenting as sudden cardiac arrest [6]. ACS are nearly always precipitated by acute thrombosis induced by a ruptured or eroded atherosclerotic coronary plaque with or without concomitant vasospasm, causing a sudden and critical reduction in blood flow [7].

16.3.2.1 Plaque Rupture In the complex process of plaque disruption, inflammation has been revealed as a key pathophysiological element [6, 8]. Plaque rupture occurs where the cap is thinnest and most infiltrated by foam cells (lipid-laden macrophages). In eccentric plaques, the weakest spot is often the cap margin or shoulder region, and only extremely thin fibrous caps are at risk of rupturing. Thinning of the fibrous cap probably involves two concurrent mechanisms. One is the gradual loss of vascular smooth muscle cells (SMCs) from the fibrous cap. Indeed, ruptured caps contain fewer SMCs and less collagen than intact caps, and SMCs are usually absent at the actual site of rupture. Concurrently, infiltrating macrophages in the plaque degrade the collagen-rich cap matrix via releasing matrix metalloproteinases (MMPs), especially MMP-1, MMP-8, and MMP-13 [6, 8]. With plaque rupture, cap collagen and the highly thrombogenic lipid core, enriched in tissue factor-expressing apoptotic microparticles, are exposed to the thrombogenic factors of the blood, leading to thrombogenesis [6, 8].

16.3.2.2 Plaque Erosion In addition to plaque rupture, superficial plaque erosion also causes thrombus formation, responsible for 20–25% of the cases of fatal myocardial infarctions [9, 10]. This anatomical substrate for coronary thrombosis occurs more frequently in women than in men and in persons with certain risk factors, such as hypertriglyceridemia. Many lesions, which cause coronary thrombosis because of superficial erosion, lack prominent inflammatory infiltrates, and such plaques typically exhibit proteoglycan accumulation. The mechanisms of superficial erosion are less clear than those involved in the rupture of the fibrous cap. The surface endothelium under the thrombus is usually absent, but no distinct morphological features of the underlying plaque have been identified.

Eroded and thrombosed plaques causing sudden cardiac death are often scarcely calcified, often associated with negative remodeling, and contain fewer macrophages than ruptured plaques. Apoptosis of endothelial cells could contribute to their desquamation. Reactive oxygen species and oxidative stress may play an important role in endothelial apoptosis. As endothelial cells undergo apoptosis, they produce the procoagulant tissue factor, resulting in local thrombosis in coronary arteries. Endothelial cells also express proteinases that may sever their tethers to the underlying basement membrane. In this regard, modified low-density lipoprotein (LDL) induces the expression of the enzyme MMP-14 by human endothelial cells. MMP-14 can activate MMP-2, an enzyme that degrades basement-membrane forms of nonfibrillar collagen (i.e., type IV collagen). These events collectively contribute to the superficial erosion of the atheroma, thereby promoting thrombogenesis and the formation of a thrombus that severely restricts the blood flow of the inflicted coronary artery [6, 8].

16.3.2.3 Secondary Mechanisms In rare cases, ACS may have a nonatherosclerotic etiology, such as arteritis, trauma, dissection, thromboembolism, congenital anomalies, cocaine abuse, or complications of cardiac catheterization. These are collectively known as secondary mechanisms of ACS. Hence, the key pathophysiological mechanisms, including plaque rupture and plaque erosion, as well as the secondary mechanisms of ACS, need to be understood for the correct use of the available therapeutic strategies.

16.3.3 Drug Targeting

16.3.3.1 Thrombus-Based Targeting As described earlier, the ACS spectrum includes patients with STEMI and NSTE-ACS, and NSTE-ACS is further comprised of UA and NSTEMI. The initial difference in pathophysiology and early outcomes between STEMI and NSTE-ACS leads to contrasting early treatment strategies. In STEMI, because of the complete occlusion, prompt reopening of the occluded artery is the therapeutic priority that limits the extent of myocardial injury and saves lives. In contrast, the therapeutic goal in NSTE-ACS management is to prevent progression of the thrombus to total occlusion, plaque thromboembolization, and recurrent infarction. In-hospital mortality rates for STEMI remain 50% higher than those for NSTEMI patients. However, the high rates of recurrent ischemic events in NSTE-ACS patients result in similar 1-year mortality rates in the two conditions, emphasizing the need for selecting appropriate early management strategies and secondary prevention measures.

16.3.3.2 Inflammation-Based Targeting Given the critical role of inflammation in the pathophysiological aspects of plaque rupture and thrombogenesis, multiple studies have been carried out over the past years to assess the safety and efficacy of using anti-inflammatory therapies other than statins to reduce the risk of recurrent ACS. This section introduces some of the recent trials on anti-inflammatory therapies in ACS and discusses the potential implications of such emerging modalities.

Colchicine The presence of activated neutrophils in culprit atherosclerotic plaques of patients with unstable coronary artery disease raises the possibility that inhibition of neutrophil function with colchicine may reduce the risk of plaque instability and thereby improve clinical outcomes in patients with stable coronary disease. In this regard, in a recent clinical trial with a prospective, randomized, observer-blinded endpoint design, 532 patients with stable coronary disease receiving aspirin and/or clopidogrel (93%) and statins (95%) were randomly assigned colchicine 0.5 mg/day or no colchicine and followed for a median of 3 years. The primary outcome was the composite incidence of ACS, out-of-hospital cardiac arrest, or noncardioembolic ischemic stroke. The primary outcome occurred in 15 of 282 patients (5.3%) who received colchicine and 40 of 250 patients (16.0%) assigned no colchicine (hazard ratio, 0.33; 95% confidence interval [CI], 0.18–0.59; $p < 0.001$; number needed to treat, 11). In a prespecified secondary on-treatment analysis that excluded 32 patients (11%) assigned to colchicine who withdrew within 30 days due to intestinal intolerance and a further seven patients (2%) who did not start treatment, the primary outcome occurred in 4.5% versus 16.0% (hazard ratio, 0.29; 95% CI, 0.15–0.56; $p < 0.001$). The study concluded that colchicine, 0.5 mg/day administered in addition to statins and other standard secondary prevention therapies, appeared effective for the prevention of cardiovascular events in patients with stable coronary disease [11].

The aforementioned colchicine trial was, however, relatively small (532 patients, with a total of 55 events), and the investigators did not use a double-blind design and did not report levels of inflammatory biomarkers, which might have provided a glimpse into the possible mechanisms underlying the effects of colchicine [8, 12]. Nevertheless, the encouraging results of this study should prompt a larger-scale, double-blind trial of this inexpensive agent, which has a long history of clinical use and a well-known and acceptable risk profile [8].

Darapladib and Varespladib Elevated lipoprotein-associated phospholipase A$_2$ (PLA$_2$) activity promotes the development of vulnerable atherosclerotic plaques, and elevated plasma levels of this enzyme are associated with an increased risk of coronary events. Hence, inhibition of PLA$_2$ activity may be an effective strategy for ACS intervention. In this regard, darapladib has been developed as a selective oral inhibitor of lipoprotein-associated PLA$_2$. However, a recent randomized, double-blind trial of 15,828 patients with stable coronary heart disease reported that darapladib (160 mg once daily) did not significantly reduce the risk of the primary composite endpoint of cardiovascular death, myocardial infarction, or stroke [13]. Similarly, a double-blind, randomized, multicenter trial of 5145 patients with recent ACS (VISTA-16) showed that varespladib, another inhibitor of sPLA$_2$, did not reduce the risk of recurrent cardiovascular events and instead significantly increased the risk of myocardial infarction. Hence, the sPLA$_2$ inhibition with varespladib may be harmful and is not a useful strategy to reduce adverse cardiovascular outcomes after ACS [14]. Although one cannot state with certainty whether the observed harmful effects in the VISTA-16 trial were a direct consequence of sPLA$_2$ inhibition, this specific enzyme target is unlikely to be investigated further. Nonetheless, these findings do not argue against a central role for inflammation in atherogenesis, but rather highlight that much still needs to be learned regarding the complexity of inflammatory pathways in ACS and effective targeting of the pathways with novel agents [15].

Other Emerging Anti-inflammatory Modalities Besides colchicine and PLA$_2$ inhibitors, several trials are currently ongoing to assess the safety and efficacy of other anti-inflammatory therapies [12, 16]. Among these are the Canakinumab Anti-inflammatory Thrombosis Outcomes Study (CANTOS) trial and the Cardiovascular Inflammation Reduction Trial (CIRT). The CANTOS trial evaluates the effectiveness of a human monoclonal antibody to the inflammatory cytokine interleukin-1-β in 17,200 stable patients post-myocardial infarction, randomized to receive either the subcutaneous antibody drug or placebo, and followed over 4 years. On the other hand, the CIRT study determines the effect of low-dose methotrexate (10–20 mg/week) on cardiovascular events in 7000 patients with prior acute myocardial infarction, elevated C-reactive protein levels, and diabetes. The results of such trials might lead to the development of effective anti-inflammatory therapies for ACS.

16.4 SUMMARY OF CHAPTER KEY POINTS

- The term ACS refers to a spectrum of clinical syndromes, including UA, NSTEMI, and STEMI.

- ACS are almost always associated with rupture (or erosion) of an atherosclerotic plaque and partial or complete thrombosis of the inflicted coronary artery. UA and NSTEMI are also known as non-ST-elevation ACS and STEMI as ST-elevation ACS.

- The difference between UA and myocardial infarction is whether the ischemia is severe enough to cause sufficient myocardial damage to release detectable quantities of cardiac troponins. Significant myocardial damage occurs in NSTEMI and STEMI, but not in UA.

- Since an elevation in cardiac troponins may not be detectable for hours after presentation, UA and NSTEMI are frequently indistinguishable at initial evaluation. As such, initial management is the same for these two syndromes, and they are often considered together.

- In STEMI, because of the complete occlusion, prompt reopening of the occluded artery is the therapeutic priority that limits the extent of myocardial injury and saves lives. In contrast, the therapeutic goal in UA/NSTEMI management is to prevent progression of the thrombus to total occlusion, plaque thromboembolization, and recurrent infarction.

• Given the critical role of inflammation in the pathophysiology of plaque rupture and thrombogenesis, multiple clinical trials have been carried out or planned to assess the safety and efficacy of using anti-inflammatory therapies other than statins to reduce the risk of recurrent ACS. Continued research in this area might someday lead to the development of novel therapeutic modalities specifically targeting the inflammatory component of ACS for more effective management. Until then, antithrombotic drugs remain the mainstay of pharmacological therapy for ACS. Chapter 17 considers the pharmacological basis of antithrombotic drugs, including anticoagulants, platelet inhibitors, and thrombolytic agents, in treating ACS.

16.5 SELF-ASSESSMENT QUESTIONS

16.5.1 A 55-year-old male is brought to the emergency department because of severe chest pain immediately after his dinner. ECG shows ST-segment depression and T-wave inversion. Blood chemistry reveals elevated levels of cardiac troponins and lactate dehydrogenase. Which of the following is most likely responsible for the patient's condition?
A. Coronary artery rupture
B. Coronary arteritis
C. Coronary plaque rupture
D. Coronary plaque erosion
E. Coronary vasospasm

16.5.2 Which of the following is the second most common pathophysiological mechanism of acute coronary syndromes?
A. Coronary artery rupture
B. Coronary arteritis
C. Coronary plaque rupture
D. Coronary plaque erosion
E. Coronary vasospasm

16.5.3 Cardiac troponins are elevated most likely under which of the following conditions?
A. Acute decompensated heart failure
B. Microvascular angina
C. Non-ST-elevation myocardial infarction
D. Stable ischemic heart disease
E. Unstable angina

16.5.4 Which of the following is a key feature of plaque superficial erosion?
A. Lack of prominent inflammatory infiltrates
B. Lack of proteoglycan accumulation
C. Less frequently seen in females
D. Responsible for <5% of the cases of fatal myocardial infarction
E. Less frequently observed in patients with hypertriglyceridemia

16.5.5 A 55-year-old male is brought to the emergency department because he feels like "an elephant is sitting on his chest." ECG shows ST-segment elevation. Blood chemistry reveals marked elevations of cardiac troponins and creatine phosphokinase (CPK) activity. Which of the following is most likely responsible for the patient's condition?
A. Complete coronary occlusion
B. Coronary artery rupture
C. Coronary congenital abnormalities
D. Coronary vasospasm
E. Partial coronary occlusion

REFERENCES

1   Ruff, C.T. and E. Braunwald, The evolving epidemiology of acute coronary syndromes. *Nat Rev Cardiol*, 2011. 8(3): p. 140–7.

2   Moran, A.E., et al., The global burden of ischemic heart disease in 1990 and 2010: the global burden of disease 2010 study. *Circulation*, 2014. 129(14): p. 1493–501.

3   Moran, A.E., et al., Temporal trends in ischemic heart disease mortality in 21 world regions, 1980 to 2010: the global burden of disease 2010 study. *Circulation*, 2014. 129(14): p. 1483–92.

4   Muller, J.E., Coronary artery thrombosis: historical aspects. *J Am Coll Cardiol*, 1983. 1(3): p. 893–6.

5   Jackson, S.P., Arterial thrombosis—insidious, unpredictable and deadly. *Nat Med*, 2011. 17(11): p. 1423–36.

6   Bentzon, J.F., et al., Mechanisms of plaque formation and rupture. *Circ Res*, 2014. 114(12): p. 1852–66.

7   Hamm, C.W., et al., ESC Guidelines for the management of acute coronary syndromes in patients presenting without persistent ST-segment elevation: The Task Force for the management of acute coronary syndromes (ACS) in patients presenting without persistent ST-segment elevation of the European Society of Cardiology (ESC). *Eur Heart J*, 2011. 32(23): p. 2999–3054.

8   Libby, P., Mechanisms of acute coronary syndromes and their implications for therapy. *N Engl J Med*, 2013. 368(21): p. 2004–13.

9   Arbab-Zadeh, A., et al., Acute coronary events. *Circulation*, 2012. 125(9): p. 1147–56.

10 Falk, E., et al., Update on acute coronary syndromes: the pathologists' view. *Eur Heart J*, 2013. 34(10): p. 719–28.

11 Nidorf, S.M., et al., Low-dose colchicine for secondary prevention of cardiovascular disease. *J Am Coll Cardiol*, 2013. 61(4): p. 404–10.

12 Vogel, R.A. and J.S. Forrester, Cooling off hot hearts: a specific therapy for vulnerable plaque? *J Am Coll Cardiol*, 2013. 61(4): p. 411–2.

13 Investigators, S., et al., Darapladib for preventing ischemic events in stable coronary heart disease. *N Engl J Med*, 2014. 370(18): p. 1702–11.

14 Nicholls, S.J., et al., Varespladib and cardiovascular events in patients with an acute coronary syndrome: the VISTA-16 randomized clinical trial. *JAMA*, 2014. 311(3): p. 252–62.

15 O'Donoghue, M.L., Acute coronary syndromes: targeting inflammation-what has the VISTA-16 trial taught us? *Nat Rev Cardiol*, 2014. 11(3): p. 130–2.

16 Thompson, P.L., S.M. Nidorf, and J. Eikelboom, Targeting the unstable plaque in acute coronary syndromes. *Clin Ther*, 2013. 35(8): p. 1099–107.

17

ANTICOAGULANTS, PLATELET INHIBITORS, AND THROMBOLYTIC AGENTS

17.1 OVERVIEW

As discussed in Chapter 16, acute coronary syndromes (ACS) are life-threatening manifestations of coronary atherosclerosis. ACS are typically precipitated by acute thrombosis induced by ruptured or eroded atherosclerotic coronary plaques, with or without concomitant vasoconstriction, causing a sudden and critical reduction in blood flow to myocardium. The management of ACS thus includes the use of drugs that either inhibit thrombogenesis or accelerate the lysis of the already formed thrombus. These drugs are classified, based on their primary targets along the hemostatic pathways, into anticoagulants, platelet inhibitors, and thrombolytic agents. This chapter begins with a brief introduction to the physiology and pathophysiology of hemostasis and then discusses the molecular pharmacology of the above three classes of drugs with a focus on their clinical use in the treatment of ACS. The principles and current guidelines regarding ACS management are covered in Chapters 18 and 19.

17.2 HEMOSTASIS

17.2.1 Definitions

The term hemostasis refers to the finely regulated dynamic physiological process of maintaining fluidity of the blood, repairing vascular injury, and limiting blood loss while avoiding vessel occlusion (thrombosis) and inadequate perfusion of vital organs. Either extreme, excessive bleeding or thrombosis represents a breakdown of the hemostatic mechanism and leads to disease processes [1].

Thrombosis is defined as the formation of an unwanted clot (thrombus) within a blood vessel, and it is the most common abnormality of hemostasis. Here, it is necessary to distinguish two related terms: thrombus and embolus. Thrombus refers to a clot that adheres to a vessel wall, whereas embolus refers to a detached thrombus.

17.2.2 Physiology

Although the clotting process is a dynamic, highly interwoven array of multiple steps, it can be viewed as occurring in four phases, which are outlined below and also illustrated in Figure 17.1:

1. Formation of the platelet plug: Platelets are activated at the site of vascular injury to form a platelet plug that provides the initial hemostatic response, including exposure of procoagulant phospholipids on the platelet surface and the assembly of components of the clotting cascade.
2. Clot formation: Generation or exposure of tissue factor at the wound site, its interaction with factor VII, and the subsequent generation of activated factor X are the primary physiological events of the extrinsic pathway in initiating clotting, while components of the intrinsic pathway (i.e., factors VIII, IX, XI) are responsible for amplification of this process (see Fig. 17.2).
3. Termination of clotting: The termination phase of coagulation involves antithrombin, tissue factor pathway inhibitor, and the protein C pathway. This phase is

Cardiovascular Diseases: From Molecular Pharmacology to Evidence-Based Therapeutics, First Edition. Y. Robert Li.
© 2015 John Wiley & Sons, Inc. Published 2015 by John Wiley & Sons, Inc.

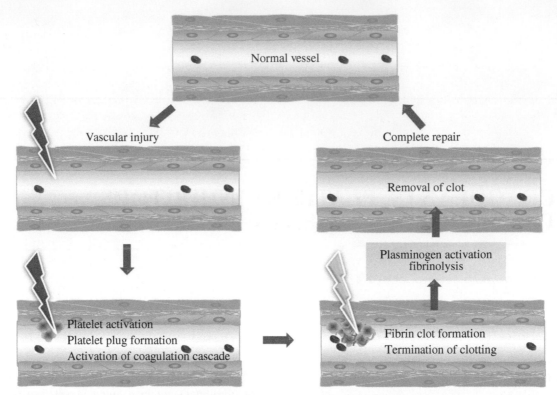

FIGURE 17.1 Physiology of coagulation and fibrinolysis in hemostasis. When a small blood vessel is injured, a localized vasoconstriction (not shown) reduces blood flow and facilitates platelet adhesion, activation, and aggregation, leading to platelet plug formation at the injured site. The vessel injury and platelet plug formation trigger activation of the coagulation cascades (see Fig. 17.2), which eventually lead to fibrin clot formation to arrest bleeding. Multiple factors in the plasma regulate the fibrin clot formation to prevent excessive clotting. The coordinated activation of fibrinolysis pathway ensures the removal of the clot once the injured vessel is repaired. Disruption of this hemostatic mechanism under pathophysiological conditions may lead to bleeding or thrombosis.

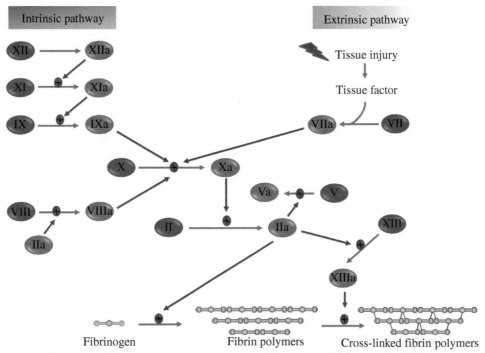

Vitamin K antagonists: (−) Posttranslational maturation of factors II, VII, IX, and X
Heparins: (−) Factors IIa and Xa
Selective factor Xa inhibitors: (−) Factor Xa
Direct thrombin inhibitors: (−) Factor IIa

FIGURE 17.2 Blood coagulation cascades and anticoagulants. As illustrated, the intrinsic and extrinsic pathways of coagulation converge with the activation of factor X and the subsequent formation of thrombin (factor IIa), which in turn catalyzes the formation of fibrin, leading to the fibrin clot formation. Vitamin K antagonists inhibit the functional maturation of factors II, VII, IX, and X. On the other hand, heparins, selective factor Xa, and direct thrombin inhibitors either indirectly or directly inhibit factor Xa and/or thrombin. For color details, please see color plate section.

critical in mediating the extent of clot formation, as demonstrated by the thrombotic disorders present in individuals with abnormalities in this pathway.

4. Clot lysis: To restore vessel patency, the clot must be properly removed. Plasminogen binds fibrin and tissue plasminogen activator (tPA), leading to the formation of active, proteolytic plasmin, which cleaves fibrin, fibrinogen, and a variety of plasma proteins and clotting factors. The plasminogen/plasminogen-activator system is also tightly regulated and, together with the parallel coagulation cascade, ensures the fluidity of the blood. Abnormal bleeding can result from diminished clot generation or enhanced clot lysis, while excessive clot formation or reduced clot lysis can lead to excessive thrombosis.

17.2.3   Disorders of Hemostasis and Drug Therapy

The disorders of hemostasis are typically classified into thromboembolic disorders and bleeding disorders. Common thromboembolic disorders include (i) ACS, (ii) deep vein thrombosis (DVT), (iii) pulmonary embolism (PE), and (iv) acute ischemic stroke. On the other hand, vitamin K deficiency, hemophilia and other coagulation factor deficiency disorders, and drug-induced bleeding are typical examples of bleeding disorders.

Management of thromboembolic disorders involves the use of anticoagulants, platelet inhibitors, and thrombolytic agents. Anticoagulants and platelet inhibitors are also collectively known as antithrombotic agents. Hence, drugs for treating thrombotic disorders, such as ACS, include antithrombotic drugs and thrombolytic agents, with the former inhibiting atherothrombogenesis and the latter causing lysis of the already formed thrombus.

17.3   ANTICOAGULANTS

Anticoagulants act on the different steps of the coagulation cascade (Fig. 17.2). They inhibit either the action of the coagulation factors or interfere with the biosynthesis of the coagulation factors. Anticoagulants are classified into four groups, as listed below:

1. Vitamin K antagonists
2. Heparins
3. Selective factor Xa (SFXa) inhibitors
4. Direct thrombin inhibitors (DTIs)

17.3.1   Vitamin K Antagonists

Listed below is the only US Food and Drug Administration (FDA)-approved vitamin K antagonist for clinical use:

• Warfarin (Coumadin)

17.3.1.1 *General Introduction to Drug Class*   Vitamin K is a critical cofactor in the posttranslational modification and functionality of coagulation factors II, VII, IX, and X. Vitamin K antagonists include the commonly used warfarin (structure shown in Fig. 17.3) and the rarely used dicumarol. Initially used as a rodenticide, warfarin is now widely used clinically as an oral anticoagulant.

17.3.1.2 *Chemistry and Pharmacokinetics*   Warfarin is chemically known as 3-(α-acetonylbenzyl)-4-hydroxycoumarin and is a racemic mixture of the R- and S-enantiomers. Warfarin is essentially completely absorbed after oral administration. It is extensively metabolized by hepatic cytochrome P450 (CYP) enzymes, including CYP2C9 (major), CYP2C19, CYP2C8, CYP2C18, CYP1A2, and CYP3A4 to form inactive metabolites that are excreted primarily in the urine. The elimination half-life of warfarin is approximately 40h. Notably, the polymorphic CYP2C9 is the major enzyme in metabolizing warfarin, which explains, at least partially, the individual variations in responses to warfarin therapy.

17.3.1.3 *Molecular Mechanisms and Pharmacological Effects*   Warfarin acts by inhibiting the biosynthesis/functional maturation of vitamin K-dependent clotting factors, which include factors II, VII, IX, and X, and the anticoagulant proteins C and S. Vitamin K is an essential cofactor for the postribosomal synthesis of the vitamin K-dependent clotting factors. Vitamin K promotes the biosynthesis of γ-carboxyglutamic acid residues in the proteins that are essential for biological activity. During the carboxylation reaction, vitamin K is oxidized to the inactive form, namely, vitamin K 2,3-epoxide. The enzyme vitamin K epoxide reductase (VKOR) converts the inactive vitamin K 2,3-epoxide into the active, reduced form of vitamin K. The regeneration of reduced vitamin K is essential for the sustained synthesis of biologically functional factors II, VII, IX, and X. Warfarin is thought to interfere with clotting factor synthesis through inhibition of the C1 subunit of VKOR (VKORC1) enzyme complex, thereby reducing the regeneration of active form of vitamin K (Fig. 17.3).

Certain single nucleotide polymorphisms in the VKORC1 gene (e.g., −1639G>A) have been associated with variable warfarin dose requirements. As noted earlier, polymorphic CYP2C9 also contributes to individual's variations in warfarin pharmacokinetics. Indeed, VKORC1 and CYP2C9 gene variants generally explain the largest proportion of known variability in warfarin dosage regimen. Hence, CYP2C9 and VKORC1 genotype information, when available, may assist in selection of the initial dose of warfarin [2]. However, the exact value of genotype-guided dosing of warfarin remains controversial [3].

An anticoagulation effect generally occurs within 24h after warfarin administration. However, peak anticoagulant effect

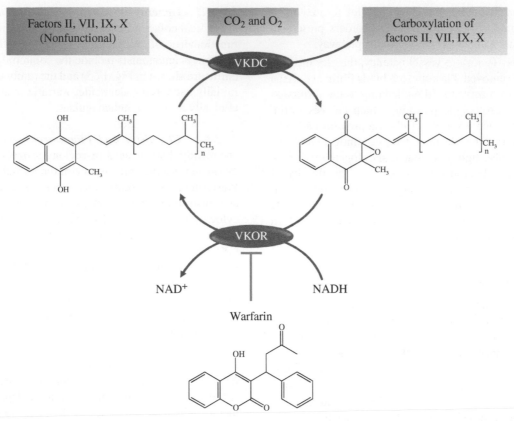

FIGURE 17.3 Molecular mechanism of action of warfarin. Warfarin inhibits vitamin K 2,3-epoxide reductase (VKOR), which is responsible for converting vitamin K epoxide to the active form of vitamin K (i.e., the reduced form of vitamin K). The reduced form of vitamin K is required for the vitamin K-dependent carboxylase (VKDC)-mediated posttranslational carboxylation and thereby functional maturation of factors II, VII, IX, and X. Inhibition of VKOR by warfarin decreases the availability of the reduced form of vitamin K, thereby inhibiting the functional maturation of the above coagulation factors. NAD^+, oxidized form of nicotinamide adenine dinucleotide; NADH, reduced form of nicotinamide adenine dinucleotide.

may be delayed 72–96 h. The duration of action of a single dose of warfarin is 2–5 days. This is consistent with the half-lives of the affected vitamin K-dependent clotting factors and anticoagulation proteins (factor II, 60 h; VII, 4–6 h; IX, 24 h; X, 48–72 h; protein C, 8 h; protein S, 30 h).

17.3.1.4 Clinical Uses Warfarin is indicated for (i) prophylaxis and treatment of venous thrombosis and its extension, PE; (ii) prophylaxis and treatment of thromboembolic complications associated with atrial fibrillation (AF) and/or cardiac valve replacement; and (iii) reduction in the risk of death, recurrent myocardial infarction (MI), and thromboembolic events, such as stroke or systemic embolization, after MI.

17.3.1.5 Therapeutic Dosages Listed below are the dosage forms and strengths of warfarin:

- Warfarin (Coumadin): Oral, 1, 2, 2.5, 3, 4, 5, 6, 7.5, and 10 mg tablets; injection, 5 mg per vial

The dosage and administration of warfarin must be individualized for each patient according to the patient's international normalized ratio (INR) response to the drug. The dose needs to be adjusted based on the patient's INR and the condition being treated. The recommended target INR is 2.5 (INR range 2.0–3.0) for venous thromboembolism, AF, certain mechanical and bioprosthetic heart valves, and post-MI.

17.3.1.6 Adverse Effects and Drug Interactions

Adverse Effects The most common adverse effects of warfarin are fatal and nonfatal hemorrhage from any tissue or organ. Other uncommon serious adverse effects include necrosis of skin and other tissues and systemic atheroemboli and cholesterol microemboli as a result of enhanced release of atheromatous plaque emboli. Other less severe adverse effects may include hypersensitivity reactions, vasculitis, liver injury, and gut disturbances.

Drug Interactions Drugs may interact with warfarin through pharmacodynamic or pharmacokinetic mechanisms. Warfarin synergizes with other anticoagulants or platelet inhibitors to cause increased risk of bleeding. As noted earlier, CYP isozymes involved in the metabolism of warfarin include CYP2C9, CYP2C19, CYP2C8, CYP2C18, CYP1A2, and CYP3A4. The more potent warfarin S-enantiomer is primarily metabolized by CYP2C9, while the R-enantiomer is mainly metabolized by CYP1A2 and CYP3A4. Hence, inhibitors of CYP2C9, CYP1A2, and/or CYP3A4 have the potential to increase the effect (increase INR) of warfarin by increasing the exposure of warfarin. On the other hand, inducers of CYP2C9, CYP1A2, and/or CYP3A4 have the potential to decrease the effect (decrease INR) of warfarin by decreasing the exposure of warfarin.

Contraindications and Pregnancy Category

- Hemorrhagic tendencies or blood dyscrasias.
- Recent or contemplated surgery of the central nervous system or eye, or traumatic surgery resulting in large open surfaces.
- Bleeding tendencies associated with certain conditions.
- Threatened abortion, eclampsia, and preeclampsia.
- Unsupervised patients with potential high levels of noncompliance.
- Spinal puncture and other diagnostic or therapeutic procedures with potential for uncontrollable bleeding
- Hypersensitivity to the drug product.
- Major regional or lumbar block anesthesia.
- Malignant hypertension.
- Warfarin is contraindicated in women who are pregnant (category X) except in pregnant women with mechanical heart valves (category D), who are at high risk of thromboembolism and for whom the benefits of warfarin may outweigh the risks.

17.3.2 Heparins

Listed below are the US FDA-approved heparin products for clinical use:

- Unfractionated heparin (UFH)
- Low-molecular-weight heparins (LMWHs):
  - Dalteparin (Fragmin)
  - Enoxaparin (Lovenox)
  - Tinzaparin (Innohep)

17.3.2.1 *General Introduction to Drug Class* Heparins, if not specified, include UFH (also known as standard heparin) and LMWHs. UFH is often prepared from bovine lung and porcine intestinal mucosa. LMWHs are derived from UFH and, as listed earlier, include dalteparin, enoxaparin, and tinzaparin, which are all available in the United States. Heparins, especially LMWHs, are widely used as anticoagulants in the management of thromboembolic disorders, including ACS.

17.3.2.2 *Chemistry and Pharmacokinetics*

Chemistry of UFH The type of UFH in clinical use is polydispersed unmodified heparin, with molecular weight ranging from 3 to 30 kDa and a mean molecular weight of approximately 15 kDa, corresponding to approximately 45 saccharide units. The polydispersed mixture consists of variably sulfated repeating disaccharide units, which are composed of a glucosamine 1,4 linked to an uronic acid, which is represented by both its epimers, namely, the glucuronic (GlcA) and the iduronic acids (IdoA) [3]. Both monosaccharides can be differently functionalized in the UFH chains: uronic acid can be 2-O-sulfated and glucosamine (GlcN) is N-sulfated (GlcNSO$_3$) or N-acetylated (GlcNAc). Moreover, GlcNSO$_3$ may be also O-sulfated in the position 3 and/or 6. The GlcNAc residues can be 6-O-sulfated or unsulfated. Hence, the possible presence of four different uronic residues and six different glucosamine derivatives makes the structure of UHF very complex and variable.

In UFH, three different domains can be identified on the basis of the abundance of GlcNSO$_3$ and GlcNAc. The first domain, namely, the NA domain, is mainly constituted of the repeating GlcA–GlcNAc disaccharides. The NA/NS domain is characterized by the succession of GlcNSO$_3$ and GlcNAc units. Finally, at the nonreducing end, a region rich in high sulfated disaccharides is found (NS domain). Notably, the NS domain contains the specific pentasaccharide unit responsible for binding to antithrombin III (see succeeding text). This interaction determines a conformational change in antithrombin III, resulting in its activation through an increase in the flexibility of its reactive site loop [3].

Chemistry of LMWHs LMWHs are obtained by chemical or enzymatic depolymerization of UFH and, depending on the production method, display different chemical structures at the reducing and/or nonreducing end of polysaccharide chains. LMWHs in clinical use have a molecular weight ranging from 2 to 9 kDa and a mean molecular weight of approximately 5 kDa, corresponding to approximately 15 saccharide units. LMWHs partially lose the ability to inactivate factor IIa as compared with UFH. Thus, for LMWHs, the ratio between anti-factor Xa and anti-factor IIa activities becomes greater than that of UHF (see section "LMWHs").

Pharmacokinetics UFH and LMWHs cannot be absorbed through the gut mucosa and thus must be administered parenterally via intravenous infusion or subcutaneous injection. The elimination half-life of UFH is dose dependent, and it appears to be primarily cleared and degraded by the reticuloendothelial system. LMWHs are primarily metabolized in the liver by desulfation and/or depolymerization to lower-molecular-weight species with much reduced biological potency. LMWHs are almost exclusively cleared by the kidney owing to their smaller sizes. Hence, renal dysfunction may result in accumulation of LMWHs, leading to bleeding. In healthy individuals, the biological half-life of LMWHs is approximately 4–6 h.

17.3.2.3   Molecular Mechanisms and Pharmacological Effects

UFH   Heparin acts at multiple sites in the normal coagulation system: (i) small amounts of heparin in combination with antithrombin III (also known as heparin cofactor) can inhibit thrombosis by inactivating factor Xa and inhibiting the conversion of prothrombin (also known as factor II) to thrombin (also known as factor IIa); (ii) once active thrombosis has developed, larger amounts of heparin can inhibit further coagulation by inactivating thrombin (IIa) and preventing the conversion of fibrinogen to fibrin; and (iii) heparin also prevents the formation of a stable fibrin clot by inhibiting the activation of factor XIII (also known as fibrin-stabilizing factor).

LMWHs   LMWHs act by primarily enhancing the inhibition of factor Xa by antithrombin III. As compared with UFH, the ability of LMWHs to enhance the inhibition of factor IIa by antithrombin III is largely lost due to their smaller sizes. Hence, as stated earlier, for LMWHs, the ratio between anti-factor Xa and anti-factor IIa activities becomes greater than that of UHF. Figure 17.4 illustrates the mechanisms of inhibition of factors Xa and IIa by UFH and LMWHs.

17.3.2.4   Clinical Uses

UFH   UFH is approved by the US FDA for clinical use in the management of the following thromboembolic conditions:

1. Anticoagulant therapy in prophylaxis and treatment of venous thrombosis and its extension
2. Low-dose regimen for prevention of postoperative DVT and PE in patients undergoing major abdomino-thoracic surgery or who, for other reasons, are at risk of developing thromboembolic disease
3. Prophylaxis and treatment of PE
4. AF with embolization
5. Treatment of acute and chronic consumptive coagu-lopathies (disseminated intravascular coagulation)
6. Prevention of clotting in arterial and cardiac surgery
7. Prophylaxis and treatment of peripheral arterial embolism

In addition to the above thromboembolic conditions, UFH may also be employed as an anticoagulant in blood transfu-sions, extracorporeal circulation, and dialysis procedures.

LMWHs   LMWHs have become more commonly used than UFH for the management of thromboembolic conditions, including DVT and ACS. Table 17.1 summarizes the US FDA-approved indications of LMWHs.

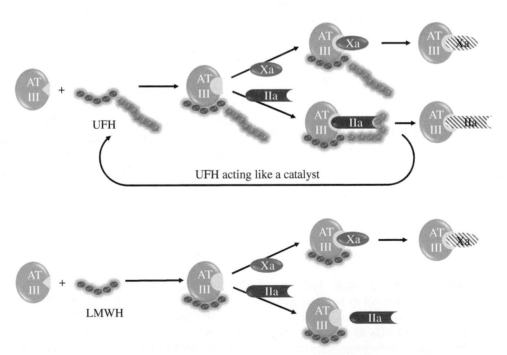

FIGURE 17.4   Molecular mechanisms of action of unfractionated heparin (UFH) and low-molecular-weight heparins (LMWHs). As illus-trated, UFH binds to antithrombin III (ATIII) and facilitates the inactivation of both factors Xa and IIa by ATIII. Due to the short chain, LMWHs only primarily facilitate ATIII-mediated inactivation of factor Xa. For color details, please see color plate section.

TABLE 17.1   The US FDA-approved indications of LMWHs

LMWH	Indication
Dalteparin	1. Prophylaxis of ischemic complications of unstable angina (UA) and non-ST-elevation myocardial infarction (NSTEMI) 2. Prophylaxis of deep vein thrombosis (DVT) in abdominal surgery, hip replacement surgery, or medical patients with severely restricted mobility during acute illness 3. Extended treatment of symptomatic venous thromboembolism to reduce the recurrence in patients with cancer
Enoxaparin	1. Prophylaxis of DVT in abdominal surgery, hip replacement surgery, knee replacement surgery, or medical patients with severely restricted mobility during acute illness; inpatient treatment of acute DVT with or without pulmonary embolism; outpatient treatment of acute DVT without pulmonary embolism 2. Prophylaxis of ischemic complications of UA and NSTEMI 3. Treatment of acute ST-elevation myocardial infarction (STEMI) managed medically or with subsequent percutaneous coronary intervention
Tinzaparin	Treatment of acute symptomatic DVT with or without pulmonary embolism when administered in conjunction with warfarin

TABLE 17.2   Dose regimens of LMWHs in patients with ACS

LMWH	Indication	Dose regimen
Dalteparin	AU/NSTEMI	For prophylaxis of ischemic complications in UA/NSTEMI, the recommended dose is 120 IU/kg (but no more than 10,000 IU/person), subcutaneously every 12 h with concurrent oral aspirin (75–165 mg once daily) therapy. Treatment should be continued until the patient is clinically stabilized. The usual duration of administration is 5–8 days. Concurrent aspirin therapy is recommended except when contraindicated
Enoxaparin	UA/NSTEMI	In patients with UA/NSTEMI, the recommended dose of enoxaparin is 1 mg/kg administered subcutaneously every 12 h in conjunction with oral aspirin therapy (100–325 mg once daily). Treatment with enoxaparin should be prescribed for a minimum of 2 days and continued until clinical stabilization. The usual duration of treatment is 2–8 days
	STEMI	In patients with STEMI, the recommended dose of enoxaparin is a single intravenous bolus of 30 mg plus a 1 mg/kg subcutaneous dose followed by 1 mg/kg administered subcutaneously every 12 h (maximum 100 mg for the first two doses only, followed by 1 mg/kg dosing for the remaining doses). All patients should receive aspirin as soon as they are identified as having STEMI and maintained with 75–325 mg once daily unless contraindicated

17.3.2.5   Therapeutic Dosages

The dosage forms and strengths of UFH and LMWHs are listed below:

- UFH: Injection, 1000, 5000, and 10,000 USP units/ml
- Dalteparin (Fragmin): Injection—single-dose prefilled syringe, 2500 IU/0.2 ml, 5000 IU/0.2 ml, 7500 IU/0.3 ml, 10,000 IU/0.4 ml, 12,500 IU/0.5 ml, 15,000 IU/0.6 ml, and 18,000 IU/0.72 ml; single-dose graduated syringe, 10,000 IU/1 ml; multiple-dose vial, 95,000 IU/9.5 ml and 95,000 IU/3.8 ml
- Enoxaparin (Lovenox): Injection—100 mg/ml concentration, prefilled syringes of 30 mg/0.3 ml and 40 mg/0.4 ml and graduated prefilled syringes of 60 mg/0.6 ml, 80 mg/0.8 ml, and 100 mg/1 ml; 150 mg/ml concentration, graduated prefilled syringes of 120 mg/0.8 ml and 150 mg/1 ml
- Tinzaparin (Innohep): Injection, 2 ml vial (20,000 anti-Xa IU/ml)

As noted earlier, UFH is generally not used in the management of ACS. The LMWHs, dalteparin and enoxaparin, but not tinzaparin, are approved for use in patients with ACS. The recommended dosage regimens of LMWHs for ACS are given in Table 17.2.

17.3.2.6   Adverse Effects and Drug Interactions

UFH   The major adverse effects of UFH therapy are hemorrhage and thrombocytopenia. Heparin-induced thrombocytopenia (HIT) is a serious immune-mediated reaction resulting from irreversible aggregation of platelets. HIT may progress to the development of venous and arterial thromboses, a condition referred to as HIT with thrombosis. Thrombotic events may also be the initial presentation for HIT. These serious thromboembolic events include DVT, PE, cerebral vein thrombosis, limb ischemia, stroke, MI, mesenteric thrombosis, renal arterial thrombosis, skin necrosis, gangrene of the extremities that may lead to amputation, and fatal outcomes. Once HIT is diagnosed or strongly suspected, UFH should be discontinued. HIT may also present as a form of delayed onset. Delayed onset HIT can occur up to several weeks after the discontinuation of heparin therapy. Patients presenting with thrombocytopenia or thrombosis after discontinuation of UFH should be evaluated for HIT. Other less severe adverse effects of UFH include skin reactions, hypersensitivity, endocrine disturbances, and elevations of hepatic aminotransferases.

Pharmacodynamic drug interactions may become clinically significant to cause increased risk of bleeding if UFH is used with oral anticoagulants or platelet inhibitors. On the other hand, digitalis, tetracyclines, nicotine, or antihistamines may partially counteract the anticoagulant action of UFH. Intravenous nitroglycerin administered to heparinized patients may result in a decrease of the partial thromboplastin time with subsequent rebound effect upon discontinuation of nitroglycerin.

UFH is contraindicated in patients (i) with severe thrombocytopenia, (ii) when suitable blood coagulation tests cannot be performed at appropriate intervals (this contraindication refers to full-dose heparin; there is usually no need to monitor coagulation parameters in patients receiving low-dose heparin), and (iii) with an uncontrolled active bleeding state, except when this is due to disseminated intravascular coagulation. The pregnancy category of UFH is C.

LMWHs The most common adverse effect of LMWHs is hematoma at the injection site. LMWHs should be used with caution in patients receiving oral anticoagulants, platelet inhibitors, and thrombolytic agents because of increased risk of bleeding. Other adverse effects of LMWHs may include anemia, thrombocytopenia, elevation of serum aminotransferases, and gut disturbance.

LMWHs are contraindicated in patients (i) with active major bleeding, (ii) with a history of HIT, (iii) undergoing epidural/neuraxial anesthesia due to increased risk of epidural or spinal hematomas, and (iv) with hypersensitivity to the drug product. The pregnancy category of LMWHs is B.

17.3.2.7 Comparison of UFH and LMWHs
While UFH and LMWHs are used in the management of various thromboembolic conditions, LMWHs have gained more popularity in recent years due to a number of advantages over UFH. For example, LMWHs have greater bioavailability than UFH when administered subcutaneously, and the duration of the anticoagulant effect of LMWHs is also greater than that of UFH owing to reduced binding to the reticuloendothelial system. Another advantage of using LMWHs is that they are much less likely to induce immune-mediated thrombocytopenia. In addition, use of LMWHs generally does not require close laboratory monitoring and can be safely administered in the outpatient setting and is thus more cost-effective than UFH.

17.3.3 SFXa Inhibitors

Listed below are the three US FDA-approved SFXa inhibitors, with fondaparinux as an indirect SFXa inhibitor and apixaban and rivaroxaban as direct SFXa inhibitors:

1. Fondaparinux (Arixtra)
2. Apixaban (Eliquis)
3. Rivaroxaban (Xarelto)

17.3.3.1 General Introduction to Drug Class
Factor Xa is essential for the propagation of the coagulation cascade (Fig. 17.2). As discussed earlier, factor Xa is a major target of UFH and LMWHs. Because of its central role in coagulation, factor Xa is an attractive target for design of new anticoagulants that selectively inhibit it. In this context, two types of SFXa inhibitors are available: direct and indirect SFXa inhibitors. Fondaparinux is a synthetic indirect SFXa inhibitor that has been in clinical use for more than a decade. A limitation of its long-term use for thromboembolic prevention is the mode of administration by subcutaneous injection. On the other hand, the recently US FDA-approved apixaban (approved in 2012) and rivaroxaban (approved in 2011) are oral agents that directly inhibit factor Xa.

17.3.3.2 Chemistry and Pharmacokinetics
Fondaparinux (structure shown in Fig. 17.5) is a synthetic pentasaccharide molecule that contains the sequence of five essential carbohydrates necessary for binding to antithrombin III and inducing the conformational change in antithrombin III required for conjugation to factor Xa. Apixaban and rivaroxaban (structures shown in Fig. 17.5) are carboxamide derivatives. The major pharmacokinetic properties of fondaparinux, apixaban, and rivaroxaban are summarized in Table 17.3.

17.3.3.3 Molecular Mechanisms and Pharmacological Effects

Fondaparinux The antithrombotic activity of fondaparinux is the result of antithrombin III-mediated selective inhibition of factor Xa. By selectively binding to antithrombin III, fondaparinux potentiates (about 300 times) the innate neutralization of factor Xa by antithrombin III. Neutralization of factor Xa interrupts the blood coagulation cascade and thus inhibits thrombin formation and thrombus development.

Apixaban and Rivaroxaban Apixaban and rivaroxaban are oral direct SFXa inhibitors. They do not require antithrombin III for antithrombotic activity. Apixaban and rivaroxaban inhibit free and clot-bound factor Xa and its prothrombinase activity. By inhibiting factor Xa, apixaban and rivaroxaban decrease thrombin generation and thrombus development.

17.3.3.4 Clinical Uses
SFXa inhibitors are primarily used in the management of DVT and PE. Apixaban and rivaroxaban are also indicated in patients with AF to reduce the risk of stroke. In this setting, use of apixaban and rivaroxaban is associated with reduced risk of intracranial hemorrhage (ICH) [4]. The use of apixaban in treating DVT is also associated with reduced risk of bleeding compared with conventional therapy (fondaparinux, followed by warfarin) [5, 6]. The US FDA-approved indications of fondaparinux, apixaban, and rivaroxaban are summarized in Table 17.4.

17.3.3.5 Therapeutic Dosages
Listed below are the dosage forms and strengths of the three US FDA-approved SFXa inhibitors:

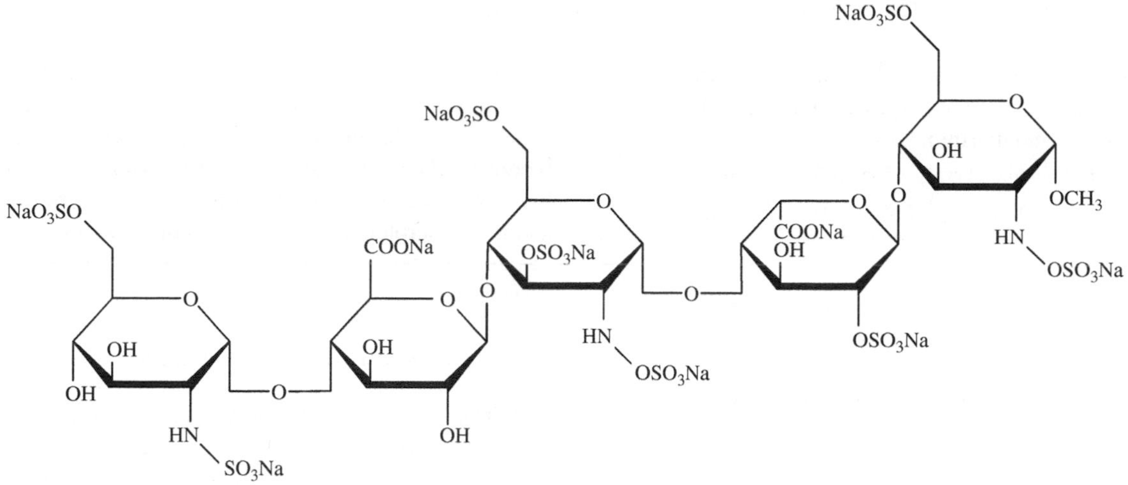

FIGURE 17.5 Structures of specific factor Xa inhibitors. As shown, fondaparinux is a pentasaccharide molecule, whereas apixaban and rivaroxaban are carboxamide derivatives.

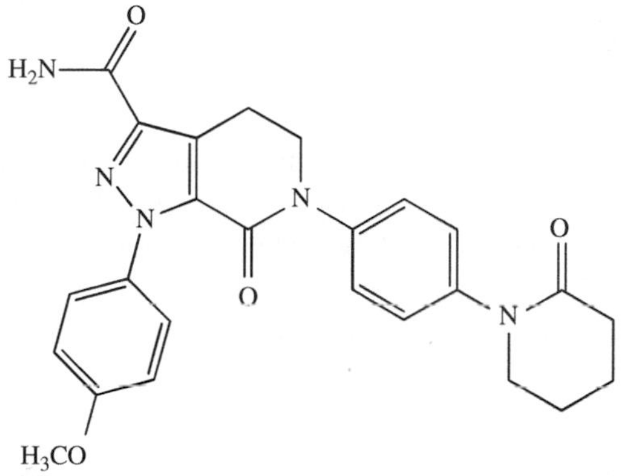

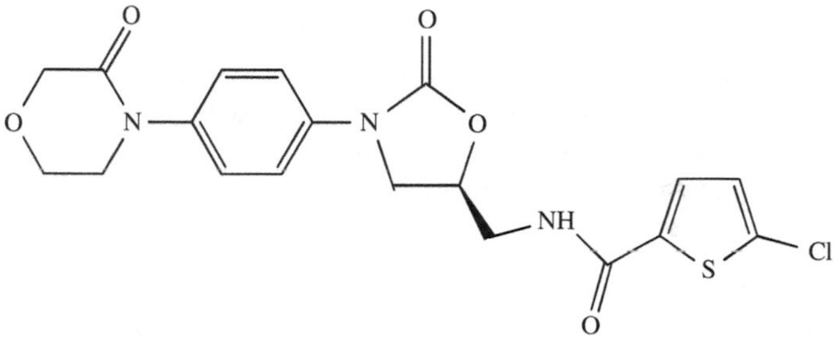

TABLE 17.3 Pharmacokinetic properties of SFXa inhibitors

SFXa inhibitor	Bioavailability (%)	Elimination half-life (h)	Metabolism and elimination
Fondaparinux	100 (subcutaneous)	17–21	Eliminated in the urine as unchanged form
Apixaban	50 (oral)	12	CYP3A4; eliminated in the urine and feces
Rivaroxaban	66–100 (oral)	5–9	CYP3A4/5, CYP2J2; eliminated in the urine and feces

Apixaban and rivaroxaban are also substrates for P-glycoprotein (ABCB1) and ABCG2.
Coadministration of rivaroxaban with food significantly increases the bioavailability, and as such, the drug is usually taken with meal.

TABLE 17.4 The US FDA-approved indications of SFXa inhibitors

SFXa inhibitor	Indication
Fondaparinux	1. Prophylaxis of DVT in patients undergoing hip fracture surgery (including extended prophylaxis), hip replacement surgery, knee replacement surgery, or abdominal surgery 2. Treatment of DVT or acute PE when administered in conjunction with warfarin
Apixaban	1. To reduce the risk of stroke and systemic embolism in patients with nonvalvular AF 2. Prophylaxis of DVT, which may lead to PE, in patients who have undergone hip or knee replacement surgery
Rivaroxaban	1. To reduce the risk of stroke and systemic embolism in patients with nonvalvular AF 2. For the treatment of DVT and PE and for the reduction in the risk of recurrence of DVT and of PE 3. Prophylaxis of DVT, which may lead to PE, in patients undergoing knee or hip replacement surgery

1. Fondaparinux (Arixtra): Subcutaneous injection, single-dose, prefilled syringes containing 2.5, 5, 7.5, or 10 mg of fondaparinux
2. Apixaban (Eliquis): Oral, 2.5 and 5 mg tablets
3. Rivaroxaban (Xarelto): Oral, 10, 15, and 20 mg tablets

The dosage regimens of SFXa inhibitors in the management of DVT, PE, and AF are given in Table 17.5.

17.3.3.6 Adverse Effects and Drug Interactions

Adverse Effects The most common adverse reactions associated with the use of SFXa inhibitors are bleeding complications. Mild local irritation (injection site bleeding, rash, and pruritus), anemia, insomnia, hypokalemia, dizziness, hypotension, and confusion may occur following subcutaneous injection of fondaparinux.

Drug Interactions Coadministration of SFXa inhibitors with other anticoagulants may increase the risk of bleeding, and hence, such a coadministration should be done with caution. Since apixaban and rivaroxaban are metabolized by CYP enzymes and substrates for P-glycoprotein, pharmacokinetic drug interactions can be significant. In this regard, strong dual inhibitors of CYP3A4 and P-glycoprotein (e.g., ketoconazole, ritonavir, clarithromycin, erythromycin, fluconazole) increase blood levels of apixaban and rivaroxaban. On the other hand, simultaneous use of strong dual inducers of CYP3A4 and P-glycoprotein (e.g., carbamazepine, phenytoin, rifampin, St. John's wort) reduces blood levels of apixaban and rivaroxaban.

Contraindications and Pregnancy Category

- The shared contraindications of the three SFXa inhibitors are presence of active major bleeding and hypersensitivity reaction to the drug product.
- The specific contraindications of fondaparinux also include (i) severe renal impairment (creatinine clearance <30 ml/min) in prophylaxis or treatment of venous thromboembolism, (ii) bacterial endocarditis, (iii) thrombocytopenia associated with a positive *in vitro* test for antiplatelet antibody in the presence of fondaparinux, and (iv) body weight <50 kg due to increased risk of bleeding (venous thromboembolism prophylaxis only).
- Pregnancy category: B for all three SFXa inhibitors.

TABLE 17.5   Dosage regimens of SFXa inhibitors in the management of DVT, PE, and AF

SFXa inhibitor	Indication	Dosage regimen
Fondaparinux	Prophylaxis of DVT	2.5 mg subcutaneously once daily after hemostasis has been established. The initial dose should be given no earlier than 6–8 h after surgery and continued for 5–9 days. For patients undergoing hip fracture surgery, extended prophylaxis up to 24 additional days is recommended
	Treatment of DVT and PE	5 mg (body weight <50 kg), 7.5 mg (50–100 kg), or 10 mg (>100 kg) subcutaneously once daily. Treatment should continue for at least 5 days until INR 2–3 achieved with warfarin (concomitant treatment with warfarin should be initiated as soon as possible, usually within 72 h)
Apixaban	Risk (stroke and systemic embolism) reduction in nonvalvular atrial fibrillation	• The typical recommended dose is 5 mg orally twice daily • In patients with at least two of the following characteristics—age ≥80 years, body weight ≤60 kg, or serum creatinine ≥1.5 mg/dl—the recommended dose is 2.5 mg orally twice daily
	Prophylaxis of DVT	2.5 mg orally twice daily
Rivaroxaban	Risk (stroke and systemic embolism) reduction in nonvalvular atrial fibrillation	• For patients with creatinine clearance >50 ml/min: 20 mg orally once daily with the evening meal • For patients with creatinine clearance of 15–50 ml/min: 15 mg orally once daily with the evening meal
	Treatment of DVT and PE and reduction in the risk of recurrence of DVT and of PE	15 mg orally twice daily with food for the first 21 days for the initial treatment of acute DVT or PE. After the initial treatment period, 20 mg orally once daily with food for the remaining treatment and the long-term reduction in the risk of recurrence of DVT and of PE
	Prophylaxis of DVT	10 mg orally once daily with or without food

17.3.4 Direct Thrombin (Factor IIa) Inhibitors

Listed below are the five US FDA-approved DTIs:

1. Bivalirudin (Angiomax)
2. Desirudin (Iprivask)
3. Lepirudin (Refludan)
4. Argatroban (Acova)
5. Dabigatran (Pradaxa)

17.3.4.1 General Introduction to Drug Class The
central role of thrombin (factor IIa) in coagulation cascade makes it an attractive target for the development of anticoagulants. As listed earlier, there are currently five DTIs approved by the US FDA for clinical use, with dabigatran being the newest member (approved in 2010). Bivalirudin, desirudin, and lepirudin, as the names indicate, are hirudin-related peptidic drugs. Hirudin is a mixture of homologous polypeptides produced by the salivary gland of *Hirudo medicinalis*, the medicinal leech. For years, surgeons had used medicinal leeches to prevent thrombosis in the fine vessels of reattached digits. Due to

the peptidic nature, the hirudin-related drugs require parenteral administration.

Argatroban is a small molecule drug that also requires parenteral administration. Dabigatran is a recently approved small molecule drug suitable for oral administration. Both argatroban and dabigatran are highly selective DTIs.

17.3.4.2 Chemistry and Pharmacokinetics Bivalirudin
is a synthetic, 20-amino-acid-long peptide, whereas desirudin and lepirudin are each a recombinant protein of 65 amino acids with structural similarity to hirudin. Argatroban and dabigatran are small molecules (structures shown in Fig. 17.6). The major pharmacokinetic properties of the DTIs are summarized in Table 17.6.

17.3.4.3 Molecular Mechanisms and Pharmacological
Effects All DTIs directly inhibit both circulating and clot-bound thrombin. As noted earlier, thrombin is a serine proteinase that plays a central role in the thrombotic process: (i) it acts to cleave fibrinogen into fibrin monomers and to activate factor XIII to factor XIIIa, allowing fibrin to develop a covalently cross-linked framework that

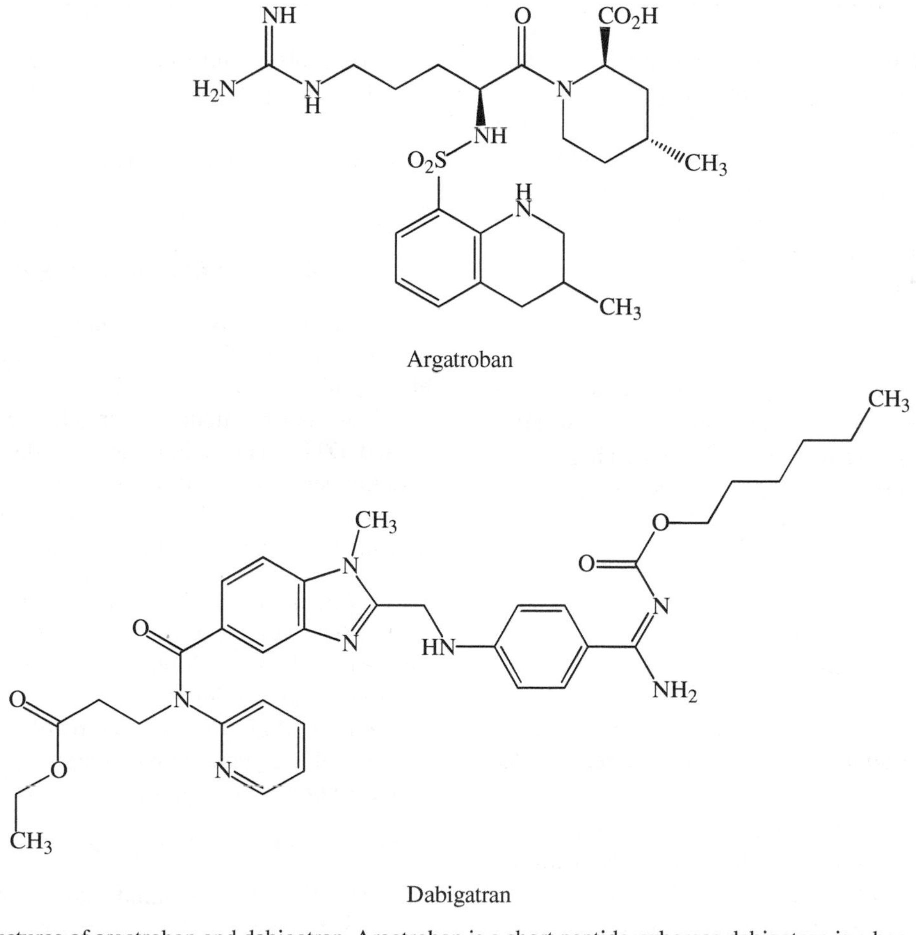

Argatroban

Dabigatran

FIGURE 17.6 Structures of argatroban and dabigatran. Argatroban is a short peptide, whereas dabigatran is a benzimidazole derivative.

TABLE 17.6   Major pharmacokinetic properties of DTIs

DTI	Oral bioavailability (%)	Elimination half-life	Metabolism and elimination
Bivalirudin		25 min	Proteolytic cleavage; renal excretion
Desirudin		2–3 h	Degradation by carboxypeptidases; renal excretion
Lepirudin		1.3 h	Hydrolysis; renal excretion
Argatroban		45 min	Possible CYP3A4/5; eliminated in the feces (major) and urine
Dabigatran	3–7	12–17 h	Glucuronidation; renal excretion

Dabigatran is a substrate of P-glycoprotein. It is not a substrate, inhibitor, or inducer of CYP enzymes. The acyl glucuronide conjugates of dabigatran are also pharmacologically active.

TABLE 17.7   The US FDA-approved indications of DTIs

DTI	Indication
Bivalirudin	It is indicated for use as an anticoagulant in patients: 1. With UA undergoing percutaneous transluminal coronary angioplasty (PTCA) 2. Undergoing PCI with provisional use of glycoprotein IIb/IIIa inhibitor as in the REPLACE-2 study [7] 3. With, or at risk of, HIT or heparin-induced thrombocytopenia and thrombosis syndrome (HITTS) undergoing PCI
Desirudin	Prophylaxis of DVT, which may lead to PE, in patients undergoing elective hip replacement surgery
Lepirudin	It is indicated for anticoagulation in patients with HIT and associated thromboembolic disease in order to prevent further thromboembolic complications
Argatroban	• For prophylaxis or treatment of thrombosis in adult patients with HIT • As an anticoagulant in adult patients with or at risk for HIT undergoing PCI
Dabigatran	• To reduce the risk of stroke and systemic embolism in patients with nonvalvular atrial fibrillation • For the treatment of DVT and PE in patients who have been treated with a parenteral anticoagulant for 5–10 days • To reduce the risk of recurrence of DVT and PE in patients who have been previously treated

stabilizes the thrombus; (ii) thrombin also activates factors V and VIII, promoting further thrombin generation; and (iii) thrombin activates platelets, stimulating aggregation and granule release. Hence, inhibition of thrombin by DTIs blocks the thrombogenic activity of thrombin. As a result, all thrombin-dependent coagulation assays are affected, for example, activated partial thromboplastin time and prothrombin time values increase in a dose-dependent fashion.

17.3.4.4 Clinical Uses
DTIs are used in the management of various thromboembolic conditions, including HIT, DVT, PE, as well as AF. Some are also indicated in patients undergoing percutaneous coronary intervention (PCI). The US FDA-approved indications of DTIs are given in Table 17.7.

17.3.4.5 Therapeutic Dosages
Listed below are the dosage forms and strengths of DRIs:

- Bivalirudin (Angiomax): Intravenous, vials containing 250 mg of bivalirudin as a sterile, lyophilized powder for reconstitution
- Desirudin (Iprivask): Subcutaneous, single-dose vials containing (15.75 mg) lyophilized powder with an accompanying sterile, nonpyrogenic diluent
- Lepirudin (Refludan): Intravenous, vials containing 50 mg

- Argatroban: Intravenous, a single-use vial containing 125 mg argatroban in 125 ml aqueous sodium chloride solution (1 mg/ml)
- Dabigatran (Pradaxa): Oral, 75 and 150 mg capsules

The dosage regimens of DTIs are summarized in Table 17.8.

17.3.4.6 Adverse Effects and Drug Interactions

Adverse Effects   The most common adverse effect shared by all DTIs is bleeding. Concomitant treatment with other anticoagulants or platelet inhibitors may increase the risk of bleeding complications. Other adverse effects of hirudin-related DTIs may include anemia, deep thrombophlebitis, injection site mass, and nausea. Argatroban may also cause hypotension, fever, headache, and diarrhea. Besides bleeding, gastritis is also a common adverse effect of oral dabigatran.

Drug Interactions   DTIs interact pharmacodynamically with other anticoagulants or platelet inhibitors to cause increased risk of bleeding, and as such, combinations should be used with caution or avoided. Since dabigatran is a substrate of P-glycoprotein, cotreatment with inhibitors or inducers of P-glycoprotein may significantly alter the plasma concentrations of dabigatran.

Contraindications and Pregnancy Category

- All DTIs are contraindicated in patients with active bleeding or a history of hypersensitivity reaction to the drug product.

TABLE 17.8 Dosage regimens of DTIs

DTI	Indication	Dosage regimen
Bivalirudin (iv)	PCI/PTCA for patients who do not have HIT/HITTS	• PCI/PTCA: 0.75 mg/kg intravenous (iv) bolus dose followed by a 1.75 mg/kg/h iv infusion for the duration of the procedure • Perform activated clotting time (ACT) test 5 min post-bolus dose. If needed, give an additional bolus of 0.3 mg/kg • After PCI/PTCA, iv infusion may be continued for up to 4 h, after which a rate of 0.2 mg/kg/h can be used for up to 20 more hours, if needed • Consider glycoprotein IIb/IIIa inhibitor administration with procedural complications
	PCI/PTCA for patients who have HIT/HITTS	• The recommended dose in patients with HIT/HITTS undergoing PCI is an iv bolus of 0.75 mg/kg. This should be followed by an infusion at a rate of 1.75 mg/kg/h for the duration of the procedure • After PCI/PTCA, iv infusion may be continued for up to 4 h, after which a rate of 0.2 mg/kg/h can be used for up to 20 more hours, if needed
	PCI/PTCA for patients with renal impairment	No reduction in bolus dose required. Consider reduction of the rate of infusion to 1 mg/kg/h for creatinine clearance <30 ml/min or 0.25 mg/kg/h if on dialysis
Desirudin (sc)	DVT	• All patients undergoing elective hip replacement surgery should be evaluated for bleeding disorder risk before prophylactic administration of desirudin • The recommended initial dose is 15 mg every 12 h administered by subcutaneous injection given up to 5–15 min prior to surgery. Administer after induction of regional block anesthesia, if used
Lepirudin (iv)	HIT	Initial dose: 0.4 mg/kg body weight (up to 110 kg) slowly intravenously (e.g., over 15–20 s) as a bolus dose, followed by 0.15 mg/kg (up to 110 kg)/h as a continuous intravenous infusion for 2–10 days or longer if clinically needed
Argatroban (iv)	HIT	2 µg (mcg)/kg/min administered as a continuous infusion
	HIT/PCI	The dose for patients with or at risk for HIT undergoing PCI is started at 25 µg (mcg)/kg/min and a bolus of 350 µg (mcg)/kg administered via a large-bore intravenous line over 3–5 min
Dabigatran (oral)	Reduction in risk of stroke and systemic embolism in nonvalvular AF	• For patients with creatinine clearance >30 ml/min: 150 mg orally twice daily • For patients with creatinine clearance of 15–30 ml/min: 75 mg orally twice daily
	Treatment and reduction in the risk of recurrence of DVT and PE	For patients with creatinine clearance >30 ml/min: 150 mg orally twice daily after 5–10 days of parenteral anticoagulation. Dose recommendation cannot be provided for creatinine clearance <30 ml/min or on dialysis

• Use of oral dabigatran is also contraindicated in patients with mechanical prosthetic heart valve due to increased risk of thromboembolic and bleeding events in such setting.
• Pregnancy category: B (bivalirudin, lepirudin, argatroban) and C (desirudin, dabigatran).

17.4 PLATELET INHIBITORS

Platelet activation and aggregation are central to atherothrombogenesis, and hence, inhibition of platelet function is an important approach to the management of atherothrombotic conditions, especially ACS. Platelet inhibitors, also called antiplatelet agents, interfere with a number of platelet functions, including aggregation, release of granule contents, and platelet-mediated vascular constriction. According to their mechanisms of action, platelet inhibitors are typically classified into the following five groups:

1. Cyclooxygenase (COX) inhibitors (aspirin)
2. P2Y$_{12}$ ADP-receptor antagonists (clopidogrel, prasugrel, ticagrelor, and ticlopidine)
3. Thrombin receptor antagonists (vorapaxar)

4. GP IIb/IIIa antagonists (abciximab, eptifibatide, tirofiban)

5. Others (dipyridamole, cilostazol)

17.4.1 COX Inhibitors

Listed below is the only COX inhibitor used for treating ACS:

- Aspirin

17.4.1.1 General Introduction to Drug Class   Aspirin, first synthesized over a century ago, is one of the most widely used medicines in the world. The drug is well known for its analgesic, antipyretic, and anti-inflammatory activities and applications in treating fever and mild pain associated with inflammatory disorders. Aspirin is a member of a group of drugs known as nonsteroidal anti-inflammatory drugs (NSAIDs) but differs from most other NSAIDs in that aspirin irreversibly inhibits COX, including COX-1 in platelets. Inhibition of COX-1 in platelets by aspirin blocks the biosynthesis of prostaglandins and the subsequent formation of thromboxanes (Tx), including TxA_2, from arachidonic acid. TxA_2 is a potent stimulator of platelet aggregation, release of active intermediates, and vasoconstriction. Aspirin is the only one among the NSAIDs that is used in the management of atherothrombotic conditions, including ACS.

17.4.1.2 Chemistry and Pharmacokinetics   Aspirin (structure shown in Fig. 17.7) is chemically described as acetylsalicylic acid. In general, immediate-release aspirin is well and completely absorbed from the gastrointestinal (GI) tract. Enteric-coated aspirin products are erratically absorbed from the GI tract. Aspirin is rapidly hydrolyzed in the plasma to salicylic acid, which then undergoes glucuronidation. Both salicylic acid and its glucuronide metabolites are eliminated via renal excretion. The elimination of aspirin follows zero-order pharmacokinetics with an elimination half-life of 6h under normal therapeutic dose. The plasma half-life may be increased to over 20h following toxic doses (10–20g).

17.4.1.3 Molecular Mechanisms and Pharmacological Effects   Aspirin causes acetylation of a serine residue at position 529 of COX, leading to irreversible enzyme inhibition. This irreversible inhibition becomes more pronounced in platelets, which, unlike other cells, do not replicate. Indeed, aspirin at low doses blocks COX-1 activity, and the effect lasts for the entire life of platelets. The potent irreversible inhibition of COX-1 (a major form of COX in platelets) in platelets blocks the formation of TxA_2 and TxA_2-mediated platelet aggregation and secretion and vasoconstriction (Fig. 17.7). At higher doses, aspirin may also inhibit the formation of prostaglandin I_2 (also known as prostacyclin) catalyzed by COX in endothelial cells. Prostaglandin I_2 is vascular protective via causing vasodilation and inhibiting

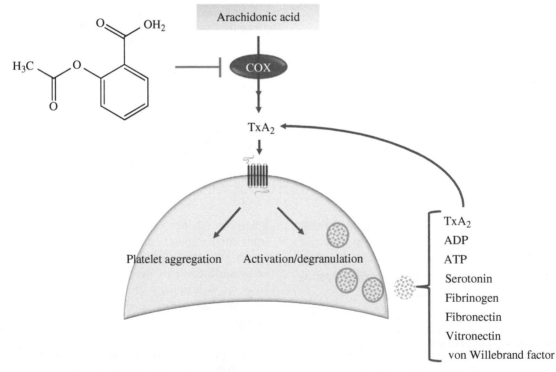

FIGURE 17.7   Molecular mechanism of action of aspirin as an antiplatelet drug. Aspirin irreversibly inhibits cyclooxygenase-1 (COX-1) in platelets. This results in decreased formation of thromboxane A_2 (TxA_2) in platelets. TxA_2 released by platelets cause platelet activation, degranulation, and aggregation. ADP, adenosine diphosphate; ATP, adenosine triphosphate.

TABLE 17.9 Cardiovascular indications of aspirin

To reduce the combined risk of death and nonfatal stroke in patients who have had ischemic stroke or transient ischemic attack due to fibrin platelet emboli (e.g., emboli formed in atrial fibrillation)

To reduce the risk of cardiovascular mortality in patients with a suspected acute myocardial infarction

To reduce the combined risk of death and nonfatal myocardial infarction in patients with a previous myocardial infarction or unstable angina

To reduce the combined risk of myocardial infarction and sudden death in patients with chronic stable angina

To reduce thrombosis in patients who have undergone revascularization procedures (e.g., PCI)

The US FDA recently issued a message to consumers stating that the evidence does not support the general use of aspirin for the primary prevention of heart attacks and strokes (http://www.fda.gov/drugs/resourcesforyou/consumers/ucm390574.htm; accessed on June 29, 2014).

platelet aggregation [8]. It is probably for this reason that low doses of aspirin are typically used for cardiovascular indications.

Noncoated aspirin acts within minutes of ingestion to stop platelets from forming blood clots. Enteric-coated aspirin takes longer to work but acts just as quickly as uncoated aspirin if chewed. Although the platelet-inhibitory effects of aspirin last for the life of the platelets (5–10 days), patients need to take aspirin every day to inhibit new platelets that are constantly being released into the circulation [9].

17.4.1.4 Clinical Uses The clinical uses of aspirin include two broad categories: (i) pain and fever relief (e.g., headache, menstrual pain, minor pain of arthritis, muscle pain, pain and fever of colds, toothache) and (ii) management of cardiovascular diseases. The cardiovascular indications of aspirin are summarized in Table 17.9.

17.4.1.5 Therapeutic Dosages Listed below are the dosage form and strengths of aspirin:

- Aspirin: Oral—low strength, 81 mg tablets; regular strength, 325 mg tablets

If rapid and complete platelet inhibition is required (e.g., if a patient is having a heart attack), the first dose of aspirin should be 160–325 mg. If noncoated aspirin is unavailable, enteric-coated tablets can be used but should be chewed to achieve a rapid effect. For long-term prevention of cardiovascular diseases, the recommended dose of aspirin is 75–325 mg once daily. Some guidelines recommend only baby aspirin (75–100 mg) for long-term prevention, based on evidence that higher doses cause more GI bleeding, but do not provide additional protection against heart attack and stroke [9].

17.4.1.6 Adverse Effects and Drug Interactions

Adverse Effects The adverse effects of aspirin are dose dependent. The most common adverse reactions are gut disturbances, including dyspepsia, heartburn, stomach pain, and gut bleeding. High doses may cause disturbances and injury of the cardiovascular, respiratory, renal, and central nervous systems. Aspirin hypersensitivity reaction is an uncommon but serious condition that can result in severe and potentially fatal anaphylaxis. Use of aspirin in children with viral infections may increase the risk of Reye's syndrome. Reye's syndrome, first described by R.D.K. Reye in 1963, is characterized by acute noninflammatory encephalopathy and fatty degenerative liver failure. The syndrome typically occurs after a viral illness, particularly an upper respiratory tract infection, influenza, varicella, or gastroenteritis, and is associated with the use of aspirin during the illness.

Drug Interactions Due to the partial involvement of prostaglandins in vasodilation and increasing renal blood flow, the concomitant use of aspirin may reduce the desired pharmacological effects (e.g., blood pressure-lowering effects) of angiotensin-converting enzyme inhibitors, diuretics, and β-blockers. Salicylic acid is highly bound to plasma proteins and as such may displace other drugs (e.g., warfarin, phenytoin, valproic acid) from their binding sites. Concomitant administration of aspirin with anticoagulants or other platelet inhibitors may increase the risk of bleeding.

Contraindications and Pregnancy Category

- Hypersensitivity reaction to the drug product.
- Children with viral infections due to increased risk of Reye's syndrome.
- Pregnancy category: D (if full-dose aspirin is taken in the third trimester).

17.4.2 P2Y$_{12}$ ADP-Receptor Antagonists

Listed below are P2Y$_{12}$ ADP-receptor antagonists currently approved by the US FDA for clinical use, with ticagrelor being the newest member (approved in 2011):

- Clopidogrel (Plavix)
- Prasugrel (Effient)
- Ticagrelor (Brilinta)
- Ticlopidine (Ticlid)

17.4.2.1 General Introduction to Drug Class P2Y$_{12}$ ADP-receptor antagonists are also simply called ADP-receptor antagonists. ADP binding to the platelet P2Y$_{12}$ ADP receptors plays an important role in platelet activation and aggregation, amplifying the initial platelet response to

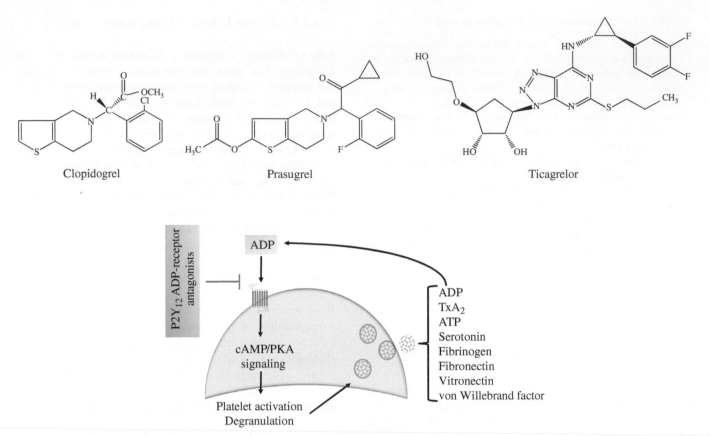

Clopidogrel   Prasugrel   Ticagrelor

FIGURE 17.8   Molecular mechanism of action of P2Y$_{12}$ ADP-receptor antagonists. Clopidogrel, prasugrel, and ticagrelor are P2Y$_{12}$ ADP-receptor inhibitors, which block ADP-induced platelet activation, degranulation, and aggregation. Clopidogrel and prasugrel are prodrugs, whose active metabolites irreversibly block P2Y$_{12}$ ADP receptors, whereas ticagrelor is a reversible receptor inhibitor. cAMP, cyclic adenosine monophosphate.

vascular damage. The antagonists of the P2Y$_{12}$ receptors are major therapeutic tools in ACS. This class includes three thienopyridine drugs, namely, clopidogrel, prasugrel, and ticlopidine, and one recently approved cyclopentyltriazolopyrimidine agent, namely, ticagrelor. Ticlopidine is not commonly used due to significant adverse effects and as such is not covered here.

17.4.2.2   Chemistry and Pharmacokinetics   As noted earlier, clopidogrel and prasugrel belong to thienopyridine compounds, whereas ticagrelor is a cyclopentyltriazolopyrimidine derivative (Fig. 17.8). Clopidogrel and prasugrel are prodrugs that undergo biotransformation to form the active metabolites that bind irreversibly to P2Y$_{12}$ ADP receptors. The major pharmacokinetic properties of the ADP-receptor antagonists are provided in Table 17.10.

17.4.2.3   Molecular Mechanisms and Pharmacological Effects   Clopidogrel and prasugrel are prodrugs, and the inhibition of platelet activation and aggregation occurs through the irreversible binding of their active metabolites to the P2Y$_{12}$ class of ADP receptors on platelets (Fig. 17.8).

On the other hand, ticagrelor and its major metabolite reversibly interact with the platelet P2Y$_{12}$ ADP receptors to prevent signal transduction and platelet activation. Ticagrelor and its active metabolite are approximately equipotent.

17.4.2.4   Clinical Uses   All P2Y$_{12}$ ADP-receptor antagonists are approved for use in the management of ACS. Clopidogrel is also approved for use in patients with a history of recent MI, recent stroke, or established peripheral arterial disease. The US FDA-approved indications of clopidogrel, prasugrel, and ticagrelor are summarized in Table 17.11.

17.4.2.5   Therapeutic Dosages   The dosage forms and strengths of the P2Y$_{12}$ ADP-receptor antagonists are given below. The recommended dosage regimens are provided in Table 17.12:

- Clopidogrel (Plavix): Oral, 75 mg tablets
- Prasugrel (Effient): Oral, 5, 10 mg tablets
- Ticagrelor (Brilinta): Oral, 90 mg tablets

TABLE 17.10   Major pharmacokinetic properties of P2Y$_{12}$ ADP-receptor antagonists

ADP-receptor antagonist	Oral bioavailability (%)	Elimination half-life (h)	Metabolism and elimination
Clopidogrel	50	6 (0.5 for the active metabolite)	Hydrolysis by esterases to form an inactive derivative; metabolized by CYP2C19, CYP3A4, CYP2B6, and CYP1A2 to form the active metabolite; eliminated in the urine and feces
Prasugrel	80	7 for the active metabolite	Rapid hydrolysis in the intestine to a thioacetone, which is then converted to the active metabolite primarily by CYP3A4 and CYP2B6 and, to a lesser extent, by CYP2C9 and CYP2C19; eliminated in the urine (major) and feces
Ticagrelor	36	7 (9 for the active metabolite)	Metabolism by CYP3A4 and the major metabolite is also pharmacologically active; eliminated in the feces (main) and urine

Ticagrelor and its major active metabolite are weak P-glycoprotein substrates and inhibitors.

TABLE 17.11   Indications of P2Y$_{12}$ ADP-receptor antagonists

P2Y$_{12}$ ADP-receptor antagonist	Indication	Benefit
Clopidogrel	ACS (UA/NSTEMI)	For patients with UA/NSTEMI, clopidogrel decreases the rate of a combined endpoint of cardiovascular death, myocardial infarction (MI), or stroke, as well as the rate of a combined endpoint of cardiovascular death, MI, stroke, or refractory ischemia
	ACS (STEMI)	For patients with STEMI, clopidogrel reduces the rate of death from any cause and the rate of a combined endpoint of death, reinfarction, or stroke. The benefit for patients who undergo primary PCI is unknown
	Recent MI, recent stroke, or established peripheral arterial disease (EPAD)	In patients with recent MI, stroke, or EPAD, clopidogrel reduces the combined endpoint of new ischemic stroke, new MI, and other vascular death
Prasugrel	ACS (UA/NSTEMI, STEMI)	Prasugrel is indicated for the reduction of thrombotic cardiovascular events (including stent thrombosis) in patients with ACS who are to be managed with PCI as follows: (i) patients with UA or NSTEMI or (ii) patients with STEMI when managed with either primary or delayed PCI
Ticagrelor	ACS (UA/NSTEMI, STEMI)	Ticagrelor is indicated to reduce the rate of thrombotic cardiovascular events in patients with ACS. It reduces the rate of a combined endpoint of cardiovascular death, MI, or stroke compared to clopidogrel. In patients treated with PCI, it also reduces the rate of stent thrombosis

TABLE 17.12   The recommended dosage regimens of P2Y$_{12}$ ADP-receptor antagonists

Drug	Indication	Dosage regimen
Clopidogrel	UA/NSTEMI	300 mg loading dose followed by 75 mg once daily, in combination with aspirin (75–325 mg once daily)
	STEMI	75 mg once daily, in combination with aspirin (75–325 mg once daily), with or without a loading dose
	Recent MI, recent stroke, or EPAD	75 mg once daily
Prasugrel	ACS	Initiate prasugrel treatment as a single 60 mg oral loading dose and then continue at 10 mg orally once daily. Patients taking prasugrel should also take aspirin (75–325 mg) daily
Ticagrelor	ACS	Initiate ticagrelor treatment with a 180 mg (two 90 mg tablets) loading dose and continue treatment with 90 mg twice daily. After the initial loading dose of aspirin (usually 325 mg), use ticagrelor with a daily maintenance dose of aspirin of 75–100 mg

17.4.2.6 Adverse Effects and Drug Interactions

Adverse Effects Bleeding, including life-threatening and fatal bleeding, is the most commonly reported adverse effect of clopidogrel, prasugrel, and ticagrelor. Dyspnea is also a common adverse effect of ticagrelor therapy.

Drug Interactions Coadministration of P2Y$_{12}$ ADP-receptor antagonists with anticoagulants and other platelet inhibitors may increase the risk of bleeding. Since ticagrelor is predominantly metabolized by CYP3A4 and, to a lesser extent, by CYP3A5 and is also a P-glycoprotein substrate, drug interactions may become clinically significant. Concomitant treatment with strong inhibitors of CYP3A4 (e.g., ketoconazole, itraconazole, voriconazole, clarithromycin, nefazodone, ritonavir, saquinavir, nelfinavir, indinavir, atazanavir, telithromycin) or potent inducers of CYP3A4 (e.g., rifampin, dexamethasone, phenytoin, carbamazepine, phenobarbital) should be avoided. Because simvastatin and lovastatin are metabolized also by CYP3A4, cotreatment with ticagrelor increases the plasma concentrations of these statins. Because of inhibition of P-glycoprotein by ticagrelor, P-glycoprotein-mediated elimination of digoxin may be reduced. As such, digoxin levels should be monitored with initiation of, or any change in, ticagrelor therapy.

Contraindications and Pregnancy Category

- Active bleeding and hypersensitivity to the drug product are contraindications shared by clopidogrel, prasugrel, and ticagrelor.
- Prasugrel is also contraindicated in patients with prior transient ischemic attack or stroke due to increased rate of stroke (ischemic and hemorrhagic).
- Ticagrelor is also contraindicated in patients with severe hepatic impairment.
- Pregnancy category: B (clopidogrel, prasugrel) and C (ticagrelor).

17.4.3 Thrombin Receptor Antagonists

Listed below is the first-in-class drug of thrombin receptor antagonists, which was approved by the US FDA in 2014:

- Vorapaxar (Zontivity)

17.4.3.1 General Introduction to Drug Class As discussed earlier, drugs have been developed to target two important platelet activation pathways, that is, (i) COX-1-mediated TxA$_2$ synthesis and the activation of platelets via TxA$_2$ receptors and (ii) ADP signaling via activating the P2Y$_{12}$ receptors. Despite the efficacy of aspirin and of a growing family of P2Y$_{12}$ ADP-receptor antagonists on the above two pathways, major cardiovascular events continue to occur in

patients with ischemic heart disease (IHD), suggesting the involvement of other platelet activation pathway(s). In this context, thrombin, a serine protease, is considered one of the most potent platelet activators and plays a central role in blood coagulation.

Platelet responses to thrombin are mediated by surface G-protein-coupled receptors, known as protease-activated receptors (PARs) or thrombin receptors. In humans, there are four known subtypes of PARs, which display wide tissue distributions. PAR-1, PAR-3, and PAR-4 are activated by thrombin, whereas PAR-2 is activated by trypsin and trypsin-like proteases and not by thrombin. Thrombin-mediated platelet activation in humans is shown to occur through PAR-1 and PAR-4, especially PAR-1. Indeed, PAR-1 acts as the principal thrombin receptor on human platelets and mediates platelet activation by sub-nanomolar thrombin concentrations, whereas PAR-4 requires higher concentrations of thrombin for activation [10, 11].

Thrombin-mediated PAR-1 cleavage results in the activation of heterotrimeric G proteins of the Gα12/13, Gαq, and Gαi/z families that interconnect several intracellular signaling pathways to the various phenotypic effects of thrombin on platelets. These include TxA$_2$ production, ADP release, serotonin and epinephrine release, activation/mobilization of P-selectin and CD40 ligand, integrin activation, and platelet aggregation. PAR-1 activation also stimulates platelet procoagulant activity, leading to enhanced thrombin formation and the consequent generation of fibrin from fibrinogen (Fig. 17.8) [10].

The advancement in understanding the critical role of PAR-1 signaling in platelet activation and atherothrombogenesis has led to the development of a novel class of antiplatelet agents able to specifically block PAR-1. Two PAR-1 antagonists, vorapaxar and atopaxar, have recently undergone clinical investigations [12–15], and vorapaxar has recently received approval by the US FDA for clinical use.

17.4.3.2 Chemistry and Pharmacokinetics Vorapaxar is a synthetic small molecule (structure shown in Fig. 17.9). The drug is readily absorbed following oral administration, with a bioavailability of approximately 100%. Vorapaxar is widely distributed and highly bound to plasma proteins. It is metabolized by CYP3A4 and CYP2J2 and eliminated mainly in the feces and, to a lesser extent, in the urine. The elimination half-life of vorapaxar is approximately 8 days.

17.4.3.3 Molecular Mechanisms and Pharmacological Effects Vorapaxar is a reversible antagonist of PAR-1 expressed on platelets, but its long half-life makes it effectively irreversible (Fig. 17.9). Vorapaxar inhibits thrombin-induced and thrombin receptor agonist peptide (TRAP)-induced platelet aggregation. Vorapaxar does not inhibit platelet

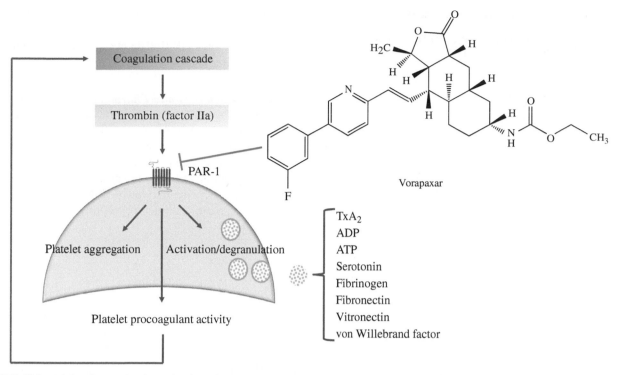

FIGURE 17.9   Molecular mechanism of action of vorapaxar. Vorapaxar is a reversible inhibitor of the thrombin receptor PAR-1 on platelets. It inhibits thrombin-induced platelet activation, degranulation, and aggregation. It also reduces thrombin-induced procoagulant activity of platelets. ADP, adenosine diphosphate; ATP, adenosine triphosphate; TxA2, thromboxane A2.

aggregation induced by ADP, collagen, or TxA_2. PAR-1 is also expressed in a wide variety of cell types, including endothelial cells, neurons, and smooth muscle cells, but the pharmacodynamic effects of vorapaxar in these cell types remain unknown.

17.4.3.4   Clinical Uses
Vorapaxar is indicated for the reduction of thrombotic cardiovascular events in patients with a history of MI or with peripheral arterial disease. It has been shown to reduce the rate of a combined endpoint of cardiovascular death, MI, stroke, and urgent coronary revascularization in this population.

17.4.3.5   Therapeutic Dosages
Listed below are the dosage form and strength of vorapaxar:

* Vorapaxar (Zontivity): Oral, 2.08 mg (equivalent to 2.5 mg vorapaxar sulfate) tablets

The recommended dosage regimen for the indication stated earlier is one tablet (2.08 mg) orally once daily, with or without food. The drug should be used in combination with aspirin and/or clopidogrel according to their indications or standard of care. There is limited clinical experience with other antiplatelet drugs or with vorapaxar as the only antiplatelet agent.

17.4.3.6   Adverse Effects and Drug Interactions

Adverse Effects   Bleeding, including life-threatening and fatal bleeding, is the most commonly reported adverse effect of vorapaxar therapy. Other rare adverse effects may include anemia, depression, rashes, eruptions, and exanthemas.

Drug Interactions   As noted earlier, vorapaxar is metabolized by CYP3A4 and CYP2J2. Drugs that are strong CYP3A4 inhibitors (e.g., ketoconazole, itraconazole, posaconazole, clarithromycin, nefazodone, ritonavir, saquinavir, nelfinavir, indinavir, boceprevir, telaprevir, telithromycin, conivaptan) or strong CYP3A4 inducers (e.g., rifampin, carbamazepine, phenytoin, St. John's wort) can significantly affect the drug disposition and its pharmacological and adverse effects. As such, concomitant use of vorapaxar with these CYP3A4-modulating drugs should be avoided.

Contraindications and Pregnancy Category

* Vorapaxar is contraindicated in patients with a history of stroke, transient ischemic attack, or ICH because of an increased risk of ICH in this population.
* Vorapaxar is contraindicated in patients with active pathological bleeding, such as ICH or peptic ulcer.
* Pregnancy category: B.

17.4.4 Glycoprotein IIb/IIIa Antagonists

Listed below are the glycoprotein (GP) IIb/IIIa antagonists approved by the US FDA for clinical use:

- Abciximab (ReoPro)
- Eptifibatide (Integrilin)
- Tirofiban (Aggrastat)

17.4.4.1 General Introduction to Drug Class

This class includes one anti-GP IIb/IIIa antibody agent, namely, abciximab, and two small molecule receptor antagonists, namely, eptifibatide and tirofiban. GP IIb/IIIa antagonists inhibit the final common pathway of platelet aggregation (the cross bridging of platelets by fibrinogen binding to the GP IIb/IIIa

receptors) and may also prevent adhesion of platelets to the vessel wall.

All of the three GP IIb/IIIa antagonists are administered intravenously.

17.4.4.2 Chemistry and Pharmacokinetics

Abciximab is the Fab fragment of the chimeric human–murine monoclonal antibody that binds to GP IIb/IIIa receptors of human platelets. As shown in Figure 17.10, eptifibatide is a cyclic heptapeptide containing six amino acids and one mercaptopropionyl (des-amino cysteinyl) residue, whereas tirofiban is a tyrosine derivative. Table 17.13 summarizes the major pharmacokinetic properties of GP IIb/IIIa antagonists.

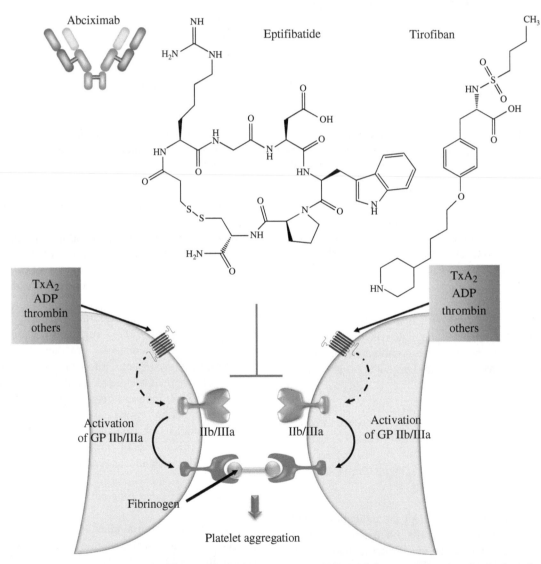

FIGURE 17.10 Molecular mechanisms of action of GP IIb/IIIa antagonists. Abciximab is a monoclonal antibody drug that directly binds (essentially irreversible) to GP IIb/IIIa, preventing fibrinogen-mediated platelet aggregation. Eptifibatide and tirofiban are small molecule drugs that reversibly antagonize fibrinogen binding to the GP IIb/IIIa receptors. Also shown in the scheme is the activation of GP IIb/IIIa receptors by TxA₂, ADP, thrombin, and other platelet activators. For color details, please see color plate section.

TABLE 17.13   Major pharmacokinetic properties of GP IIb/IIIa antagonists

GP IIb/IIIa antagonist	Elimination half-life	Metabolism and elimination
Abciximab	10–30 min (free drug)	
Eptifibatide	2.5 h	Renal excretion
Tirofiban	2 h	Eliminated in unchanged form in the urine (major) and feces

Although the half-life of free abciximab is 10–30 min, abciximab remains bound to the GP IIb/IIIa receptors and inhibits platelet aggregation as measured *in vitro* for up to 24 h after infusion is stopped.

TABLE 17.14   Indications of abciximab, eptifibatide, and tirofiban

GP IIb/IIIa antagonist	Indication
Abciximab	As an adjunct to PCI for the prevention of cardiac ischemic complications in patients undergoing PCI or in patients with UA not responding to conventional medical therapy when PCI is planned within 24 h
Eptifibatide	To decrease the rate of a combined endpoint of death or new MI in patients with UA/NSTEMI, including patients who are to be managed medically and those undergoing PCI
	To decrease the rate of a combined endpoint of death, new MI, or need for urgent intervention in patients undergoing PCI, including those undergoing intracoronary stenting
Tirofiban	To reduce the rate of thrombotic cardiovascular events (combined endpoint of death, myocardial infarction, or refractory ischemia/repeat cardiac procedure) in patients with UA/NSTEMI

17.4.4.3   Molecular Mechanisms and Pharmacological Effects

Abciximab   Abciximab binds to the intact platelet GP IIb/IIIa receptor, which is a member of the integrin family of adhesion receptors and the major platelet surface receptor involved in platelet aggregation. Abciximab inhibits platelet aggregation by preventing the binding of fibrinogen, von Willebrand factor, and other adhesive molecules to GP IIb/IIIa receptor sites on activated platelets (Fig. 17.10). The mechanism of action is thought to involve steric hindrance and/or conformational effects to block access of large molecules to the receptor rather than direct interaction with the RGD (arginine–glycine–aspartic acid) binding site of GP IIb/IIIa receptor.

Abciximab binds with similar affinity to the vitronectin receptor, also known as the $\alpha v \beta 3$ integrin. The vitronectin receptor mediates the procoagulant properties of platelets and the proliferative properties of vascular endothelial and smooth muscle cells. In addition, abciximab also binds to the activated Mac-1 receptor on monocytes and neutrophils. However, the relationship of these *in vitro* data to the clinical efficacy of abciximab is unknown.

Eptifibatide and Tirofiban   Eptifibatide and tirofiban reversibly inhibit platelet aggregation by preventing the binding of fibrinogen, von Willebrand factor, and other adhesive ligands to GP IIb/IIIa receptors. When administered intravenously, they inhibit *ex vivo* platelet aggregation in a dose- and concentration-dependent manner. Platelet aggregation inhibition is reversible following cessation of the drug infusion; this is thought to result from dissociation of the drugs from the platelets (Fig. 17.10).

17.4.4.4   Clinical Uses   The US FDA-approved clinical indications of the GP IIb/IIIa antagonists are given in Table 17.14.

17.4.4.5   Therapeutic Dosages   The dosage forms and strengths of the GP IIb/IIIa antagonists are as follows:

- Abciximab (ReoPro): Intravenous, 2 mg/ml supplied in 5 ml vials containing 10 mg
- Eptifibatide (Integrilin): Intravenous, 20 mg/10 ml (2 mg/ml) in a single-use vial for bolus injection, 75 mg/100 ml (0.75 mg/ml) in a single-use vial for infusion, and 200 mg/100 ml (2 mg/ml) in a single-use vial for infusion
- Tirofiban (Aggrastat): Intravenous, 5 mg/100 ml (0.05 mg/ml) and 12.5 mg/250 ml (0.05 mg/ml)

Abciximab   The recommended dosage of abciximab in adults is a 0.25 mg/kg intravenous bolus administered 10–60 min before the start of PCI, followed by a continuous intravenous infusion of 0.125 µg (mcg)/kg/min (to a maximum of 10 µg/min) for 12 h.

Patients with unstable angina (UA) not responding to conventional medical therapy and who are planned to undergo PCI within 24 h may be treated with an abciximab 0.25 mg/kg intravenous bolus followed by an 18–24 h intravenous infusion of 10 µg (mcg)/min, concluding 1 h after the PCI.

Eptifibatide   For patients with UA/NSTEMI, 180 µg (mcg)/kg intravenous (iv) bolus should be administered as soon as possible after diagnosis, followed by continuous infusion of 2 µg (mcg)/kg/min. Infusion should continue until hospital discharge or initiation of coronary artery bypass grafting (CABG) surgery, up to 72 h. If a patient is to undergo PCI, the infusion should be continued until hospital discharge or for up to 18–24 h after the procedure, whichever comes first, allowing for up to 96 h of therapy. Aspirin, 160–325 mg, should be given daily.

For patients with PCI, 180 μg (mcg)/kg iv bolus should be given immediately before PCI followed by continuous infusion of 2 μg (mcg)/kg/min and a second bolus of 180 μg (mcg)/kg (given 10 min after the first bolus). Infusion should be continued until hospital discharge or for up to 18–24 h, whichever comes first. A minimum of 12 h of infusion is recommended. In patients who undergo CABG surgery, eptifibatide infusion should be discontinued prior to surgery. Aspirin, 160–325 mg, should be given 1–24 h prior to PCI and daily thereafter.

Tirofiban   The recommended regimen of tirofiban therapy for patients with UA/NSTEMI is to administer intravenously 25 μg (mcg)/kg over 3 min and then 0.15 μg (mcg)/kg/min (or 0.075 μg/kg/min for patients with serum creatinine <60 ml/min) for up to 18 h.

17.4.4.6   Adverse Effects and Drug Interactions

Bleeding is the most common adverse effects of GP IIb/IIIa antagonist therapy. These drugs may also cause thrombocytopenia. Coadministration of other antiplatelet agents, thrombolytics, and heparin increases the risk of bleeding associated with GP IIb/IIIa antagonist therapy. GP IIb/IIIa antagonists are contraindicated in patients with active bleeding or conditions that may predispose to bleeding complications (e.g., severe hypertension, major surgical procedures, thrombocytopenia) and with known hypersensitivity to the drug product. All three drugs belong to pregnancy category B.

17.4.5   Other Platelet Inhibitors

Listed below are two miscellaneous platelet inhibitors:

1. Cilostazol (Pletal)
2. Dipyridamole (Persantine)

Both drugs are miscellaneous platelet inhibitors, available in oral formulations, and dipyridamole is also for intravenous injection. These drugs are mainly used for non-ACS conditions.

17.4.5.1   Cilostazol

Cilostazol inhibits phosphodiesterases (PDEs) (more specific for PDE3). This results in increased levels of cyclic-3′,5′-adenosine monophosphate (cAMP) in platelets and vascular smooth muscle cells, leading to inhibition of platelet aggregation and vasodilation, respectively. Cilostazol reversibly inhibits platelet aggregation induced by a variety of stimuli, including thrombin, and ADP.

Following oral administration, cilostazol is readily absorbed and extensively metabolized by hepatic CYP3A4 and, to a lesser extent, CYP2C19, with metabolites largely excreted in the urine. The elimination half-life of cilostazol is 12 h.

Cilostazol is indicated for the reduction of symptoms of intermittent claudication, as indicated by an increased walking distance. It is contraindicated in patients with congestive heart failure (due to decreased survival) and hemostatic disorders or active bleeding. Common adverse effects of cilostazol include headache, palpation, and diarrhea.

17.4.5.2   Dipyridamole

Dipyridamole inhibits the uptake of adenosine into platelets, endothelial cells, and erythrocytes. This inhibition results in an increase in local concentrations of adenosine that acts on the platelet A_2 receptors, thereby stimulating platelet adenylate cyclase and increasing platelet cAMP levels. Via this mechanism, platelet aggregation is inhibited in response to various stimuli, such as platelet-activating factor (PAF), collagen, and ADP.

Dipyridamole also inhibits PDEs in various tissues. While the inhibition of cAMP-PDE is weak, therapeutic levels of dipyridamole inhibit cyclic-3′,5′-guanosine monophosphate-PDEs (cGMP-PDEs), thereby augmenting the increase in cGMP produced by nitric oxide. Through this mechanism, dipyridamole lowers systemic blood pressure and improves coronary blood flow.

Dipyridamole is metabolized in the liver via glucuronidation and the metabolites are excreted in the bile. The drug has an initial half-life of 40 min, and the terminal half-life is approximately 10 h.

Oral dipyridamole is indicated as an adjunct to warfarin in the prevention of postoperative thromboembolic complications of cardiac valve replacement. Dipyridamole injection is indicated as an alternative to exercise in thallium myocardial perfusion imaging for the evaluation of coronary artery disease in patients who cannot exercise adequately. Dipyridamole is generally well tolerated, and adverse effects may include headache, dizziness, and gut disturbances.

17.5   THROMBOLYTIC AGENTS

Listed below are 6 thrombolytic agents for clinical use. Streptokinase is currently not available in the United States:

- Recombinant forms of human tPA:
  - Alteplase (Activase)
  - Reteplase (Retavase)
  - Tenecteplase (TNKase)
- Urokinase (Abbokinase)
- Streptokinase (Streptase)
- Anistreplase (Streptokinase + plasminogen)

17.5.1   General Introduction to Drug Class

As stated in Chapter 16, plaque rupture/erosion and thrombus formation are key pathophysiological processes of ACS. The introduction of thrombolytic (also known as fibrinolytic) therapy was a major advance in the treatment of acute

STEMI resulting from complete coronary occlusion. PCI is now preferred for most patients if it can be performed by an experienced operator with <2h delay from presentation to the emergency department. However, due in part to limited availability of primary PCI, thrombolytic therapy remains an important therapeutic modality for STEMI as well as ischemic stroke (see Unit VIII).

Currently used thrombolytic drugs are intravenously infused plasminogen activators that activate the blood fibrinolytic system. These agents have a high specificity for their substrate plasminogen, hydrolyzing a peptide bond to yield the active enzyme plasmin. Free plasmin is rapidly neutralized by the serine proteinase inhibitor α-antiplasmin, whereas fibrin-bound plasmin is protected from rapid inhibition, thereby promoting thrombus lysis. The clinically available thrombolytic agents are classified into two categories: (i) recombinant forms of human tPA (a serine protease), which include alteplase, reteplase, and tenecteplase, and (ii) non-tPA agents, including urokinase, streptokinase, and anistreplase. Notably, streptokinase, a protein obtained from streptococci, is much less costly than other thrombolytic agents, but outcomes are inferior and the drug is associated with high risk of hypersensitivity reaction. This drug is extensively used in other parts of the world, but not available in the United States. Anistreplase is a preformed complex of streptokinase and plasminogen. Hence, streptokinase and anistreplase are not considered here.

TABLE 17.15   Major pharmacokinetic properties of thrombolytic agents

Thrombolytic drug	Elimination half-life	Metabolism and elimination
Alteplase	<5 min (initial)	Eliminated by the liver
Reteplase	13–16 min	Eliminated by the liver and kidney
Tenecteplase	20–24 min (initial) 90–130 min (terminal)	Eliminated by the liver
Urokinase	6–19 min	Eliminated by the liver

17.5.2   Chemistry and Pharmacokinetics

Alteplase is a GP of 527 amino acids synthesized using the cDNA for natural human tPA. Reteplase is a nonglycosylated deletion mutein of tPA, containing the kringle 2 and the protease domains of human tPA. Reteplase contains 355 of the 527 amino acids of native tPA (amino acids 1–3 and 176–527). Tenecteplase is a 527-amino-acid GP developed by introducing the following modifications to the cDNA for natural human tPA: a substitution of threonine 103 with asparagine and a substitution of asparagine 117 with glutamine, both within the kringle 1 domain, and a tetra-alanine substitution at amino acids 296–299 in the protease domain.

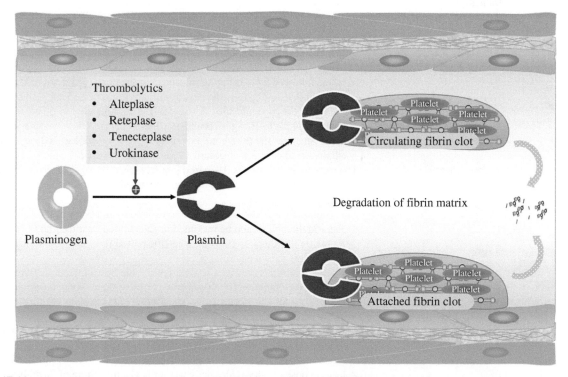

FIGURE 17.11   Molecular mechanism of action of thrombolytic drugs. Thrombolytic drugs stimulate the conversion of plasminogen to plasmin, which then degrades fibrin, leading to thrombus lysis. For color details, please see color plate section.

Urokinase is a thrombolytic agent obtained from human neonatal kidney cells grown in tissue culture. The principal active ingredient of the agent is the low-molecular-weight form of urokinase and consists of an A chain of 2 kDa linked by a sulfhydryl bond to a B chain of 30.4 kDa. The major pharmacokinetic properties of the aforementioned thrombolytic agents are given in Table 17.15.

17.5.3 Molecular Mechanisms and Pharmacological Effects

All recombinant tPA agents catalyze the cleavage of endogenous plasminogen to generate plasmin. Plasmin in turn degrades the fibrin matrix of the thrombus, thereby exerting thrombolytic action (Fig. 17.11).

Alteplase is a recombinant form of human native tPA without molecular modifications. It has the property of fibrin-enhanced conversion of plasminogen to plasmin. It produces limited conversion of plasminogen to plasmin in the absence of fibrin. When introduced into the systemic circulation at pharmacological concentrations, alteplase binds to fibrin in a thrombus and converts the entrapped plasminogen to plasmin.

Tenecteplase is a modified form of human tPA that binds to fibrin and converts plasminogen to plasmin. *In vitro* studies demonstrate that, in the presence of fibrin, conversion of plasminogen to plasmin by tenecteplase is also increased relative to its conversion in the absence of fibrin. This fibrin specificity decreases systemic activation of plasminogen and the resulting degradation of circulating fibrinogen as compared to a molecule lacking this property.

Urokinase is an enzyme produced by the kidney and found in the urine. There are two forms of urokinase that differ in molecular weight but have similar clinical effects. Abbokinase is the low-molecular-weight form. It converts plasminogen to plasmin.

17.5.4 Clinical Uses

All three recombinant tPA agents are indicated for use in the management of acute MI for the reduction of mortality and/or improvement of ventricular function. Alteplase is also indicated for the management of acute ischemic stroke in adults for improving neurological recovery and reducing the incidence of disability (see Unit VIII) and in the management of acute massive PE in adults.

Urokinase is only approved for use in the management of PE. It is indicated in adults (i) for the lysis of acute massive pulmonary emboli, defined as obstruction of blood flow to a lobe or multiple segments, or (ii) for the lysis of pulmonary emboli accompanied by unstable hemodynamics, that is, failure to maintain blood pressure without supportive measures.

TABLE 17.16 Dosage regimens of recombinant tPA agents

Recombinant tPA	Indication	Dosage regimen
Alteplase	Acute MI	Accelerated infusion: for patients weighing >67 kg, the recommended dose administered is 100 mg as a 15 mg intravenous bolus, followed by 50 mg infused over the next 30 min and then 35 mg infused over the next 60 min; for patients weighing ≤ 67 kg, the recommended dose is administered as a 15 mg intravenous bolus, followed by 0.75 mg/kg infused over the next 30 min not to exceed 50 mg and then 0.50 mg/kg over the next 60 min not to exceed 35 mg 3 h infusion: the recommended dose is 100 mg administered as 60 mg in the first hour (of which 6–10 mg is administered as a bolus), 20 mg over the second hour, and 20 mg over the third hour. For smaller patients (<65 kg), a dose of 1.25 mg/kg administered over 3 h, as described earlier, may be used
	Acute ischemic stroke	The recommended dose is 0.9 mg/kg (not to exceed 90 mg total dose) infused over 60 min with 10% of the total dose administered as an initial intravenous bolus over 1 min
	PE	The recommended dose is 100 mg administered by intravenous infusion over 2 h. Heparin therapy should be instituted or reinstituted near the end of or immediately following the alteplase infusion when the partial thromboplastin time or thrombin time returns to twice normal or less
Reteplase	Acute MI	Reteplase is administered as a 10 + 10 unit double-bolus injection. Two 10 unit bolus injections are required for a complete treatment. Each bolus is administered as an intravenous injection over 2 min. The second bolus is given 30 min after initiation of the first bolus injection
Tenecteplase	Acute MI	The recommended total dose should not exceed 50 mg and is based upon patient weight: 30 mg for <60 kg, 35 mg for ≥60 to <70 kg, 40 mg for ≥70 to <80 kg, 45 mg for ≥80 to <90 kg, 50 mg for ≥90 kg
Urokinase	PE	A loading dose of 4400 IU/kg body weight is given at a rate of 90 ml/h over a period of 10 min. This is followed by a continuous infusion of 4400 IU/kg/h at a rate of 15 ml for 12 h

17.5.5 Therapeutic Dosages

The dosage forms and strengths of the thrombolytic agents are given below:

- Alteplase (Activase): Intravenous, 50 and 100 mg vials
- Reteplase (Retavase): Intravenous, 10.4 units (18.1 mg) vials
- Tenecteplase (TNKase): Intravenous, 50 mg vials
- Urokinase (Abbokinase): 250,000 IU vials

The dosage regimens of the recombinant tPA agents in treating ACS and other conditions are given in Table 17.16.

17.5.6 Adverse Effects and Drug Interactions

The most common adverse effect of thrombolytic therapy is bleeding. Cotreatment with anticoagulants or other platelet inhibitors may increase the risk of bleeding complications. Thrombolytic agents may also cause cholesterol embolism, a rare but serious condition. They may provoke cardiac arrhythmias associated with reperfusion. Other adverse effects may include hypersensitivity reaction, hypotension, and gut disturbances. Contraindications of thrombolytic therapy include active bleeding and conditions that increase the risk of bleeding (e.g., severe uncontrolled hypertension, recent trauma, major surgical procedures). Alteplase, reteplase, and tenecteplase all belong to pregnancy category C, whereas urokinase is in B category.

17.6 SUMMARY OF CHAPTER KEY POINTS

- Atherothrombogenesis is a key pathophysiological process of ACS. Anticoagulants, platelet inhibitors, and thrombolytic agents are widely used in the management of ACS as well as other thromboembolic disorders, including DVT and PE.
- Anticoagulants act on the different steps of the coagulation cascades. They inhibit either the action of the coagulation factors or interfere with their biosynthesis. Commonly used anticoagulants include warfarin, heparins (UFH, LMWHs), selective factor Xa inhibitors (fondaparinux, apixaban, rivaroxaban), and direct thrombin inhibitors (bivalirudin, desirudin, lepirudin, argatroban, dabigatran).
- Platelet inhibitors block a number of platelet functions, including aggregation, release of granule contents, and platelet-mediated vascular constriction. Based on their mechanisms of action, platelet inhibitors are classified into five groups: (i) COX inhibitors (aspirin), (ii) $P2Y_{12}$ ADP-receptor antagonists (clopidogrel,

prasugrel, ticagrelor, and ticlopidine), (iii) thrombin receptor antagonists (vorapaxar), (iv) GP IIb/IIIa antagonists (abciximab, eptifibatide, tirofiban), and others (cilostazol, dipyridamole). Platelet inhibitor therapy has become a cornerstone of the medical management of ACS.

- The clinically available thrombolytic agents (also called fibrinolytic agents) are classified into two categories: (i) recombinant forms of human tPA (a serine protease), which include alteplase, reteplase, and tenecteplase, and (ii) non-tPA agents, including urokinase, streptokinase, and anistreplase. These agents cause the conversion of plasminogen to plasmin, which then degrades the fibrin matrix of the thrombus, thereby exerting thrombolytic action.
- Although PCI is now preferred for most patients with ACS, due in part to limited availability of primary PCI, thrombolytic therapy with recombinant tPA agents remains as an important treatment option for acute MI as well as ischemic stroke.

17.7 SELF-ASSESSMENT QUESTIONS

17.7.1. Following successful acute management of a non-ST-elevation acute coronary syndrome, a 55-year-old male is prescribed a drug that interferes with $P2Y_{12}$ ADP-receptor-mediated signaling in platelets. Which of the following is most likely prescribed?
A. Clopidogrel
B. Dabigatran
C. Dipyridamole
D. Enoxaparin
E. Eptifibatide

17.7.2. A 60-year-old female with deep vein thrombosis is prescribed a drug that inhibits the enzyme that converts the inactive vitamin K 2,3-epoxide into the active, reduced form of vitamin K. Which of the following drugs is she most likely given?
A. Aspirin
B. Lepirudin
C. Prasugrel
D. Tinzaparin
E. Warfarin

17.7.3. A 65-year-old male develops severe thrombocytopenia in response to treatment with an anticoagulant. Which of the following drugs is most likely responsible for this patient's thrombocytopenia?
A. Dabigatran
B. Dipyridamole
C. Prasugrel
D. Unfractionated heparin
E. Warfarin

17.7.4. A 60-year-old male is brought to the emergency department of a rural hospital 90 min after the onset of severe chest pain during yard work. He has a history of poorly controlled stage 1 hypertension and elevated blood cholesterol, but does not smoke and drink alcohol. ECG changes confirm the diagnosis of an acute ST-elevation myocardial infarction. Which of the following should be given to open his occluded coronary artery?
   A. Abciximab
   B. Alteplase
   C. Dalteparin
   D. Fondaparinux
   E. Prasugrel

17.7.5. A 50-year-old male presenting with an ST-elevation acute coronary syndrome is given reteplase to open his occluded coronary artery. Which of the following adverse effects is the patient most likely to experience?
   A. Acute kidney injury
   B. Bleeding
   C. Hepatotoxicity
   D. Neutropenia
   E. Thrombocytopenia

17.7.6. A 45-year-old male, presenting with a non-ST-elevation acute coronary syndrome, is given prasugrel following a successful percutaneous coronary intervention. Which of the following best describes the rationale of the antiplatelet therapy?
   A. Preventing cerebral hemorrhage
   B. Preventing long QT syndrome
   C. Preventing reocclusion of the coronary artery
   D. Preventing thrombocytopenia
   E. Preventing transient ischemic attack

17.7.7. A 62-year-old female is brought to the emergency department of a rural hospital 90 min after the onset of severe chest pain. Elevated cardiac troponin T and ECG changes confirm the diagnosis of ST-elevation myocardial infarction. Due to the lack of percutaneous coronary intervention facility, she is given a drug to accelerate the conversion of plasminogen to plasmin so as to restore myocardial perfusion. Which of the following drugs is most likely administered to the patient for the above purpose?
   A. Clopidogrel
   B. Aspirin
   C. Heparin
   D. Tenecteplase
   E. Warfarin

17.7.8. Following a successful percutaneous coronary intervention with stent placement, a 48-year-old male with stable ischemic heart disease is prescribed ticagrelor to reduce the rate of stent thrombosis. What is the mechanism of action of ticagrelor for the above stated purpose of use?
   A. Blocking GP IIb/IIIa receptors
   B. Blocking P2Y$_{12}$ ADP receptors
   C. Decreasing cAMP degradation
   D. Decreasing thromboxane A$_2$ formation
   E. Inhibiting factor Xa

17.7.9. A 62-year-old female undergoes an abdominal surgical procedure. She is admitted for postoperative observation and started on subcutaneous heparin treatment to prevent formation of deep vein thrombosis, a major risk factor for pulmonary embolism. Which of the following adverse reactions is the patient likely to experience?
   A. Hyperthyroidism
   B. Liver injury
   C. Pulmonary fibrosis
   D. Thrombocytopenia
   E. Ventricular arrhythmias

17.7.10. A patient develops severe thrombocytopenia in response to the treatment with unfractionated heparin and still requires parenteral anticoagulation. Which of the following drugs would be most suitable for this patient?
   A. Abciximab
   B. Clopidogrel
   C. Lepirudin
   D. Streptokinase
   E. Tirofiban

REFERENCES

1 Borissoff, J.I., H.M. Spronk, and H. ten Cate, The hemostatic system as a modulator of atherosclerosis. *N Engl J Med*, 2011. 364(18): p. 1746–60.

2 Pirmohamed, M., et al., A randomized trial of genotype-guided dosing of warfarin. *N Engl J Med*, 2013. 369(24): p. 2294–303.

3 Kimmel, S.E., et al., A pharmacogenetic versus a clinical algorithm for warfarin dosing. *N Engl J Med*, 2013. 369(24): p. 2283–93.

4 Chatterjee, S., et al., New oral anticoagulants and the risk of intracranial hemorrhage: traditional and Bayesian meta-analysis and mixed treatment comparison of randomized trials of new oral anticoagulants in atrial fibrillation. *JAMA Neurol*, 2013. 70(12): p. 1486–90.

5 Agnelli, G., et al., Apixaban for extended treatment of venous thromboembolism. *N Engl J Med*, 2013. 368(8): p. 699–708.

6 Agnelli, G., et al., Oral apixaban for the treatment of acute venous thromboembolism. *N Engl J Med*, 2013. 369(9): p. 799–808.

7 Lincoff, A.M., et al., Bivalirudin and provisional glycoprotein IIb/IIIa blockade compared with heparin and planned glycoprotein IIb/IIIa blockade during percutaneous coronary intervention: REPLACE-2 randomized trial. *JAMA*, 2003. 289(7): p. 853–63.

8 Mitchell, J.A. and T.D. Warner, COX isoforms in the cardiovascular system: understanding the activities of non-steroidal anti-inflammatory drugs. *Nat Rev Drug Discov*, 2006. 5(1): p. 75–86.

9 Paikin, J.S. and J.W. Eikelboom, Cardiology patient page: aspirin. *Circulation*, 2012. 125(10): p. e439–42.

10 Angiolillo, D.J., D. Capodanno, and S. Goto, Platelet thrombin receptor antagonism and atherothrombosis. *Eur Heart J*, 2010. 31(1): p. 17–28.

11 de Souza Brito, F. and P. Tricoci, Novel anti-platelet agents: focus on thrombin receptor antagonists. *J Cardiovasc Transl Res*, 2013. 6(3): p. 415–24.

12 Morrow, D.A., et al., Vorapaxar in the secondary prevention of atherothrombotic events. *N Engl J Med*, 2012. 366(15): p. 1404–13.

13 Scirica, B.M., et al., Vorapaxar for secondary prevention of thrombotic events for patients with previous myocardial infarction: a prespecified subgroup analysis of the TRA 2 degrees P-TIMI 50 trial. *Lancet*, 2012. 380(9850): p. 1317–24.

14 Tricoci, P., et al., Thrombin-receptor antagonist vorapaxar in acute coronary syndromes. *N Engl J Med*, 2012. 366(1): p. 20–33.

15 Rollini, F., A. Tello-Montoliu, and D.J. Angiolillo, Atopaxar: a review of its mechanism of action and role in patients with coronary artery disease. *Future Cardiol*, 2012. 8(4): p. 503–11.

18

MANAGEMENT OF UNSTABLE ANGINA AND NON-ST-ELEVATION MYOCARDIAL INFARCTION: PRINCIPLES AND GUIDELINES

18.1 OVERVIEW

Unstable angina (UA) and non-ST-elevation myocardial infarction (NSTEMI) constitute a clinical syndrome subset of the acute coronary syndromes (ACS). In the spectrum of ACS, UA/NSTEMI is defined by electrocardiographic (ECG) ST-segment depression or prominent T-wave inversion and/or positive biomarkers of necrosis (e.g., increased plasma levels of cardiac troponins for NSTEMI, but not for UA) in the absence of ST-segment elevation and in an appropriate clinical setting (chest discomfort or anginal equivalent). UA/NSTEMI often results from the disruption or erosion of an atherosclerotic plaque and a subsequent cascade of pathophysiological coagulation, leading to platelet-enriched thrombus formation and occlusion of coronary blood flow. This pathophysiological mechanism, as discussed in Chapters 16 and 17, serves as the foundation for the use of anticoagulants, platelet inhibitors, and thrombolytic agents in the management of ACS.

This chapter discusses the general principles and current guidelines for the management of patients with UA/NSTEMI, focusing on pharmacological therapies. The chapter begins with an introduction to some most widely cited guidelines, including the American College of Cardiology/American Heart Association (ACC/AHA) 2007 guideline and its 2011 and 2012 focused updates, as well as the European Society of Cardiology (ESC) 2011 guideline. The chapter then considers the general principles of management of UA/NSTEMI followed by a detailed survey of the current guideline-based recommendations. Management of ST-elevation myocardial infarction (STEMI) is covered in Chapter 19.

18.2 INTRODUCTION TO EVIDENCE-BASED GUIDELINES

Over the past several years, guidelines from various organizations in the United States and other nations have been developed to provide evidence-based recommendations on the management of UA/NSTEMI. Among them, the ACC/AHA 2007 guideline with focused updates in 2011 and 2012 and the ESC 2011 guideline have been most widely cited. This section provides a brief introduction to these guidelines so as to set a stage for the subsequent consideration of guideline-based management of UA/NSTEMI. It also provides the reader a source of references for detailed information on UA/NSTEMI, including both diagnosis and management.

18.2.1 The ACC/AHA 2007 Guideline and Its Focused Updates in 2011 and 2012

The ACC/AHA 2007 guideline represents the current full recommendations on the management of UA/NSTEMI. The executive summary is published in the August 14, 2007, issue of the *Journal of the American College of Cardiology* and the August 14, 2007, issue of *Circulation*. The full-text guideline is e-published in the same issue of the journals noted earlier [1, 2], as well as posted on the websites of ACC (www.acc.org) and AHA (www.americanheart.org). The full guideline document is divided into seven sections: (i) introduction; (ii) initial evaluation and management; (iii) early hospital care; (iv) coronary revascularization; (v) late hospital care, hospital discharge, and posthospital discharge care; (vi) special groups; and (vii) conclusion and future directions.

Since the publication of the 2007 full guideline, there have been two focused updates by the American College of Cardiology Foundation (ACCF)/AHA, one in 2011 [3] and the other in 2012 [4]. The biggest change of the 2012 focused update is the recommendation to consider ticagrelor, a drug approved by the US Food and Drug Administration (FDA) in 2011 (see Chapter 17) as a treatment option in addition to clopidogrel and prasugrel. The panel's report highlights both the benefits (anticlotting action) and risks (bleeding) of the new drug.

18.2.2   The ESC 2011 Guideline

The European task force for the management of ACS patients presenting with NSTEMI released a new guideline in 2011 [5]. This guideline updates a previous version issued in 2007 [6] and contains a number of important new recommendations (see Section 18.2.3). The full guideline document consists of six sections: (i) introduction to epidemiology and pathophysiology; (ii) diagnosis; (iii) prognosis assessment; (iv) treatment, which further includes anti-ischemic agents, antiplatelet agents, anticoagulants, and coronary revascularization, special population and conditions, and long-term management; (v) performance measures; and (vi) management strategy.

18.2.3   Major New Changes of the ACCF/AHA 2011/2012 Focused Updates and the ESC 2011 Guideline

The 2011/2012 ACCF/AHA focused updates and the 2011 ESC guideline for the management of UA/NSTEMI include several new changes compared to the previous guidelines issued in 2007. Among the most important new recommendations for clinicians are those regarding the use of the newest $P2Y_{12}$ ADP-receptor antagonists, prasugrel and ticagrelor, genotyping and platelet function testing in appropriate patients on clopidogrel, and concomitant use of clopidogrel with percutaneous coronary intervention (PCI). The new updates and guideline also stress the importance of estimating the risk of cardiovascular and bleeding events using various risk assessment models, support coronary computed tomography angiogram (cCTA) in low- to intermediate-risk patients, recommend against aggressive glycemic control during hospitalization with ACS, and advocate measures to prevent contrast-induced nephropathy. Finally, the updates and guideline emphasize the importance of quality improvement and performance assessment in the management of patients with UA/NSTEMI [7].

18.2.4   The NICE 2010 Guideline

The British National Institute for Health and Care Excellence (NICE) 2010 clinical guideline for the early management of UA/NSTEMI offers evidence-based advice on the care and treatment of adults with UA/NSTEMI from the time of diagnosis to leaving hospital. The full guideline is available at www.nice.org.uk, and the summary is published in the April 2010 issue of *British Medical Journal* [8]. The 2010 NICE guideline covers seven areas: (i) risk assessment, (ii) antiplatelet treatment, (iii) antithrombotic treatment, (iv) invasive management, (v) testing for ischemia, (vi) assessing left ventricular function, and (vii) rehabilitation and discharge planning.

18.2.5   Other Guidelines

In addition to the United States and Europe, countries, such as Canada, Australia, and New Zealand, have also released updated guidelines for the management of UA/NSTEMI. With regard to general principles of management, these guidelines from Canada [9] and Australia/New Zealand [10] are largely consistent with those from the ACCF/AHA, the ESC, and the NICE, as described earlier.

18.3   GENERAL PRINCIPLES OF MANAGEMENT OF UA/NSTEMI

UA, NSTEMI, and STEMI are the three presentations of ACS. The first step in the management of patients with ACS is prompt recognition, since the beneficial effects of therapy are greatest when performed soon after hospital presentation. For patients presenting to the emergency department with chest pain suspicious for an ACS, the diagnosis of myocardial infarction (MI) can be confirmed by the ECG and serum cardiac biomarker elevation; the history is relied upon heavily to make the diagnosis of UA.

Once the diagnosis of either UA or an acute NSTEMI is made, the management of the patient involves the simultaneous achievement of several goals. These include (i) relief of ischemic pain; (ii) assessment of the patient's hemodynamic status and correction of abnormalities; (iii) estimation of the level of risk of death and nonfatal cardiac ischemic events; (iv) selection and initiation of a treatment strategy to minimize myocardial injury and reduce mortality, that is, an early invasive strategy (with angiography and intent for revascularization with PCI or coronary artery bypass grafting [CABG] as defined by the anatomy) versus a conservative strategy with pharmacological therapy; and (v) long-term care to prolong survival, including lifestyle modifications and drug therapy. The following section discusses current guideline-based recommendations on the management of patients with UA/NSTEMI.

18.4   GUIDELINE-BASED RECOMMENDATIONS ON THE MANAGEMENT OF UA/NSTEMI

Treatment of patients with UA/NSTEMI involves the use of four therapeutic modalities: (i) anti-ischemic and analgesic therapy, (ii) antiplatelet therapy, (iii) anticoagulant therapy, and (iv) coronary revascularization. This section discusses the recommendations from the ACC/AHA and ESC guidelines on each of the aforementioned treatment modalities. The section also considers the guideline-based recommendations for treating UA/NSTEMI in special patient groups.

18.4.1   Anti-ischemic and Analgesic Therapy

For about 100 years, inhaled oxygen has been administered to all patients suspected of having an acute MI. The basis for this practice was the belief that oxygen supplementation raised often-deficient arterial oxygen content to improve myocardial oxygenation, thereby reducing infarct size. This assumption is conditional and not evidence based [11–13]. While such physiological changes may pertain in some patients who are hypoxemic, considerable data suggest that oxygen therapy may be detrimental in others, particularly in patients with normoxia. Hyperoxia, which might occur with the administration of oxygen to normoxic individuals, has been shown to have a direct vasoconstricting effect on the coronary arteries, which may actually worsen myocardial ischemia. There are no large, contemporary, randomized studies that examine clinical outcomes after this intervention. Hence, this long-accepted but potentially harmful tradition urgently needs reevaluation. In this context, several randomized controlled trials, including the AVOID study [14] and DETO$_2$X-AMI trial [15], are currently underway to examine the role of oxygen in ACS. Results of these trials may have widespread implications on the use of oxygen therapy in patients with ACS. In fact, as described in the following text, clinical guidelines appear to be changing, favoring use of oxygen only in hypoxemic patients, and then cautiously titrating to individual oxygen tensions.

Anti-ischemic drugs either decrease myocardial oxygen demand (by decreasing heart rate, lowering blood pressure, reducing preload, or decreasing myocardial contractility) or increase myocardial oxygen supply (by inducing coronary vasodilatation). Anti-ischemic agents for UA/NSTEMI typically include nitrates (Chapter 11), β-blockers (Chapter 8), calcium channel blockers (CCBs) (Chapter 10), and ranolazine (Chapter 11). Besides the aforementioned typical anti-ischemic agents, inhibitors of the renin–angiotensin–aldosterone system (RAAS) (Chapter 9) may also improve myocardial ischemia via mechanisms including vasodilation, decreased water and salt retention, and reduced afterload. In addition, morphine has potent analgesic and anxiolytic effects, as well as hemodynamic effects that are potentially beneficial in UA/NSTEMI. Morphine causes venodilation and can produce modest reductions in heart rate (through increased vagal tone) and systolic blood pressure to further reduce myocardial oxygen demand. Hence, in a broad sense, both RAAS inhibitors and morphine are considered anti-ischemic agents in patients with UA/NSTEMI [16]. In contrast, analgesic therapy with

selective cyclooxygenase (COX)-2 inhibitors and nonsteroidal anti-inflammatory drugs (NSAIDs) (except for low-dose aspirin) may increase the risk of cardiovascular events in patients with UA/NSTEMI. Hence, patients taking these drugs at the time of UA/NSTEMI should discontinue them immediately [16].

The guideline recommendations from the ACC/AHA and the ESC on the use of anti-ischemic agents in UA/NSTEMI are summarized in Tables 18.1 and 18.2, respectively.

18.4.2   Antiplatelet Therapy

Platelet activation and subsequent aggregation play a dominant role in the propagation of arterial thrombosis and consequently are the key therapeutic targets in the management of ACS. Antiplatelet therapy is essential for modifying the disease process and its progression to death, MI, or recurrent MI in the majority of patients who have ACS due to thrombosis on a plaque. Antiplatelet therapy should be instituted as early as possible when the diagnosis of UA/NSTEMI is made in order to reduce the risk of both acute ischemic complications and recurrent atherothrombotic events. The molecular pharmacology of antiplatelet drugs is covered in Chapter 17. Tables 18.3 and 18.4 summarize the guideline recommendations on antiplatelet therapy in patients with UA/NSTEMI.

18.4.3   Anticoagulant Therapy

Anticoagulants are used in the treatment of UA/NSTEMI to inhibit thrombin generation and/or activity, thereby reducing thrombus-related events. There is evidence that anticoagulation is effective in addition to platelet inhibition and that the combination of the two is more effective than either treatment alone. Guideline recommendations for the use of anticoagulant therapy in patients with UA/NSTEMI are available from various organizations, including the ACC/AHA and the ESC. The major recommendations are summarized in Tables 18.5 and 18.6.

18.4.4   Additional Considerations on Antithrombotic (Antiplatelet and Anticoagulant) Therapy

As noted earlier, antithrombotic therapy is essential for improving prognosis in patients with ACS due to thrombosis on a ruptured or eroded plaque. A combination of aspirin, an anticoagulant, and an additional antiplatelet drug represents the most effective therapy. The intensity of treatment is tailored to individual risk, and such a triple-antithrombotic treatment is used in patients with continuing ischemia or with other high-risk features and in patients oriented to an early invasive strategy. In this context, the ACCF/AHA 2012 focused update provides recommendations for additional management of antiplatelet and anticoagulant therapy [19]. These include 8 class I, 4 class IIa, 2 class IIb, and 1 class III

TABLE 18.1   The ACC/AHA 2007 guideline recommendations for anti-ischemic therapy in UA/NSTEMI [16]

COR	Recommendation	LOE
I	Supplemental oxygen should be administered to patients with UA/NSTEMI with an arterial saturation <90%, respiratory distress, or other high-risk features for hypoxemia	B
	Patients with UA/NSTEMI with ongoing ischemic discomfort should receive sublingual nitroglycerin (0.4 mg) every 5 min for a total of 3 doses, after which assessment should be made about the need for intravenous nitroglycerin, if not contraindicated	C
	Intravenous nitroglycerin is indicated in the first 48 h after UA/NSTEMI for treatment of persistent ischemia, heart failure (HF), or hypertension. The decision to administer intravenous nitroglycerin and the dose used should not preclude therapy with other proven mortality-reducing interventions such as β-blockers or angiotensin-converting enzyme inhibitors (ACEIs)	B
	Oral β-blocker therapy should be initiated within the first 24 h for patients who do not have one or more of the following: (i) signs of HF, (ii) evidence of a low-output state, (iii) increased risk for cardiogenic shock, or (iv) other relative contraindications to β-blockade (PR interval >0.24 s, second- or third-degree heart block, active asthma, or reactive airway disease)	B
	In UA/NSTEMI patients with continuing or frequently recurring ischemia and in whom β-blockers are contraindicated, a nondihydropyridine CCB (e.g., verapamil or diltiazem) should be given as initial therapy in the absence of clinically significant left ventricular (LV) dysfunction or other contraindications	B
	An ACEI should be administered orally within the first 24 h to UA/NSTEMI patients with pulmonary congestion or LV ejection fraction (LVEF) ≤40%, in the absence of hypotension (systolic blood pressure <100 mm Hg or <30 mm Hg below baseline) or known contraindications to that class of medications	A
	An angiotensin receptor blocker (ARB) should be administered to UA/NSTEMI patients who are intolerant of ACEIs and have either clinical or radiological signs of HF or LVEF ≤40%	A
IIa	It is reasonable to administer supplemental oxygen to all patients with UA/NSTEMI during the first 6 h after presentation	C
	In the absence of contradictions to its use, it is reasonable to administer morphine sulfate intravenously to UA/NSTEMI patients if there is uncontrolled ischemic chest discomfort despite nitroglycerin, provided that additional therapy is used to manage the underlying ischemia	B
	It is reasonable to administer intravenous β-blockers at the time of presentation for hypertension to UA/NSTEMI patients who do not have one or more of the following: (i) signs of HF, (ii) evidence of low-output state, (iii) increased risk for cardiogenic shock, or (iv) other relative contraindications to β-blockade (PR interval >0.24 s, second- or third-degree heart block, active asthma, or reactive airway disease)	B
	Oral long-acting nondihydropyridine CCBs are reasonable for use in UA/NSTEMI patients for recurrent ischemia in the absence of contraindications after β-blockers and nitrates have been fully used	C
	An ACEI administered orally within the first 24 h of UA/NSTEMI can be useful in patients without pulmonary congestion or LVEF ≤40% in the absence of hypotension (systolic blood pressure <100 or <30 mm Hg below baseline) or known contraindications to that class of medications	B
	Intra-aortic balloon pump (IABP) counterpulsation is reasonable in UA/NSTEMI patients for severe ischemia that is continuing or recurs frequently despite intensive medical therapy, for hemodynamic instability in patients before or after coronary angiography, and for mechanical complications of myocardial infarction	C
IIb	The use of extended-release forms of nondihydropyridine CCBs instead of β-blockers may be considered in patients with UA/NSTEMI	B
	Immediate-release dihydropyridine CCBs in the presence of adequate β-blockade may be considered in patients with UA/NSTEMI with ongoing ischemic symptoms or hypertension	B
III	Nitrates should not be administered to UA/NSTEMI patients with systolic blood pressure <90 or ≥30 mm Hg below baseline, severe bradycardia (<50 beats/min), tachycardia (>100 beats/min) in the absence of symptomatic HF, or right ventricular infarction	C
	Nitroglycerin or other nitrates should not be administered to patients with UA/NSTEMI who had received a phosphodiesterase inhibitor for erectile dysfunction within 24 h of sildenafil or 48 h of tadalafil use. The suitable time for the administration of nitrates after vardenafil has not been determined	C
	Immediate-release dihydropyridine CCBs should not be administered to patients with UA/NSTEMI in the absence of a β-blocker	A
	An intravenous ACEI should not be given to patients within the first 24 h of UA/NSTEMI because of the increased risk of hypotension. A possible exception may be patients with refractory hypertension	B
	It may be harmful to administer intravenous β-blockers to UA/NSTEMI patients who have contraindications to β-blockade, signs of HF or low-output state, or other risk factors for cardiogenic shock	A
	Nonsteroidal anti-inflammatory drugs (except for aspirin), whether nonselective or COX-2-selective agents, should not be administered during hospitalization for UA/NSTEMI because of the increased risks of mortality, reinfarction, hypertension, HF, and myocardial rupture associated with their use	C

COR and LOE denote class of recommendation and level of evidence, respectively (see Chapter 5 for description).

TABLE 18.2   The ESC 2011 guideline recommendations for anti-ischemic therapy in UA/NSTEMI [17]

COR	Recommendation	LOE
I	Oral or intravenous nitrate treatment is indicated to relieve angina; intravenous nitrate treatment is recommended in patients with recurrent angina and/or signs of HF	C
	Patients on chronic β-blocker therapy admitted with ACS should be continued on β-blocker therapy if not in Killip class[a] ≥III	B
	Oral β-blocker treatment is indicated in all patients with LV dysfunction without contraindications	B
	CCBs are recommended for symptom relief in patients already receiving nitrates and β-blockers (dihydropyridine CCBs) and in patients with contraindications to β-blockade (nondihydropyridine CCBs)	B
	CCBs are recommended in patients with vasospastic angina	C
IIa	Intravenous β-blocker treatment at the time of admission should be considered for patients in a stable hemodynamic condition (Killip class <III) with hypertension and/or tachycardia	C
III	Nifedipine or other dihydropyridine CCBs are not recommended unless combined with β-blocker	B

[a]The Killip classification is a system used in individuals with an acute myocardial infarction for risk stratification. First published in 1967, the system focuses on physical examination and the development of HF to predict risk as described below: class I, no evidence of heart failure; class II, findings of mild-to-moderate HF; class III, pulmonary edema; and class IV, cardiogenic shock defined as systolic blood pressure <90 mm Hg and signs of hypoperfusion. COR and LOE denote class of recommendation and level of evidence, respectively (see Chapter 5 for description).

TABLE 18.3   The ACCF/AHA 2012 focused update recommendations for antiplatelet therapy of UA/NSTEMI [18, 19]

COR	Recommendation	LOE
I	Aspirin should be administered to UA/NSTEMI patients as soon as possible after hospital presentation and continued indefinitely in patients who tolerate it	A
	A loading dose followed by daily maintenance dose of either clopidogrel (LOE: B), prasugrel (in PCI-treated patients) (LOE: C), or ticagrelor (LOE: C) should be administered to UA/NSTEMI patients who are unable to take aspirin because of hypersensitivity or major gastrointestinal intolerance	
	Patients with definite UA/NSTEMI at medium or high risk and in whom an initial invasive strategy is selected should receive dual antiplatelet therapy on presentation (LOE: A). Aspirin should be initiated on presentation (LOE: A). The choice of a second antiplatelet therapy to be added to aspirin on presentation includes one of the following (note that there are no data for therapy with two concurrent P2Y$_{12}$ ADP-receptor inhibitors, and this is not recommended in the case of aspirin allergy): • Before PCI: (i) clopidogrel (LOE: B), or (ii) ticagrelor (LOE: B), or (iii) an intravenous (iv) GP IIb/IIIa inhibitor (LOE: A). Intravenous eptifibatide and tirofiban are the preferred GP IIb/IIIa inhibitors (LOB: B) • At the time of PCI: (i) clopidogrel if not started before PCI (LOE: A), or (ii) prasugrel (LOE: B), or (iii) ticagrelor (LOE: B), or (iv) an iv GP IIb/IIIa inhibitor (LOE: A)	
	For UA/NSTEMI patients in whom an initial conservative (i.e., noninvasive) strategy is selected, clopidogrel or ticagrelor (loading dose followed by daily maintenance dose) should be added to aspirin and anticoagulant therapy as soon as possible after admission and administered for up to 12 months	B
	For UA/NSTEMI patients in whom an initial conservative strategy is selected, if recurrent symptoms/ischemia, HF, or serious arrhythmias subsequently appear, then diagnostic angiography should be performed (LOE: A). Either an iv GP IIb/IIIa inhibitor (eptifibatide or tirofiban [LOE: A]), clopidogrel (loading dose followed by daily maintenance dose [LOE: B]), or ticagrelor (loading dose followed by daily maintenance dose [LOE: B]) should be added to aspirin and anticoagulant therapy before diagnostic angiography (upstream) (LOE: C)	
	A loading dose of P2Y$_{12}$ ADP-receptor inhibitor therapy is recommended for UA/NSTEMI patients for whom PCI is planned. One of the following regimens should be used: (i) clopidogrel 600 mg should be given as early as possible before or at the time of PCI (LOE: B); or (ii) prasugrel 60 mg should be given promptly and no later than 1 h after PCI once coronary anatomy is defined and a decision is made to proceed with PCI (LOE: B); or (iii) ticagrelor 180 mg should be given as early as possible before or at the time of PCI (LOE: B)	
	The duration and maintenance dose of P2Y$_{12}$ ADP-receptor inhibitor therapy should be as follows: (i) In UA/NSTEMI patients undergoing PCI, either clopidogrel 75 mg daily, prasugrel 10 mg daily, or ticagrelor 90 mg twice daily should be given for at least 12 months (LOE: B). (ii) If the risk of morbidity because of bleeding outweighs the anticipated benefits afforded by P2Y$_{12}$ ADP-receptor inhibitor therapy, earlier discontinuation should be considered (LOE: C)	

(Continued)

TABLE 18.3 (*Continued*)

COR	Recommendation	LOE
IIa	For UA/NSTEMI patients in whom an initial conservative strategy is selected and who have recurrent ischemic discomfort with aspirin, a P2Y$_{12}$ ADP-receptor inhibitor (clopidogrel or ticagrelor), and anticoagulant therapy, it is reasonable to add a GP IIb/IIIa inhibitor before diagnostic angiography	C
	For UA/NSTEMI patients in whom an initial invasive strategy is selected, it is reasonable to omit administration of an iv GP IIb/IIIa inhibitor if bivalirudin is selected as the anticoagulant and at least 300 mg of clopidogrel was administered at least 6 h earlier than planned catheterization or PCI	B
IIb	For UA/NSTEMI patients in whom an initial conservative (i.e., noninvasive) strategy is selected, it may be reasonable to add eptifibatide or tirofiban to anticoagulant and oral antiplatelet therapy	B
	Prasugrel 60 mg may be considered for administration promptly upon presentation in patients with UA/NSTEMI for whom PCI is planned, before definition of coronary anatomy if both the risk for bleeding is low and the need for CABG is considered unlikely	C
	The use of upstream GP IIb/IIIa inhibitors may be considered in high-risk UA/NSTEMI patients already receiving aspirin and a P2Y$_{12}$ ADP-receptor inhibitor (clopidogrel or ticagrelor) who are selected for an invasive strategy, such as those with elevated troponin levels, diabetes, or significant ST-segment depression, and who are not otherwise at high risk for bleeding	B
	In patients with definite UA/NSTEMI undergoing PCI as part of an early invasive strategy, the use of a loading dose of clopidogrel of 600 mg, followed by a higher maintenance dose of 150 mg daily for 6 days, and then 75 mg daily may be reasonable in patients not considered at high risk for bleeding	B
III (no benefit)	Abciximab should not be administered to patients in whom PCI is not planned	A
	In UA/NSTEMI patients who are at low risk for ischemic events (e.g., TIMI risk score[a] ≤2) or at high risk of bleeding and who are already receiving aspirin and a P2Y$_{12}$ ADP-receptor inhibitor, upstream GP IIb/IIIa inhibitors are not recommended	B
III (harm)	In UA/NSTEMI patients with a prior history of stroke and/or TIA for whom PCI is planned, prasugrel is potentially harmful as part of a dual antiplatelet therapy regimen	B

[a]TIMI risk score: In patients with UA/NSTEMI, the TIMI risk score is a simple prognostication scheme that categorizes a patient's risk of death and ischemic events and provides a basis for therapeutic decision making [20].

COR and LOE denote class of recommendation and level of evidence, respectively (see Chapter 5 for description).

TABLE 18.4 **The ESC 2011 guideline recommendations for antiplatelet agents [17]**

COR	Recommendation	LOE
I	Aspirin should be given to all patients without contraindications at an initial loading dose of 150–300 mg and at a maintenance dose of 75–100 mg daily long term regardless of treatment strategy	A
	A P2Y$_{12}$ ADP-receptor inhibitor should be added to aspirin as soon as possible and maintained over 12 months, unless there are contraindications, such as excessive risk of bleeding	A
	A proton pump inhibitor (preferably not omeprazole) in combination with dual antiplatelet therapy is recommended in patients with a history of gastrointestinal hemorrhage or peptic ulcer and appropriate for patients with multiple other risk factors (*Helicobacter pylori* infection; age ≥65 years; concurrent use of anticoagulants or steroids)	A
	Prolonged or permanent withdrawal of P2Y$_{12}$ ADP-receptor inhibitors within 12 months after the index event is discouraged unless clinically indicated	C
	Ticagrelor (180 mg loading dose, 90 mg twice daily) is recommended for all patients at moderate to high risk of ischemic events (e.g., elevated cardiac troponins), regardless of initial treatment strategy and including those pretreated with clopidogrel (which should be discontinued when ticagrelor is commenced)	B
	Prasugrel (60 mg loading dose, 10 mg daily dose) is recommended for P2Y$_{12}$ ADP-receptor inhibitor-naïve patients (especially diabetics) in whom coronary anatomy is known and who are proceeding to PCI unless there is a high risk of life-threatening bleeding or other contraindications	B
	Clopidogrel (300 mg loading dose, 75 mg daily dose) is recommended for patients who cannot receive ticagrelor or prasugrel	A
	A 600 mg loading dose of clopidogrel (or a supplementary 300 mg dose at PCI following an initial 300 mg loading dose) is recommended for patients scheduled for an invasive strategy when ticagrelor or prasugrel is not an option	B
	The choice of combination of oral antiplatelet agents, a GP IIb/IIIa receptor inhibitor, and anticoagulants should be made in relation to the risk of ischemic and bleeding events	C
	Among patients who are already treated with dual antiplatelet therapy, the addition of a GP IIb/IIIa receptor inhibitor for high-risk PCI (elevated cardiac troponins, visible thrombus) is recommended if the risk of bleeding is low	B

(*Continued*)

TABLE 18.4 (*Continued*)

COR	Recommendation	LOE
IIa	A higher maintenance dose of clopidogrel 150 mg daily should be considered for the first 7 days in patients managed with PCI and without an increased risk of bleeding	B
	In patients pretreated with $P2Y_{12}$ ADP-receptor inhibitors who need to undergo nonemergent major surgery (including CABG), postponing surgery at least for 5 days after cessation of ticagrelor or clopidogrel, and 7 days for prasugrel, if clinically feasible and unless the patient is at high risk of ischemic events should be considered	C
	Ticagrelor or clopidogrel should be considered to be (re)started after CABG surgery as soon as considered safe	B
	Eptifibatide or tirofiban added to aspirin should be considered prior to angiography in high-risk patients not preloaded with $P2Y_{12}$ ADP-receptor inhibitors	C
IIb	Increasing the maintenance dose of clopidogrel based on platelet function testing is not advised as routine but may be considered in selected cases	B
	Genotyping and/or platelet function testing may be considered in selected cases when clopidogrel is used	B
	In high-risk patients, eptifibatide or tirofiban may be considered prior to early angiography in addition to dual antiplatelet therapy, if there is ongoing ischemia and the risk of bleeding is low	C
III	The combination of aspirin with another NSAID (selective COX-2 inhibitors and nonselective NSAIDs) is not recommended	C
	GP IIb/IIIa receptor inhibitors are not recommended routinely before angiography in an invasive treatment strategy	A
	GP IIb/IIIa receptor inhibitors are not recommended for patients on dual antiplatelet therapy who are treated conservatively	A

COR and LOE denote class of recommendation and level of evidence, respectively (see Chapter 5 for description).

TABLE 18.5 The ACC/AHA 2007 guideline recommendations for anticoagulant therapy [16]

COR	Recommendation	LOE
I	Anticoagulant therapy should be added to antiplatelet therapy in UA/NSTEMI patients as soon as possible after presentation:	
	1. For patients in whom an invasive strategy is selected, regimens with established efficacy at a LOE of A include enoxaparin and unfractionated heparin (UFH), and those with established efficacy at a LOE of B include bivalirudin and fondaparinux	
	2. For patients in whom a conservative strategy is selected, regimens using either enoxaparin or UFH (LOE: A) or fondaparinux (LOE: B) have established efficacy	
	3. In patients in whom a conservative strategy is selected and who have an increased risk of bleeding, fondaparinux is preferable (LOE: B)	
IIa	For UA/NSTEMI patients in whom an initial conservative strategy is selected, enoxaparin or fondaparinux is preferable to UFH as anticoagulant therapy, unless CABG is planned within 24 h	B

COR and LOE denote class of recommendation and level of evidence, respectively (see Chapter 5 for description).

TABLE 18.6 The ESC 2011 guideline recommendations for anticoagulant therapy [17]

COR	Recommendations	LOE
I	Anticoagulation is recommended for all patients in addition to antiplatelet therapy	A
	The anticoagulation should be selected according to both ischemic and bleeding risks and according to the efficacy–safety profile of the chosen agent	C
	Fondaparinux (2.5 mg subcutaneously daily) is recommended as having the most favorable efficacy–safety profile with respect to anticoagulation	A
	If the initial anticoagulant is fondaparinux, a single bolus of UFH (85 IU/kg adapted to activated clotting time or 60 IU in the case of concomitant use of GP IIb/IIIa receptor inhibitors) should be added at the time of PCI	B
	Enoxaparin (1 mg/kg twice daily) is recommended when fondaparinux is not available	B
	If fondaparinux or enoxaparin is not available, UFH with a target activated partial thromboplastin time (aPTT) of 50–70 s or other low-molecular-weight heparins (LMWHs) at the specific recommended doses are indicated	C
	Bivalirudin plus provisional GP IIb/IIIa receptor inhibitors are recommended as an alternative to UFH plus GP IIb/IIIa receptor inhibitors in patients with an intended urgent or early invasive strategy, particularly in patients with a high risk of bleeding	B
	In a purely conservative strategy, anticoagulation should be maintained up to hospital discharge	A
IIa	Discontinuation of anticoagulation should be considered after an invasive procedure unless otherwise indicated	C
III	Crossover of heparins (UFH and LMWH) is not recommended	B

COR and LOE denote class of recommendation and level of evidence, respectively (see Chapter 5 for description).

recommendations. Notably, the class III (no benefit) recommendation states that intravenous fibrinolytic therapy (also known as thrombolytic therapy) is not indicated in patients without an acute ST-segment elevation, a true posterior MI, or a presumed new left bundle-branch block (LOE: A). Hence, fibrinolytic agents are contraindicated in patients with UA/NSTEMI. Summarized below are the 8 class I recommendations. The reader is suggested to refer to the full guideline update [19] for more details:

1. For UA/NSTEMI patients in whom an initial conservative strategy is selected and no subsequent features appear that would necessitate diagnostic angiography (recurrent symptoms/ischemia, heart failure, or serious arrhythmias), a stress test should be performed (LOE: B). If, after stress testing, the patient is classified as not at low risk, diagnostic angiography should be performed (LOE: A). If, after stress testing, the patient is classified as being at low risk, the instructions noted below should be followed in preparation for discharge: (i) continue aspirin indefinitely (LOE: A); (ii) continue clopidogrel or ticagrelor for up to 12 months (LOE: B); (iii) discontinue iv GP IIb/IIIa inhibitor if started previously (LOE: A); and (iv) continue UFH for 48 h (LOE: A) or administer enoxaparin (LOE: A) or fondaparinux (LOE: B) for the duration of hospitalization, up to 8 days, and then discontinue anticoagulant therapy.

2. For UA/NSTEMI patients in whom CABG is selected as a postangiography management strategy, the instructions noted below should be followed: (i) continue aspirin (LOE: A); (ii) see (3) below in this section; (iii) discontinue iv GP IIb/IIIa inhibitor (eptifibatide or tirofiban) 4 h before CABG (LOE: B); and (iv) anticoagulant therapy should be managed as follows—continue UFH (LOE: B), discontinue enoxaparin 12–24 h before CABG and dose with UFH per institutional practice (LOE: B), discontinue fondaparinux 24 h before CABG and dose with UFH per institutional practice (LOE: B), and discontinue bivalirudin 3 h before CABG and dose with UFH per institutional practice (LOE: B).

3. In patients taking a $P2Y_{12}$ ADP-receptor inhibitor in whom CABG is planned and can be delayed, it is recommended that the drug be discontinued to allow for dissipation of the antiplatelet effect (LOE: B). The period of withdrawal should be at least 5 days in patients receiving clopidogrel (LOE: B) or ticagrelor (LOE: C) and at least 7 days in patients receiving prasugrel (LOE: C) unless the need for revascularization and/or the net benefit of the $P2Y_{12}$ ADP-receptor inhibitor therapy outweighs the potential risks of excess bleeding (LOE: C).

4. For UA/NSTEMI patients in whom PCI has been selected as a postangiography management strategy, the instructions noted below should be followed: (i) continue aspirin (LOE: A); (ii) administer a loading dose of a $P2Y_{12}$ ADP-receptor inhibitor if not started before diagnostic angiography (LOE: A); and (iii) discontinue anticoagulant therapy after PCI for uncomplicated cases (LOE: B).

5. For UA/NSTEMI patients in whom medical therapy is selected as a management strategy and in whom no significant obstructive coronary artery disease (CAD) on angiography was found, antiplatelet and anticoagulant therapy should be administered at the discretion of the clinician (LOE: C). For patients in whom evidence of coronary atherosclerosis is present (e.g., luminal irregularities or intravascular ultrasound-demonstrated lesions), albeit without flow-limiting stenoses, long-term treatment with aspirin and other secondary prevention measures should be prescribed (LOE: C).

6. For UA/NSTEMI patients in whom medical therapy is selected as a management strategy and in whom CAD was found on angiography, the following approach is recommended: (i) continue aspirin (LOE: A); (ii) administer a loading dose of clopidogrel or ticagrelor if not given before diagnostic angiography (LOE: B); (iii) discontinue intravenous GP IIb/IIIa inhibitor if started previously (LOE: B); and (iv) anticoagulant therapy should be managed as follows—continue intravenous UFH for at least 48 h or until discharge if given before diagnostic angiography (LOE: A), continue enoxaparin for duration of hospitalization, up to 8 days, if given before diagnostic angiography (LOE: A), continue fondaparinux for duration of hospitalization, up to 8 days, if given before diagnostic angiography (LOE: B), and either discontinue bivalirudin or continue at a dose of 0.25 mg/kg/h for up to 72 h at the physician's discretion if given before diagnostic angiography (LOE: B).

7. For UA/NSTEMI patients in whom a conservative strategy is selected and who do not undergo angiography or stress testing, the instructions noted below should be followed: (i) continue aspirin indefinitely (LOE: A); (ii) continue clopidogrel or ticagrelor for up to 12 months (LOE: B); (iii) discontinue intravenous GP IIb/IIIa inhibitor if started previously (LOE: A); and (iv) continue UFH for 48 h (LOE: A) or administer enoxaparin (LOE: A) or fondaparinux (LOE: B) for the duration of hospitalization, up to 8 days, and then discontinue anticoagulant therapy.

8. For UA/NSTEMI patients in whom an initial conservative strategy is selected and in whom no subsequent features appear that would necessitate diagnostic angiography (recurrent symptoms/ischemia, heart failure, or serious arrhythmias), LVEF should be measured (LOE: B).

TABLE 18.7   The ACCF/AHA 2012 update recommendations for revascularization in patients with UA/NSTEMI [19]

COR	Recommendation	LOE
I	An early invasive strategy (i.e., diagnostic angiography with intent to perform revascularization) is indicated in UA/NSTEMI patients who have refractory angina or hemodynamic or electrical instability (without serious comorbidities or contraindications to such procedures)	B
	An early invasive strategy (i.e., diagnostic angiography with intent to perform revascularization) is indicated in initially stabilized UA/NSTEMI patients (without serious comorbidities or contraindications to such procedures) who have an elevated risk for clinical events	A
IIa	It is reasonable to choose an early invasive strategy (within 12–24 h of admission) over a delayed invasive strategy for initially stabilized high-risk patients with UA/NSTEMI. For patients not at high risk, a delayed invasive approach is also reasonable	B
	In initially stabilized patients, an initially conservative (i.e., a selectively invasive) strategy may be considered as a treatment strategy for UA/NSTEMI patients (without serious comorbidities or contraindications to such procedures) who have an elevated risk for clinical events, including those who are troponin positive (LOE: B). The decision to implement an initial conservative (vs. initial invasive) strategy in these patients may be made by considering physician and patient preference (LOE: C)	B, C
III (no benefit)	An early invasive strategy (i.e., diagnostic angiography with intent to perform revascularization) is not recommended in patients with extensive comorbidities (e.g., liver or pulmonary failure, cancer), in whom the risks of revascularization and comorbid conditions are likely to outweigh the benefits of revascularization	C
	An early invasive strategy (i.e., diagnostic angiography with intent to perform revascularization) is not recommended in patients with acute chest pain and a low likelihood of ACS	C
	An early invasive strategy (i.e., diagnostic angiography with intent to perform revascularization) should not be performed in patients who will not consent to revascularization regardless of the findings	C

COR and LOE denote class of recommendation and level of evidence, respectively (see Chapter 5 for description).

18.4.5   Coronary Revascularization

At present, most patients with UA/NSTEMI in the United States, and many patients elsewhere, undergo coronary arteriography and revascularization during the index hospitalization. Revascularization for UA/NSTEMI relieves symptoms, shortens hospital stay, and improves prognosis. The indications and timing for myocardial revascularization and choice of preferred approach (PCI or CABG) depend on many factors including the patient's condition, the presence of risk features, comorbidities, and the extent and severity of the lesions as identified by coronary angiography [17].

Risk stratification should be performed as early as possible to identify high-risk individuals rapidly and reduce the delay to an early invasive approach. However, patients with UA/NSTEMI represent a heterogeneous population in terms of risk and prognosis. This extends from low-risk patients who benefit from conservative treatment and a selective invasive approach to patients at high risk for death and cardiovascular events who should be rapidly referred for angiography and revascularization. Therefore, risk stratification is critical for selection of the optimal management strategy. Analysis of the patient risk profile may be performed by assessment of generally accepted high-risk criteria and/or applying predefined risk scores such as the GRACE risk score or TIMI risk score [17].

The major recommendations for coronary vascularization in patients with UA/NSTEMI are summarized in Tables 18.7 and 18.8.

18.4.6   Recommendations for Special Patient Groups

The previous sections have considered the management of UA/NSTEMI in the general patient populations. Patients with UA/NSTEMI are frequently also inflicted with other disorders. For example, ~20–30% of patients with UA/NSTEMI have known diabetes, and at least as many have undiagnosed diabetes or impaired glucose tolerance. Hence, management of UA/NSTEMI in patients with concomitant diseases should also consider the appropriate treatment of the comorbidities. In addition, other factors, including age, gender, and extreme body weight, should also be considered for the proper management of UA/NSTEMI. This section discusses the current guideline recommendations in special patient groups, including the elderly and those with diabetes, chronic kidney disease (CKD), or heart failure.

18.4.6.1   The Elderly

General Considerations   Older adults represent a group of patients in whom baseline risk is higher and who have more comorbidities but who derive equivalent or greater benefit (e.g., invasive vs. conservative strategy) compared to younger patients. Although a precise definition of "older patients" or "elderly" has not been established in the medical literature, many studies have used this term to refer to those who are 75 years and older. On the basis of a large national ACS registry, older patients make up a substantial portion of those presenting with UA/NSTEMI, with 35% older than 75

TABLE 18.8   The ESC 2011 guideline recommendations for revascularization in patients with UA/NSTEMI [17]

COR	Recommendation	LOE
I	An invasive strategy (within 72 h after first presentation) is indicated in patients with (i) at least one high-risk criterion[a] and/or (ii) recurrent symptoms	A
	Urgent coronary angiography (<2 h after first presentation) is recommended in patients at very high ischemic risk (refractory angina, with associated heart failure, life-threatening ventricular arrhythmias, or hemodynamic instability)	C
	An early invasive strategy (<24 h after first presentation) is recommended in patients with a GRACE score[b] >140 or with at least one primary high-risk criterion	A
	Noninvasive documentation of inducible ischemia is recommended in low-risk patients without recurrent symptoms before deciding for invasive evaluation	A
	The revascularization strategy (ad hoc culprit lesion PCI/multivessel PCI/CABG) should be based on the clinical status as well as the disease severity, that is, distribution and angiographic lesion characteristics (e.g., SYNTAX score[c]), according to the local "Heart Team" protocol	C
	As there are no safety concerns related to the use of drug-eluting stents (DESs) in ACS, such stents are indicated based on an individual basis taking into account baseline characteristics, coronary anatomy, and bleeding risk	A
III	PCI of nonsignificant lesions is not recommended	C
	Routine invasive evaluation of low-risk patients is not recommended	A

[a]The criteria for high risk regarding indication for coronary revascularization include relevant rise or fall in cardiac troponins, dynamic ST- or T-wave changes (symptomatic or silent), diabetes, renal insufficiency (estimated glomerular filtration rate <60 ml/min/1.73 m^2), reduced left ventricular function (ejection fraction <40%), early postinfarction angina, recent PCI, prior CABG, and intermediate to high Global Registry of Acute Coronary Events (GRACE) risk score.
[b]GRACE risk score: The GRACE risk score, along with the TIMI risk score, is widely used for the assessment of the risk of death among patients with ACS [21].
[c]SYNTAX score: See note to Table 15.6 for description.
COR and LOE denote class of recommendation and level of evidence, respectively (see Chapter 5 for description).

years and 11% aged more than 85 years [2]. Hence, management of AU/NSTEMI in the elderly represents a significant issue in cardiovascular medicine.

Characteristics of the Elderly with UA/NSTEMI   Older persons also present with a number of special and complex challenges with regard to the management of UA/NSTEMI [2]. These include the following:

1. Older persons who develop UA/NSTEMI are more likely to present with atypical symptoms, including dyspnea and confusion, rather than with the chest pain typically experienced by younger patients with acute myocardial ischemia. Conversely, noncardiac comorbidities, such as chronic obstructive pulmonary disease, gastroesophageal reflux disease, upper-body musculoskeletal symptoms, pulmonary embolism, and pneumonia, also are more frequent and may be associated with chest pain at rest that can mimic classic symptoms of UA/NSTEMI. Hence, successful recognition of true myocardial ischemia in the elderly is often more difficult than in younger patients.

2. Older patients are more likely than younger patients to have altered or abnormal cardiovascular anatomy and physiology, including a diminished β-sympathetic response, increased cardiac afterload due to decreased arterial compliance and arterial hypertension, orthostatic hypotension, cardiac hypertrophy, and ventricular dysfunction, especially diastolic dysfunction.

3. Older patients typically have developed significant cardiac comorbidities and risk factors, such as hypertension, prior MI, heart failure, cardiac conduction abnormalities, prior CABG, peripheral and cerebrovascular disease, diabetes, renal insufficiency, and stroke.

4. Because of the larger burden of concomitant diseases, older patients tend to be treated with a greater number of medications and are at higher risk for drug interactions. Hence, among an already high-risk population, older age is associated with higher disease severity and higher disease and treatment risk at presentation.

5. Due to altered drug metabolism and other physiological changes, older patients are at higher risk of drug toxicity. Older persons are particularly vulnerable to adverse events from cardiovascular drugs due to altered drug metabolism and distribution, as well as to exaggerated drug effects. This is particularly true for the risk of bleeding with antiplatelet agents and anticoagulants but also for hypotension, bradycardia, and renal failure. Reductions in cardiac output and in renal and hepatic perfusion and function decrease the rate of elimination of drugs in the elderly. Additionally, older patients typically have lower drug distribution volumes (due to a lower body mass). As a result, drugs need to be carefully selected and individually adjusted. Current evidence demonstrates that older adults are frequently excessively dosed. Hence, precautions need

TABLE 18.9   The ACC/AHA 2007 guideline recommendations for management of UA/NSTEMI in older adults [2]

COR	Recommendation	LOE
I	Older patients with UA/NSTEMI should be evaluated for appropriate acute and long-term therapeutic interventions in a similar manner as younger patients with UA/NSTEMI	A
I	Decisions on management of older patients with UA/NSTEMI should not be based solely on chronologic age, but should be patient centered, with consideration given to general health, functional and cognitive status, comorbidities, life expectancy, and patient preferences and goals	B
I	Attention should be given to appropriate dosing (i.e., adjusted by weight and estimated creatinine clearance) of pharmacological agents in older patients with UA/NSTEMI, because they often have altered pharmacokinetics (due to reduced muscle mass, renal and/or hepatic dysfunction, and reduced volume of distribution) and pharmacodynamics (increased risks of hypotension and bleeding)	B
I	Older UA/NSTEMI patients face increased early procedural risks with revascularization relative to younger patients, yet the overall benefits from invasive strategies are equal to or perhaps greater in older adults and are recommended	B
I	Consideration should be given to patient and family preferences, quality of life issues, end-of-life preferences, and sociocultural differences in older patients with UA/NSTEMI	C

COR and LOE denote class of recommendation and level of evidence, respectively (see Chapter 5 for description).

TABLE 18.10   The ESC 2011 guideline recommendations for management of UA/NSTEMI in older adults [17]

COR	Recommendation	LOE
I	Because of the frequent atypical presentation, elderly patients (>75 years) should be investigated for UA/NSTEMI at low level of suspicion	C
	Treatment decisions in the elderly (>75 years) should be made in the context of estimated life expectancy, comorbidities, quality of life, and patient wishes and preferences	C
	Choice and dosage of antithrombotic drugs should be tailored in elderly patients to prevent the occurrence of adverse effects	C
IIa	Elderly patients should be considered for an early invasive strategy with the option of possible revascularization, after careful weighing up of the risks and benefits	B

COR and LOE denote class of recommendation and level of evidence, respectively (see Chapter 5 for description).

to be taken to personalize these therapies (i.e., beginning with lower doses than in younger patients, whenever appropriate, and providing careful observation for adverse effects) [2].

Recommendations   Both the ACC/AHA and the ESC guidelines provide recommendations and detailed rationales for particular management of UA/NSTEMI in the elderly. Some of the major recommendations are provided in Tables 18.9 and 18.10.

18.4.6.2   Diabetes

General Considerations   Diabetes occurs perhaps in more than 50% of patients with UA/NSTEMI and is an independent predictor of adverse outcomes. Patients with diabetes have a twofold higher risk of death. In addition, patients with impaired glucose tolerance or impaired fasting blood glucose have a worse prognosis than patients with normal glucose metabolism but a better prognosis than patients with confirmed diabetes.

A consensus statement by the American Association of Clinical Endocrinologists and the American Diabetes Association summarized that "although hyperglycemia is associated with adverse outcomes after acute MI, reduction of glycemia per se and not necessarily the use of insulin is associated with improved outcomes. It remains unclear, however, whether hyperglycemia is a marker of underlying health status or is a mediator of complications after acute MI. Noniatrogenic hypoglycemia has also been associated with adverse outcomes and is a predictor of higher mortality" [22].

Diabetes is associated with more extensive CAD, unstable lesions, frequent comorbidities, and less favorable long-term outcomes with coronary revascularization, especially with percutaneous transluminal coronary angioplasty (PTCA). Given the diffuse nature of diabetic coronary disease, the relative benefits of CABG over PCI may well persist for diabetic patients, even in the era of drug-eluting stents.

Although the benefit of appropriate glycemic control in ACS has been widely recognized, the efficacy of intensified insulin-based glycemic control in ACS remains controversial [23]. For example, a recent randomized trial reported that intensive glucose management strategy, aiming at a plasma glucose level of 85–110 mg/dl by using intravenous insulin, did not reduce infarct size in hyperglycemic patients

TABLE 18.11   The ACCF/AHA 2012 update recommendations for management of UA/NSTEMI in diabetics [19]

COR	Recommendation	LOE
I	Medical treatment in the acute phase of UA/NSTEMI and decisions on whether to perform stress testing, angiography, and revascularization should be similar in patients with and without diabetes	A
IIa	For patients with UA/NSTEMI and multivessel disease, CABG with use of the internal mammary arteries can be beneficial over PCI in patients being treated for diabetes	B
	PCI is reasonable for UA/NSTEMI patients with diabetes with single-vessel disease and inducible ischemia	B
	It is reasonable to use an insulin-based regimen to achieve and maintain glucose levels less than 180 mg/dl while avoiding hypoglycemia for hospitalized patients with UA/NSTEMI with either a complicated or an uncomplicated course	B

COR and LOE denote class of recommendation and level of evidence, respectively (see Chapter 5 for description).

TABLE 18.12   The ESC 2011 guideline recommendations for management of UA/NSTEMI in diabetics [17]

COR	Recommendation	LOE
I	All patients with UA/NSTEMI should be screened for diabetes. Blood glucose levels should be monitored frequently in patients with known diabetes or admission hyperglycemia	C
	Treatment of elevated blood glucose should avoid both excessive hyperglycemia (10–11 mM or 180–200 mg/dl) and hypoglycemia (<5 mM or <90 mg/dl)	B
	Antithrombotic treatment is indicated as in nondiabetic patients	C
	Renal function should be closely monitored following contrast exposure	C
	An early invasive strategy is recommended	A
	Drug-eluting stents are recommended to reduce rates of repeat revascularization	A
	CABG surgery should be favored over PCI in diabetic patients with main stem lesions and/or advanced multivessel disease	B

COR and LOE denote class of recommendation and level of evidence, respectively (see Chapter 5 for description).

with ACS treated with PCI and was associated with harm [24]. On the other hand, a long-term (20 years) follow-up trial demonstrated that intensified insulin-based glycemic control after acute MI in patients with diabetes and hyperglycemia at admission had a long-lasting beneficial effect on longevity [25]. Although the effect of glucose lowering might be less apparent with presently available, more effective lipid-lowering and blood pressure-lowering drugs, improved glycemic control might still be important for longevity after acute MI [25].

Recommendations   The guideline recommendations for management of UA/NSTEMI in diabetic patients are summarized in Tables 18.11 and 18.12.

18.4.6.3   CKD

General Considerations   Renal dysfunction is present in 30–40% of patients with UA/NSTEMI. Patients with CKD more frequently present with heart failure and without typical chest pain. Patients with UA/NSTEMI and CKD often do not receive guideline-recommended therapy. CKD constitutes a major risk factor for adverse outcomes after MI, including NSTEMI and other coronary patient subsets, and is an independent predictor of short- and long-term mortality and of major bleeding in patients with UA/NSTEMI [17]. In addition to higher risk of antithrombotic

drug-associated bleeding in CKD, angiography also carries an increased risk of contrast-induced nephropathy, and the usual benefits of percutaneous interventions can be lessened or abolished. Indeed, PCI in patients with CKD is associated with a higher rate of early and late complications of bleeding, restenosis, and death [2]. Hence, the identification of CKD is important in that it represents an ACS subgroup with a far more adverse prognosis but for whom interventions have less certain benefits.

Recommendations   To guide the proper management of UA/NSTEMI in patients with CKD, both the ACCF/AHA and the ESC have published guideline-based recommendations (Tables 18.13 and 18.14). The ESC also has provided recommendations specific for the use of antithrombotic drugs in UA/NSTEMI patients with CKD (Table 18.15).

18.4.6.4   Heart Failure   Heart failure is one of the most frequent and deadly complications of ACS including UA/NSTEMI. Both left ventricular dysfunction and heart failure are independent predictors of mortality and other major adverse cardiac events in UA/NSTEMI. Patients with UA/NSTEMI and heart failure less frequently receive evidence-based therapies, including β-blockers and ACEIs, or ARBs, coronary angiography, and revascularization [17]. In its 2011 guideline, the ESC provides recommendations for management of UA/NSTEMI in patients with heart failure

TABLE 18.13   The ACCF/AHA 2012 update recommendations for management of UA/NSTEMI in patients with CKD [19]

COR	Recommendation	LOE
I	Creatinine clearance should be estimated in UA/NSTEMI patients and the doses of renally cleared medications should be adjusted according to the pharmacokinetic data for specific medications	B
	Patients undergoing cardiac catheterization with receipt of contrast media should receive adequate preparatory hydration	B
	Calculation of the contrast volume to creatinine clearance ratio is useful to predict the maximum volume of contrast media that can be given without significantly increasing the risk of contrast-associated nephropathy	B
IIa	An invasive strategy is reasonable in patients with mild (stage 2) and moderate (stage 3) CKD. There are insufficient data on benefit–risk of invasive strategy in UA/NSTEMI patients with advanced CKD (stages 4 and 5)	B

COR and LOE denote class of recommendation and level of evidence, respectively (see Chapter 5 for description).

TABLE 18.14   The ESC 2011 guideline recommendations for management of UA/NSTEMI in patients with CKD [17]

COR	Recommendation	LOE
I	Kidney function should be assessed by creatinine clearance (CrCl) or estimated glomerular filtration rate (eGFR) in patients with UA/NSTEMI, with special attention to elderly people, women, and patients with low body weight, as near-normal serum creatinine levels may be associated with lower than expected CrCl and eGFR levels	C
	Patients with UA/NSTEMI and CKD should receive the same first-line antithrombotic treatment as patients devoid of CKD, with appropriate dose adjustments according to the severity of renal dysfunction	B
	Depending on the degree of renal dysfunction, dose adjustment or switch to UFH with fondaparinux, enoxaparin, and bivalirudin and dose adjustment with small molecule GP IIb/IIIa receptor inhibitors are indicated	B
	UFH infusion adjusted to aPTT is recommended when CrCl is <30 ml/min or eGFR is <30 ml/min/1.73 m^2 with most anticoagulants (fondaparinux <20 ml/min)	C
	In patients with UA/NSTEMI and CKD considered for invasive strategy, hydration and low- or iso-osmolar contrast medium at low volume (<4 ml/kg) are recommended	B
	CABG or PCI is recommended in patients with CKD amenable to revascularization after careful assessment of the risk–benefit ratio in relation to the severity of renal dysfunction	B

COR and LOE denote class of recommendation and level of evidence, respectively (see Chapter 5 for description).

TABLE 18.15   The ESC 2011 guideline recommendations for use of antithrombotic drugs in UA/NSTEMI patients with CKD [17]

Antithrombotic	Recommendation
Clopidogrel	No information in patients with renal dysfunction
Prasugrel	No dose adjustment necessary, including in patients with end-stage disease
Ticagrelor	No dose reduction required; no information in dialysis patients
Enoxaparin	Dose reduction to 1 mg/kg once daily in the case of severe renal failure (CrCl <30 ml/min). Consider monitoring of anti-factor Xa activity
Fondaparinux	Contraindicated in severe renal failure (CrCl <20 ml/min); drug of choice in patients with moderately reduced renal function (CrCl 30–60 ml/min)
Bivalirudin	Patients with moderate renal impairment (30–59 ml/min) should receive an infusion of 1.75 mg/kg/h. If the creatinine clearance is <30 ml/min, reduction of the infusion rate to 1 mg/kg/h should be considered. No reduction in the bolus dose is needed. If a patient is on hemodialysis, the infusion rate should be reduced to 0.25 mg/kg/h
Abciximab	No specific recommendations for the use of abciximab or for dose adjustment in the case of renal failure. Careful evaluation of hemorrhagic risk is needed before using the drug in the case of renal failure
Eptifibatide	The infusion dose should be reduced to 1 µg (mcg)/kg/min in patients with CrCl <50 ml/min. The dose of the bolus remains unchanged at 180 µg (mcg)/kg. Eptifibatide is contraindicated in patients with CrCl <30 ml/min
Tirofiban	Dose adaptation is required in patients with renal failure; 50% of the bolus dose and infusion if CrCl is <30 ml/min

COR and LOE denote class of recommendation and level of evidence, respectively (see Chapter 5 for description).

TABLE 18.16   The ESC 2011 guideline recommendations for management of UA/NSTEMI in patients with heart failure [17]

COR	Recommendation	LOE
I	β-Blockers and ACEIs/ARBs appropriately titrated are indicated in patients with UA/NSTEMI and left ventricular dysfunction with or without signs of heart failure	A
	Aldosterone receptor antagonists, preferably eplerenone, are indicated in patients with UA/NSTEMI, left ventricular dysfunction, and heart failure	A
	Patients with UA/NSTEMI and LV dysfunction or heart failure are recommended to undergo coronary revascularization, if amenable to it	A
IIa	Patients with UA/NSTEMI and severe LV dysfunction should be considered after 1 month for device therapy (cardiac resynchronization therapy and/or implantable cardioverter defibrillator) in addition to optimal medical therapy whenever indicated	B

COR and LOE denote class of recommendation and level of evidence, respectively (see Chapter 5 for description).

TABLE 18.17   The ACCF/AHA 2012 update recommendations for convalescent and long-term antiplatelet therapy [19]

COR	Recommendation	LOE
I	For UA/NSTEMI patients treated medically without stenting, aspirin should be prescribed indefinitely (LOE: A); clopidogrel (75 mg/day) or ticagrelor (90 mg twice daily) should be prescribed for up to 12 months (LOE: B)	
	For UA/NSTEMI patients treated with a stent (bare-metal stent [BMS] or DES), aspirin should be continued indefinitely (LOE: A). The duration and maintenance dose of $P2Y_{12}$ ADP-receptor inhibitor therapy should be as follows: (i) Clopidogrel 75 mg daily, prasugrel 10 mg daily, or ticagrelor 90 mg twice daily should be given for at least 12 months in patients receiving DES and up to 12 months for patients receiving BMS (LOE: B). (ii) If the risk of morbidity, because of bleeding, outweighs the anticipated benefits afforded by $P2Y_{12}$ ADP-receptor inhibitor therapy, earlier discontinuation should be considered (LOE: C)	
	Clopidogrel 75 mg daily (LOE: B), prasugrel 10 mg daily (in PCI-treated patients) (LOE: C), or ticagrelor 90 mg twice daily (LOE: C) should be given to patients recovering from UA/NSTEMI when aspirin is contraindicated or not tolerated because of hypersensitivity or gastrointestinal intolerance (despite use of gastroprotective agents, such as proton pump inhibitors)	
IIa	After PCI, it is reasonable to use 81 mg/day of aspirin in preference to higher maintenance doses	B
IIb	For UA/NSTEMI patients who have an indication for anticoagulation, the addition of warfarin may be reasonable to maintain an international normalized ratio (INR) of 2.0–3.0	B
	Continuation of a $P2Y_{12}$ ADP-receptor inhibitor beyond 12 months may be considered in patients following DES placement	C
III (no benefit)	Dipyridamole is not recommended as an antiplatelet agent in post-UA/NSTEMI patients because it has not been shown to be effective	B

COR and LOE denote class of recommendation and level of evidence, respectively (see Chapter 5 for description).

[17], and the key points are summarized in Table 18.16. However, this topic is not addressed in the ACC/AHA 2007 guideline or its 2011 and 2012 updates.

18.4.7   Long-Term Management

18.4.7.1   General Considerations   The acute phase of UA/ NSTEMI is usually over within 2 months. The risk of progression to MI or the development of recurrent MI or death is highest during that period. At 1–3 months after the acute phase, most patients resume a clinical course similar to that in patients with chronic stable ischemic heart disease (SIHD).

The broad goals during the hospital discharge phase are twofold: (i) to prepare the patient for normal activities to the extent possible and (ii) to use the acute event as an opportunity to reevaluate the plan of care, particularly lifestyle and risk factor modifications. Aggressive risk factor modifications that can prolong survival should be the main goals of long-term management of SIHD (see Chapter 15). Patients who have undergone successful PCI with an uncomplicated course are usually discharged the next day, and patients who undergo uncomplicated CABG are generally discharged 4–7 days after CABG.

Medical management of low-risk patients after noninvasive stress testing and coronary angiography can typically be accomplished rapidly, with discharge soon after testing. Medical management of a high-risk group of patients who are unsuitable for or unwilling to undergo revascularization could require vigilant inpatient monitoring in order to achieve adequate ischemic symptom control with medical therapy that will minimize future morbidity and mortality and improve quality of life [19].

This section focuses on discussing the general principles and current guideline-based recommendations regarding long-term pharmacological therapy and secondary prevention in patients following the acute phase of UA/NSTEMI. The reader may refer to the ACCF/AHA 2007 guideline and its 2011 and 2012 updates for recommendations on other aspects of the long-term management, including postdischarge follow-up and cardiac rehabilitation, among others [19].

18.4.7.2 Long-Term Pharmacological Therapy and Secondary Prevention Secondary prevention is of paramount importance since ischemic events continue to accrue at a high rate after the acute phase. In a database of 16,321 ACS patients, 20% of all patients were rehospitalized and 18% of the men and 23% of the women >40 years of age died during the first year following the acute ischemic event [17]. In this context, secondary prevention has a major impact on long-term outcome in patients who have survived the acute phase. It is worth repeating here that all measures and treatments with proven efficacy in secondary prevention should be implemented. These include lifestyle modifications, control of risk factors, and prescription of the drug classes with proven efficacy, such as antiplatelet agents (aspirin, $P2Y_{12}$ ADP-receptor inhibitors), β-blockers, statins, ACE inhibitors or ARBs, and aldosterone receptor antagonists. Since ACS patients without release of cardiac biomarkers (i.e., patients with UA) are less likely to receive guideline-oriented pharmacological secondary prevention as compared with NSTEMI patients, it should be stressed that all ACS patients do benefit from comprehensive secondary prevention, as outlined earlier.

Guidelines on secondary prevention of cardiovascular diseases, as covered elsewhere in the book (Chapters 5, 12, and 15), are of relevance, and the reader may refer to the aforementioned chapters for detailed information. This section considers only the ACCF/AHA 2012 update recommendations on long-term antiplatelet therapy in patients who have survived the acute phase of UA/NSTEMI (Table 18.17).

18.5 SUMMARY OF CHAPTER KEY POINTS

- UA, NSTEMI, and STEMI are the three presentations of ACS. The first step in the management of patients with ACS is prompt recognition and diagnosis of the syndromes.
- Once the diagnosis of either UA or an acute NSTEMI is made, the management of the patient involves the simultaneous achievement of several goals, including relief of ischemic pain, risk stratification, treatment to minimize myocardial injury and reduce mortality, and long-term care to prolong survival.

- Multiple guidelines have been developed to provide evidence-based recommendations on the management of UA/NSTEMI. Among them, the ACC/AHA 2007 guideline with focused updates in 2011 and 2012 and the ESC 2011 guideline are extensively cited and widely adopted.
- Guideline-based management of patients with UA/NSTEMI involves the use of four therapeutic modalities, including anti-ischemic and analgesic therapy, antiplatelet therapy, anticoagulant therapy, and coronary revascularization.
- Anti-ischemic drugs either decrease myocardial oxygen demand or increase myocardial oxygen supply so as to improve myocardial oxygen imbalance. Anti-ischemic agents for UA/NSTEMI typically include nitrates, β-blockers, CCBs, and ranolazine.
- Antiplatelet therapy is essential to modify the disease process and its progression to death, MI, or recurrent MI in the majority of patients who have ACS due to thrombosis on a ruptured plaque.
- Antiplatelet therapy should be instituted as early as possible when the diagnosis of UA/NSTEMI is made in order to reduce the risk of both acute ischemic complications and recurrent atherothrombotic events.
- Anticoagulants are used to inhibit thrombin generation and/or activity, thereby reducing thrombus-related events. The combination of anticoagulation and antiplatelet therapy is more effective than either treatment alone.
- Revascularization for UA/NSTEMI relieves symptoms, shortens hospital stay, and improves prognosis. The indications and timing for myocardial revascularization and choice of preferred approach (PCI or CABG) depend on the patient's condition, the presence of risk features and comorbidities, and the extent and severity of the lesions as identified by coronary angiography.
- The noninvasive intravenous fibrinolytic therapy is not indicated in patients without an acute ST-segment elevation, a true posterior MI, or a presumed new left bundle-branch block. Hence, fibrinolytic agents are contraindicated in patients with UA/NSTEMI.
- Management of UA/NSTEMI in patients with concomitant diseases (e.g., diabetes, CKD, heart failure) should also consider the appropriate treatment of the comorbidities. In addition, other factors, including age, gender, and extreme body weight, should also be considered for the proper management of UA/NSTEMI.
- Long-term management of SIHD should focus on aggressive risk factor modifications to prolong survival. Secondary prevention is of paramount importance since ischemic events continue to accrue at a high rate after the acute phase. All measures and treatments with

proven efficacy in secondary prevention should be implemented. These include lifestyle modifications, control of risk factors, and prescription of the drug classes with proven efficacy, such as antiplatelet agents (aspirin, $P2Y_{12}$ ADP-receptor inhibitors), β-blockers, statins, ACEIs or ARBs, and aldosterone receptor antagonists.

18.6 SELF-ASSESSMENT QUESTIONS

18.6.1 A 60-year-old male presents with unstable angina. Which of the following drugs is most likely given to the patient to reduce myocardial oxygen demand and decrease the risk of cardiac death?
  A. Carvedilol
  B. Digoxin
  C. Dobutamine
  D. Prazosin
  E. Verapamil

18.6.2 A 65-year-old female is brought to the emergency department because of severe chest pain. ECG and blood chemistry reveal that she is having a non-ST-elevation myocardial infarction. Her blood pressure is 135/75 mm Hg and heart rate 80 beats/min. Which of the following should be initiated within the first 24 h?
  A. Intravenous dobutamine
  B. Intravenous enalapril
  C. Intravenous nicardipine
  D. Oral ibuprofen
  E. Oral metoprolol

18.6.3 A 54-year-old patient presents to the emergency room with severe "heartburn," anxiety, and sweating. Immediate ECG suggests an acute coronary syndrome. Which of the following should be administered as soon as possible and continued indefinitely if the patient is diagnosed with an ACS?
  A. Amlodipine
  B. Aspirin
  C. Clopidogrel
  D. Nitroglycerin
  E. Warfarin

18.6.4 A 60-year-old patient presents to the emergency department due to shortness of breath and chest tightness. ECG and blood chemistry show that he is having a non-ST-elevation acute coronary syndrome. History reveals that he had a stroke 2 years ago. In designing a treatment regimen for this patient who is scheduled to undergo percutaneous coronary intervention, which of the following drugs should be avoided?
  A. Aspirin
  B. Clopidogrel
  C. Enoxaparin
  D. Eptifibatide
  E. Prasugrel

18.6.5 A 46-year-old male with non-ST-elevation myocardial infarction has undergone a successful percutaneous coronary intervention with stent placement. The patient is put on dual antiplatelet therapy with aspirin (81 mg daily) and clopidogrel (75 mg daily). How long should clopidogrel be prescribed in the absence of contraindications?
  A. 1 month
  B. 2 months
  C. 3 months
  D. 6 months
  E. 12 months

REFERENCES

1 Anderson, J.L., et al., ACC/AHA 2007 guidelines for the management of patients with unstable angina/non-ST-elevation myocardial infarction: a report of the American College of Cardiology/American Heart Association Task Force on Practice Guidelines (Writing Committee to Revise the 2002 Guidelines for the Management of Patients With Unstable Angina/Non-ST-Elevation Myocardial Infarction) developed in collaboration with the American College of Emergency Physicians, the Society for Cardiovascular Angiography and Interventions, and the Society of Thoracic Surgeons endorsed by the American Association of Cardiovascular and Pulmonary Rehabilitation and the Society for Academic Emergency Medicine. *J Am Coll Cardiol*, 2007. 50(7): p. e1–157.

2 Anderson, J.L., et al., ACC/AHA 2007 guidelines for the management of patients with unstable angina/non ST-elevation myocardial infarction: a report of the American College of Cardiology/American Heart Association Task Force on Practice Guidelines (Writing Committee to Revise the 2002 Guidelines for the Management of Patients With Unstable Angina/Non ST-Elevation Myocardial Infarction): developed in collaboration with the American College of Emergency Physicians, the Society for Cardiovascular Angiography and Interventions, and the Society of Thoracic Surgeons: endorsed by the American Association of Cardiovascular and Pulmonary Rehabilitation and the Society for Academic Emergency Medicine. *Circulation*, 2007. 116(7): p. e148–304.

3 Wright, R.S., et al., 2011 ACCF/AHA focused update of the guidelines for the management of patients with unstable angina/non-ST-elevation myocardial infarction (updating the 2007 guideline): a report of the American College of Cardiology Foundation/American Heart Association Task Force on Practice Guidelines. *Circulation*, 2011. 123(18): p. 2022–60.

4 Writing Committee Members, et al., 2012 ACCF/AHA focused update of the guideline for the management of patients with unstable angina/non-ST-elevation myocardial infarction (updating the 2007 guideline and replacing the 2011 focused update): a report of the American College of Cardiology Foundation/American

Heart Association Task Force on practice guidelines. *Circulation*, 2012. 126(7): p. 875–910.

5   Hamm, C.W., et al., ESC guidelines for the management of acute coronary syndromes in patients presenting without persistent ST-segment elevation: The Task Force for the management of acute coronary syndromes (ACS) in patients presenting without persistent ST-segment elevation of the European Society of Cardiology (ESC). *Eur Heart J*, 2011. 32(23): p. 2999–3054.

6   Bassand, J.P., et al., Guidelines for the diagnosis and treatment of non-ST-segment elevation acute coronary syndromes. *Eur Heart J*, 2007. 28(13): p. 1598–660.

7   Thomas, D. and R.P. Giugliano, Management of non-ST-segment elevation acute coronary syndrome: comparison of the updated guidelines from North America and Europe. *Crit Pathw Cardiol*, 2012. 11(2): p. 62–73.

8   Crowe, E., et al., Early management of unstable angina and non-ST segment elevation myocardial infarction: summary of NICE guidance. *BMJ*, 2010. 340: p. c1134.

9   Fitchett, D.H., et al., Assessment and management of acute coronary syndromes (ACS): a Canadian perspective on current guideline-recommended treatment—part 1: non-ST-segment elevation ACS. *Can J Cardiol*, 2011. 27 Suppl A: p. S387–401.

10 Chew, D.P., et al., 2011 Addendum to the National Heart Foundation of Australia/Cardiac Society of Australia and New Zealand Guidelines for the management of acute coronary syndromes (ACS) 2006. *Heart Lung Circ*, 2011. 20(8): p. 487–502.

11 Moradkhan, R. and L.I. Sinoway, Revisiting the role of oxygen therapy in cardiac patients. *J Am Coll Cardiol*, 2010. 56(13): p. 1013–6.

12 Bennett, M.H., J.P. Lehm, and N. Jepson, Hyperbaric oxygen therapy for acute coronary syndrome. *Cochrane Database Syst Rev*, 2011(8): p. CD004818.

13 Kones, R., Oxygen therapy for acute myocardial infarction-then and now. A century of uncertainty. *Am J Med*, 2011. 124(11): p. 1000–5.

14 Stub, D., et al., A randomized controlled trial of oxygen therapy in acute myocardial infarction Air Verses Oxygen In myocarDial infarction study (AVOID Study). *Am Heart J*, 2012. 163(3): p. 339–345 e1.

15 Hofmann, R., et al., DETermination of the role of OXygen in suspected Acute Myocardial Infarction trial. *Am Heart J*, 2014. 167(3): p. 322–8.

16 Anderson, J.L., et al., ACC/AHA 2007 guidelines for the management of patients with unstable angina/non-ST-elevation myocardial infarction: a report of the American College of Cardiology/American Heart Association Task Force on Practice Guidelines (Writing Committee to Revise the 2002 Guidelines for the Management of Patients With Unstable Angina/Non-ST-Elevation Myocardial Infarction) developed

in collaboration with the American College of Emergency Physicians, the Society for Cardiovascular Angiography and Interventions, and the Society of Thoracic Surgeons endorsed by the American Association of Cardiovascular and Pulmonary Rehabilitation and the Society for Academic Emergency Medicine. *J Am Coll Cardiol*, 2007. 50(7): p. e1–157.

17 Hamm, C.W., et al., ESC guidelines for the management of acute coronary syndromes in patients presenting without persistent ST-segment elevation: The Task Force for the management of acute coronary syndromes (ACS) in patients presenting without persistent ST-segment elevation of the European Society of Cardiology (ESC). *Eur Heart J*, 2011. 32(23): p. 2999–3054.

18 Jneid, H., et al., 2012 ACCF/AHA focused update of the guideline for the management of patients with unstable angina/non-ST-elevation myocardial infarction (updating the 2007 guideline and replacing the 2011 focused update): a report of the American College of Cardiology Foundation/American Heart Association Task Force on Practice Guidelines. *J Am Coll Cardiol*, 2012. 60(7): p. 645–81.

19 Anderson, J.L., et al., 2012 ACCF/AHA focused update incorporated into the ACCF/AHA 2007 guidelines for the management of patients with unstable angina/non-ST-elevation myocardial infarction: a report of the American College of Cardiology Foundation/American Heart Association Task Force on Practice Guidelines. *J Am Coll Cardiol*, 2013. 61(23): p. e179–347.

20 Antman, E.M., et al., The TIMI risk score for unstable angina/non-ST elevation MI: a method for prognostication and therapeutic decision making. *JAMA*, 2000. 284(7): p. 835–42.

21 Granger, C.B., et al., Predictors of hospital mortality in the global registry of acute coronary events. *Arch Intern Med*, 2003. 163(19): p. 2345–53.

22 Moghissi, E.S., et al., American Association of Clinical Endocrinologists and American Diabetes Association consensus statement on inpatient glycemic control. *Endocr Pract*, 2009. 15(4): p. 353–69.

23 Dandona, P. and W.E. Boden, Intensive glucose control in hyperglycemic patients with acute coronary syndromes: still smoke, but no fire. *JAMA Intern Med*, 2013. 173(20): p. 1905–6.

24 de Mulder, M., et al., Intensive glucose regulation in hyperglycemic acute coronary syndrome: results of the randomized BIOMarker study to identify the acute risk of a coronary syndrome-2 (BIOMArCS-2) glucose trial. *JAMA Intern Med*, 2013. 173(20): p. 1896–904.

25 Ritsinger, V., et al., Intensified insulin-based glycaemic control after myocardial infarction: mortality during 20 year follow-up of the randomised Diabetes Mellitus Insulin Glucose Infusion in Acute Myocardial Infarction (DIGAMI 1) trial. *Lancet Diabetes Endocrinol*, 2014. 2(8): p. 627–33.

19

MANAGEMENT OF ST-ELEVATION MYOCARDIAL INFARCTION: PRINCIPLES AND GUIDELINES

19.1 OVERVIEW

Acute ST-elevation myocardial infarction (STEMI) accounts for ~30% of all cases of acute coronary syndromes (ACS). The high early mortality for patients with STEMI is largely due to the significant myocardial ischemic injury. However, immediate reperfusion either pharmacologically with fibrinolysis or mechanically by primary percutaneous coronary intervention (PCI) or coronary artery bypass grafting (CABG) limits the size of the infarction and reduces mortality. Reperfusion therapy by primary PCI reduces mortality and the risk of reinfarction beyond the benefits achieved by fibrinolysis, especially when the primary PCI is initiated within 90 min of first medical contact. The use of adjuvant therapy with platelet inhibitors and anticoagulants is essential to enhance the results of reperfusion and/or maintain vessel patency following either mode of reperfusion.

Despite the recent success in reducing the mortality of STEMI, the management of STEMI continues to undergo major changes. Good practice should be based on sound evidence, derived from well-conducted randomized clinical trials. This chapter considers the principles of management of patients with STEMI based primarily on the recommendations of the most current guideline from the American College of Cardiology Foundation (ACCF) and American Heart Association (AHA). The chapter begins with a brief introduction to definition and epidemiology of STEMI and then provides an overview on the ACCF/AHA guideline development as well as related guidelines developed in other nations. The chapter focuses on discussing the general principles and the 2013 ACCF/AHA guideline-based recommendations with regard to both acute and nonacute managements of patients with STEMI. Special attention is given to drug therapy, which is an important component of STEMI treatment.

19.2 DEFINITION AND EPIDEMIOLOGY

STEMI is a clinical syndrome defined by characteristic symptoms of myocardial ischemia in association with persistent electrocardiographic (ECG) ST-segment elevation and subsequent release of biomarkers of myocardial necrosis [1]. Diagnostic ST-segment elevation in the absence of left ventricular hypertrophy or left bundle-branch block is defined by the European Society of Cardiology (ESC)/ACCF/AHA/World Heart Federation Task Force for the "Universal Definition of Myocardial Infarction" as new ST-segment elevation at the J point in at least two contiguous leads of ≥2 mm (0.2 mV) in men or ≥1.5 mm (0.15 mV) in women in leads V_2–V_3 and/or of ≥1 mm (0.1 mV) in other contiguous chest leads or the limb leads [2].

Community incidence rates for STEMI have declined over the past decade, whereas those for non-ST-segment elevation ACS have increased (also see Chapter 18). At present, STEMI comprises ~25–40% of myocardial infarction (MI) presentations. The major epidemiological data on STEMI are summarized in Box 19.1.

Cardiovascular Diseases: From Molecular Pharmacology to Evidence-Based Therapeutics, First Edition. Y. Robert Li.
© 2015 John Wiley & Sons, Inc. Published 2015 by John Wiley & Sons, Inc.

BOX 19.1   EPIDEMIOLOGY OF STEMI

The percentage of ACS or MI cases with ST-segment elevation varies in different registries/databases and depends heavily on the age of patients included and the type of surveillance used. According to the National Registry of Myocardial Infarction 4 (NRMI-4), ~29% of patients with MI are patients with STEMI. The AHA's "Get With The Guidelines (GWTG)" project found that 32% of the patients with MI in the coronary artery disease (CAD) module are patients with STEMI (personal communication from AHA GWTG staff, October 1, 2007). The Global Registry of Acute Coronary Events (GRACE) study, which includes US patient populations, found that 38% of ACS patients have STEMI, whereas the second Euro Heart Survey on ACS (EHS-ACS-II) reported that ~47% of patients with ACS have STEMI.

In addition, the percentage of ACS or MI cases with ST-segment elevation appears to be declining. In an analysis of 46,086 hospitalizations for ACS in the Kaiser Permanente Northern California study, the percentage of MI cases with ST-segment elevation decreased from 48.5 to 24% between 1999 and 2008.

Approximately 30% of patients with STEMI are women. Female sex was a strong independent predictor of failure to receive reperfusion therapy among patients who had no contraindications in the Can Rapid Risk Stratification of Unstable Angina Patients Suppress Adverse Outcomes with Early Implementation of the ACC/AHA Guidelines (CRUSADE) registry.

Approximately 23% of patients with STEMI in the United States have diabetes, and three quarters of all deaths among patients with diabetes are related to coronary artery disease. Diabetes is associated with higher short- and long-term mortality after STEMI, and in patients with diabetes, both hyperglycemia and hypoglycemia are associated with worse outcomes.

The elderly comprise a growing segment of the population and present special challenges for diagnosis and management that may lead to disparities in care and delays in treatment [3, 4].

19.3   INTRODUCTION TO RECENT GUIDELINES ON THE MANAGEMENT OF STEMI

Several guidelines on STEMI from professional organizations in both the United States and other countries have been published over the last several years. This section provides an overview on the development of the ACCF/AHA guidelines for the management of STEMI. STEMI guidelines from other countries or organizations are also introduced here.

19.3.1   The ACCF/AHA Guidelines

The American College of Cardiology (ACC) and the AHA have jointly produced guidelines in the area of cardiovascular disease since 1980. The ACC/AHA Task Force on Practice Guidelines (Task Force), charged with developing, updating, and revising practice guidelines for cardiovascular diseases and procedures, directs and oversees this effort. Writing committees are charged with regularly reviewing and evaluating all available evidence to develop balanced, patient-centric recommendations for clinical practice. In this

regard, a notable development was the ACC/AHA 2004 guideline for the management of patients with STEMI. The 2004 guideline resulted from the revision of the 1999 guideline for the management of patients with AMI. The 2004 guideline was subsequently updated in 2007 and 2009. In 2013, the ACCF/AHA published the 2013 guideline for the management of STEMI. Table 19.1 summaries the recent development of ACC/AHA guidelines on the management of STEMI.

19.3.2   Other Guidelines

In addition to the ACCF/AHA, many international organizations have also published guidelines on the management of STEMI. Among them, the most notable one is the ESC guidelines for the management of STEMI. The ESC guidelines represent the official position of the ESC on a given topic and are regularly updated. The current ESC guideline on STEMI was published in 2012, and this guideline is largely consistent with the 2013 ACCF/AHA guideline. Table 19.2 lists the current guidelines on STEMI from the ESC and other organizations/countries.

TABLE 19.1   The ACC/AHA STEMI guideline development in the past decade

Year	Title of the guideline	Description
2004	ACC/AHA guidelines for the management of patients with ST-elevation myocardial infarction: a report of the American College of Cardiology/American Heart Association Task Force on Practice Guidelines [5]	This 2004 guideline resulted from the revision of the 1999 guidelines for the management of patients with acute myocardial infarction [6]
2007	2007 focused update of the ACC/AHA 2004 guidelines for the management of patients with ST-elevation myocardial infarction: a report of the American College of Cardiology/American Heart Association Task Force on Practice Guidelines	
2009	2009 focused updates: ACC/AHA guidelines for the management of patients with ST-elevation myocardial infarction (updating the 2004 guideline and 2007 focused update) and ACC/AHA/SCAI guidelines on percutaneous coronary intervention (updating the 2005 guideline and 2007 focused update): a report of the American College of Cardiology Foundation/American Heart Association Task Force on Practice Guidelines [7]	
2013	2013 ACCF/AHA guideline for the management of ST-elevation myocardial infarction: a report of the American College of Cardiology Foundation/American Heart Association Task Force on Practice Guidelines [1]	Particular emphasis of the 2013 guideline is placed on advances in reperfusion therapy, organization of regional systems of care, transfer algorithms, evidence-based antithrombotic and medical therapies, and secondary prevention strategies to optimize patient-centered care. By design, the document is narrower in scope than the 2004 STEMI guideline, in an attempt to provide a more focused tool for practitioners. References related to management guidelines are provided whenever appropriate, including those pertaining to PCI, CABG, heart failure, cardiac devices, and secondary prevention

TABLE 19.2   Recent guidelines on STEMI from other nations and organizations

Nation/organization	Title	Year
ESC	ESC guidelines for the management of acute myocardial infarction in patients presenting with ST-segment elevation [8]	2012
Canada	Assessment and management of acute coronary syndromes (ACS): a Canadian perspective on current guideline-recommended treatment—Part 2: ST-segment elevation myocardial infarction [9]	2011
National Heart Foundation of Australia and Cardiac Society of Australia and New Zealand	2011 Addendum to the National Heart Foundation of Australia/Cardiac Society of Australia and New Zealand Guidelines for the management of acute coronary syndromes (ACS) 2006 [10]	2011

19.4   PRINCIPLES AND GUIDELINE RECOMMENDATIONS FOR THE MANAGEMENT OF STEMI

This section first introduces the general principles of acute management of STEMI. It then describes the current recommendations from the 2013 ACCF/AHA guideline [1].

19.4.1   General Principles

The management of STEMI consists of both diagnosis and treatment. Treatment includes both pharmacological and nonpharmacological treatments. The management of STEMI is conventionally divided into two phases: acute management and nonacute management.

Once the diagnosis of an acute STEMI is made, the early management of the patient involves the simultaneous achievement of several goals, including (i) relief of ischemic pain, (ii) assessment of the hemodynamic state and correction of abnormalities that are present, (iii) initiation of reperfusion therapy with primary PCI or fibrinolysis, and (iv) initiation of antithrombotic therapy.

The above acute management is followed by the initiation of short- and long-term interventions aimed at improving in-hospital and long-term outcomes. This nonacute management in the hours and days following the very early decision-making period includes use of various types of cardiovascular drugs as well as risk assessment and posthospitalization plan of care.

The general principles of STEMI management are considered in the context of the five steps of management, as outlined below [1].

19.4.1.1 Step 1: Timely Initial Diagnosis
The first step in the management of patients with an acute STEMI is timely diagnosis, since the beneficial effects of therapy with reperfusion are greatest when performed soon after presentation. Management, including both diagnosis and treatment, of acute MI starts at the point of first medical contact, defined as the point at which the patient is either initially assessed by a paramedic or physician or other medical personnel in the prehospital setting or the patient arrives at the hospital emergency department and therefore often in the outpatient setting. A working diagnosis of MI must first be made. This is usually based on a history of chest pain lasting for 20 min or more, not responding to nitroglycerine. Important clues are a history of CAD and radiation of the pain to the neck, lower jaw, or left arm. The pain may not be severe. It is important to note that ~30% of the STEMI patients present with less-typical symptoms, such as nausea/vomiting, shortness of breath, fatigue, palpitations, or syncope. Timely diagnosis is essential to successful management. ECG monitoring should be initiated as soon as possible in all patients with suspected STEMI to detect life-threatening arrhythmias and allow prompt defibrillation if indicated.

19.4.1.2 Step 2: Simultaneous Achievement of Multiple Goals of Alleviating Symptoms and Restoring Blood Flow
Once the diagnosis of an acute STEMI is made, the early management of the patients involves the simultaneous achievement of the following several goals:

- Relief of ischemic pain
- Assessment of the hemodynamic state and correction of abnormalities that are present
- Initiation of reperfusion therapy with primary PCI or fibrinolysis
- Antithrombotic therapy to prevent rethrombosis or acute stent thrombosis
- β-Blocker therapy to prevent recurrent ischemia and life-threatening ventricular arrhythmias

19.4.1.3 Step 3: In-Hospital Initiation of Multiple Drugs to Improve the Long-Term Prognosis
Step 2 is followed by the in-hospital initiation of different drugs that may improve the long-term prognosis. These drugs include the following:

- Antiplatelet therapy to reduce the risk of recurrent coronary artery thrombosis or, with PCI, coronary artery stent thrombosis

- Angiotensin-converting enzyme inhibitor (ACEI) therapy to prevent remodeling of the left ventricle
- Statin therapy to reduce mortality and prolong survival
- Anticoagulation in the presence of left ventricular thrombus or chronic atrial fibrillation to prevent embolization

19.4.1.4 Step 4: Risk Assessment after STEMI
Prior to discharge, the patients should undergo risk stratification, including noninvasive measurement of resting left ventricular function and possibly exercise testing.

19.4.1.5 Step 5: Posthospitalization Plan of Care
Medications upon discharge from the hospital should include those important for risk reduction, including platelet inhibitors, a β-blocker, a statin, and possibly an ACEI or an angiotensin receptor blocker (ARB). Posthospitalization plan of care should also include exercise-based cardiac rehabilitation and secondary prevention program and risk factor modifications as discussed elsewhere in the book.

19.4.2 The 2013 ACCF/AHA Guideline Recommendations

The current recommendations of the 2013 ACCF/AHA guideline on the management of STEMI cover nine areas, ranging from acute management to nonacute management [1]. These nine areas are outlined below. This section considers the major 2013 ACCF/AHA guideline recommendations on each of the areas. Wherever pertinent, recommendations of the 2012 ESC guideline [8] are also covered:

1. Recommendations pertinent to onset of MI
2. Recommendations pertinent to reperfusion at a PCI-capable hospital
3. Recommendations pertinent to reperfusion at a non-PCI-capable hospital
4. Recommendations pertinent to delayed invasive management
5. Recommendations pertinent to CABG surgery
6. Recommendations pertinent to routine drug therapies
7. Recommendations pertinent to complications after STEMI
8. Recommendations pertinent to risk assessment after STEMI
9. Recommendations pertinent to posthospitalization plan of care

19.4.2.1 Recommendations Pertinent to Onset of MI
The 2013 ACCF/AHA guideline recommendations pertinent to onset of MI cover two aspects: (i) regional systems of STEMI care, reperfusion therapy, and time-to-treatment

TABLE 19.3   The 2013 ACCF/AHA guideline recommendations pertinent to regional systems of STEMI care, reperfusion therapy, and time-to-treatment goals [1]

COR	Recommendations	LOE
I	All communities should create and maintain a regional system of STEMI care that includes assessment and continuous quality improvement of emergency medical services and hospital-based activities. Performance can be facilitated by participating in programs, such as "Mission: Lifeline and the Door-to-Balloon Alliance"	B
	Performance of a 12-lead ECG by emergency medical services personnel at the site of first medical contact (FMC) is recommended in patients with symptoms consistent with STEMI	B
	Reperfusion therapy should be administered to all eligible patients with STEMI with symptom onset within the prior 12 h	A
	Primary PCI is the recommended method of reperfusion when it can be performed in a timely fashion by experienced operators	A
	Emergency medical services transport directly to a PCI-capable hospital for primary PCI is the recommended triage strategy for patients with STEMI, with an ideal FMC-to-device time system goal of 90 min or less	B
	Immediate transfer to a PCI-capable hospital for primary PCI is the recommended triage strategy for patients with STEMI who initially arrive at or are transported to a non-PCI-capable hospital, with an FMC-to-device time system goal of 120 min or less	B
	In the absence of contraindications, fibrinolytic therapy should be administered to patients with STEMI at non-PCI-capable hospitals when the anticipated FMC-to-device time at a PCI-capable hospital exceeds 120 min because of unavoidable delays	B
	When fibrinolytic therapy is indicated or chosen as the primary reperfusion strategy, it should be administered within 30 min of hospital arrival	B
IIa	Reperfusion therapy is reasonable for patients with STEMI and symptom onset within the prior 12–24 h who have clinical and/or ECG evidence of ongoing ischemia. Primary PCI is the preferred strategy in this population	B

COR and LOE denote class of recommendation and level of evidence, respectively (see Chapter 5 for description).

TABLE 19.4   The 2013 ACCF/AHA guideline recommendations pertinent to evaluation and management of patients with STEMI and out-of-hospital cardiac arrest [1]

COR	Recommendations	LOE
I	Therapeutic hypothermia should be started as soon as possible in comatose patients with STEMI and out-of-hospital cardiac arrest caused by ventricular fibrillation or pulseless ventricular tachycardia, including patients who undergo primary PCI	B
	Immediate angiography and PCI when indicated should be performed in resuscitated out-of-hospital cardiac arrest patients whose initial ECG shows STEMI	B

COR and LOE denote class of recommendation and level of evidence, respectively (see Chapter 5 for description).

goals (Table 19.3) and (ii) evaluation and management of patients with STEMI and out-of-hospital cardiac arrest (Table 19.4). Relief of pain is of paramount importance, not only for humane reasons, but because the pain is associated with sympathetic activation that causes vasoconstriction and increases the workload of the heart. Titrated intravenous opioids (e.g., morphine) are the analgesics most commonly used in this context. Recommendations for relief pain, breathlessness, and anxiety are provided in the 2012 ESC guideline [8] (Table 19.5).

19.4.2.2   Recommendations Pertinent to Reperfusion at a PCI-Capable Hospital   The 2013 ACCF/AHA guideline recommendations pertinent to reperfusion at a PCI-capable hospital cover five major areas: (i) primary PCI in STEMI, (ii) aspiration thrombectomy, (iii) use of stents in patients with STEMI, (iv) antiplatelet therapy to support primary PCI

TABLE 19.5   The 2012 ESC guideline recommendations for relieving pain, breathless, and anxiety [8]

COR	Recommendations	LOE
I	Titrated intravenous opioids are indicated to relieve pain	C
	Oxygen is indicated in patients with hypoxia (saturated oxygen <95%), breathlessness, or acute heart failure	C
IIa	Tranquilizer may be considered in very anxious patients	C

COR and LOE denote class of recommendation and level of evidence, respectively (see Chapter 5 for description).

for STEMI, and (v) anticoagulant therapy to support primary PCI [1]. The pertinent recommendations are summarized in Tables 19.6, 19.7, 19.8, and 19.9.

TABLE 19.6   The 2013 ACCF/AHA guideline recommendations pertaining to primary PCI in STEMI at a PCI-capable hospital [1]

COR	Recommendations	LOE
I	Primary PCI should be performed in patients with STEMI and ischemic symptoms of <12 h of duration	A
	Primary PCI should be performed in patients with STEMI and ischemic symptoms of <12 h of duration who have contraindications to fibrinolytic therapy, irrespective of the time delay from first medical contact	B
	Primary PCI should be performed in patients with STEMI and cardiogenic shock or acute severe heart failure, irrespective of time delay from myocardial infarction onset	B
IIa	Primary PCI is reasonable in patients with STEMI if there is clinical and/or ECG evidence of ongoing ischemia between 12 and 24 h after symptom onset	B
III (harm)	PCI should not be performed in a noninfarct artery at the time of primary PCI in patients with STEMI who are hemodynamically stable	B

COR and LOE denote class of recommendation and level of evidence, respectively (see Chapter 5 for description).

TABLE 19.7   The 2013 ACCF/AHA guideline recommendations pertaining to use of stents in patients with STEMI at a PCI-capable hospital [1]

COR	Recommendations	LOE
I	Placement of a stent (bare-metal stent [BMS] or drug-eluting stent [DES]) is useful in primary PCI for patients with STEMI	A
	BMS should be used in patients with high bleeding risk, inability to comply with 1 year of dual antiplatelet therapy (DAPT), or anticipated invasive or surgical procedures in the next year	C
III (harm)	DES should not be used in primary PCI for patients with STEMI who are unable to tolerate or comply with a prolonged course of DAPT because of the increased risk of stent thrombosis with premature discontinuation of one or both agents	B

COR and LOE denote class of recommendation and level of evidence, respectively (see Chapter 5 for description).

TABLE 19.8   The 2013 ACCF/AHA guideline recommendations pertaining to antiplatelet therapy to support primary PCI for STEMI at a PCI-capable hospital [1]

COR	Recommendations	LOE
I	Aspirin (162–325 mg) should be given before primary PCI	B
	After PCI, aspirin should be continued indefinitely	A
	A loading dose of a $P2Y_{12}$ ADP-receptor inhibitor should be given as early as possible or at time of primary PCI to patients with STEMI. Options include (i) clopidogrel 600 mg, or (ii) prasugrel 60 mg, or (iii) ticagrelor 180 mg	B
	$P2Y_{12}$ ADP-receptor inhibitor therapy should be given for 1 year to patients with STEMI who receive a stent (BMS or DES) during primary PCI using the following maintenance doses: (i) clopidogrel 75 mg daily, or (ii) prasugrel 10 mg daily, or (iii) ticagrelor 90 mg twice a day	B
IIa	It is reasonable to use 81 mg of aspirin in preference to higher maintenance doses after primary PCI	B
	It is reasonable to begin treatment with an intravenous GP IIb/IIIa receptor antagonist, such as abciximab (LOE: A), high-bolus-dose tirofiban (LOE: B), or double-bolus eptifibatide (LOE: B) at the time of primary PCI (with or without stenting or clopidogrel pretreatment) in selected patients with STEMI who are receiving unfractionated heparin (UFH)	
IIb	It may be reasonable to administer an intravenous GP IIb/IIIa receptor antagonist in the precatheterization laboratory setting (e.g., ambulance, emergency department) to patients with STEMI for whom primary PCI is intended	B
	It may be reasonable to administer intracoronary abciximab to patients with STEMI undergoing primary PCI	B
	Continuation of a $P2Y_{12}$ ADP-receptor inhibitor beyond 1 year may be considered in patients undergoing DES placement	C
III (harm)	Prasugrel should not be administered to patients with a history of prior stroke or transient ischemic attack due to increased risk of hemorrhagic stroke	B

COR and LOE denote class of recommendation and level of evidence, respectively (see Chapter 5 for description).

TABLE 19.9   The 2013 ACCF/AHA guideline recommendations pertaining to anticoagulant therapy to support primary PCI for STEMI at a PCI-capable hospital [1]

COR	Recommendation	LOE
I	For patients with STEMI undergoing primary PCI, the following supportive anticoagulant regimens are recommended: (i) UFH, with additional boluses administered as needed to maintain therapeutic activated clotting time levels, taking into account whether a GP IIb/IIIa receptor antagonist has been administered (LOE: C), or (ii) bivalirudin with or without prior treatment with UFH (LOE: B)	C, B
IIa	In patients with STEMI undergoing PCI who are at high risk of bleeding, it is reasonable to use bivalirudin monotherapy in preference to the combination of UFH and a GP IIb/IIIa receptor antagonist	B
III (harm)	Fondaparinux should not be used as the sole anticoagulant to support primary PCI because of the risk of catheter thrombosis	B

COR and LOE denote class of recommendation and level of evidence, respectively (see Chapter 5 for description).

TABLE 19.10   The 2013 ACCF/AHA guideline recommendations pertaining to fibrinolytic therapy when there is an anticipated delay to performing primary PCI within 120 min of first medical contact [1]

COR	Recommendation	LOE
I	In the absence of contraindications, fibrinolytic therapy should be given to patients with STEMI and onset of ischemic symptoms within the previous 12 h when it is anticipated that primary PCI cannot be performed within 120 min of first medical contact	A
IIa	In the absence of contraindications and when PCI is not available, fibrinolytic therapy is reasonable for patients with STEMI if there is clinical and/or ECG evidence of ongoing ischemia within 12–24 h of symptom onset and a large area of myocardium at risk or hemodynamic instability	C
III (harm)	Fibrinolytic therapy should not be administered to patients with ST depression except when a true posterior (inferobasal) MI is suspected or when associated with ST elevation in lead aVR	B

COR and LOE denote class of recommendation and level of evidence, respectively (see Chapter 5 for description).

19.4.2.3   Recommendations Pertinent to Reperfusion at a Non-PCI-Capable Hospital

The 2013 ACCF/AHA guideline recommendations pertinent to reperfusion at a non-PCI-capable hospital cover three major areas: (i) fibrinolytic therapy when there is an anticipated delay to performing primary PCI within 120 min of first medical contact, (ii) adjunctive antithrombotic therapy with fibrinolysis, and (iii) transfer to a PCI-capable hospital after fibrinolytic therapy. The major recommendations are summarized in Tables 19.10, 19.11, 19.12, and 19.13 [1].

19.4.2.4   Recommendations Pertinent to Delayed Invasive Management

The 2013 ACCF/AHA guideline recommendations for delayed invasive management cover four major areas: (i) coronary angiography in patients who initially were managed with fibrinolytic therapy or who did not receive reperfusion, (ii) PCI of an infarct artery in patients who initially were managed with fibrinolysis or who did not receive reperfusion therapy, (iii) PCI of a non-infarct artery before hospital discharge, and (iv) adjunctive antithrombotic therapy to support delayed PCI after fibrinolytic therapy. Major recommendations are listed in Tables 19.13, 19.14, 19.15, and 19.16 [1].

19.4.2.5   Recommendations Pertinent to CABG

The 2013 ACCF/AHA guideline recommendations pertinent to

CABG surgery cover two major aspects: (i) CABG in patients with STEMI and (ii) timing of urgent CABG in patients with STEMI in relation to use of antiplatelet agents. The recommendations on the above two aspects are summarized in Tables 19.17 and 19.18 [1].

19.4.2.6   Recommendations Pertinent to Routine Drug Therapies

Drug therapies are a key component of the management of STEMI. They are not only important for the acute management of STEMI but also essential for the non-acute management of STEMI to improve long-term patient survival and quality of life. This section describes the current recommendations from the 2013 ACCF/AHA guideline with regard to routine drug therapies in the management of patients with STEMI, including both acute and nonacute managements. The routine drug therapies include the use of the following classes of drugs: (i) β-blockers, (ii) inhibitors of the renin–angiotensin–aldosterone system (RAAS), (iii) lipid management with statins, and (iv) other drugs. Table 19.19 summarizes the guideline recommendations pertaining to the use of the first 3 classes of drugs, and the guideline comments on other drugs, including nitrates, calcium channel blockers (CCBs), oxygen, and analgesics, are given in Table 19.20 [1].

19.4.2.7   Recommendations Pertinent to Complications after STEMI

The 2013 ACCF/AHA guideline

TABLE 19.11   The 2013 ACCF/AHA guideline recommendations pertaining to adjunctive antithrombotic therapy with fibrinolysis [1]

COR	Recommendation	LOE
I	Aspirin (162–325 mg loading dose) and clopidogrel (300 mg loading dose for ≤75 years of age, 75 mg dose for patients >75 years of age) should be administered to patients with STEMI who receive fibrinolytic therapy	A
	Aspirin should be continued indefinitely (LOE: A) and clopidogrel (75 mg daily) should be continued for at least 14 days (LOE: A) and up to 1 year (LOE: C) in patients with STEMI who receive fibrinolytic therapy	
	Patients with STEMI undergoing reperfusion with fibrinolytic therapy should receive anticoagulant therapy for a minimum of 48 h, and preferably for the duration of the index hospitalization, up to 8 days or until revascularization is performed (LOE: A)	
	Recommended regimens include (i) UFH administered as a weight-adjusted intravenous bolus and infusion to obtain an activated partial thromboplastin time of 1.5–2.0 times control, for 48 h or until revascularization (LOE: C); (ii) enoxaparin administered according to age, weight, and creatinine clearance, given as an intravenous bolus, followed in 15 min by subcutaneous injection for the duration of the index hospitalization, up to 8 days or until revascularization (LOE: A); or (iii) fondaparinux administered with initial intravenous dose, followed in 24 h by daily subcutaneous injections if the estimated creatinine clearance is >30 ml/min, for the duration of the index hospitalization, up to 8 days or until revascularization (LOE: B)	
IIa	It is reasonable to use aspirin 81 mg/day in preference to higher maintenance doses after fibrinolytic therapy	B

COR and LOE denote class of recommendation and level of evidence, respectively (see Chapter 5 for description).

TABLE 19.12   The 2013 ACCF/AHA guideline recommendations pertaining to transfer to a PCI-capable hospital after fibrinolytic therapy [1]

COR	Recommendation	LOE
I	Immediate transfer to a PCI-capable hospital for coronary angiography is recommended for suitable patients with STEMI who develop cardiogenic shock or acute severe heart failure, irrespective of the time delay from MI onset	B
IIa	Urgent transfer to a PCI-capable hospital for coronary angiography is reasonable for patients with STEMI who demonstrate evidence of failed reperfusion or reocclusion after fibrinolytic therapy	B
	Transfer to a PCI-capable hospital for coronary angiography is reasonable for patients with STEMI who have received fibrinolytic therapy even when hemodynamically stable and with clinical evidence of successful reperfusion. Angiography can be performed as soon as logistically feasible at the receiving hospital, and ideally within 24 h, but should not be performed within the first 2–3 h after administration of fibrinolytic therapy	B

COR and LOE denote class of recommendation and level of evidence, respectively (see Chapter 5 for description).

TABLE 19.13   The 2013 ACCF/AHA guideline recommendations pertaining to coronary angiography in patients who initially were managed with fibrinolytic therapy or who did not receive reperfusion [1]

COR	Recommendation	LOE
I	Cardiac catheterization and coronary angiography with intent to perform revascularization should be performed after STEMI in patients with any of the following: (i) cardiogenic shock or acute severe heart failure that develops after initial presentation (LOE: B), (ii) intermediate- or high-risk findings on predischarge noninvasive ischemia testing (LOE: B), or (iii) myocardial ischemia that is spontaneous or provoked by minimal exertion during hospitalization (LOE: C)	
IIa	Coronary angiography with intent to perform revascularization is reasonable for patients with evidence of failed reperfusion or reocclusion after fibrinolytic therapy. Angiography can be performed as soon as logistically feasible	B
	Coronary angiography is reasonable before hospital discharge in stable patients with STEMI after successful fibrinolytic therapy. Angiography can be performed as soon as logistically feasible, and ideally within 24 h, but should not be performed within the first 2–3 h after administration of fibrinolytic therapy	B

COR and LOE denote class of recommendation and level of evidence, respectively (see Chapter 5 for description).

recommendations pertaining to complications after STEMI cover four major areas: (i) treatment of cardiogenic shock, (ii) management of cardiac arrhythmias, (iii) management of pericarditis after STEMI, and (iv) management of thromboembolic complications. The main recommendations regarding the above areas are summarized in Table 19.21. The reader is advised to refer to the full guideline [1] for comments on the management of other complications,

TABLE 19.14 The 2013 ACCF/AHA guideline recommendations pertaining to PCI of an infarct artery in patients who initially were managed with fibrinolysis or who did not receive reperfusion therapy [1]

COR	Recommendations	LOE
I	PCI of an anatomically significant stenosis in the infarct artery should be performed in patients with suitable anatomy and any of the following: (i) cardiogenic shock or acute severe HF (LOE: B), (ii) intermediate- or high-risk findings on predischarge noninvasive ischemia testing (LOE: C), or (iii) myocardial ischemia that is spontaneous or provoked by minimal exertion during hospitalization (LOE: C)	
IIa	Delayed PCI is reasonable in patients with STEMI and evidence of failed reperfusion or reocclusion after fibrinolytic therapy. PCI can be performed as soon as logistically feasible at the receiving hospital	B
	Delayed PCI of a significant stenosis in a patent infarct artery is reasonable in stable patients with STEMI after fibrinolytic therapy. PCI can be performed as soon as logistically feasible at the receiving hospital, and ideally within 24 h, but should not be performed within the first 2–3 h after administration of fibrinolytic therapy	B
IIb	Delayed PCI of a significant stenosis in a patent infarct artery >24 h after STEMI may be considered as part of an invasive strategy in stable patients	B
III (no benefit)	Delayed PCI of a totally occluded infarct artery >24 h after STEMI should not be performed in asymptomatic patients with 1- or 2-vessel disease if they are hemodynamically and electrically stable and do not have evidence of severe ischemia	B

COR and LOE denote class of recommendation and level of evidence, respectively (see Chapter 5 for description).

TABLE 19.15 The 2013 ACCF/AHA guideline recommendations pertaining to PCI of a noninfarct artery before hospital discharge [1]

COR	Recommendation	LOE
I	PCI is indicated in a noninfarct artery at a time separate from primary PCI in patients who have spontaneous symptoms of myocardial ischemia	C
IIa	PCI is reasonable in a noninfarct artery at a time separate from primary PCI in patients with intermediate- or high-risk findings on noninvasive testing	B

COR and LOE denote class of recommendation and level of evidence, respectively (see Chapter 5 for description).

TABLE 19.16 The 2013 ACCF/AHA guideline recommendations pertaining to adjunctive antithrombotic therapy to support delayed PCI after fibrinolytic therapy [1]

COR	Recommendation	LOE
I	After PCI, aspirin should be continued indefinitely	A
	Clopidogrel should be provided as follows: (i) a 300 mg loading dose should be given before or at the time of PCI to patients who did not receive a previous loading dose and who are undergoing PCI within 24 h of receiving fibrinolytic therapy; (ii) a 600 mg loading dose should be given before or at the time of PCI to patients who did not receive a previous loading dose and who are undergoing PCI more than 24 h after receiving fibrinolytic therapy; and (iii) a dose of 75 mg daily should be given after PCI	C
	For patients with STEMI undergoing PCI after receiving fibrinolytic therapy with intravenous UFH, additional boluses of intravenous UFH should be administered as needed to support the procedure, taking into account whether GP IIb/IIIa receptor antagonists have been administered	C
	For patients with STEMI undergoing PCI after receiving fibrinolytic therapy with enoxaparin, if the last subcutaneous dose was administered within the prior 8 h, no additional enoxaparin should be given; if the last subcutaneous dose was administered between 8 and 12 h earlier, enoxaparin 0.3 mg/kg IV should be given	B
IIa	After PCI, it is reasonable to use 81 mg of aspirin per day in preference to higher maintenance doses	B
	Prasugrel, in a 60 mg loading dose, is reasonable once the coronary anatomy is known in patients who did not receive a previous loading dose of clopidogrel at the time of administration of a fibrinolytic agent, but prasugrel should not be given sooner than 24 h after administration of a fibrin-specific agent[a] or 48 h after administration of a non-fibrin-specific agent[b]	B
	Prasugrel, in a 10 mg daily maintenance dose, is reasonable after PCI	B
III (harm)	Prasugrel should not be administered to patients with a history of prior stroke or transient ischemic attack	B
	Fondaparinux should not be used as the sole anticoagulant to support PCI (an additional anticoagulant with anti-factor IIa activity should be administered because of the risk of catheter thrombosis)	C

[a] Fibrin-specific fibrinolytic agents include alteplase, reteplase, and tenecteplase.
[b] Streptokinase is a non-fibrin-specific fibrinolytic agent.
COR and LOE denote class of recommendation and level of evidence, respectively (see Chapter 5 for description).

TABLE 19.17   The 2013 ACCF/AHA guideline recommendations pertaining to CABG in patients with STEMI [1]

COR	Recommendation	LOE
I	Urgent CABG is indicated in patients with STEMI and coronary anatomy not amenable to PCI who have ongoing or recurrent ischemia, cardiogenic shock, severe heart failure, or other high-risk features	B
	CABG is recommended in patients with STEMI at time of operative repair of mechanical defects	B
IIa	The use of mechanical circulatory support is reasonable in patients with STEMI who are hemodynamically unstable and require urgent CABG	C
IIb	Emergency CABG within 6 h of symptom onset may be considered in patients with STEMI who do not have cardiogenic shock and are not candidates for PCI or fibrinolytic therapy	C

COR and LOE denote class of recommendation and level of evidence, respectively (see Chapter 5 for description).

TABLE 19.18   The 2013 ACCF/AHA guideline recommendations pertaining to timing of urgent CABG in patients with STEMI in relation to use of antiplatelet agents [1]

COR	Recommendation	LOE
I	Aspirin should not be withheld before urgent CABG	C
	Clopidogrel or ticagrelor should be discontinued at least 24 h before urgent on-pump CABG, if possible	B
	Short-acting intravenous GP IIb/IIIa receptor antagonists (eptifibatide, tirofiban) should be discontinued at least 2–4 h before urgent CABG	B
	Abciximab should be discontinued at least 12 h before urgent CABG	B
IIb	Urgent off-pump CABG within 24 h of clopidogrel or ticagrelor administration might be considered, especially if the benefits of prompt revascularization outweigh the risks of bleeding	B
	Urgent CABG within 5 days of clopidogrel or ticagrelor administration or within 7 days of prasugrel administration might be considered, especially if the benefits of prompt revascularization outweigh the risks of bleeding	C

The purpose of the timing management is to reduce the risk of major bleeding due to antiplatelet therapy without compromising the benefit of urgent CABG, especially when the benefits of revascularization outweigh the risks of bleeding, as often may be the case among patients with ACS.
COR and LOE denote class of recommendation and level of evidence, respectively (see Chapter 5 for description).

TABLE 19.19   The 2013 ACCF/AHA guideline recommendations pertaining to the use of beta-blockers, RAAS inhibitors, and statins in the management of STEMI [1]

Drug class	Recommendation	COR	LOE	Rationale
β-Blockers	Oral β-blockers should be initiated in the first 24 h in patients with STEMI who do not have any of the following: signs of heart failure, evidence of a low-output state, increased risk for cardiogenic shock, or other contraindications to use of oral β-blockers (PR interval >0.24 s, second- or third-degree heart block, active asthma, or reactive airway disease)	I	B	β-Blockers reduce the risks of recurrent MI and ventricular fibrillation but increase the risk of cardiogenic shock [11]
	β-Blockers should be continued during and after hospitalization for all patients with STEMI and with no contraindications to their use	I	B	
	Patients with initial contraindications to the use of β-blockers in the first 24 h after STEMI should be reevaluated to determine their subsequent eligibility	I	C	
	It is reasonable to administer intravenous β-blockers at the time of presentation to patients with STEMI and no contraindications to their use who are hypertensive or have ongoing ischemia	IIa	B	

(Continued)

TABLE 19.19   (*Continued*)

Drug class	Recommendation	COR	LOE	Rationale
RAAS inhibitors	An ACEI should be administered within the first 24 h to all patients with STEMI with anterior location, heart failure, or ejection fraction ≤40%, unless contraindicated	I	A	Oral ACEIs reduce both fatal and nonfatal cardiovascular events; early initiation of eplerenone (<7 days) reduces the rates of all-cause mortality, sudden cardiac death, and cardiovascular mortality/hospitalization in patients with STEMI [12]
	An ARB should be given to patients with STEMI who have indications for, but are intolerant of, ACEIs	I	B	
	An aldosterone receptor antagonist should be given to patients with STEMI and no contraindications who are already receiving an ACEI and β-blocker and who have an ejection fraction ≤40% and either symptomatic heart failure or diabetes	I	B	
	ACEIs are reasonable for all patients with STEMI and no contraindications to their use	IIa	A	
Lipid management	High-intensity statin therapy[a] should be initiated or continued in all patients with STEMI and no contraindications to its use	I	A	Treatment with statins in patients stabilized after an ACS, including STEMI, lowers the risk of IHD death, recurrent MI, stroke, and the need for coronary revascularization
	It is reasonable to obtain a fasting lipid profile in patients with STEMI, preferably within 24 h of presentation	IIa	C	

[a]High-intensity stain therapy: See Chapter 5 for description.
COR and LOE denote class of recommendation and level of evidence, respectively (see Chapter 5 for description).

TABLE 19.20   The 2013 ACCF/AHA guideline comments pertaining to the use of other drugs in the management of STEMI [1]

Drug class	Comments
Nitrates	• Although nitroglycerin can ameliorate symptoms and signs of myocardial ischemia by reducing left ventricular preload and increasing coronary blood flow, it generally does not attenuate the myocardial injury associated with epicardial coronary artery occlusion unless vasospasm plays a significant role • Intravenous nitroglycerin may be useful to treat patients with STEMI and hypertension or heart failure • Nitrates should not be given to patients with hypotension, marked bradycardia or tachycardia, right ventricular infarction, or phosphodiesterase inhibitor (e.g., sildenafil) use within the previous 24–48 h • There is no role for the routine use of oral nitrates in the convalescent phase of STEMI
CCBs	• CCBs have no beneficial effect on infarct size or the rate of reinfarction when CCB therapy is initiated during either the acute or convalescent phase of STEMI • CCBs may be useful, however, to relieve ischemia, lower blood pressure, or control the ventricular response rate to atrial fibrillation in patients who are intolerant of β-blockers • Caution is advised in patients with left ventricular systolic dysfunction • The use of the immediate-release nifedipine is contraindicated in patients with STEMI because of hypotension and reflex sympathetic activation with tachycardia
Oxygen	• Few data exist to support or refute the value of the routine use of oxygen in the acute phase of STEMI, and more research is needed • A pooled Cochrane analysis of three trials showed a threefold higher risk of death for patients with confirmed acute MI treated with oxygen than for patients with acute MI managed on room air • Oxygen therapy is appropriate for patients who are hypoxemic (oxygen saturation <90%) and may have a salutary placebo effect in others. Supplementary oxygen may, however, increase coronary vascular resistance • Oxygen should be administered with caution to patients with chronic obstructive pulmonary disease and carbon dioxide retention
Analgesics	• In the absence of a history of hypersensitivity, morphine sulfate is the drug of choice for pain relief in patients with STEMI, especially those whose course is complicated by acute pulmonary edema. It can alleviate the work of breathing, reduce anxiety, and favorably affect ventricular loading conditions • Nonsteroidal anti-inflammatory drugs (except for aspirin) and COX-2 inhibitors are contraindicated in patients with STEMI. They should not be initiated in the acute phase and should be discontinued in patients using them before hospitalization. Use of these drugs may be associated with an increased risk of death, reinfarction, cardiac rupture, hypertension, renal insufficiency, and heart failure

TABLE 19.21   The 2013 ACCF/AHA guideline recommendations pertaining to the management of complications after STEMI [1]

Complication	Recommendation	COR	LOE
Cardiogenic shock	Emergency revascularization with either PCI or CABG is recommended in suitable patients with cardiogenic shock due to pump failure after STEMI irrespective of the time delay from MI onset	I	B
	In the absence of contraindications, fibrinolytic therapy should be administered to patients with STEMI and cardiogenic shock who are unsuitable candidates for either PCI or CABG	I	B
	The use of intra-aortic balloon pump counterpulsation can be useful for patients with cardiogenic shock after STEMI who do not quickly stabilize with pharmacological therapy	IIa	B
	Alternative left ventricular assist devices for circulatory support may be considered in patients with refractory cardiogenic shock	IIb	C
Cardiac arrhythmias	Implantable cardioverter-defibrillator therapy is indicated before discharge in patients who develop sustained ventricular tachycardia/ventricular fibrillation more than 48 h after STEMI, provided the arrhythmia is not due to transient or reversible ischemia, reinfarction, or metabolic abnormalities	I	B
	Temporary pacing is indicated for symptomatic bradyarrhythmias unresponsive to medical treatment	I	C
Pericarditis	Aspirin is recommended for treatment of pericarditis after STEMI (Note: Glucocorticoids and nonsteroidal anti-inflammatory drugs are potentially harmful for treatment of pericarditis after STEMI due to possible aneurysm and delayed healing)	I	B
	Administration of acetaminophen, colchicine, or narcotic analgesics may be reasonable if aspirin, even in higher doses, is not effective	IIb	C
	Glucocorticoids and nonsteroidal anti-inflammatory drugs are potentially harmful for treatment of pericarditis after STEMI	III (harm)	B
Thromboembolic complications	Anticoagulant therapy with a vitamin K antagonist should be provided to patients with STEMI and atrial fibrillation with CHADS$_2$ score[a] ≥2, mechanical heart valves, venous thromboembolism, or hypercoagulable disorder	I	C
	The duration of triple-antithrombotic therapy with a vitamin K antagonist, aspirin, and a P2Y$_{12}$ ADP-receptor inhibitor should be minimized to the extent possible to limit the risk of bleeding	I	C
	Anticoagulant therapy with a vitamin K antagonist is reasonable for patients with STEMI and asymptomatic LV mural thrombi	IIa	C
	Anticoagulant therapy may be considered for patients with STEMI and anterior apical akinesis or dyskinesis	IIb	C
	Targeting vitamin K antagonist therapy to a lower international normalized ratio (INR) (e.g., 2.0–2.5) might be considered in patients with STEMI who are receiving dual antiplatelet therapy	IIb	C

[a] CHADS$_2$ score: First published in 2001, the CHADS$_2$ score system based on the individual's conditions (congestive heart failure, hypertension, age, diabetes, stroke) has been widely used to predict the risk of stroke in patients with atrial fibrillation [13].
COR and LOE denote class of recommendation and level of evidence, respectively (see Chapter 5 for description).

including severe heart failure (also see Unit VI), right ventricular infarction, mechanical complications, acute kidney injury, and hyperglycemia.

19.4.2.8   Recommendations Pertinent to Risk Assessment after STEMI

Initial risk stratification should be performed early with the use of information available at the time of presentation. However, risk assessment is a continuous process that requires recalibration on the basis of data obtained during the hospital stay. Such data include the success of reperfusion therapy, the events that occur during the hospital course (such as hemorrhagic complications), and the findings from noninvasive and invasive testing, particularly as they relate to the assessment of left ventricular systolic function. Table 19.22 summarizes the 2013 ACCF/AHA guideline recommendations

pertaining to (i) the use of noninvasive testing for ischemia before discharge, (ii) assessment of left ventricular function, and (iii) assessment of risk for sudden cardiac death (SCD).

19.4.2.9   Recommendations Pertinent to Posthospitalization Plan of Care

The continued survival and improvement of life quality of the patients with STEMI depend heavily on the availability and execution of a comprehensive posthospitalization plan of care. The posthospitalization plan of care for patients with STEMI should address in detail several complex issues, including the following eight aspects [1]:

1. Medication adherence and titration (antithrombotic agents, β-blockers, RAAS inhibitors, statins)
2. Physical activity/cardiac rehabilitation

TABLE 19.22   The 2013 ACCF/AHA guideline recommendations pertaining to risk assessment after STEMI [1]

Assessment	Recommendation	COR	LOE
Use of noninvasive testing for ischemia	Noninvasive testing for ischemia should be performed before discharge to assess the presence and extent of inducible ischemia in patients with STEMI who have not had coronary angiography and do not have high-risk clinical features for which coronary angiography would be warranted	I	B
	Noninvasive testing for ischemia might be considered before discharge to evaluate the functional significance of a noninfarct artery stenosis previously identified at angiography	IIb	C
	Noninvasive testing for ischemia might be considered before discharge to guide the postdischarge exercise prescription	IIb	C
Left ventricular function	Left ventricular ejection fraction (LVEF) should be measured in all patients with STEMI	I	C
Risk for SCD	Patients with an initially reduced LVEF who are possible candidates for implantable cardioverter-defibrillator therapy should undergo reevaluation of LVEF 40 or more days after discharge	I	B

COR and LOE denote class of recommendation and level of evidence, respectively (see Chapter 5 for description).

TABLE 19.23   The 2013 ACCF/AHA guideline class I recommendations pertaining to posthospitalization plan of care [1]

Recommendation	LOE
Posthospital systems of care designed to prevent hospital readmissions should be used to facilitate the transition to effective, coordinated outpatient care for all patients with STEMI	B
Exercise-based cardiac rehabilitation/secondary prevention programs are recommended for patients with STEMI	B
A clear, detailed, and evidence-based plan of care that promotes medication adherence, timely follow-up with the healthcare team, appropriate dietary and physical activities, and compliance with interventions for secondary prevention should be provided to patients with STEMI	C
Encouragement and advice to stop smoking and to avoid secondhand smoke should be provided to patients with STEMI	A

LOE denotes class of recommendation and level of evidence, respectively (see Chapter 5 for description of LOE and class of recommendation).

3. Risk factor modifications and lifestyle interventions including smoking cessation, diet, and nutrition

4. Management of comorbidities including overweight/obesity, dyslipidemias, diabetes, hypertension, heart failure, and cardiac arrhythmias

5. Psychosocial factors, such as sexual activity, gender-specific issues, depression, stress, alcohol use, and culture-sensitive issues

6. Provider follow-up including follow-up by the cardiologist and primary care provider, as well as influenza vaccination

7. Patient/family education including plan of care for acute MI, recognizing symptoms of MI, and cardiopulmonary resuscitation training for family members

8. Socioeconomic factors such as access to health insurance coverage, access to healthcare providers, disability, and social service

Execution of the posthospitalization plan of care is critical and often challenging. Failure to understand and comply with the plan of care may account for the high rate of rehospitalization and increased mortality. The 2013 ACCF/AHA guideline provides four class I recommendations regarding posthospitalization plan of care (Table 19.23).

19.5   SUMMARY OF CHAPTER KEY POINTS

- STEMI is a clinical syndrome defined by characteristic symptoms of myocardial ischemia in association with ST elevation and subsequent release of biomarkers of myocardial necrosis. STEMI accounts for ~30% of all ACS and is associated with high early mortality.

- Despite the recent success in reducing the mortality of STEMI, the management of STEMI continues to undergo major changes. Good clinical practice should be based on solid evidence, derived from well-conducted randomized clinical trials. Such trials also provide a foundation for developing management guidelines.

- Several guidelines on STEMI from professional organizations in both the United States and other countries have been developed over the past few years. Among them, the most notable one is the 2013 ACCF/AHA guideline for the management of STEMI.

- The 2013 ACCF/AHA guideline places particular emphasis on advances in reperfusion therapy, organization of regional systems of care, transfer algorithms, evidence-based antithrombotic and other drug therapies, and secondary prevention strategies to optimize patient-centered care.

- The general principles of STEMI management are embedded in the context of the five steps of management: (i) timely initial diagnosis, (ii) simultaneous achievement of multiple goals of alleviating symptoms and restoring blood flow, (iii) in-hospital initiation of multiple drugs to improve the long-term prognosis, (iv) risk assessment after STEMI, and (v) posthospitalization plan of care. The 2013 ACCF/AHA guideline provides detailed evidence-based recommendations for the management in each of the above five steps.

19.6 SELF-ASSESSMENT QUESTIONS

19.6.1 A 55-year-old female is brought to the emergency department 1 h after the onset of severe chest pain during dinner. Cardiac enzyme test and ECG changes confirm the diagnosis of acute ST-elevation myocardial infarction. She undergoes percutaneous coronary intervention and placement of 3 drug-eluting stents and is also given a chimeric monoclonal antibody, in combination with aspirin and heparin, to prevent platelet aggregation. Which of the following is the molecular target of this monoclonal antibody drug?
   A. Antithrombin III
   B. Coagulation factor IIa
   C. GP IIb/IIIa receptor
   D. $P2Y_{12}$ ADP receptor
   E. PAR-1 receptor

19.6.2 A 56-year-old male presents to the emergency department of a rural hospital with chest tightness, anxiety, and sweating, first noticed 45 min ago while he was doing yard work. He is obese and has hypercholesterolemia for 10 years. ECG and cardiac enzyme test confirm the diagnosis of an acute ST-elevation myocardial infarction. It is decided to put him on fibrinolytic therapy. Which of the following should be given to lyse the coronary thrombus?
   A. Alteplase
   B. Aspirin
   C. Clopidogrel
   D. Unfractionated heparin
   E. Warfarin

19.6.3 A 55-year-old patient is diagnosed with acute ST-elevation myocardial infarction with left ventricular dysfunction (LVEF=39%). Which of the following should be administered within the first 24 h to reduce cardiovascular death?
   A. Ibuprofen
   B. Lisinopril
   C. Morphine
   D. Nitroglycerin
   E. Verapamil

19.6.4 A 65-year-old male develops pericarditis following an ST-elevation acute coronary syndrome. Which of the following would be most suitable for treating his pericarditis?
   A. Aspirin
   B. Colchicine
   C. Ibuprofen
   D. Indomethacin
   E. Prednisone

19.6.5 Which of the following should be initiated or continued in all patients with STEMI and no contraindications to its use in order to lower the risk of ischemic cardiac death, recurrent myocardial infarction, and stroke?
   A. High-oxygen therapy
   B. High-dose calcium antagonist therapy
   C. High-dose colchicine therapy
   D. High-dose ibuprofen therapy
   E. High-intensity statin therapy

REFERENCES

1 O'Gara, P.T., et al., 2013 ACCF/AHA guideline for the management of ST-elevation myocardial infarction: a report of the American College of Cardiology Foundation/American Heart Association Task Force on Practice Guidelines. *Circulation*, 2013. 127(4): p. e362–425.

2 Thygesen, K., et al., Third universal definition of myocardial infarction. *Nat Rev Cardiol*, 2012. 9(11): p. 620–33.

3 Go, A.S., et al., Heart disease and stroke statistics—2013 update: a report from the American Heart Association. *Circulation*, 2013. 127(1): p. e6–245.

4 Yeh, R.W., et al., Population trends in the incidence and outcomes of acute myocardial infarction. *N Engl J Med*, 2010. 362(23): p. 2155–65.

5 Antman, E.M., et al., ACC/AHA guidelines for the management of patients with ST-elevation myocardial infarction: a report of the American College of Cardiology/American Heart Association Task Force on Practice Guidelines (Committee to Revise the 1999 Guidelines for the Management of Patients with Acute Myocardial Infarction). *Circulation*, 2004. 110(9): p. e82–292.

6 Ryan, T.J., et al., 1999 update: ACC/AHA guidelines for the management of patients with acute myocardial infarction: executive summary and recommendations: a report of the American College of Cardiology/American Heart Association Task Force on Practice Guidelines (Committee on Management of Acute Myocardial Infarction). *Circulation*, 1999. 100(9): p. 1016–30.

7 Kushner, F.G., et al., 2009 focused updates: ACC/AHA guidelines for the management of patients with ST-elevation myocardial infarction (updating the 2004 guideline and 2007 focused update) and ACC/AHA/SCAI guidelines on percutaneous coronary intervention (updating the 2005 guideline and 2007

focused update): a report of the American College of Cardiology Foundation/American Heart Association Task Force on Practice Guidelines. *Circulation*, 2009. 120(22): p. 2271–306.

8   Task Force on the management of ST-segment elevation acute myocardial infarction of the European Society of Cardiology (ESC), et al., ESC guidelines for the management of acute myocardial infarction in patients presenting with ST-segment elevation. *Eur Heart J*, 2012. 33(20): p. 2569–619.

9   Fitchett, D.H., et al., Assessment and management of acute coronary syndromes (ACS): a Canadian perspective on current guideline-recommended treatment—part 2: ST-segment elevation myocardial infarction. *Can J Cardiol*, 2011. 27 Suppl A: p. S402–12.

10 Chew, D.P., et al., 2011 addendum to the National Heart Foundation of Australia/Cardiac Society of Australia and New Zealand guidelines for the management of acute coronary syndromes (ACS) 2006. *Heart Lung Circ*, 2011. 20(8): p. 487–502.

11 Chen, Z.M., et al., Early intravenous then oral metoprolol in 45,852 patients with acute myocardial infarction: randomised placebo-controlled trial. *Lancet*, 2005. 366(9497): p. 1622–32.

12 Adamopoulos, C., et al., Timing of eplerenone initiation and outcomes in patients with heart failure after acute myocardial infarction complicated by left ventricular systolic dysfunction: insights from the EPHESUS trial. *Eur J Heart Fail*, 2009. 11(11): p. 1099–105.

13 Gage, B.F., et al., Validation of clinical classification schemes for predicting stroke: results from the National Registry of Atrial Fibrillation. *JAMA*, 2001. 285(22): p. 2864–70.

UNIT VI

HEART FAILURE

20

OVERVIEW OF HEART FAILURE AND DRUG THERAPY

20.1 INTRODUCTION

Heart failure (HF) is a common clinical syndrome representing the end-stage of a number of different cardiac diseases. It can result from any structural or functional cardiac disorder that impairs the ability of the ventricle to fill with or eject blood. HF is a serious condition, and currently there is no cure. Many people with HF may live a reasonably good quality of life when the condition is appropriately managed with HF medications and healthy lifestyle changes. This chapter provides an overview of HF, including its definition, classification, and epidemiology. The chapter also examines the pathophysiology of HF which serves as a basis for drug targeting and pharmacotherapy of this clinical syndrome. The drug classes for treating HF and the underlying pharmacological basis are covered in Chapter 21. Chapter 22 considers the principles and current guidelines for HF management.

20.2 DEFINITION, CLASSIFICATION, AND EPIDEMIOLOGY

20.2.1 Definition

HF is a complex clinical syndrome that can result from any structural or functional impairment of ventricular filling or ejection of blood [1]. The cardinal manifestations of HF are dyspnea and fatigue, which may limit exercise tolerance, and fluid retention, which may then lead to pulmonary congestion and peripheral edema. Both abnormalities can impair the functional capacity and quality of life of affected individuals, but they do not necessarily dominate the clinical picture at the same time. It is important to note that HF is not equivalent to cardiomyopathy or to left ventricular (LV) dysfunction; these latter terms describe possible structural or functional reasons for the development of HF [1]. Instead, HF is defined as a clinical syndrome that is characterized by specific symptoms (dyspnea and fatigue) in the medical history and signs (edema, rales) on the physical examination. There is no single diagnostic test for HF because it is largely a clinical diagnosis that is based on a careful history and physical examination.

20.2.2 Classification

HF can be classified in various ways [2, 3]. The classification schemes may be based on the time-course of the syndrome, the systolic or diastolic function, the side of the heart involved, the left ventricular ejection fraction, and the severity of the symptoms (Table 20.1).

In addition to the above terminology on HF, another term that is sometimes still used, particularly in the United States, is congestive HF. This term may describe acute or chronic HF with evidence of congestion (i.e., sodium and water retention). Because of HF, blood flow out of the heart slows and blood returning to the heart through the veins backs up, causing congestion in the body's tissues, which is manifested as peripheral edema (e.g., edema in the legs and ankles). Sometimes fluid collects in the lungs and interferes with breathing, causing shortness of breath, especially when a person is lying down. This is known as pulmonary edema, and if left untreated, can cause respiratory distress. HF also affects the kidneys' ability to dispose of sodium and water.

TABLE 20.1   Classification of HF[a]

Classification basis	Type	Description
Time course	Acute HF and chronic HF	Patients who have had HF for some time are often said to have chronic HF. A treated patient with symptoms and signs, which have remained generally unchanged for at least a month, is said to be stable. If chronic stable HF deteriorates, the patient may be described as decompensated and this may happen suddenly, that is, acutely, usually leading to hospital admission, an event of considerable prognostic importance. New (*de novo*) HF may present acutely, for example, as a consequence of acute myocardial infarction or in a subacute (gradual) fashion, for example in a patient who has had asymptomatic cardiac dysfunction, often for an indeterminate period, and may persist or resolve (patients may become compensated). Acute HF is not a single disease, but rather a family of related disorders. Hence, the term "acute HF syndromes" (AHFS) is commonly used to describe the acute conditions. AHFS can be defined as the new onset or recurrence of gradually or rapidly developing syndromes and signs of HF requiring urgent or emergent therapy and resulting in hospitalization.
Side of the heart	Left-sided HF and right-sided HF	In left-sided or left ventricular HF, the ability of the left ventricle to fill with or eject blood is impaired. There are two types of left-sided HF: systolic and diastolic HF (see below). Right-sided or right ventricular HF usually occurs as a result of left-sided failure. When the left ventricle fails, increased fluid pressure is, in effect, transferred back through the lungs, ultimately damaging the heart's right side. When the right side loses pumping power, blood backs up in the body's veins. This usually causes swelling in the legs and ankles.
Systolic or diastolic	Systolic HF and diastolic HF	In systolic HF, the left ventricle loses its ability to contract normally. The heart can't pump with enough force to push enough blood into circulation, and hence the left ventricular ejection fraction is reduced. In diastolic HF, the left ventricle loses its ability to relax normally because the muscle has become stiff. The heart can't properly fill with blood during the resting period between each beat, and the left ventricular ejection fraction (LVEF) is preserved or normal.
Ejection fraction	HF with reduced ejection fraction (HF-REF) and HF with preserved ejection fraction (HF-PEF)	HF-REF is typically defined as the clinical diagnosis of HF and LVEF less than or equal to 40% (or 0.4). HF-REF is seen in systolic HF, and thus also loosely referred to as systolic HF. HF-PEF is typically defined as clinical diagnosis of HF and LVEF ≥50% (or 0.5). However, it has also been variably classified as LVEF >40, >45, >50, and ≥55%. The ACC/AHA classifies HF-PEF with LVEF of 41–49% as "borderline," and HF-PEF with EF >40% as "improved" [1]. HF-PEF is seen in diastolic HF, and as such is also loosely referred to as diastolic HF. It accounts for approximately 50% of all HF cases.
Severity and disease development	New York Heart Association (NYHA) functional classification and the ACC/AHA stage classification	The symptomatic severity is graded by NYHA into four classes. See Table 20.2 for NYHA functional classification. The stage system recognizes that HF, like coronary artery disease, has established risk factors and structural prerequisites; that the development of HF has asymptomatic and symptomatic phases; and that specific treatments targeted at each stage can reduce the morbidity and mortality of HF. See Table 20.3 for ACC/AHA stage classification.

[a] Modified from Ref. [2, 3].

TABLE 20.2   The NYHA functional classification of HF

Class	Description
I	No limitation of physical activity
	Ordinary physical activity does not cause undue fatigue, palpitation, or dyspnea
II	Slight limitation of physical activity
	Comfortable at rest, but ordinary physical activity results in fatigue, palpitation, or dyspnea
III	Marked limitation of physical activity
	Comfortable at rest, but less than ordinary activity results in fatigue, palpitation, or dyspnea
IV	Unable to carry on any physical activity without discomfort
	Symptoms present at rest, and if any physical activity is undertaken, discomfort is increased

TABLE 20.3   The ACC/AHA stage classification of HF and its relationship to NYHA functional classification

ACC/AHA stage	Description	Equivalent to NYHA classification
A	At high risk for HF	None
	No identified structural or functional abnormality	
	No signs or symptoms	
B	Structural heart disease but without signs or symptoms of HF	I
C	Structural heart disease with prior or current signs or symptoms of HF	I–IV
D	Advanced structural heart disease and marked symptoms of HF at rest despite maximal medical therapy	IV
	Refractory HF requiring specialized interventions	

Modified from Ref. [1].

This retained water further aggravates edema in the body's tissues. Because some patients present without signs or symptoms of volume overload, the term "heart failure" is preferred over "congestive HF" [1].

20.2.3 Epidemiology

HF is a major public health issue that affects as many as 23 million people worldwide [4]. Approximately 1–2% of the population in developed countries has HF with the prevalence rising to 10% or more among persons 70 years of age or order [2]. There are many causes of HF, and these vary in different parts of the world. The most common cause of HF is ischemic heart disease (myocardial infarction). Other causes include dilated cardiomyopathies, familial cardiomyopathies, diabetic cardiomyopathy, and toxic cardiomyopathy. Major risk factors for HF include hypertension, diabetes, metabolic syndrome, and atherosclerotic diseases [1, 5].

At least half of patients with HF have a low LVEF (i.e., HF-REF). HF-REF is the best understood type of HF in terms of pathophysiology and treatment, and is the focus of current guidelines on HF management. Coronary artery disease (CAD) is the cause of approximately two-thirds of cases of systolic HF, although hypertension and diabetes are probable contributing factors in many cases.

HF-PEF seems to have a different epidemiological and etiological profile from HF-REF. Patients with HF-PEF are older and more often female and obese than those with HF-REF. They are less likely to have CAD and more likely to have hypertension and atrial fibrillation. Indeed, hypertension remains the most important cause of HF-PEF, with a prevalence of 60–89% from large controlled trials, epidemiological studies, and HF registries [1].

In the United States, it is estimated that approximately 5.1 million Americans ≥20 years of age have HF. Projections show that the prevalence of HF will increase 46% from 2012 to 2030, resulting in >8 million people ≥18 years of age with HF [6]. Currently, there are 825,000 new HF cases annually in the United States, and the incidence approaches 1% of the population after 65 years of age. At 40 years of age, the lifetime risk of developing HF for both men and women is 1 in 5 (20%). Survival after HF diagnosis has improved over time. However, the death rate remains high: approximately 50% of people diagnosed with HF will die within 5 years. In 2012, total cost for HF was estimated to be $30.7 million. Projections show that by 2030, the total cost of HF will increase almost 127% to approximately $70 billion from 2012 [7]. Hence, HF is a major issue of public health in the United States as well as worldwide.

20.3 PATHOPHYSIOLOGY AND DRUG TARGETING

20.3.1 Pathophysiology of HF-REF and Drug Targeting

The mechanisms of HF, especially systolic HF (or HF-REF) have been investigated from a variety of perspectives during the past several decades [8]. The pathophysiological process of diastolic HF is initiated by a primary myocardial injury, most commonly myocardial infarction. This injury results in left ventricular (LV) systolic dysfunction. In patients with LV systolic dysfunction, the maladaptive changes occurring in surviving myocytes and extracellular matrix after myocardial injury lead to pathological remodeling of the ventricle with dilatation and impaired contractility, one measure of

which is a reduced LVEF [2, 9]. What characterizes untreated systolic dysfunction is progressive worsening of these changes over time, with increasing enlargement of the left ventricle and decline in LVEF, even though the patient may be symptomless initially. Two mechanisms are thought to account for this progression. The first is occurrence of further events leading to additional cardiomyocyte death (e.g., recurrent myocardial infarction). The other is the systemic responses induced by the decline in systolic function (i.e., a decreased cardiac output), especially the activation of the sympathetic nervous system and the renin–angiotensin–aldosterone system (RAAS) (Fig. 20.1).

The above neurohormonal adaptations, in the form of activation of the RAAS and sympathetic nervous system by decreased cardiac output, can temporarily contribute to

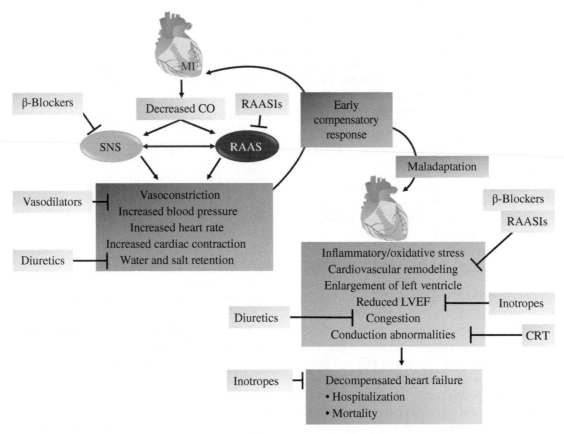

FIGURE 20.1 Pathophysiology of heart failure with reduced ejection fraction (HF-REF) and drug targeting mechanisms. As illustrated, HF is typically initiated by a primary myocardial injury, most commonly myocardial infarction (MI). This injury results in left ventricular dysfunction and decreased cardiac output (CO). The reduced CO in turn sets off an initially compensatory response and subsequently progressive maladaptation, leading to myocardial remodeling and deterioration of left ventricular dysfunction. This vicious cycle is perpetuated primarily via sustained activation of the renin–angiotensin–aldosterone system (RAAS) and the sympathetic nervous system (SNS). Drugs that target the RAAS and SNS retard cardiac remodeling and slow disease progression, and as such, have become the cornerstone of HF management. Diuretics, vasodilators, and positive inotropic agents act at various steps of the HF pathophysiology to improve symptoms associated with left ventricular dysfunction. On the other hand, cardiac-resynchronization therapy (CRT) improves left ventricular function via correcting the conduction abnormalities. RAASIs denotes RAAS inhibitors, which include angiotensin-converting enzyme inhibitors, angiotensin receptor blockers, and aldosterone receptor antagonists. LVEF denotes left ventricular ejection fraction.

maintenance of perfusion of vital organs in two ways: (1) maintenance of systemic pressure by vasoconstriction, resulting in redistribution of blood flow to vital organs; and (2) restoration of cardiac output by increasing myocardial contractility and heart rate and by expansion of the extracellular fluid volume. Although some of the responses are initially compensatory, eventually they contribute to the further worsening of heart function and disease progression, creating a pathophysiological vicious cycle, accounting for many of the clinical features of the HF syndrome [2]. Interruption of the above processes, especially the sympathetic nervous system and the RAAS, is thus the basis of much of the effective treatment of systolic HF [8, 10]. In this context, as discussed in Chapter 22, the mainstay of therapy of systolic HF includes the use of β-blocker and the RAAS inhibitors (Fig. 20.1).

Clinically, the aforementioned changes are associated with the development of symptoms and worsening of the symptoms over time. In this regard, the elevation in diastolic pressures of the failing heart is transmitted to the atria and to the pulmonary and systemic venous circulations, and the ensuing elevation in capillary pressures promotes the development of pulmonary congestion and peripheral edema. On the other hand, the increase in left ventricular afterload induced by the rise in peripheral resistance can both directly depress cardiac function and enhance the rate of deterioration of myocardial function via pathological remodeling. Moreover, activation of the RAAS and the sympathetic nervous system results in inflammatory and oxidative stress, contributing to pathological cardiac remodeling. Collectively, these deleterious effects due to maladaptation cause diminished quality of life, declining functional capacity, episodes of frank decompensation leading to hospital admission (which is often recurrent and costly to health services), and premature death, usually due to pump failure or a ventricular arrhythmia. The acute decompensated HF necessitates the use of (i) positive inotropic agents to augment myocardial contractility and hence improve cardiac output and the perfusion of vital organs, and (ii) diuretics to alleviate edema, especially pulmonary edema, to reduce pulmonary congestion (Fig. 20.1).

The limited cardiac reserve of such patients is also dependent on atrial contraction, synchronized contraction of the left ventricle, and a normal interaction between the right and left ventricles. Intercurrent events affecting any of these including the development of atrial fibrillation or conduction abnormalities (such as left bundle branch block) or imposing an additional hemodynamic load on the failing heart (e.g., anemia) can lead to acute decompensation. Indeed, HF patients, frequently have conduction abnormalities which further aggravate LV dysfunction. The efficacy of cardiac-resynchronization therapy (CRT) in HF is consistent with the above electro-pathophysiological concept [11–13] (Fig. 20.1).

20.3.2 Pathophysiology of HF-PEF and Drug Targeting

HF-PEF is estimated to account for approximately 50% of all HF cases, and the prevalence and hospitalization related to HF-PEF are rising. Indeed, HF-PEF is increasing out of proportion to HF-REF, and its prognosis is worsening while that of HF-REF is improving. The health and economic impact of HF-PEF is at least as great as that of HF-REF, with similar severity of chronic exercise intolerance, acute hospitalization rates, and substantial mortality [14, 15].

In contrast to HF-REF, to date, there are no approved therapies to reduce hospitalization or mortality for HF-PEF, largely due to the pathophysiological heterogeneity that exists within the broad spectrum of HF-PEF. This syndrome was historically considered to be caused exclusively by left ventricular diastolic dysfunction due to hypertrophy and other remodeling processes that result in LV stiffness. However, recent research has identified multiple other contributory factors, including: (i) decreased left ventricular systolic reserve; (ii) systemic and pulmonary vascular dysfunction; (iii) endothelial dysfunction, decreased nitric oxide bioavailability, and increased arterial stiffness; and (iv) decreased right heart function and left atrial function [16, 17].

Multiple individual mechanisms frequently coexist within the same patient to cause symptomatic HF, but between patients with HF-PEF, the extent to which each component is operative can differ widely, further confounding treatment approaches. In line with this notion, clinical trials on HF-PEF therapies have been largely unsuccessful. These include the trials on RAAS inhibitors, digoxin, β-blockers, and phosphodiesterase-5 (PDE5) inhibitors (e.g., sildenafil) [15]. Because of the lack of specific treatment, the management of HF-PEF is currently limited to diuretics and treatment of comorbidities [15].

20.4 SUMMARY OF CHAPTER KEY POINTS

- HF is a major public health issue that affects as many as 23 million people worldwide. It is a complex clinical syndrome that can result from any structural or functional impairment of ventricular filling or ejection of blood. The cardinal manifestations of HF are dyspnea and fatigue (which may limit exercise tolerance), and fluid retention (which may lead to pulmonary congestion and peripheral edema).

- HF can be classified in various ways. For example, based on time course, HF is classified into acute and chronic HF. Based on LVEF, HF is divided into HF-REF and HF-PEF, with each accounting for approximately 50% of all cases. HF is also classified into different classes (I–IV) or stages (A–D) according to diseases severity and development.

- The pathophysiological process of HF-REF is initiated by a primary myocardial injury, most commonly myocardial infarction, and the disease progression is catalyzed by the initially compensatory activation of both the RAAs and the sympathetic nervous system. Interruption of the above processes by β-blockers and RAAS inhibitors is the basis of much of the effective treatment of HF-REF.

- In contrast to HF-REF, to date, there are no approved therapies to reduce hospitalization or mortality for HF-PEF. This is largely due to the pathophysiological heterogeneity that exists within the broad spectrum of HF-PEF. Because of the lack of specific treatment, the management of HF-PEF is currently limited to diuretics and treatment of comorbidities.

20.5   SELF-ASSESSMENT QUESTIONS

20.5.1. According to the 2013 ACCF/AHA guideline for the management of heart failure, HF-REF is defined as the clinical diagnosis of HF and reduced LVEF of which of the following?
   A. ≤25
   B. ≤30
   C. ≤35
   D. ≤40
   E. ≤45

20.5.2. Heart failure is a common clinical syndrome that affects over 5 million Americans and as many as 23 million people worldwide. Which of the following is the etiology responsible for the majority of heart failure cases?
   A. Atrial fibrillation
   B. Diabetic cardiomyopathy
   C. Hypertension
   D. Myocardial infarction
   E. Toxic cardiomyopathy

20.5.3. Based on the changes in left ventricular ejection fraction, patients with heart failure are divided into two categories: those with HF-REF and those with HF-PEF. Which of the following is most likely the ratio of the prevalence of the two types of heart failure?
   A. 0.1
   B. 0.25
   C. 0.5
   D. 1.0
   E. 2.0

20.5.4. A 60-year-old male is brought to the emergency department because of severe shortness of breath upon minimum exertion. History reveals that he has chronic heart failure of 2 years of duration. If the patient's symptom is associated with his heart failure, which of the following best describes the stage of the patient's heart failure?
   A. ACC/AHA stage A
   B. ACC/AHA stage B
   C. ACC/AHA stage C
   D. NYHA class I
   E. NYHA class II

20.5.5. A 62-year-old female with HF-REF has been managed successfully with a treatment regimen that includes a diuretic, an angiotensin-converting enzyme inhibitor, and a β-blocker. She is free of heart failure symptoms at rest and even able to carry out ordinary physical activity without experiencing fatigue, palpitation, or dyspnea. Which of the following best describes the patient's condition?
   A. ACC/AHA stage A
   B. ACC/AHA stage B
   C. ACC/AHA stage C
   D. ACC/AHA stage D

REFERENCES

1 Yancy, C.W., et al., 2013 ACCF/AHA guideline for the management of heart failure: a report of the American College of Cardiology Foundation/American Heart Association Task Force on Practice Guidelines. *J Am Coll Cardiol*, 2013. 62(16): p. e147–239.

2 McMurray, J.J., et al., ESC Guidelines for the diagnosis and treatment of acute and chronic heart failure 2012: the Task Force for the Diagnosis and Treatment of Acute and Chronic Heart Failure 2012 of the European Society of Cardiology. Developed in collaboration with the Heart Failure Association (HFA) of the ESC. *Eur Heart J*, 2012. 33(14): p. 1787–847.

3 Hunt, S.A., ACC/AHA 2005 guideline update for the diagnosis and management of chronic heart failure in the adult: a report of the American College of Cardiology/American Heart Association Task Force on Practice Guidelines (Writing Committee to Update the 2001 Guidelines for the Evaluation and Management of Heart Failure). *J Am Coll Cardiol*, 2005. 46(6): p. e1–82.

4 Bui, A.L., T.B. Horwich, and G.C. Fonarow, Epidemiology and risk profile of heart failure. *Nat Rev Cardiol*, 2011. 8(1): p. 30–41.

5 Mudd, J.O. and D.A. Kass, Tackling heart failure in the twenty-first century. *Nature*, 2008. 451(7181): p. 919–28.

6 Heidenreich, P.A., et al., Forecasting the impact of heart failure in the United States: a policy statement from the American Heart Association. *Circ Heart Fail*, 2013. 6(3): p. 606–19.

7 Go, A.S., et al., Heart disease and stroke statistics—2014 update: a report from the American Heart Association. *Circulation*, 2014. 129(3): p. e28–292.

8 Braunwald, E., Heart failure. *JACC Heart Fail*, 2013. 1(1): p. 1–20.

9 Heusch, G., et al., Cardiovascular remodelling in coronary artery disease and heart failure. *Lancet*, 2014. 383(9932): p. 1933–43.

10 Lymperopoulos, A., G. Rengo, and W.J. Koch, Adrenergic nervous system in heart failure: pathophysiology and therapy. *Circ Res*, 2013. 113(6): p. 739–53.

11 Goldenberg, I., et al., Survival with cardiac-resynchronization therapy in mild heart failure. *N Engl J Med*, 2014. 370(18): p 1694–701.

12 Tang, A.S., et al., Cardiac-resynchronization therapy for mild-to-moderate heart failure. *N Engl J Med*, 2010. 363(25): p 2385–95.

13 Holzmeister, J. and C. Leclercq, Implantable cardioverter defibrillators and cardiac resynchronisation therapy. *Lancet*, 2011. 378(9792): p. 722–30.

14 Kitzman, D.W. and B. Upadhya, Heart failure with preserved ejection fraction: a heterogenous disorder with multifactorial pathophysiology. *J Am Coll Cardiol*, 2014. 63(5): p. 457–9.

15 Butler, J., et al., Developing therapies for heart failure with preserved ejection fraction: current state and future directions. *JACC Heart Fail*, 2014. 2(2): p. 97–112.

16 Sharma, K. and D.A. Kass, Heart failure with preserved ejection fraction: mechanisms, clinical features, and therapies. *Circ Res*, 2014. 115(1): p. 79–96.

17 Borlaug, B.A., The pathophysiology of heart failure with preserved ejection fraction. *Nat Rev Cardiol*, 2014. 11(9): p. 507–15.

21

DRUGS FOR HEART FAILURE

21.1 OVERVIEW

As introduced in Chapter 20, treatment of heart failure (HF) involves the use of various classes of drugs that target the different pathophysiological pathways underlying this syndrome. The major drug classes for treating HF include diuretics (Chapter 7), β-blockers (Chapter 8), inhibitors of the renin–angiotensin–aldosterone system (RAAS) (Chapter 9), vasodilators (Chapter 11), and positive inotropic agents. This chapter discusses the pharmacological basis of using the above drug classes in treating HF. Since inotropic drugs have not been covered elsewhere in the book, this chapter examines the molecular pharmacology of this drug class in HF treatment. The chapter also introduces emerging therapeutics for HF. The principles and guidelines regarding management of HF are considered in Chapter 22.

21.2 DIURETICS FOR HEART FAILURE

Diuretics inhibit the reabsorption of sodium or chloride at specific sites in the renal tubules (see Chapter 7). Loop diuretics (e.g., bumetanide, furosemide, and torsemide), as indicated by the name, act at the loop of Henle, whereas thiazides and thiazide-type drugs (e.g., chlorthalidone, hydrochlorothiazide, and indapamide) and potassium-sparing agents (e.g., eplerenone and spironolactone) act in the distal portion of the tubule. Loop diuretics are the most potent diuretic drugs among the above three drug classes and have emerged as the preferred diuretic agents for use in most patients with heart failure (HF). Thiazide and thiazide-type

diuretics may be considered in hypertensive patients with HF and mild fluid retention because they confer more persistent antihypertensive effects [1]. In contrast, potassium-sparing diuretics are not used for the purpose of diuresis due to their limited ability to directly reduce edema. However, these agents have been shown to retard the disease progression and reduce mortality of systolic HF independent of their diuretic activity [2, 3].

Clinical trials have demonstrated the ability of diuretic drugs to increase urinary sodium excretion and decrease physical signs of fluid retention in patients with HF. In intermediate-term studies, diuretics have been shown to improve symptoms and exercise tolerance in patients with HF. However, unlike β-blockers and RAAS inhibitors, the effects of diuretic effects on morbidity and mortality of patients with systolic HF are not known [1, 4]. Diuretics are the only drugs used for the treatment of HF that can adequately control the fluid retention of HF. Appropriate use of diuretics is a key element in the success of other drugs used for the treatment of HF. The use of inappropriately low doses of diuretics will result in fluid retention. Conversely, the use of inappropriately high doses of diuretics will lead to volume contraction, which can increase the risk of symptomatic hypotension and worsen renal function [1, 4].

21.3 β-BLOCKERS FOR HF

β-Blocker therapy (see Chapter 8), advocated for HF by some investigators since the 1970s, represents a major advance in the treatment of patients with systolic HF [5].

Cardiovascular Diseases: From Molecular Pharmacology to Evidence-Based Therapeutics, First Edition. Y. Robert Li.
© 2015 John Wiley & Sons, Inc. Published 2015 by John Wiley & Sons, Inc.

Multiple large-scale clinical trials have provided unequivocal evidence for the efficacy of β-blocker therapy in systolic HF. Long-term treatment with β-blockers can lessen the symptoms of systolic HF, improve the patient's clinical status, and enhance the patient's overall sense of well-being. In addition, like RAAS inhibitors, β-blockers can reduce the risk of death and the combined risk of death or hospitalization of patients with systolic HF. These benefits of β-blocker therapy have been observed in patients with or without coronary artery disease (CAD) and in patients with or without diabetes, as well as in women and blacks. Moreover, the favorable effects of β-blocker therapy have also been reported in patients already taking angiotensin-converting enzyme inhibitors (ACEIs) [1].

Three β-blockers have been shown to be effective in reducing the risk of death in patients with chronic systolic HF. They are bisoprolol, carvedilol, and sustained-release metoprolol succinate. However, the positive findings with these three β-blockers should not be considered a β-blocker class effect. In this context, bucindolol has been shown to lack uniform effectiveness across different populations, and short-acting metoprolol tartrate has been found to be less effective in HF clinical trials [1]. Hence, for β-blocker treatment of systolic HF, bisoprolol, carvedilol, or sustained-release metoprolol succinate should be chosen.

21.4 INHIBITORS OF THE RAAS FOR HF

As described in Chapter 9, inhibitors of RAAS include ACEIs, angiotensin receptor blockers (ARBs), aldosterone receptor antagonists (e.g., eplerenone and spironolactone; these are also known as potassium-sparing diuretics), and direct renin inhibitors (e.g., aliskiren). This section summarizes the clinical trial evidence on the efficacy of each of the above drug classes in treating systolic HF.

21.4.1 ACEIs

A number of large clinical trials have demonstrated that ACEIs can decrease the risk of death and reduce hospitalization of patients with systolic HF [6, 7]. Although the mortality benefit of ACEIs is strongest across New York Heart Association (NYHA) class II–IV HF, the benefits have been observed in patients with mild, moderate, or severe symptoms of systolic HF and in patients with or without CAD. Hence, ACEIs should be used in all HF patients with reduced left ventricular ejection fraction (LVEF), whether or not they are symptomatic. Unless there is a contraindication, ACEIs are used together with a β-blocker to achieve additive benefits. However, patients should not be given an ACEI if they have experienced life-threatening adverse reactions, such as angioedema during previous medication exposure,

or if they are pregnant or plan to become pregnant (also see Chapter 9) [1]. If the patients cannot tolerate ACEIs, ARBs may be considered.

21.4.2 ARBs

In several placebo-controlled studies, long-term therapy with ARBs produced hemodynamic, neurohormonal, and clinical effects consistent with those expected after interference with the RAAS in systolic HF [8–10]. Reduced hospitalization and mortality of patients with systolic HF by ARB therapy have been demonstrated. ACEIs remain the first choice for inhibition of the RAAS in systolic HF, but ARBs can now be considered a reasonable alternative [1]. In this context, ARBs are used in patients with systolic HF who are ACEI intolerant; an ACE-inhibition intolerance primarily related to persistent dry cough is the most common indication. In addition, an ARB may be used as an alternative to an ACEI in patients who are already taking an ARB for another reason, such as hypertension, and who subsequently develop systolic HF [1].

21.4.3 Aldosterone Receptor Antagonists

Spironolactone and eplerenone (also see Chapters 7 and 9) block receptors that bind aldosterone and other corticosteroids and are also known as mineralocorticoid receptor antagonists. These drugs are also classified as potassium-sparing diuretics though they are not used as direct diuretic agents due to limited ability of diuresis. Likely via blocking aldosterone-mediated inflammation and cardiovascular remodeling, both spironolactone and eplerenone exert beneficial effects in systolic HF. The landmark RALES trial demonstrated that blockade of aldosterone receptors by spironolactone, in addition to standard therapy, substantially reduced the risk of both morbidity and death among patients with severe diastolic HF; a remarkable 30% reduction in the risk of death among patients in the spironolactone group was attributed to a lower risk of both death from progressive HF and sudden death from cardiac causes. The frequency of hospitalization for worsening HF was also 35% lower in the spironolactone group than in the placebo group. In addition, patients who received spironolactone had a significant improvement in the symptoms of HF, as assessed on the basis of the NYHA functional class [3]. Similarly, eplerenone has been shown to reduce all-cause deaths, cardiovascular deaths, or HF hospitalizations in a wider range of patients with systolic HF [2, 11, 12]. In view of the substantial benefits of aldosterone receptor antagonist therapy, clinicians should strongly consider the addition of spironolactone or eplerenone for all patients with HF with reduced ejection fraction (HF-REF) who are already on ACEIs (or ARBs) and β-blockers [1].

21.4.4   Direct Renin Inhibitors

Aliskiren is a first-in-class drug that directly inhibits renin, a key enzyme of the RAAS (see Chapter 9). The ASTRONAUT trial is an international double-blind, placebo-controlled study involving 1639 patients with HF-REF, and the study concluded that among patients hospitalized for HF-REF, initiation of aliskiren in addition to standard therapy did not reduce cardiovascular death or HF rehospitalization at 6 or 12 months after discharge [13]. A subsequent prespecified subgroup analysis from the ASTRONAUT trial suggested that the addition of aliskiren to standard HF therapy in nondiabetic patients was generally well tolerated and improved postdischarge outcomes and biomarker profiles. In contrast, diabetic patients receiving aliskiren appeared to have worse postdischarge outcomes [14]. In view of the above data, aliskiren is not presently recommended as an alternative to an ACEI or ARB in treating systolic HF.

21.5   VASODILATORS FOR HF

21.5.1   Vasodilators in Acute HF

Vasodilators used in the management of HF include nitrates (e.g., nitroglycerin, isosorbide dinitrate), hydralazine, and nesiritide (see Chapter 11). They are often used in treating acute HF. Although vasodilators such as nitrates reduce preload and afterload and increase stroke volume, there is no solid evidence that they relieve dyspnea or improve other clinical outcomes [4]. Vasodilators are probably most useful in patients with hypertension and should be avoided in patients with a systolic blood pressure less than 110 mmHg. Excessive falls in blood pressure should also be avoided because hypotension is associated with higher mortality in patients with acute HF [4].

Nesiritide (a recombinant form of human B-type natriuretic peptide that causes vasodilation) is approved in the United States for early relief of dyspnea in patients with acute HF. A recent randomized trial involving 7141 patients who were hospitalized with acute HF reported that nesiritide therapy was not associated with an increase or a decrease in the rate of death and rehospitalization and had a small, nonsignificant effect on dyspnea when used in combination with other therapies. Although nesiritide was not associated with a worsening of renal function, it was associated with an increase in rates of hypotension. On the basis of these results, the trial concluded that nesiritide cannot be recommended for routine use in the broad population of patients with acute HF [15]. In line with the above notion, the more recent Renal Optimization Strategies Evaluation (ROSE) study, a multicenter, double-blind, placebo-controlled trial with 360 hospitalized patients with acute HF and renal dysfunction, concluded that in participants with acute HF and renal dysfunction, neither low-dose dopamine nor low-dose nesiritide

enhanced decongestion or improved renal function when added to diuretic therapy [16].

21.5.2   Vasodilators in Chronic HF

The combination of isosorbide dinitrate and hydralazine has been used for treating chronic HF, especially in African Americans [1, 17]. An early randomized trial involving 642 men with impaired cardiac function and reduced exercise tolerance reported that addition of hydralazine and isosorbide dinitrate to the therapeutic regimen of digoxin and diuretics in patients with chronic congestive HF exerted a favorable effect on left ventricular function and mortality. Notably, the mortality-risk reduction in the group treated with hydralazine and isosorbide dinitrate was 36% by 3 years [18]. However, as summarized elsewhere [1], in two other trials that compared the vasodilator combination with an ACEI, the ACEI produced more favorable effects on survival in patients with chronic congestive HF. A post hoc retrospective analysis of these vasodilator trials demonstrated particular efficacy of isosorbide dinitrate and hydralazine in the African American cohort [1]. In a subsequent large-scale trial (A-HeFT) involving a total of 1050 black patients who had NYHA class III or IV HF with dilated ventricles demonstrated that the addition of a fixed-dose combination of hydralazine and isosorbide dinitrate to standard therapy with an ACEI or ARB, a β-blocker, and an aldosterone receptor antagonist offered significant benefits. Notably, the study was terminated early owing to a significantly higher mortality rate in the placebo group than in the group given isosorbide dinitrate plus hydralazine. Addition of the fixed-dose combination of hydralazine and isosorbide dinitrate resulted in a 43% relative reduction in mortality, 33% relative reduction in the rate of first hospitalization for HF, and a significant improvement in the quality of life in black patients with advanced HF [19]. The favorable results of the A-HeFT study provide the foundation for guideline-based class I recommendations on using the combination of hydralazine and isosorbide dinitrate to reduce morbidity or mortality in African Americans with systolic HF [1, 4] (see Chapter 22).

21.6   POSITIVE INOTROPIC AGENTS FOR HF

Listed below are the US Food and Drug Administration (FDA)-approved positive inotropic agents (PIAs) for use in the management of HF. Digoxin is a digitalis glycoside. Dobutamine and dopamine are β-adrenergic receptor agonists, whereas inamrinone and milrinone are phosphodiesterase 3 (PDE3) inhibitors:

- Digoxin (Digox)
- Dobutamine (Dobutrex)

- Dopamine (Intropin)
- Inamrinone (Inocor)
- Milrinone (Primacor)

21.6.1 Introduction

In cardiovascular medicine, inotropic agents (also known as inotropes) refer to the drugs that alter the contractility of the myocardium. Negative inotropic agents weaken the force of myocardial contraction, whereas PIAs increase the strength of myocardial contraction. PIAs are typically classified into the following four classes: (1) digitalis (e.g., digoxin), (2) β-adrenergic receptor agonists (e.g., dobutamine and dopamine), (3) PDE3 inhibitors (e.g., inamrinone and milrinone), and (4) calcium-sensitizing agents (e.g., levosimendan, not approved for use in the United States).

PIAs have beneficial hemodynamic effects in patients with acute systolic HF due primarily to a direct increase in cardiac output. While the efficacy of acute PIA therapy is well established in patients with acute decompensated HF, long-term use of these agents (except for digitoxin) is associated with increased mortality.

21.6.2 Digitalis (Digoxin)

21.6.2.1 Chemistry and Pharmacokinetics
Digoxin (structure shown in Fig. 21.1) is one of the cardiac (or digitalis) glycosides, a closely related group of drugs having in common specific effects on the myocardium. These drugs are found in a number of plants. Digoxin is extracted from the leaves of *Digitalis lanata*.

The term "digitalis" is used to designate the whole group of glycosides. The glycosides are composed of two portions: a sugar and a cardenolide (hence "glycosides"). Digoxin is the only digitalis approved by the US FDA for clinical use.

Digoxin is readily absorbed following oral administration with oral bioavailability ranging from 60 to 80%. Digoxin is concentrated in tissues and therefore has a large apparent volume of distribution. Only a small percentage (16%) of a dose of digoxin is metabolized via a cytochrome P450-independent mechanism. A major portion of the drug is excreted unchanged in the urine with an elimination half-life of 1.5–2 days.

21.6.2.2 Molecular Mechanisms and Pharmacological Effects
Digoxin inhibits Na$^+$/K$^+$-ATPase. Inhibition of the enzyme on the plasma membrane of cardiomyocytes leads to an increase in the intracellular concentration of Na$^+$, which in turn stimulates Na$^+$/Ca^{2+}exchange, leading to an increase in the intracellular concentration of Ca^{2+} and thereby increased myocardial contraction (Fig. 21.2).

The beneficial effects of digoxin result from its direct action on cardiac muscle as mentioned earlier, as well as an indirect action mediated by the autonomic nervous system. The autonomic effects of digoxin include (i) a vagomimetic action, which is responsible for the effects of the drug on the

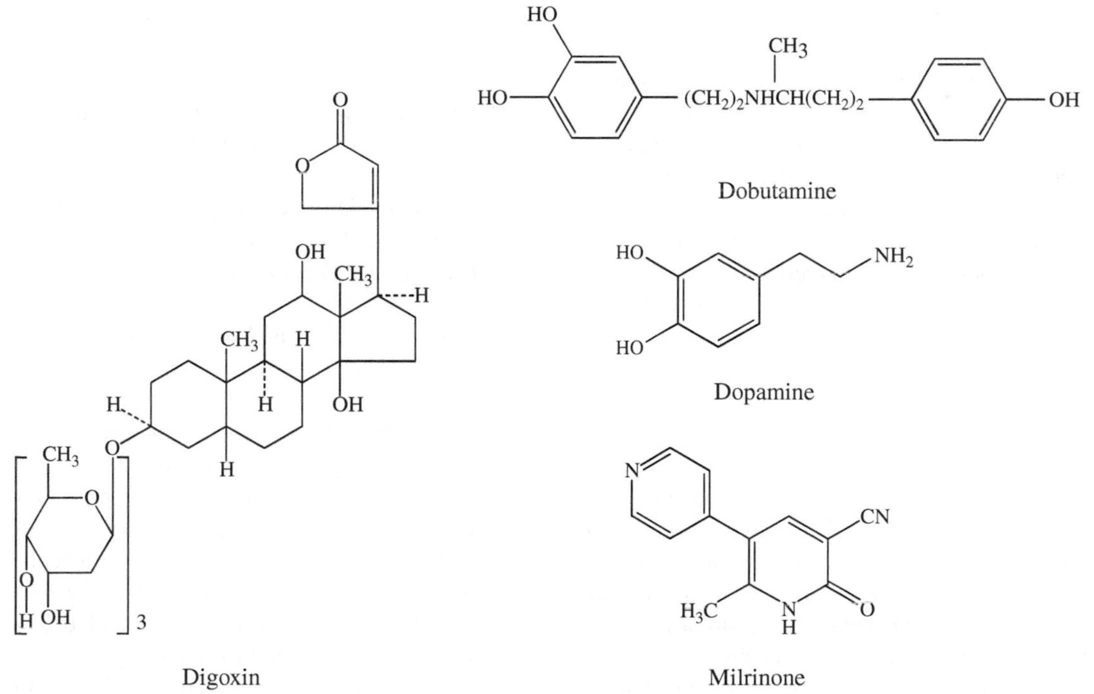

FIGURE 21.1 Structures of positive inotropic agents. Digoxin is a digitalis glycoside that is present naturally in the leaves of foxglove (*Digitalis lanata*). Dobutamine and dopamine are catecholamines, whereas milrinone is a bipyridine derivative.

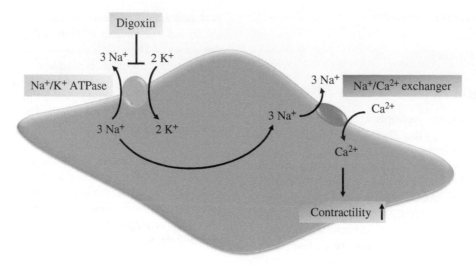

FIGURE 21.2 Molecular mechanisms of action of digoxin as a positive inotrope in treating heart failure. Digoxin inhibits Na^+/K^+-ATPase. Inhibition of this enzyme on the plasma membrane of cardiomyocytes leads to an increase in the intracellular concentration of Na^+, which in turn stimulates Na^+/Ca^{2+}exchange, leading to an increase in the intracellular concentration of Ca^{2+} and thereby increased myocardial contraction.

sinoatrial and atrioventricular (AV) nodes, and (ii) baroreceptor sensitization, which results in increased afferent inhibitory activity and reduced activity of the sympathetic nervous system and RAAS for any given increment in mean arterial pressure.

The pharmacological consequences of the above direct and indirect effects are (i) an increase in the force and velocity of myocardial systolic contraction (positive inotropic action), (ii) a decrease in the degree of activation of the sympathetic nervous system and RAAS (neurohormonal deactivating effect), and (iii) slowing of the heart rate and decreased conduction velocity through the AV node (vagomimetic effect).

The beneficial effects of digoxin in HF are mediated by its positive inotropic and neurohormonal deactivating effects, whereas the effects of the drug in atrial arrhythmias are related, at least partly, to its vagomimetic actions. However, in high doses, digoxin increases sympathetic outflow from the central nervous system (CNS). This increase in sympathetic activity may be an important factor in digoxin toxicity.

21.6.2.3 Clinical Uses

HF Digoxin is indicated for the treatment of mild-to-moderate systolic HF. Digoxin increases LVEF and improves HF symptoms as evidenced by exercise capacity and reduced HF-related hospitalizations and emergency care [20, 21]. However, the effect of digoxin on mortality remains controversial. While the initial DIG trial reported no effect of digoxin on the mortality of patients with systolic HF [20], a prespecified subgroup analysis of the DIG trial reported that digoxin therapy also exerted survival benefits in addition to decreased hospitalization in high-risk chronic HF

patients with NYHA classes III–IV, LVEF less than 25%, or cardiothoracic ratio greater than 55% [22]. Regardless of the controversies on its survival benefits, where possible, digoxin should be used with a diuretic and an ACEI (or ARB).

Chronic Atrial Fibrillation Digoxin is indicated for the control of ventricular response rate in patients with chronic atrial fibrillation. In these patients, digoxin slows rapid ventricular response rate in a linear dose–response fashion from 0.25 to 0.75 mg/day. However, digoxin should not be used for the treatment of multifocal atrial tachycardia.

21.6.2.4 Therapeutic Dosages
Listed below are the dosage forms and strengths of digoxin:

- Digoxin (Digox): Oral, 125 and 250 µg (mcg) tablets; injection (intravenous (iv) or intramuscular (im)), 500 µg (mcg)/2 ml ampules

In selecting a digoxin dosing regimen, it is important to consider factors that affect digoxin blood levels (e.g., body weight, age, renal function, concomitant drugs) since toxic levels of digoxin are only slightly higher than therapeutic levels (narrow therapeutic index). Dosing can be either initiated with a loading dose followed by maintenance dosing if rapid titration is desired or initiated with maintenance dosing without a loading dose.

Table 21.1 summarizes the recommended dosage regimens for both oral and parenteral administration. It should be noted that parenteral administration of digoxin should be used only when the need for rapid digitalization is urgent or when the drug cannot be taken orally. IM injection can lead to severe pain at the injection site; thus, IV administration is preferred. If the drug must be administered by the IM route,

TABLE 21.1   Digoxin dosage regimens for treating HF and chronic atrial fibrillation

Route	Loading dose	Maintenance dose
Oral	8–12 µg (mcg): administer half the total loading dose initially, then 25% the loading dose every 6–8 h twice	125–500 µg (mcg) once daily
IV	8–12 µg (mcg): administer half the total loading dose initially, then 25% the loading dose every 6–8 h twice	Starting maintenance dose: 2.4–3.6 µg (mcg)/kg/day, given once daily. Doses may be increased every 2 weeks according to clinical response, plasma drug levels, and toxicity

TABLE 21.2   Major pharmacokinetic properties of dobutamine and dopamine

Drug	Onset of action	Elimination half-life	Metabolism and elimination
Dobutamine	1–2 min	2 min	Catechol-O-methyltransferase (COMT); eliminated in the urine
Dopamine	<5 min	2 min	Monoamine oxidase; COMT; eliminated in the urine

it should be injected deep into the muscle followed by massage. For adults, no more than 500 µg (mcg) of digoxin should be injected into a single site. It cannot be overemphasized that the recommended dosage regimens are based upon average patient response, and substantial individual variation can be expected. Accordingly, ultimate dosage selection must be based upon clinical assessment of the individual patient.

21.6.2.5 Adverse Effects and Drug Interactions

Adverse Effects   Digoxin adverse effects mainly involve the heart, the gut, and the CNS. Therapeutic doses of digoxin can cause heart block in patients with preexisting sinoatrial or AV conduction disorders, and heart block can be avoided by adjusting the dose of digoxin. Digoxin causes gut disturbance, leading to anorexia, nausea, vomiting, and diarrhea. The CNS effects of digoxin include visual disturbances (blurred or yellow vision), headache, weakness, dizziness, apathy, confusion, and mental disturbances (e.g., anxiety, depression, delirium, and hallucination). In addition, gynecomastia has been occasionally observed following the prolonged use of digoxin. Thrombocytopenia and maculopapular rash and other skin reactions have been rarely observed.

Drug Interactions   Potassium-depleting diuretics including thiazides and loop diuretics are a major contributing factor to digitalis toxicity. Calcium, particularly if administered rapidly by the IV route, may produce serious arrhythmias in digitalized patients. Quinidine, verapamil, amiodarone, propafenone, indomethacin, itraconazole, alprazolam, and spironolactone raise the plasma digoxin concentration due to a reduction in clearance and/or in volume of distribution of the drug. Erythromycin and clarithromycin (and possibly other macrolide antibiotics) and tetracycline may increase digoxin absorption in patients who inactivate digoxin by bacterial metabolism in the lower intestine, so that digitalis intoxication may result.

Contraindications and Pregnancy Category

- Digoxin is contraindicated in patients with ventricular fibrillation or in patients with a known hypersensitivity to digoxin. A hypersensitivity reaction to other digitalis preparations usually constitutes a contraindication to digoxin.
- Pregnancy category: C.

21.6.3   Dobutamine and Dopamine

21.6.3.1 Chemistry and Pharmacokinetics   Dobutamine and dopamine are β-adrenergic receptor agonists that stimulate cardiac contractility. Dopamine is a natural catecholamine formed by the decarboxylation of 3,4-dihydroxyphenylalanine, whereas dobutamine is a synthetic catecholamine (structures shown in Fig. 21.1). The major pharmacokinetic properties of these two drugs are given in Table 21.2.

21.6.3.2 Molecular Mechanisms and Pharmacological Effects

Dobutamine   Dobutamine is a direct-acting inotropic agent whose primary activity results from stimulation of the β-adrenergic receptors of the heart while producing comparatively mild chronotropic, hypertensive, arrhythmogenic, and vasodilative effects. It does not cause the release of endogenous norepinephrine. In animal studies, dobutamine produces less increase in heart rate and less decrease in peripheral vascular resistance for a given inotropic effect than isoproterenol.

In patients with depressed cardiac function, both dobutamine and isoproterenol increase the cardiac output to a similar degree. In the case of dobutamine, this increase is usually not accompanied by marked increases in heart rate (although tachycardia is occasionally observed), and the cardiac stroke volume is usually increased. In contrast, isoproterenol increases the cardiac index primarily by increasing the heart rate, while stroke volume changes little or declines.

Dopamine Dopamine produces positive chronotropic and inotropic effects on the myocardium, resulting in increased heart rate and cardiac contractility. This is accomplished directly by exerting an agonist action on β-adrenergic receptors and indirectly by causing release of norepinephrine from storage sites in sympathetic nerve endings.

The cardiovascular effects of dopamine are dose related: (i) at low rates of infusion (0.5–2 μg/kg/min), it causes vasodilation possibly due to specific activation of dopamine receptors in the renal, mesenteric, coronary, and intracerebral vascular beds. The vasodilation in renal vascular beds is accompanied by increased glomerular filtration rate, renal blood flow, sodium excretion, and urine flow. (ii) At intermediate rates of infusion (2–10 μg/kg/min), dopamine acts to stimulate the β_1-adrenergic receptors, resulting in augmented myocardial contractility, increased sinus rate, and enhanced impulse conduction in the heart. There is little, if any, stimulation of the β_2-adrenergic receptors (activation of β_2-adrenergic receptors causes peripheral vasodilation). (iii) At higher rates of infusion (10–20 μg/kg/min), dopamine causes activation of α-adrenergic receptors, leading to vasoconstriction and increased blood pressure.

21.6.3.3   Clinical Uses

Dobutamine Dobutamine injection is indicated when parenteral therapy is necessary for inotropic support in the short-term treatment of adults with cardiac decompensation due to depressed contractility resulting either from organic heart disease or from cardiac surgical procedures. In patients who have atrial fibrillation with rapid ventricular response, a digitalis preparation should be used prior to institution of therapy with dobutamine. Recent studies suggested that dobutamine is the most commonly used PIA in patients hospitalized with HF, followed by dopamine and milrinone [23, 24].

Dopamine Dopamine is indicated for the correction of hemodynamic imbalances present in the shock syndrome due to myocardial infarctions, trauma, endotoxic septicemia, open heart surgery, renal failure, and chronic cardiac decompensation as in congestive HF. However, a recent multicenter, double-blind, placebo-controlled clinical trial (ROSE) of 360 hospitalized patients with acute HF and renal dysfunction concluded that in patients with acute HF and renal dysfunction, low-dose dopamine did not enhance decongestion or improve renal function when added to diuretic therapy [16].

21.6.3.4   Therapeutic Dosages
Listed below are the dosage forms and strengths of dobutamine and dopamine. The recommended dosage regimens in treating acute HF are summarized in Table 21.3:

TABLE 21.3   Recommended dosage regimens of dobutamine and dopamine

Drug	Dosage regimen
Dobutamine	Typical dose range: 2.5–15 μg (mcg)/kg/min. On rare occasions, infusion rates up to 40 μg (mcg)/kg/min are required to obtain the desired effect
Dopamine	• Initial dosage rate: 2–5 μg (mcg)/kg/min in patients who are likely to respond to modest increments of heart force and renal perfusion • In more seriously ill patients, begin administration at a dose rate of 5 μg (mcg)/kg/min and increase gradually, using 5–10 μg (mcg)/kg/min increments, up to 20–50 μg (mcg)/kg/min as needed

- Dobutamine (Dobutrex): IV, 12.5 mg/ml available as 20 ml single-dose vials containing 250 mg dobutamine or 40 ml single-dose vials containing 500 mg dobutamine
- Dopamine: IV, 800 mg/5 ml (160 mg/ml) single-dose vials

21.6.3.5   Adverse Effects and Drug Interactions

Adverse Effects The major adverse effects of dobutamine and dopamine are related to the cardiovascular responses, including increased heart rate and blood pressure, and ventricular arrhythmias. Precipitous decreases in blood pressure have occasionally been described in association with these drugs. Other uncommon effects may include nausea, headache, anginal pain, nonspecific chest pain, palpitations, and shortness of breath.

Drug Interactions Dobutamine may be ineffective if the patient has recently received a β-blocker. In such a case, the peripheral vascular resistance may increase. Because dopamine is metabolized by monoamine oxidase (MAO), inhibition of this enzyme prolongs and potentiates the effect of dopamine. Patients who have been treated with MAO inhibitors within 2–3 weeks prior to the administration of dopamine should receive initial doses of dopamine not greater than one tenth (1/10) of the usual dose. Dopamine may also interact with cyclopropane or halogenated hydrocarbon anesthetics to cause arrhythmias. Administration of phenytoin to patients receiving dopamine has been reported to lead to hypotension and bradycardia.

Contraindications and Pregnancy Category

- Dobutamine is contraindicated in patients with idiopathic hypertrophic subaortic stenosis and in patients who have shown previous manifestations of hypersensitivity to the drug.

- Dopamine should not be used in patients with pheochromocytoma. It should not be administered in the presence of uncorrected tachyarrhythmias or ventricular fibrillation.
- Pregnancy category: B (dobutamine) and C (dopamine).

21.6.4   PDE3 Inhibitors

21.6.4.1   Chemistry and Pharmacokinetics   Inamrinone (also called amrinone) and milrinone (structure shown in Fig. 21.1) represent a new class of PIAs distinct from digitalis glycosides or β-adrenergic receptor agonists. These drugs are relatively selective inhibitors of PDE3 and as such are commonly known as PDE3 inhibitors. Since inamrinone is rarely used, this section focuses on milrinone. Milrinone is a bipyridine derivative. Following IV administration, milrinone is approximately 70% bound to plasma proteins, and the elimination half-life is about 2 h. The drug is primarily eliminated unchanged in the urine.

21.6.4.2   Molecular Mechanisms and Pharmacological Effects   Milrinone is a positive inotrope and vasodilator, with little chronotropic activity. At therapeutic doses, milrinone is a selective inhibitor of PDE3 in both myocardium and vascular smooth muscle. Inhibition of PDE3 in cardiomyocytes causes increased levels of 3′-5′-cyclic adenosine monophosphate (cAMP), which in turn cause increases in intracellular Ca^{2+}, thereby augmenting contractile force of the myocardium. On the other hand, increased levels of cAMP in smooth muscle cells augment cAMP-dependent contractile protein phosphorylation, thereby leading to smooth muscle relaxation and vasodilation.

21.6.4.3   Clinical Uses   Milrinone is indicated for the short-term IV treatment of patients with acute decompensated HF. However, the exact benefit of short-term IV milrinone in treating patients hospitalized for an exacerbation of chronic HF remains questionable [25, 26].

21.6.4.4   Therapeutic Dosages   Listed below are the dosage forms and strengths of milrinone:

- Milrinone (Primacor): IV, 1 mg/ml supplied in 10 ml (10 mg/10 ml) and 20 ml (20 mg/20 ml) single-dose vials

Milrinone should be administered with a loading dose followed by a continuous infusion (maintenance dose). The recommended loading dose is 50 μg (mcg)/kg, administered slowly over 10 min. The standard maintenance dose is 0.5 μg (mcg)/kg/min with a total daily dose of 0.77 mg/kg.

21.6.4.5   Adverse Effects and Drug Interactions   The most common adverse effects of milrinone therapy are ventricular arrhythmias. Other cardiovascular adverse effects include hypotension and angina. The drug may also cause headaches, usually mild to moderate in severity. Drug interactions seem to be insignificant for milrinone therapy. Milrinone is contraindicated in patients who are hypersensitive to it. The drug is in pregnancy category C.

21.6.5   Calcium-Sensitizing Agents

Calcium-sensitizing agents are those that increase the sensitivity of the cardiac myocardial contractile apparatus to calcium, causing an increase in tension development and myocardial contractility. They represent a novel class of investigational PIAs for treating HF. One of them, namely, levosimendan, has undergone multiple clinical trials for treating acute decompensated HF [27–29]. Overall, these trials demonstrated that IV levosimendan seemed to improve hemodynamic performance and relief symptoms in patients with acute decompensated HF [27–29]. Levosimendan is available in a number of countries. However, it is not a US FDA-approved drug.

21.7   EMERGING DRUGS FOR HF

As discussed earlier, the management of systolic HF has been advanced substantially over the past decades due to better understanding of the underlying pathophysiology and the mechanistically based therapies with β-blockers and RAAS inhibitors as well as diuretics and inotropes. However, the 5-year morality rate of HF remains unacceptably high. This has necessitated continued search for more effective therapeutic modalities. In this context, a number of novel therapeutic agents for HF have emerged and undergone randomized clinical trials with promising results. Summarized below are some of the promising agents;

1. *Ivabradine*: This novel drug inhibits sinus rate by specifically blocking I_f current of the pacemaker cells (see Chapter 14). It was shown to improve the clinical outcomes in patients with chronic HF-REF with a heart rate greater than or equal to 70 beats/min [30, 31].
2. *Sildenafil*: This PDE5 inhibitor is also effective in treating pulmonary hypertension (see Chapter 12). It was found to improve hemodynamic parameters in patients with HF-REF [32, 33].
3. *Serelaxin*: It is a recombinant form of human relaxin-2, a vasoactive peptide hormone with many biological and hemodynamic effects, including vasodilation and stimulation of myocardial contraction [34]. Serelaxin treatment of patients with acute HF (with either preserved or reduced ejection fraction) was well tolerated and associated with dyspnea relief and improvement in other clinical outcomes, including reduced mortality irrespective of ejection fraction [35, 36];

4. *Cardiac stem cell and gene therapy* [37, 38]: The promise of cardiac gene therapy is exemplified with the notable success of a phase 1/phase 2 first-in-human clinical gene therapy trial (CUPID) using an adeno-associated virus serotype 1 (AAV1) vector carrying the sarcoplasmic reticulum calcium ATPase gene (AAV1/SERCA2a) in patients with advanced HF [39]. The CUPID trial reported that the risk of pre-specified recurrent cardiovascular events was reduced by 82% in the high-dose AAV1/SERCA2a versus placebo group and no safety concerns were noted during a 3-year follow-up [39]. Deficiency/dysfunction of SERCA2a, a protein involved in reloading sarcoplasmic reticulum with Ca^{2+} during systole, is an important mechanism of systolic HF [40]. The success of the CUPID study has prompted a phase 2b trial to further investigate the efficacy and safety of intracoronary administration of AAV1/SERCA2a in patients with advanced HF [41].

21.8 SUMMARY OF CHAPTER KEY POINTS

- Treatment of HF involves the use of various classes of drugs, including diuretics, β-blockers, RAAS inhibitors, vasodilators, and PIAs.

- Diuretic drugs increase urinary sodium excretion and decrease physical signs of fluid retention in patients with HF. Loop diuretics are preferred drugs for fluid volume reduction in most patients with HF. The effects of diuretic effects on morbidity and mortality of patients with systolic HF are not known.

- β-Blocker therapy represents a major advance in the treatment of patients with systolic HF. Long-term treatment with β-blockers relieves the symptoms of systolic HF, improves the patient's clinical status, and enhances the patient's overall sense of well-being. Like RAAS inhibitors, β-blockers can reduce the risk of death and the combined risk of death or hospitalization of patients with systolic HF.

- ACEIs reduce the risk of death and decrease hospitalization of patients with systolic HF of various severities. ACEIs should be used in all HF patients with reduced LVEF and be used together with a β-blocker to achieve additive benefits.

- Long-term therapy with ARBs reduces hospitalization and mortality of patients with systolic HF. ACEIs remain the first choice for inhibition of the RAAS in systolic HF, but ARBs can now be considered a reasonable alternative in patients with systolic HF who are ACEI intolerant.

- Both spironolactone and eplerenone exert beneficial effects in systolic HF, including improving symptoms and reducing mortality and HF hospitalizations. Clinicians should strongly consider the addition of spironolactone or eplerenone for all patients with HF-REF who are already on ACEIs (or ARBs) and β-blockers.

- Vasodilators used in the management of HF include nitrates (e.g., nitroglycerin, isosorbide dinitrate), hydralazine, and nesiritide. They are often used in treating acute HF though the efficacy in this clinical setting remains unclear.

- Addition of a fixed-dose combination of hydralazine and isosorbide dinitrate to standard therapy with an ACEI or ARB, a β-blocker, and an aldosterone receptor antagonist can offer significant benefits in black patients with advanced systolic HF. These benefits include remarkable reduction in mortality and the rate of first hospitalization for heart failure and a significant improvement in the quality of life.

- PIAs include digitalis (e.g., digoxin), β-adrenergic receptor agonists (e.g., dobutamine and dopamine), PDE3 inhibitors (e.g., inamrinone and milrinone), and calcium-sensitizing agents (e.g., levosimendan). In general, these drugs have beneficial hemodynamic effects in patients with acute systolic HF due primarily to a direct increase in cardiac output. While the efficacy of acute PIA therapy is well established in patients with acute decompensated HF, long-term use of many of these agents (except for digoxin) is associated with increased mortality.

- A number of novel therapeutic agents for HF have emerged and undergone randomized clinical trials with promising results. These include the sinus rate inhibitor ivabradine, the PDE5 inhibitor sildenafil, the recombinant form of human relaxin-2 serelaxin, as well as gene therapy. Further development of these and other novel therapeutic modalities will contribute to the effective management of HF.

21.9 SELF-ASSESSMENT QUESTIONS

21.9.1. Following an acute myocardial infarction, a patient develops signs of hypoperfusion requiring drug management. What effects would milrinone, digoxin, and dobutamine have in common if each were administered individually to the patient?
   A. Decrease of atrioventricular conduction
   B. Decrease of cAMP metabolism
   C. Decrease of venous return
   D. Increase of cardiac afterload
   E. Increase of ventricular contractility

21.9.2. A 60-year-old male being treated with a positive inotropic agent is brought to the emergency department due to nausea and vomiting, blurred and abnormally colored vision, and palpitations. Which of the following would be the most suitable antidote if a drug overdosage is suspected?

A. Abciximab
B. Digoxin immune fab
C. Insulin
D. Magnesium sulfate
E. *N*-Acetylcysteine

21.9.3. A 71-year-old male comes to his physician complaining of severe shortness of breath at night and swelling of the ankles. He has reduced his activity over the past 6 months because of chest pain when he exerts himself. Physical examination reveals rales over both lungs, enlargement of the liver, and pitting edema of the ankles. Blood pressure is 140/90 mmHg and heart rate is 95 beats/min. His ECG is normal at rest. Cardiac enzymes and blood electrolytes are normal. Because the major immediate problem in this patient is heart failure, which of the following drug therapies should be initiated right away for his condition?

A. Atenolol and quinidine
B. Furosemide and lisinopril
C. Nitroglycerin and spironolactone
D. Prazosin and losartan
E. Verapamil and digoxin

21.9.4. A 55-year-old male is having acute decompensated heart failure following a myocardial infarction. His acute condition requires drug treatment to increase cardiac output. Which of the following would be most suitable for improving hypoperfusion in this patient?

A. Dobutamine
B. Lisinopril
C. Losartan
D. Metoprolol
E. Nitroprusside

21.9.5. A 65-year-old male is diagnosed with advanced chronic systolic heart failure. Which of the following drugs would be most likely prescribed to prolong his life?

A. Digoxin
B. Furosemide
C. Hydralazine
D. Nesiritide
E. Spironolactone

21.9.6. A 60-year-old African American male with chronic systolic heart failure is being treated with a combination regimen consisting of hydrochlorothiazide, lisinopril, and metoprolol. Which of the following, if added to the above regimen, would most likely further prolong his life?

A. Amlodipine plus nitroglycerin
B. Dobutamine plus digoxin
C. Furosemide plus milrinone
D. Hydralazine plus isosorbide dinitrate
E. Nesiritide plus dopamine

21.9.7. A 65-year-old male presenting with an ST-elevation acute coronary syndrome develops cardiogenic shock. He is given a positive inotrope to improve cardiac output. This inotrope is a substrate of monoamine oxidase. Which of the following drugs is most likely used?

A. Digoxin
B. Dobutamine
C. Dopamine
D. Milrinone
E. Serelaxin

21.9.8. A 72-year-old female with chronic systolic heart failure of 3 years of duration is being placed on spironolactone in addition to standard therapy to improve symptoms and prolong survival. What is the most likely mechanism of action for spironolactone to prolong this patient's life?

A. Decreased cardiac remodeling
B. Decreased nitric oxide bioavailability
C. Increased diuresis
D. Inhibition of Na^+/K^+-ATPase
E. Positive inotropic effect

21.9.9. A 65-year-old male develops angioedema following a multidrug treatment regimen for his heart failure. Which of the following drugs is most likely responsible for causing his angioedema?

A. Carvedilol
B. Chlorthalidone
C. Eplerenone
D. Irbesartan
E. Lisinopril

21.9.10. A 62-year-old male with a history of myocardial infarction presents to his physician complaining of shortness of breath. On examination, his heart rate is 110 beats/min and respiratory rate is 22 min. He has rales in both lung fields, a normal sinus rhythm, and a mild pitting ankle edema. A chest X-ray reveals cardiomegaly, and his ejection fraction on echocardiogram is 39%. Which of the following drugs would alleviate this patient's symptoms by reducing both the preload and afterload on the heart without affecting its inotropic state?

A. Digoxin
B. Diltiazem
C. Enalapril
D. Milrinone
E. Propranolol

REFERENCES

1 Yancy, C.W., et al., 2013 ACCF/AHA guideline for the management of heart failure: a report of the American College of Cardiology Foundation/American Heart Association Task

Force on Practice Guidelines. *J Am Coll Cardiol*, 2013. 62(16): p. e147–239.

2   Zannad, F., et al., Eplerenone in patients with systolic heart failure and mild symptoms. *N Engl J Med*, 2011. 364(1): p. 11–21.

3   Pitt, B., et al., The effect of spironolactone on morbidity and mortality in patients with severe heart failure. Randomized Aldactone Evaluation Study Investigators. *N Engl J Med*, 1999. 341(10): p. 709–17.

4   McMurray, J.J., et al., ESC guidelines for the diagnosis and treatment of acute and chronic heart failure 2012: the Task Force for the Diagnosis and Treatment of Acute and Chronic Heart Failure 2012 of the European Society of Cardiology. Developed in collaboration with the Heart Failure Association (HFA) of the ESC. *Eur Heart J*, 2012. 33(14): p. 1787–847.

5   Bristow, M.R., Treatment of chronic heart failure with beta-adrenergic receptor antagonists: a convergence of receptor pharmacology and clinical cardiology. *Circ Res*, 2011. 109(10): p. 1176–94.

6   Krum, H. and J.R. Teerlink, Medical therapy for chronic heart failure. *Lancet*, 2011. 378(9792): p. 713–21.

7   Lang, C.C. and A.D. Struthers, Targeting the renin-angiotensin-aldosterone system in heart failure. *Nat Rev Cardiol*, 2013. 10(3): p. 125–34.

8   Cohn, J.N., G. Tognoni, and Valsartan Heart Failure Trial Investigators, A randomized trial of the angiotensin-receptor blocker valsartan in chronic heart failure. *N Engl J Med*, 2001. 345(23): p. 1667–75.

9   Demers, C., et al., Impact of candesartan on nonfatal myocardial infarction and cardiovascular death in patients with heart failure. *JAMA*, 2005. 294(14): p. 1794–8.

10 Konstam, M.A., et al., Effects of high-dose versus low-dose losartan on clinical outcomes in patients with heart failure (HEAAL study): a randomised, double-blind trial. *Lancet*, 2009. 374(9704): p. 1840–8.

11 Pitt, B., et al., Eplerenone, a selective aldosterone blocker, in patients with left ventricular dysfunction after myocardial infarction. *N Engl J Med*, 2003. 348(14): p. 1309–21.

12 Eschalier, R., et al., Safety and efficacy of eplerenone in patients at high risk for hyperkalemia and/or worsening renal function: analyses of the EMPHASIS-HF study subgroups (Eplerenone in Mild Patients Hospitalization and SurvIval Study in Heart Failure). *J Am Coll Cardiol*, 2013. 62(17): p. 1585–93.

13 Gheorghiade, M., et al., Effect of aliskiren on postdischarge mortality and heart failure readmissions among patients hospitalized for heart failure: the ASTRONAUT randomized trial. *JAMA*, 2013. 309(11): p. 1125–35.

14 Maggioni, A.P., et al., Effect of aliskiren on post-discharge outcomes among diabetic and non-diabetic patients hospitalized for heart failure: insights from the ASTRONAUT trial. *Eur Heart J*, 2013. 34(40): p. 3117–27.

15 O'Connor, C.M., et al., Effect of nesiritide in patients with acute decompensated heart failure. *N Engl J Med*, 2011. 365(1): p. 32–43.

16 Chen, H.H., et al., Low-dose dopamine or low-dose nesiritide in acute heart failure with renal dysfunction: the ROSE acute heart failure randomized trial. *JAMA*, 2013. 310(23): p. 2533–43.

17 Gupta, D., et al., Nitrate therapy for heart failure: benefits and strategies to overcome tolerance. *JACC Heart Fail*, 2013. 1(3): p. 183–91.

18 Cohn, J.N., et al., Effect of vasodilator therapy on mortality in chronic congestive heart failure. Results of a Veterans Administration Cooperative Study. *N Engl J Med*, 1986. 314(24): p. 1547–52.

19 Taylor, A.L., et al., Combination of isosorbide dinitrate and hydralazine in blacks with heart failure. *N Engl J Med*, 2004. 351(20): p. 2049–57.

20 Digitalis Investigation Group, The effect of digoxin on mortality and morbidity in patients with heart failure. *N Engl J Med*, 1997. 336(8): p. 525–33.

21 Ambrosy, A.P., et al., The use of digoxin in patients with worsening chronic heart failure: reconsidering an old drug to reduce hospital admissions. *J Am Coll Cardiol*, 2014. 63(18): p. 1823–32.

22 Gheorghiade, M., et al., Effect of oral digoxin in high-risk heart failure patients: a pre-specified subgroup analysis of the DIG trial. *Eur J Heart Fail*, 2013. 15(5): p. 551–9.

23 Partovian, C., et al., Hospital patterns of use of positive inotropic agents in patients with heart failure. *J Am Coll Cardiol*, 2012. 60(15): p. 1402–9.

24 Allen, L.A., et al., Hospital variation in intravenous inotrope use for patients hospitalized with heart failure: insights from Get With The Guidelines. *Circ Heart Fail*, 2014. 7(2): p. 251–60.

25 Cuffe, M.S., et al., Short-term intravenous milrinone for acute exacerbation of chronic heart failure: a randomized controlled trial. *JAMA*, 2002. 287(12): p. 1541–7.

26 Felker, G.M., et al., Heart failure etiology and response to milrinone in decompensated heart failure: results from the OPTIME-CHF study. *J Am Coll Cardiol*, 2003. 41(6): p. 997–1003.

27 Follath, F., et al., Efficacy and safety of intravenous levosimendan compared with dobutamine in severe low-output heart failure (the LIDO study): a randomised double-blind trial. *Lancet*, 2002. 360(9328): p. 196–202.

28 Mebazaa, A., et al., Levosimendan vs dobutamine for patients with acute decompensated heart failure: the SURVIVE Randomized Trial. *JAMA*, 2007. 297(17): p. 1883–91.

29 Packer, M., et al., Effect of levosimendan on the short-term clinical course of patients with acutely decompensated heart failure. *JACC Heart Fail*, 2013. 1(2): p. 103–11.

30 Swedberg, K., et al., Ivabradine and outcomes in chronic heart failure (SHIFT): a randomised placebo-controlled study. *Lancet*, 2010. 376(9744): p. 875–85.

31 Borer, J.S., et al., Efficacy and safety of ivabradine in patients with severe chronic systolic heart failure (from the SHIFT study). *Am J Cardiol*, 2014. 113(3): p. 497–503.

32 Zhuang, X.D., et al., PDE5 inhibitor sildenafil in the treatment of heart failure: a meta-analysis of randomized controlled trials. *Int J Cardiol*, 2014. 172(3): p. 581–7.

33 Wu, X., et al., Additional use of a phosphodiesterase 5 inhibitor in patients with pulmonary hypertension secondary to chronic systolic heart failure: a meta-analysis. *Eur J Heart Fail*, 2014 p. 16: 444–53.

34 Bathgate, R.A., et al., Relaxin family peptides and their receptors. *Physiol Rev*, 2013. 93(1): p. 405–80.

35 Teerlink, J.R., et al., Serelaxin, recombinant human relaxin-2, for treatment of acute heart failure (RELAX-AHF): a randomised, placebo-controlled trial. *Lancet*, 2013. 381(9860): p. 29–39.

36 Filippatos, G., et al., Serelaxin in acute heart failure patients with preserved left ventricular ejection fraction: results from the RELAX-AHF trial. *Eur Heart J*, 2014. 35(16): p. 1041–50.

37 Braunwald, E., Heart failure. *JACC Heart Fail*, 2013. 1(1): p. 1–20.

38 Pleger, S.T., et al., Heart failure gene therapy: the path to clinical practice. *Circ Res*, 2013. 113(6): p. 792–809.

39 Zsebo, K., et al., Long-term effects of AAV1/SERCA2a gene transfer in patients with severe heart failure: analysis of recurrent cardiovascular events and mortality. *Circ Res*, 2014. 114(1): p. 101–8.

40 Kho, C., A. Lee, and R.J. Hajjar, Altered sarcoplasmic reticulum calcium cycling—targets for heart failure therapy. *Nat Rev Cardiol*, 2012. 9(12): p. 717–33.

41 Greenberg, B., et al., Design of a phase 2b trial of intracoronary administration of AAV1/SERCA2a in patients with advanced heart failure: the CUPID 2 trial (calcium up-regulation by percutaneous administration of gene therapy in cardiac disease phase 2b). *JACC Heart Fail*, 2014. 2(1): p. 84–92.

22

MANAGEMENT OF HEART FAILURE: PRINCIPLES AND GUIDELINES

22.1 OVERVIEW

As described in Chapter 20, heart failure (HF) is a clinical syndrome that can be manifested as acute and chronic HF, or HF with reduced ejection fraction (HF-REF) and HF with preserved ejection fraction (HF-PEF). HF-REF and HF-PEF are also loosely known as systolic HF and diastolic HF, respectively. In chronic HF, about half of the patients have HF-REF, which is the best understood HF with regard to pathophysiology and treatment. This chapter thus focuses on the management of HF-REF. It also covers the management of HF-PEF and acute HF syndromes.

22.2 MANAGEMENT OF HF-REF

HF-REF is characterized by abnormalities in systolic function, that is, reduced left ventricular ejection fraction (LVEF) usually with progressive chamber dilation and eccentric remodeling. Because the dominant abnormality is with systolic function, this syndrome, as noted earlier, is also known as systolic heart failure (HF). As noted in Chapter 20, HF-REF has been defined by a variety of LVEF partition values ranging from <35 to 50%. This book adopts the definition of the American College of Cardiology Foundation/American Heart Association (ACCF/AHA) [1], defining HF-REF as the clinical diagnosis of HF with LVEF ≤40%. This section first introduces current guidelines on the management of HF-REF including those from both the United States and other nations. The section then considers the 2013 ACCF/AHA guideline recommendations for the management of HF-REF with a focus on pharmacological treatment.

22.2.1 Introduction to Current Guidelines on the Management of HF-REF

The fundamental goals of HF treatment are to improve symptoms (including risk of hospitalization), slow or reverse deterioration in myocardial function, and reduce mortality. Several major organizations have published extensive guidelines for the management of HF (primarily HF-REF). These include the 2013 ACCF/AHA guideline for the management of heart failure [1], the 2012 European Society of Cardiology (ESC) guideline for the diagnosis and treatment of acute and chronic heart failure [2], and the 2010 Heart Failure Society of America (HFSA) comprehensive heart failure practice guideline [3]. With few exceptions, these organizations make similar recommendations regarding the treatment of HF-REF. Notably, the 2013 ACCF/AHA guideline provides a detailed coverage of all key areas of HF management, and the document is divided into many sections, including (1) introduction, (2) definition of HF, (3) HF classifications, (4) epidemiology, (5) cardiac structural abnormalities and other causes of HF, (6) initial and serial evaluation of the HF patient, (7) treatment of stages A to D, (8) the hospitalized patient, (9) important comorbidities in HF, (10) surgical/percutaneous/transcatheter interventional treatments of HF, (11) coordinating care for patients with chronic HF, (12) quality metrics/performance measures, (13) evidence gaps and future research directions, and (14) appendices and other online supplementary materials. The main document contains 924 references. Section 2.2 focuses on describing the guideline recommendations for treatment of HF stages A to D. The reader is advised to refer to the full guideline document for details.

Cardiovascular Diseases: From Molecular Pharmacology to Evidence-Based Therapeutics, First Edition. Y. Robert Li.
© 2015 John Wiley & Sons, Inc. Published 2015 by John Wiley & Sons, Inc.

22.2.2 The 2013 ACCF/AHA Guideline Recommendations for Treatment of HF Stages A to D

22.2.2.1 Recommendations for Treatment of HF Stage A

As discussed in Chapter 20, inclusion of stage A in HF classification by the ACC/AHA emphasizes the importance of treating the risk factors of HF, especially hypertension and dyslipidemias, to prevent or slow the subsequent development of HF. Table 22.1 summarizes the 2013 ACCF/AHA guideline recommendations for treatment of stage A HF.

22.2.2.2 Recommendations for Treatment of HF Stage B

Stage B HF patients have structural abnormalities and reduced LVEF, but do not have HF symptoms. These patients are most often identified during an evaluation for other disorders, such as abnormal heart sounds, abnormal electrocardiogram (ECG),

abnormal chest X-ray, hypertension or hypotension, arrhythmias, acute myocardial infarction (MI), or pulmonary or systemic thromboembolic events [1]. In general, all recommendations for patients with stage A HF also apply to those with stage B HF, particularly with respect to control of blood pressure in the patient with LV hypertrophy and the optimization of lipids with statins. CAD is a major risk factor for the development of symptomatic HF and a key target of evidence-based therapy for preventing the progression into clinical HF [1]. Table 22.2 outlines the 2013 ACCF/AHA guideline recommendations for the treatment of stage B HF patients.

22.2.2.3 Recommendations for Treatment of HF Stage C

Stage C HF patients have structural heart disease with prior or current signs or symptoms of HF. Treatment of these patients involves the implementation of a multicomponent approach that includes (i) nonpharmacological interventions (e.g., education, social support, sodium restriction, treatment of sleep disorders, weight loss, physical activity, exercise prescription, and cardiac rehabilitation), (ii) pharmacological treatment (e.g., diuretics, ACEIs, ARBs, β-blockers, aldosterone receptor antagonists, combination of hydralazine and isosorbide dinitrate, digoxin, and others), and (iii) device therapy (e.g., implantable cardioverter-defibrillator and cardiac resynchronization therapy). Notably, a number of evidence-based drug therapies have become available for the treatment of patients with stage C HF-REF. The 2013 ACCF/AHA guideline recommendations on pharmacological treatment of stage C HF-REF are given in Table 22.3.

TABLE 22.1 The 2013 ACCF/AHA guideline recommendations for treatment of stage A HF[a]

COR	Recommendation	LOE
I	Hypertension and lipid disorders should be controlled in accordance with contemporary guidelines to lower the risk of HF	A
I	Other conditions that may lead to or contribute to HF, such as obesity, diabetes, tobacco use, and known cardiotoxic agents, should be controlled or avoided	C

COR and LOE stand for class of recommendation and level of evidence, respectively (see Chapter 5 for description).
[a]Ref. 1.

TABLE 22.2 The 2013 ACCF/AHA guideline recommendations for treatment of stage B HF[a]

COR	Recommendation	LOE
I	In all patients with a recent or remote history of MI or acute coronary syndromes (ACS) and reduced LVEF, angiotensin-converting enzyme inhibitors (ACEIs) should be used to prevent symptomatic HF and reduce mortality. In patients intolerant of ACEIs, angiotensin receptor blockers (ARBs) are appropriate unless contraindicated	A
	In all patients with a recent or remote history of MI or ACS and reduced LVEF, evidence-based β-blockers (e.g., carvedilol) should be used to reduce mortality	B
	In all patients with a recent or remote history of MI or ACS, statins should be used to prevent symptomatic HF and cardiovascular events	A
	In patients with structural cardiac abnormalities, including LV hypertrophy, in the absence of a history of MI or ACS, blood pressure should be controlled in accordance with clinical practice guidelines for hypertension to prevent symptomatic HF	A
	ACEIs should be used in all patients with a reduced LVEF to prevent symptomatic HF, even if they do not have a history of MI	A
	β-Blockers should be used in all patients with a reduced LVEF to prevent symptomatic HF, even if they do not have a history of MI	C
IIa	To prevent sudden death, placement of an implantable cardioverter defibrillator (ICD) is reasonable in patients with asymptomatic ischemic cardiomyopathy who are at least 40 days post-MI, have an LVEF of ≤30%, are on appropriate medical therapy, and have reasonable expectation of survival with a good functional status for >1 year	B
III (harm)	Nondihydropyridine calcium channel blockers (e.g., verapamil, diltiazem) with negative inotropic effects may be harmful in asymptomatic patients with low LVEF and no symptoms of HF after MI	C

COR and LOE stand for class of recommendation and level of evidence, respectively (see Chapter 5 for description).
[a]Ref. 1.

TABLE 22.3   The 2013 ACCF/AHA guideline recommendations for pharmacological treatment of stage C HF[a]

Drug class	Recommendation	COR	LOE
Diuretics	Diuretics are recommended in patients with HF-REF who have evidence of fluid retention, unless contraindicated, to improve symptoms	I	C
ACEIs	ACEIs are recommended in patients with HF-REF and current or prior symptoms, unless contraindicated, to reduce morbidity and mortality	I	A
ARBs	ARBs are recommended in patients with HF-REF with current or prior symptoms who are ACEI intolerant, unless contraindicated, to reduce morbidity and mortality	I	A
	ARBs are reasonable to reduce morbidity and mortality as alternatives to ACEIs as first-line therapy for patients with HF-REF, especially for patients already taking ARBs for other indications, unless contraindicated	IIa	A
	Addition of an ARB may be considered in persistently symptomatic patients with HF-REF who are already being treated with an ACE inhibitor and a β-blocker in whom an aldosterone antagonist is not indicated or tolerated	IIb	A
	Routine combined use of an ACEI, ARB, and aldosterone receptor antagonist is potentially harmful for patients with HF-REF	III (harm)	C
β-Blockers	Use of 1 of the 3 β-blockers proven to reduce mortality (e.g., bisoprolol, carvedilol, and sustained-release metoprolol succinate) is recommended for all patients with current or prior symptoms of HF-REF, unless contraindicated, to reduce morbidity and mortality	I	A
Aldosterone receptor antagonists	Aldosterone receptor antagonists are recommended in patients with NYHA class II–IV HF and who have LVEF ≤35%, unless contraindicated, to reduce morbidity and mortality. Patients with NYHA class II HF should have a history of prior cardiovascular hospitalization or elevated plasma natriuretic peptide levels to be considered for aldosterone receptor antagonists. Creatinine should be 2.5 mg/dl or less in men or 2.0 mg/dl or less in women (or estimated glomerular filtration rate >30 ml/min/1.73 m^2), and potassium should be less than 5.0 mM. Careful monitoring of potassium, renal function, and diuretic dosing should be performed at initiation and closely followed thereafter to minimize risk of hyperkalemia and renal insufficiency	I	A
	Aldosterone receptor antagonists are recommended to reduce morbidity and mortality following an acute MI in patients who have LVEF ≤40% who develop symptoms of HF or who have a history of diabetes, unless contraindicated	I	B
	Inappropriate use of aldosterone receptor antagonists is potentially harmful because of life-threatening hyperkalemia or renal insufficiency when serum creatinine is greater than 2.5 mg/dl in men or greater than 2.0 mg/dl in women (or estimated glomerular filtration rate <30 ml/min/1.73 m^2) and/or potassium greater than 5.0 mM	III (harm)	B
Hydralazine and isosorbide dinitrate	The combination of hydralazine and isosorbide dinitrate is recommended to reduce morbidity and mortality for patients self-described as African Americans with NYHA class III–IV HF-REF receiving optimal therapy with ACEIs and β-blockers, unless contraindicated	I	A
	A combination of hydralazine and isosorbide dinitrate can be useful to reduce morbidity or mortality in patients with current or prior symptomatic HF-REF who cannot be given an ACEI or ARB because of drug intolerance, hypotension, or renal insufficiency, unless contraindicated	IIa	B
Digoxin	Digoxin can be beneficial in patients with HF-REF, unless contraindicated, to decrease hospitalizations for HF	IIa	B
Anticoagulants	Patients with chronic HF with permanent/persistent/paroxysmal atrial fibrillation (AF) and an additional risk factor for cardioembolic stroke (history of hypertension, diabetes, previous stroke or transient ischemic attack, or ≥75 years of age) should receive chronic anticoagulant therapy	I	A
	The selection of an anticoagulant agent (warfarin, dabigatran, apixaban, or rivaroxaban) for permanent/persistent/paroxysmal AF should be individualized on the basis of risk factors, cost, tolerability, patient preference, potential for drug interactions, and other clinical characteristics, including time in the international normalized ratio therapeutic range if the patient has been taking warfarin	I	C
	Chronic anticoagulation is reasonable for patients with chronic HF who have permanent/persistent/paroxysmal AF but are without an additional risk factor for cardioembolic stroke	IIa	B
	Anticoagulation is not recommended in patients with chronic HF-REF without AF, a prior thromboembolic event, or a cardioembolic source	III (harm)	B
Statins[b]	Statins are not beneficial as adjunctive therapy when prescribed solely for the diagnosis of HF in the absence of other indications for their use	III (no benefit)	A
Omega-3 fatty acid	Omega-3 polyunsaturated fatty acid supplementation is reasonable to use as adjunctive therapy in patients with NYHA class II–IV symptoms and HF-REF (or HF-PEF), unless contraindicated, to reduce mortality and cardiovascular hospitalizations	IIa	B

TABLE 22.3   *(Continued)*

Drug class	Recommendation	COR	LOE
Nutritional supplements[c]	Nutritional supplements as treatment for HF are not recommended in patients with current or prior symptoms of HF-*R*EF	III (no benefit)	B
Hormonal therapies	Hormonal therapies other than to correct deficiencies are not recommended for patients with current or prior symptoms of HF-REF	III (no benefit)	C
Positive inotropic agents	Long-term use of infused positive inotropic drugs is potentially harmful for patients with HF-REF, except as palliation for patients with end-stage disease who cannot be stabilized with standard medical treatment	III (harm)	C
Others	Drugs known to adversely affect the clinical status of patients with current or prior symptoms of HF-REF are potentially harmful and should be avoided or withdrawn whenever possible. These include most antiarrhythmic drugs, most calcium channel-blocking drugs (except for amlodipine), nonsteroidal anti-inflammatory drugs, and thiazolidinediones	III (harm)	B

COR and LOE stand for class of recommendation and level of evidence, respectively (see Chapter 5 for description).
[a]Ref. 1.
[b]A recent analysis of the CORONA trial reported, however, when repeat events were included, rosuvastatin was shown to reduce the risk of heart failure hospitalizations by approximately 15–20%, equating to approximately 76 fewer admissions per 1000 patients treated over a median follow-up of 33 months [4].
[c]This ACCF/AHA recommendation is further supported by a recent trial showing no effects of micronutrient supplementation on LVEF in patients with chronic stable HF [5].

TABLE 22.4   The 2013 ACCF/AHA guideline recommendations for inotropic support of stage D HF[a]

COR	Recommendation	LOE
I	Until definitive therapy (e.g., coronary revascularization, MCS, heart transplantation) or resolution of the acute precipitating problem, patients with cardiogenic shock should receive temporary intravenous inotropic support to maintain systemic perfusion and preserve end-organ performance	C
IIa	Continuous intravenous inotropic support is reasonable as "bridge therapy" in patients with stage D HF refractory to guideline-directed medical therapy (GDMT) and device therapy who are eligible for and awaiting MCS or cardiac transplantation	B
IIb	Short-term, continuous intravenous inotropic support may be reasonable in those hospitalized patients presenting with documented severe systolic dysfunction who present with low blood pressure and significantly depressed cardiac output to maintain systemic perfusion and preserve end-organ performance	B
	Long-term, continuous intravenous inotropic support may be considered as palliative therapy for symptom control in select patients with stage D HF despite optimal GDMT and device therapy who are not eligible for either MCS or cardiac transplantation	B
III (harm)	Long-term use of either continuous or intermittent, intravenous parenteral positive inotropic agents, in the absence of specific indications or for reasons other than palliative care, is potentially harmful in the patient with HF	B
	Use of parenteral inotropic agents in hospitalized patients without documented severe systolic dysfunction, low blood pressure, or impaired perfusion and evidence of significantly depressed cardiac output, with or without congestion, is potentially harmful	B

COR and LOE stand for class of recommendation and level of evidence, respectively (see Chapter 5 for description).
[a]Ref. 1.

The reader is suggested to refer to the full guideline document for recommendations on nonpharmacological interventions and device therapy in patients with stage C HF.

22.2.2.4   Recommendations for Treatment of HF Stage D

Stage D HF is defined as patients with truly refractory HF who might be eligible for specialized, advanced treatment strategies, such as mechanical circulatory support (MCS), procedures to facilitate fluid removal, continuous inotropic infusions, or cardiac transplantation or other innovative or experimental surgical procedures, or for end-of-life care, such as hospice [6]. The 2013 ACCF/AHA guideline recommendations for treating stage D HF address issues including

(i) thorough evaluation to ascertain the diagnosis of stage D, (ii) water restriction, (iii) inotropic support, (iv) MCS, and (v) cardiac transplantation. Table 22.4 summarizes the recommendations for inotropic support of stage D HF.

22.3   MANAGEMENT OF HF-PEF

22.3.1   General Considerations

HF-PEF accounts for approximately 50% of HF cases and the prevalence is increasing (see Chapter 20). This syndrome is characterized by a normal or near normal (or "preserved")

LVEF, normal LV end-diastolic volume, and abnormal diastolic function, usually with concentric remodeling or hypertrophy. Although the dominant abnormality resides in diastole, systolic dysfunction may also occur. In patients with HF-PEF, certain types of hemodynamic stress including atrial fibrillation, tachycardia, elevated blood pressure, and myocardial ischemia are associated with worsening of diastolic dysfunction and clinical symptoms.

In contrast to HF-REF, currently, there are no specific therapies for HF-PEF. Randomized trials using comparable and efficacious agents for HF-REF have generally been disappointing when used in patients with HF-PEF [7–10]. Current treatment of HF-REF remains largely empiric and focuses on control of symptoms and comorbidities as well as risk factor modifications, including (i) control of systolic and diastolic hypertension, (ii) control of ventricular rate in patients with atrial fibrillation, (iii) control of pulmonary congestion and peripheral edema with diuretics and nitrates, and (iv) treatment of myocardial ischemia (including coronary revascularization) and control of diabetes to prevent left ventricular remodeling that predisposes to diastolic dysfunction and increased ventricular stiffness [7, 11, 12].

22.3.2 Guideline Recommendations

Several guidelines have been developed to address the management of HF-PEF, including the aforementioned 2013 ACC/AHA [1], the 2012 ESC [2], and the 2010 HFSA [3] guidelines. The major recommendations of the 2013 ACC/AHA and the 2010 HFSA guidelines for the treatment of patients with HF-PEF are summarized in Tables 22.5 and 22.6.

22.4 MANAGEMENT OF ACUTE HEART FAILURE SYNDROMES

22.4.1 Definition and Precipitating Factors

The terms acute heart failure (AHF), acute decompensated heart failure (ADHF), and acute heart failure syndromes (AHFS) are frequently used interchangeably. Because patients with AHFS usually require hospitalization, the term "hospitalized patient with HF" has been used in recent ACCF/AHA guidelines since 2009 [1, 13].

AHFS can be defined as new-onset or gradual or rapidly worsening HF signs and symptoms requiring urgent therapy [14]. Irrespective of the underlying cause or precipitating factors, pulmonary and systemic congestion due to elevated ventricular filling pressures is a nearly universal finding in AHFS [14]. Common factors that precipitate AHFS include (i) nonadherence with medication regimen and sodium and/or fluid restriction, (ii) acute myocardial ischemia (e.g., ACS), (iii) uncorrected high blood pressure, (iv) atrial fibrillation and other arrhythmias, (v) recent addition of negative inotropic drugs (e.g., verapamil, nifedipine, diltiazem, β-blockers), (vi) pulmonary embolus, (vii) initiation of drugs that increase salt retention (e.g., steroids, thiazolidinediones, nonsteroidal anti-inflammatory drugs), (viii) excessive alcohol or illicit drug use, (ix) endocrine abnormalities (e.g., diabetes, hyperthyroidism, hypothyroidism), and (x) concurrent infections (e.g., pneumonia) [1].

The majority of AHFS patients have worsening chronic HF. However, after initial management resulting in stabilization, these patients should no longer be considered acute but chronic HF [14].

TABLE 22.5 The 2013 ACCF/AHA guideline recommendations for pharmacological treatment for stage C HF-PEF[a]

COR	Recommendation	LOE
I	Systolic and diastolic blood pressure should be controlled in patients with HF-PEF in accordance with published clinical practice guidelines to prevent morbidity	B
	Diuretics should be used for relief of symptoms due to volume overload in patients with HF-PEF	C
IIb	Coronary revascularization is reasonable in patients with CAD in whom symptoms (angina) or demonstrable myocardial ischemia is judged to be having an adverse effect on symptomatic HF-PEF despite GDMT	C
	Management of AF according to published clinical practice guidelines in patients with HF-PEF is reasonable to improve symptomatic HF	C
	The use of β-blocking agents, ACEIs, and ARBs in patients with hypertension is reasonable to control blood pressure in patients with HF-PEF	C
	The use of ARBs might be considered to decrease hospitalizations for patients with HF-PEF	B
III (no benefit)	Routine use of nutritional supplements is not recommended for patients with HF-PEF	C

COR and LOE stand for class of recommendation and level of evidence, respectively (see Chapter 5 for description).
[a]Ref. 1.

TABLE 22.6   The HFSA 2010 treatment recommendations for HF-PEF[a]

Recommendation[b]	SOE[c]
Careful attention to differential diagnosis is recommended in patients with HF and preserved LVEF to distinguish among a variety of cardiac disorders, because treatments may differ. These various entities may be distinguished based on echocardiography, electrocardiography, and stress imaging (via exercise or pharmacological means, using myocardial perfusion or echocardiographic imaging) and cardiac catheterization	C
Evaluation for ischemic heart disease and inducible myocardial ischemia is recommended in patients with HF-PEF	C
Blood pressure monitoring is recommended in patients with HF-PEF	C
Counseling on the use of a low-sodium diet is recommended for all patients with HF, including those with preserved LVEF	C
Diuretic treatment is recommended in all patients with HF and clinical evidence of volume overload, including those with preserved LVEF. Treatment may begin with either a thiazide or loop diuretic. In more severe volume overload or if response to a thiazide is inadequate, treatment with a loop diuretic should be implemented. Excessive diuresis, which may lead to orthostatic changes in blood pressure and worsening renal function, should be avoided	C
In the absence of other specific indications for these drugs, ARBs or ACEIs may be considered in patients with HF-PEF	C
ACEIs should be considered in all patients with HF-PEF who have symptomatic atherosclerotic cardiovascular disease or diabetes and one additional risk factor. In patients who meet these criteria but are intolerant to ACEIs, ARBs should be considered	C
β-Blocker treatment is recommended in patients with HF-PEF who have (i) prior myocardial infarction (SOE: A), (ii) hypertension (SOE: B), and (iii) atrial fibrillation requiring control of ventricular rate (SOE: B)	
Calcium channel blockers should be considered in patients with HF-PEF and (i) atrial fibrillation requiring control of ventricular rate and intolerance to β-blockers (in these patients, diltiazem or verapamil should be considered) (SOE: C), (ii) symptom-limiting angina (SOE: A), and (iii) hypertension (SOE: C)	
Measures to restore and maintain sinus rhythm may be considered in patients who have symptomatic atrial flutter/fibrillation and preserved LVEF, but this decision should be individualized	C

[a]Ref. 3.
[b]HFSA system for classifying the strength of recommendations: (i) "is recommended" (part of routine care and exceptions to therapy should be minimized), (ii) "should be considered" (majority of patients should receive the intervention and some discretion in application to individual patients should be allowed), (iii) "may be considered" (individualization of therapy is indicated), and (iv) "is not recommended" (therapeutic intervention should not be used).
[c]HFSA system for classifying the strength of evidence (SOE): (i) "level A," randomized controlled clinical trials (may be assigned based on results of a single trial); (ii) "level B," cohort and case control studies, post hoc analysis, subgroup analysis, and meta-analysis, and prospective observational studies or registries; and (iii) "level C," expert opinion, observational studies/epidemiological findings, and safety reporting from large-scale use in practice.

22.4.2   Principles of Management

22.4.2.1   *General Considerations*   AHFS is a life-threatening condition that requires immediate medical attention and usually leads to urgent admission to hospital. In most cases, AHF arises as a result of deterioration in patients with a previous diagnosis of HF (either HF-REF or HF-PEF), and all of the aspects of chronic management described earlier in this chapter apply to these patients.

Patients with AHFS may present with a spectrum of conditions ranging from life-threatening pulmonary edema or cardiogenic shock to a condition characterized, predominantly, by worsening peripheral edema. Diagnosis and treatment are usually carried out in parallel, especially in patients who are particularly unwell, and management must be initiated promptly.

22.4.2.2   *Treatment Goals*   Although the immediate goals of treatment are to improve symptoms and stabilize the patient's hemodynamic condition, longer-term management, including postdischarge care, is also particularly important to prevent recurrences and improve prognosis in AHFS [2]. To help guide the management, the 2010 HFSA guideline recommends 12 treatment goals for patients admitted for AHFS [3]. These include (1) improve symptoms, especially congestion and low-output symptoms; (2) restore normal oxygenation; (3) optimize volume status; (4) identify etiology; (5) identify and address precipitating factors; (6) optimize chronic oral therapy; (7) minimize side effects; (8) identify patients who might benefit from revascularization; (9) identify patients who might benefit from device therapy; (10) identify risk of thromboembolism and need

for anticoagulant therapy; (11) educate patients concerning medications and self-management of HF; and (12) consider and, where possible, initiate a disease-management program.

22.4.2.3 Key Medical Therapies
Although not strictly "evidence based" in the same way as treatments for chronic HF-REF described earlier, the key medical therapies for AHFS are oxygen, diuretics, and vasodilators. Positive inotropes are used more selectively for those with severely compromised cardiac output, and mechanical support of the circulation is required only rarely [2]. In patients with hemodynamic instability, β-blockers should not be used. However, β-blockers should be initiated or titrated upward later in stable patients.

22.4.3 Guideline Recommendations

Multiple guidelines have recently been published to provide evidence-based recommendations for the treatment of AHFS. In the 2013 ACCF/AHA guideline [1], management of AHFS is addressed in a separate section, entitled "The Hospitalized Patient," which covers various aspects, including recommendations on pharmacological therapies (Table 22.7). The 2012 ESC guideline provides recommendations for the treatment of AHFS under various settings, including (i) patients with pulmonary congestion/edema without shock (Table 22.8); (ii) patients with hypotension, hypoperfusion, or shock (Table 22.9); (iii) patients with an ACS (Table 22.10); (iv) patients with atrial fibrillation and a rapid ventricular rate (Table 22.11); and (v) patients with severe bradycardia or heart block (Table 22.11).

TABLE 22.7   The 2013 ACCF/AHA guideline recommendations on drug therapies in patients with AHFS[a]

Drug therapy	Recommendation	COR	LOE
GDMT	In patients with HF-REF experiencing a symptomatic exacerbation of HF requiring hospitalization during chronic maintenance treatment with GDMT, it is recommended that GDMT be continued in the absence of hemodynamic instability or contraindications	I	B
	Initiation of β-blocker therapy is recommended after optimization of volume status and successful discontinuation of intravenous diuretics, vasodilators, and inotropic agents. β-Blocker therapy should be initiated at a low dose and only in stable patients. Caution should be used when initiating β-blockers in patients who have required positive inotropes during their hospital course	I	B
Diuretics	Patients with HF admitted with evidence of significant fluid overload should be promptly treated with intravenous loop diuretics to reduce morbidity	I	B
	If patients are already receiving loop diuretic therapy, the initial intravenous dose should equal or exceed their chronic oral daily dose and should be given as either intermittent boluses or continuous infusion. Urine output and signs and symptoms of congestion should be serially assessed, and the diuretic dose should be adjusted accordingly to relieve symptoms, reduce volume excess, and avoid hypotension	I	B
	The effect of HF treatment should be monitored with careful measurement of fluid intake and output, vital signs, body weight that is determined at the same time each day, and clinical signs and symptoms of systemic perfusion and congestion. Daily serum electrolytes, urea nitrogen, and creatinine concentrations should be measured during the use of intravenous diuretics or active titration of HF medications	I	C
	When diuresis is inadequate to relieve symptoms, it is reasonable to intensify the diuretic regimen using either (i) higher doses of intravenous loop diuretics or (ii) addition of a second (e.g., thiazide) diuretic drug	IIa	B
	Low-dose dopamine infusion may be considered in addition to loop diuretic therapy to improve diuresis and better preserve renal function and renal blood flow	IIb	B
Vasodilators	If symptomatic hypotension is absent, intravenous nitroglycerin, sodium nitroprusside, or nesiritide may be considered an adjuvant to diuretic therapy for relief of dyspnea in patients admitted with acutely decompensated HF	IIb	A
Anticoagulants	Patients admitted to the hospital with decompensated HF should receive venous thromboembolism prophylaxis with an anticoagulant medication if the risk–benefit ratio is favorable	I	B
Arginine vasopressin antagonists	In patients hospitalized with volume overload, including HF, who have persistent severe hyponatremia and are at risk for or having active cognitive symptoms despite water restriction and maximization of GDMT, vasopressin antagonists may be considered in the short term to improve serum sodium concentration in hypervolemic, hyponatremic states with either a V_2 receptor selective or a nonselective vasopressin antagonist	IIb	B

COR and LOE stand for class of recommendation and level of evidence, respectively (see Chapter 5 for description).
[a]Ref. 1.

TABLE 22.8   The 2012 ESC guideline recommendations for the treatment of AHFS patients with pulmonary congestion/edema, but without shock[a]

COR	Recommendation	LOE
I	An intravenous (iv) loop diuretic is recommended to improve breathlessness and relieve congestion. Symptoms, urine output, renal function, and electrolytes should be monitored regularly during use of iv diuretic	B
	High-flow oxygen is recommended in patients with a capillary oxygen saturation <90% or PaO_2 <60 mm Hg (8.0 kPa) to correct hypoxemia	D
	Thromboembolism prophylaxis (e.g., with low-molecular-weight heparin) is recommended in patients not already anticoagulated and with no contraindication to anticoagulation to reduce the risk of deep venous thrombosis and pulmonary embolism	A
IIa	Noninvasive ventilation (e.g., continuous positive airway pressure) should be considered in dyspnea patients with pulmonary edema and a respiratory rate >20 breaths per minute to improve breathlessness and reduce hypercapnia and acidosis. Noninvasive ventilation can reduce blood pressure and should not generally be used in patients with a systolic blood pressure <85 mm Hg (and blood pressure should be monitored regularly when this treatment is used)	B
	An iv opiate (along with an antiemetic) should be considered in particularly anxious, restless, or distressed patients to relieve these symptoms and improve breathlessness. Alertness and ventilatory effort should be monitored frequently after administration because opiates can depress respiration	C
	An iv infusion of a nitrate should be considered in patients with pulmonary congestion/edema and a systolic blood pressure >110 mm Hg, who do not have severe mitral or aortic stenosis, to reduce pulmonary capillary wedge pressure and systemic vascular resistance. Nitrates may also relieve dyspnea and congestion. Symptoms and blood pressure should be monitored frequently during administration of iv nitrates	B
IIb	An iv infusion of sodium nitroprusside may be considered in patients with pulmonary congestion/edema and a systolic blood pressure >110 mm Hg, who do not have severe mitral or aortic stenosis, to reduce pulmonary capillary wedge pressure and systemic vascular resistance. Caution is recommended in patients with acute myocardial infarction. Sodium nitroprusside may also relieve dyspnea and congestion. Symptoms and blood pressure should be monitored frequently during administration of iv sodium nitroprusside	B
III	Inotropic agents are not recommended unless the patient is hypotensive (systolic blood pressure <85 mm Hg), hypoperfused, or shocked because of safety concerns (atrial and ventricular arrhythmias, myocardial ischemia, and death)	C

COR and LOE stand for class of recommendation and level of evidence, respectively (see Chapter 5 for description).
[a]Ref. 2.

TABLE 22.9   The 2012 ESC guideline recommendations for the treatment of AHFS patients with hypotension, hypoperfusion, or shock[a]

COR	Recommendation	LOE
I	Electrical cardioversion is recommended if an atrial or ventricular arrhythmia is thought to be contributing to the patient's hemodynamic compromise in order to restore sinus rhythm and improve the patient's clinical condition	C
IIa	An iv infusion of an inotrope (e.g., dobutamine) should be considered in patients with hypotension (systolic blood pressure <85 mm Hg) and/or hypoperfusion to increase cardiac output, increase blood pressure, and improve peripheral perfusion. The ECG should be monitored continuously because inotropic agents can cause arrhythmias and myocardial ischemia	C
	Short-term mechanical circulatory support should be considered (as a "bridge to recovery") in patients remaining severely hypoperfused despite inotropic therapy and with a potentially reversible cause (e.g., viral myocarditis) or a potentially surgically correctable cause (e.g., acute interventricular septal rupture)	C
IIb	An iv infusion of levosimendan (or a phosphodiesterase inhibitor) may be considered to reverse the effect of β-blockers if β-blockade is thought to be contributing to hypoperfusion. The ECG should be monitored continuously because inotropic agents can cause arrhythmias and myocardial ischemia. Because these agents are also vasodilators, blood pressure should be monitored carefully	C
	A vasopressor (e.g., dopamine or norepinephrine) may be considered in patients who have cardiogenic shock, despite treatment with an inotrope, to increase blood pressure and vital organ perfusion. The ECG should be monitored as these agents can cause arrhythmias and/or myocardial ischemia. Intra-arterial blood pressure measurement should be considered	C
	Short-term mechanical circulatory support may be considered (as a "bridge to decision") in patients deteriorating rapidly before a full diagnostic and clinical evaluation can be made	C

COR and LOE stand for class of recommendation and level of evidence, respectively (see Chapter 5 for description).
[a]Ref. 2.

TABLE 22.10   The 2012 ESC guideline recommendations for the treatment of AHFS patients with an ACS[a]

COR	Recommendation	LOE
I	Immediate primary percutaneous coronary intervention (PCI) or coronary artery bypass grafting (CABG) in selected cases is recommended if there is an ST-segment elevation or a new left bundle-branch block (LBBB) ACS in order to reduce the extent of myocyte necrosis and reduce the risk of premature death	A
	Alternative to PCI or CABG: intravenous thrombolytic therapy is recommended, if PCI/CABG cannot be performed, or if there is ST-segment elevation or new LBBB, to reduce the extent of myocyte necrosis and the risk of premature death	A
	Early PCI (or CABG in selected patients) is recommended if there is non-ST-segment elevation ACS in order to reduce the risk of recurrent ACS. Urgent revascularization is recommended if the patient is hemodynamically unstable	A
	Eplerenone is recommended to reduce the risk of death and subsequent cardiovascular hospitalization in patients with an LVEF ≤40%	B
	An ACEI (or ARB) is recommended in patients with an LVEF ≤40%, after stabilization, to reduce the risk of death, recurrent myocardial infarction, and hospitalization for HF	A
	A β-blocker is recommended in patients with an LVEF ≤40%, after stabilization, to reduce the risk of death and recurrent myocardial infarction	B
IIa	An iv opiate (along with an antiemetic) should be considered in patients with ischemic chest pain to relieve this symptom (and improve breathlessness). Alertness and ventilatory effort should be monitored frequently after administration because opiates can depress respiration	C

COR and LOE stand for class of recommendation and level of evidence, respectively (see Chapter 5 for description).
[a]Ref. 2.

TABLE 22.11   The 2012 ESC guideline recommendations for the treatment of AHFS patients with atrial fibrillation and a rapid ventricular rate and patients with severe bradycardia or heart block[a]

COR	Recommendation	LOE
Patients with atrial fibrillation (AF)		
I	Patients should be fully anticoagulated (e.g., with iv heparin), if not already anticoagulated and with no contraindication to anticoagulation, as soon as AF is detected to reduce the risk of systemic arterial embolism and stroke	A
	Electrical cardioversion is recommended in patients hemodynamically compromised by AF and in whom urgent restoration of sinus rhythm is required to improve the patient's clinical condition rapidly	C
	Electrical cardioversion or pharmacological cardioversion with amiodarone should be considered in patients when a decision is made to restore sinus rhythm nonurgently ("rhythm control" strategy). This strategy should only be employed in patients with a first episode of AF of <48 h duration (or in patients with no evidence of left atrial appendage thrombus on transesophageal echocardiography)	C
	Intravenous administration of a cardiac glycoside should be considered for rapid control of the ventricular rate	C
III	Dronedarone is not recommended because of safety concerns (increased risk of hospital admission for cardiovascular causes and an increased risk of premature death), particularly in patients with an LVEF ≤40%	A
	Class I antiarrhythmic agents are not recommended because of safety concerns (increased risk of premature death), particularly in patients with LV systolic dysfunction	A
Patients with bradycardia or heart block		
I	Pacing is recommended in patients hemodynamically compromised by severe bradycardia or heart block to improve the patient's clinical condition	C

COR and LOE stand for class of recommendation and level of evidence, respectively (see Chapter 5 for description).
[a]Ref. 2.

22.5   SUMMARY OF CHAPTER KEY POINTS

- This chapter describes the general principles and guidelines regarding management of HF-REF, HF-PEF, and AHFS.
- The fundamental goals of treatment of HF are to improve symptoms (including risk of hospitalization), slow or reverse deterioration in myocardial function, and reduce mortality. Management of HF involves implementation of a multicomponent approach, and pharmacological therapy is an essential component of HF management, especially for treatment of HF-REF.
- Several major organizations have published detailed guidelines for the management of HF. These include the 2013 ACCF/AHA guideline for the management of heart failure, the 2012 ESC guideline for the diagnosis and treatment of acute and chronic heart failure, and the 2010 HFSA comprehensive heart failure practice guideline.

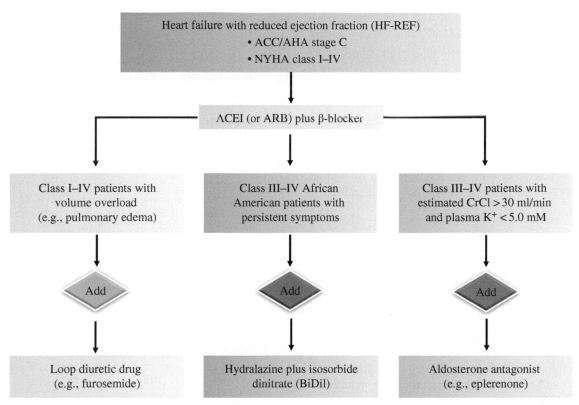

FIG 22.1 2013 ACCF/AHA guideline flowchart illustration of evidence-based pharmacological therapies in the management of stage C HF-REF. This scheme is based on the flowchart of the guideline document (*Circulation* 2013, 128:e240–e327) with modifications.

- The above guidelines are essentially similar with regard to the general approach to treating HF-REF. In this context, these guidelines recommend evidence-based specific pharmacological therapies, including β-blockers and RAAS inhibitors to improve symptoms, retard disease progression, and prolong survival in patients with HF-REF. The 2013 ACCF/AHA guideline's flowchart (Fig. 22.1) summarizes the key steps of evidence-based, guideline-directed medical (pharmacological) therapy of stage C HF-REF, highlighting the critical importance of drug therapy in the management of HF-REF.

- In contrast to the GDMT for HF-REF, treatment of HF-PEF remains largely empiric and focuses on control of symptoms and comorbidities as well as risk factor modifications. This represents an unmet area in cardiovascular medicine, which warrants more research to develop effective therapies.

- AHFS can be defined as new-onset or gradual or rapidly worsening HF signs and symptoms requiring urgent therapy. Although the immediate goals of treatment are to improve symptoms and stabilize the patient's hemodynamic condition, longer-term management, including postdischarge care, is particularly important to prevent recurrences and improve prognosis in AHFS.

- Although not strictly "evidence based" in the same way as treatments for chronic HF-REF, the key medical therapies for AHFS are oxygen, diuretics, and vasodilators. Positive inotropes are used more selectively for those with severely compromised cardiac output and hypoperfusion.

22.6 SELF-ASSESSMENT QUESTIONS

22.6.1 The 2013 ACCF/AHA guideline for the management of heart failure recommends use of 1 of the 3 β-blockers proven to reduce mortality for all patients with current or prior symptoms of HF-REF, unless contraindicated, to reduce morbidity and mortality. Which of the following are the three recommended β-blockers for treating HF-REF?
A. Atenolol, carvedilol, and metoprolol
B. Bisoprolol, carvedilol, and metoprolol
C. Bisoprolol, carvedilol, and propranolol
D. Bisoprolol, labetalol, and metoprolol
E. Labetalol, metoprolol, and propranolol

22.6.2 The combination of hydralazine and isosorbide dinitrate is recommended by the 2013 ACCF/AHA guideline to reduce morbidity and mortality for patients self-described as which of the following ethical groups with NYHA class III–IV HF-REF receiving optimal therapy with ACEIs and β-blockers?

A. African Americans
B. Asian Americans
C. Caucasians
D. Hispanic Americans

22.6.3 A 63-year-old female, with chronic systolic heart failure, has been managed with a multidrug treatment regimen. Which of the following agents is most likely able to prolong the survival of this patient?

A. Carvedilol
B. Digoxin
C. Dopamine
D. Milrinone
E. Nesiritide

22.6.4 A 65-year-old female has been admitted to the coronary care unit with an acute left ventricular myocardial infarction. She develops acute decompensated heart failure with marked pulmonary edema, but no evidence of peripheral edema or weight gain. Which of the following drugs would be most useful for alleviating her pulmonary congestion?

A. Digoxin
B. Furosemide
C. Metoprolol
D. Spironolactone
E. Verapamil

22.6.5 A 69-year-old female with chronic systolic heart failure is treated with a regimen that includes lisinopril. She subsequently develops persistent dry cough. Which of the following drugs is most suitable to replace lisinopril in the regimen?

A. Hydralazine
B. Hydrochlorothiazide
C. Losartan
D. Nesiritide
E. Nitroglycerin

REFERENCES

1   Yancy, C.W., et al., 2013 ACCF/AHA guideline for the management of heart failure: a report of the American College of Cardiology Foundation/American Heart Association Task Force on Practice Guidelines. *J Am Coll Cardiol*, 2013. 62(16): p. e147–239.

2   McMurray, J.J., et al., ESC guidelines for the diagnosis and treatment of acute and chronic heart failure 2012: the Task Force for the Diagnosis and Treatment of Acute and Chronic Heart Failure 2012 of the European Society of Cardiology. Developed in collaboration with the Heart Failure Association (HFA) of the ESC. *Eur Heart J*, 2012. 33(14): p. 1787–847.

3   Heart Failure Society of America, et al., HFSA 2010 comprehensive heart failure practice guideline. *J Card Fail*, 2010. 16(6): p. e1–194.

4   Rogers, J.K., et al., Effect of rosuvastatin on repeat heart failure hospitalizations: the CORONA Trial (Controlled Rosuvastatin Multinational Trial in Heart Failure). *JACC Heart Fail*, 2014. 2(3): p. 289–97.

5   McKeag, N.A., et al., The effect of multiple micronutrient supplementation on left ventricular ejection fraction in patients with chronic stable heart failure: a randomized, placebo-controlled trial. *JACC Heart Fail*, 2014. 2(3): p. 308–17.

6   Hunt, S.A., et al., 2009 focused update incorporated into the ACC/AHA 2005 guidelines for the Diagnosis and Management of Heart Failure in Adults: a report of the American College of Cardiology Foundation/American Heart Association Task Force on Practice Guidelines: developed in collaboration with the International Society for Heart and Lung Transplantation. *Circulation*, 2009. 119(14): p. e391–479.

7   Butler, J., et al., Developing therapies for heart failure with preserved ejection fraction: current state and future directions. *JACC Heart Fail*, 2014. 2(2): p. 97–112.

8   Edelmann, F., et al., Effect of spironolactone on diastolic function and exercise capacity in patients with heart failure with preserved ejection fraction: the Aldo-DHF randomized controlled trial. *JAMA*, 2013. 309(8): p. 781–91.

9   Shah, A.M., et al., Cardiac structure and function in heart failure with preserved ejection fraction: baseline findings from the echocardiographic study of the Treatment of Preserved Cardiac Function Heart Failure with an Aldosterone Antagonist trial. *Circ Heart Fail*, 2014. 7(1): p. 104–15.

10   Redfield, M.M., et al., Effect of phosphodiesterase-5 inhibition on exercise capacity and clinical status in heart failure with preserved ejection fraction: a randomized clinical trial. *JAMA*, 2013. 309(12): p. 1268–77.

11   Garg, N., et al., Heart failure with a normal left ventricular ejection fraction: epidemiology, pathophysiology, diagnosis and management. *Am J Med Sci*, 2013. 346(2): p. 129–36.

12   Sharma, K. and D.A. Kass, Heart failure with preserved ejection fraction: mechanisms, clinical features, and therapies. *Circ Res*, 2014. 115(1): p. 79–96.

13   Hunt, S.A., et al., 2009 Focused update incorporated into the ACC/AHA 2005 guidelines for the Diagnosis and Management of Heart Failure in Adults A Report of the American College of Cardiology Foundation/American Heart Association Task Force on Practice Guidelines Developed in Collaboration With the International Society for Heart and Lung Transplantation. *J Am Coll Cardiol*, 2009. 53(15): p. e1–e90.

14   Gheorghiade, M. and P.S. Pang, Acute heart failure syndromes. *J Am Coll Cardiol*, 2009. 53(7): p. 557–73.

UNIT VII

CARDIAC ARRHYTHMIAS

23

OVERVIEW OF CARDIAC ARRHYTHMIAS AND DRUG THERAPY

23.1 INTRODUCTION

The human heart is both a mechanical and an electronic organ, with the mechanical component pumping the blood and the electronic component controlling the rhythm of the pump. A cardiac arrhythmia is an abnormal rhythm of the heart and is caused by problems with the heart's electrical system, also known as the conduction system. The electrical impulses may happen too fast, too slowly, or erratically, causing the heart to beat too fast, too slowly, or erratically. There are two broad classes of cardiac arrhythmias: bradycardias (also called bradyarrhythmias) and tachycardias (also known as tachyarrhythmias). A bradycardia occurs when the heart rate is <60 beats/min, whereas a tachycardia happens when the heart rate is >100 beats/ min. Each class consists of multiple disorders. This chapter introduces the ICD-10 classification of cardiac arrhythmias and describes the major types of cardiac arrhythmias and their epidemiology. The chapter also considers the pathophysiological mechanisms of cardiac arrhythmias so as to provide a basis for understanding the drug targeting. The major classes of antiarrhythmic agents are covered in Chapter 24. Chapter 25 describes the general principles and current guidelines with regard to the management of the major types of cardiac arrhythmias.

23.2 DEFINITION, CLASSIFICATION, AND EPIDEMIOLOGY

23.2.1 Definition and General Considerations

The term cardiac arrhythmia may be broadly defined as any abnormality or perturbation of the cardiac conduction system, including the sinoatrial (SA) node, atrioventricular (AV) node,

bundle of His, and Purkinje fibers. The SA node, displaying properties of automaticity, spontaneously depolarizes, sending a depolarization wave in the form of cardiac action potential over the atrium, depolarizing the AV node, propagating over the His–Purkinje system, and depolarizing the ventricle in a systematic and coordinated fashion (Fig. 23.1). A cardiac arrhythmia arises when there is an abnormality in either the initiation or conduction of the electrical pulses.

There are many different types of cardiac arrhythmias, which are classified in various ways. Section 23.2.2 introduces the ICD-10 classification of cardiac arrhythmias, which represents the most comprehensive classification of cardiac arrhythmias. The conventional classification of cardiac arrhythmias and the common forms of cardiac arrhythmias are described in Section 23.2.3.

The normal sinus rhythm of the heart can be disturbed through failure of automaticity, such as sick sinus syndrome, or through overactivity, such as inappropriate sinus tachycardia. Ectopic foci prematurely exciting the myocardium on a single or continuous basis result in premature atrial contractions and premature ventricular complexes. Sustained tachyarrhythmias in the atria, such as atrial fibrillation (AF), paroxysmal atrial tachycardia, and supraventricular tachycardia, originate because of micro- or macro-reentry. In general, the severity of cardiac arrhythmias depends on the presence or absence of structural heart diseases.

23.2.2 The ICD-10 Classification

The ICD-10 is the 10th revision of the International Statistical Classification of Diseases and Related Health Problems (ICD) (also see Chapter 1). The ICD-10 defines

Cardiovascular Diseases: From Molecular Pharmacology to Evidence-Based Therapeutics, First Edition. Y. Robert Li.
© 2015 John Wiley & Sons, Inc. Published 2015 by John Wiley & Sons, Inc.

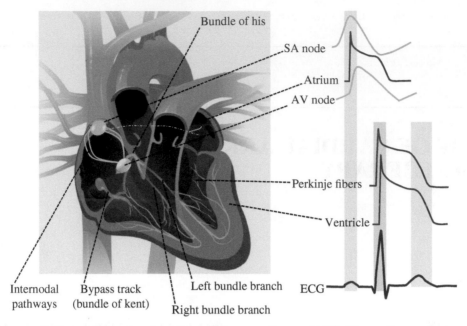

FIGURE 23.1   Cardiac conduction system, action potential, and electrocardiogram (ECG). The cardiac conduction system consists of the sinoatrial (SA) node, atrial ventricular (AV) node, bundle of His, left and right Bundle branches, and Purkinje fibers. In addition, internodal pathways connect SA and AV nodes. Conduction fibers also travel from the SA node to the left ventricle. In some individual, a bypass track connects the right atrium to the right ventricle. This accessory pathway also known as bundle of Kent predisposes the individual to the development of Wolff–Parkinson–White syndrome. For color details, please see color plate section.

the universe of diseases, disorders, injuries, and other related health conditions. These entities are listed in a comprehensive way so that everything is covered. ICD-10 organizes information into standard groupings of diseases. Cardiac arrhythmias are listed under Chapter IX of ICD-10—Diseases of the Circulatory System (I00–I99) within Other Forms of Heart Diseases (I30–I52)—and span from I44 to I49 (Table 23.1). As shown in the table, disorders within I44–I46 are all bradycardias except for I45.6 (pre-excitation syndrome). Likewise, disorders within I47–I49 are all tachyarrhythmias except for I49.5 (sick sinus syndrome), which is classified under bradycardia although over 50% patients with sick sinus syndrome have alteration of paroxysms of atrial tachyarrhythmias and periods of slow atrial and ventricular rates (also see Table 23.2).

23.2.3   Conventional Classification

The ICD-10 is the most comprehensive scheme for classifying cardiac arrhythmias. While the ICD-10 classification is used as the international standard, the detailed list oftentimes causes confusion, especially among the non-cardiologists. As such, cardiac arrhythmias are frequently classified in conventional ways. Based on the changes in heart rate, they are broadly classified into bradyarrhythmias and tachyarrhythmias. Based on the origin, they are classified into supraventricular arrhythmias and ventricular

arrhythmias. Arrhythmias that start in the atria or AV node are supraventricular (above the ventricles) arrhythmias. Ventricular arrhythmias begin in the ventricles. Figure 23.2 illustrates a conventional classification scheme of cardiac arrhythmias, and some of the individual disorders in each group are described in Tables 23.2, 23.3, and 23.4.

23.2.4   Epidemiology

Cardiac arrhythmias are common disorders. It is estimated that over 5 million Americans have one or more forms of cardiac arrhythmias, most over the age of 50 years. Some also have CAD or other forms of cardiac disorders, but many don't. The general risk factors for cardiac arrhythmias include CAD, hypertension, metabolic disorders, smoking, consumption of alcohol, and use of certain medications. Genetic factors also contribute to certain arrhythmias, such as LQTS. While in the vast majority of cases cardiac arrhythmias themselves are not life threatening, some, such as AF, may predispose the individuals to more severe thromboembolic conditions, such as ischemic stroke. Life-threatening cardiac arrhythmias include sustained VT and VF. The prevalence of cardiac arrhythmias increases with age, even when there's no clear sign of CAD. AF is the most common form of cardiac arrhythmias. The prevalence and incidence of common cardiac arrhythmias are summarized in Table 23.5.

TABLE 23.1 The ICD-10 classification of cardiac arrhythmias[a]

ICD-10 code	ICD-10 subcode and disorder of heart rhythm
I44: Atrioventricular and left bundle-branch block	I44.0. Atrioventricular block, first degree
	I44.1. Atrioventricular block, second degree
	Atrioventricular block, types I and II
	Mobitz block, types I and II
	Second-degree block, types I and II
	Wenckebach block
	I44.2. Atrioventricular block, complete
	Complete heart block NOS[b]
	Third-degree block
	I44.3. Other and unspecified atrioventricular blocks
	Atrioventricular block NOS
	I44.4. Left anterior fascicular block
	I44.5. Left posterior fascicular block
	I44.6. Other and unspecified fascicular blocks
	Left bundle-branch hemiblock NOS
	I44.7. Left bundle-branch block, unspecified
I45: Other conduction disorders	I45.0. Right fascicular block
	I45.1. Other and unspecified right bundle-branch blocks
	Right bundle-branch block NOS
	I45.2. Bifascicular block
	I45.3. Trifascicular block
	I45.4. Nonspecific intraventricular block
	Bundle-branch block NOS
	I45.5. Other specified heart blocks
	Sinoatrial block
	Sinoauricular block
	Excl.[c]: Heart block NOS (I45.9)
	I45.6. Preexcitation syndrome
	Anomalous atrioventricular excitation
	Atrioventricular conduction:
	• Accelerated
	• Accessory
	• Preexcitation
	Lown–Ganong–Levine syndrome
	Wolff–Parkinson–White syndrome
	I45.8. Other specified conduction disorders
	Atrioventricular [AV] dissociation
	Interference dissociation
	Long QT syndrome
	Excl.: Prolongation of QT interval (R94.3)[d]
	I45.9. Conduction disorder, unspecified
	Heart block NOS
	Stokes–Adams syndrome
I46: Cardiac arrest	I46.0. Cardiac arrest with successful resuscitation
Excl.:	
Cardiogenic shock (R57.0)	
Complicating:	
• Abortion or ectopic or molar pregnancy (O00–O07, O08.8)	
• Obstetric surgery and procedures (O75.4)	
	I46.1. Sudden cardiac death, so described
	Excl.: Sudden death:
	• NOS (R96.-)
	• With:
	Conduction disorder (I44–I45)
	Myocardial infarction (I21–I22)
	I46.9. Cardiac arrest, unspecified

(Continued)

TABLE 23.1 (*Continued*)

ICD-10 code	ICD-10 subcode and disorder of heart rhythm
I47: Paroxysmal tachycardia	I47.0. Reentry ventricular arrhythmia
Excl.:	I47.1. Supraventricular tachycardia
Complicating:	Paroxysmal tachycardia:
• Abortion or ectopic or molar pregnancy (O00–O07, O08.8)	• Atrial
• Obstetric surgery and procedures (O75.4)	• Atrioventricular [AV]
Tachycardia:	• Junctional
• NOS (R00.0)[e]	• Nodal
• Sinoauricular NOS (R00.0)	I47.2. Ventricular tachycardia
• Sinus [sinusal] NOS (R00.0)	I47.9. Paroxysmal tachycardia, unspecified
	Bouveret(–Hoffmann) syndrome
I48: Atrial fibrillation and flutter	
I49: Other cardiac arrhythmias	I49.0. Ventricular fibrillation and flutter
Excl.:	
Bradycardia:	
• NOS (R00.1)[f]	I49.1. Atrial premature depolarization
	Atrial premature beats
• Sinoatrial (R00.1)	I49.2. Junctional premature depolarization
• Sinus (R00.1)	I49.3. Ventricular premature depolarization
• Vagal (R00.1)	I49.4. Other and unspecified premature depolarization
Complicating:	Ectopic beats
• Abortion or ectopic or molar pregnancy (O00–O07, O08.8)	Extrasystoles
• Obstetric surgery and procedures (O75.4)	Extrasystolic arrhythmias
Neonatal cardiac dysrhythmia (P29.1)[g]	Premature:
	• Beats NOS
	• Contractions
	I49.5. Sick sinus syndrome
	Tachycardia–bradycardia syndrome
	I49.8. Other specified cardiac arrhythmias
	Rhythm disorder:
	• Coronary sinus
	• Ectopic
	• Nodal
	I49.9. Cardiac arrhythmia, unspecified
	Arrhythmia (cardiac) NOS

[a] Adapted from http://apps.who.int/classifications/icd10/browse/2010/en.
[b] NOS, not otherwise specified.
[c] Excl., excluding;
[d] R94.3: Abnormal results of cardiovascular function studies within "Abnormal Findings on Diagnostic Imaging and in Function Studies, without Diagnosis" under Chapter XVIII: Symptoms, Signs and Abnormal Clinical and Laboratory Findings, Not Elsewhere Classified.
[e] R00.0: Tachycardia, unspecified.
[f] R00.1: Bradycardia, unspecified. Both R00.0 and R00.1 are under R00: Abnormalities of Heart Rate within the "Symptoms and Signs Involving the Circulatory and Respiratory Systems" under Chapter XVIII: Symptoms, Signs and Abnormal Clinical and Laboratory Findings, Not Elsewhere Classified.
[g] P29.1: Neonatal cardiac dysrhythmia within "Respiratory and Cardiovascular Disorders Specific to the Perinatal Period" under Chapter XVI: Certain Conditions Originating in the Perinatal Period.

23.3 PATHOPHYSIOLOGY AND DRUG TARGETING

23.3.1 Pathophysiology

The cardiac conduction system is designed for electrical impulse creation and propagation. It allows for initiation of impulses in the atrium, slowed conduction in the AV node, and rapid propagation through the His–Purkinje system to allow synchronous contraction in the ventricles. Cardiac pacemaker cells have the inherent property of spontaneous depolarization, which creates the cardiac impulse. Cells within the sinus node have the fastest rate of spontaneous depolarization, and therefore, the sinus node is the main pacemaker region of the heart. The AV node has the second fastest rate of spontaneous depolarization, which allows it to create an escape rhythm if the sinus node is diseased. Abnormalities or disturbances of the cardiac conduction systems result in a number of cardiac arrhythmias (Fig. 23.2).

The mechanisms of arrhythmias are generally divided into two categories: (1) abnormalities in impulse formation and (2) abnormalities in impulse conduction. As illustrated

TABLE 23.2   Bradycardias

Bradycardia	Description
SA node dysfunction (SND)	SND and AV conduction block are the most common causes of pathological bradycardia. SND can result from either extrinsic factors (e.g., drugs, hypothyroidism, sleet apnea) or intrinsic causes (e.g., sick sinus syndrome[a], coronary artery disease [CAD], inflammation)
AV conduction block	AV conduction block is defined as a delay or interruption in the transmission of an impulse, either transient or permanent, from the atria to the ventricles due to an anatomic or functional impairment in the conduction system. The conduction can be delayed, intermittent, or absent. Conduction block from the atrium to the ventricle can occur for various reasons, including hypothyroidism, drugs, infections, inflammation, heritable/congenital diseases, and CAD. The AV conduction block can be classified based on the severity of block into first-, second-, and third-degree block
	First-degree AV block: PR interval >0.21 s with all atrial impulses conducted. In other words, first-degree block is slowed conduction without missed beats
	Second-degree AV block: missed beats, often in a regular pattern, for example, 2:1, 3:2, or higher degrees of block
	Third-degree AV block: complete heart block, in which no supraventricular impulses are conducted to the ventricles

[a] Sick sinus syndrome: The term "sick sinus syndrome" was first used in 1967 to describe the sluggish return of SA nodal activity in some patients following electrical electroversion. It is now a term applied to a syndrome encompassing multiple sinus node abnormalities, including (i) persistent spontaneous sinus bradycardia not caused by drugs and inappropriate for the physiological circumstance, (ii) sinus arrest or exit block, (iii) combination of SA and AV conduction disturbances, and (iv) alteration of paroxysms of rapid regular or irregular atrial tachyarrhythmias and periods of slow atrial and ventricular rates, as part of the bradycardia–tachycardia syndrome (also known as tachycardia–bradycardia syndrome) (also see Table 23.1).

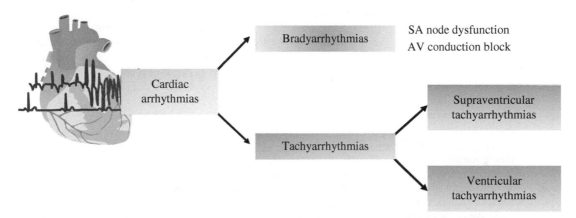

FIGURE 23.2   Conventional classification scheme of cardiac arrhythmias. Based on the heart rate, cardiac arrhythmias are classified into bradyarrhythmias (heart rate <60 beats/min) and tachyarrhythmias (heart rate >100 beats/min). Based on the origin, cardiac arrhythmias are classified into (i) supraventricular arrhythmias, which originate in tissues above the ventricles, including the SA node, atria, and AV node, and (ii) ventricular arrhythmias, which originate in the ventricles. Bradyarrhythmias are primarily supraventricular, whereas tachyarrhythmias can be either supraventricular or ventricular.

in Figure 23.3, abnormalities in impulse formation include altered automaticity and triggered activity. Altered automaticity can be manifested as either depressed automaticity or enhanced automaticity. Triggered activity is caused by afterdepolarization, which is further divided into early afterdepolarization and delayed afterdepolarization. Abnormalities in impulse conduction include reentry and conduction blockage. This section briefly reviews the electrophysiology and electropathophysiology of the cardiac conduction system to help the reader understand the pathophysiological mechanisms of cardiac arrhythmias and the basis of drug targeting in treating these disorders.

23.3.1.1   Altered   Automaticity   Normal automaticity involves the slow, progressive depolarization of the membrane potential (spontaneous diastolic depolarization or phase 4 depolarization) until a threshold potential is reached, at which point an action potential is initiated. The spontaneous discharge rate of the SA nodal complex (70–80 beats/min under resting conditions) exceeds that of all other subsidiary or latent pacemakers in the AV node and His bundle (50–60 beats/min), and Purkinje fibers (30–40 beats/min). As a result, the impulse initiated by the SA node depresses the activity of subsidiary pacemaker sites before they can spontaneously depolarize to threshold. However, the lower pacemaker

TABLE 23.3   Supraventricular tachyarrhythmias

Supraventricular tachyarrhythmia	Description
Atrial premature complexes (APCs)	APCs, also known as atrial premature beats (APBs), atrial premature depolarizations (APDs), premature atrial beats (PABs), or premature atrial complexes (PACs), result from premature activation of the atria arising from a site other than the sinus node. They are observed on the surface electrocardiogram (ECG) as a P wave that occurs relatively early in the cardiac cycle (i.e., prematurely before the next sinus P wave should occur) and has a different morphology from the sinus P wave. APCs are the most common arrhythmia identified during extended ECG monitoring. APCs typically are asymptomatic although some patients experience palpitations or an irregularity of the pulse
Junctional premature complexes (JPCs)	JPCs, also known as junctional premature beats (JPBs) or junctional premature depolarizations (JPDs), arise within the AV junction (e.g., AV node and His bundle). JPCs are very uncommon. They may cause no symptoms or lead to symptoms, such as palpitations
Sinus tachycardia	Sinus tachycardia is defined as an increase in sinus rate to >100 beats/min. Sinus tachycardia is classified into physiological sinus tachycardia and inappropriate sinus tachycardia (IST). Physiological sinus tachycardia occurs in keeping with the level of physical, emotional, pathological, or pharmacological stress. These include exercise, fever, and hyperthyroidism. IST, also called chronic nonparoxysmal sinus tachycardia, is an unusual condition that occurs in individuals without apparent heart disease or other causes for sinus tachycardia, such as hyperthyroidism or fever. Affected patients have an elevated resting heart rate and/or an exaggerated heart rate response to exercise; many patients have both
	The underlying pathological basis for IST is likely to be multifactorial, but two main mechanisms have been proposed: (1) enhanced automaticity of the sinus node and (2) abnormal autonomic regulation of the sinus node with excess sympathetic and reduced parasympathetic tone. Sinus tachycardia is often asymptomatic, although the patient may complain of a rapid heartbeat
Atrial fibrillation (AF)	AF is the most common chronic arrhythmia, which is characterized by disorganized, rapid, and irregular atrial activation. The ventricular response to the rapid atrial activation is also irregular, and the ventricular rate in untreated patients also tends to be rapid. Many patients with AF are asymptomatic. In some patients, it causes adverse consequences related to a reduction in cardiac output (e.g., palpitation, exercise intolerance, hypotension) and to atrial and atrial appendage thrombus formation (e.g., ischemic stroke and peripheral embolization)
Atrial flutter (AFL)	Macro-reentrant atrial tachycardias involving the atrial myocardium are collectively referred to as AFL. AFL and macro-reentrant atrial tachycardia are frequently used interchangeably, with both denoting a nonfocal source of an atrial arrhythmia. There are two forms of AFL: typical and atypical. Typical (also known as counterclockwise) right AFL represents approximately 80% of all AFL. Typical right AFL has an atrial rate of 260–300 beats/min with a ventricular response that tends to be 2:1 or typically 130–150 beats/min. Typical AFL is caused by a macro-reentrant right atrial circuit around the tricuspid annulus, and as such, it is also known as cavotricuspid isthmus-dependent flutter. AFL is less common than AF. AFL occurs most often in patients with chronic obstructive pulmonary disease (COPD) but may be seen also in those with rheumatic or CAD, congestive heart failure, atrial septal defect, or surgically repaired congenital heart disease. AFL is often associated with palpitation (acute) and fatigue (chronic). Similar to AF, AFL also predisposes the individual to thromboembolism
Atrial tachycardias	**Multifocal atrial tachycardia (MAT)** MAT, also called chaotic atrial tachycardia, is characterized by at least three distinct P wave morphologies and often at least three different PR intervals, and the associated atrial and ventricular rates are typically between 100 and 150 beats/min. MAT is the signature tachycardia of patients with severe pulmonary diseases, such as COPD. It can also occur in the presence of coronary, valvular, hypertensive, and other types of heart diseases, particularly when associated with heart failure and/or underlying lung disease. In most cases, the clinical manifestations of MAT differ from those of other tachyarrhythmias in that symptoms predominantly relate to the underlying precipitating illness rather than the arrhythmia. Patients rarely present with symptoms, such as palpitations
	Focal atrial tachycardias (FATs) FATs are caused by regular atrial activation from atrial areas with centrifugal spread. FATs are usually manifest by atrial rates between 150 and 250 beats/min, with a P wave contour different from that of the sinus P wave. Neither the sinus nor the AV node plays a role in the initiation or perpetuation of the tachycardia. FATs do not occur randomly throughout the atria, but rather cluster at predefined anatomic locations. Most patients with FAT report palpitations associated with their episodes of tachycardia. Rarely, patients may present with syncope or exacerbation of an underlying cardiac condition (e.g., angina)

(Continued)

TABLE 23.3 (*Continued*)

Supraventricular tachyarrhythmia		Description
AV nodal tachycardias	AV nodal reentrant tachycardia (AVNRT)	AVNRT is the most common form of paroxysmal supraventricular tachycardias (PSVTs) and is more prevalent in females. AVNRT is associated with palpitations, dizziness, and neck pulsations. It is not usually associated with structural heart disease. Rates of tachycardia are often between 140 and 250 beats/min. Although the reentrant circuit was initially thought to be confined to the compact AV node, a more contemporary view recognizes the usual participation of perinodal atrial tissue as the most common component of the reentrant circuit
	AV junctional tachycardias	Junctional tachycardias are further classified into different forms, including focal junctional tachycardia and nonparoxysmal junctional tachycardia. Focal junctional tachycardia, also known as automatic or paroxysmal junctional tachycardia, is a very uncommon arrhythmia. It usually presents in young adulthood. Nonparoxysmal junctional tachycardia is a benign arrhythmia that is characterized by a narrow complex tachycardia with rates of 70–120 beats/min. The arrhythmogenic mechanism is thought to be associated with an enhanced automaticity arising from a high junctional focus or in response to a triggered mechanism. It shows a typical "warm-up" and "cool-down" pattern and cannot be terminated by pacing maneuvers. The most important feature about this tachycardia is that it may be a marker for a serious underlying condition, such as digitalis toxicity, postcardiac surgery, hypokalemia, or myocardial ischemia
Tachycardias associated with accessory AV pathways		They refer to reentrant tachycardias with an anatomically defined circuit that consists of two distinct pathways, the normal AV conduction system and an AV accessory pathway, linked by common proximal (the atria) and distal (the ventricles) tissues. If sufficient differences in conduction time and refractoriness exist between the normal conduction system and the accessory pathway, a properly timed premature impulse of atrial, junctional, or ventricular origin can initiate reentry
		Two terms, distinguished by the presence or absence of arrhythmias, have been used to describe individuals with accessory pathways: (i) the Wolff–Parkinson–White (WPW) pattern is applied to the patient with preexcitation manifest on an ECG in the absence of symptomatic arrhythmias, and (ii) the WPW syndrome is applied to the patient with both preexcitation manifest on an ECG and symptomatic arrhythmias (e.g., atrioventricular reentrant [or reciprocating] tachycardia [AVRT] and AF) involving the accessory pathway
		The two major types of arrhythmias in persons with an AV accessory pathway are orthodromic AVRT (including permanent junctional reciprocating tachycardia [PJRT]) and antidromic AVRT
		Orthodromic AVRT comprises 90–95% of the reentrant tachycardias associated with the WPW syndrome
		The diagnosis of WPW syndrome is reserved for patients who have both preexcitation and tachyarrhythmias

may take over the pacemaking function of the heart if the faster pacemaker fails or slows.

Variations in autonomic tone may have a major effect on normal automaticity. The automaticity can also be either suppressed or enhanced under pathogenic conditions. Under pathogenic conditions that depolarize cells, myocardial cells outside the specialized conduction system also may acquire automaticity, a phenomenon termed abnormal automaticity. Hence, the term enhanced cardiac automaticity refers to the accelerated generation of an action potential by either normal pacemaker tissue (enhanced normal automaticity) or by nonpacemaker tissue within the myocardium (abnormal automaticity). Enhanced normal automaticity accounts for the occurrence of sinus tachycardia, while abnormal automaticity may result in various atrial or ventricular tachyarrhythmias.

23.3.1.2 Triggered Activity Triggered activity refers to impulse initiation that is dependent on afterdepolarizations. Afterdepolarizations are membrane voltage oscillation during (early afterdepolarization) or after (delayed afterdepolarization) an action potential (Fig. 23.4). Hence, triggered activity or responses can be viewed as early depolarizations

that reach threshold potential, depolarize cell membranes, and result in additional action potentials. Propagation of these triggered responses produces ventricular premature depolarizations that may initiate polymorphic VT, such as TdP, in susceptible subjects. The development of afterdepolarization is potentiated by bradycardia, hypokalemia, hypomagnesemia, and many drugs, especially antiarrhythmic agents.

23.3.1.3 Conduction Block Conduction delay and block can result in bradyarrhythmias or tachyarrhythmias. Bradyarrhythmias occur when the propagating action potential (impulse) is blocked and is followed by asystole or a slow escape rhythm. On the other hand, tachyarrhythmias occur when the delay and block produce reentrant excitation (see Section 23.3.1.4.). Conduction block usually is due to fibrosis or calcification of the AV node, His bundle, or right and left bundle branches. Conduction block also may result from elevated parasympathetic tone, as seen during sleep or in well-trained athletes, or may be due to agents that act on the AV node, such as digitalis, β-blockers, or non-dihydropyridine calcium channel blockers.

TABLE 23.4 Ventricular tachyarrhythmias

Ventricular tachyarrhythmia		Description
Premature ventricular complexes (PVCs)		PVCs, also known as ventricular premature beats, ventricular premature complexes, premature ventricular beats, or ventricular extrasystoles, are triggered from the ventricular myocardium in a variety of situations. PVCs are common and occur in a broad spectrum of the population. This includes patients without structural heart disease and those with any form of cardiac disease, independent of severity. PVCs produce few or no symptoms in the vast majority of patients. Some patients may experience palpitations or dizziness
Ventricular tachycardia (VT)	Nonsustained VT	VT is defined as a cardiac arrhythmia of three or more consecutive ventricular beats in duration at a rate of >100 beats/min (cycle length <600 ms). VT is further classified into nonsustained VT and sustained VT. As described below, torsades de pointes (TdP) is also a form of VT
		Nonsustained VT is defined as a VT terminating spontaneously in <30 s. Nonsustained VT can be monomorphic (with a single QRS morphology) or polymorphic (with a changing QRS morphology at cycle length between 600 and 180 ms). Nonsustained VT is often asymptomatic, but some patients experience palpitations, light-headedness, presyncope, or dyspnea
	Sustained VT	Sustained VT is defined as a VT >30 s in duration and/or requiring termination due to hemodynamic compromise in <30 s. Sustained VT can be monomorphic or polymorphic. Sustained VT is potentially life threatening due to hemodynamic instability
	TdP	TdP is a form of polymorphic VT that occurs in the setting of acquired or congenital QT interval prolongation [1] (see long QT syndrome (LQTS)). In the specific case of TdP, these variations take the form of a progressive, sinusoidal, cyclic alteration of the QRS axis. The peaks of the QRS complexes appear to "twist" around the isoelectric line of the recording, hence the name torsades de pointes or "twisting of the points." Typical features of TdP include an antecedent prolonged QT interval, particularly in the last sinus beat preceding the onset of the arrhythmia; a ventricular rate of 160–250 beats/min; irregular RR intervals; and a cycling of the QRS axis through 180° every 5–20 beats. TdP is usually short-lived and terminates spontaneously. However, most patients experience multiple episodes of the arrhythmia, and episodes can recur in rapid succession, potentially degenerating to ventricular fibrillation (VF) and sudden cardiac death
Ventricular flutter		Ventricular flutter is a regular (cycle length variability 30 ms or less) monomorphic VT that occurs at a very rapid rate of 150–300 beats/min (usually around 200 beats/min). Often, no distinctive T waves are discernible. There is no isoelectric interval between QRS complexes. P waves or evidence of atrial activity is absent since the ventricular rate is rapid
VF		Rapid, usually >300 beats/min, grossly irregular ventricular rhythm with marked variability in QRS cycle length, morphology, and amplitude. Both ventricular flutter and VF represent severe derangements of the heartbeat that usually terminate fatally within 3–5 min unless corrective measures are undertaken promptly
LQTS		The LQTS is a disorder of myocardial repolarization characterized by a prolonged QT interval on the ECG [2]. As noted earlier, this syndrome is associated with an increased risk of TdP, a characteristic life-threatening cardiac arrhythmia. The LQTS may be either congenital or acquired. Congenital LQTS consists of defects in cardiac ion channels that are responsible for cardiac depolarization. Defects that enhance sodium or calcium inward currents or inhibit outward potassium currents during the plateau phase of the action potential prolong action potential duration and hence the QT interval. Acquired LQTS usually results from drug therapy, hypokalemia, or hypomagnesemia. The primary symptoms in patients with LQTS include palpitations, syncope, seizures, and sudden cardiac death

23.3.1.4 Reentry

Definition of Reentry Electric activity in the form of action potential during each normal cardiac cycle begins in the SA node and continues until the entire heart has been activated. Each cell becomes activated in turn, and the cardiac impulse dies out when all fibers have been discharged and are completely refractory. During this absolute refractory period, the cardiac impulse has no place to go. It must be extinguished and restarted by the next sinus impulse. However, this well-controlled conduction of impulses may be altered under pathogenic conditions, leading to a electropathophysiological process, namely, reentry. This term is alternatively known as reentrant excitation, circus movement, reciprocal beat, or echo beat. Reentry occurs when a propagating impulse fails to die out after normal activation of the heart and persists to reexcite the heart after the refractory period has ended.

Reentry as the Most Important Electropathophysiological Mechanism of Arrhythmias Reentry is the electropathophysiological mechanism responsible for the majority of clinically important arrhythmias. These include AF, AFL, AVNRT, tachycardias associated with accessory AV pathways, VT after myocardial infarction with the presence of left ventricular scar, and VF. Patients who develop reentrant arrhythmias usually have an anatomic or electrical (functional)

TABLE 23.5 Prevalence and incidence of common cardiac arrhythmias

Cardiac arrhythmia	Prevalence and/or incidence
SND	The prevalence of SND has been estimated to be between 403 and 666 per million; this would translate into 125,000–207,000 people in the United States. The estimated incidence rate for SND is 63 per million per year [3]
AF	Estimates of the prevalence of AF in the United States ranged from approximately 2.7 to 6.1 million in 2010, and the AF prevalence is expected to rise to between approximately 5.6 and 12 million by 2050 [3]. In a Medicare sample, the incidence of AF was approximately 28 per 1000 person-years and did not change substantially between 1993 and 2007 [4]
Atrial tachycardias (ATs) (focal AT and multifocal AT)	Repetitive focal AT was estimated to account for between 5 and 15% of arrhythmias in adults undergoing study for PSVTs [5]. Early studies suggested that on the ECG multifocal AT occurred in 0.05–0.32% of electrocardiograms interpreted in general hospitals and in 0.37% of hospitalized patients [6]
AV nodal tachycardias (AVNRT)	AVNRT is categorized as one of the PSVTs and is the most common of these arrhythmias, accounting for nearly two thirds of all cases. The term PSVT is applied to intermittent supraventricular tachycardias with abrupt onset and offset other than AF, AFL, and multifocal AT. Because they have distinct clinical characteristics, these narrow complex tachycardias are usually considered separately. PSVT occurs with an incidence of 35 per 100,000 person-years [7]
Tachycardias associated with accessory AV pathways	The prevalence of a WPW pattern on the surface ECG is estimated at 0.13–0.25% in the general population. The prevalence of the WPW syndrome (WPW pattern plus tachyarrhythmia) is substantially lower than that of the WPW pattern alone, perhaps as low as 2% of patients with the WPW pattern on surface ECG [8]
PVCs	PVCs are one of the most common arrhythmias and can occur in individuals with or without heart diseases. The prevalence of PVCs varies greatly, with estimates of <3% to >60% in asymptomatic individuals A cross-sectional analysis of the 15,792 individuals (aged 45–65 years) from the four US communities participating at visit 1 of the Atherosclerosis Risk In Communities (ARIC) study showed that based on a 2 min ECG, PVCs are present in >6% of middle-aged adults. Increasing age, the presence of heart diseases, faster sinus rates, African American ethnicity, male sex, lower educational attainment, and lower serum magnesium or potassium levels are directly related to PVC prevalence. Independently of these factors, hypertension is associated with a 23% increase in the prevalence of PVCs [9]
VT (nonsustained VT, sustained VT, and TdP)	The estimated prevalence of nonsustained VT in the general population (both with and without heart disease) is between 0 and 4%, although this is most likely an underestimate [10]. The prevalence of nonsustained VT becomes more common as age increases The incidence of sustained VT in the United States is not well quantified because of the clinical overlap of VT with ventricular fibrillation (VF) As many as 300,000 persons die annually in the United States of sudden cardiac death (SCD) [11, 12]. Along with VF, sustained VT is responsible for nearly all of the arrhythmic SCD The true incidence and prevalence of drug-induced TdP in the US general population is largely unknown. By extrapolating data from non-US registries, it has been estimated that 12,000 cases of drug-induced TdP occur annually in the United States [3]
VF	VF is the most commonly identified arrhythmia in cardiac arrest patients. SCD accounts for approximately 300,000 deaths per year in the United States, of which 75–80% are due to VF. This represents an incidence of 0.08–0.16% per year in the adult population. VF is commonly the first expression of CAD and is responsible for approximately 50% of deaths from CAD, often within the first hour after the onset of an acute myocardial infarction
LQTS	The prevalence of LQTS is difficult to estimate. However, given the currently increasing frequency of diagnosis, LQTS may be expected to occur in 1–4 in 10,000 individuals [13]

abnormality, which could be caused by an accessory pathway, by an abnormal separation of adjacent fibers that may form two limbs of a reentrant circuit, or by juxtaposed fibers that possess different electrophysiological characteristics, often resulting from abnormalities of the myocardium and Purkinje fibers as the result of a disease process.

Models of Reentry Several models of reentry have been proposed, and among them, the circus movement reentry model has been well described [14]. In the simplest form of the circus movement model, as illustrated in Figure 23.5, reentry requires the presence of a unidirectional block (anterograde conduction is blocked, and retrograde conduction is permitted but with slowed conduction velocity), in which the action potential can travel only in one direction, and a long enough circuit, in which recovery from refractoriness occurs before the approach of the leading edge of depolarization. The unidirectional block is usually caused by a disease processes, such as ischemia. Consequently, the retrograde conduction through the diseased tissue is also slowed, which promotes reentry.

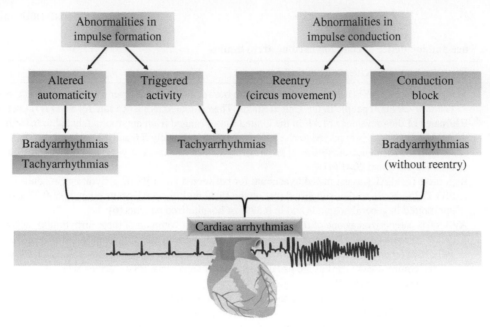

FIGURE 23.3 General mechanisms of cardiac arrhythmias. Cardiac arrhythmias stem from abnormalities in either impulse formation or impulse conduction or both.

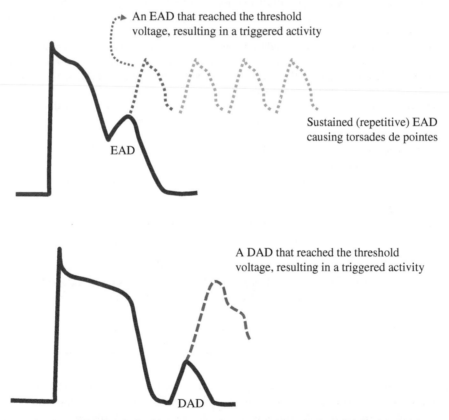

FIGURE 23.4 Afterdepolarization. Afterdepolarizations occur when a normal action potential triggers abnormal depolarizations. The abnormal depolarizations may occur during the plateau or repolarizing phase of the normal action potential, which is known as early afterdepolarization (EAD), or shortly after depolarization of the normal action potential, which is called delayed afterdepolarization (DAD). EAD during the plateau phase (phase 2) results from an inward calcium ion current, and during the plateau phase, sodium ion channels are inactivated. On the other hand, EAD during depolarization phase (phase 3) results from an inward sodium ion current mediated by partially recovered sodium ion channels. EAD mainly occurs during phase 3, and if an EAD is sustained, it may cause torsades de pointes. In contrast to that of EAD, the mechanism of DAD remains to be elucidated. It has been suggested that DAD may be caused by intracellular calcium ion accumulation. The higher intracellular calcium ion concentrations lead to activation of Na^+/Ca^{2+} exchanges, and the resulting electrogenic influx of three sodium ions for each extruded calcium ion causes membrane depolarization. The significance of DAD in cardiac arrhythmias is less clear compared with that of EAD.

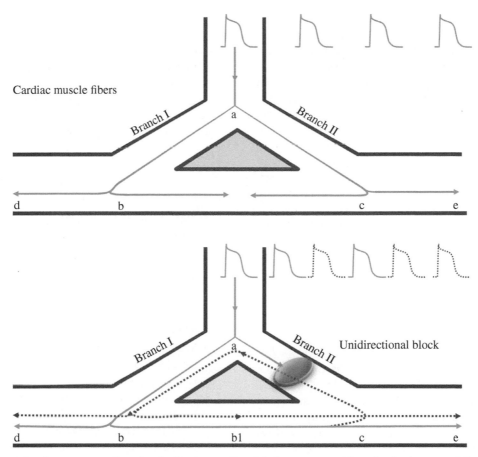

FIGURE 23.5 Normal and reentrant cardiac electrical pathways. Panel A shows a normal impulse (action potential) conduction through a bifurcating pathway to depolarize the different areas of myocardium. Panel B shows formation of a reentrant circuit (dashed lines) due to unidirectional block of one of the branch of the bifurcation pathway leading to tachyarrhythmias. As illustrated in panel A, a normal impulse arrives at point a, where it travels through the bifurcating pathway (branches I and II). When the impulses travel down branches I and II to arrive at points b and c, respectively, the impulses are able to once again travel in two opposite directions. The impulses between b and c canceled each other, whereas as the rest travel to points d and e respectively, they cause excitation of the myocardium in different areas. As depicted in panel B, a unidirectional block occurs in branch II due to pathologic factors, such as ischemia. Hence, in branch II, the impulse from point a cannot travel to point c, whereas impulses are able to travel from point c to a. Because of this unidirectional block in branch II, when the impulse from branch I arrives at point b, it will travel in two directions: one is toward point d as in normal conduction (panel A), and the other toward point b1 because of no impulse from phase II to cancel it. When the impulse from point b1 arrives at point c, it will travel in two directions: one through the retrograde conduction along branch II toward point a (reentry), and the other continues toward point e. Once the reentered impulse arrives at point a, it will continue to travel along branch I to point b, forming a circuit. As illustrated, reentrant conduction results in extra beats, thereby tachyarrhythmias. For color details, please see color plate section.

Conditions That Affect Reentry Since the length of the pathway is fixed and determined by the anatomy, the conduction velocity of the action potential and the duration of the refractory periods are the two main factors that influence reentry. Conditions that depress the conduction velocity or abbreviate the refractory period promote the development of reentry (see previous explanation). On the other hand, conditions that speed conduction velocity and prolong the refractory period hinder reentry. Although the aforementioned notion is valid as a general rule, dramatic suppression of the already slowed retrograde conduction by antiarrhythmic drugs, such as the class I antiarrhythmic agents, may lead to complete extinguishment of the retrograde

impulse, thereby the termination of the reentry. Indeed, blocking conduction velocity is one of the main mechanisms for terminating reentry by antiarrhythmic drugs, especially the class I agents (see Chapter 24).

23.3.2 Drug Targeting

23.3.2.1 General Considerations Drug therapy for cardiac arrhythmias has a long history. For example, digitalis has been used to treat patients with AF for >100 years, and the value of quinidine for terminating AF first became apparent at the beginning of the twentieth century. Drug therapy used to be the mainstay of treatment for both ventricular and supraventricular arrhythmias. However, increasing knowledge about the potentially significant adverse effects of these medications, together with the emergence of new and effective nonpharmacological approaches to the treatment of arrhythmias, has led to rapidly diminished use of antiarrhythmic agents over the past two to three decades. Despite the decline in their use, antiarrhythmic agents are still useful for treating specific arrhythmic disorders and play a supporting role in many cases. Currently available antiarrhythmic drug options include roughly a dozen of rather old agents. These drugs have been traditionally classified into four classes: I, II, III, and IV. These agents target on specific ion channels to interfere with the initiation and conduction of cardiac action potential to achieve the desired therapeutic effects. Due to the complex nature of the membrane ion channels and the nonselectivity of inhibition by the antiarrhythmic agents, use of the currently available antiarrhythmic agents is associated with significant adverse effects, including the paradoxical capacity to induce arrhythmic disorders, a phenomenon known as proarrhythmic activity (see Chapter 24).

23.3.2.2 Drug Targeting on Reentry

23.3.2.2 Drug Targeting on Reentry As discussed earlier, reentry is the most important electropathophysiological mechanism for cardiac arrhythmias. Many antiarrhythmic agents act via suppressing reentry. This occurs mainly via two mechanisms: (1) prolongation of refractory period and (2) decrease of conduction velocity. This notion seems to be contradictory with the earlier statement that conditions that depress the conduction velocity promote the development of reentry. How does slowing conduction velocity by antiarrhythmic agents stop reentry? This is due to the further suppression of the already slowed conduction velocity of the retrograde conduction by antiarrhythmic drugs (especially the class I drugs) that leads to a complete extinguishment of the retrograde impulse and thereby the termination of the reentry. This would not affect significantly the impulse conduction in the normal pathway because the drugs preferentially act on the diseased tissue (see Chapter 24 for more discussion).

23.4 SUMMARY OF CHAPTER KEY POINTS

- Cardiac arrhythmia is broadly defined as any abnormality or perturbation of the cardiac conduction system, including the SA node, AV node, bundle of His, and Purkinje fibers. A cardiac arrhythmia arises when there is an abnormality in either the initiation or conduction of the electrical pulses.

- The ICD-10 represents the most comprehensive classification scheme of cardiac arrhythmias. Cardiac arrhythmias are listed under Chapter IX of ICD-10—Diseases of the Circulatory System (I00–I99) within

Other Forms of Heart Diseases (I30–I52)—and span from I44 to I49.

- Conventionally, cardiac arrhythmias are broadly classified into bradyarrhythmias and tachyarrhythmias based on the changes in heart rates. Based on the origin, they are classified into supraventricular arrhythmias and ventricular arrhythmias.

- Bradycardias include SA node dysfunction and AV conduction block. Bradycardias are hence primarily supraventricular as the origin of the arrhythmias is above the ventricles. In contrast, tachyarrhythmias can be supraventricular or ventricular.

- Supraventricular tachyarrhythmias include atrial premature complexes, junctional premature complexes, sinus tachycardia, atrial fibrillation (AF), AFL, atrial tachycardias (multifocal atrial tachycardia and focal atrial tachycardia), AV nodal tachycardias (AVNRT and AV junctional tachycardias), and tachycardias associated with accessory AV pathways (e.g., WPW syndrome).

- Ventricular tachyarrhythmias include PVCs, VT (nonsustained VT, sustained VT, and TdP), ventricular flutter, VF, and LQTS.

- The term PSVT is applied to intermittent supraventricular tachycardias with abrupt onset and offset other than AF, AFL, and multifocal atrial tachycardia. Because they have distinct clinical characteristics, these narrow complex tachycardias are usually considered separately.

- While in the vast majority of cases the cardiac arrhythmia itself is not life threatening; some, such as AF, may predispose the individual to more severe problems (e.g., ischemic stroke). Life-threatening cardiac arrhythmias include sustained VT and VF.

- The mechanisms of arrhythmias are generally divided into two categories: (1) abnormalities in impulse formation and (2) abnormalities in impulse conduction, including reentry.

- The term reentry is alternatively known as reentrant excitation, circus movement, reciprocal beat, or echo beat. Reentry occurs when a propagating impulse fails to die out after normal activation of the heart and persists to reexcite the heart after the refractory period has ended.

- Reentry is the electropathophysiological mechanism responsible for the majority of clinically important arrhythmias. These include AF, AFL, AVNRT, tachycardias associated with accessory AV pathways, VT after myocardial infarction with the presence of left ventricular scar, and VF.

- Drug therapy used to be the mainstay of treatment for cardiac arrhythmias. However, use of antiarrhythmic agents has declined over the past decades due to the serious adverse effects associated with their use as well as the availability of more effective nonpharmacological modalities.

- Many antiarrhythmic agents act via suppressing reentry. This occurs mainly through two mechanisms: (1) prolongation of refractory period and (2) decrease of conduction velocity.

23.5 SELF-ASSESSMENT QUESTIONS

23.5.1 A 28-year-old woman suddenly has rapid palpitations accompanied by chest pain and dizziness while playing her cello. She is brought to an emergency department. She has a faint regular pulse of 190 beats/min. Her blood pressure is 82/54 mmHg. Cardiovascular examination reveals no signs of heart failure. An electrocardiogram shows a regular tachycardia with a narrow QRS complex and no apparent P waves (from Ref. [15]). Further electrophysiological studies supported a diagnosis of AV nodal reentrant tachycardia. Which of the following best describes the patient's arrhythmia?
    A. The patient's arrhythmia is also known as the Wolff–Parkinson–White syndrome.
    B. The patient's arrhythmia is an atypical type of atrial flutter.
    C. The patient's arrhythmia is frequently associated with long QT syndrome.
    D. The patient's arrhythmia is the most common form of PSVTs.

23.5.2 The long QT syndrome is a familial, usually autosomal dominant disease characterized by an abnormally prolonged ventricular repolarization phase and a propensity toward polymorphic ventricular tachycardia (often termed torsades de pointes) and sudden cardiac death. At least 10 different forms of the long QT syndrome have been described, but in approximately 45% of genotyped patients, the underlying causes are mutations in the KCNQ1 (also known as KVLQT1 or Kv7.1) gene, which encodes the pore-forming alpha subunits of the channels generating I_{Ks}, an adrenergic-sensitive, slow outward potassium current. This form of the long QT syndrome is designated as long QT syndrome type 1. In type 1 long QT syndrome, mutations in the KCNQ1 gene may result in which of the following?
    A. Diminished early afterdepolarization
    B. Increased conduction velocity of action potential
    C. Prolongation of the refractory period of action potential
    D. Shortening of refractory period of action potential

23.5.3 After the sudden death of a 13-year-old girl while she was playing basketball, her family comes to the clinic for medical evaluation. Her parents' resting electrocardiograms (ECGs) are normal, but her 9-year-old sister's ECG shows an abnormally long QT interval. There is a history of recurrent syncope in female relatives of the maternal grandmother, but there is no family history of other sudden deaths, the sudden infant death syndrome, drowning, or death from a motor vehicle accident (from Ref. [2]). If this 13-year-old girl had long QT syndrome, what would be the most likely cause of her death?
    A. Atrial fibrillation
    B. Premature ventricular complexes
    C. Torsades de pointes
    D. Wolff–Parkinson–White syndrome

23.5.4 Recently published in the *New England Journal of Medicine* (from Ref. [16]), the Race Associated Cardiac Arrest Event Registry (RACER) study assessed the incidence, clinical profile, and outcomes of cardiac arrest in 10.9 million participants of marathons and half marathons held between 2000 and 2010 in the United States. Findings from this large registry study indicate that marathons and half marathons are associated with a low overall risk of cardiac arrest and sudden death. Cardiac arrest, most commonly attributable to hypertrophic cardiomyopathy or atherosclerotic coronary disease, occurs primarily among male marathon participants; the incidence rate in this group increased during the past decade. On the basis of their findings, the RACER investigators conclude that "the risk associated with long-distance running events is equivalent to or lower than the risk associated with other vigorous physical activity." Which of the following dysrhythmias is the most common cause of sudden cardiac arrest and death in the marathon runners as well as the general public?
    A. Atrial fibrillation
    B. Torsades de pointes
    C. Ventricular fibrillation
    D. Wolff–Parkinson–White syndrome

23.5.5 Postoperative atrial fibrillation or flutter is one of the most common complications of cardiac surgery and significantly increases morbidity and healthcare utilization. A recent large-scale, double-blind, placebo-controlled, randomized clinical trial involving a total of 1516 patients scheduled for cardiac surgery in 28 centers in the United States, Italy, and Argentina concluded that among patients undergoing cardiac surgery, perioperative supplementation with fish oil (omega-3 fatty acids), compared with placebo, did not reduce the risk of postoperative atrial fibrillation (from Ref. [17]). Which of the following is the primary electropathophysiological mechanism of the arrhythmia investigated in the OPERA trial?

A. AV conduction block
B. Enhanced normal automaticity
C. Prolongation of refractory period
D. Reentry in the atrium
E. Triggered activity in the AV node

REFERENCES

1   Drew BJ, et al. Prevention of torsade de pointes in hospital settings: a scientific statement from the American Heart Association and the American College of Cardiology Foundation. *Circulation* 2010;121: p. 1047–60.

2   Roden DM. Clinical practice. Long-QT syndrome. *N Engl J Med* 2008;358: p. 169–76.

3   Go AS, et al. Heart disease and stroke statistics—2013 update: a report from the American Heart Association. *Circulation* 2013;127: p. e6–245.

4   Piccini JP, et al. Incidence and prevalence of atrial fibrillation and associated mortality among medicare beneficiaries, 1993–2007. *Circ Cardiovasc Qual Outcomes* 2012;5: p. 85–93.

5   Chen SA, et al. Sustained atrial tachycardia in adult patients. Electrophysiological characteristics, pharmacological response, possible mechanisms, and effects of radiofrequency ablation. *Circulation* 1994;90: p. 1262–78.

6   Kastor JA. Multifocal atrial tachycardia. *N Engl J Med* 1990;322: p. 1713–7.

7   Orejarena LA, et al. Paroxysmal supraventricular tachycardia in the general population. *J Am Coll Cardiol* 1998;31: p. 150–7.

8   Cohen MI, et al. PACES/HRS expert consensus statement on the management of the asymptomatic young patient with a Wolff-Parkinson-White (WPW, ventricular preexcitation)

9   Simpson RJ, Jr., et al.. Prevalence of premature ventricular contractions in a population of African American and white men and women: the Atherosclerosis Risk in Communities (ARIC) study. *Am Heart J* 2002;143: p. 535–40.

10 Kinder C, et al.. The clinical significance of nonsustained ventricular tachycardia: current perspectives. *Pacing Clin Electrophysiol* 1994;17: p. 637–64.

11 Kong MH, et al. Systematic review of the incidence of sudden cardiac death in the United States. *J Am Coll Cardiol* 2011;57: p. 794–801.

12 Fishman GI, et al. Sudden cardiac death prediction and prevention: report from a National Heart, Lung, and Blood Institute and Heart Rhythm Society Workshop. *Circulation* 2010;122: p. 2335–48.

13 Earle N, et al. Community detection of long QT syndrome with a clinical registry: an alternative to ECG screening programs? *Heart Rhythm* 2013;10: p. 233–8.

14 Keating MT, and Sanguinetti MC. Molecular and cellular mechanisms of cardiac arrhythmias. *Cell* 2001;104:569–80.

15 Delacretaz, E., Clinical practice: supraventricular tachycardia. *N Engl J Med*, 2006. 354: p. 1039–1051

16 Kim, J.H., et al., Cardiac arrest during long-distance running races. *N Engl J Med*, 2012. 366: p. 130–140.

17 Mozaffarian, D., et al., Fish oil and postoperative atrial fibrillation: the Omega-3 Fatty Acids for Prevention of Post-operative Atrial Fibrillation (OPERA) randomized trial. *JAMA*, 2012. 308: p. 2001–2011.

electrocardiographic pattern: developed in partnership between the Pediatric and Congenital Electrophysiology Society (PACES) and the Heart Rhythm Society (HRS). Endorsed by the governing bodies of PACES, HRS, the American College of Cardiology Foundation (ACCF), the American Heart Association (AHA), the American Academy of Pediatrics (AAP), and the Canadian Heart Rhythm Society (CHRS). *Heart Rhythm* 2012;9: p. 1006–24.

24

DRUGS FOR CARDIAC ARRHYTHMIAS

24.1 OVERVIEW

Pharmacological therapy had traditionally been the mainstay of treatment for cardiac arrhythmias, including supraventricular and ventricular tachycardias. However, increasing knowledge about the potentially significant adverse effects of the antiarrhythmic agents, together with the emergence of new and more effective nonpharmacological approaches of good safety profile to the treatment of cardiac arrhythmias, has led to a marked decline of antiarrhythmic drug therapy in the past two decades. Nevertheless, some of the available antiarrhythmic drugs when used appropriately are still effective options for the management of certain cardiac arrhythmic conditions. This chapter discusses the molecular pharmacology of the classical antiarrhythmic drugs with a focus on those whose efficacy is supported by recent clinical research. The chapter begins with a brief introduction to cardiac electrophysiology and the Vaughan–Williams classification of antiarrhythmic drugs into four classes and then focuses on discussing the molecular mechanisms of action for each of the drug classes and their clinical uses. The chapter also considers antiarrhythmic agents that do not fall into the above classification scheme as well as emerging drugs with promising results from randomized clinical trials.

24.2 CLASSIFICATION OF ANTIARRHYTHMIC DRUGS

As stated in Chapter 23, antiarrhythmic drugs act primarily by suppressing or preventing abnormal formation or conduction of cardiac action potential. Because sodium channels are involved in the spontaneous depolarization of phase 4 of action potential in nodal cells and in the phase 0 depolarization of nonpacemaker cells, drugs that block sodium channels can reduce abnormal automaticity and slow conduction of the cardiac impulse. Also as noted in Chapter 23, the automaticity is determined by the rate of phase 4 spontaneous depolarization in pacemaker cells, whereas the conduction velocity of action potential in non-pacemaker cells is determined by the rate of phase 0 depolarization. Similarly, since calcium channels are involved in the phase 0 depolarization of pacemaker cells, calcium channel blockers (CCBs) inhibit the conduction of action potential through the nodal cells or tissues. Likewise, drugs that block repolarizing potassium channels can prolong repolarization and the action potential duration (APD) and thereby increase the refractory period of cardiac tissue and hinder reentry. Drugs that block β-adrenergic receptors attenuate the sympathetic stimulation of cardiac automaticity and conduction velocity and thereby prevent the overstimulation of sympathetic tone that contributes to some forms of cardiac arrhythmias.

On the basis of the abovementioned mechanisms, E.M. Vaughan–Williams has divided the antiarrhythmic drugs into four main classes, with class I consisting of sodium channel blockers, class II consisting of β-blockers, class III consisting of potassium channel blockers and other drugs that prolong the APD, and class IV consisting of CCBs. The Vaughan–Williams classification has proven useful in helping understand the clinical effects of the antiarrhythmic agents. However, it represents an oversimplification of the electrophysiological events that occur with each of the drug classes. In addition, a few drugs (e.g., adenosine)

Cardiovascular Diseases: From Molecular Pharmacology to Evidence-Based Therapeutics, First Edition. Y. Robert Li.
© 2015 John Wiley & Sons, Inc. Published 2015 by John Wiley & Sons, Inc.

TABLE 24.1   Vaughan–Williams classification of antiarrhythmic drugs and overview of their major electrophysiological effects

Class	Basic mechanisms	Subclass and/or representative drugs		Major electrophysiological effects
I	Sodium channel blockage	IA—moderate sodium channel blockers; also cause potassium channel blockage	Disopyramide Procainamide Quinidine	Moderate reduction in phase 0 slope; increase of action potential duration (APD); increase of effective refractory period (ERP)
		IB—weak sodium channel blockers	Lidocaine Mexiletine	Small reduction in phase 0 slope; decrease of APD; decrease of ERP
		IC—strong sodium channel blockers	Flecainide Propafenone	Pronounced reduction in phase 0 slope; no effect on APD or ERP
II	β-Adrenergic receptor blockage	Esmolol Metoprolol Propranolol Other β-blockers		Blockage of sympathetic activity; decrease of cardiac automaticity and conduction; increase of ERP
III	Potassium channel blockage	Amiodarone Dofetilide Dronedarone Ibutilide Sotalol		Delay of repolarization (phase 3) and increase of APD and ERP
IV	Calcium channel blockage	Diltiazem Verapamil		Blockage of L-type calcium channels; most effective at SA and AV nodes; decrease of AV conduction velocity

do not fit into any of these categories, and some drugs (e.g., amiodarone) could be included in more than one category. Table 24.1 shows the Vaughan–Williams classification and the basic mechanism of action associated with each class of the antiarrhythmic drugs.

24.3   CLASS I ANTIARRHYTHMIC DRUGS

24.3.1   General Aspects of Class I Drugs

24.3.1.1   Sodium Channel-Binding and Dissociation Properties   Class I drugs act by blocking the sodium channels, thereby inhibiting phase 0 depolarization in non-pacemaker cells and phase 4 spontaneous depolarization in pacemaker cells. These drugs are positively charged and may interact with specific amino acid residues in the internal pore of the sodium channels to cause inhibition. As summarized in Table 24.1, class I drugs are further divided into three subgroups. This subclassification is based on their different effects on sodium channels, which in turn result from the different binding and dissociation properties. Table 24.2 summarizes the sodium channel-binding and dissociation properties of class I drugs. The effects of these drugs on action potential are illustrated in Figure 24.1.

24.3.1.2   Use-Dependent Blockage   As shown in Table 24.2, antiarrhythmic drugs preferentially bind to sodium channels when the channels are at active (open) and/or inactive states, and consequently, dissociation occurs when the channels are at rest. During faster heart rates, less time exists for the drugs to dissociate from the

TABLE 24.2   Sodium channel-binding and dissociation properties of class I drugs

Subclass	Sodium channel blockage	Binding and dissociation from sodium channels	Affinity for sodium channels at open or inactivated state
IA[a]	Moderate	IA drugs are intermediate in terms of the rapidity of binding and dissociation	High affinity for open state
IB	Weak	IB drugs have the most rapid binding and dissociation from the receptor	High affinity for both open and inactivated states
IC	Strong	IC drugs have the slowest binding and dissociation from the receptor	High affinity for open state

[a]IA drugs also inhibit repolarizing potassium channels, leading to prolongation of ERP.

channels due to less time for the channels to be at rest, thereby resulting in an increased number of blocked channels and enhanced blockade. These pharmacological effects may cause a progressive decrease in conduction velocity of action potential and a widening of the QRS complex in electrocardiogram (ECG). This property is known as use-dependent blockage. This explains why antiarrhythmic drugs are more effective on tissue that is firing rapidly, that is, tachyarrhythmias.

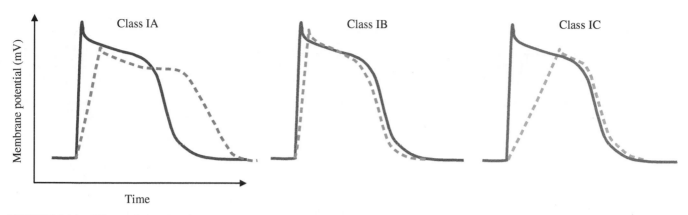

FIGURE 24.1 Effects of class I antiarrhythmic drugs on action potential of nonpacemaker cells. Of note is the ability of class IA drugs to prolong repolarization. Solid lines indicate normal action potentials and dashed lines show the drug effects.

24.3.2 Specific Class IA Drugs

- Disopyramide (Norpace)
- Procainamide (Pronestyl)
- Quinidine

24.3.2.1 Chemistry and Pharmacokinetics The structures of the three class IA drugs are shown in Figure 24.2. Table 24.3 summarizes the major pharmacokinetic properties of these drugs.

24.3.2.2 Molecular Mechanisms and Pharmacological Effects

Quinidine Quinidine depresses the rapid inward depolarizing sodium current, thereby slowing phase 0 depolarization and reducing the amplitude of the action potential without affecting the resting potential. In pacemaker cells and Purkinje fibers, it reduces the slope of phase 4 depolarization and shifts the threshold voltage upward toward zero. The result is slowed conduction and reduced automaticity in all parts of the heart, with increase of the effective refractory period (ERP) relative to the duration of the action potential in the atria, ventricles, and Purkinje tissues. Quinidine also directly prolongs the refractory period by blocking potassium channels.

Quinidine increases the fibrillation thresholds of the atria and ventricles, and it raises the ventricular defibrillation threshold as well. By slowing conduction and prolonging the ERP, quinidine can interrupt or prevent reentrant arrhythmias and arrhythmias due to increased automaticity, including atrial flutter (AFL), atrial fibrillation (AF), and paroxysmal supraventricular tachycardia (PSVT). In patients with the sick sinus syndrome, quinidine can cause marked sinus node depression and bradycardia. In most patients, however, use of quinidine is associated with an increase in the sinus rate.

Quinidine prolongs the QT interval in a dose-related fashion. This may lead to increased ventricular automaticity and polymorphic ventricular tachycardias, including torsades de pointes. In addition, quinidine has anticholinergic activity and negative inotropic activity, and it acts peripherally as an α-adrenergic antagonist, causing vasodilation.

Recently, the use of quinidine in treating cardiac arrhythmias has received renewed attention. Quinidine may be particularly effective for the management of ventricular arrhythmias in Brugada syndrome, a genetic disorder caused by abnormalities in sodium channels. However, quinidine is inaccessible in many countries due to its limited use in cardiovascular medicine and, consequently, a lack of incentives for drug manufacturing [1, 2].

Disopyramide and Procainamide Disopyramide and procainamide are similar to quinidine with regard to electrophysiological effects. The anticholinergic activity of disopyramide is more potent than that of quinidine. Among the IA drugs, procainamide has the weakest anticholinergic activity.

24.3.2.3 Clinical Uses and Therapeutic Dosages Listed below are the dosage forms and strengths of the three class IA drugs:

1. Disopyramide (Norpace): Oral, 100 and 150 mg capsules; 150 mg extended-release capsules
2. Procainamide (Pronestyl): Oral, 250, 375, and 500 mg tablets and capsules; 250, 500, 750, and 1000 mg sustained-release tablets; injection, 100 mg/ml in multiple-dose 10 ml vials; 500 mg/ml in 2 ml vials
3. Quinidine: Oral, 200 and 300 mg quinidine sulfate tablets; 300 mg extended-release quinidine sulfate tablets; 324 mg extended-release quinidine gluconate tablets; injection, 80 mg/ml in 10 ml multiple-dose vials

The cardiovascular indications of class IA drugs and the dosage regimens are given in Table 24.4.

24.3.2.4 Adverse Effects and Drug Interactions The major adverse effects and drug interactions for IA drugs are summarized in Table 24.5.

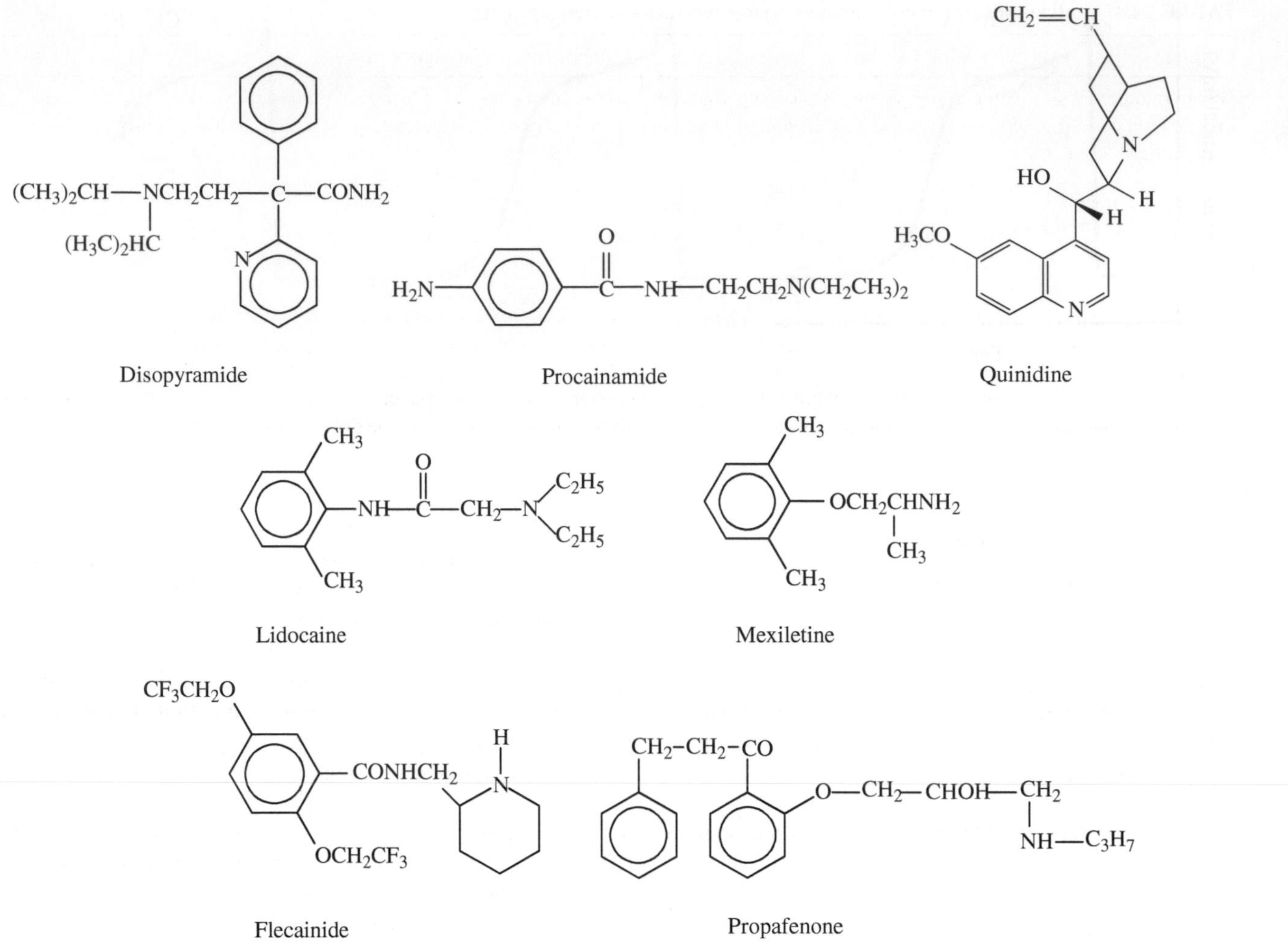

FIGURE 24.2 Structures of class I drugs. Disopyramide, procainamide, and quinidine are class IA drugs; lidocaine and mexiletine are class IIB drugs; and flecainide and propafenone are class IC drugs.

TABLE 24.3 **Major pharmacokinetic properties of class IA drugs**

IA drug	Oral bioavailability (%)	Elimination half-life (h)	Metabolism and elimination
Disopyramide	100	4–10	CYP3A4; 50% eliminated unchanged and 30% as metabolites in the urine
Procainamide	85	3–4	Acetylation by N-acetyltransferases; eliminated unchanged and as metabolites in the urine
Quinidine	70 (40–100)	6–8	CYP3A4; 20% eliminated unchanged in the urine

24.3.3 Specific Class IB Drugs

- Lidocaine (Xylocaine)
- Mexiletine (Mexitil)

24.3.3.1 Chemistry and Pharmacokinetics Lidocaine and mexiletine are structurally related drugs (structures shown in Fig. 24.2). The major pharmacokinetic properties of these two class IB drugs are summarized in Table 24.6.

24.3.3.2 Molecular Mechanisms and Pharmacological Effects Lidocaine and mexiletine inhibit the inward sodium current, thus reducing the rate of rise of the action potential (phase 0). They decrease the ERP in Purkinje fibers. The decrease in ERP is of lesser magnitude than the decrease in APD, with a resulting increase in the ERP/APD ratio. This increased ratio appears to be responsible for the inhibition of reentry and thereby raise the ventricular fibrillation threshold.

TABLE 24.4   Indications and recommended dosage regimens of class IA drugs

IA drug	Indication	Dosage regimen
Disopyramide	Suppression of life-threatening ventricular arrhythmias	400–800 mg/day in divided doses
Procainamide (injection)	Suppression of life-threatening ventricular arrhythmias	Initial loading infusion: 20 mg/ml, 1 ml/min for up to 25–30 min Maintenance infusion: 2 mg/ml, 1–3 ml/min
Quinidine sulfate (tablets)	Conversion of atrial fibrillation/flutter to sinus rhythm	400 mg every 6 h
	Reduction in the frequency of relapse into atrial fibrillation/flutter	200 mg every 6 h
	Suppression of life-threatening ventricular arrhythmias	Not available
Quinidine gluconate (extended-release tablets)	Conversion of atrial fibrillation/flutter to sinus rhythm	648 mg (two tablets) every 8 h
	Reduction in the frequency of relapse into atrial fibrillation/flutter	324 mg (one tablet) every 8–12 h
	Suppression of life-threatening ventricular arrhythmias	Not available
Quinidine gluconate (injection)	Conversion of atrial fibrillation/flutter to sinus rhythm	Infusion rate: no faster than 0.25 mg/kg/min. Most arrhythmias that will respond to intravenous quinidine will respond to a total dose of <5 mg/kg, but some patients may require as much as 10 mg/kg
	Suppression of life-threatening ventricular arrhythmias	Not available

TABLE 24.5   Adverse effects and drug interactions for class IA drugs

IA drug	Adverse effects	Drug interactions	Contraindications
Disopyramide	The most serious adverse reactions are hypotension and congestive heart failure. The most common adverse reactions are associated with the anticholinergic properties of the drug. Urinary retention is the most serious anticholinergic effect	Inhibitors of CYP3A4 may increase plasma levels of disopyramide	It is contraindicated in the presence of cardiogenic shock, preexisting second- or third-degree AV block (if no pacemaker is present), congenital QT prolongation, or known hypersensitivity to the drug. The drug is in pregnancy category C
Procainamide	Adverse effects include (i) hypotension and serious disturbances of cardiac rhythm, (ii) a lupus erythematosus-like syndrome, (iii) hematological toxicity, (iv) gut disturbances and liver injury, and (v) nervous system disturbances	Patients taking procainamide who require neuromuscular blocking agents, such as succinylcholine, may require less than usual doses of the latter, due to procainamide effects on reducing acetylcholine release	Contraindications include complete heart block, idiosyncratic hypersensitivity, lupus erythematosus, and torsades de pointes. The drug is in pregnancy category C
Quinidine	Most common adverse effects are gut disturbances. The most serious adverse effects are proarrhythmic actions, including prolongation of QT interval, which can lead to torsades de pointes. Other effects include cinchonism (a syndrome that may include tinnitus, reversible high-frequency hearing loss, deafness, vertigo, blurred vision, diplopia, photophobia, headache, confusion, and delirium), liver injury, and autoimmunity	Drug interactions are very significant. For example, quinidine levels are increased by coadministration of amiodarone or cimetidine. Drugs that affect CYP3A4 may affect quinidine metabolism. Quinidine inhibits the elimination of digoxin. It potentiates the anticoagulatory action of warfarin. Quinidine inhibits CYP2D6 and, as such, affects the metabolism of drugs that are substrates of CYP2D6 (e.g., flecainide, codeine, hydrocodone)	Quinidine is contraindicated in patients who are known to be allergic to it or who have developed thrombocytopenic purpura during prior therapy with quinidine or quinine. In the absence of a functioning artificial pacemaker, quinidine is contraindicated in any patient whose cardiac rhythm is dependent upon a junctional or idioventricular pacemaker, including patients in complete atrioventricular block. Quinidine is contraindicated in patients who, like those with myasthenia gravis, might be adversely affected by an anticholinergic agent. The drug is in pregnancy category C

Because lidocaine and mexiletine have greater affinity for both open and inactivated states of sodium channels (Table 24.2), they are particularly effective in blocking rapidly firing tissues, such as ischemic tissue, where there is a higher likelihood of sodium channels being in the open or inactivated state. This explains their efficacy in suppressing ventricular arrhythmias resulting from myocardial ischemia (e.g., myocardial infarction).

24.3.3.3 Clinical Uses and Therapeutic Dosages

Listed below are the dosage forms and strengths of the two class IB drugs:

- Lidocaine (Xylocaine): Intravenous (iv) injection, 100 mg in 5 ml ampules
- Mexiletine (Mexitil): Oral, 150 mg capsules

The cardiovascular indications of class IB drugs and the dosage regimens are given in Table 24.7.

24.3.3.4 Adverse Effects and Drug Interactions

The major adverse effects and drug interactions for class IB drugs are summarized in Table 24.8.

TABLE 24.6   Major pharmacokinetic properties of class IB drugs

IB drug	Oral bioavailability (%)	Elimination half-life (h)	Metabolism and elimination
Lidocaine (iv)		1.5–2	CYP1A2 (major); CYP3A4; eliminated in the urine
Mexiletine	90	10–12	CYP2D6 (major); CYP1A2; eliminated in the urine

24.3.4 Specific Class IC Drugs

- Flecainide (Tambocor)
- Propafenone (Rythmol)

24.3.4.1 Chemistry and Pharmacokinetics

Both flecainide and propafenone are oral antiarrhythmic agents (structures shown in Fig. 24.2). Their major pharmacokinetic properties are summarized in Table 24.9.

24.3.4.2 Molecular Mechanisms and Pharmacological Effects

Among class I antiarrhythmic drugs, class IC drugs are the most potent sodium channel blockers. Both flecainide and propafenone dramatically reduce the rate of phase 0 of action potential, thereby slowing conduction velocity and inhibiting reentry in all parts of the heart with the greatest effect on the His–Purkinje system. Blockage of sodium channels also slows down phase 4 depolarization in pacemaker cells, resulting in decreased spontaneous automaticity. Class IC drugs depress triggered activity likely due to inhibition of inward sodium ion current mediated by partially recovered sodium ion channels during depolarization phase (phase 3) (see Chapter 23). Class IB drugs, especially propafenone, exert cardiac depressing effects likely due to β-blocking activity.

24.3.4.3 Clinical Uses and Therapeutic Dosages

Listed below are the dosage forms and strengths of the two class IC drugs:

- Flecainide (Tambocor): 50, 100, and 150 mg tablets
- Propafenone (Rythmol): Oral, 150, 225, and 300 mg tablets; 225, 325, and 425 mg extended-release capsules

The cardiovascular indications of class IC drugs and the recommended dosage regimens are given in Table 24.10.

TABLE 24.7   Indications and recommended dosage regimens of class IB drugs

IB drug	Indication	Dosage regimen
Lidocaine[a]	Intravenous administration for acute management of (i) ventricular arrhythmias occurring during cardiac manipulation such as cardiac surgery and (ii) life-threatening arrhythmias, particularly those that are ventricular in origin, such as those that occur during acute myocardial infarction	The usual dose is 50–100 mg administered intravenously under ECG monitoring. This dose may be administered at the rate of approximately 25 mg/min. If the initial injection of 50–100 mg does not produce a desired response, a second dose may be repeated after 5 min. No >200–300 mg should be administered during a 1 h period. Following bolus administration, intravenous infusions of lidocaine may be initiated at the rate of 1–4 mg/min
Mexiletine	Life-threatening ventricular arrhythmias	200–300 mg given every 8 h

[a] The role of lidocaine in preventing life-threatening ventricular arrhythmias in myocardial infarction has recently been questioned due to mixed results in clinical trials. The majority of the evidence suggests that prophylactic use of lidocaine in this clinical setting is not associated with improved clinical outcomes, and as such, the prophylactic use of lidocaine is not currently recommended by the American College of Cardiology Foundation/American Heart Association (ACCF/AHA) [3, 4]. Most physicians administer iv lidocaine only to patients with life-threatening arrhythmias secondary to acute myocardial infarction. It may be considered as an important drug for termination of ventricular tachycardia and prevention of ventricular fibrillation after cardioversion in the setting of acute coronary syndromes (ACS).

TABLE 24.8   Major adverse effects and drug interactions of class IB drugs

IB drug	Adverse effect	Drug interaction	Contraindication
Lidocaine	The major adverse effects are related to the central nervous system (CNS) and cardiovascular system. CNS reactions are excitatory and/or depressant and may be characterized by light-headedness, nervousness, apprehension, euphoria, confusion, dizziness, drowsiness, tinnitus, and blurred or double vision. Cardiovascular reactions may include bradycardia, hypotension, and cardiovascular collapse, which may lead to cardiac arrest	Lidocaine should be used with caution in patients with digitalis toxicity accompanied by atrioventricular (AV) block	Lidocaine is contraindicated in patients with a known history of hypersensitivity to local anesthetics of the amide type. It should not be used in patients with Stokes–Adams syndrome, with Wolff–Parkinson–White syndrome, or with severe degrees of sinoatrial, AV, or intraventricular block in the absence of an artificial pacemaker. The drug is in pregnancy category B
Mexiletine	The drug is relatively well tolerated. The most common adverse effects are related to gut disturbance. The drug may also affect the cardiovascular system (e.g., syncope and hypotension) and CNS (e.g., short-term memory loss and psychological changes)	Since mexiletine is a substrate for the metabolic pathways involving CYP2D6 and CYP1A2 enzymes, inhibition or induction of either of these enzymes would be expected to alter mexiletine plasma concentrations	It is contraindicated in the presence of cardiogenic shock or preexisting second- or third-degree AV block (if no pacemaker is present). The drug is in pregnancy category C

TABLE 24.9   Major pharmacokinetic properties of class IC drugs

IC drug	Oral bioavailability (%)	Elimination half-life (h)	Metabolism and elimination
Flecainide	~100	12–27	CYP2D6; eliminated in unchanged form and metabolites in the urine
Propafenone	3–21 (dose dependent)	2–10	CYP2D6, CYP3A4, and CYP1A2; eliminated in the urine and feces

TABLE 24.10   Cardiovascular indications and dosage regimens of class IC drugs

IC drug	Indication	Dosage regimen
Flecainide	In patients without structural heart disease, flecainide is indicated for the prevention of (i) paroxysmal supraventricular tachycardias (PSVT), including atrioventricular (AV) nodal reentrant tachycardia, AV reentrant tachycardia, and other supraventricular tachycardias of unspecified mechanism associated with disabling symptoms; and (ii) paroxysmal atrial fibrillation/flutter (PAF) associated with disabling symptoms	For patients with PSVT and patients with PAF, the recommended starting dose is 50 mg every 12 h. The doses may be increased in increments of 50 mg b.i.d. every 4 days until efficacy is achieved. The maximum recommended dose for patients with paroxysmal supraventricular arrhythmias is 300 mg/day
	Prevention of documented ventricular arrhythmias, such as sustained ventricular tachycardia (VT), that are life-threatening. It should not be used in patients with recent myocardial infarction due to increased mortality	For sustained VT, the recommended starting dose is 100 mg every 12 h. This dose may be increased in increments of 50 mg b.i.d. every 4 days until efficacy is achieved. The maximum dose recommended is 400 mg/day
Propafenone	In patients without structural heart disease, propafenone is indicated to prolong the time to recurrence of (i) PAF associated with disabling symptoms and (ii) PSVT associated with disabling symptoms	It is recommended that therapy be initiated with 150 mg propafenone given every 8 h (450 mg/day). Dosage may be increased at a minimum of 3–4-day intervals to 225 mg every 8 h (675 mg/day) and, if necessary, to 300 mg every 8 h (900 mg/day)
	Treatment of documented ventricular arrhythmias, such as sustained VT, that are life-threatening	

TABLE 24.11 Major adverse effects, drug interactions, and contraindications of class IC drugs

IC drug	Adverse effect	Drug interaction	Contraindication
Flecainide[a]	In post-myocardial infarction patients with asymptomatic premature ventricular complexes and nonsustained ventricular tachycardia, flecainide therapy was found to be associated with a 5.1% rate of death and nonfatal cardiac arrest, compared with a 2.3% rate in a matched placebo group. This is largely due to its proarrhythmic effects. Common noncardiovascular effects may include dizziness, visual disturbance, headache, nausea, and malaise	Drugs that inhibit CYP2D6, such as quinidine, might increase the plasma concentrations of flecainide in patients that are on chronic flecainide therapy, especially if these patients are extensive metabolizers (i.e., patients with high CYP2D6 activity). When amiodarone is added to flecainide therapy, plasma flecainide levels may increase at least twofold	Flecainide is contraindicated in patients with preexisting second- or third-degree atrioventricular (AV) block or with right bundle branch block when associated with a left hemiblock (bifascicular block), unless a pacemaker is present to sustain the cardiac rhythm should complete heart block occur. Flecainide is also contraindicated in the presence of cardiogenic shock or known hypersensitivity to the drug. This drug is in pregnancy category C
Propafenone	Adverse reactions associated with propafenone occur most frequently in the gastrointestinal, cardiovascular, and central nervous systems. Most common events are unusual taste, dizziness, first-degree AV block, intraventricular conduction delay, nausea and/or vomiting, and constipation	Propafenone increases plasma levels of drugs, including digoxin and warfarin, if coadministrated. Drugs that inhibit CYP2D6, CYP1A2, and CYP3A4 might lead to increased plasma levels of propafenone	It is contraindicated in the presence of uncontrolled congestive heart failure; cardiogenic shock; sinoatrial, AV, and intraventricular disorders of impulse generation and/or conduction (e.g., sick sinus node syndrome, AV block) in the absence of an artificial pacemaker; bradycardia; marked hypotension; bronchospastic disorders; manifest electrolyte imbalance; and known hypersensitivity to the drug. It is a pregnancy category C drug

[a]Flecainide was included in the National Heart, Lung, and Blood Institute's Cardiac Arrhythmia Suppression Trial (CAST), a long-term, multicenter, randomized, double-blind study in patients with asymptomatic non-life-threatening ventricular arrhythmias who had a myocardial infarction >6 days but <2 years previously. An excessive mortality or nonfatal cardiac arrest rate was seen in patients treated with flecainide compared with that seen in patients assigned to a carefully matched placebo-treated group. This rate was 16/315 (5.1%) for flecainide and 7/309 (2.3%) for the matched placebo. The average duration of treatment with flecainide in this study was 10 months. The applicability of the CAST results to other populations (e.g., those without recent myocardial infarction) is uncertain, but at present, it is prudent to consider the risks of class IC agents (including flecainide), coupled with the lack of any evidence of improved survival, generally unacceptable in patients without life-threatening ventricular arrhythmias, even if the patients are experiencing unpleasant, but not life-threatening, symptoms or signs.

24.3.4.4 Adverse Effects and Drug Interactions As with other antiarrhythmic drugs, use of class IC drugs may be associated with a number of significant adverse effects. Table 24.11 summarizes the major adverse effects, drug interactions, and contraindications of flecainide and propafenone.

24.4 CLASS II ANTIARRHYTHMIC DRUGS

24.4.1 General Aspects

Class II antiarrhythmic drugs are β-blockers (see Chapter 8). These drugs block the proarrhythmic effects of β_1-adrenergic receptor activation resulting from sympathetic stimulation. Sympathetic activation of β_1-adrenergic receptors in sinoatrial (SA) nodal cells accelerates phase 4 depolarization, thereby increasing automaticity. Activation of β_1-adrenergic receptors in atrioventricular (AV) nodal cells augments phase 0 Ca^{2+} currents, resulting in increased conduction velocity and elevated risk of reentry. Activation of β_1-adrenergic receptors in AV nodal cells also increases phase 3 K^+ currents, leading to shortened refractory period. These β_1-adrenergic receptor-mediated effects collectively contribute to the development of cardiac arrhythmias associated with sympathetic hyperactivation.

By blocking the aforementioned β_1-adrenergic receptor-mediated actions, class II antiarrhythmic drugs reduce heart rate, slow AV conduction velocity, and prolong AV refractory period. Slowing AV nodal conduction velocity and prolongation of AV nodal refractory period result in decreased reentry. As noted in Chapter 23, AV nodal reentry is a fundamental mechanism of supraventricular tachyarrhythmias.

24.4.2 β-Blockers Commonly Used for Treating Arrhythmias

Although many β-blockers have been approved for use in the United States, metoprolol, carvedilol, atenolol, propranolol, and esmolol have been most widely used to treat cardiac arrhythmias. It is generally considered that β-blockers possess class effects and that if titrated to the proper dose, all can be used effectively to treat cardiac arrhythmias as well as other cardiovascular diseases. However, differences in pharmacokinetic or pharmacodynamic properties influence the choice of the particular β-blockers (see Chapter 8).

24.5   CLASS III ANTIARRHYTHMIC DRUGS

Listed below are the US Food and Drug Administration (FDA)-approved class III antiarrhythmic drugs. Among them, dronedarone is the newest member of this drug class. Sotalol is a β-blocker that also blocks potassium channels to prolong ERP, and as such, it is usually classified as a class III drug. It is sometimes also classified as a class II drug:

- Amiodarone (Cordarone)
- Dronedarone (Multaq)
- Dofetilide (Tikosyn)
- Ibutilide (Corvert)
- Sotalol (Betapace)

24.5.1   General Aspects

Class III drugs block the repolarizing potassium channels, thereby prolonging repolarization and ERP. Prolongation of ERP reduces reentry. On the other hand, blockage of ventricular potassium currents results in prolongation of the QT interval on ECG. Prolongation of QT interval may predispose to torsades de pointes, a polymorphic ventricular tachycardia.

In addition to the blockage of potassium channels, amiodarone and dronedarone also inhibit sodium channels in depolarized tissues, a characteristic effect of class IB drugs. Amiodarone and dronedarone also block calcium channels as well as β-adrenergic receptors.

24.5.2   Specific Class III Drugs

24.5.2.1   *Amiodarone and Dronedarone*

Chemistry and Pharmacokinetics   Both amiodarone and dronedarone are benzofuran derivatives (structures shown in Fig. 24.3), which are chemically described as 2-butyl-3-benzofuranyl 4-[2-(diethylamino)-ethoxy]-3,5-diiodophenyl ketone and *N*-{2-butyl-3-[4-(3-dibutylaminopropoxy)benzoyl] benzofuran-5-yl} methanesulfonamide, respectively. In contrast to amiodarone, dronedarone does not contain the iodine moiety and, as such, lacks thyroid effects (see the following text). The major pharmacokinetic properties of amiodarone and dronedarone are summarized in Table 24.12.

Molecular Mechanisms and Pharmacological Effects Amiodarone is generally considered a class III antiarrhythmic drug, but it possesses electrophysiological characteristics of all four classes of antiarrhythmics, including blockage of sodium channels, β-adrenergic receptors (as well as α-adrenergic receptors), potassium channels, and calcium channels.

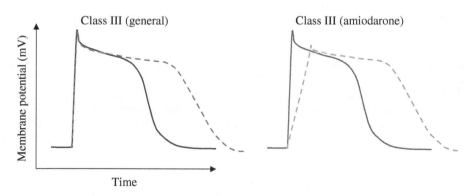

FIGURE 24.3   Effects of class III antiarrhythmic drugs on action potential of nonpacemaker cells. Of note is the ability of amiodarone (also dronedarone) to inhibit phase 0 upstroke. Solid lines indicate normal action potentials and dashed lines show the drug effects.

TABLE 24.12   Major pharmacokinetic properties of amiodarone and dronedarone

Drug	Oral bioavailability (%)	Elimination half-life	Metabolism and elimination
Amiodarone	50	2.5–10 days (initial phase); 40–55 days (terminal phase)	CYP3A4, CYP2C8; eliminated in the feces via biliary excretion
Dronedarone	4–15	13–19 h	CYP3A4; eliminated in the feces via biliary excretion (major) and also in the urine (minor)

400 DRUGS FOR CARDIAC ARRHYTHMIAS

Like class I drugs, amiodarone blocks sodium channels at rapid pacing frequencies, and like class II drugs, it exerts a noncompetitive antisympathetic action (blockage of both β- and α-adrenergic receptors). One of its main effects is to lengthen the cardiac action potential, a class III effect. This results in prolongation of ERP. The negative chronotropic effect of amiodarone in nodal tissues is similar to the effect of class IV drugs. The antisympathetic action and the block of calcium and potassium channels are responsible for the negative dromotropic effects on the sinus node and for the slowing of conduction and prolongation of ERP in the AV node. Its vasodilatory action can decrease cardiac workload and consequently myocardial oxygen consumption.

As an amiodarone derivative, dronedarone is believed to possess similar mechanisms of actions of amiodarone, that is, dronedarone has antiarrhythmic properties belonging to all four Vaughan–Williams classes. However, the contribution of each of these activities to the clinical effect is not clear.

Clinical Uses and Therapeutic Dosages The dosage forms and strengths of amiodarone and dronedarone are listed below. The clinical indications and dosage regimens of these two drugs are summarized in Table 24.13:

- Amiodarone (Cordarone): Oral, 200 and 400 mg tablets; iv, 50 mg/ml in 3 ml (150 mg) single-dose vials; 450 mg/9 ml and 900 mg/18 ml in multiple-dose vials
- Dronedarone (Multaq): Oral, 400 mg tablets

Adverse Effects and Drug Interactions Both amiodarone and dronedarone can cause significant adverse effects, including pulmonary toxicity and cardiovascular adverse

reactions. Because they are metabolized by CYP enzymes, especially CYP3A4, drug interactions are also remarkable. Table 24.14 summarizes the major adverse effects, drug interactions, and contraindications of these two drugs.

24.5.2.2 Ibutilide and Dofetilide

Chemistry and Pharmacokinetics Ibutilide and dofetilide are sulfanilide derivatives (structures shown in Fig. 24.4). Their major pharmacokinetic properties are given in Table 24.15.

Molecular Mechanisms and Pharmacological Effects Ibutilide and dofetilide, like other class III drugs, are traditionally thought to block outward potassium currents, thereby resulting in prolongation of action potential and ERP. While dofetilide at clinically relevant concentrations is known to block the cardiac ion channel carrying the rapid component of the delayed rectifier potassium current and have no effects on sodium channels, ibutilide may primarily affect sodium channels. In this regard, voltage clamp studies indicate that ibutilide, at nanomolar concentrations, delays repolarization mainly by activation of a slow, inward current (predominantly sodium), rather than by blocking outward potassium currents. This also leads to prolongation of atrial and ventricular APD and refractoriness.

Ibutilide produces mild slowing of the sinus rate and AV conduction. In contrast, dofetilide does not influence cardiac conduction velocity and sinus node function. Like other class III drugs, ibutilide and dofetilide cause prolongation of the QT interval, which may be associated with torsades de pointes.

Clinical Uses and Therapeutic Dosages Listed below are the dosage forms and strengths of ibutilide and dofetilide.

TABLE 24.13 Clinical indications and dosage regimens of amiodarone and dronedarone

Drug	Indication	Dosage regimen
Amiodarone (iv)	Amiodarone injection is indicated for initiation of treatment and prophylaxis of frequently recurring ventricular fibrillation (VF) and hemodynamically unstable ventricular tachycardia (VT) in patients refractory to other therapy. Amiodarone (iv) also can be used to treat patients with VT/VF for whom oral amiodarone is indicated but who are unable to take oral medication	The recommended starting dose of amiodarone is about 1000 mg over the first 24 h of therapy, delivered by the following infusion regimen: (i) loading infusions, 50 mg over the first 10 min (15 mg/min), followed by 360 mg over the next 6 h (1 mg/min), and (ii) maintenance infusion, 540 mg over the remaining 18 h (0.5 mg/min). After the first 24 h, continue the maintenance infusion rate of 0.5 mg/min (720 mg/24 h) utilizing a concentration of 1–6 mg/ml. The rate of the maintenance infusion may be increased to achieve effective arrhythmia suppression
Amiodarone (tablet)	Treatment of recurrent VF and recurrent hemodynamically unstable VT	The recommended oral dosage regimen is as follows: (i) loading doses of 800–1600 mg/day for 1–3 weeks, (ii) adjusted doses of 600–800 mg/day for 1 month, and (iii) maintenance dose of 400 mg/day
Dronedarone	It is indicated to reduce the risk of hospitalization for atrial fibrillation (AF) in patients in sinus rhythm with a history of paroxysmal or persistent AF	The recommended dosage is 400 mg twice daily. It should be taken as one tablet with the morning meal and one tablet with the evening meal

TABLE 24.14 Adverse effects, drug interactions, and contraindications of amiodarone and dronedarone

Drug	Adverse effect	Drug interaction	Contraindication
Amiodarone	The most serious adverse effects of amiodarone include pulmonary toxicity (e.g., pulmonary fibrosis), arrhythmias, and severe liver injury. Common adverse effects include gut disturbances, solar dermatitis/photosensitivity, visual and neurological disturbances, elevated liver enzymes, and pulmonary inflammation or fibrosis. Amiodarone may also cause thyroid abnormalities (hypothyroidism or hyperthyroidism) and congestive heart failure. The adverse reactions most frequently requiring discontinuation of amiodarone include pulmonary infiltrates or fibrosis, paroxysmal ventricular tachycardia, congestive heart failure, and elevation of liver enzymes. Other symptoms causing drug discontinuations less often include visual disturbances, solar dermatitis, blue skin discoloration, hyperthyroidism, and hypothyroidism	Since amiodarone is a substrate for CYP3A4, drugs/substances that inhibit CYP3A4 may decrease the metabolism and increase plasma concentrations of amiodarone. Amiodarone inhibits P-glycoprotein and certain CYP enzymes, including CYP1A2, CYP2C9, CYP2D6, and CYP3A4. This inhibition can result in unexpectedly high plasma levels of other drugs that are metabolized by those CYP enzymes or are substrates of P-glycoprotein	It is contraindicated in patients with cardiogenic shock; severe sinus node dysfunction, causing marked sinus bradycardia; second- or third-degree atrioventricular (AV) block; and when episodes of bradycardia have caused syncope (except when used in conjunction with a pacemaker). It is also contraindicated in patients with a known hypersensitivity to the drug or to any of its components, including iodine. Amiodarone is in pregnancy category D
Dronedarone	The most common adverse effects of dronedarone are gut disturbances. The drug can also cause new or worsening heart failure, liver injury, pulmonary toxicity, and electrolyte disturbances. Dronedarone may increase cardiovascular mortality in patients with permanent atrial fibrillation who are at risk for major vascular events	Dronedarone is metabolized by CYP3A4 and is a moderate inhibitor of CYP3A4 and CYP2D6, as well as P-glycoprotein, and has potentially important drug interactions when coadministered with drugs that are substrates, inhibitors, or inducers of the above enzymes. Dronedarone may also have significant pharmacodynamics interactions. For example, digoxin can potentiate the electrophysiological effects of dronedarone. Coadministration of drugs prolonging the QT interval (such as certain phenothiazines, tricyclic antidepressants, certain macrolide antibiotics, and class I and class III antiarrhythmics) is contraindicated because of the potential risk of torsades de pointes-type ventricular tachycardia	Dronedarone has a number of contraindications, including permanent AF, recently decompensated heart failure requiring hospitalization or NYHA class IV heart failure, second- or third-degree AV block or sick sinus syndrome (except when used in conjunction with a functioning pacemaker), bradycardia <50 beats/min, concomitant use of a strong CYP3A inhibitor, concomitant use of drugs or herbal products that prolong the QT interval and may induce torsades de pointes, liver or lung toxicity related to the previous use of amiodarone, severe hepatic impairment, QTc Bazett interval ≥500 ms, and hypersensitivity to the drug product. Dronedarone is also contraindicated in pregnancy and nursing mothers

Their clinical indications and dosage regimens are summarized in Table 24.16:

- Dofetilide (Tikosyn): Oral, 0.125, 0.25, and 0.5 mg capsules
- Ibutilide (Corvert): Injection, 0.1 mg/ml in 10 ml (1 mg/10 ml) single-dose vials

Adverse Effects and Drug Interactions Although both dofetilide and ibutilide can be associated with increased risk of cardiac arrhythmias, including torsades de pointes, ibutilide is generally considered well tolerated. While ibutilide appears to be free of drug interactions, drug interactions for dofetilide are extensive. The adverse effects, drug interactions, and contraindications of dofetilide and ibutilide are summarized in Table 24.17.

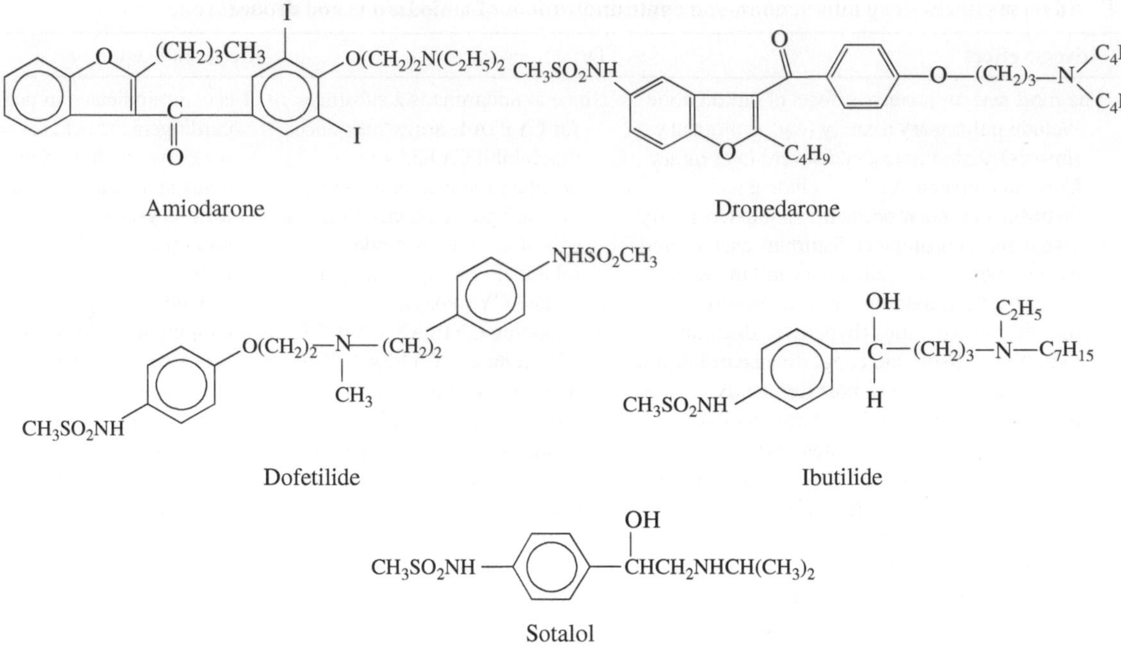

FIGURE 24.4 Structure of class III drugs. Notably, dronedarone is devoid of iodine, which accounts for its inability to cause thyroid abnormalities.

TABLE 24.15 Major pharmacokinetic properties of ibutilide and dofetilide

Drug	Oral bioavailability (%)	Elimination half-life (h)	Metabolism and elimination
Ibutilide (iv)		6 (2–12)	Oxidation of the heptyl side chain; eliminated in the urine (major) and feces (minor)
Dofetilide	>90	10	Weak metabolism by CYP3A4; eliminated in the urine in unchanged form (~80% of administered dose)

TABLE 24.16 Indications and dosage regimens of dofetilide and ibutilide

Drug	Indication	Dosage regimen
Dofetilide	The drug is indicated for (i) maintenance of normal sinus rhythm (delay in time to recurrence of atrial fibrillation/atrial flutter) in patients with atrial fibrillation/atrial flutter of >1-week duration who have been converted to normal sinus rhythm and (ii) the conversion of atrial fibrillation and atrial flutter to normal sinus rhythm	The usual recommended dose is 0.5 mg twice daily. The dosage must be individualized according to calculated creatinine clearance and QTc
Ibutilide	Ibutilide injection is indicated for the rapid conversion of atrial fibrillation or atrial flutter of recent onset to sinus rhythm	For patients whose body weight ≥60 kg, initial infusion of 1 mg over 10 min. For patients whose body weight <60 kg, initial infusion of 0.01 mg/kg over 10 min. For either regimen, if the arrhythmia does not terminate within 10 min after the end of the initial infusion, a second 10 min infusion of equal strength may be administered 10 min after completion of the first infusion

24.5.2.3 Sotalol

Chemistry and Pharmacokinetics Sotalol, a sulfanilide compound (structure shown in Fig. 24.4), is an antiarrhythmic drug with class II (β-adrenoreceptor blocking) and class III (cardiac APD prolongation) properties. Following oral administration, sotalol is readily absorbed with an oral bioavailability of 90–100%. Sotalol does not undergo significant metabolism and is excreted predominantly via the kidney in the unchanged form with a mean elimination half-life of 12 h.

TABLE 24.17   Adverse effects, drug interactions, and contraindications of dofetilide and ibutilide

Drug	Adverse effect	Drug interaction	Contraindication
Dofetilide	The most frequent adverse events include headache, chest pain, and dizziness. Dofetilide can cause torsades de pointes in a dose-dependent manner	Clinical studies show that coadministration of a number of drugs can significantly increase dofetilide exposure. These include cimetidine, verapamil, ketoconazole, trimethoprim alone or in combination with sulfamethoxazole, and hydrochlorothiazide alone or in combination with triamterene. Dofetilide is eliminated in the kidney by cationic secretion. Inhibitors of renal cationic secretion can increase its plasma concentration. In addition, drugs that are actively secreted via this route (e.g., triamterene, metformin, and amiloride) should be coadministered with care as they might increase dofetilide levels. In addition, dofetilide is metabolized, to a small extent, by the CYP3A4. Inhibitors of the CYP3A4 could increase systemic dofetilide exposure	Dofetilide is contraindicated in patients with congenital or acquired long QT syndromes. It is also contraindicated in patients with severe renal impairment (calculated creatinine clearance <20 ml/min) The concomitant use of verapamil or the cation transport system inhibitors cimetidine, trimethoprim (alone or in combination with sulfamethoxazole), and ketoconazole with dofetilide is contraindicated as each of these drugs causes a substantial increase in dofetilide plasma concentrations. In addition, other known inhibitors of the renal cation transport system such as prochlorperazine, dolutegravir, and megestrol should not be used in patients on dofetilide The concomitant use of hydrochlorothiazide (alone or in combination such as with triamterene) with dofetilide is contraindicated because this has been shown to significantly increase dofetilide plasma concentrations and QT interval prolongation Dofetilide is contraindicated in patients with a known hypersensitivity to the drug. Dofetilide belongs to pregnancy category C
Ibutilide	Ibutilide is generally well tolerated and the major adverse effects are related to cardiac arrhythmias, including torsades de pointes. Nausea is another adverse reaction	Drug interactions of ibutilide have not been systemically studied. Ibutilide appears to have no significant interactions with drugs, including calcium channel blockers, digoxin, and β-blockers	Ibutilide is contraindicated in patients who have previously demonstrated hypersensitivity to the drug product. Ibutilide belongs to pregnancy category C

Molecular Mechanisms and Pharmacological Effects Sotalol blocks both β-androgenic receptors and repolarizing potassium channels and thus has both class II and class III antiarrhythmic properties. The class II (β-blockade) electrophysiological effects of sotalol are manifested by increased sinus cycle length (slowed heart rate), decreased AV nodal conduction, and increased AV nodal refractoriness. The class III electrophysiological effects include prolongation of the atrial and ventricular monophasic action potentials and ERP prolongation of atrial muscle, ventricular muscle, and AV accessory pathways (where present) in both the anterograde and retrograde directions.

Clinical Uses and Therapeutic Dosages Listed below are the dosage forms and strengths of sotalol. The clinical indications and dosage regimens are provided in Table 24.18:

- Sotalol (Betapace): Oral, 80, 120, and 160 mg tablets; injection, 150 mg in 10 ml (15 mg/ml) vials

Adverse Effects and Drug Interactions Adverse reactions that are clearly related to sotalol are those that are typical of its class II (β-blocking) and class III (cardiac APD prolongation) effects. The common documented β-blocking adverse reactions (e.g., bradycardia, dyspnea, and fatigue) and class III effects (e.g., QT interval prolongation) are dose dependent. Sotalol can cause serious ventricular arrhythmias, including torsades de pointes. Factors, such as reduced creatinine clearance, gender (female), and larger doses, increase the risk of torsades de pointes.

Since sotalol is primarily eliminated by renal excretion in unchanged form, drugs that are metabolized by CYP enzymes do not alter the pharmacokinetics of sotalol. However, sotalol may interact pharmacodynamically with some drugs. For example, sotalol should be administered with caution in conjunction with nondihydropyridine CCBs because of possible additive effects on AV conduction or ventricular function. Additionally, concomitant use of these drugs may have additive effects on blood pressure, possibly leading to hypotension.

TABLE 24.18   Clinical indications and dosage regimens of sotalol

Drug dosage form	Indication	Dosage regimen
Sotalol (tablet)	Maintenance of normal sinus rhythm (delay in time to recurrence of atrial fibrillation/atrial flutter [AF/AFL]) in patients with symptomatic AF/AFL who are currently in sinus rhythm	The dose of sotalol must be individualized according to calculated creatinine clearance. The recommended initial dose of sotalol is 80 mg twice daily
	Treatment of documented ventricular arrhythmias, such as sustained ventricular tachycardia, that are life-threatening	The recommended initial dose is 80 mg twice daily. This dose may be increased, if necessary, after appropriate evaluation to 240 or 320 mg/day (120–160 mg twice daily)
Sotalol (iv)	Sotalol (iv) can substitute for oral sotalol in patients who are unable to take sotalol orally	The bioavailability of oral sotalol is between 90 and 100%. The corresponding dose of iv sotalol is, therefore, slightly less than that of the oral dose. The starting dose of iv sotalol is 75 mg infused over 5 h once or twice daily based on the creatinine clearance
	Sotalol (iv) is indicated for the maintenance of normal sinus rhythm (delay in time to recurrence of AF/AFL) in patients with symptomatic AF/AFL who are currently in sinus rhythm	The recommended initial dose is 112.5 mg iv sotalol twice a day. It may be increased to 150 mg once or twice a day
	Sotalol (iv) is indicated for the treatment of documented life-threatening ventricular arrhythmias	The recommended initial dose of iv sotalol is 75 mg infused over 5 h once or twice daily based on creatinine clearance. The dose may be increased in increments of 75 mg/day every 3 days

Sotalol is contraindicated in patients with sinus brady-cardia, sick sinus syndrome or second- and third-degree AV block (unless a functioning pacemaker is present), congenital or acquired long QT syndromes, baseline QT interval >450 ms, cardiogenic shock, uncontrolled heart failure, hypokalemia (<4 mM), creatinine clearance <40 ml/min, asthma, and previous evidence of hypersensitivity to sotalol. Sotalol is a pregnancy category B drug.

24.6   CLASS IV ANTIARRHYTHMIC DRUGS

- Diltiazem (Cardizem)
- Verapamil (Calan, Isoptin)

24.6.1   General Aspects

Class IV drugs are nondihydropyridine CCBs and include diltiazem and verapamil (also see Chapter 10). By blocking L-type calcium channels, class IV drugs can slow the sinus rate, increase the refractoriness of and prolong conduction through the AV node, and depress left ventricular contractility. Verapamil has a more pronounced effect on SA and AV nodes than diltiazem.

24.6.2   Chemistry and Pharmacokinetics

Diltiazem is a benzothiazepine derivative, whereas verapamil is a phenethylamine compound. The major pharmacokinetic properties of diltiazem and verapamil are given in Table 24.19.

TABLE 24.19   Major pharmacokinetic properties of diltiazem and verapamil

Drug	Oral bioavailability (%)	Elimination half-life (h)	Metabolism and elimination
Diltiazem[a]	40	3.0–4.5	CYP3A4; eliminated in the feces via biliary excretion and in the urine
Verapamil	20–35	4.5–12	CYP3A4; eliminated in the urine (major) and feces (minor)

[a] Diltiazem is also an inhibitor of CYP3A4.

24.6.3   Molecular Mechanisms and Pharmacological Effects

Diltiazem and verapamil inhibit the influx of calcium ions during membrane depolarization of cardiac and vascular smooth muscle. The therapeutic efficacy of the drugs in supraventricular tachycardias is related to their ability to slow AV nodal conduction time and prolong AV nodal refractoriness. Diltiazem and verapamil exhibit frequency (use)-dependent effects on AV nodal conduction such that they may selectively reduce the heart rate during tachycardias involving the AV node with little or no effect on normal AV nodal conduction at normal heart rates. These drugs slow the ventricular rate in patients with a rapid ventricular response during AF or AFL. They convert PSVT to normal sinus rhythm by interrupting the reentry circuit in AV nodal reentrant tachycardias.

Like other CCBs, because of their effect on vascular smooth muscle, diltiazem and verapamil reduce peripheral resistance, thereby decreasing blood pressure. These drugs relax coronary artery and prevent coronary artery spasm and reduce myocardial oxygen demand. As such, they are used to treat angina (see Chapters 14 and 15).

24.6.4 Clinical Uses and Therapeutic Dosages

The dosage forms and strengths of diltiazem and verapamil for cardiac arrhythmias are listed below. Diltiazem

and verapamil can be administered both orally and parenterally. Both oral and injection dosage forms of verapamil are indicated for treating arrhythmias, whereas only diltiazem injection is used for treating arrhythmias. The use of these drugs in treating arrhythmias and the corresponding dosage regimens are summarized in Table 24.20:

- Diltiazem (Cardizem): Injection, 0.5% (5 mg/ml) in 5, 10, and 25 ml single-use vials
- Verapamil (Calan, Isoptin): Oral, 40, 80, and 120 mg tablets; injection, 2.5 mg/ml in 2 ml (5 mg/ml) single-dose vials

TABLE 24.20 Indications and dosage regimens of diltiazem and verapamil in treating cardiac arrhythmias

Drug	Indication	Dosage regimen
Diltiazem (intravenous (iv))	Temporary control of rapid ventricular rate in AF or AFL. It should not be used in patients with AF or AFL associated with an accessory bypass tract such as in Wolff–Parkinson–White (WPW) syndrome or short PR syndrome	For iv bolus injection, the initial dose of diltiazem should be 0.25 mg/kg body weight as a bolus administered over 2 min (20 mg is a reasonable dose for the average patient). If response is inadequate, a second dose may be administered after 15 min. The second bolus dose of diltiazem should be 0.35 mg/kg administered over 2 min (25 mg is a reasonable dose for the average patient). Subsequent iv bolus doses should be individualized for each patient
		For continued reduction of the heart rate (up to 24 h) in patients with AF or AFL, an iv infusion of diltiazem may be administered. Immediately following bolus administration of diltiazem and reduction of heart rate, begin an iv infusion of diltiazem. The recommended initial infusion rate of diltiazem is 10 mg/h. Some patients may maintain response to an initial rate of 5 mg/h. The infusion rate may be increased in 5 mg/h increments up to 15 mg/h as needed, if further reduction in heart rate is required. The infusion may be maintained for up to 24 h. Infusion duration exceeding 24 h and infusion rates exceeding 15 mg/h are not recommended
	Rapid conversion of PSVT to sinus rhythm. This includes AV nodal reentrant tachycardias and reciprocating tachycardias associated with an extranodal accessory pathway such as the WPW syndrome or short PR syndrome	Use the same bolus iv injection as described above for temporary control of rapid ventricular rate in AF or AFL. Typically, PSVT can be converted to normal sinus rhythm in most patients within 3 min of the first or second bolus dose
Verapamil (tablet)	Oral verapamil is indicated for (i) treatment of arrhythmias in association with digitalis for the control of ventricular rate at rest and during stress in patients with chronic AF and/or AFL and (ii) prophylaxis of repetitive PSVT	The dosage in digitalized patients with chronic AF ranges from 240 to 320 mg/day in divided (t.i.d. or q.i.d.) doses. The dosage for prophylaxis of PSVT (nondigitalized patients) ranges from 240 to 480 mg/day in divided (t.i.d. or q.i.d.) doses. In general, maximum effects for any given dosage will be apparent during the first 48 h of therapy
Verapamil (iv)	Temporary control of rapid ventricular rate in AF or AFL except when the AF and/or AFL are associated with accessory bypass tracts such as in WPW and Lown–Ganong–Levine (LGL) syndromes. Rapid conversion to sinus rhythm of PSVT, including those associated with accessory bypass tracts such as in WPW and LGL syndromes	The initial dose can be 5–10 mg (0.075–0.15 mg/kg body weight) given as an iv bolus over at least 2 min. If the initial response is not adequate, a repeat dose of 10 mg (0.15 mg/kg) can be given 30 min after the first dose

24.6.5 Adverse Effects and Drug Interactions

24.6.5.1 Diltiazem The adverse effects of diltiazem are generally mild and transient. Hypotension is the most commonly reported adverse event. Other adverse effects include injection site reactions, flushing, junctional rhythm or isorhythmic dissociation, and AV block.

Diltiazem is both a substrate and an inhibitor of CYP3A4. Other drugs that are specific substrates, inhibitors, or inducers of this enzyme may have a significant impact on the efficacy and adverse effect profile of diltiazem. Patients taking other drugs that are substrates of CYP3A4, especially patients with renal and/or hepatic impairment, may require dosage adjustment when starting or stopping concomitantly administered diltiazem in order to maintain optimum therapeutic blood levels.

Diltiazem is contraindicated in patients (i) with sick sinus syndrome except in the presence of a functioning ventricular pacemaker, (ii) with second- or third-degree AV block except in the presence of a functioning ventricular pacemaker, (iii) with severe hypotension or cardiogenic shock, and (iv) who have demonstrated hypersensitivity to the drug. In addition, iv diltiazem and iv β-blockers should not be administered together or in close proximity (within a few hours) due to synergistic cardiac suppression. The drug is also contraindicated in patients with AF or AFL associated with an accessory bypass tract, such as in Wolff–Parkinson–White (WPW) syndrome or short PR syndrome. As with other agents that slow AV nodal conduction and do not prolong the refractoriness of the accessory pathway (e.g., verapamil, digoxin), in rare instances, patients in AF or AFL associated with an accessory bypass tract may experience a potentially life-threatening increase in heart rate accompanied by hypotension when treated with diltiazem. Diltiazem is also contraindicated in patients with wide complex ventricular tachycardia because of increased risk of hemodynamic deterioration and ventricular fibrillation. Hence, it is important to distinguish wide complex QRS tachycardia of supraventricular origin from that of ventricular origin prior to administration of diltiazem. Diltiazem is a drug in pregnancy category C.

24.6.5.2 Verapamil Serious adverse effects of verapamil are uncommon when treatment follows the recommended dosage regimens. Adverse reactions are related to (i) cardiovascular effects (e.g., symptomatic hypotension, bradycardia, severe tachycardia), (ii) central nervous system (CNS) effects (e.g., dizziness, headache), and (iii) gastrointestinal disturbances (e.g., nausea, abdominal discomfort). Constipation is the most common adverse effect of oral verapamil therapy.

Clinically significant drug interactions have been reported with CYP3A4 inhibitors (e.g., erythromycin, ritonavir) or CYP3A4 inducers (e.g., rifampin). The use of statin drugs that are CYP3A4 substrates (e.g., atorvastatin, simvastatin) in combination with verapamil may increase the risk of statin-induced myopathy (see Chapter 4).

Verapamil is contraindicated in (i) severe left ventricular dysfunction, (ii) hypotension or cardiogenic shock, (iii) sick sinus syndrome (except in patients with a functioning artificial ventricular pacemaker), (iv) second- or third-degree AV block (except in patients with a functioning artificial ventricular pacemaker), (v) patients with AF or AFL and an accessory bypass tract (e.g., WPW and Lown–Ganong–Levine (LGL) syndromes) due to increased risk of developing ventricular tachyarrhythmia including ventricular fibrillation, and (vi) patients with known hypersensitivity to verapamil. In addition, verapamil injection is also contraindicated in patients receiving iv β-adrenergic blocking drugs (e.g., propranolol) since both may have a depressant effect on myocardial contractility and AV conduction. Administration of iv verapamil to patients with wide complex ventricular tachycardia can result in marked hemodynamic deterioration and ventricular fibrillation and as such should be avoided. Verapamil is a drug in pregnancy category C.

24.7 OTHER ANTIARRHYTHMIC DRUGS

- Adenosine (Adenocard)
- Digoxin (Digox) (see Chapter 21)
- Magnesium
- Vernakalant (Brinavess)

24.7.1 Adenosine

24.7.1.1 Chemistry and Pharmacokinetics Adenosine (structure shown in Fig. 24.5) is an endogenous nucleoside occurring in all cells of the body. Intravenously administered adenosine is rapidly cleared from the circulation via cellular uptake, primarily by erythrocytes and vascular endothelial cells. Intracellular adenosine is rapidly metabolized either via phosphorylation to adenosine monophosphate by adenosine kinase or via deamination to inosine by adenosine deaminase in the cytosol. The plasma elimination half-life of adenosine is only about 10 s.

24.7.1.2 Molecular Mechanisms and Pharmacological Effects The antiarrhythmic effects of adenosine are mediated by specific G-protein-coupled receptor signaling (Fig. 24.5). Adenosine activates acetylcholine (Ach)-sensitive potassium channels in the atrium and SA and AV nodes, resulting in membrane hyperpolarization and the subsequent inhibition of voltage-dependent calcium channels. Adenosine slows AV conduction time and inhibits reentry pathways through the AV node. The drug can restore normal sinus rhythm in patients with PSVT, including PSVT associated with WPW syndrome.

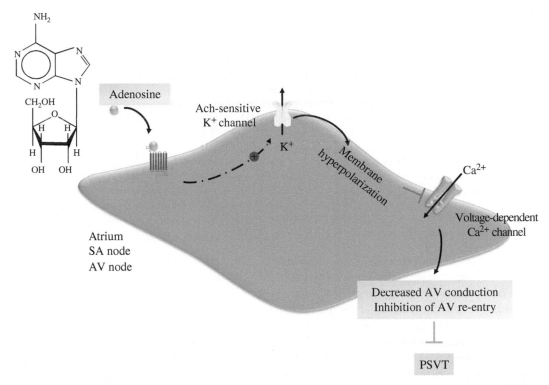

FIGURE 24.5 Molecular mechanism of action of adenosine in suppressing paroxysmal supraventricular tachycardias (PSVT). Adenosine via activating its receptor (G-protein-coupled receptor) causes activation of acetylcholine (Ach)-sensitive potassium channels, thereby leading to membrane hyperpolarization. As a result, voltage-dependent calcium channels are inhibited, leading to decreased atrioventricular (AV) conduction and reduced reentry, key mechanisms responsible for PSVT.

24.7.1.3 Clinical Uses and Therapeutic Dosages Listed below are the dosage form and strengths of adenosine in treating cardiac arrhythmias:

- Adenosine (Adenocard): iv, 6 mg/2 ml (3 mg/ml) in 2 ml (fill volume) syringe; 12 mg/4 ml (3 mg/ml) in 4 ml (fill volume) syringe

iv adenosine is indicated for the conversion to sinus rhythm of PSVT, including that associated with accessory bypass tracts (WPW syndrome). Adenosine does not convert AF, AFL, or ventricular tachycardia to normal sinus rhythm.

Adenosine should be given as a rapid bolus by the peripheral iv route. The recommended dosage regimens for adults and pediatric patients are given in Table 24.21.

24.7.1.4 Adverse Effects and Drug Interactions The adverse effects are usually transient as adenosine is rapidly metabolized. The adverse reactions are related to (i) vasodilation (e.g., facial flushing, headache, sweating, palpitations, chest pain, hypotension), (ii) respiratory effects (e.g., shortness of breath/dyspnea, chest pressure, hyperventilation), (iii) CNS reactions (e.g., light-headedness, dizziness, tingling in arms, numbness), and (iv) gastrointestinal disturbances (e.g., nausea).

Adenosine is relatively free of pharmacokinetic drug interactions. Pharmacodynamic drug interactions may occur. For example, the effects of adenosine are antagonized by methylxanthines, such as caffeine and theophylline. In the presence of these methylxanthines, larger doses of adenosine may be required or adenosine may not be effective. Adenosine effects are potentiated by dipyridamole. Thus, smaller doses of adenosine may be effective in the presence of dipyridamole.

Adenosine is contraindicated in patients with (i) second- or third-degree AV block (except in patients with a functioning artificial pacemaker); (ii) sinus node disease, such as sick sinus syndrome or symptomatic bradycardia (except in patients with a functioning artificial pacemaker); and (iii) known hypersensitivity to adenosine. Adenosine is in pregnancy category C.

24.7.2 Magnesium

Magnesium ion is the fourth most abundant cation in the human body. It is involved in multiple essential physiological, biochemical, and cellular processes regulating cardiovascular function. Magnesium plays a critical role in modulating vascular smooth muscle tone, endothelial cell function, and myocardial excitability, and as such, dysregulation of magnesium may play an important part in the pathogenesis of cardiovascular

TABLE 24.21 Dosage regimens of adenosine in converting PSVT to sinus rhythm in adult and pediatric patients

Patient	Dosage regimen
Adult	• Initial dose: 6 mg given as a rapid iv bolus (administered over a 1–2 s period) • Repeated administration: If the first dose does not result in elimination of the PSVT within 1–2 min, 12 mg should be given as a rapid iv bolus. This 12 mg dose may be repeated a second time if required
Pediatric	• The dosages used in neonates, infants, children, and adolescents were equivalent to those administered to adults on a weight basis. For pediatric patients with a body weight ≥50 kg, administer the adult dose • For pediatric patients with a body weight <50 kg, initial dose: Give 0.05–0.1 mg/kg as a rapid iv bolus given either centrally or peripherally. A saline flush should follow • For pediatric patients with a body weight <50 kg, repeat administration: If conversion of PSVT does not occur within 1–2 min, additional bolus injections of adenosine can be administered at incrementally higher doses, increasing the amount given by 0.05–0.1 mg/kg. Follow each bolus with a saline flush. This process should continue until sinus rhythm is established or a maximum single dose of 0.3 mg/kg is used

disorders including hypertension, atherosclerosis, coronary artery disease, congestive heart failure, and cardiac arrhythmias [5]. Probably the most widely accepted and practiced use of magnesium in cardiovascular medicine is for the prevention and/or treatment of cardiac arrhythmias. Multiple randomized clinical trials and/or meta-analyses of the randomized trials suggest a role for magnesium supplementation in the management of postoperative arrhythmias, AF, and ventricular arrhythmias (including torsades de pointes) [5–7]. However, the exact efficacy of magnesium in the prevention and treatment of cardiac arrhythmias remains to be established in large-scale, randomized controlled trials. Likewise, basic research is warranted to further elucidate the underlying molecular mechanisms of magnesium-mediated antiarrhythmic activity.

24.7.3 Vernakalant

Vernakalant is a novel and relatively atrium-selective drug, which inhibits atrium-selective potassium ion currents with only a small inhibitory effect on the rapidly activating delayed rectifier potassium ion current in the ventricle. This results in prolongation of atrial refractoriness and slowing of atrial impulse conduction. The drug has been recently approved in Europe as an iv therapy for rapid conversion of recent-onset AF to sinus rhythm in adults [8, 9]. Vernakalant is currently not a US FDA-approved drug for clinical use.

24.8 SUMMARY OF CHAPTER KEY POINTS

• Antiarrhythmic drugs are conventionally classified into four classes with class I consisting of sodium channel blockers, class II consisting of β-blockers, class III consisting of potassium channel blockers and other drugs that prolong the APD, and class IV consisting of CCBs.
• Class I drugs act by blocking the sodium channels, thereby inhibiting phase 0 depolarization in nonpacemaker cells and phase 4 spontaneous depolarization in pacemaker cells. These drugs are further classified into three sub-

classes (IA, IB, IC) based on their sodium channel-binding and dissociation properties. Class I drugs are used for treating both supraventricular and ventricular arrhythmias, including life-threatening ventricular tachycardias.
• Class II antiarrhythmic drugs are β-blockers (see Chapter 8). These drugs block the proarrhythmic effects of β_1-adrenergic receptor activation resulting from sympathetic stimulation. By blocking β_1-adrenergic receptor-mediated actions, class II antiarrhythmic drugs reduce heart rate, slow AV conduction velocity, and prolong AV refractory period. Slowing AV nodal conduction velocity and prolongation of AV nodal refractory period collectively result in decreased reentry, a fundamental mechanism of supraventricular arrhythmias. Class II drugs are useful in treating supraventricular and ventricular arrhythmias.
• Class III drugs block the potassium channels, thereby prolonging repolarization and the refractory period. Prolongation of refractory period reduces reentry. On the other hand, blockage of ventricular potassium currents results in prolongation of the QT interval on ECG. Prolongation of QT interval may predispose to torsades de pointes, a polymorphic ventricular tachycardia. In addition to the blockage of potassium channels, amiodarone and dronedarone also block sodium channels in depolarized tissues, a characteristic effect of class IB drugs. Amiodarone and dronedarone block calcium channels, as well as β-adrenergic receptors. Sotalol is also a β-blocker. Class III drugs are effective in treating supraventricular and ventricular arrhythmias, including life-threatening ventricular tachycardias.
• Class IV drugs are nondihydropyridine CCBs and include diltiazem and verapamil (also see Chapter 10). By blocking L-type calcium channels, class IV drugs can slow the sinus rate, increase the refractoriness of and prolong conduction through the AV node, and depress left ventricular contractility. Class IV drugs are used for rapid conversion of PSVT to sinus rhythm and

for temporary control of rapid ventricular rate in atrial fibrillation (AF) or AFL.

- In addition to the above four classes of drugs, intravenous adenosine is indicated for the conversion to sinus rhythm of PSVT. Digoxin is used for controlling ventricular responses in patients with chronic AF. Clinical studies suggest a role for magnesium supplementation in the management of postoperative arrhythmias, AF, and ventricular arrhythmias (including torsades de pointes). Vernakalant, a novel atrium-selective drug recently approved in Europe, represents a new therapy for rapid conversion of recent-onset AF to sinus rhythm in adults. These drugs along with those in the Vaughan–Williams classes, when used appropriately, are effective options for the management of cardiac arrhythmic conditions. Chapter 25 considers the current guideline recommendations on cardiac arrhythmia management, including drug therapy.

24.9 SELF-ASSESSMENT QUESTIONS

24.9.1 A 25-year-old hospitalized patient develops paroxysmal supraventricular tachycardia (PSVT). A calcium channel blocker is given as a bolus administered over 2 min. Immediately after the injection, the patient's PSVT is converted to sinus rhythm. Which of the following drugs is most likely prescribed?
A. Amlodipine
B. Clevidipine
C. Diltiazem
D. Nifedipine
E. Nisoldipine

24.9.2 A 58-year-old female with cardiac arrhythmia is being treated with a sodium channel blocker in a rural clinic. After a few weeks of treatment, she develops a lupus-like syndrome. She was previously identified as a "slow acetylator." Which of the following drugs has been most likely prescribed to treat the arrhythmia in this patient?
A. Amiodarone
B. Metoprolol
C. Procainamide
D. Quinidine
E. Verapamil

24.9.3 A 20-year-old female is admitted to the emergency room because of palpitations and syncopal episodes. She is found to be hypotensive and her ECG shows a very rapid AV nodal reentrant tachycardia. Which of the following drugs would be most appropriate as a rapid iv bolus for her condition?
A. Adenosine
B. Bethanechol
C. Dobutamine

D. Metoprolol
E. Procainamide

24.9.4 A 60-year-old female, diagnosed with dilated cardiomyopathy with atrial fibrillation and a rapid ventricular rate, is treated with a drug to control her ventricular rate. The drug also affects cardiac contractility, placing the patient in pulmonary edema. Which of the following drugs is most likely used?
A. Amiodarone
B. Digoxin
C. Lidocaine
D. Nifedipine
E. Verapamil

24.9.5 After beginning drug therapy to suppress ventricular tachycardia, a man complains of cold intolerance and of being tired all the time. His thyroid-stimulating hormone is found to be elevated. Which of the following drugs is most likely responsible for the patient's complaints?
A. Adenosine
B. Amiodarone
C. Dronedarone
D. Procainamide
E. Verapamil

24.9.6 A 35-year-old female with a history of poorly controlled thyrotoxicosis has recurrent episodes of tachycardia with severe short of breath. When she is admitted to the emergency department with one of these episodes, which of the following drugs would be most suitable for treating her condition?
A. Amiodarone
B. Diltiazem
C. Disopyramide
D. Esmolol
E. Quinidine

24.9.7 A 60-year-old male is admitted to the emergency department because of an acute myocardial infarction. Which of the following would be the most appropriate prophylactic antiarrhythmic therapy for this patient?
A. Diltiazem
B. Lidocaine
C. Metoprolol
D. Procainamide
E. Quinidine

24.9.8 A 55-year-old male living in a rural area is being treated for atrial fibrillation. He complains of headache, dizziness, and tinnitus during the treatment. Which of the following antiarrhythmic drugs is the patient most likely taking?
A. Amiodarone
B. Dronedarone
C. Propranolol
D. Quinidine
E. Verapamil

24.9.9   A 19-year-old female presents to the emergency department with a racing heart that began while she was playing football 45 min ago. Her ECG shows a rapid and regular heart rate. She is given an intravenous bolus of a drug that activates G-protein-coupled receptors, and her heart rate is converted back to the sinus rhythm. Which of the following best explains the mechanism of action of this drug in terminating her tachyarrhythmia?

A. Increased potassium ion efflux
B. Increased sodium ion efflux
C. Inhibition of Ach-sensitive K^+ channel
D. Inhibition of membrane Na^+/K^+-ATPase
E. Inhibition of Na^+/Ca^{2+} exchanger

24.9.10   A 55-year-old male with atrial fibrillation accompanied by a rapid ventricular rate has been managed successfully with a drug that exerts a positive inotropic effect. Which of the following is most likely the primary molecular target of the drug?

A. Acetylcholine-sensitive K^+ channel
B. β_1-Adrenergic receptor
C. Na^+/Ca^{2+} exchanger
D. Na^+/K^+-ATPase
E. Voltage-dependent Na^+ channel

REFERENCES

1 Viskin, S., et al., Quinidine, a life-saving medication for Brugada syndrome, is inaccessible in many countries. *J Am Coll Cardiol*, 2013. 61(23): p. 2383–7.

2 Viskin, S., et al., Inaccessibility to quinidine therapy is about to get worse. *J Am Coll Cardiol*, 2013. 62(4): p. 355.

3 O'Connor, R.E., et al., Part 9: Acute coronary syndromes: 2010 international consensus on cardiopulmonary resuscitation and emergency cardiovascular care science with treatment recommendations. *Circulation*, 2010. 122(16 Suppl 2): p. S422–65.

4 American College of Cardiology Foundation/American Heart Association Task Force on Practice Guidelines, et al., 2013 ACCF/AHA guideline for the management of ST-elevation myocardial infarction: a report of the American College of Cardiology Foundation/American Heart Association Task Force on Practice Guidelines. *J Am Coll Cardiol*. 61(4): p. e78–140.

5 Kolte, D., et al., Role of magnesium in cardiovascular diseases. *Cardiol Rev*, 2014. 22(4): p. 182–92.

6 Lee, H.Y., S. Ghimire, and E.Y. Kim, Magnesium supplementation reduces postoperative arrhythmias after cardiopulmonary bypass in pediatrics: a metaanalysis of randomized controlled trials. *Pediatr Cardiol*, 2013. 34(6): p. 1396–403.

7 Onalan, O., et al., Meta-analysis of magnesium therapy for the acute management of rapid atrial fibrillation. *Am J Cardiol*, 2007. 99(12): p. 1726–32.

8 Guerra, F., et al., Intravenous vernakalant for the rapid conversion of recent onset atrial fibrillation: systematic review and meta-analysis. *Expert Rev Cardiovasc Ther*, 2014: p. 1–9.

9 Vizzardi, E., et al., A new antiarrhythmic drug in the treatment of recent-onset atrial fibrillation: vernakalant. *Cardiovasc Ther*, 2013. 31(5): p. e55–62.

25

MANAGEMENT OF CARDIAC ARRHYTHMIAS: PRINCIPLES AND GUIDELINES

25.1 OVERVIEW

Most arrhythmias are considered harmless and are left untreated. Some arrhythmias, such as ventricular tachycardias, can be life-threatening medical emergencies and may result in cardiac arrest and sudden cardiac death. Some others, such as atrial fibrillation (AF), may not be associated with any significant symptoms, but may dispose the patients to potentially life-threatening stroke or embolism.

Treatment of arrhythmias has been evolving from the primary use of drugs to the use of nondrug devices in combination with drug therapy when necessary. This chapter discusses the general principles and guidelines for the management of arrhythmias with an emphasis on pharmacological therapy. Since arrhythmias are a large group of disorders, the chapter focuses on those that have the most significant clinical and public health impact, including atrial fibrillation and certain forms of ventricular tachyarrhythmias.

25.2 GENERAL PRINCIPLES OF MANAGEMENT

The fundamental goal of therapy for cardiac arrhythmias is twofold: (1) to improve symptoms of the patients and (2) to prevent potentially serious outcomes, especially life-threatening ventricular arrhythmias and sudden death. Stroke prevention in patients with AF and atrial flutter (AFL) is also an important treatment objective. Hence, selection of the treatment approach (including therapeutic modality and duration of treatment) should take into consideration both the severity and frequency of arrhythmia-related symptoms and the risks associated with the therapy itself, no matter if it is a device- or drug-based modality.

25.3 MANAGEMENT OF SUPRAVENTRICULAR ARRHYTHMIAS

As described in Chapter 23, supraventricular arrhythmias comprise a wide spectrum of disorders including, in descending order of frequency, AF, AFL, atrial tachycardias, atrioventricular (AV) nodal tachycardias, and Wolff–Parkinson–White (WPW) syndrome, among others. While being not life threatening in most cases, they may cause important symptoms, such as palpitations, chest discomfort, breathlessness, anxiety, and syncope, which significantly impair quality of life. Medical therapy has variable efficacy, and most patients are not rendered free of symptoms. Research over the past several decades has revealed fundamental mechanisms involved in the initiation and maintenance of supraventricular arrhythmias. Knowledge of mechanisms has in turn led to the development of highly effective surgical and catheter-based treatments. Nevertheless, drug therapy is still an effective option for patients under certain conditions [1–4].

Due to the high prevalence of AF and its role in ischemic stroke, management of this particular supraventricular arrhythmia is often addressed separately. Hence, this section is divided into guidelines on supraventricular arrhythmias excluding AF and guidelines specifically on AF.

Cardiovascular Diseases: From Molecular Pharmacology to Evidence-Based Therapeutics, First Edition. Y. Robert Li.
© 2015 John Wiley & Sons, Inc. Published 2015 by John Wiley & Sons, Inc.

25.3.1 Guidelines on Supraventricular Arrhythmias Excluding AF

The 2003 American College of Cardiology/American Heart Association/European Society of Cardiology (ACC/AHA/ESC) guideline for the management of patients with supraventricular arrhythmias is the most comprehensive guidance that provides treatment recommendations for various forms of supraventricular arrhythmias (excluding AF) [5]. These include (i) sinus tachyarrhythmias (physiological sinus tachycardia and inappropriate sinus tachycardia), (ii) AV nodal reciprocating tachycardia, (iii) focal and nonparoxysmal junctional tachycardia, (iv) AV reciprocating tachycardia (extranodal accessory pathways), (v) focal atrial tachycardias, and (vi) macro-reentrant atrial tachycardias (isthmus-dependent AFL and noncavotricuspid isthmus-dependent AFL). The major guideline recommendations are summarized in Table 25.1.

25.3.2 Current Guidelines on the Management of AF

25.3.2.1 General Considerations

25.3.2.1.1 Definition and Classification AF is a supraventricular tachyarrhythmia characterized by uncoordinated atrial activation with consequent deterioration of atrial mechanical function. On the electrocardiogram (ECG), AF is characterized by the replacement of consistent P waves by rapid oscillations or fibrillatory waves that vary in amplitude, shape, and timing, associated with an irregular, frequently rapid ventricular response when AV conduction is intact [6].

Traditionally, the terms acute AF and chronic AF have been used to describe the temporal nature of a patient's AF. These two terms have been largely replaced with a classification scheme proposed by the American Heart Association/American College of Cardiology/Heart Rhythm Society (AHA/ACC/HRS) [7] (Table 25.2).

TABLE 25.1 The 2003 ACC/AHA/ESC recommendations on the management of supraventricular arrhythmias [5]

Supraventricular arrhythmia	Management
Sinus tachyarrhythmias (physiological and inappropriate sinus tachycardia)	β-Blockers may be considered as first-line therapy; nondihydropyridine calcium channel blockers (CCBs), such as verapamil and diltiazem, may also be used if β-blockers are contraindicated
Atrioventricular nodal reciprocating tachycardia (AVNRT)	Acute management: (i) the most effective and rapid means of terminating any hemodynamically unstable narrow or wide QRS complex tachycardia is direct-current (DC) cardioversion; (ii) iv adenosine or DC cardioversion for acute management of regular narrow QRS complex tachycardia (the nondihydropyridine CCBs verapamil/diltiazem or β-blocker metoprolol may be used if adenosine is contraindicated); (iii) DC cardioversion is the first choice for unstable wide QRS complex tachycardia; (iv) for pharmacologic termination of a stable wide QRS complex tachycardia, iv procainamide and/or sotalol is recommended. Amiodarone is also acceptable
	Chronic management: (i) catheter ablation is recommended for poorly tolerated AVNRT with hemodynamic intolerance; (ii) for patients with frequent, recurrent sustained episodes of AVNRT who prefer long-term oral therapy instead of catheter ablation, a spectrum of antiarrhythmic agents is available. Standard therapy includes nondihydropyridine CCBs, β-blockers, and digoxin. In patients without structural heart disease who do not respond to AV nodal-blocking agents, the class IC drugs flecainide and propafenone have become the preferred choice
Focal and nonparoxysmal junctional tachycardia	For focal junctional tachycardia, drug therapy (β-blocker, flecainide) is only variably successful, and ablative techniques have been introduced to cure tachycardia. Catheter ablation can be curative
	The mainstay of managing nonparoxysmal junctional tachycardia is to correct the underlying abnormality. β-Blockers and CCBs may be used
Atrioventricular (AV) reciprocating tachycardia (extranodal accessory pathways)	Acute management: for patients with wide complex (preexcited) tachycardias, adenosine should be used with caution because it may produce AF with a rapid ventricular rate in preexcited tachycardias. Ibutilide, procainamide, and flecainide, which are capable of slowing the conduction through the pathway, are preferred
	Chronic management: antiarrhythmic drugs represent one therapeutic option for the management of accessory pathway-mediated arrhythmias, but they have been increasingly replaced by catheter ablation. Antiarrhythmic drugs that primarily modify conduction through the AV node include digoxin, verapamil, β-blockers, adenosine, and diltiazem. Antiarrhythmic drugs that depress conduction across the accessory pathway include class I drugs, such as procainamide, disopyramide, propafenone, and flecainide, as well as class III antiarrhythmic drugs, such as ibutilide, sotalol, and amiodarone
Focal atrial tachycardias (AT)	Catheter ablation: regardless of whether the arrhythmia is due to abnormal automaticity, triggering, or micro-reentry, focal AT is ablated by targeting the site of origin of the AT
	Focal AT is difficult to treat with drugs
AFL and macro-re-entrant AT	Acute treatment of AFL might include the initial use of electrical pacing, DC or chemical cardioversion (iv ibutilide), or AV nodal-blocking agents (β-blockers, verapamil, or diltiazem)

TABLE 25.2   The AHA/ACC/HRS classification scheme of AF [7]

Classification	Definition
Paroxysmal AF	AF that terminates spontaneously or with intervention within 7 days of onset. Episodes may recur with variable frequency
Persistent AF	Continuous AF that is sustained for >7 days
Long-standing persistent AF	Continuous AF of >12 months of duration
Permanent AF	Permanent AF is used when there has been a joint decision by the patient and clinician to cease further attempts to restore and/or maintain sinus rhythm. Acceptance of AF represents a therapeutic attitude on the part of the patient and clinician rather than an inherent pathophysiological attribute of the AF. Acceptance of AF may change as symptoms, the efficacy of therapeutic interventions, and patient and clinician preferences evolve
Nonvalvular AF	It refers to AF in the absence of rheumatic mitral stenosis, a mechanical or bioprosthetic heart valve, or mitral valve repair

Epidemiology and Pathophysiology AF is the most common pathological supraventricular tachycardia. In the United States, estimates of the prevalence of AF ranged from approximately 2.7 to 6.1 million in 2010, and AF prevalence is expected to rise to between approximately 5.6 and 12 million in 2050 [8]. In 2010, AF was mentioned on 107,335 US death certificates and was the underlying cause in 16,454 of those deaths. AF is associated with a four- to fivefold increased risk of ischemic stroke. It is also an independent risk factor for ischemic stroke severity, recurrence, and mortality. Individuals with AF have a twofold increased risk of dementia. Notably, AF and heart failure share many antecedent risk factors, and approximately 40% individuals with either AF or heart failure will develop the other condition [8].

Risk factors for AF include advancing age, European ancestry, body size (greater height and body mass index), electrocardiography features (left ventricular hypertrophy, left atrial enlargement), diabetes, hypertension, and presence of coronary heart disease, heart failure, and valvular heart disease. Additional risk factors for AF include clinical and subclinical hyperthyroidism, chronic kidney disease (CKD), and heavy alcohol consumption, as well as family history and genetic variations [3, 8].

AF is caused by multiple electrical wavelets appearing in the atria simultaneously, resembling the waves that would be produced if one dropped several pebbles in a bucket of water at the same time [3]. Available data support a "focal" triggering mechanism involving automaticity or multiple reentrant wavelets, but these mechanisms are not mutually exclusive and may coexist. The important observation that a focal source for AF could be identified and ablation of this source could extinguish AF supported a focal origin. While pulmonary veins are the most frequent source of these rapidly atrial impulses, foci have also been found in the superior vena cava, ligament of Marshall, left posterior free wall, crista terminalis, and coronary sinus [7].

25.3.2.2   General Principles of Management of AF

Management Goals Management of patients with AF involves three goals: (1) prevention of thromboembolism, (2) heart rate control, and (3) correction of the rhythm disturbance. These three goals are not mutually exclusive.

Once prevention of thromboembolism is addressed, the initial management decision involves primarily a rate-control or rhythm-control strategy. Under the rate-control strategy, the ventricular rate is controlled with no commitment to restore or maintain sinus rhythm. On the other hand, the rhythm-control strategy attempts restoration and/or maintenance of sinus rhythm, and it also requires attention to rate control.

Depending on the patient's course, the strategy initially chosen may prove unsuccessful and the alternate strategy is then adopted. Regardless of whether the rate-control or rhythm-control strategy is pursued, attention must also be directed to antithrombotic therapy for prevention of thromboembolism [7].

Determinants of Management Strategy At the initial encounter, an overall management strategy should be discussed with the patient, considering multiple factors [7]. These include the following:

1. Type and duration of AF
2. Severity and type of symptoms
3. Associated cardiovascular disease
4. Patient age
5. Associated medical conditions
6. Short-term and long-term treatment goals
7. Pharmacological and nonpharmacological therapeutic options

25.3.2.3   Current Guidelines New guidelines for the management of AF have recently been published by the ACCF/AHA/HRS [6, 7], the Canadian Cardiovascular Society (CCS) [9], and the ESC [10, 11]. In general, these guidelines

are similar with regard to management goals and evidence-based recommendations. This section focuses on describing the major treatment recommendations from the 2014 AHA/ACC/HRS guideline for the management of AF [7].

The 2014 AHA/ACC/HRS guideline for the management of AF is a comprehensive document that consists of eight sections: (1) introduction, (2) background and pathophysiology, (3) clinical evaluation, (4) prevention of thromboembolism, (5) rate control, (6) rhythm control, (7) specific patient groups and AF, and (8) evidence gaps and future research directions. Summarized in the following text are the major guideline recommendations on the prevention of thromboembolism, rate control, and rhythm control. The reader is advised to refer to the full document for other aspects.

Recommendations on Prevention of Thromboembolism   As noted earlier, AF is associated with a four- to fivefold increased risk of ischemic stroke. Hence, prevention of thromboembolism and ischemic stroke is a top priority in the management of AF. Summarized in Table 25.3 are the key recommendations on prevention of thromboembolism in patients with AF.

Recommendations on Rate Control   Rate control in AF is an important component of AF management. It impacts quality of life, reduces morbidity, and decreases the potential for developing tachycardia-induced cardiomyopathy. Multiple drugs, including β-blockers, nondihydropyridine CCBs (e.g., diltiazem, verapamil), digoxin, and certain antiarrhythmic drugs (e.g., amiodarone, sotalol), have been evaluated with regard to efficacy in attaining rate control in patients with AF. In general, β-blockers are the most common agents utilized for rate control, followed by nondihydropyridine CCBs, digoxin, and amiodarone. For effective rate control, patient comorbidities must be understood in order to avoid medications that may precipitate adverse events, such as decompensation of HF, exacerbation of chronic obstructive pulmonary disease (COPD), or acceleration of conduction in patients with preexcitation [7]. Table 25.4 summarizes the major guideline recommendations on rate control of AF.

Recommendations on Rhythm Control   The last goal of AF management is to restore and maintain sinus rhythm. This is typically achieved by utilizing a combination of approaches, including cardioversion, antiarrhythmic drugs, and radiofrequency catheter ablation in the setting of appropriate anticoagulation and rate control. Although an initial rate-control strategy is reasonable for many patients, several considerations favor pursuing a rhythm-control strategy. Successful sinus rhythm maintenance is associated with improvements in symptoms and quality of life for some patients. Persistent symptoms associated with AF remain the most compelling indication for a rhythm-control strategy. Other factors that may favor attempts at rhythm control

include difficulty in achieving adequate rate control, younger patient age, tachycardia-mediated cardiomyopathy, first episode of AF, AF that is precipitated by an acute illness, and patient preference [7]. With regard to rhythm control, the 2014 AHA/ACC/HRS guideline covers five areas, including (1) electrical and pharmacological conversion of AF and AFL, (2) pharmacological agents for preventing AF and maintaining sinus rhythm, (3) AF catheter ablation to maintain sinus rhythm, (4) pacemaker and implantable cardioverter defibrillators (ICDs) for the prevention of AF, and (5) surgery maze procedures. This section only considers recommendations on pharmacological cardioversion (Table 25.5) and antiarrhythmic drugs for maintaining sinus rhythms (Table 25.6).

Management of AF in Specific Patient Groups   AF occurs frequently in patients with other concomitant disorders, either as a consequence of the comorbidity or as an accompanying disorder. These include hypertrophic cardiomyopathy, acute coronary syndromes (ACS), hyperthyroidism, pulmonary diseases (e.g., COPD), WPW and preexcitation syndromes, heart failure, and postoperative cardiac and thoracic surgery. Table 25.7 summarizes the major guideline recommendations for the management of AF in patients with the above concomitant disorders.

25.4   MANAGEMENT OF VENTRICULAR ARRHYTHMIAS

25.4.1   General Considerations

The term "ventricular arrhythmias" encompasses a wide spectrum of dysrhythmias, ranging from single premature ventricular complexes (PVCs) to nonsustained ventricular tachycardia (VT), sustained VT, ventricular flutter, and ventricular fibrillation (VF). These arrhythmias predominantly occur in patients with structural heart diseases, such as ischemic and dilated cardiomyopathies. However, benign forms of VT can also occur among individuals without evidence of cardiac diseases. Sustained VTs/VF are an important cause of morbidity and the most common cause of sudden cardiac death (SCD), accounting for 75–80% of cases [15, 16].

Sudden cardiac arrest (SCA) and SCD refer to the sudden cessation of cardiac activity with hemodynamic collapse, typically due to sustained VT/VF. These events mostly occur in patients with structural heart diseases (which may not have been previously diagnosed), particularly coronary heart disease. The event is referred to as SCA (or aborted SCD) if an intervention (e.g., defibrillation) or spontaneous reversion restores circulation. The event is called SCD if the patient dies.

The geographic incidence of SCD varies as a function of coronary heart disease prevalence in different regions. Estimates for the United States range from <200,000 to >450,000 SCDs

TABLE 25.3   The 2014 AHA/ACC/HRS guideline recommendations on prevention of thromboembolism in patients with AF [7]

COR	Recommendation	LOE
I	In patients with AF, antithrombotic therapy should be individualized based on shared decision making after discussion of the absolute and relative risks of stroke and bleeding and the patient's values and preferences	C
	Selection of antithrombotic therapy should be based on the risk of thromboembolism irrespective of whether the AF pattern is paroxysmal, persistent, or permanent	B
	In patients with nonvalvular AF, the CHA_2DS_2-VASc score[a] is recommended for assessment of stroke risk	B
	For patients with AF who have mechanical heart valves, warfarin is recommended and the target international normalized ratio (INR) intensity (2.0–3.0 or 2.5–3.5) should be based on the type and location of the prosthesis	B
	For patients with nonvalvular AF with prior stroke, transient ischemic attack (TIA), or a CHA_2DS_2-VASc score of ≥2, oral anticoagulants are recommended. Options include warfarin (INR, 2.0–3.0) (LOE, A), dabigatran (LOE, B), rivaroxaban (LOE, B), or apixaban (LOE, B)	
	Among patients treated with warfarin, the INR should be determined at least weekly during initiation of antithrombotic therapy and at least monthly when anticoagulation (INR in range) is stable	A
	For patients with nonvalvular AF unable to maintain a therapeutic INR level with warfarin, use of a direct thrombin or factor Xa inhibitor (dabigatran, rivaroxaban, or apixaban) is recommended	C
	Reevaluation of the need for and choice of antithrombotic therapy at periodic intervals is recommended to reassess stroke and bleeding risks	C
	Bridging therapy with unfractionated heparin (UFH) or low-molecular-weight heparin (LMWH) is recommended for patients with AF and a mechanical heart valve undergoing procedures that require interruption of warfarin. Decisions regarding bridging therapy should balance the risks of stroke and bleeding	C
	For patients with AF without mechanical heart valves who require interruption of warfarin or newer anticoagulants for procedures, decisions about bridging therapy (LMWH or UFH) should balance the risks of stroke and bleeding and the duration of time a patient will not be anticoagulated	C
	Renal function should be evaluated prior to initiation of direct thrombin or factor Xa inhibitors and should be reevaluated when clinically indicated and at least annually	B
	For patients with atrial flutter, antithrombotic therapy is recommended according to the same risk profile used for AF	C
IIa	For patients with nonvalvular AF and a CHA_2DS_2-VASc score of 0, it is reasonable to omit antithrombotic therapy	B
	For patients with nonvalvular AF with a CHA_2DS_2-VASc score of ≥2 and who have end-stage CKD (creatinine clearance [CrCl] <15 ml/min) or are on hemodialysis, it is reasonable to prescribe warfarin (INR, 2.0–3.0) for oral anticoagulation	B
IIb	For patients with nonvalvular AF and a CHA_2DS_2-VASc score of 1, no antithrombotic therapy or treatment with an oral anticoagulant or aspirin may be considered	C
	For patients with nonvalvular AF and moderate-to-severe CKD with CHA_2DS_2-VASc scores of ≥2, treatment with reduced doses of direct thrombin or factor Xa inhibitors may be considered (e.g., dabigatran, rivaroxaban, or apixaban), but safety and efficacy have not been established	C
	In patients with AF undergoing percutaneous coronary intervention, bare-metal stents may be considered to minimize the required duration of dual antiplatelet therapy. Anticoagulation may be interrupted at the time of the procedure to reduce the risk of bleeding at the site of peripheral arterial puncture	C
	Following coronary revascularization (percutaneous or surgical) in patients with AF and a CHA_2DS_2-VASc score of ≥2, it may be reasonable to use clopidogrel (75 mg once daily) concurrently with oral anticoagulants but without aspirin	B
III (no benefit)	The direct thrombin inhibitor dabigatran and the factor Xa inhibitor rivaroxaban are not recommended in patients with AF and end-stage CKD or on hemodialysis because of the lack of evidence from clinical trials regarding the balance of risks and benefits	C
III (harm)	The direct thrombin inhibitor dabigatran should not be used in patients with AF and a mechanical heart valve due to increased risk of thromboembolic and bleeding events in such setting	B

[a] CHA_2DS_2-VASc score is a commonly used score system for estimating the risk of stroke by considering the risk factors of stroke including *c*ongestive heart failure/left ventricular dysfunction, *h*ypertension, *a*ge ≥75 (doubled), *d*iabetes, *s*troke (doubled)–*v*ascular disease, *a*ge 65–74, and *s*ex category (female) [12]. COR and LOE stand for class of recommendation and level of evidence, respectively (see Chapter 5 for definitions).

TABLE 25.4   The 2014 AHA/ACC/HRS guideline recommendations on rate control in patients with AF [7]

COR	Recommendation	LOE
I	Control of the ventricular rate using a β-blocker or nondihydropyridine CCB is recommended for patients with paroxysmal, persistent, or permanent AF	B
	Intravenous administration of a β-blocker or nondihydropyridine CCB is recommended to slow the ventricular heart rate in the acute setting in patients without preexcitation. In hemodynamically unstable patients, electrical cardioversion is indicated	B
	In patients who experience AF-related symptoms during activity, the adequacy of heart rate control should be assessed during exertion, adjusting pharmacological treatment as necessary to keep the ventricular rate within the physiological range	C
IIa	A heart rate control (resting heart rate <80 beats/min) strategy is reasonable for symptomatic management of AF	B
	Intravenous amiodarone can be useful for rate control in critically ill patients without preexcitation	B
	AV nodal ablation with permanent ventricular pacing is reasonable to control the heart rate when pharmacological therapy is inadequate and rhythm control is not achievable	B
IIb	A lenient rate-control strategy (resting heart rate <110 beats/min) may be reasonable as long as patients remain asymptomatic and left ventricular (LV) systolic function is preserved	B
	Oral amiodarone may be useful for ventricular rate control when other measures are unsuccessful or contraindicated	C
III (harm)	AV nodal ablation with permanent ventricular pacing should not be performed to improve rate control without prior attempts to achieve rate control with medications	C
	Nondihydropyridine CCBs should not be used in patients with decompensated HF as these may lead to further hemodynamic compromise	C
	In patients with preexcitation and AF, digoxin, nondihydropyridine CCBs, or intravenous amiodarone should not be administered as they may increase the ventricular response and may result in ventricular fibrillation	B
	Dronedarone should not be used to control the ventricular rate in patients with permanent AF as it increases the risk of the combined endpoint of stroke, myocardial infarction, systemic embolism, or cardiovascular death	B

COR and LOE stand for class of recommendation and level of evidence, respectively (see Chapter 5 for description).

TABLE 25.5   The 2014 AHA/ACC/HRS guideline recommendations on pharmacological cardioversion in patients with AF [7]

COR	Recommendation	LOE
I	Flecainide, dofetilide, propafenone, and intravenous ibutilide are useful for pharmacological cardioversion of AF or AFL provided contraindications to the selected drug are absent	A
IIa	Administration of oral amiodarone is a reasonable option for pharmacological cardioversion of AF	A
	Propafenone or flecainide ("pill in the pocket") in addition to a β-blocker or nondihydropyridine CCB is reasonable to terminate AF outside the hospital once this treatment has been observed to be safe in a monitored setting for selected patients	B
III (harm)	Dofetilide therapy should not be initiated out of hospital owing to the risk of excessive QT prolongation that can cause torsades de pointes	B

COR and LOE stand for class of recommendation and level of evidence, respectively (see Chapter 5 for description).

annually, with the most widely used estimates in the range of 300,000–350,000 SCDs annually. The variation is due, in part, to the inclusion criteria used in individual studies. Overall, event rates in Europe are similar to those in the United States, also with significant geographic variations reported [17, 18].

25.4.2   General Principles of Management of Ventricular Arrhythmias

25.4.2.1   Selection of Management Strategies   The selection of appropriate therapy for the management of ventricular arrhythmias (e.g., PVCs, nonsustained VT, sustained mono-morphic and polymorphic VT, and ventricular flutter/VF)

requires an understanding of the etiology and mechanism of the arrhythmias, an appreciation of the associated medical conditions that may contribute to and/or exacerbate the arrhythmias, the risk posed by the arrhythmias, and risk-to-benefit aspects of the selected therapy. Management of symptomatic ventricular arrhythmias may involve (i) discontinuation of offending proarrhythmic drugs and other eliciting agents, (ii) pharmacological therapy with antiarrhythmic agents, and (iii) nonpharmacological approaches, such as implantable devices, ablation, and surgery. Due to the limited efficacy as well as significant adverse effects of antiarrhythmic drugs, in recent years, the nonpharmacological approaches have been becoming the mainstay of therapy, especially for severe forms of ventricular arrhythmias.

TABLE 25.6   The 2014 AHA/ACC/HRS guideline recommendations on antiarrhythmic drugs to maintain sinus rhythms in patients with AF [7]

COR	Recommendation	LOE
I	Before initiating antiarrhythmic drug therapy, treatment of precipitating or reversible causes of AF is recommended	C
	The following antiarrhythmic drugs are recommended in patients with AF to maintain sinus rhythm, depending on underlying heart disease and comorbidities:	A
	1. Amiodarone	
	2. Dofetilide	
	3. Dronedarone	
	4. Flecainide	
	5. Propafenone	
	6. Sotalol	
	The risks of the antiarrhythmic drug, including proarrhythmias, should be considered before initiating therapy with each drug	C
	Owing to its potential toxicities, amiodarone should only be used after consideration of risks and when other agents have failed or are contraindicated	C
IIa	A rhythm-control strategy with pharmacological therapy can be useful in patients with AF for the treatment of tachycardia-induced cardiomyopathy	C
IIb	It may be reasonable to continue current antiarrhythmic drug therapy in the setting of infrequent, well-tolerated recurrences of AF, when the drug has reduced the frequency or symptoms of AF	C
III (harm)	Antiarrhythmic drugs for rhythm control should not be continued when AF becomes permanent (LOE, C), including dronedarone (LOE, B)	
	Dronedarone should not be used for treatment of AF in patients with New York Heart Association (NYHA) class III and IV HF or patients who have had an episode of decompensated HF in the past 4 weeks	B

COR and LOE stand for class of recommendation and level of evidence, respectively (see Chapter 5 for description).

25.4.2.2   General Principles of Antiarrhythmic Drug Therapy

25.4.2.2.1   Potential Proarrhythmic Activity of Antiarrhythmic Drugs
With the exception of β-blockers, the currently available antiarrhythmic drugs have not been shown in randomized clinical trials to be effective in the primary management of patients with life-threatening ventricular arrhythmias or in the prevention of SCD. As a general rule, antiarrhythmic agents may be effective as adjunctive therapy in the management of arrhythmia-prone patients under special circumstances. Because of potential adverse effects (proarrhythmic activity) of the available antiarrhythmic drugs, these agents must be used with caution. In this context, as discussed in Chapter 24, many of the antiarrhythmic agents have the potential to precipitate life-threatening ventricular tachyarrhythmias.

β-Blockers as Effective Antiarrhythmic Drugs   β-Blockers are effective in suppressing ventricular ectopic beats and arrhythmias as well as in reducing SCD in a spectrum of cardiac disorders in patients with and without HF. In patients with acute myocardial infarction, early (within 24 h of presentation) administration of β-blockers is associated with a reduction in the incidence of VF and is recommended for all patients without contraindications [19]. β-Blockers are safe and effective antiarrhythmic agents that can be considered the mainstay of antiarrhythmic drug therapy. The mechanism of antiarrhythmic efficacy of this class of drugs involves competitive inhibition of β-adrenergic receptor activation-induced increases in automaticity and conduction velocity, as well as shortening of refractory period, thereby leading to reduced reentry (see Chapters 8 and 24).

Amiodarone as a Major Antiarrhythmic Drug   Amiodarone has a spectrum of actions that include blockage of potassium repolarization currents that can inhibit or terminate ventricular arrhythmias by suppressing reentry. The overall long-term survival benefit from amiodarone is controversial, with most studies showing no clear advantage over placebo. Chronic administration of amiodarone is associated with complex drug interactions and a number of adverse effects involving the lung, liver, thyroid, and skin. As a general rule, the longer the therapy and the higher dose of amiodarone, the greater is the likelihood that adverse effects will require discontinuance of the drug. A meta-analysis of 15 randomized controlled trials involving 8522 patients suggests that amiodarone reduces the risk of SCD by 29% and may represent a viable alternative in patients who are not eligible for or who do not have access to ICD therapy for the prevention of SCD. However, the meta-analysis also reveals that amiodarone therapy is neutral with respect to all-cause mortality and is associated with a two- and fivefold increased risk of pulmonary and thyroid toxicity, respectively [20]. Nevertheless, amiodarone is still a viable option for the management of life-threatening VTs and cardiac arrest [21].

TABLE 25.7　The 2014 AHA/ACC/HRS guideline recommendations on the management of AF in patients with additional specific conditions [7]

COR	Recommendation	LOE
Patients with hypertrophic cardiomyopathy (HCM)		
I	Anticoagulation is indicated in patients with HCM with AF independent of the CHA_2DS_2-VASc score[a]	B
IIa	Antiarrhythmic medications can be useful to prevent recurrent AF in patients with HCM. Amiodarone or disopyramide combined with a β-blocker or an nondihydropyridine CCB is a reasonable therapy	C
	AF catheter ablation can be beneficial in patients with HCM in whom a rhythm-control strategy is desired when antiarrhythmic drugs fail or are not tolerated	B
IIb	Sotalol, dofetilide, and dronedarone may be considered for a rhythm-control strategy in patients with HCM	C
Patients with ACS		
I	Urgent direct-current cardioversion of new-onset AF in the setting of ACS is recommended for patients with hemodynamic compromise, ongoing ischemia, or inadequate rate control	C
	Intravenous β-blockers are recommended to slow a rapid ventricular response to AF in patients with ACS who do not display HF, hemodynamic instability, or bronchospasm	C
	For patients with ACS and AF with CHA_2DS_2-VASc score of ≥2, anticoagulation with warfarin is recommended unless contraindicated	C
IIb	Administration of amiodarone or digoxin may be considered to slow a rapid ventricular response in patients with ACS and AF associated with severe LV dysfunction and heart failure (HF) or hemodynamic instability	C
	Administration of nondihydropyridine CCBs might be considered to slow a rapid ventricular response in patients with ACS and AF only in the absence of significant HF or hemodynamic instability	C
Patients with hyperthyroidism		
I	β-Blockers are recommended to control ventricular rate in patients with AF complicating thyrotoxicosis unless contraindicated	C
	In circumstances in which a β-blocker cannot be used, a nondihydropyridine CCB is recommended to control the ventricular rate	C
Patients with pulmonary diseases		
I	A nondihydropyridine CCB is recommended to control the ventricular rate in patients with AF and chronic obstructive pulmonary disease	C
	Direct-current cardioversion should be attempted in patients with pulmonary disease who become hemodynamically unstable as a consequence of new-onset AF	C
Patients with WPW and preexcitation syndromes		
I	Prompt direct-current cardioversion is recommended for patients with AF, WPW, and rapid ventricular response who are hemodynamically compromised	C
	Intravenous procainamide or ibutilide to restore sinus rhythm or slow the ventricular rate is recommended for patients with preexcited AF and rapid ventricular response who are not hemodynamically compromised	C
	Catheter ablation of the accessory pathway is recommended in symptomatic patients with preexcited AF, especially if the accessory pathway has a short refractory period that allows rapid antegrade conduction	C
III (harm)	Administration of intravenous amiodarone, adenosine, digoxin (oral or intravenous), or nondihydropyridine CCBs (oral or intravenous) in patients with WPW syndrome who have preexcited AF is potentially harmful as they accelerate the ventricular rate	B
Patients with HF		
I	Control of resting heart rate using either a β-blocker or a nondihydropyridine CCB is recommended for patients with persistent or permanent AF and compensated HF with preserved ejection fraction (HF-PEF)	B
	In the absence of preexcitation, intravenous β-blocker administration (or a nondihydropyridine CCB in patients with HF-PEF) is recommended to slow the ventricular response to AF in the acute setting, with caution needed in patients with overt congestion, hypotension, or HF with reduced ejection fraction (HF-REF)	B
	In the absence of preexcitation, intravenous digoxin or amiodarone is recommended to control heart rate acutely in patients with HF	B
	Assessment of heart rate control during exercise and adjustment of pharmacological treatment to keep the rate in the physiological range is useful in symptomatic patients during activity	C
	Digoxin is effective to control resting heart rate in patients with HF-REF	C

TABLE 25.7 (*Continued*)

COR	Recommendation	LOE
IIa	A combination of digoxin and a β-blocker (or a nondihydropyridine CCB for patients with HF-PEF) is reasonable to control resting and exercise heart rate in patients with AF	B
	It is reasonable to perform AV node ablation with ventricular pacing to control heart rate when pharmacological therapy is insufficient or not tolerated	B
	Intravenous amiodarone can be useful to control the heart rate in patients with AF when other measures are unsuccessful or contraindicated	C
	For patients with AF and rapid ventricular response causing or suspected of causing tachycardia-induced cardiomyopathy, it is reasonable to achieve rate control by either AV nodal blockade or a rhythm-control strategy	B
	For patients with chronic HF who remain symptomatic from AF despite a rate-control strategy, it is reasonable to use a rhythm-control strategy	C
IIb	Oral amiodarone may be considered when resting and exercise heart rate cannot be adequately controlled using a β-blocker (or a nondihydropyridine CCB in patients with HF-PEF) or digoxin, alone or in combination	C
	AV node ablation may be considered when the rate cannot be controlled and tachycardia-mediated cardiomyopathy is suspected	C
III (harm)	AV node ablation should not be performed without a pharmacological trial to achieve ventricular rate control	C
	For rate control, intravenous nondihydropyridine CCBs, intravenous β-blockers, and dronedarone should not be administered to patients with decompensated HF	C
Patients of postoperative cardiac and thoracic surgery		
I	Treating patients who develop AF after cardiac surgery with a β-blocker is recommended unless contraindicated	A
	A nondihydropyridine CCB is recommended when a β-blocker is inadequate to achieve rate control in patients with postoperative AF	B
IIa	Preoperative administration of amiodarone reduces the incidence of AF in patients undergoing cardiac surgery and is reasonable as prophylactic therapy for patients at high risk for postoperative AF	A
	It is reasonable to restore sinus rhythm pharmacologically with ibutilide or direct-current cardioversion in patients who develop postoperative AF, as advised for nonsurgical patients	B
	It is reasonable to administer antiarrhythmic medications in an attempt to maintain sinus rhythm in patients with recurrent or refractory postoperative AF, as advised for other patients who develop AF	B
	It is reasonable to administer antithrombotic medication in patients who develop postoperative AF, as advised for nonsurgical patients	B
	It is reasonable to manage well-tolerated, new-onset postoperative AF with rate control and anticoagulation with cardioversion if AF does not revert spontaneously to sinus rhythm during follow-up	C
IIb	Prophylactic administration of sotalol may be considered for patients at risk of developing AF following cardiac surgery	B
	Administration of colchicine[a] may be considered for patients postoperatively to reduce AF following cardiac surgery	B

[a] CHA$_2$DS$_2$-VASc score: see note to Table 25.3.
[b] A randomized controlled trial suggested an efficacy for colchicine in reducing postoperative atrial fibrillation likely due to its anti-inflammatory activity [13]. In this regard, inflammation plays an important role in the pathogenesis of postoperative atrial fibrillation [14].
COR and LOE stand for class of recommendation and level of evidence, respectively (see Chapter 5 for description).

Indications for Antiarrhythmic Drug Therapy Antiarrhythmic drugs may be indicated under various conditions, including (i) patients with ventricular tachyarrhythmias who do not meet criteria for an ICD, (ii) patients with ICDs who have recurrent VT/VF with frequent appropriate ICD firing, and (iii) patients with ICDs who have paroxysmal or chronic AF with rapid rates and inappropriate ICD [17].

For patients with ventricular tachyarrhythmias who do not meet criteria for an ICD, β-blockers are the first-line therapy. If the β-blocker therapy at full therapeutic dose is not effective, amiodarone or sotalol can be selected with monitoring for adverse effects during administration.

For patients with ICDs who have recurrent VT/VF with frequent appropriate ICD firing, if the condition is in its extreme (also known as defibrillator storm or tachycardia storm), it requires the addition of antiarrhythmic drugs and/or catheter ablation for control of the recurrent VT and associated ICD shocks. The class III agent sotalol is effective in suppressing atrial and ventricular arrhythmias. The combination of β-blockers and amiodarone is an alternative approach. Because many such patients have low LVEF and poor renal function, amiodarone and β-blockers rather than sotalol can be the first-line therapy for defibrillator storm. Sotalol should be avoided in patients with severely depressed LV function or significant HF due to its negative inotropic effect.

For patients with ICDs who have paroxysmal or chronic AF with rapid rates and inappropriate ICD, it is essential to control the rapid ventricular response to atrial tachyarrhythmias. To this end, combination therapy with a β-blocker and a nondihydropyridine CCB is useful. Amiodarone can be used for rate control if other therapies are contraindicated, not tolerated, or ineffective. Ablation of the AV node may be required when pharmacological therapy fails.

25.4.3   Guideline Recommendations for the Management of Ventricular Arrhythmias

The ACC/AHA/ESC 2006 guideline for the management of patients with ventricular arrhythmias and the prevention of SCD provides a comprehensive coverage on the management of various forms of ventricular arrhythmias. In particular, the guideline addresses acute management of specific ventricular arrhythmias, including (i) ventricular tachyarrhythmias (e.g., sustained monomorphic VT, repetitive monomorphic VT,

polymorphic VT, incessant VT, and torsades de pointes) and (ii) VF/SCA. In addition, the guideline also addresses general management of ventricular arrhythmias associated with specific diseases and conditions, including cardiomyopathies, HF, genetic disorders, and drug toxicities. This section considers primarily the guideline recommendations on drug therapy for the above conditions. It is worth repeating that nonpharmacological approaches take a more important part in the management of cardiac arrhythmias, including ventricular dysrhythmias.

25.4.3.1   Acute Management of Specific Ventricular Arrhythmias

25.4.3.1.1 Acute Management of Ventricular Tachyarrhythmias   The guideline recommendations on drug therapy in the acute management of various forms of

TABLE 25.8   The 2006 ACC/AHA/ESC guideline recommendations on drug therapy in the acute management of various forms of ventricular tachyarrhythmias [17]

COR	Recommendation	LOE
Sustained monomorphic VT		
IIa	Intravenous procainamide is reasonable for initial treatment of patients with stable sustained monomorphic VT	B
	Intravenous amiodarone is reasonable in patients with sustained monomorphic VT that is hemodynamically unstable, refractory to conversion with countershock, or recurrent despite treatment with procainamide or other agents	C
IIb	Intravenous lidocaine might be reasonable for the initial treatment of patients with stable sustained monomorphic VT specifically associated with acute myocardial ischemia or infarction	C
III	CCBs, such as verapamil and diltiazem, should not be used in patients to terminate wide QRS complex tachycardia of unknown origin, especially in patients with a history of myocardial dysfunction	C
Repetitive monomorphic VT		
IIa	Intravenous amiodarone, β-blockers, and intravenous procainamide can be useful for treating repetitive monomorphic VT in the context of coronary artery disease and idiopathic VT	C
Polymorphic VT		
I	Intravenous β-blockers are useful for patients with recurrent polymorphic VT, especially if ischemia is suspected or cannot be excluded	B
	Intravenous loading with amiodarone is useful for patients with recurrent polymorphic VT in the absence of abnormal repolarization related to congenital or acquired long QT syndrome (LQTS)	C
IIb	Intravenous lidocaine may be reasonable for treatment of polymorphic VT specifically associated with acute myocardial ischemia or infarction	C
Incessant VT		
I	Revascularization and β-blockade followed by intravenous antiarrhythmic drugs, such as procainamide or amiodarone, are recommended for patients with recurrent or incessant polymorphic VT due to acute myocardial ischemia	C
IIa	Intravenous amiodarone or procainamide followed by VT ablation can be effective in the management of patients with frequently recurring or incessant monomorphic VT	B
IIb	Intravenous amiodarone and intravenous β-blockers separately or together may be reasonable in patients with VT storm	C
Torsades de pointes		
IIa	Management with intravenous magnesium sulfate is reasonable for patients who present with LQTS and few episodes of torsades de pointes. Magnesium is not likely to be effective in patients with a normal QT interval	B
	β-Blockade combined with pacing is reasonable acute therapy for patients who present with torsades de pointes and sinus bradycardia	C
	Isoproterenol is reasonable as temporary treatment in acute patients who present with recurrent pause-dependent torsades de pointes who do not have congenital LQTS	B
IIb	Potassium repletion to 4.5–5 mM may be considered for patients who present with torsades de pointes	B
	Intravenous lidocaine or oral mexiletine may be considered in patients who present with LQT3 and torsades de pointes. LQT3 is one of the three most common forms of LQTS (LQT1, LQT2, and LQT3)	C

COR and LOE stand for class of recommendation and level of evidence, respectively (see Chapter 5 for description).

ventricular tachyarrhythmias are summarized in Table 25.8. The reader is advised to refer to the full guideline document for details on nonpharmacological treatment approaches [17].

Acute Management of SCA Cardiac arrest can be caused by VF, pulseless VT, pulseless electric activity, and ventricular asystole, with VF and pulseless VT being the most common causes. Survival from these cardiac arrest rhythms requires both basic life support and a system of advanced cardiovascular life support (ACLS) with integrated post-cardiac arrest care. The foundation of successful ACLS is high-quality cardiopulmonary resuscitation (CPR) and, for VF/pulseless VT, attempted defibrillation within minutes of collapse. For victims of witnessed VF arrest, early CPR and rapid defibrillation can significantly increase the chance for survival to hospital discharge. On the other hand, other ACLS therapies, such as drug therapy and advanced airway management, although associated with an increased rate of return of spontaneous circulation (ROSC), have not been conclusively shown to increase the rate of survival to hospital discharge [22].

To facilitate evidence-based management of cardiac arrest, the AHA has published its 2010 guideline for CPR and emergency cardiovascular care [22]. As compared with the 2005 ACLS guideline [23], the 2010 guideline incorporates several key changes, including that (i) continuous quantitative waveform capnography is recommended for confirmation and monitoring of endotracheal tube placement; (ii) cardiac arrest algorithms are simplified and redesigned to emphasize the importance of high-quality CPR (including chest compressions of adequate rate and depth, allowing complete chest recoil after each compression, minimizing interruptions in chest compressions, and avoiding excessive ventilation); (iii) atropine is no longer recommended for routine use in the management of pulseless electrical activity and ventricular asystole; (iv) there is an increased emphasis on physiologic monitoring to optimize CPR quality and detect ROSC; (v) chronotropic drug (e.g., epinephrine) infusions are recommended as an alternative to pacing in symptomatic and unstable bradycardia; and (vi) adenosine is recommended as a safe and potentially effective therapy in the initial management of stable undifferentiated regular monomorphic wide complex tachycardia [22].

This section considers primarily the guideline recommendations on drug therapy in VF/pulseless VT. In this context, the 2010 AHA guideline recommendations include that (i) when VF/pulseless VT persists after at least 1 shock and a 2 min CPR period, a vasopressor can be given with the primary goal of increasing myocardial blood flow during CPR and achieving ROSC (class IIb; LOE, A); (ii) amiodarone is the first-line anti-arrhythmic agent given during cardiac arrest because it has been clinically demonstrated to improve the rate of ROSC and hospital admission in adults with refractory VF/pulseless VT— amiodarone may be considered when VF/VT is unresponsive to CPR, defibrillation, and vasopressor therapy (class IIb; LOE, A);

(iii) if amiodarone is unavailable, lidocaine may be considered, but in clinical studies, lidocaine has not been demonstrated to improve rates of ROSC and hospital admission compared with amiodarone (class IIb; LOE, B); and (iv) magnesium sulfate should be considered only for torsades de pointes associated with a long QT interval (class IIb; LOE, B) [23].

25.4.3.2 *Management of Ventricular Arrhythmias Associated with Specific Diseases and Conditions* In addition to the acute management of ventricular arrhythmias, the ACC/AHA/ESC 2006 guideline also provides recommendations on the management of ventricular arrhythmias associated with cardiomyopathies (e.g., nonischemic dilated, hypertrophic, and arrhythmogenic right ventricular cardiomyopathies), heart failure, genetic syndromes (e.g., LQTS, Brugada syndrome, and catecholaminergic polymorphic VT), idiopathic VT, and drug toxicities (e.g., digitalis toxicity, drug-induced LQTS, and sodium channel blocker-related toxicity). The major guideline recommendations on drug therapy for the above diseases and conditions are summarized in Table 25.9.

25.5 SUMMARY OF CHAPTER KEY POINTS

- Treatment of arrhythmias has been evolving from the primary use of drugs to the use of nonpharmacological devices in combination with drug therapy when necessary.

- The fundamental goal of therapy for cardiac arrhythmias is twofold: to improve symptoms of the patients and to prevent potentially serious outcomes, especially life-threatening ventricular arrhythmias and sudden death.

- Stroke prevention in patients with AF and AFL is an important treatment objective.

- The 2003 ACC/AHA/ESC guideline for the management of patients with supraventricular arrhythmias is the most comprehensive guidance that provides treatment recommendations for various forms of supraventricular arrhythmias (excluding AF).

- Due to the high prevalence of AF and its role in ischemic stroke, management of this particular supraventricular arrhythmia is often addressed separately.

- Management of patients with AF involves three goals: prevention of thromboembolism, heart rate control, and correction of the rhythm disturbance. These three goals are not mutually exclusive.

- The 2014 AHA/ACC/HRS guideline on the management of AF is a comprehensive document that provides evidence-based recommendations for prevention of thromboembolism, rate control, and rhythm control. It also addresses management of AF in specific patient groups, including HCM, ACS, hyperthyroidism, COPD, WPW syndrome, heart failure, and postoperative cardiac and thoracic surgery.

TABLE 25.9   The ACC/AHA/ESC 2006 guideline recommendations on drug therapy of ventricular arrhythmias associated with specific diseases and conditions [16]

COR	Recommendation	LOE
Dilated cardiomyopathy (nonischemic)		
IIb	Amiodarone may be considered for sustained ventricular tachycardia (VT) or VF in patients with nonischemic cardiomyopathy	C
Hypertrophic cardiomyopathy (HCM)		
IIa	Amiodarone therapy can be effective for treatment in patients with HCM with a history of sustained VT and/or VF when an ICD is not feasible	C
IIb	Amiodarone may be considered for primary prophylaxis against SCD in patients with HCM who have 1 or more major risk factor for SCD if ICD implantation is not feasible. The major risk factors for SCD include cardiac arrest, spontaneous sustained VT, family history of premature sudden death, unexplained syncope, left ventricular thickness ≥30 mm, abnormal exercise blood pressure, and nonsustained spontaneous VT	C
Arrhythmogenic right ventricular cardiomyopathy (ARVC)		
IIa	Amiodarone or sotalol can be effective for treatment of sustained VT or VF in patients with ARVC when ICD implantation is not feasible	C
Heart failure (HF)		
I	Amiodarone, sotalol, and/or other β-blockers are recommended pharmacological adjuncts to ICD therapy to suppress symptomatic ventricular tachyarrhythmias (both sustained and nonsustained) in otherwise optimally treated patients with HF	C
	Amiodarone is indicated for the suppression of acute hemodynamically compromising ventricular or supraventricular tachyarrhythmias when cardioversion and/or correction of reversible causes has failed to terminate the arrhythmia or prevent its early recurrence	B
IIb	Amiodarone, sotalol, and/or β-blockers may be considered as pharmacological alternatives to ICD therapy to suppress symptomatic ventricular tachyarrhythmias (both sustained and nonsustained) in optimally treated patients with HF for whom ICD therapy is not feasible	C
Long QT syndrome (LQTS)		
I	β-Blockers are recommended for patients with an LQTS clinical diagnosis (i.e., in the presence of prolonged QT interval)	B
	Implantation of an ICD along with the use of β-blockers is recommended for LQTS patients with previous cardiac arrest and who have reasonable expectation of survival with a good functional status or >1 year	B
IIa	β-Blockers can be effective to reduce SCD in patients with a molecular LQTS analysis and normal QT interval	B
	Implantation of an ICD with continued use of β-blockers can be effective to reduce SCD in LQTS patients experiencing syncope and/or VT while receiving β-blockers and who have reasonable expectation of survival with a good functional status for >1 year	b
IIb	Implantation of an ICD with the use of β-blockers may be considered for prophylaxis of SCD for patients in categories possibly associated with higher risk of cardiac arrest such as LQT2 and LQT3 and who have reasonable expectation of survival with a good functional status for >1 year	B
Brugada syndrome		
IIb	Quinidine might be reasonable for the treatment of electrical storm in patients with Brugada syndrome	C
Catecholaminergic polymorphic ventricular tachycardia (CPVT)		
I	β-Blockers are indicated for patients who are clinically diagnosed with CPVT on the basis of the presence of spontaneous or documented stress-induced ventricular arrhythmias	C
	Implantation of an ICD with the use of β-blockers is indicated for patients with CPVT who are survivors of cardiac arrest and who have reasonable expectation of survival with a good functional status for >1 year	C
IIa	β-Blockers can be effective in patients without clinical manifestations when the diagnosis of CPVT is established during childhood based on genetic analysis	C
	Implantation of an ICD with the use of β-blockers can be effective for affected patients with CPVT with syncope and/or documented sustained VT while receiving β-blockers and who have reasonable expectation of survival with a good functional status for >1 year	C
IIb	β-Blockers may be considered for patients with CPVT who were genetically diagnosed in adulthood and never manifested clinical symptoms of tachyarrhythmias	C
Idiopathic VT		
IIa	Drug therapy with β-blockers and/or calcium channel blockers (and/or IC agents in right ventricular outflow tract VT) can be useful in patients with structurally normal hearts with symptomatic VT arising from the right ventricle	C
Digitalis toxicity		
I	An antidigitalis antibody is recommended for patients who present with sustained ventricular arrhythmias, advanced AV block, and/or asystole that are considered due to digitalis toxicity	A

TABLE 25.9 (*Continued*)

COR	Recommendation	LOE
IIa	Magnesium or pacing is reasonable for patients who take digitalis and present with severe toxicity (sustained ventricular arrhythmias, advanced AV block, and/or asystole)	C
III	Management by lidocaine or phenytoin is not recommended for patients taking digitalis and who present with severe toxicity (sustained ventricular arrhythmias, advanced AV block, and/or asystole)	C
Drug-induced LQTS		
I	In patients with drug-induced LQTS, removal of the offending agent is indicated	A
IIa	Management with intravenous magnesium sulfate is reasonable for patients who take QT-prolonging drugs and present with few episodes of torsades de pointes in which the QT remains long	B
	Atrial or ventricular pacing or isoproterenol is reasonable for patients taking QT-prolonging drugs who present with recurrent torsades de pointes	B
IIb	Potassium ion repletion to 4.5–5 mM may be reasonable for patients who take QT-prolonging drugs and present with few episodes of torsades de pointes in whom the QT remains long	C
Sodium channel blocker-related toxicity		
I	In patients with sodium channel blocker-related toxicity, removal of the offending agent is indicated	A
IIa	Stopping the drug, reprogramming the pacemaker, or repositioning leads can be useful in patients taking sodium channel blockers who present with elevated defibrillation thresholds or pacing requirement	C
	In patients taking sodium channel blockers who present with atrial flutter with 1:1 AV conduction, withdrawal of the offending agent is reasonable. If the drug needs to be continued, additional AV nodal blockade with diltiazem, verapamil, or β-blocker or AFL ablation can be effective	C
IIb	Administration of a β-blocker and a sodium bolus may be considered for patients taking sodium channel blockers if the tachycardia becomes more frequent or more difficult to cardiovert	C

COR and LOE stand for class of recommendation and level of evidence, respectively (see Chapter 5 for description).

- Management of symptomatic ventricular arrhythmias may involve discontinuation of offending proarrhythmic drugs and other eliciting agents, pharmacological therapy with antiarrhythmic agents, and nonpharmacological approaches, such as implantable devices, ablation, and surgery.

- Due to the limited efficacy in treating ventricular arrhythmias as well as significant adverse effects of antiarrhythmic drugs, in recent years, the nonpharmacological approaches have been becoming the mainstay of therapy, especially for severe forms of ventricular arrhythmias.

- With the exception of β-blockers, the currently available antiarrhythmic drugs have not been shown in randomized clinical trials to be effective in the primary management of patients with life-threatening ventricular arrhythmias or in the prevention of SCD. As a general rule, antiarrhythmic agents may be effective as adjunctive therapy in the management of arrhythmia-prone patients under special circumstances.

- β-Blockers are effective in suppressing ventricular ectopic beats and arrhythmias as well as in reducing SCD in a spectrum of cardiac disorders in patients with and without HF. In patients with acute myocardial infarction, early (within 24 h of presentation) administration of β-blockers is associated with a reduction in the incidence of VF and is thus recommended for all patients without contraindications.

- Amiodarone reduces the risk of SCD and may represent a viable alternative in patients who are not eligible for or who do not have access to ICD therapy for the prevention of SCD. The 2010 AHA guideline for CPR and emergency cardiovascular care suggests that amiodarone be used when VF/VT is unresponsive to CPR, defibrillation, and vasopressor therapy. Amiodarone may also be useful for other forms of ventricular arrhythmias associated with specific diseases and conditions.

25.6 SELF-ASSESSMENT QUESTIONS

25.6.1. A 46-year-old male is brought to the emergency department because of palpitation. Physical exam reveals a heart rate of 105 beats/min. Electrocardiography indicates that he is having atrial fibrillation. He is otherwise healthy. Which of the following electrophysiological changes is most likely responsible for his condition?
A. Early and late afterdepolarization in AV node
B. QT prolongation in atrial tissue
C. Focal triggering and reentry in the atrial tissue
D. Reentry in the AV node
E. Reentry in the ventricular tissue

25.6.2. A 66-year-old diabetic female is diagnosed with persistent atrial fibrillation (AF). This diagnosis is based on which of the following criteria?
A. Continuous AF that is sustained for over 12 h
B. Continuous AF that is sustained for over 24 h
C. Continuous AF that is sustained for over 3 days
D. Continuous AF that is sustained for over 5 days
E. Continuous AF that is sustained for over 7 days

25.6.3. A 70-year-old female presents to the physician's office complaining of frequent episodes of palpitations. History reveals that she has hypertension and chronic kidney disease of 10 and 6 years of duration, respectively. She is diagnosed with paroxysmal atrial fibrillation with a CHA_2DS_2-VASc score of 2. Which of the following drugs would be most appropriately prescribed to prevent stroke in this patient?

A. Amodarine
B. Aspirin
C. Digoxin
D. Diltiazem
E. Warfarin

25.6.4. A 58-year-old patient with a mechanical heart valve and atrial fibrillation requires anticoagulation therapy to prevent thromboembolism. Which of the following drugs would be most suitable?

A. Aspirin
B. Clopidogrel
C. Dabigatran
D. Unfractionated heparin
E. Warfarin

25.6.5. A 54-year-old male develops ventricular fibrillation following an acute myocardial infarction. He is unresponsive to cardiopulmonary resuscitation, defibrillation, and vasopressor therapy. Which of the following drugs would be most suitable for treating his condition?

A. Amiodarone
B. Diltiazem
C. Ibutilide
D. Lidocaine
E. Quinidine

REFERENCES

1   Whinnett, Z.I., S.M. Sohaib, and D.W. Davies, Diagnosis and management of supraventricular tachycardia. *BMJ*, 2012. 345: p. e7769.

2   Marine, J.E., Catheter ablation therapy for supraventricular arrhythmias. *JAMA*, 2007. 298(23): p. 2768–78.

3   Link, M.S., Clinical practice. Evaluation and initial treatment of supraventricular tachycardia. *N Engl J Med*, 2012. 367(15): p. 1438–48.

4   Lee, G., P. Sanders, and J.M. Kalman, Catheter ablation of atrial arrhythmias: state of the art. *Lancet*, 2012. 380(9852): p. 1509–19.

5   Blomstrom-Lundqvist, C., et al., ACC/AHA/ESC guidelines for the management of patients with supraventricular arrhythmias—executive summary. a report of the American college of cardiology/American heart association task force on practice guidelines and the European society of cardiology committee for practice guidelines (writing committee to develop guidelines for the management of patients with supraventricular arrhythmias)

developed in collaboration with NASPE-heart rhythm society. *J Am Coll Cardiol*, 2003. 42(8): p. 1493–531.

6   Fuster, V., et al., 2011 ACCF/AHA/HRS focused updates incorporated into the ACC/AHA/ESC 2006 guidelines for the management of patients with atrial fibrillation: a report of the American college of cardiology foundation/American heart association task force on practice guidelines developed in partnership with the European society of cardiology and in collaboration with the European heart rhythm association and the heart rhythm society. *J Am Coll Cardiol*, 2011. 57(11): p. e101–98.

7   January, C.T., et al., 2014 AHA/ACC/HRS guideline for the management of patients with atrial fibrillation: a report of the American College of Cardiology/American Heart Association Task Force on Practice Guidelines and the Heart Rhythm Society. J Am Coll Cardiol, 2014. 64(21): p. e1-e76.

8   Go, A.S., et al., Heart disease and stroke statistics—2014 update: a report from the American Heart Association. *Circulation*, 2014. 129(3): p. e28-292.

9   Skanes, A.C., et al., Focused 2012 update of the Canadian cardiovascular society atrial fibrillation guidelines: recommendations for stroke prevention and rate/rhythm control. *Can J Cardiol*, 2012. 28(2): p. 125–36.

10 Camm, A.J., et al., 2012 focused update of the ESC guidelines for the management of atrial fibrillation: an update of the 2010 ESC guidelines for the management of atrial fibrillation. Developed with the special contribution of the European heart rhythm association. *Eur Heart J*, 2012. 33(21): p. 2719–47.

11 European Heart Rhythm Association, et al., Guidelines for the management of atrial fibrillation: the Task Force for the Management of Atrial Fibrillation of the European Society of Cardiology (ESC). *Eur Heart J*, 2010. 31(19): p. 2369–429.

12 Boriani, G., et al., Improving stroke risk stratification using the CHADS2 and CHA2DS2-VASc risk scores in patients with paroxysmal atrial fibrillation by continuous arrhythmia burden monitoring. *Stroke*, 2011. 42(6): p. 1768–70.

13 Imazio, M., et al., Colchicine reduces postoperative atrial fibrillation: results of the Colchicine for the Prevention of the Postpericardiotomy Syndrome (COPPS) atrial fibrillation substudy. *Circulation*, 2011. 124(21): p. 2290–5.

14 Jacob, K.A., et al., Inflammation in new-onset atrial fibrillation after cardiac surgery: a systematic review. *Eur J Clin Invest*, 2014. 44(4): p. 402–28.

15 Roberts-Thomson, K.C., D.H. Lau, and P. Sanders, The diagnosis and management of ventricular arrhythmias. *Nat Rev Cardiol*, 2011. 8(6): p. 311–21.

16 John, R.M., et al., Ventricular arrhythmias and sudden cardiac death. *Lancet*, 2012. 380(9852): p. 1520–9.

17 European Heart Rhythm Association, et al., ACC/AHA/ESC 2006 guidelines for management of patients with ventricular arrhythmias and the prevention of sudden cardiac death: a report of the American College of Cardiology/American Heart Association Task Force and the European Society of Cardiology Committee for Practice Guidelines (Writing Committee to Develop Guidelines for Management of Patients With Ventricular Arrhythmias and the Prevention of

Sudden Cardiac Death). *J Am Coll Cardiol*, 2006. 48(5): p. e247–346.

18 Deo, R. and C.M. Albert, Epidemiology and genetics of sudden cardiac death. *Circulation*, 2012. 125(4): p. 620–37.

19 American College of Emergency Physicians, et al., 2013 ACCF/AHA guideline for the management of ST-elevation myocardial infarction: a report of the American College of Cardiology Foundation/American Heart Association Task Force on Practice Guidelines. *J Am Coll Cardiol*, 2013. 61(4): p. e78–140.

20 Piccini, J.P., J.S. Berger, and C.M. O'Connor, Amiodarone for the prevention of sudden cardiac death: a meta-analysis of randomized controlled trials. *Eur Heart J*, 2009. 30(10): p. 1245–53.

21 Field, J.M., et al., Part 1: executive summary: 2010 American Heart Association Guidelines for Cardiopulmonary Resuscitation and Emergency Cardiovascular Care. *Circulation*, 2010. 122(18 Suppl 3): p. S640–56.

22 Neumar, R.W., et al., Part 8: adult advanced cardiovascular life support: 2010 American Heart Association Guidelines for Cardiopulmonary Resuscitation and Emergency Cardiovascular Care. *Circulation*, 2010. 122(18 Suppl 3): p. S729–67.

23 ECC Committee and Task Forces of the American Heart A, 2005 American Heart Association guidelines for cardiopulmonary resuscitation and emergency cardiovascular care. *Circulation*, 2005. 112(24 Suppl): p. IV1–203.

UNIT VIII

ISCHEMIC STROKE

26

OVERVIEW OF ISCHEMIC STROKE AND DRUG THERAPY

26.1 INTRODUCTION

Stroke, also known as cerebrovascular accident, is arguably the most feared cardiovascular event among healthy subjects and those with cardiovascular diseases. Stroke as one of the major causes of mortality, morbidity, and long-term disability imposes an enormous economic burden both in the United States and worldwide. Since stroke is part of cerebrovascular diseases, this chapter first defines cerebrovascular diseases and introduces the ICD-10 classification scheme of cerebrovascular diseases. The chapter then examines the definition and classification of stroke. This chapter also considers the epidemiology and risk factors of stroke, which provide a rationale for evidence-based management of this dread cardiovascular disorder. The chapter ends with a brief overview of the pathophysiology of ischemic stroke (the predominant form of stroke) and the pathophysiological basis of drug targeting in ischemic stroke. Summary of the key points of the chapter and case-based self-assessment questions are provided to facilitate the reader's understanding of the topic and knowledge retention as well as critical thinking.

26.2 DEFINITION AND CLASSIFICATION OF CEREBROVASCULAR DISEASES

26.2.1 Definition of Cerebrovascular Diseases

The term cerebrovascular diseases refers to a group of disorders of the vasculature that affect the circulation of blood to the brain, causing limited or no blood flow to affected areas of the brain. As described next, the ICD-10 classifies cerebrovascular diseases into 10 subgroups: ICD-10 I60–I69.

26.2.2 The ICD-10 Classification of Cerebrovascular Diseases

The ICD-10 is the 10th revision of the International Statistical Classification of Diseases and Related Health Problems (ICD) (also see Chapter 1). The ICD-10 defines the universe of diseases, disorders, injuries, and other related health conditions. These entities are listed in a comprehensive way so that everything is covered. The ICD-10 organizes information into standard groupings of diseases. Cerebrovascular diseases are listed under Chapter 9 of ICD-10: Diseases of the circulatory system (I00–I99) and span from I60 to I69 (Table 26.1).

26.3 DEFINITION AND CLASSIFICATION OF STROKE

A stroke or cerebrovascular accident is defined as an abrupt onset of a neurological deficit that occurs as a result of either inadequate focal blood flow (ischemic stroke) or hemorrhage into the brain tissue (parenchymal or intracerebral hemorrhage) or surrounding subarachnoid space (subarachnoid hemorrhage) (Table 26.1). Focal ischemic stroke is usually caused by thrombosis of the cerebral vessels themselves or by emboli from a proximal arterial source or the heart. Focal ischemic stroke (often simply referred to as ischemic stroke) accounts for 87% of all cases. A generalized reduction in cerebral blood flow due to systemic hypotension (e.g., cardiogenic shock) usually produces syncope and, in more severe instances, results in hypoxic–ischemic encephalopathy. This condition is, however, not classified as stroke.

Cardiovascular Diseases: From Molecular Pharmacology to Evidence-Based Therapeutics, First Edition. Y. Robert Li.
© 2015 John Wiley & Sons, Inc. Published 2015 by John Wiley & Sons, Inc.

TABLE 26.1   ICD-10 classification of cerebrovascular diseases[a]

ICD-10 code	Grouping of diseases
I60	Subarachnoid hemorrhage
I61	Intracerebral hemorrhage
I62	Other nontraumatic intracranial hemorrhage
I63	Cerebral infarction
I64	Stroke, not specified as hemorrhage or infarction
I65	Occlusion and stenosis of precerebral arteries, not resulting in cerebral infarction
I66	Occlusion and stenosis of cerebral arteries, not resulting in cerebral infarction
I67	Other cerebrovascular diseases
I68	Cerebrovascular disorders in diseases classified elsewhere
I69	Sequelae of cerebrovascular disease

[a] ICD-10 cerebrovascular diseases do not include transient cerebral ischemic attacks and related syndromes, traumatic intracranial hemorrhage, and vascular dementia. These disorders are classified into other disease groups. For example, transient cerebral ischemic attack is included in G45 "Transient cerebral ischemic attacks and related syndromes" under "Episodic and paroxysmal disorders" of ICD-10 Chapter 6—"Diseases of the nervous system" (G00–G99).

> **BOX 26.1   STATISTICS ON STROKE IN THE UNITED STATES [4]**
>
> - On average, a stroke occurs every 40 s.
> - Stroke is currently the no. 4 cause of death, accounting for one of every 19 deaths.
> - On average, every 4 min, someone dies of stroke.
> - Encouragingly, from 1999 to 2009, the annual stroke death rate decreased 36.9% and the actual number of stroke deaths declined 22.9%.
> - Approximately 56% of stroke deaths occur out of the hospital.
> - More women than men die of stroke each year because of the large number of elderly women. Women accounted for almost 60% of stroke deaths in 2009.
> - Stroke patients >85 years of age (the very elderly) make up 17% of all stroke patients.
> - The direct and indirect cost of stroke in 2009 was $38.6 billion.

Another condition related to cerebral ischemia is transient ischemic attack (TIA). A TIA is a transient episode of neurological dysfunction caused by focal brain, spinal cord, or retinal ischemia, without acute infarction [1]. Although TIA is not classified as stroke, patients with TIAs are at high risk of early stroke [2]. Because of this, TIA and stroke are frequently considered together in treatment guidelines from professional organizations, such as the American Heart Association/American Stroke Association [3].

26.4   EPIDEMIOLOGY OF STROKE

Based on the American Heart Association Heart Disease and Stroke Statistics, 2013 Update [4], an estimated 6,800,000 Americans at the age of 20 and over have had a stroke, and each year, about 790,500 Americans experience a new or recurrent stroke. Approximately 610,000 of these are first attacks and 185,000 are recurrent attacks. When considered separately from other cardiovascular diseases, stroke ranks no. 4 among all causes of death, behind diseases of the heart, cancer, and chronic lower respiratory diseases in the United States. Stroke mortality as an underlying cause of death in 2009 was 128,824; any-mention mortality in 2009 was 215,864, and the death rate was 38.9 per 100,000. In the United States, as with other industrialized countries, stroke rates, adjusted for age, declined over the last 30 years. However, the aging population implies that absolute numbers of stroke may stabilize or increase over the next two decades. Indeed, assuming no changes in current trends, by 2030, the prevalence of stroke is projected to increase by

25%, and the economic costs of stroke will nearly triple [5]. Because improvements in medical care are reducing stroke mortality even further, the prevalence of adult stroke-related disability is also likely to increase [6]. Additional statistics on stroke in the United States are listed in Box 26.1.

In global perspective, continuing industrialization of Asia and other regions is increasing unhealthy lifestyles, which promote stroke and other cardiovascular diseases. As a result, the highest rates of stroke mortality and disability-adjusted life years lost occur in Asia, Russia, and Eastern Europe. Stroke is increasing rapidly in Eastern Europe and Central Asia compared with Western Europe and the United States. In China and other developing countries, rates of stroke and other cardiovascular diseases are projected to increase dramatically due to combination of an aging population and the high prevalence of smoking and hypertension. The types of stroke are also changing in rapidly developing Asian countries such as China, with an increase in ischemic stroke and a decline in hemorrhagic stroke to approach patterns seen in industrialized countries [7]. Worldwide, approximately 16 million first-ever strokes occur annually, with a death toll of approximately 5.7 million people per year. Due to the increasing prevalence of stroke as a result of poorly controlled risk factors in low- and middle-income countries, the global death toll from stroke is projected to reach approximately 8 million per year by 2030 [8]. The increasing global burden of stroke due to medical care costs and disability warrants continued efforts to implement systematic approaches including community education on risk factors for stroke and their reduction. The next section describes common and emerging risk factors for stroke. Understanding stroke risk

TABLE 26.2 Risk factors for stroke

	Risk factor	Description
Modifiable	Hypertension	Hypertension is by far the most potent risk factor for stroke including both ischemic and hemorrhagic strokes. Hypertension causes a two- to fourfold increase in the risk of stroke before age 80. Of note, prehypertension is also associated with incident stroke
	Cigarette smoking	Cigarette smoking causes about a twofold increase in the risk of ischemic stroke and up to a fourfold increase in the risk of subarachnoid hemorrhagic stroke. Smoking is perhaps the most important modifiable risk factor in preventing subarachnoid hemorrhagic stroke. Discontinuation of smoking reduces stroke risk across sex, race, and age groups
	Heart diseases	Common heart disorders such as coronary artery disease, valve defects, irregular heart beat (atrial fibrillation), and enlargement of one of the heart's chambers can result in blood clots that may break loose and block vessels in or leading to the brain. Atrial fibrillation is a powerful risk factor for stroke, independently increasing risk approximately fivefold throughout all ages. Atrial fibrillation, which is more prevalent in older people, is responsible for one in four strokes after age 80 and is associated with higher mortality and disability
	Warning signs or history of TIA or stroke	If you've previously had a TIA or stroke, your risk of having a stroke is many times greater than someone who has never had one. The prevalence of TIA in the United States is estimated to be at least 5 million people. Approximately 15% of all strokes are heralded by a TIA. The short-term risk of stroke after TIA is approximately 3–10% at 2 days and 9–17% at 90 days
	Diabetes	In terms of stroke and cardiovascular diseases, having diabetes is the equivalent of aging 15 years. Diabetes increases ischemic stroke incidence at all ages. In people with a history of TIA or minor stroke, impaired glucose tolerance nearly doubles the stroke risk compared with those with normal glucose levels and triples the risks for those with diabetes
	Cholesterol imbalance	People with high blood cholesterol have an increased risk for stroke. Also, it appears that a low level of high-density lipoprotein (HDL) cholesterol is a risk factor for stroke in men, but more data are needed to verify its effect in women
	Physical inactivity and obesity	Obesity and inactivity are associated with hypertension, diabetes, and heart disease. Waist circumference to hip circumference ratio equal to or above the midvalue for the population increases the risk of ischemic stroke threefold. Moderate to vigorous physical activity is associated with an overall 35% reduction in risk of ischemic stroke compared with no physical activity
	Chronic kidney disease	People with creatinine ≥1.5 mg/dl show an increased risk for stroke. A meta-analysis of >280,000 patients showed a 43% increased incident stroke risk among patients with a glomerular filtration rate (GFR) <60 ml/min/1.73 m² [9]
	Sleep apnea	Sleep apnea is an independent risk factor for stroke, increasing the risk of stroke or stroke mortality twofold
	Depression	A meta-analysis of 28 prospective cohort studies (comprising 317,540 participants) showed that depression is associated with a 45% increase for total stroke and 55% increase for fatal stroke [10]
	Air pollution	Exposure to ambient air pollution, including airborne particulate matter, has been shown to be associated with increased mortality of stroke [11]
Nonmodifiable	Age	Stroke occurs in all age groups. The risk of stroke doubles for each decade between the ages of 55 and 85. But strokes also can occur in childhood or adolescence
	Gender	Men have a higher risk for stroke, but more women die from stroke. Men generally do not live as long as women, so men are usually younger when they have their strokes and therefore have a higher rate of survival
	Race	People from certain ethnic groups have a higher risk of stroke. For African Americans, stroke is more common and more deadly, even in young and middle-aged adults, than for any ethnic or other racial groups in the United States. The age-adjusted incidence of stroke is about twice as high in African Americans and Hispanic Americans as in Caucasians
	Family history of stroke	Stroke seems to run in some families. Several factors may contribute to familial stroke. Members of a family might have a genetic tendency for stroke risk factors, such as an inherited predisposition for hypertension or diabetes. The influence of a common lifestyle among family members also could contribute to familial stroke
	Genetic loci	Genome-wide association studies have revealed the association of certain genetic loci with an increased risk of stroke [12, 13]

factors provides a rationale for "stroke systems of care," a comprehensive approach to the care of patients with stroke or at high risk of developing stroke (see Chapter 28 for discussion of "stroke systems of care").

26.5 RISK FACTORS OF STROKE

The main risk factors for ischemic and hemorrhagic stroke, respectively, are well known and can be differentiated into nonmodifiable (age, sex, genetic predisposition, ethnicity) and modifiable ones (smoking, hypertension, lifestyle factors, diabetes) with variations in the impact of specific risk factors between hemorrhagic stroke and ischemic stroke [4]. For example, atrial fibrillation (AF) is a risk factor for ischemic stroke but not for hemorrhagic stroke. The risk factors for stroke are similar to those for coronary heart disease. Hypertension is by far the most potent risk factor for stroke. Table 26.2 summarizes the major risk factors for stroke as well as the emerging conditions associated with stroke, such as depression [10] and air pollution [14].

26.6 PATHOPHYSIOLOGY OF ISCHEMIC STROKE AND DRUG TARGETING

26.6.1 Pathophysiology of Ischemic Stroke

Acute occlusion of an intracranial vessel causes reduction of blood flow to the brain region that it supplies. A complete blockage of blood flow to the region causes cell death within a few minutes. If blood flow is restored before significant cell death occurs, the patients may experience only transient symptoms (i.e., a TIA). Conversely, infarction occurs upon prolonged occlusion. Tissue surrounding the core region of infarction is ischemic but without irreversible cell injury and is referred to as ischemic penumbra. This region is functionally impaired but potentially salvageable. The penumbra may be detected by magnetic resonance imaging [15]. The ischemic penumbra will progress into infarction unless reperfusion is improved or cells made relatively more resistant to injury.

With time, the infarct core expands into the entire ischemic penumbra, and the therapeutic opportunity to prevent cell death is lost. Therefore, detecting a penumbra in patients can help identify those who might benefit most from acute treatments that restore blood flow (e.g., thrombolytic therapy) or treatments for the future that render viable brain cells more resistant to ischemic injury [16, 17] (Fig. 26.1).

Several modes of cell death occur during ischemia. In the area of severely limited blood supply, adenosine triphosphate (ATP) depletion leads to acidosis and loss of ionic homeostasis. As a consequence, cells swell and membrane ruptures, and necrosis occurs. Within the ischemic penumbra, multiple mechanisms have been identified to contribute to the irreversible injury of the cells, which usually die of apoptosis. These pathophysiological mechanisms include excitotoxicity, calcium dysregulation, mitochondrial dysfunction, formation of reactive oxygen/nitrogen species (ROS/RNS), and inflammation [16, 17]. It should be mentioned that these mechanisms are not mutually exclusive, and oxidative stress appears to play a significant role as the other mechanistic pathways all lead to the increased formation of ROS/RNS. Although restoration of blood flow is essential for salvage of the ischemic penumbra, reperfusion also causes additional tissue injury beyond that generated by ischemia alone. Reperfusion results in production of a much larger amount of ROS/RNS than ischemia, and as such, oxidative stress plays a more pronounced part in reperfusion injury (Fig. 26.2). The oxidative stress mechanism of ischemic stroke provides a basis for developing antioxidant-based modalities for stroke intervention.

26.6.2 Drug Targeting in Ischemic Stroke

The current framework of therapeutic options for the prevention and treatment of stroke comprises a spectrum of five fields of management: (1) primary prevention, (2) recanalization and thrombolysis, (3) neuroprotection, (4) secondary prevention, and (5) neurorepair [18]. As illustrated in Figure 26.3, pharmacological therapies are a major component of stroke intervention. The pharmacological agents used in the prevention and treatment of ischemic stroke are discussed in Chapter 27.

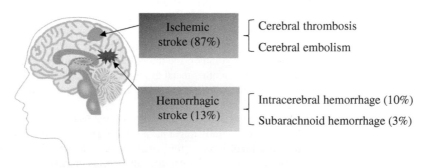

Ischemic stroke (87%)
— Cerebral thrombosis
— Cerebral embolism

Hemorrhagic stroke (13%)
— Intracerebral hemorrhage (10%)
— Subarachnoid hemorrhage (3%)

FIGURE 26.1 Classification of stroke. Stroke is conventionally classified into two groups, ischemic stroke and hemorrhagic stroke, with the former accounting for 87% of all stroke cases.

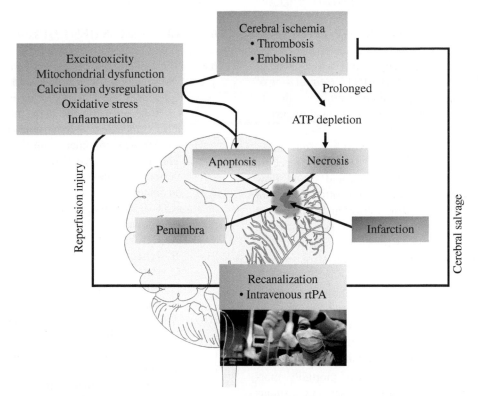

FIGURE 26.2 Pathophysiology of ischemic stroke. Prolonged cerebral ischemia leads to infarction due to necrotic cell death of brain tissue as a result of adenosine triphosphate (ATP) depletion. Cells in the penumbra primarily die of apoptosis due to less severe ischemia. Multiple pathways are implicated in cell death in penumbra, including excitotoxicity, mitochondrial dysfunction, calcium ion dysregulation, oxidative stress, and inflammation. Restoration of blood flow through the use of thrombolytic agents is essential for salvage of the cells in penumbra and thereby preventing the growth of the cerebral infarction. Although recanalization is crucial for acute management of ischemic stroke, restoration of blood flow also causes reperfusion injury with oxidative stress and inflammation as an important mechanism.

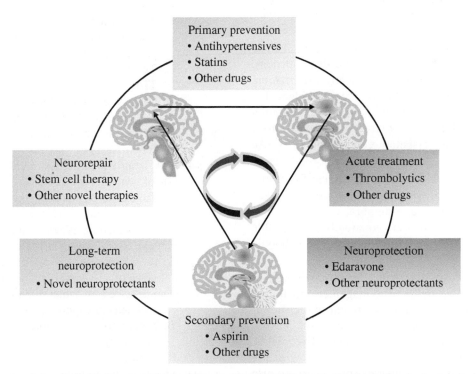

FIGURE 26.3 Management options for the preventive and therapeutic intervention of ischemic stroke. As depicted, pharmacological therapies play an important part in the overall management of ischemic stroke. Of note is the fact that at present neuroprotection and neurorepair are largely experimental approaches whose clinical efficacy remains to be established. Edaravone (structure shown in Fig. 27.1) is a free radical scavenging agent that is approved for treating ischemic stroke in Japan. For color details, please see color plate section.

26.7   SUMMARY OF CHAPTER KEY POINTS

- The term cerebrovascular diseases refers to a group of disorders of the vasculature that affect the circulation of blood to the brain, causing limited or no blood flow to affected areas of the brain. Stroke is the most important form of cerebrovascular diseases.

- A stroke or cerebrovascular accident is defined as an abrupt onset of a neurological deficit that occurs as a result of either inadequate focal blood flow (ischemic stroke) or hemorrhage into the brain tissue (parenchymal or intracerebral hemorrhage) or surrounding subarachnoid space (subarachnoid hemorrhage). Ischemic stroke and hemorrhagic stroke account for 87 and 13% of all stroke cases, respectively.

- Transient ischemic attack (TIA) is another condition related to cerebral ischemia. It is defined as a transient episode of neurological dysfunction caused by focal brain, spinal cord, or retinal ischemia, without acute infarction. TIA is not stroke, but a risk factor for stroke.

- Stroke as one of the major causes of mortality, morbidity, and long-term disability imposes an enormous economic burden both in the United States and worldwide. Although the stroke rates, adjusted for age, declined over the past three decades in the United States and other developed countries, in low- and middle-income nations, rates of stroke are projected to increase dramatically due to combination of an aging population and the high prevalence of smoking, hypertension, diabetes, and dyslipidemias.

- The major modifiable risk factors for stroke include hypertension, smoking, diabetes, and dyslipidemias. Risk factors for stroke also include atrial fibrillation, TIA, physical inactivity, obesity, chronic kidney disease, and sleep apnea. Depression and air pollution are emerging risk factors for stroke. Control of stroke risk factors is an essential component of stroke management.

- Two critical pathophysiological processes of ischemic stroke are cerebral infarction due to prolonged occlusion and ischemic penumbra, a region that is functionally impaired but potentially salvageable upon restoration of blood flow via using thrombolytic agents.

- Current framework of stroke intervention comprises a spectrum of five fields of management, including primary prevention, acute treatment with thrombolytic agents, neuroprotection, secondary prevention, and neurorepair. Among these five fields, neuroprotection and neurorepair are largely at the experimental stage although they show great promise (Chapter 27).

26.8   SELF-ASSESSMENT QUESTIONS

26.8.1. The United States is required to use the ICD for the classification of diseases and injuries under an agreement with the World Health Organization (WHO). By using the ICD, the United States collects, processes, and presents mortality data in a similar way to other countries around the world. This permits comparison of data across countries. Periodically, new revisions are developed to reflect advances in medical science. According to the ICD-10, which of the following is not classified as a cerebrovascular disease?
   A. Cerebral infarction
   B. Intracerebral hemorrhage
   C. Parenchymal hemorrhage
   D. Subarachnoid hemorrhage
   E. Transient cerebral ischemic attack

26.8.2. A 65-year-old patient with atrial fibrillation is prescribed aspirin (81 mg, once daily) by a cardiologist. What is the purpose of treating the patient with aspirin?
   A. To augment neurorepair
   B. To prevent ischemic stroke
   C. To prevent myocardial infarction
   D. To reduce fever associated with atrial fibrillation
   E. To stop atrial fibrillation

26.8.3. Transient ischemic attack (TIA) is also known as mini stroke. This notion implies that:
   A. The cerebral infarction in TIA is much smaller than that in regular stroke.
   B. The infarction in TIA is much smaller than the penumbra in typical stroke.
   C. The penumbra is much smaller than infarction in TIA.
   D. TIA is a minor form of stroke caused by moderate cerebral ischemia.
   E. TIA is not stroke, but an independent risk factor for stroke.

26.8.4. M.A. Ikram and associates recently carried out an analysis of genome-wide association data generated from four large cohorts composing the Cohorts for Heart and Aging Research in Genomic Epidemiology consortium, including 19,602 white persons (mean [+/− SD] age, 63+/−8 years) in whom 1544 incident strokes (1164 ischemic strokes) developed over an average follow-up of 11 years [12]. The main purpose of such genome-wide association studies of stroke is to:
   A. Delineate the family pedigree of stroke patients
   B. Figure out the molecular basis of stroke pathology
   C. Find out the family history of stroke patients
   D. Study the impact of aging on stroke
   E. Understand the genetic component of stroke

26.8.5. Despite the increase in the global burden of stroke, advances are being made. In 2008, after years of being the third leading cause of death in the United States, stroke dropped to fourth. Which of the following is currently the third leading cause of death in the United States?

A. Accidents (unintentional injuries)
B. Cancer
C. Chronic lower respiratory diseases
D. Diabetes
E. Heart diseases

REFERENCES

1 Easton, J.D., et al., Definition and evaluation of transient ischemic attack: a scientific statement for healthcare professionals from the American Heart Association/American Stroke Association Stroke Council; Council on Cardiovascular Surgery and Anesthesia; Council on Cardiovascular Radiology and Intervention; Council on Cardiovascular Nursing; and the Interdisciplinary Council on Peripheral Vascular Disease. The American Academy of Neurology affirms the value of this statement as an educational tool for neurologists. *Stroke*, 2009. 40(6): p. 2276–93.

2 Siket, M.S. and J. Edlow, Transient ischemic attack: an evidence-based update. *Emerg Med Pract*, 2013. 15(1): p. 1–26.

3 Furie, K.L., et al., Guidelines for the prevention of stroke in patients with stroke or transient ischemic attack: a guideline for healthcare professionals from the American Heart Association/American Stroke Association. *Stroke*, 2011. 42(1): p. 227–76.

4 Lloyd-Jones, D., et al., Heart disease and stroke statistics—2010 update: a report from the American Heart Association. *Circulation*, 2010. 121(7): p. e46–215.

5 Heidenreich, P.A., et al., Forecasting the future of cardiovascular disease in the United States: a policy statement from the American Heart Association. *Circulation*, 2011. 123(8): p. 933–44.

6 Sacco, R.L., et al., What the million hearts initiative means for stroke: a presidential advisory from the American Heart Association/American Stroke Association. *Stroke*, 2012. 43(3): p. 924–8.

7 Kinlay, S., Changes in stroke epidemiology, prevention, and treatment. *Circulation*, 2011. 124(19): p. e494–6.

8 Strong, K., C. Mathers, and R. Bonita, Preventing stroke: saving lives around the world. *Lancet Neurol*, 2007. 6(2): p. 182–7.

9 Lee, M., et al., Low glomerular filtration rate and risk of stroke: meta-analysis. *BMJ*, 2010. 341: p. c4249.

10 Pan, A., et al., Depression and risk of stroke morbidity and mortality: a meta-analysis and systematic review. *JAMA*, 2011. 306(11): p. 1241–9.

11 Chen, R., et al., Acute effect of ambient air pollution on stroke mortality in the china air pollution and health effects study. *Stroke*, 2013. 44(4): p. 954–60.

12 Ikram, M.A., et al., Genomewide association studies of stroke. *N Engl J Med*, 2009. 360(17): p. 1718–28.

13 Anderson, C.D., et al., Chromosome 9p21 in ischemic stroke: population structure and meta-analysis. *Stroke*, 2010. 41(6): p. 1123–31.

14 Mateen, F.J. and R.D. Brook, Air pollution as an emerging global risk factor for stroke. *JAMA*, 2011. 305(12): p. 1240–1.

15 Asdaghi, N. and S.B. Coutts, Stroke: neuroimaging in acute stroke-where does MRI fit in? *Nat Rev Neurol*, 2011. 7(1): p. 6–7.

16 Moskowitz, M.A., E.H. Lo, and C. Iadecola, The science of stroke: mechanisms in search of treatments. *Neuron*, 2010. 67(2): p. 181–98.

17 Chen, R.L., et al., Ischemic stroke in the elderly: an overview of evidence. *Nat Rev Neurol*, 2010. 6(5): p. 256–65.

18 Blanco, M. and J. Castillo, Stroke in 2012: major advances in the treatment of stroke. *Nat Rev Neurol*, 2013. 9(2): p. 68–70.

27

DRUGS FOR ISCHEMIC STROKE

27.1 OVERVIEW

As described in Chapter 26, the preventive and therapeutic intervention of ischemic stroke involves five areas of management: (1) primary prevention, (2) early treatment of acute ischemic stroke, (3) neuroprotection, (4) secondary prevention, and (5) neurorepair. Although neuroprotection and neurorepair are promising strategies, presently, they are primarily experimental approaches. Drug therapies remain as a major component of the effective intervention of ischemic stroke. This chapter discusses the major classes of drugs that are used in each of the above areas. Since most of the drugs discussed in this chapter are covered in the preceding chapters, this chapter summarizes the current evidence-based consensus statements on their clinical efficacy in preventive and therapeutic intervention of ischemic stroke.

27.2 DRUGS FOR PRIMARY PREVENTION OF ISCHEMIC STROKE

Primary prevention of stroke refers to the treatment of individuals with no previous history of stroke. The goal of primary prevention by pharmacological therapies is to treat the modifiable risk factors of ischemic stroke. As described in Chapter 26, hypertension, dyslipidemias, obesity, diabetes, and atrial fibrillation (AF) are among the chief modifiable risk factors of ischemic stroke that are relatively amenable to drug therapy [1–3]. As such, the pharmacological approaches to stroke risk reduction include the use of

(i) antihypertensive medications to control high blood pressure (see Chapter 12), (ii) statins (and other lipid-lowering drugs) to treat dyslipidemias (see Chapter 4), (iii) drugs to treat comoribund conditions in diabetic patients, and (iv) anticoagulants and antiplatelet agents (see Chapter 17) to reduce the risk of ischemic stroke in patients with AF.

27.2.1 Treatment of Hypertension

27.2.1.1 Hypertension Hypertension remains the most important well-documented, modifiable risk factor for stroke, and treatment of hypertension is among the most effective strategies for preventing both ischemic and hemorrhagic stroke [1, 4, 5]. In a meta-analysis of 23 randomized trials on antihypertensive medication compared with no drug therapy, a 32% reduction in stroke risk was found with pharmacological treatment [6]. The benefit of hypertension treatment in preventing stroke is well established across the spectrum of age groups, including adults of ≥80 years of age [7]. Reduction in blood pressure is generally more important than the specific agents used to achieve this goal. The management of hypertension is covered in Chapter 12.

27.2.1.2 Prehypertension As noted in Chapter 26, prehypertension (blood pressure levels: 120–139/80–89 mm Hg) is also a risk factor of stroke. A recent meta-analysis of 16 randomized trials involving 70,664 individuals with prehypertension showed that antihypertensive therapy in cohorts with prehypertensive blood pressure levels reduced the risk of stroke by 22% compared with placebo [8].

Cardiovascular Diseases: From Molecular Pharmacology to Evidence-Based Therapeutics, First Edition. Y. Robert Li.
© 2015 John Wiley & Sons, Inc. Published 2015 by John Wiley & Sons, Inc.

This finding can have important clinical and public health implications as nearly 30% of American adults of ≥20 years of age have prehypertension [9]. The global prevalence of prehypertension is estimated to be >30% of the adult population, and over 40% of adults and 15% of children in some areas of Asian countries including China [10, 11].

27.2.2 Treatment of Dyslipidemias with Statins

27.2.2.1 Statins for Ischemic Stroke
Statins lower LDL cholesterol by 30–50%, depending on the formulation and dosage (see Chapter 4). Treatment with statins reduces the risk of stroke in patients with atherosclerosis or at high risk for atherosclerosis [1]. One meta-analysis of 26 trials that included >90,000 patients found that statins reduced the risk of all strokes by approximately 21% [12]. The most likely mechanism underlying stroke prevention by statins is the retardation or regression of atherosclerosis. In this context, a meta-analysis of randomized trials of statins in combination with other preventive strategies, involving 165,792 individuals, showed that each 1 mM (39 mg/dl) decrease in LDL cholesterol equated to a reduction in relative risk for stroke of 21.1% [13]. Although statin therapy conclusively reduces the risk of ischemic stroke, the efficacy of lipid-modifying therapies other than statins on the risk of ischemic stroke has not been well established.

27.2.2.2 Statins and Intracerebral Hemorrhagic Stroke
The efficacy of statins in hemorrhagic stroke is controversial, and current evidence suggests that statin therapy neither decreases nor increases the risk of intracerebral hemorrhagic stroke. Early studies reported an increased risk of intracerebral hemorrhagic stroke associated with statin therapy. A recent meta-analysis of 31 randomized trials involving 91,588 subjects showed that active statin therapy was not associated with significant increase in intracerebral hemorrhage [14]. Likewise, a meta-analysis of 42 trials involving 121,000 patients did not show a preventive effect for statin therapy in hemorrhagic stroke either [15].

27.2.3 Treatment of Comoribund Conditions of Diabetics: Antihypertensives and Statins

Persons with diabetes have both an elevated susceptibility to atherosclerosis and an increased prevalence of proatherogenic risk factors, notably hypertension and dyslipidemias [1]. A comprehensive program that includes tight control of hypertension with angiotensin-converting enzyme (ACE) inhibitors or angiotensin receptor blockers treatment reduces risk of stroke in persons with diabetes.

Glycemic control reduces microvascular complications, but there is no evidence that improved glycemic control reduces the risk of incident stroke. Adequately powered studies and meta-analyses show that statin treatment of patients with diabetes but not with cardiovascular diseases decreases risk of a first stroke as well as other cardiovascular events [16, 17]. It is necessary also to mention that multiple meta-analyses of randomized trials show an increased risk of new onset diabetes associated with statin therapy, especially intensive dose statin therapy [18, 19]. This finding should not discourage the use of statin therapy in stroke prevention in diabetic patients considering the well-established benefit of statin therapy in reducing stroke risk in diabetics.

27.2.4 Treatment of AF with Anticoagulants and Antiplatelet Agents

27.2.4.1 Anticoagulants
Worldwide, AF is the most common arrhythmia with a 25% lifetime risk in adults [20]. AF, even in the absence of cardiac valvular disease, is associated with a four- to fivefold increased risk of ischemic stroke due to embolism of stasis-induced thrombus formation in the left atrial appendage. AF is responsible for approximately 15% of all ischemic stroke cases.

Warfarin, the standard oral anticoagulation therapy for patients with AF, is highly efficacious in reducing stroke and death in high-risk patients with this condition [1]. Novel oral anticoagulants, including apixaban, dabigatran, and rivaroxaban, have been developed as alternatives to warfarin [21]. These new oral anticoagulants are as efficacious as warfarin for the prevention of stroke and systemic embolism in patients with AF. With a decreased risk for intracranial bleeding, they appear to have a favorable safety profile, making them promising alternatives to warfarin [22–25]. A recent (2012) Scientific Advisory for Health Professionals from the American Heart Association/American Stroke Association recommends that the novel oral anticoagulants dabigatran and apixaban be efficacious alternatives to warfarin for the primary prevention of ischemic stroke in patient with nonvalvular AF [26].

27.2.4.2 Antiplatelet Agents
The antiplatelet agent aspirin is effective for the prevention of stroke in patients with AF but less effective than warfarin and the new oral anticoagulants [27]. Dual antiplatelet therapy (aspirin plus clopidogrel) is more effective than aspirin alone for preventing stroke in patients with AF, but the risk of bleeding is also augmented [1]. Antiplatelet therapy with aspirin is recommended for low-risk and some moderate-risk patients with AF, based on patient preference, estimated bleeding risk if anticoagulated, and access to high-quality anticoagulation monitoring [1].

27.3   DRUGS FOR EARLY TREATMENT OF ACUTE ISCHEMIC STROKE

27.3.1   Thrombolytic Drugs

27.3.1.1   Intravenous Tissue Plasminogen Activator Intravenous administration of recombinant tissue plasminogen activator (tPA) remains the only US Food and Drug Administration (FDA)-approved pharmacological therapy for the treatment of patients with acute ischemic stroke [28] (also see Chapter 17). As discussed in Chapter 17, there are currently three forms of recombinant tPA: (1) alteplase (synthesized using the cDNA for natural human tPA), (2) reteplase (genetically altered), and (3) tenecteplase (genetically altered). Among the three forms, only alteplase is approved by the US FDA for treating ischemic stroke. As such, in this chapter, the term tPA, if not specified, refers to only alteplase.

The use of intravenous tPA is associated with improved outcomes for a broad spectrum of patients who can be treated within 3 h of the last known well time before symptom onset and a mildly more selective spectrum of patients who can be treated between 3 and 4.5 h of the last known well time [29, 30]. Most importantly, earlier treatment is more likely to result in a favorable outcome. Patients within 3 h of onset with major strokes (NIHSS score >22) (see Box 27.1 for NIHSS score) have a very poor prognosis, but some positive treatment effect with intravenous tPA remains [28]. Treatment with intravenous tPA is associated with increased rates of intracranial hemorrhage, which may be fatal. However, the benefit of intravenous tPA outweighs this risk in acute ischemic patients.

BOX 27.1   THE NATIONAL INSTITUTES OF HEALTH STROKE SCALE

The National Institutes of Health Stroke Scale (NIHSS) is a standardized method used by physicians and other healthcare professionals to measure the level of impairment caused by a stroke. The NIHSS measures several aspects of brain function, including consciousness, vision, sensation, movement, speech, and language. A certain number of points are given for each impairment uncovered during a focused neurological examination. A maximal score of 42 represents the most severe and devastating stroke:

- 0: No stroke
- 1–4: Minor stroke
- 5–15: Moderate stroke
- 16–20: Moderate/severe stroke
- 21–42: Severe stroke

Although current guidelines recommend use of intravenous tPA within 4.5 h of stroke onset, evidence suggests that the benefit of tPA treatment may probably extend beyond 4.5 h. In this context, an updated systemic review and meta-analysis showed that the benefit of intravenous tPA may extend up to 6 h after stroke onset in some patients [31].

27.3.1.2   Endovascular Intervention: Intra-arterial tPA The term endovascular intervention deals with endovascular techniques that were originally pioneered for diagnostic purposes by radiologists. Basic techniques involve the introduction of a catheter percutaneously into a large blood vessel. Endovascular intervention has become an important approach to disease management. The number of options for endovascular treatment of ischemic stroke has increased substantially over the past decade to include intra-arterial fibrinolysis, mechanical clot retrieval with the Mechanical Embolus Removal in Cerebral Ischemia (Merci) Retrieval System (Concentric Medical, Inc., Mountain View, CA), mechanical clot aspiration with the Penumbra System (Penumbra, Inc., Alameda, CA), and acute angioplasty and stenting [28]. However, the clinical efficacy of these endovascular interventions remains to be established [32–35]. In this context, a recent randomized trial studied 362 patients with acute ischemic stroke randomly assigned, within 4.5 h after onset, to endovascular therapy (intra-arterial thrombolysis with tPA, mechanical clot disruption or retrieval, or a combination of these approaches) or intravenous tPA. Treatments were given as soon as possible after randomization. The results of this trial in patients with acute ischemic stroke indicate that endovascular therapy is not superior to standard treatment with intravenous tPA [33].

27.3.1.3   Combination of Intravenous tPA and Intra-arterial tPA Endovascular therapy is increasingly used after the administration of intravenous tPA for patients with moderate-to-severe acute ischemic stroke. However, such a combination approach may not produce better outcomes than intravenous tPA. Indeed, J.P. Broderick and associates recently reported a randomized trial involving 656 patients with ischemic stroke who had received intravenous tPA within 3 h after symptom onset. The patients were randomized to receive additional endovascular therapy (intra-arterial tPA) or intravenous tPA alone. The trial showed similar safety outcomes and no significant difference in functional independence with endovascular therapy after intravenous tPA, as compared with intravenous tPA alone [34].

27.3.1.4   Intravenous tPA for Warfarin-Treated Patients Intravenous tPA is known to improve outcomes in ischemic stroke; however, patients receiving long-term chronic warfarin therapy may face an increased risk for intracranial hemorrhage when treated with tPA. Although early studies suggested an increased risk, a recent large observational study using data from the American Heart Association Get With The Guidelines-Stroke Registry, of 23,437 patients with ischemic stroke and with international normalized ratio (INR) ≤1.7, treated with intravenous tPA in 1203 registry hospitals from April 2009 through June 2011 didn't show an increased risk of intracranial hemorrhage associated with intravenous tPA in warfarin-treated patients (INR ≤1.7) as compared with non-warfarin-treated patients [36]. Although the true risk of intracranial hemorrhage associated with intravenous tPA in warfarin-treated patients (INR ≤1.7) remains controversial, current guidelines endorse administering intravenous tPA to warfarin-treated patients if their INR is ≤1.7. This guideline is consistent with the notion that intravenous tPA remains substantially underutilized in the United States for patients with stroke; with few if any alternate therapies, medical professionals should ensure that eligible patients receive this important therapy and keep in mind that the real risk is in not treating otherwise eligible patients, who may then have prolonged morbidity from their stroke [37].

27.3.1.5 Intravenous Tenecteplase As noted earlier, among the various forms of tPA (e.g., alteplase, reteplase, tenecteplase), intravenous alteplase is the only US FDA-approved treatment for acute ischemic stroke (also see Table 17.16 of Chapter 17). Tenecteplase, a genetically engineered mutant of human tPA, is an alternative thrombolytic agent. A recent phase 2B trial involved 75 patients randomized to receive alteplase (0.9 mg/kg body weight) or tenecteplase (0.1 or 0.25 mg/kg body weight) <6 h after the onset of ischemic stroke. The trial showed that tenecteplase was associated with significantly better reperfusion and clinical outcomes than alteplase in patients with stroke who were selected on the basis of computed tomography (CT) perfusion imaging [38]. The clinical efficacy of tenecteplase remains to be established by large-scale trials.

27.3.2 Antiplatelet Agents

Currently available data demonstrate a small but statistically significant decline in mortality and unfavorable outcomes with the administration of aspirin within 48 h after stroke. It appears that the primary effects of aspirin are attributable to a reduction in early recurrent stroke. Data regarding the utility of other antiplatelet agents, including clopidogrel or dipyridamole alone or in combination with aspirin, for the treatment of acute ischemic stroke are not well established. In addition, data on the safety of antiplatelet agents when given within 24 h of intravenous fibrinolysis are lacking [28]. Nevertheless, a meta-analysis of 12 randomized trials involving 3766 patients with acute cerebral ischemia (≤3 days) reported that dual antiplatelet therapy (aspirin + dipyridamole and aspirin + clopidogrel) appeared to be safe and effective in reducing stroke recurrence and combined vascular events in patients with acute ischemic stroke or transient ischemic attack (TIA) as compared with monotherapy [39].

Due to relatively small numbers of outcome events included in the aforementioned meta-analysis, the clinical efficacy and safety of dual antiplatelet therapy in early management of acute ischemic stroke remain unclear. The ongoing Platelet-Oriented Inhibition in New TIA and Minor Ischemic Stroke (POINT; www.pointtrial.org) trial, which is comparing aspirin plus clopidogrel versus aspirin alone, will add more data to this question. Moreover, the ongoing Triple Antiplatelets for Reducing Dependency after Ischemic Stroke (TARDIS; www.tardistrial.org/) trial further addresses the question by comparing the efficacy of more intensive antiplatelet therapy (aspirin + clopidogrel + dipyridamole) with that of the guideline-based dual antiplatelet therapy (aspirin + dipyridamole).

27.3.3 Anticoagulants

The results of several clinical trials demonstrate an increased risk of bleeding complications with early administration of either unfractionated heparin (UFH) or low-molecular-weight heparin (LMWH). Early administration of anticoagulants does not lessen the risk of early neurological worsening. Data indicate that early administration of UFH or LMWH does not lower the risk of early recurrent stroke, including among people with cardioembolic sources [28]. The usefulness of the new oral anticoagulants in the early management of acute ischemic stroke has not been established.

27.4 DRUGS FOR NEUROPROTECTION

27.4.1 Definition of Neuroprotection

Neuroprotection refers to the concept of applying a therapy that directly affects the brain tissue to salvage or delay the infarction of the still-viable ischemic penumbra, rather than reperfusing the tissue. Because many potential neuroprotective therapies are likely safe and potentially efficacious in hemorrhagic as well as ischemic stroke, the ideal neuroprotective therapy would be initiated as early as possible in the course of therapy, including in the prehospital setting, and be continued while other measures are instituted, such as brain imaging followed by fibrinolytic or endovascular revascularization [28].

Pharmacological agents that limit the cellular effects of acute ischemia or reperfusion may limit neurological injury after stroke. Potential therapeutic strategies include curbing the effects of excitatory amino acids, such as glutamate, transmembrane fluxes of calcium, intracellular activation of proteases, apoptosis, free radical damage, and inflammatory responses. More than one thousand published reports of various experimental neuroprotective treatments for acute stroke exist, culminating in well over one hundred clinical trials. Although most clinical trials testing these therapies have produced disappointing outcomes [28], some have shown potentially promising results.

27.4.2   Edaravone

Due to favorable clinical outcomes, the free radical scavenger edaravone (structure shown in Fig. 27.1) has been approved by the regulatory authority in Japan for the treatment of acute ischemic stroke in that country [40–42]. However, it should be mentioned that the regulatory process for drugs in Japan is different from that in the United States and Europe, and the development of this compound outside Japan is unclear.

27.4.3   Citicoline

Citicoline (structure shown in Fig. 27.1) is an exogenous form of cytidine-5′-diphosphocholine (CDP-choline) used in membrane biosynthesis. Citicoline may reduce ischemic injury by stabilizing cell membranes and decreasing free radical formation in experimental models. Early clinical studies also showed a clinical efficacy in treating acute

ischemic stroke [43, 44]. However, the clinical efficacy of citicoline in the treatment of acute ischemic stroke was not supported by a recent large-scale international, randomized, multicenter, placebo-controlled study [45]. This trial involved 2298 patients with moderate-to-severe ischemic stroke who all received stroke care according to local treatment practice, including tPA for eligible patients presenting within 4.5 h after the onset of the stroke. Patients were randomly assigned in a 1:1 ratio to receive citicoline or placebo within 24 h after the onset of symptoms (1000 mg every 12 h intravenously during the first 3 days and orally thereafter for a total of 6 weeks [2 × 500 mg oral tablets given every 12 h]). Global recovery was similar in both groups, and no significant differences were reported in the safety variables or in the rate of adverse events. The trial concluded that citicoline is not efficacious in the treatment of moderate-to-severe acute ischemic stroke [45].

27.4.4   Statins

As stated earlier, the benefit of statins in primary prevention of ischemic stroke has been well established [14, 15]. Because of their pluripotent effects, statins have also been tested for their efficacy in the management of acute ischemic stroke. A meta-analysis of 27 studies (largely observational studies) involving 113,148 subjects showed that statin

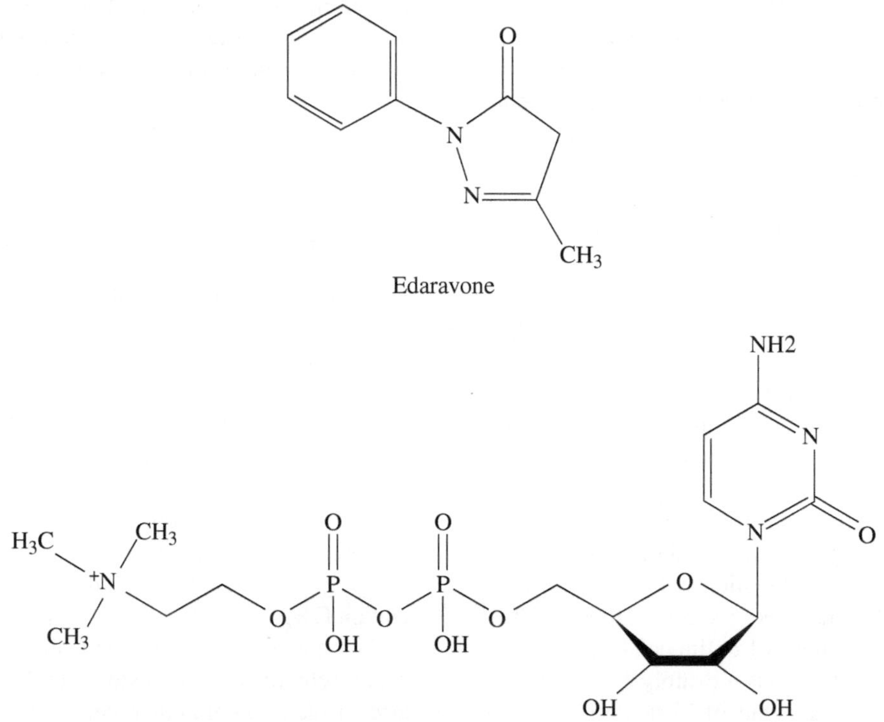

FIGURE 27.1   Structures of edaravone and citicoline. Edaravone and citicoline are potential neuroprotectants. Edaravone, a free radical scavenging compound, has been approved for clinical use in the treatment of ischemic stroke in Japan.

therapy at stroke onset was associated with improved outcome [46]. A recent multicenter study on prospectively collected data of 2,072 stroke patients treated with intravenous thrombolysis indicated that statin use in the acute phase was associated with neurological improvement, favorable functional outcome, and a reduced risk of neurological deterioration and death [47]. However, insufficient data were available from randomized trials to establish if statins are safe and effective in cases of acute ischemic stroke and TIA [48].

27.5 DRUGS FOR SECONDARY PREVENTION OF ISCHEMIC STROKE

Stroke is the leading cause of death and disability worldwide. In the United States, nearly 800,000 strokes occur annually, with almost one in four being a recurrent event [9]. Hence, prevention of recurrence, that is, secondary prevention, is crucial. Indeed, secondary prevention can reduce risk of recurrent stroke by 80–90% [49, 50]. As illustrated in Figure 27.2, secondary prevention of stroke involves (i) general risk factor reduction (hypertension, diabetes, smoking, and dyslipidemias, among others), (ii) medical treatments for patients with cardiogenic embolism, and (iii) antithrombotic therapy for noncardiogenic stroke or TIA [49–51]. Secondary preventive strategies may also include carotid revascularization for high-grade carotid stenosis [49, 50].

27.5.1 Drugs for General Risk Factor Reduction

The risk of recurrent stroke can be reduced greatly with the implementation of secondary stroke preventive measures to control the risk factors of stroke. These include (i) pharmacological treatment of hypertension, diabetes, and dyslipidemias, (ii) smoking cessation, (iii) dietary modifications, (iv) weight control, and (v) regular exercise [49, 52].

27.5.2 Drugs for Patients with Cardiogenic Embolism

Cardiogenic cerebral embolism is responsible for approximately 20% of ischemic strokes [52]. There is a history of nonvalvular AF in about one half of the cases, valvular heart disease in one fourth, and left ventricular mural thrombus in almost one third. Both persistent and paroxysmal AF are potent predictors of first as well as recurrent stroke. In the United States, >75,000 cases of stroke per year are attributed to AF. It has been estimated that AF affects over 2 million Americans and becomes more frequent with age, ranking as the leading cardiac arrhythmia in the elderly (see Chapter 25). Of all AF patients, those with a prior stroke or TIA have the highest relative risk of stroke.

Anticoagulants are preferred medications for secondary prevention of stroke in patients with AF or other cardiogenic embolism conditions (e.g., acute myocardial infarction and left ventricular thrombus, native valvular heart disease, and prosthetic heart valves). Although warfarin is the anticoagulant of choice in many current guidelines, recent studies also support the clinical efficacy of the new oral anticoagulants in the secondary prevention of stroke in patients with cardiogenic embolism [53]. In this regard, a recent (2012) Scientific Advisory for Health Professionals from the American Heart Association/American Stroke Association recommends the novel oral anticoagulant dabigatran as an efficacious alternative to warfarin for the prevention of recurrent ischemic stroke in patients with nonvalvular AF [26]. Evidence supporting the efficacy for the antiplatelet drug aspirin in AF is weaker than for anticoagulants [52].

Secondary prevention

• General risk reduction

• Medical treatment of cardiogenic embolism

• Antithrombotic therapy of noncardiogenic stroke or TIA

• Carotid revascularization for high-grade carotid stenosis

FIGURE 27.2 Secondary prevention of ischemic stroke. As illustrated, secondary prevention of ischemic stroke involves multiple strategies, ranging from treating risk factors of stroke to antithrombotic therapy. Secondary prevention may also include carotid revascularization for high-grade carotid stenosis.

27.5.3 Antithrombotic Therapy for Noncardioembolic Stroke (Specifically Atherosclerotic, Lacunar, or Cryptogenic Infarcts) and TIA

27.5.3.1 Antiplatelet Agents Four antiplatelet drugs have been approved by the US FDA for prevention of vascular events among patients with a stroke or TIA: (i) aspirin, (ii) combination aspirin/dipyridamole, (iii) clopidogrel, and (iv) ticlopidine. On average, these agents reduce the relative risk of stroke, myocardial infarction, or death by about 22%, but important differences exist between agents that have direct implications for therapeutic selection. Selection among these four agents should be based on relative effectiveness, safety, cost, patient characteristics, and patient preference [52, 54].

Notably, a recent double-blind, multicenter trial involving 3020 patients with recent symptomatic lacunar infarcts identified by magnetic resonance imaging showed that among patients with recent lacunar strokes, the addition of clopidogrel to aspirin did not significantly reduce the risk of recurrent stroke and did significantly increase the risk of bleeding and death [55]. It is also worth mentioning that patients who present with a first or recurrent stroke are commonly already on antiplatelet therapy. Unfortunately, up to now, there have been no clinical trials to indicate that switching antiplatelet agents reduces the risk for subsequent events.

27.5.3.2 Oral Anticoagulants Randomized trials have addressed the use of oral anticoagulants to prevent recurrent stroke among patients with noncardioembolic stroke. Overall, they do not show a favorable effect for anticoagulants in the secondary prevention of noncardioembolic ischemic stroke [52, 54, 56]. Current guidelines do not recommend the use of anticoagulants in the secondary prevention of noncardioembolic ischemic stroke [52].

27.6 STEM CELL THERAPY FOR NEUROREPAIR

Over the past two decades, stem cell-based neurorepair has emerged as a promising therapeutic option for ischemic stroke. Studies in animal models show that stem cell transplantation can improve function by replacing neurons or by trophic actions, modulation of inflammation, promotion of angiogenesis, remyelination and axonal plasticity, and neuroprotection. Endogenous neural stem cells are also potential therapeutic targets because they produce new neurons after stroke [57].

Despite the supposed immunoprivileged status of the brain, allogeneic grafts of stem cell-derived neurons and glia remain susceptible to rejection. A novel alternative strategy to avoid graft rejection and immunosuppression is to generate induced pluripotent stem cells (iPSCs) from somatic cells [58]. In this context, K. Oki and associates transplanted long-term self-renewing neuroepithelial-like stem cells, generated from adult human fibroblast-derived iPSCs, into stroke-damaged mouse and rat striatum or cortex [59]. The transplanted cells stopped proliferating, could survive without forming tumors for at least 4 months, and differentiated into morphologically mature neurons. Grafted cells exhibited electrophysiological properties of mature neurons and received synaptic input from host neurons. Most importantly, recovery of forepaw movements was observed by 1 week after transplantation. This study provides the first evidence that transplantation of human iPSC-derived cells may be a safe and efficient approach to promoting repair and recovery after stroke [58, 59].

Clinical trials of stem cell therapy in stroke patients are still in their infancy. There have been a number of recent phase I/II trials, primarily investigating the role in ischemic stroke patients [60–62]. The completed trials have in general demonstrated the safety and feasibility of stem cell therapy in patients with stroke. Ongoing and future trials need to define the best stem cell type, route of delivery, and timing of therapy [58, 60, 63]. In this context, the close collaborations among the US FDA, National Institutes of Health (NIH), industries, and academia are crucial for the advancement of stem cell therapy of ischemic stroke [64].

27.7 SUMMARY OF CHAPTER KEY POINTS

- Intervention of ischemic stroke involves primary prevention, early treatment of acute stroke, neuroprotection, secondary prevention, and neurorepair. At present, neuroprotection and neurorepair are largely experimental approaches, and their clinical efficacy remains to be established.

- Drug therapies remain as a major component of the effective intervention of ischemic stroke.

- Primary prevention of stroke refers to the treatment of individuals with no previous history of stroke. The goal of primary prevention by pharmacological therapies is to treat the modifiable risk factors of ischemic stroke, including hypertension, dyslipidemias, diabetes, and AF. Treatment of these risk factors greatly reduces the risk of ischemic stroke.

- Drugs for early treatment of acute ischemic stroke include intravenous tPA for recanalization and use of antiplatelet agents for possibly reducing early recurrent stroke.

- tPA remains the only US FDA-approved pharmacological therapy for treatment of patients with acute ischemic stroke.

- Neuroprotection refers to the concept of applying a therapy that directly affects the brain tissue to salvage or delay the infarction of the still-viable ischemic penumbra, rather than reperfusing the tissue. The potential pharmacological neuroprotectants include edaravone, citicoline, and statins; however, their clinical efficacy in treating acute ischemic stroke remains controversial.

- Secondary prevention of ischemic stroke, that is, prevention of stroke recurrence, can reduce the risk of stroke recurrence by 80–90%. The pharmacological measures for secondary prevention include drugs for treating general risk factors (hypertension, dyslipidemias, and diabetes), anticoagulants for treating patients with cardiogenic embolism, and antiplatelet agents for treating noncardioembolic stroke and transient ischemic attack.

- Stem cell therapy has emerged as a promising therapeutic option for ischemic stroke. Although studies in animal models show great promise, the clinical efficacy of stem cell therapy remains to be established via large-scale clinical trials.

27.8 SELF-ASSESSMENT QUESTIONS

27.8.1. An 81-year-old man arrived at the emergency department at 9:15 A.M. with speech difficulty and weakness on the right side. He had awakened that morning without symptoms. During breakfast at 8 A.M., his wife saw him slumped over and fell from the chair to the floor. He was unable to speak and could not move his right arm or leg. She called 911, and he was transported to the emergency department. He made a few attempts to speak, but his speech was unintelligible. He could move his right arm and leg but could not lift either limb off the bed. Computed tomography (CT) of the brain showed no hemorrhage and no early ischemic changes. Blood pressure was 160/90 mm Hg. His platelet count, glucose level, and prothrombin time were all normal. After the patient returned from imaging at 10 A.M., a neurologist was consulted, who confirmed the presumptive diagnosis of acute ischemic stroke and recommended immediate initiation of intravenous thrombolytic therapy (from Ref. [65]). Which of the following was most likely given for the thrombolytic therapy as recommended by the neurologist?
   A. Intravenous dabigatran
   B. Intravenous heparin
   C. Intravenous tenecteplase
   D. Intravenous alteplase
   E. Intravenous warfarin

27.8.2. Which of the following scenarios would not justify recanalization by intravenous thrombolytic therapy in the patient described in 27.8.1?
   A. If everything remained the same except that patient was treated with warfarin (INR ≤ 1.7)
   B. If the patient arrived at 10:45 A.M. and the presumptive diagnosis of acute ischemic stroke was confirmed at 11:00 A.M.
   C. If the patient arrived at 11:45 A.M. and the presumptive diagnosis of acute ischemic stroke was confirmed at 12:00 P.M.
   D. If the patient arrived at 2:45 P.M. and the presumptive diagnosis of acute ischemic stroke was confirmed at 3:00 P.M.

27.8.3. A 77-year-old woman with a history of hypertension treated with metoprolol presents for her annual examination. She reports no new symptoms. The examination is remarkable only for an irregular heart rate. Electrocardiographic testing reveals atrial fibrillation at an average rate of 75 beats/min. She has no history of arrhythmia, coronary disease, valvular disease, diabetes, alcohol abuse, or transient ischemic attack or stroke. She reports a family history of ischemic stroke (modified from Ref. [66]). Which of the following should be recommended to prevent stroke in this patient?
   A. Intra-arterial tPA
   B. Intravenous warfarin
   C. Oral atorvastatin
   D. Oral dabigatran
   E. Oral lisinopril

27.8.4. A 62-year-old woman is seen 1 week after an ischemic stroke. She had presented to another hospital with dysphasia and right-sided weakness; magnetic resonance imaging (MRI) showed a recent infarction in the left parietal cortex, and computed tomographic angiography (CTA) showed a high-grade stenosis in the left proximal internal carotid artery with normal intracranial vessels. She was treated with intravenous recombinant tissue plasminogen activator and discharged home, taking aspirin and a statin. She stopped smoking 12 years ago. On examination, the blood pressure is 145/90 mm Hg. She reports some mild residual clumsiness of her right hand (from Ref. [67]). Which of the following is an inappropriate recommendation for preventing stroke recurrence?
   A. Continuing her statin therapy and providing low-dose aspirin (e.g., 81 mg daily)
   B. Informing the patient about lifestyle factors and the importance of avoiding smoking and obesity and exercising regularly
   C. Lowering her blood pressure with an angiotensin-converting enzyme (ACE) inhibitor and a thiazide diuretic
   D. Referring this patient for prompt carotid endarterectomy, although carotid stenting would also be reasonable, given her age

    E. Treating the patient with a novel oral anticoagulant to prevent the recurrence of the noncardio-embolic ischemic stroke

27.8.5. Mr. V, a man with severe coronary, aortic, and peripheral artery disease, had an episode of brain ischemia caused by severe preocclusive carotid artery disease in the neck. The major treatment options for his symptomatic carotid artery disease are optimizing medical treatment, carotid endarterectomy, and carotid artery stenting. Selection of treatment must take into consideration his severe symptomatic coronary artery disease as well as Mr. V's concerns about surgery (from Ref. [68]). Which of the following drug therapies is appropriate for preventing an ischemic stroke in this patient?

    A. Aspirin
    B. Citicoline
    C. Dabigatran
    D. Warfarin
    E. tPA

REFERENCES

1 Goldstein, L.B., et al., Guidelines for the primary prevention of stroke: a guideline for healthcare professionals from the American Heart Association/American Stroke Association. *Stroke*, 2011. 42(2): p. 517–84.

2 O'Donnell, M.J., et al., Risk factors for ischaemic and intracerebral haemorrhagic stroke in 22 countries (the INTERSTROKE study): a case-control study. *Lancet*, 2010. 376(9735): p. 112–23.

3 Tu, J.V., Reducing the global burden of stroke: INTERSTROKE. *Lancet*, 2010. 376(9735): p. 74–5.

4 Endres, M., et al., Primary prevention of stroke: blood pressure, lipids, and heart failure. *Eur Heart J*, 2011. 32(5): p. 545–52.

5 Soros, P., et al., Antihypertensive treatment can prevent stroke and cognitive decline. *Nat Rev Neurol*, 2013. 9(3): p. 174–8.

6 Psaty, B.M., et al., Health outcomes associated with various antihypertensive therapies used as first-line agents: a network meta-analysis. *JAMA*, 2003. 289(19): p. 2534–44.

7 Beckett, N.S., et al., Treatment of hypertension in patients 80 years of age or older. *N Engl J Med*, 2008. 358(18): p. 1887–98.

8 Sipahi, I., et al., Effect of antihypertensive therapy on incident stroke in cohorts with prehypertensive blood pressure levels: a meta-analysis of randomized controlled trials. *Stroke*, 2012. 43(2): p. 432–40.

9 Go, A.S., et al., Heart disease and stroke statistics—2013 update: a report from the American Heart Association. *Circulation*, 2013. 127(1): p. e6–245.

10 Zhang, W. and N. Li, Prevalence, risk factors, and management of prehypertension. *Int J Hypertens*, 2011. 2011: p. 605359.

11 Guo, X., et al., Gender-specific prevalence and associated risk factors of prehypertension among rural children and adolescents in Northeast China: a cross-sectional study. *Eur J Pediatr*, 2013. 172(2): p. 223–30.

12 Amarenco, P., et al., Statins in stroke prevention and carotid atherosclerosis: systematic review and up-to-date meta-analysis. *Stroke*, 2004. 35(12): p. 2902–9.

13 Amarenco, P. and J. Labreuche, Lipid management in the prevention of stroke: review and updated meta-analysis of statins for stroke prevention. *Lancet Neurol*, 2009. 8(5): p. 453–63.

14 McKinney, J.S. and W.J. Kostis, Statin therapy and the risk of intracerebral hemorrhage: a meta-analysis of 31 randomized controlled trials. *Stroke*, 2012. 43(8): p. 2149–56.

15 O'Regan, C., et al., Statin therapy in stroke prevention: a meta-analysis involving 121,000 patients. *Am J Med*, 2008. 121(1): p. 24–33.

16 Cholesterol Treatment Trialists Collaborators, et al., Efficacy of cholesterol-lowering therapy in 18,686 people with diabetes in 14 randomised trials of statins: a meta-analysis. *Lancet*, 2008. 371(9607): p. 117–25.

17 Taylor, F., et al., Statins for the primary prevention of cardiovascular disease. *Cochrane Database Syst Rev*, 2013. 1: p. CD004816.

18 Preiss, D., et al., Risk of incident diabetes with intensive-dose compared with moderate-dose statin therapy: a meta-analysis. *JAMA*, 2011. 305(24): p. 2556–64.

19 Navarese, E.P., et al., Meta-analysis of impact of different types and doses of statins on new-onset diabetes mellitus. *Am J Cardiol*, 2013. 111(8): p. 1123–30.

20 Lloyd-Jones, D.M., et al., Lifetime risk for development of atrial fibrillation: the Framingham Heart Study. *Circulation*, 2004. 110(9): p. 1042–6.

21 Cabral, K.P. and J. Ansell, Oral direct factor Xa inhibitors for stroke prevention in atrial fibrillation. *Nat Rev Cardiol*, 2012. 9(7): p. 385–91.

22 Miller, C.S., et al., Meta-analysis of efficacy and safety of new oral anticoagulants (dabigatran, rivaroxaban, apixaban) versus warfarin in patients with atrial fibrillation. *Am J Cardiol*, 2012. 110(3): p. 453–60.

23 Dogliotti, A., E. Paolasso, and R.P. Giugliano, Novel oral anticoagulants in atrial fibrillation: a meta-analysis of large, randomized, controlled trials vs warfarin. *Clin Cardiol*, 2013. 36(2): p. 61–7.

24 Dentali, F., et al., Efficacy and safety of the novel oral anticoagulants in atrial fibrillation: a systematic review and meta-analysis of the literature. *Circulation*, 2012. 126(20): p. 2381–91.

25 Granger, C.B., et al., Apixaban versus warfarin in patients with atrial fibrillation. *N Engl J Med*, 2011. 365(11): p. 981–92.

26 Furie, K.L., et al., Oral antithrombotic agents for the prevention of stroke in nonvalvular atrial fibrillation: a science advisory for healthcare professionals from the American Heart Association/American Stroke Association. *Stroke*, 2012. 43(12): p. 3442–53.

27 Lip, G.Y., et al., Modification of outcomes with aspirin or apixaban in relation to CHADS(2) and CHA(2)DS(2)-VASc scores in patients with atrial fibrillation: a secondary analysis

of the AVERROES study. *Circ Arrhythm Electrophysiol*, 2013. 6(1): p. 31–8.

28 Jauch, E.C., et al., Guidelines for the early management of patients with acute ischemic stroke: a guideline for healthcare professionals from the American Heart Association/American Stroke Association. *Stroke*, 2013. 44(3): p. 870–947.

29 Hacke, W., et al., Thrombolysis with alteplase 3 to 4.5 hours after acute ischemic stroke. *N Engl J Med*, 2008. 359(13): p. 1317–29.

30 Lees, K.R., et al., Time to treatment with intravenous alteplase and outcome in stroke: an updated pooled analysis of ECASS, ATLANTIS, NINDS, and EPITHET trials. *Lancet*, 2010. 375(9727): p. 1695–703.

31 Wardlaw, J.M., et al., Recombinant tissue plasminogen activator for acute ischaemic stroke: an updated systematic review and meta-analysis. *Lancet*, 2012. 379(9834): p. 2364–72.

32 Chimowitz, M.I., Endovascular treatment for acute ischemic stroke—still unproven. *N Engl J Med*, 2013. 368(10): p. 952–5.

33 Ciccone, A., et al., Endovascular treatment for acute ischemic stroke. *N Engl J Med*, 2013. 368(10): p. 904–13.

34 Broderick, J.P., et al., Endovascular therapy after intravenous t-PA versus t-PA alone for stroke. *N Engl J Med*, 2013. 368(10): p. 893–903.

35 Kidwell, C.S., et al., A trial of imaging selection and endovascular treatment for ischemic stroke. *N Engl J Med*, 2013. 368(10): p. 914–23.

36 Xian, Y., et al., Risks of intracranial hemorrhage among patients with acute ischemic stroke receiving warfarin and treated with intravenous tissue plasminogen activator. *JAMA*, 2012. 307(24): p. 2600–8.

37 Alberts, M.J., Cerebral hemorrhage, warfarin, and intravenous tPA: the real risk is not treating. *JAMA*, 2012. 307(24): p. 2637–9.

38 Parsons, M., et al., A randomized trial of tenecteplase versus alteplase for acute ischemic stroke. *N Engl J Med*, 2012. 366(12): p. 1099–107.

39 Geeganage, C.M., et al., Dual or mono antiplatelet therapy for patients with acute ischemic stroke or transient ischemic attack: systematic review and meta-analysis of randomized controlled trials. *Stroke*, 2012. 43(4): p. 1058–66.

40 EAISG, Effect of a novel free radical scavenger, edaravone (MCI-186), on acute brain infarction. Randomized, placebo-controlled, double-blind study at multicenters. *Cerebrovasc Dis*, 2003. 15(3): p. 222–9.

41 Nakase, T., S. Yoshioka, and A. Suzuki, Free radical scavenger, edaravone, reduces the lesion size of lacunar infarction in human brain ischemic stroke. *BMC Neurol*, 2011. 11: p. 39.

42 Feng, S., et al., Edaravone for acute ischaemic stroke. *Cochrane Database Syst Rev*, 2011(12): p. CD007230.

43 Davalos, A., et al., Oral citicoline in acute ischemic stroke: an individual patient data pooling analysis of clinical trials. *Stroke*, 2002. 33(12): p. 2850–7.

44 Saver, J.L., Citicoline: update on a promising and widely available agent for neuroprotection and neurorepair. *Rev Neurol Dis*, 2008. 5(4): p. 167–77.

45 Davalos, A., et al., Citicoline in the treatment of acute ischaemic stroke: an international, randomised, multicentre, placebo-controlled study (ICTUS trial). *Lancet*, 2012. 380(9839): p. 349–57.

46 Ni Chroinin, D., et al., Statin therapy and outcome after ischemic stroke: systematic review and meta-analysis of observational studies and randomized trials. *Stroke*, 2013. 44(2): p. 448–56.

47 Cappellari, M., et al., The THRombolysis and STatins (THRaST) study. *Neurology*, 2013. 80(7): p. 655–61.

48 Elkind, M.S., Stroke: a step closer to statin therapy for stroke? *Nat Rev Neurol*, 2013. 9(5): p. 242–4.

49 Spence, J.D., Secondary stroke prevention. *Nat Rev Neurol*, 2010. 6(9): p. 477–86.

50 Davis, S.M. and G.A. Donnan, Clinical practice. Secondary prevention after ischemic stroke or transient ischemic attack. *N Engl J Med*, 2012. 366(20): p. 1914–22.

51 Rothwell, P.M., A. Algra, and P. Amarenco, Medical treatment in acute and long-term secondary prevention after transient ischaemic attack and ischaemic stroke. *Lancet*, 2011. 377(9778): p. 1681–92.

52 Furie, K.L., et al., Guidelines for the prevention of stroke in patients with stroke or transient ischemic attack: a guideline for healthcare professionals from the American Heart Association/American Stroke Association. *Stroke*, 2011. 42(1): p. 227–76.

53 Rasmussen, L.H., et al., Primary and secondary prevention with new oral anticoagulant drugs for stroke prevention in atrial fibrillation: indirect comparison analysis. *BMJ*, 2012. 345: p. e7097.

54 Warden, B.A., A.M. Willman, and C.D. Williams, Antithrombotics for secondary prevention of noncardioembolic ischaemic stroke. *Nat Rev Neurol*, 2012. 8(4): p. 223–35.

55 Benavente, O.R., et al., Effects of clopidogrel added to aspirin in patients with recent lacunar stroke. *N Engl J Med*, 2012. 367(9): p. 817–25.

56 Berlie, H.D., A. Hammad, and L.A. Jaber, Health issues in the Arab American community. The use of glucose-lowering agents and aspirin among Arab Americans with diabetes. *Ethn Dis*, 2007. 17(2 Suppl 3): p. S3-42–45.

57 Lindvall, O. and Z. Kokaia, Stem cell research in stroke: how far from the clinic? *Stroke*, 2011. 42(8): p. 2369–75.

58 Blanco, M. and J. Castillo, Stroke in 2012: major advances in the treatment of stroke. *Nat Rev Neurol*, 2013. 9(2): p. 68–70.

59 Oki, K., et al., Human-induced pluripotent stem cells form functional neurons and improve recovery after grafting in stroke-damaged brain. *Stem Cells*, 2012. 30(6): p. 1120–33.

60 Banerjee, S., et al., The potential benefit of stem cell therapy after stroke: an update. *Vasc Health Risk Manag*, 2012. 8: p. 569–80.

61 Lemmens, R. and G.K. Steinberg, Stem cell therapy for acute cerebral injury: what do we know and what will the future bring? *Curr Opin Neurol*, 2013. 26(6): p. 617–25.

62 Hess, D.C., et al., A double-blind placebo-controlled clinical evaluation of MultiStem for the treatment of ischemic stroke. *Int J Stroke*, 2014. 9(3): p. 381–6.

63 Mir, O. and S.I. Savitz, Stem cell therapy in stroke treatment: is it a viable option? *Expert Rev Neurother*, 2013. 13(2): p. 119–21.

64 Kleitman, N., M.S. Rao, and D.F. Owens, Pluripotent stem cells in translation: a Food and Drug Administration-National Institutes of Health collaboration. *Stem Cells Transl Med*, 2013. 2(7): p. 483–7.

65 Wechsler, L.R., et al., Intravenous thrombolytic therapy for acute ischemic stroke. *N Engl J Med*, 2011. 364: p. 2138–46.

66 Page, R.L., Newly diagnosed atrial fibrillation. *N Engl J Med*, 2004. 351: p. 2408–2416.

67 Stephen, M.D., Donnan, G.A., Secondary prevention after ischemic stroke or transient ischemic attack. *N Engl J Med*, 2012. 366: p. 1914–22.

68 Caplan, L.R., A 70-year-old man with a transient ischemic attack: review of internal carotid artery stenosis. *JAMA*, 2008. 300: p. 81–90.

28

MANAGEMENT OF ISCHEMIC STROKE: PRINCIPLES AND GUIDELINES

28.1 OVERVIEW

As noted in Chapter 1, to focus on reducing mortality attributable to cardiovascular diseases, including stroke, and improving overall cardiovascular health, including brain health, the American Heart Association/American Stroke Association (AHA/ASA) established a 2020 Health Impact Goal to improve the cardiovascular health of all Americans by 20% while reducing deaths related to cardiovascular diseases and stroke by 20%. Similarly, the US Department of Health and Human Services (DHHS) has established a Healthy People 2020 target of reducing the rate of death attributable to stroke by 20% from a 2007 baseline [1]. Achieving the above goals relies on understanding and implementation of a comprehensive "stroke systems of care" consisting of primary prevention, community education, acute management, secondary prevention, and rehabilitation, among others. This chapter first introduces the concept of stroke systems of care and then describes the current guidelines on the early management of acute ischemic stroke. The chapter focuses on the principles of pharmacological management of acute ischemic stroke. The primary and secondary prevention of ischemic stroke are discussed in Chapter 27.

28.2 INTRODUCTION TO STROKE SYSTEMS OF CARE

The Institute of Medicine of the US National Academy of Sciences has concluded that the fragmentation of the delivery of healthcare services frequently results in suboptimal treatment, safety concerns, and inefficient use of healthcare resources. To ensure that scientific knowledge is translated into practice, the Institute of Medicine has recommended the establishment of coordinated systems of care that integrate preventive and treatment services and promote patient access to evidence-based care [2]. In general, the fragmented approach to stroke care that exists in most regions of the United States has failed to provide an effective integrated system for stroke prevention, treatment, and rehabilitation because of inadequate linkages and coordination among the fundamental components of stroke care. Although individual components of a stroke system may be well developed, these components often operate in isolation. The problem of access to coordinated stroke care may be exacerbated in rural or neurologically underserved (inadequate access to neurological expertise) areas [3].

In view of the aforementioned considerations and in line with the recommendations by the Institute of Medicine [2], the American Stroke Association (ASA) in 2005 proposed a systems approach known as "stroke systems of care" and recommended that a stroke system should coordinate and promote patient access to the full range of activities and services associated with stroke prevention, treatment, and rehabilitation. The ASA "stroke systems of care" includes the following seven key components (Fig. 28.1) [3]:

1. Primordial and primary prevention: Primordial prevention refers to strategies designed to decrease the development of disease risk factors (e.g., efforts to decrease the development of obesity, increase exercise,

---

Cardiovascular Diseases: From Molecular Pharmacology to Evidence-Based Therapeutics, First Edition. Y. Robert Li.
© 2015 John Wiley & Sons, Inc. Published 2015 by John Wiley & Sons, Inc.

447

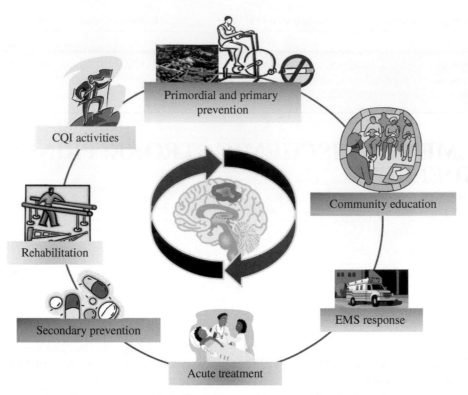

FIGURE 28.1 Schematic illustration of the American Stroke Association "stroke systems of care." As illustrated, the "stroke systems of care" consists of seven interrelated components of care, ranging from primary prevention to acute management and secondary prevention and finally to rehabilitation and continued quality improvement (CQI) activities. Hence, the "stroke systems of care" emphasizes the holistic approach to preventing and treating stroke.

and provide a well-balanced diet). Thus, primordial prevention encompasses the entire population and is not limited to individuals with recognized risk factors for stroke or other cardiovascular diseases. Primary prevention refers to the treatment of established disease risk factors.

2. Community education: Despite numerous efforts to increase awareness, overall knowledge among the public remains poor with regard to stroke risk factors, the signs and symptoms of stroke, and the availability of a time-sensitive therapy, especially among groups at the highest risk for stroke. Improving the public's knowledge of the risk factors, signs, and symptoms of stroke is critical to improving the quality of stroke care.

3. Notification and response of emergency medical services (EMS): The effective notification and response of EMS for stroke involve a complex interaction among the public, the applicable EMS programs, and the relevant hospital emergency departments. Stroke patients or a bystander witnessing a stroke must recognize the signs and symptoms of stroke and the importance of calling an emergency response telephone number immediately to help initiate effective therapy as rapidly as possible.

4. Acute stroke treatment, including the hyperacute and emergency department phases: One critical element of the multidisciplinary stroke system is the hospital-based acute stroke team. This is the component of the stroke system that is prepared to handle the hyperacute phase of diagnosis and treatment of acute stroke events.

5. Subacute stroke treatment and secondary prevention: The treatment of stroke patients during the subacute phase, including the early implementation of secondary prevention regimens, is critical to optimizing patient outcomes.

6. Rehabilitation: Stroke rehabilitation involves a combined and coordinated use of medical, social, educational, and vocational measures for retraining individuals to reach their maximal physical, psychological, social, vocational, and avocational potential.

7. Continuous quality improvement (CQI) activities: A critical function of a systems approach to stroke care is the use of CQI strategies to ascertain whether and to what extent various efforts are succeeding in improving patient care. CQI relies on data accessibility and transfer among all appropriate facilities and providers.

TABLE 28.1   AHA/ASA current guidelines on the management of ischemic stroke

Title of guideline (note)	Year	Journal of publication
Guidelines for the early management of patients with acute ischemic stroke: A guideline for healthcare professionals from the American Heart Association/American Stroke Association (early management of acute ischemic stroke)	2013	*Stroke* 2013; 44:870–947
Guidelines for the primary prevention of stroke: A guideline for healthcare professionals from the American Heart Association/American Stroke Association (primary prevention of stroke)	2011	*Stroke* 2011; 42:517–584
Guidelines for the prevention of stroke in patients with stroke or transient ischemic attack: A guideline for healthcare professionals from the American Heart Association/American Stroke Association (secondary prevention of stroke)	2011	*Stroke* 2011; 42:227–276

TABLE 28.2   International guidelines on the management of ischemic stroke

Country/organization	Title of guideline	Year	Publication
European Stroke Organisation (ESO)	Guidelines for the management of ischemic stroke and transient ischemic attack 2008	2008	*Cerebrovasc Dis* 2008; 25:457–507
National Institute for Health and Care Excellence (NICE) (United Kingdom)	Diagnosis and initial management of acute stroke and transient ischemic attack (TIA)	2008	http://www.nice.org.uk/CG68
Canadian Stroke Network (Canada)	Canadian best practice recommendations for stroke care	2010 update	http://www.strokebestpractices.ca/

The "stroke systems of care" represents the complete spectrum of stroke management in the United States. Multiple evidence-based guidelines have been published over the past few years to address the various components of the stroke systems of care [4–9] (Table 28.1). In addition, guidelines from Europe and Canada and other countries are also available [10–14] (Table 28.2). This chapter focuses on the recommendations from the current AHA/ASA guidelines on the early management of acute ischemic stroke, a crucial component of stroke intervention and stroke systems of care.

28.3   CURRENT AHA/ASA GUIDELINES ON EARLY MANAGEMENT OF ACUTE ISCHEMIC STROKE

28.3.1   General Principles of Early Management of Acute Ischemic Stroke

28.3.1.1   Main Goals of Early Management of Acute Ischemic Stroke   There are four main goals in the initial phase of acute stroke management. They are (i) to ensure medical stability, (ii) to quickly reverse conditions that are contributing to the patient's problem, (iii) to determine if patients with acute ischemic stroke are candidates for thrombolytic therapy and screen for potential contraindications to thrombolysis, and (iv) to begin to uncover the pathophysiological basis of the neurological symptoms.

28.3.1.2   Components of Early Management of Acute Ischemic Stroke   The early management of acute ischemic stroke involves multiple areas of care, which are intimately linked in a timely manner. The 2010 AHA guidelines on advanced cardiac life support stroke chain of survival describe the critical links to the process of moving a patient from stroke ictus through recognition, transport, triage, early diagnosis and treatment, and finally hospital disposition (Fig. 28.2) [15]. The 2013 AHA/ASA guidelines draw on the above stroke chain of survival and recommend 12 components of the early management of acute ischemic stroke [4]. These 12 components are outlined in the following text, and the reader may refer to the full guidelines [4] for detailed description of each of the 12 components:

1. Prehospital stroke management
2. Emergency evaluation and diagnosis of acute ischemic stroke
3. General supportive care and treatment of acute complications
4. Intravenous fibrinolysis
5. Endovascular interventions
6. Anticoagulants
7. Antiplatelet agents
8. Volume expansion, vasodilators, and induced hypertension
9. Neuroprotective modalities

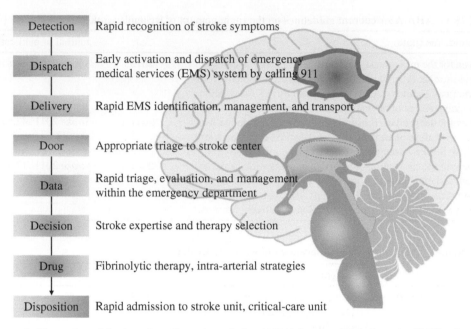

Detection — Rapid recognition of stroke symptoms

Dispatch — Early activation and dispatch of emergency medical services (EMS) system by calling 911

Delivery — Rapid EMS identification, management, and transport

Door — Appropriate triage to stroke center

Data — Rapid triage, evaluation, and management within the emergency department

Decision — Stroke expertise and therapy selection

Drug — Fibrinolytic therapy, intra-arterial strategies

Disposition — Rapid admission to stroke unit, critical-care unit

FIGURE 28.2 Schematic illustration of the American Heart Association (AHA) "stroke chain of survival." Effective management of acute stroke depends on the timely diagnosis and treatment. The AHA "7 D's of Stroke Care" remains the major steps in diagnosis and treatment of stroke and identifies the key points at which delays can occur [15].

10. Surgical intervention
11. Admission to hospital and general acute treatment (after hospitalization)
12. Treatment of acute neurological complications (ischemic brain edema, hemorrhagic transformation)

28.3.2 Current AHA/ASA Guidelines on Early Management of Acute Ischemic Stroke

This section focuses on describing the 2013 AHA/ASA guideline recommendations related to pharmacological therapy, which are mainly included in components (3–9) listed in Section 28.3.1.2. The recommendations and the corresponding class and level of evidence for each of the components (3–9) are summarized in Tables 28.3, 28.4, 28.5, 28.6, 28.7, 28.8, and 28.9. The reader may refer to the full guidelines [4] for recommendations on other components of the early management of acute ischemic stroke.

28.4 SUMMARY OF CHAPTER KEY POINTS

- The current framework for the management of stroke is conceptualized in the "stroke systems of care," which specifies seven domains of care: primary prevention, community education, notification and response of emergency medical services, acute stroke treatment, subacute stroke treatment and secondary prevention, rehabilitation, and continuous quality improvement activities.

- Multiple evidence-based guidelines have been published over the past few years to address the various components of the stroke systems of care. Most notably, among them are the guidelines from the American Heart Association/American Stroke Association (AHA/ASA).

- The AHA/ASA recently released the 2013 guidelines for the early management of patients with acute ischemic stroke: A guideline for health professionals from the American Heart Association/American Stroke Association.

- The above 2013 AHA/ASA guideline provides recommendations on 12 areas of early management of acute ischemic stroke, ranging from prehospital stroke management, emergency evaluation and diagnosis, and intravenous thrombolysis to surgical intervention and treatment of acute neurological complications.

- Intravenous administration of tPA is the only US FDA-approved pharmacological therapy for the treatment of patients with acute ischemic stroke within 4.5 h after stroke onset. Intravenous tPA for acute ischemic stroke is now widely accepted and recommended (class I, level of evidence A) by the AHA/ASA and many other international organizations.

- According to the 2013 AHA/ASA guideline, intra-arterial fibrinolysis is beneficial for treatment of carefully selected patients with major ischemic strokes of <6 h duration caused by occlusions of the MCA who are not otherwise candidates for intravenous tPA.

TABLE 28.3 The 2013 AHA/ASA recommendations on general supportive care and treatment of acute complications [4]

COR	Recommendation (explanation)	LOE
I	Cardiac monitoring is recommended to screen for atrial fibrillation and other potentially serious cardiac arrhythmias that would necessitate emergency cardiac interventions. Cardiac monitoring should be performed for at least the first 24 h	B
	Patients who have elevated blood pressure and are otherwise eligible for treatment with intravenous recombinant tissue plasminogen activator (tPA) should have their blood pressure carefully lowered so that their systolic blood pressure is <185 mm Hg and their diastolic blood pressure is <110 mm Hg before fibrinolytic therapy is initiated. If medications are given to lower blood pressure, the clinician should be sure that the blood pressure is stabilized at the lower level before beginning treatment with intravenous tPA and maintained below 180/105 mm Hg for at least the first 24 h after intravenous tPA treatment (blood pressure is usually elevated in acute ischemic stroke)	B
	Airway support and ventilatory assistance are recommended for the treatment of patients with acute stroke who have decreased consciousness or who have bulbar dysfunction that causes compromise of the airway	C
	Supplemental oxygen should be provided to maintain oxygen saturation >94%	C
	Sources of hyperthermia (temperature >38°C) should be identified and treated, and antipyretic medications should be administered to lower temperature in hyperthermic patients with stroke	C
	Until other data become available, consensus exists that the previously described blood pressure recommendations should be followed in patients undergoing other acute interventions to recanalize occluded vessels, including intra-arterial fibrinolysis	C
	In patients with markedly elevated blood pressure who do not receive fibrinolysis, a reasonable goal is to lower blood pressure by 15% during the first 24 h after onset of stroke. The level of blood pressure that would mandate such treatment is not known, but consensus exists that medications should be withheld unless the systolic blood pressure is >220 mm Hg or the diastolic blood pressure is >120 mm Hg	C
	Hypovolemia should be corrected with intravenous normal saline, and cardiac arrhythmias that might be reducing cardiac output should be corrected	C
	Hypoglycemia (blood glucose <60 mg/dl) should be treated in patients with acute ischemic stroke. The goal is to achieve normoglycemia	C
IIa	Evidence from one clinical trial indicates that initiation of antihypertensive therapy within 24 h of stroke is relatively safe. Restarting antihypertensive medications is reasonable after the first 24 h for patients who have preexisting hypertension and are neurologically stable unless a specific contraindication to restarting treatment is known	B
	No data are available to guide selection of medications for the lowering of blood pressure in the setting of acute ischemic stroke. The antihypertensive medications and doses described in the full guidelines [4] are reasonable choices based on general consensus	C
	Evidence indicates that persistent in-hospital hyperglycemia during the first 24 h after stroke is associated with worse outcomes than normoglycemia, and thus, it is reasonable to treat hyperglycemia to achieve blood glucose levels in a range of 140–180 mg/dl and to closely monitor to prevent hypoglycemia in patients with acute ischemic stroke	C
IIb	The management of arterial hypertension in patients not undergoing reperfusion strategies remains challenging. Data to guide recommendations for treatment are inconclusive or conflicting. Many patients have spontaneous declines in blood pressure during the first 24 h after onset of stroke. Until more definitive data are available, the benefit of treating arterial hypertension in the setting of acute ischemic stroke is not well established. Patients who have malignant hypertension or other medical indications for aggressive treatment of blood pressure should be treated accordingly	C
III	Supplemental oxygen is not recommended in nonhypoxic patients with acute ischemic stroke	B

COR and LOE denote class of recommendation and level of evidence, respectively; see Chapter 5 for description.

TABLE 28.4   The 2013 AHA/ASA recommendations on intravenous fibrinolysis[a]

COR	Recommendation	LOE
I	Intravenous tPA (0.9 mg/kg, maximum dose 90 mg) is recommended for selected patients who may be treated within 3 h of onset of ischemic stroke. Physicians should review the criteria outlined in the full guidelines [4] to determine the eligibility of the patient. A recommended regimen for observation and treatment of patients who receive intravenous tPA is described in the full guidelines [4]	A
	In patients eligible for intravenous tPA, benefit of therapy is time dependent, and treatment should be initiated as quickly as possible. The door-to-needle time (time of bolus administration) should be within 60 min from hospital arrival	A
	Intravenous tPA (0.9 mg/kg, maximum dose 90 mg) is recommended for administration to eligible patients who can be treated in the time period of 3–4.5 h after stroke onset. The eligibility criteria for treatment in this time period are similar to those for people treated at earlier time periods within 3 h, with the following additional exclusion criteria: patients >80 years old, those taking oral anticoagulants regardless of international normalized ratio (INR), those with a baseline National Institutes of Health Stroke Scale (NIHSS) score >25, those with imaging evidence of ischemic injury involving more than one third of the middle cerebral artery (MCA) territory, or those with a history of both stroke and diabetes mellitus	B
	Intravenous tPA is reasonable in patients whose blood pressure can be lowered safely (to below 185/110 mm Hg) with antihypertensive agents, with the physician assessing the stability of the blood pressure before starting intravenous tPA	B
	In patients undergoing fibrinolytic therapy, physicians should be aware of and prepared to emergently treat potential side effects, including bleeding complications and angioedema that may cause partial airway obstruction	B
IIa	Intravenous tPA is reasonable in patients with a seizure at the time of onset of stroke if evidence suggests that residual impairments are secondary to stroke and not a postictal phenomenon	C
IIb	The effectiveness of sonothrombolysis for treatment of patients with acute stroke is not well established	B
	The usefulness of the intravenous administration of tenecteplase, reteplase, desmoteplase, urokinase, or other fibrinolytic agents and the intravenous administration of ancrod or other defibrinogenating agents is not well established, and they should only be used in the setting of a clinical trial	B
	The effectiveness of intravenous treatment with tPA is not well established and requires further study for patients who can be treated in the time period of 3–4.5 h after stroke but have one or more of the following exclusion criteria: (i) patients >80 years old; (ii) those taking oral anticoagulants, even with INR ≤1.7; (iii) those with a baseline NIHSS score >25; or (iv) those with a history of both stroke and diabetes mellitus	C
	Use of intravenous fibrinolysis in patients with conditions of mild stroke deficits, rapidly improving stroke symptoms, major surgery in the preceding 3 months, and recent myocardial infarction may be considered, and potential increased risk should be weighed against the anticipated benefits. These circumstances require further study	C
III	The intravenous administration of streptokinase for treatment of stroke is not recommended	A
	The use of intravenous tPA in patients taking direct thrombin inhibitors or direct factor Xa inhibitors may be harmful and is not recommended unless sensitive laboratory tests such as activated partial thromboplastin time (aPTT), INR, platelet count, ecarin clotting time (ECT), and thrombin time or appropriate direct factor Xa activity assays are normal or the patient has not received a dose of these agents for >2 days (assuming normal renal metabolizing function). Similar consideration should be given to patients being considered for intra-arterial tPA. Further study is required	C

[a]Ref. 4.

COR and LOE denote class of recommendation and level of evidence, respectively; see Chapter 5 for description.

TABLE 28.5   The 2013 AHA/ASA recommendations on endovascular interventions[a]

COR	Recommendation	LOE
I	Patients eligible for intravenous tPA should receive intravenous tPA even if intra-arterial treatments are being considered	A
	Intra-arterial fibrinolysis is beneficial for the treatment of carefully selected patients with major ischemic strokes of <6h duration caused by occlusions of the MCA who are not otherwise candidates for intravenous tPA. The optimal dose of intra-arterial tPA is not well established, and tPA does not have US Food and Drug Administration (FDA) approval for intra-arterial use	B
	As with intravenous fibrinolytic therapy, reduced time from symptom onset to reperfusion with intra-arterial therapies is highly correlated with better clinical outcomes, and all efforts must be undertaken to minimize delays to definitive therapy	B
	Intra-arterial treatment requires the patient to be at an experienced stroke center with rapid access to cerebral angiography and qualified interventionalists. An emphasis on expeditious assessment and treatment should be made. Facilities are encouraged to define criteria that can be used to credential individuals who can perform intra-arterial revascularization procedures. Outcomes on all patients should be tracked	C
	When mechanical thrombectomy is pursued, stent retrievers such as Solitaire FR and Trevo are generally preferred to coil retrievers such as Merci. The relative effectiveness of the Penumbra System versus stent retrievers is not yet characterized	A
IIa	The Merci, Penumbra System, Solitaire FR, and Trevo thrombectomy devices can be useful in achieving recanalization alone or in combination with pharmacological fibrinolysis in carefully selected patients. Their ability to improve patient outcomes has not yet been established. These devices should continue to be studied in randomized controlled trials to determine the efficacy of such treatments in improving patient outcomes	B
	Intra-arterial fibrinolysis or mechanical thrombectomy is reasonable in patients who have contraindications to the use of intravenous fibrinolysis	C
IIb	Rescue intra-arterial fibrinolysis or mechanical thrombectomy may be reasonable approaches to recanalization in patients with large-artery occlusion who have not responded to intravenous fibrinolysis. Additional randomized trial data are needed	B
	The usefulness of mechanical thrombectomy devices other than the Merci retriever, the Penumbra System, Solitaire FR, and Trevo is not well established. These devices should be used in the setting of clinical trials	C
	The usefulness of emergent intracranial angioplasty and/or stenting is not well established. These procedures should be used in the setting of clinical trials	C
	The usefulness of emergent angioplasty and/or stenting of the extracranial carotid or vertebral arteries in unselected patients is not well established. Use of these techniques may be considered in certain circumstances, such as in the treatment of acute ischemic stroke resulting from cervical atherosclerosis or dissection. Additional randomized trial data are needed	C

[a]Ref. 4.
COR and LOE denote class of recommendation and level of evidence, respectively; see Chapter 5 for description.

TABLE 28.6   The 2013 AHA/ASA recommendations on treatment with anticoagulants[a]

COR	Recommendation	LOE
IIb	At present, the usefulness of argatroban or other thrombin inhibitors for treatment of patients with acute ischemic stroke is not well established. These agents should be used in the setting of clinical trials	B
	The usefulness of urgent anticoagulation in patients with severe stenosis of an internal carotid artery ipsilateral to an ischemic stroke is not well established	B
III	Urgent anticoagulation, with the goal of preventing early recurrent stroke, halting neurological worsening, or improving outcomes after acute ischemic stroke, is not recommended for treatment of patients with acute ischemic stroke	A
	Urgent anticoagulation for the management of noncerebrovascular conditions is not recommended for patients with moderate-to-severe strokes because of an increased risk of serious intracranial hemorrhagic complications	A
	Initiation of anticoagulant therapy within 24h of treatment with intravenous tPA is not recommended	B

[a]Ref. 4.
COR and LOE denote class of recommendation and level of evidence, respectively; see Chapter 5 for description.

TABLE 28.7   The 2013 AHA/ASA recommendations on treatment with antiplatelet agents[a]

COR	Recommendation (explanation)	LOE
I	Oral administration of aspirin (initial dose is 325 mg) within 24–48 h after stroke onset is recommended for treatment of most patients	A
IIb	The usefulness of clopidogrel for the treatment of acute ischemic stroke is not well established. Further research testing the usefulness of the emergency administration of clopidogrel in the treatment of patients with acute stroke is required	C
	The efficacy of intravenous tirofiban and eptifibatide is not well established, and these agents should be used only in the setting of clinical trials	C
III	Aspirin is not recommended as a substitute for other acute interventions for treatment of stroke, including intravenous tPA	B
	The administration of other intravenous antiplatelet agents that inhibit the glycoprotein IIb/IIIa receptor is not recommended. Further research testing the usefulness of emergency administration of these medications as a treatment option in patients with acute ischemic stroke is required	B
	The administration of aspirin (or other antiplatelet agents) as an adjunctive therapy within 24 h of intravenous fibrinolysis is not recommended (*due to increased intracranial hemorrhage*)	C

[a]Ref. 4.
COR and LOE denote class of recommendation and level of evidence, respectively; see Chapter 5 for description.

TABLE 28.8   The 2013 AHA/ASA recommendations on interventions involving volume expansion, use of vasodilators, and induced hypertension[a]

COR	Recommendation	LOE
I	In exceptional cases with systemic hypotension producing neurological sequelae, a physician may prescribe vasopressors to improve cerebral blood flow. If drug-induced hypertension is used, close neurological and cardiac monitoring is recommended	C
IIb	The administration of high-dose albumin is not well established as a treatment for most patients with acute ischemic stroke until further definitive evidence regarding efficacy becomes available	B
	At present, use of devices to augment cerebral blood flow for the treatment of patients with acute ischemic stroke is not well established. These devices should be used in the setting of clinical trials	B
	The usefulness of drug-induced hypertension in patients with acute ischemic stroke is not well established. Induced hypertension should be performed in the setting of clinical trials	B
III	Hemodilution by volume expansion is not recommended for treatment of patients with acute ischemic stroke	A
	The administration of vasodilatory agents, such as pentoxifylline, is not recommended for treatment of patients with acute ischemic stroke	A

[a]Ref. 4.
COR and LOE denote class of recommendation and level of evidence, respectively; see Chapter 5 for description.

TABLE 28.9   The 2013 AHA/ASA recommendations on neuroprotective modalities[a]

COR	Recommendation	LOE
IIa	Among patients already taking statins at the time of onset of ischemic stroke, continuation of statin therapy during the acute period is reasonable	B
IIb	The utility of induced hypothermia for the treatment of patients with ischemic stroke is not well established, and further trials are recommended	B
	At present, transcranial near-infrared laser therapy is not well established for the treatment of acute ischemic stroke, and further trials are recommended	B
III	At present, no pharmacological agents with putative neuroprotective actions have demonstrated efficacy in improving outcomes after ischemic stroke, and therefore, other neuroprotective agents are not recommended	A
	Data on the utility of hyperbaric oxygen are inconclusive, and some data imply that the intervention may be harmful. Thus, with the exception of stroke secondary to air embolization, this intervention is not recommended for treatment of patients with acute ischemic stroke	B

[a]Ref. 4.
COR and LOE denote class of recommendation and level of evidence, respectively; see Chapter 5 for description.

- Urgent anticoagulation, with the goal of preventing early recurrent stroke, halting neurological worsening, or improving outcomes after acute ischemic stroke, is not recommended for treatment of patients with acute ischemic stroke based on the 2013 AHA/ASA guideline.

- Oral administration of aspirin (initial dose is 325 mg) within 24–48 h after stroke onset is recommended for treatment of most patients by the 2013 AHA/ASA guideline. The administration of aspirin (or other antiplatelet agents) as an adjunctive therapy within 24 h of intravenous fibrinolysis is not recommended (*due to increased intracranial hemorrhage*).

- At present, no pharmacological agents with putative neuroprotective actions have demonstrated efficacy in improving outcomes after ischemic stroke, and therefore, pharmacological neuroprotective agents are not recommended by the 2013 AHA/ASA guideline.

28.5 SELF-ASSESSMENT QUESTIONS

28.5.1. In his recent article on a modern history of stroke, M.A. Kelly stated that "Prevention and treatment of stroke has changed substantially since the time of Franklin Delano Roosevelt who died of an intracerebral hemorrhage in 1945. As the understanding of stroke pathophysiology advanced, the beneficial effects of antiplatelet and anticoagulant drugs were recognized. Imaging of blood vessels by angiography made surgical therapies possible. Later noninvasive computerized tomography and magnetic resonance imaging distinguished hemorrhagic from ischemic stroke and gave new insight into stroke mechanisms. Stroke prevention became possible by selective management of stroke risk factors. Thrombolytics introduced in 1996 provided the first actual treatment of ischemic stroke. The field of stroke continues to advance as medical and surgical treatments are refined and indications made clear, organized systems of care become standard, and new imaging techniques and endovascular therapies are developed" ([16]). According to the 2013 AHA/ASA guidelines on the early management of acute ischemic stroke, which of the following is a class I (LOE: A) recommendation on thrombolytic therapy in patients with acute ischemic stroke?

A. In patients eligible for intravenous tPA, benefit of therapy is time dependent, and treatment should be initiated as quickly as possible. The door-to-needle time (time of bolus administration) should be within 60 min from hospital arrival.

B. Intra-arterial fibrinolysis is beneficial for treatment of carefully selected patients with major ischemic strokes of <6 h duration caused by occlusions of the MCA who are not otherwise candidates for intravenous tPA.

C. Intravenous tPA (0.9 mg/kg, maximum dose 90 mg) is recommended for selected patients who may be treated within 4.5 h of onset of ischemic stroke.

D. The intravenous administration of streptokinase for treatment of stroke is recommended.

28.5.2. Which of the following is a class I (LOE: A) recommendation on endovascular interventions from the 2013 AHA/ASA guidelines on early management of acute ischemic stroke?

A. As with intravenous fibrinolytic therapy, reduced time from symptom onset to reperfusion with intra-arterial therapies is highly correlated with better clinical outcomes, and all efforts must be undertaken to minimize delays to definitive therapy.

B. Intra-arterial fibrinolysis or mechanical thrombectomy is reasonable in patients who have contraindications to the use of intravenous fibrinolysis.

C. Patients eligible for intravenous tPA should receive intravenous tPA even if intra-arterial treatments are being considered.

D. The usefulness of emergent intracranial angioplasty and/or stenting is not well established. These procedures should be used in the setting of clinical trials.

28.5.3. A 42-year-old woman with a history of exertional dyspnea, severe mitral stenosis (mitral valve area, 0.9 cm²) secondary to rheumatic heart disease, and atrial fibrillation was admitted for percutaneous mitral balloon valvotomy to be performed with the MULTI-TRACK (NuMED) single-wire, double-balloon device. After the patient underwent transseptal puncture during cardiac catheterization, consciousness became impaired, vertigo developed, and binocular visual loss occurred. The doctors become concerned about a possible acute stroke, and cerebral angiography was performed, revealing total occlusion of the distal basilar artery, which suggested acute thromboembolism. Selective intra-arterial thrombolysis was initiated 40 min after the ictus. Follow-up cerebral angiography was performed and showed resolution of the occlusion. Within 12 h of the event, all neurologic abnormalities had resolved. The patient underwent mitral valve replacement at a later date (slightly modified from [17]). Which of the following was most likely used for the intra-arterial thrombolysis in this patient?

A. Alteplase
B. Desmoteplase

C. Reteplase

D. Tenecteplase

E. Urokinase

28.5.4. An 18-year-old man was transferred to the hospital because of blurred vision, dysarthria, and ataxia, which had reportedly progressed to coma. Upon analyzing the clinical and radiological findings as well as the patient's history, the patient was diagnosed with ischemic stroke from occlusion of the midbasilar artery (although it seems surprising that such a young and previously healthy person can have a stroke, up to 14% of all strokes occur in children and young adults). Intravenous thrombolysis with tissue plasminogen activator appears safe in young patients with stroke, but unfortunately, this young man's stroke was diagnosed 5h after onset (modified from [18]). Which of the following would be recommended?

A. Hemodilution by volume expansion

B. Immediate oral administration of aspirin

C. Intra-arterial tPA

D. Urgent anticoagulation

28.5.5. For the patient described in 5.4, which of the following would be a reasonable recommendation?

A. Initiation of warfarin therapy within 12h of stroke onset

B. Oral administration of aspirin (initial dose is 325mg) within 24–48h after stroke onset

C. Supplementation of oxygen even the patient was nonhypoxic

D. The administration of pentoxifylline

REFERENCES

1 Heidenreich, P.A., et al., Forecasting the future of cardiovascular disease in the United States: a policy statement from the American Heart Association. *Circulation*, 2011. 123(8): p. 933–44.

2 Washburn, E.R., Fast forward: a blueprint for the future from the Institute of Medicine. *Physician Exec*, 2001. 27(3): p. 8–14.

3 Schwamm, L.H., et al., Recommendations for the establishment of stroke systems of care: recommendations from the American Stroke Association's Task Force on the Development of Stroke Systems. *Stroke*, 2005. 36(3): p. 690–703.

4 Jauch, E.C., et al., Guidelines for the early management of patients with acute ischemic stroke: a guideline for healthcare professionals from the American Heart Association/American Stroke Association. *Stroke*, 2013. 44(3): p. 870–947.

5 Goldstein, L.B., et al., Guidelines for the primary prevention of stroke: a guideline for healthcare professionals from the American Heart Association/American Stroke Association. *Stroke*, 2011. 42(2): p. 517–84.

6 Furie, K.L., et al., Guidelines for the prevention of stroke in patients with stroke or transient ischemic attack: a guideline for healthcare professionals from the American Heart Association/American Stroke Association. *Stroke*, 2011. 42(1): p. 227–76.

7 Connolly, E.S., Jr., et al., Guidelines for the management of aneurysmal subarachnoid hemorrhage: a guideline for healthcare professionals from the American Heart Association/American Stroke Association. *Stroke*, 2012. 43(6): p. 1711–37.

8 Saposnik, G., et al., Diagnosis and management of cerebral venous thrombosis: a statement for healthcare professionals from the American Heart Association/American Stroke Association. *Stroke*, 2011. 42(4): p. 1158–92.

9 Broderick, J., et al., Guidelines for the management of spontaneous intracerebral hemorrhage in adults: 2007 update: a guideline from the American Heart Association/American Stroke Association Stroke Council, High Blood Pressure Research Council, and the Quality of Care and Outcomes in Research Interdisciplinary Working Group. *Circulation*, 2007. 116(16): p. e391–413.

10 European Stroke Organisation (ESO) Executive Committee, ESO Writing Committee, Guidelines for management of ischaemic stroke and transient ischaemic attack 2008. *Cerebrovasc Dis*, 2008. 25(5): p. 457–507.

11 Werring, D.J. and M.M. Brown, New NICE guideline on acute stroke and TIA: need for major changes in delivery of stroke treatment. *Heart*, 2009. 95(10): p. 841–3.

12 Bayley, M., et al., Balancing evidence and opinion in stroke care: the 2008 best practice recommendations. *CMAJ*, 2008. 179(12): p. 1247–9.

13 Wright, L., et al., Stroke management: updated recommendations for treatment along the care continuum. *Intern Med J*, 2012. 42(5): p. 562–9.

14 Wang, Y.J., et al., Chinese guidelines for the secondary prevention of ischemic stroke and transient ischemic attack 2010. *CNS Neurosci Ther*, 2012. 18(2): p. 93–101.

15 Jauch, E.C., et al., Part 11: adult stroke: 2010 American Heart Association Guidelines for Cardiopulmonary Resuscitation and Emergency Cardiovascular Care. *Circulation*, 2010. 122(18 Suppl 3): p. S818–28.

16 Kelly, M.A., Stroke: a modern history. *Am J Ther*, 2011. 18: p. 51–6.

17 Duraes, A.R., and J.C., Brito, Selective intraarterial thrombolysis for cardioembolic stroke. *N Engl J Med*, 2012. 367: p. e24.

18 Yager, P.H., A.B. Singhal, and R.G. Nogueria, Case 31-2012: an 18-year-old man with blurred vision, dysarthria, and ataxia. *N Engl J Med*, 2012. 367: p. 1450–60.

INDEX

Note: f: figure; t: table; b: box

Cardiovascular Diseases: From Molecular Pharmacology to Evidence-Based Therapeutics, First Edition. Y. Robert Li.
© 2015 John Wiley & Sons, Inc. Published 2015 by John Wiley & Sons, Inc.